Design of Ship Hull Structu

Yasuhisa Okumoto · Tetsuo Okada · Yu Takeda ·
Masaki Mano · Masanobu Toyoda

Design of Ship Hull Structures

A Practical Guide for Engineers

Second Edition

Springer

Yasuhisa Okumoto
Higashihiroshima, Japan

Yu Takeda (Deceased)
Yokohama, Japan

Masanobu Toyoda
Marine Data Solution Group
Japan Marine United Corporation
Yokohama, Japan

Tetsuo Okada
Faculty of Engineering
Yokohama National University
Yokohama, Japan

Masaki Mano
Japan Ship Design Council
Yokohama, Japan

ISBN 978-3-032-11789-2 ISBN 978-3-032-12500-2 (eBook)
https://doi.org/10.1007/978-3-032-12500-2

This Springer imprint is published by the registered company Springer Nature Switzerland AG
The registered company address is: Gewerbestrasse 11, 6330 Cham, Switzerland

Yu Takeda has died before the publication of this book.

Preface to the Second Edition

Sixteen years have passed since the original edition of the book was published, and since then, the field of ship structural design has undergone significant developments, including advanced direct load and structural analyses, new structural design methods, and the rapid increase in the size of container ships. To reflect such new developments and to make the book more suitable to modern professionals and practitioners, the authors undertook a significant revision of the book.

Throughout the book, outdated contents have been renewed with the latest information and data. Most significant revisions are as follows:

Part I

- Chapter 4: A new section titled "Design by Direct Load and Structural Analysis" was added, where new approach for rational structural design using high fidelity numerical simulations of loads and structural responses is discussed.
- Chapter 5: A new section titled "Increase in Container Ship Size" was added, touching upon rapid increase in the size of container ships due to the high growth of worldwide seaborne trade, especially after 1990, and challenges to achieve such a rapid size increase successfully.

Part II

- Chapter 8: This chapter has been completely rearranged, introducing virtual fixation method to solve continuous beams and many examples of continuous beams encountered in ship structural design, such as irregular span, simply supported end by a header, partially distributed or concentrated load, and forced displacement. Comprehensive formulation and discussions were included also about the effect of asymmetrical (L-shaped) stiffener, effect of axial stress, and effect of shear stress.

- Chapter 9: A new section titled "Large Opening on Girder Web Plate" was added to give simple formulae expressing the secondary bending stresses induced on both the narrow sides of a large opening on a girder web plate.
- Chapter 10: A comprehensive formulation of slot strength calculation was newly introduced.
- Chapter 12: Sect. 12.3 "Plates Supporting Bending and Axial Load Simultaneously" was completely revised to account for additional lateral force due to the axial load and the aspect ratio.

Part III

- Chapter 20: A new section titled "Review of Break-in-Two Accidents" was newly added to review past important break-in-two accidents of ships.
- Chapter 21: A new subsection titled "Arrangement of Primary Supporting Members" was newly added to explain the history and principles of the arrangement of primary supporting members of tankers.
- Chapter 23: A new section "Bilge Shell and Arch Effect" was added to discuss broader range of topics on the arch effect, including formulation of strength evaluation formulae for unstiffened and stiffened bilge shell plating, and benefits and drawbacks of arch effect in ship structures.
- Chapter 24: Significant revision was made to the section "Horizontal Girder Arrangement on Bulkheads" to explain optimum arrangement of horizontal girders on a transverse bulkhead of tankers.

Dr. Masanobu Toyoda joined our author team for this second edition, resulting in the author team spanning five generations from the dawn period of modern ship hull design (from the 1960s) to the most recent rapid size increase in container ships and development of structural design methodologies.

Many new formulae introduced in the second edition, especially those in Chap. 8, Part II were developed in the lecture "Structural Mechanics of Ships" and "Ship and Marine Structural Design Methodologies" given by one of the authors, Prof. Tetsuo Okada, at Yokohama National University.

Last but not least, we would like to express our sincere gratitude to the late Mr. Masahiro Fujiwara, whose personal communications with us provided us with great inspiration and support, particularly in the development of mathematical formulae in Chaps. 8–10 in Part II and Chap. 23 in Part III.

Higashihiroshima, Japan — Yasuhisa Okumoto
Yokohama, Japan — Tetsuo Okada
Yokohama, Japan — Yu Takeda
Yokohama, Japan — Masaki Mano
Yokohama, Japan — Masanobu Toyoda
August 2025

Preface to the First Edition

The ship design is divided generally into four parts, hull form design, arrangement design, hull structure design, and fitting design (hull fitting and machinery fitting). The design of merchant ships starts with the owner's requirements such as kind and volume of cargo, transportation route and time generally. Sometimes the owner has a special requirement such as no bulkhead in hold.

Based on the above requirements, a general arrangement plan is roughly designed, and the studies are to be done from stability, strength, operation, and habitability viewpoints. Thus, the general arrangement plan is finally decided with correction if necessary. Referring to the general arrangement plan and the lines plan which shows the hull form, the hull structure design is carried out to decide the size, position, and materials of the structural members, including the fabrication and assembly methods.

The most important duty of the hull structure design is to provide a hull structure strong enough against internal and external loads. Textbooks or handbooks of hull strength are helpful to the hull structure designer. However, these books are generally written from the viewpoint of the strength theory and do not seem to be sufficient from the design viewpoint.

The authors are hull structure designers in four generations, from the developing era of structural design by large increase of the ship size and increase of ship production to establishing era of the design technology using computer: CAD and CAE. In this book the experiences of the authors in the above generations are condensed from the design viewpoint. The authors are convinced that this book will be strong weapon for designers to design the hull structure as well as for students to understand the hull structure design in the world.

Higashihiroshima, Japan — Yasuhisa Okumoto
Yokohama, Japan — Yu Takeda
Yokohama, Japan — Masaki Mano
Yokohama, Japan — Tetsuo Okada
August 2008

Competing Interests The authors have no competing interests to declare that are relevant to the content of this manuscript.

About the Authors

Yasuhisa Okumoto, Dr. Eng.
1965–1991: Ishikawajima-Harima Heavy Industries Co., Ltd. (IHI)
Shipbuilding and Offshore Division
1992–2010: Professor of Faculty of Engineering, Kinki University
2010–2017: Professor of JWRI, Osaka University

Tetsuo Okada, Dr. Eng.
1986–2002: Ishikawajima-Harima Heavy Industries Co., Ltd. (IHI)
Shipbuilding and Offshore Division
2002–2012: IHI Marine United Inc.
2013–2014: Japan Marine United Corporation
2015–: Professor of Faculty of Engineering, Yokohama National University

Yu Takeda, Dr. Eng., Deceased
1981–1997: Ishikawajima-Harima Heavy Industries Co., Ltd. (IHI)
Shipbuilding and Offshore Division
1997–2008: Ishikawajima-Harima Heavy Industries Co., Ltd. (IHI)
Manager, Research Institute

Masaki Mano, Dr. Eng., FRINA
1952–1986: Ishikawajima-Harima Heavy Industries Co., Ltd. (IHI)
Shipbuilding and Offshore Division and Research Institute
1987–1999: Professor of Faculty of Engineering, Kinki University
Consulting Naval Architect

Masanobu Toyoda, Dr. Eng.
1996–2002: Ishikawajima-Harima Heavy Industries Co., Ltd. (IHI)
Shipbuilding and Offshore Division
2002–2012: IHI Marine United Inc.
2013–: Japan Marine United Corporation
General Manager, Marine Data Solution Group

Contents

List of Figures

List of Tables

Part I
Fundamentals

Chapter 1
Philosophy of Hull Structure Design

1.1 Importance of Hull Structure Design

Hull structure design has recently become more and more important. Ship hull structure design has always been considered, technically as well as economically, to be a very important part of the shipyard management, since the strength of a ship is concern and the cost of the hull steel is about 20% of total ship's cost. Therefore, a good executive needs many talented designers to assist especially in hull structure design.

Between 1950 and 1970, due to economical demand, ship dimensions had to increase, ship speed had to be higher, and the specialization of ships was promoted. Worldwide economies developed quickly, due to the consumption of cheap fuel oil, and a mass consumption era commenced. Consequently cargo movement increased. In order to satisfy this social requirement, large, high speed and specialized ships were designed and constructed. In Fig. 1.1 worldwide seaborne cargo movement is shown.

The hull structure designers played a major role in this era. They produced reliable ships and simultaneously increased the size of the ships step by step, based on past experience and immediate feeding back of performance results of newly completed ships. In Fig. 1.2 the transition of scaling-up of ships is shown. It goes without saying that the ship dimensions increased step by step. The largest ship actually constructed was the 565,000 DWT (dead weight tons) tanker *Seawise Giant*. However, a preliminary design of a 1,000,000 DWT tanker (Million tonner) was carried out also, and building docks, to make the construction of that ship possible, were actually constructed. However, after the oil crisis, no bigger tanker than *Seawise Giant* was built.

The main engine output increased in order to obtain higher vessel speeds, and vibration problems became apparent due to a rigid propeller shaft in a flexible hull structure of the increasing in vessel dimensions. However these problems were solved by the joint efforts of hull structure designers.

Y. Okumoto et al., *Design of Ship Hull Structures*,
https://doi.org/10.1007/978-3-032-12500-2_1

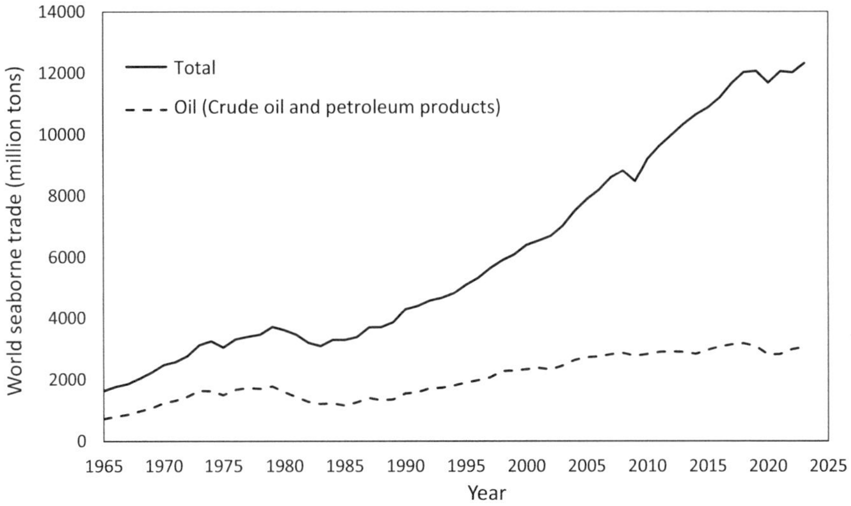

Fig. 1.1 World seaborne trade

The specialization of ships occurred in response to the need for effective transportation. In addition to the tankers previously in service, ore and bulk carriers, container ships, designated car carriers etc. appeared. In these specialized ships many special requirements of hull structure, that were not necessary before, had to be incorporated in order to increase transportation efficiency. Hull structure designers overcame these problems due to these special requirements when designing these ships.

The oil crises of 1973 and 1979 halted the size increase of ships and further development of higher vessel speeds. At the same time worldwide shipbuilding and shipping recessions started. As the price of crude oil increased from U.S. $1.6 to U.S. $12 and then sky-rocketing to U.S. $34, energy saving trends spread all over the world. The earlier increase in cargo movement reversed and gradually decreased. The surplus of shipbuilding industry facilities and the over-tonnage of ships accelerated the recession.

The most important topic in the energy saving era was to reduce fuel consumption. So the utmost was done on hull form improvement, propulsion (propeller) efficiency, main engine fuel consumption and efficiency of the propulsion plant. The main actors in this period were hull form designers, machinery designers, and main engine builders.

Regarding energy saving, hull structure designers could not contribute a great deal. They could only reduce hull steel weight to have smaller ships which have less resistance. Application of higher strength steels expanded in this period, especially thanks to newly developed TMCP technology, which is explained in Sect. 6.3.

Shipbuilding was depressed to 1990, and hull structure designers have been the main players once again in the recovering period after that. After 1990, economic

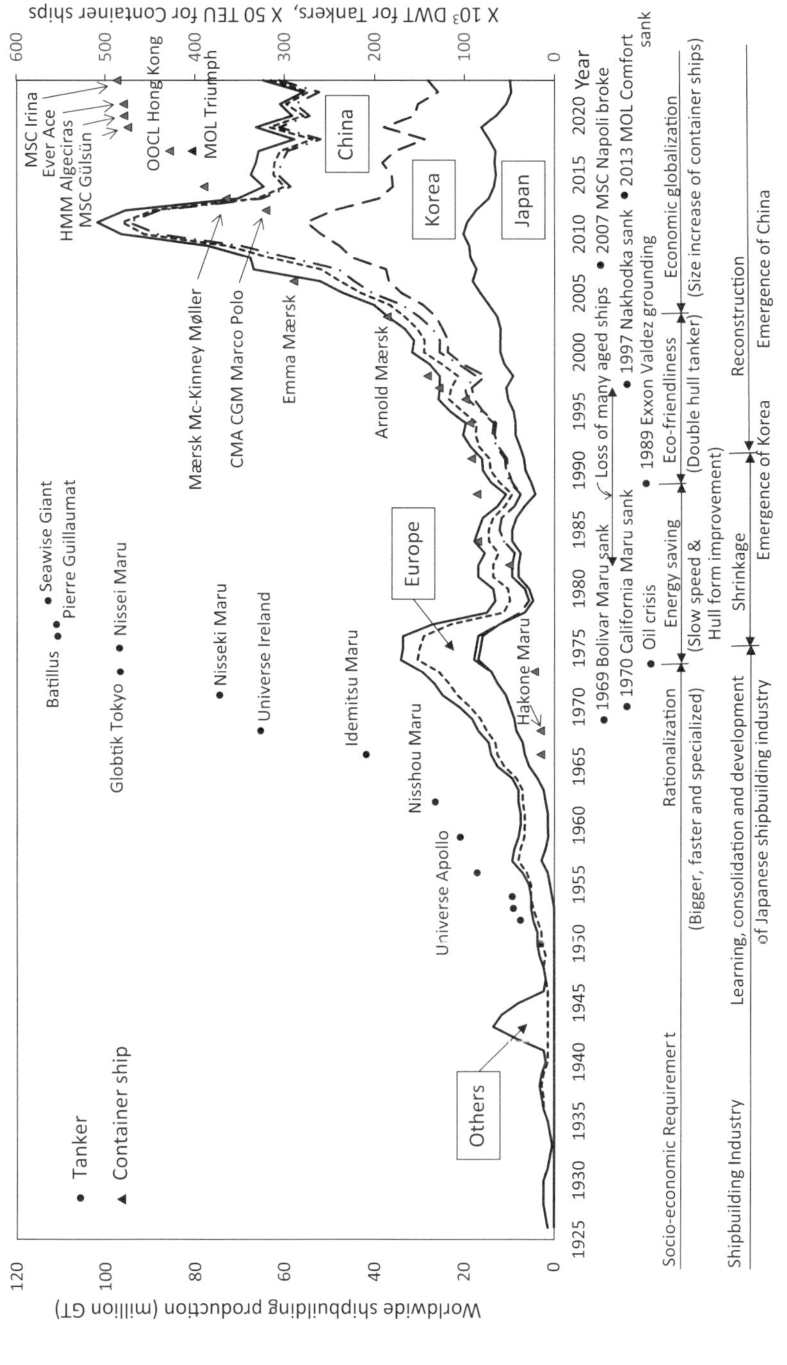

Fig. 1.2 Environments, worldwide shipbuilding production and size increase of ships

globalization progressed and worldwide seaborne trade other than oil expanded rapidly. China and other emerging countries produced and exported a vast and various kinds of products to the world market, and as a result, container ships became larger and larger in this period.

Due to competition among ship owners themselves and also among shipyards it was hard to survive in these rapidly changing times. Under these circumstances the specialization of ships was promoted to achieve a more profitable cargo transportation unit. Consequently the demands on the hull structure were greater and greater, and some of them seemed to be impossible to carry out.

Whether a new structure can be realized or not, depends on what the boundary conditions are, and this is up to the hull structure designer to decide. It has been said for a long time that shipbuilding is "experience engineering" and ships are to be designed and built, based on past experience. However, the situation has changed as the hull structure designer has to design new and unusual structures which have never been built before, so there is not much experience available. Survival of the shipyard will depend on the success of the hull structure design team.

1.2 Design Procedure of Structures

Not only the hull structure design but also general structural design is carried out in accordance with the procedure shown in Fig. 1.3. It is a well known fact in the shipbuilding field that design starts with the requirements of the type and volume of cargoes, routes of transportation, speed (from where to where in how many days), or for new types of ship, e.g. the million tonner, whether it will be available or not.

Then a preliminary general arrangement is to be prepared based on these requirements. The purpose of a ship is to carry cargo so that the size, shape and the arrangement of cargo holds are decided in the first stage. The results of hull structure research are important in the decision-making process regarding the shape and arrangement of cargo holds.

The estimation of the forces applied to the structures shown in Fig. 1.3 means the establishment of static loads such as buoyancy, weight of ship, cargo weight and the estimation of dynamic loads such as wave load and exciting force. The wave load is a kind of natural phenomenon and is to be estimated based on accumulated observed data.

In Fig. 1.3 the design flow, shown in vertical arrows, is decided by the structural designer, and the horizontal arrows show the items to be based on social standards and research results. These concepts are supplied to the designers by outsiders. Because the setting of the forces applied to the hull structure affects the design very much, hull structure designers should always pay attention to research in this subject.

The estimation of structural failure mode due to such forces is important for a reliable design. Not only theoretical knowledge but also observations experienced on actual failures are necessary in estimating the failure modes correctly. In the

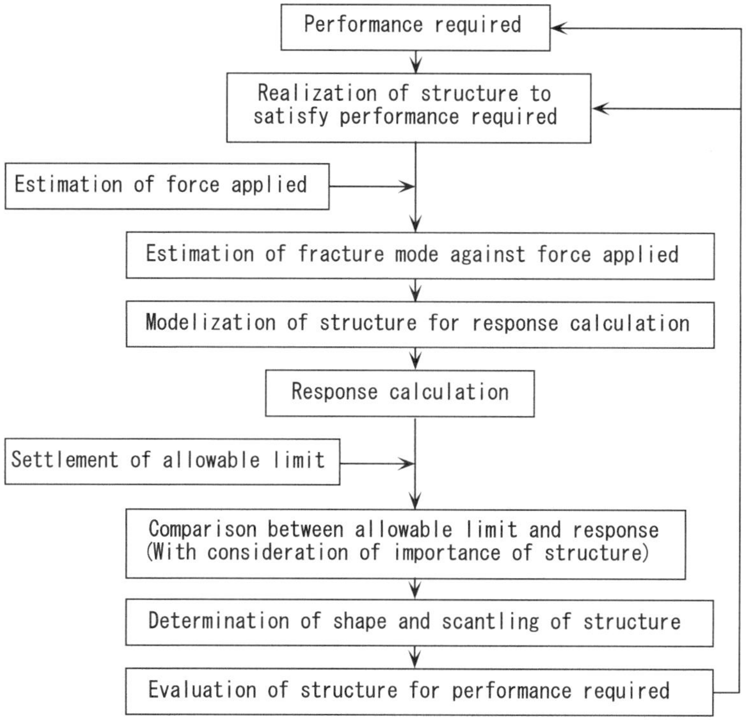

Fig. 1.3 Sequence of structure design

modelization of strength response calculations the correct failure mode should be applied.

Around 1965 many buckling and cracks occurred on the web plates of big girders in newly built large ships because designers designed the girder based on bending moment without considering shear force, which is important for big girders.

It is easy to understand the phenomena after solving the problem when it happened for the first time, but it is difficult to visualize and predict it beforehand. And even if it can be predicted, it is very hard for designers to apply countermeasures in advance, which require time, money and labor. With the correct estimation of failure mode and the effective response calculation, hull structure designers can design satisfactory structures.

The next step is to calculate responses for the forces applied. In the old style response calculation, each structural member is assumed to be a simple beam fixed at both ends. Recently, with the help of a computer whole structures can be modeled and analyzed for the estimated forces. However, for the reliable design, considering statistical distribution of forces and deterioration of materials, more research will be required.

The stochastic finite element analysis is attracting attention to obtain damage probability under the statistical force data and scantling data. A pioneering work in

1982 was applied to estimate remaining life of a land-based oil storage tank under an uncertain corrosion on the bottom plate [1]. The application is expanding to deal with the shape uncertainty [2] and to design a dynamic damper to reduce vibration of a superstructure under stochastic mass and stiffness [3].

For the allowable limit, there are allowable stress, allowable deformation, allowable amplitudes of vibration and allowable reliability. These limits are usually stipulated by external and social requirements such as the classification society's rules. The hull structure designer has no decisive right but should pay special attention to them. And here there are very important items, which include: load estimations, response calculations and allowable limit. They should be investigated together and simultaneously. Any changes in the method of load estimation must be followed by modifying the allowable limit.

The importance of the structural members is based on the damage caused by the failure of the member. It is reasonable to apply a different safety factor, corresponding to the importance of the member.

Where the calculated response is within the allowable limit, the shape and the scantling of the member are to be accepted and decided upon, provided that the values are moderate. If these values are not moderate, further options must be considered. For example, with the existing material and facilities, the million tonner cannot be built and the maximum available tanker will be one of 700,000 DWT. Additionally a 5-hold arrangement gives problems, while a 6-holds one would be preferable from a structural viewpoint.

In Fig. 1.3 there is no description of the material, which is one of the most important items in the structural design, because mild steel, which has been a major shipbuilding material, has splendid properties, tough, strong, high weldability and low price. The hull structure designer can use it without any special attention. However, nowadays high tensile steel is becoming more common as a shipbuilding material, especially after the TMCP steel was invented in 1980's. When the high tensile steel and other special materials are used as shipbuilding material, the hull structure designer must pay careful attention to these materials.

1.3 Hull Structure Design Policy

Prior to designing a ship, a design policy must be established and confirmed. The principle of hull structure design is illustrated in Fig. 1.4; high reliability, good performance, and easy maintenance, same as in any other design process. These three elements are to be established from the user's point of view, who operates ships and contributes to worldwide economy.

High reliability can be achieved by reliable design stated after. Good performance means a well-balanced structure which can carry out the given duty during the given period. Good performance can be obtained by rationalization. Reliability concerns the safety of ship, crew member, and cargo, so as to prevent any failures which

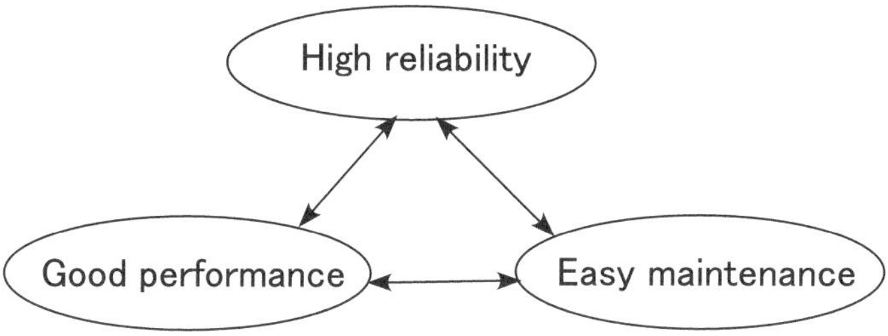

Fig. 1.4 Principle of hull structure design

could jeopardize the ship's safety. Easy maintenance means less repair cost and good accessibility for inspection.

To realize a good balance in the above policy; high reliability, good performance, and easy maintenance, a rational design system must be established, based on the theoretical approach and plentiful experience.

1.4 Basic Idea of Hull Structure Design

The basic idea of hull structure design is: "A ship will break when the load applied is bigger than the ship's strength". It is a simple but grim reality. Once the ship is broken the safety of the ship will be in jeopardy, then the crew's lives and the owner's property (ship and cargo) are at stake.

Even today when science and technology have progressed remarkably, there still exists an unknown part in the forces applied on the ship and the strength of the ship's structure. The hull structure designer should design the structure very carefully, based on the updated knowledge.

Actually hull structures are designed using the classification society's rules; however, the rules are not 100% perfect. The designer must do his/her utmost in the design.

When a failure such as a crack, buckling, collapse, etc. arises in a ship, it is regarded as a good chance to gain engineering knowledge, though the damage is not pleasing in itself;

1. to go to the spot on board
2. to see the appearance of the failure
3. to consider the phenomenon at that locality.

It is important to observe carefully the detail such as figure, size, direction, etc., and to estimate the magnitude, direction, source, and transmitting route of the forces, by seeing the actual damage to the structure. If possible, learn of the operating conditions at the time of the event from the crew concerned.

After that, experiments are also carried out if necessary. Next, the cause of the damage and failures are predicted by analysis, and countermeasures are determined in accordance with the calculation. The worst case is to calculate using a complex model such as the FEM program without seeing the actual damage. Recently, this tendency is increasing due to the rise in computer simulations. Structural designers tend to rely on the result of FEM calculations without considering the principle of strength.

1.5 Studies on Loads Applied

Before 1960, the wave bending moment was calculated by assuming that the wave had the same length as the ship and the height was 1/20 of this length. After intense and numerous observations, research of ocean waves, and research on ship motion in waves, the wave bending moment can be estimated rather accurately nowadays. In Fig. 1.5 the reduction of the midship section modulus required by Nippon Kaiji Kyokai (NK) is shown. In the figure the level of 1.0 shows the guidelines for the 1930 load line regulations [4, 5].

Figure 1.5 shows that the required section modulus has reduced year by year especially for larger ships. These are the fruits of the above research. Due to this research, large ships can be satisfactorily constructed.

Also before 1960, the tank length of a tanker was limited to 15 m. It was too limiting in order to build large tankers economically from the hull steel weight,

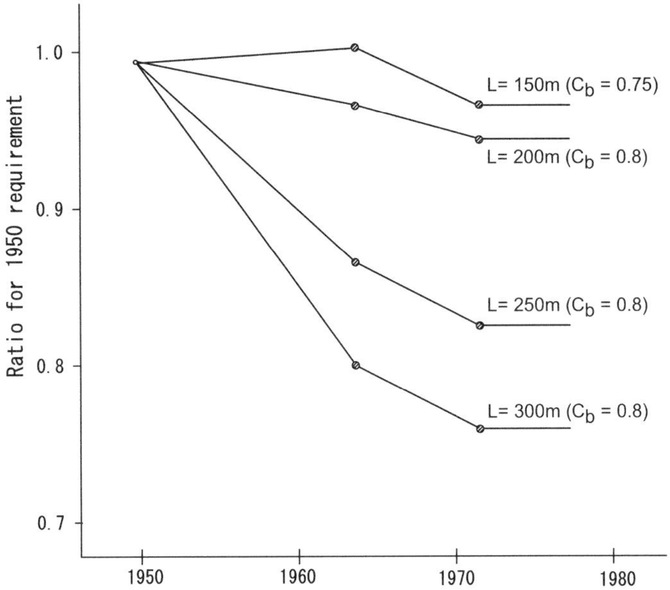

Fig. 1.5 Reduction of section modulus required (tanker)

fitting weight and cargo operation viewpoints. However, the designer did not have any reason to increase the tank length over 15 m.

Much research was carried out on cargo oil movement in tanks caused by ship motion and sloshing pressure by the cargo oil motion. And as a result tank length was extended up to a value of 20% of the total ship length. According to the old regulation, a 250,000 DWT VLCC tanker must have 40 or 50 tanks of 20 m breadth, 35 m depth and 15 m length which is uneconomical and inconvenient for operation.

As the above examples show, rationalization can be achieved by research, thus discarding the obsolete traditional ways and methods.

1.6 Reliable Design

It is stressed in Sect. 1.4 that the structural design is based on the forces applied and the strength of the ship. In this chapter how to maintain the ship's safety and reliability, for the applied forces and the strength of the ship itself, is introduced [6].

There are three important procedures in the hull structure design:

1. The forces applied are to be estimated
2. The structural response is calculated applying by these forces on the hull structure
3. The obtained responses are assessed against allowable values.

To maintain safety and reliability it is important to have good accuracy in these three values. In the design of ship structures the force applied is divided into two, static loads which are the difference between weight and buoyancy, and dynamic loads such as fluctuating forces and impact forces from waves. The former can be calculated with sufficient accuracy, but the latter is difficult to estimate because it has many variations depending on the geographical area and the season.

Response calculations can now be done with sufficient accuracy by means of a framework or Finite Element Method thanks to the development of computers. The calculated results agree well with the results of model tests and full-scale tests.

The allowable limits can be decided, in the case of ordinary ships for which past experience has accumulated, by comparing results of response calculations for estimated applied forces with damage data of the ship in service. This is an empirical reliable design.

For new types of ship and for special ships which have no damage history, the empirical method cannot be applied. In this case response calculations will be done with applied forces which are estimated as accurately as possible. And the allowable limit is to be decided based on the property of materials to be used and the theory of fracture mechanics. This is a theoretical reliable design.

A flowchart of a reliable design is shown in Fig. 1.6 in which "theoretical failure probability flow" corresponds to the above theoretical reliable design and "empirical failure probability flow" corresponds to the empirical reliable design. The values in Fig. 1.6 are variable and statistical. Many parameters in ship design are not definite but take statistical values which have a distribution around representative values. For

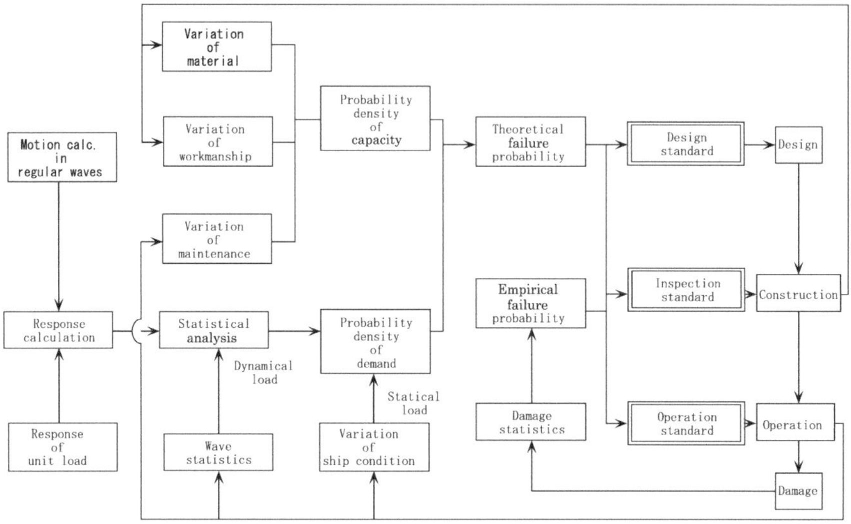

Fig. 1.6 Flow diagram of reliable design

example, concerning the forces applied and the strength of the structure, the former changes depending on loading and wave conditions, and the latter changes depending on tolerance of materials, accuracy of manufacturing, and design standards.

Theoretical reliable design aims to keep reliability by gathering data on statistical parameters and calculating failure probabilities theoretically for many kinds of failure modes.

The failure probability can be obtained by considering the distribution patterns of the forces applied and the strength of the structure [7]. As shown in Fig. 1.7, if probability density functions of forces applied D (hereafter forces applied is to be called demand) and strength of structure C (hereafter strength of structure is to be called capacity) are given with the horizontal axis indicating stress X, the failure probability is given by the following equation:

$$P_f = \int_0^\infty Q_d\{X\}P_c\{X\}dX = \int_0^\infty (1 - Q_c\{X\})P_d\{X\}dX \tag{1.6.1}$$

where $P_d\{X\}$: probability density function of demand

$P_c\{X\}$: probability density function of capacity

$Q_d\{X\}$: probability of demand exceeding certain value

$Q_c\{X\}$: probability of capacity exceeding certain value.

P_f can be estimated approximately by the overlapped area of $P_d\{X\}$ and $P_c\{X\}$ in Fig. 1.7. A ship's failure is divided into plastic collapse, buckling, fatigue fracture and brittle fracture. For plastic collapse and buckling the maximum demand is most important, and the maximum demand caused by the biggest wave once per 10^8 cycles

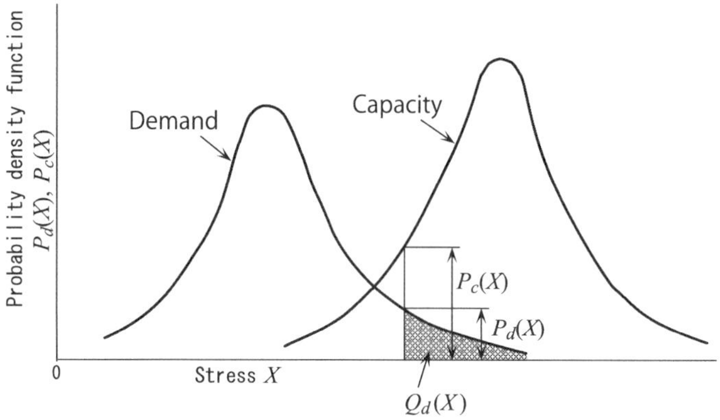

Fig. 1.7 Probability of fracture

is applied for plastic collapse and buckling analysis. Assuming the wave period is about 7 s and there are 320 sailing days in a year, a ship encounters 10^8 waves in 25 years of her life. For fatigue fracture analysis a concept of cumulative damage intensity is applied using distributions of demand in these 10^8 cycle of waves. Brittle fracture can be avoided by selecting a grade of materials and welding condition to meet the ship's service environment.

To reduce the probability of failure, it is effective to reduce the overlapping area of curve D (demand) and curve C (capacity) in Fig. 1.7. For that there are two ways; one is to have a bigger ratio between the mean values of C and D which is called safety factor, and the other is to make both curves sharper, which means more accurate estimation of demand and more accurate control of capacity. Treating this matter theoretically, Fig. 1.8 is obtained [8]. Here, the distributions of demand and capacity are assumed to be normal.

In Fig. 1.8 the horizontal axis is the mean safety factor N, which is the ratio $\overline{C}/\overline{D}$ of the mean value of the demand $\overline{D}$ and mean value of the capacity $\overline{C}$. The failure probability is plotted for various combinations of the coefficient of variation V (standard deviation / mean value) of demand and capacity. From Fig. 1.8, it can be seen that the failure probability changes remarkably with the coefficient V, and in case of smaller V, failure probability can be reduced considerably by a small increase in the safety factor. In the case of a bigger V, failure probability can be reduced a little by increasing the safety factor.

To achieve reliable design it is important first to determine the variation in the necessary parameters. Standardization and systematization will reduce these variables and the probability of failure. Design standards, inspection standards, and operation standards play very important roles in keeping a ship's reliability by limiting the variation in every parameter which affects a ship's safety over her life.

On the other hand, reliable design based on the theoretical failure probabilities as stated above can be gained by using empirical failure probability. In the case of ordinary ships which have accumulated past experiences, allowable limits are to be

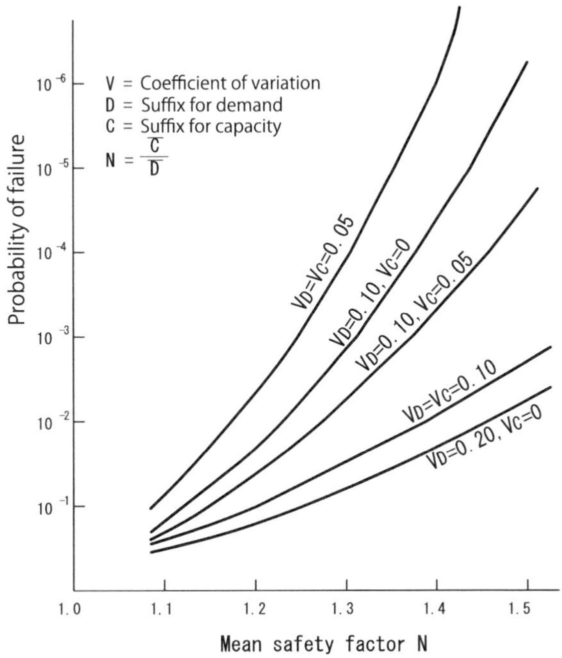

Fig. 1.8 Probability of fracture and mean safety factor

decided based on the past experiences. And ships built with similar design, material, and workmanship, and operated with similar operation standards will have similar reliability as the previous ship.

For new types and special ships never built before, reliable design with theoretical failure probability is the surest way at this moment. However, for ordinary ships which have previous experience, reliable design using empirical failure probability is most applicable.

In addition to the above it is important for the designer to keep in mind that the reliability, rationalization, and economy have to be well balanced.

Apart from the above explanation for new buildings, the failure probability of ships in service changes year by year depending on the maintenance. This is shown in Fig. 1.9 [1]. Hatched part shows failure probability after aging.

Given the distributions of demand and capacity, failure probability can be calculated according to a text book [9]. The important point in reliable design is the criterion under which the failure probability should be maintained.

There are many kinds of structural members, for example longitudinal strength members, transverse strength members, and local strength members, in the hull structure. How much reliability to be given to each member is to be decided by the designer according to his/her design policy.

In this procedure the degree of importance stated in Fig. 1.3 can be applied. The degree of importance will be proportional to the damage suffered from the failure of the member, and a bigger degree of importance corresponds to bigger damage. A

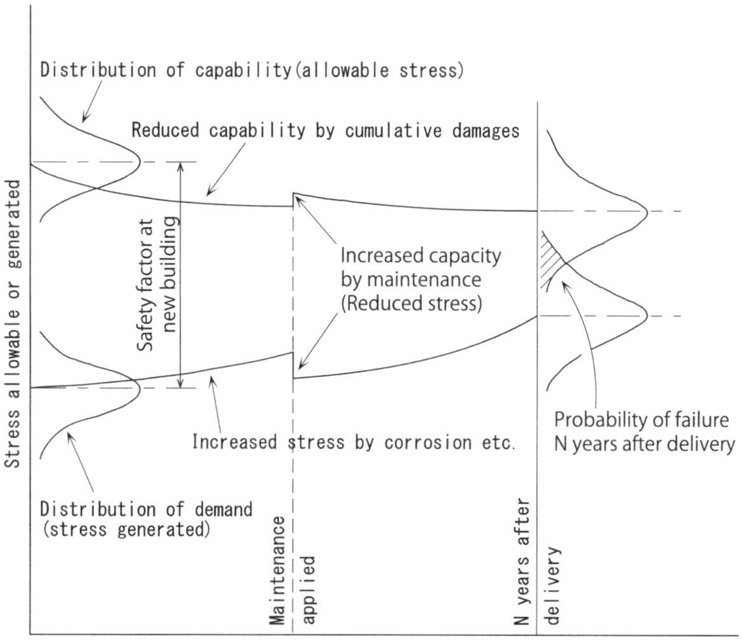

Fig. 1.9 Probability of fracture with consideration of reduced strength and increases stress by aging

smaller degree of importance is for a member that can be repaired before the failure escalates into serious damage. There are three kinds of damage suffered as follows:

(1) how dangerous and how extensive
(2) how much will social matters be affected
(3) how big the economic loss.

In the case of land structures with hazardous materials such as a nuclear power stations, chemical plants, etc. where people live who have no direct connection with these companies, the failure probability should not be decided only by the designer but should be decided with social agreement, because once failure happens the local inhabitants will suffer from the damage. In the case of a ship it is decided usually without any social agreement.

Transportation is also a part of social welfare and a ship will be one of the lifelines where there is no alternative transportation, for example supplying electricity, gas and water.

The optimum criterion of reliability is shown in Fig. 1.10. The horizontal axis is ship's total reliability in which each member of the hull structure is evaluated by a rational reliable design. The curve [A] is initial investment which increases with the improvement of the ship's total reliability. The curve [B] is the maintenance cost including repair cost due to the damage suffered, which decreases with the improvement of the reliability. The optimum point will be the minimum point of the curve [A] + [B].

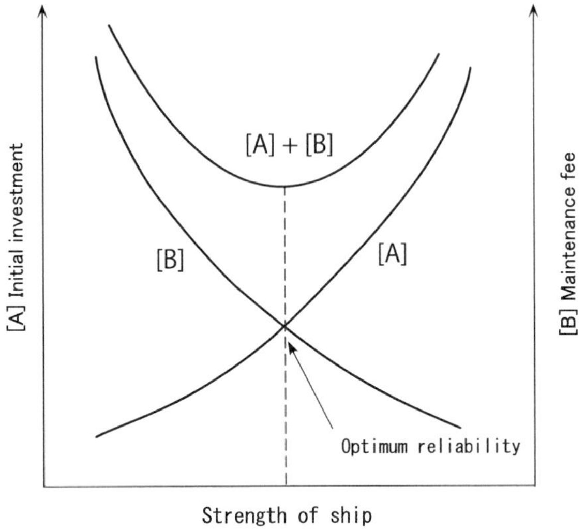

Fig. 1.10 How to decide optimum reliability

Chapter 2
Structural Design Loads

2.1 Introduction

When a ship is sailing at sea, it is subjected to various load patterns with many magnitudes which cause deformation of its structure, as well as stresses. The structural designer needs to know the hull structure load features, as accurately as possible: direction of the working load, frequency of occurrence, distribution pattern on the hull structure and behavior in the time domain, etc. The first design step is to assume exact loads acting on the structure concerned, in order to estimate the structural strength in a reasonable way and consequently to develop the design.

In this chapter, the classification of loads being applied to a hull structure will be explained, then the features of typical load components will be described, and finally the method of estimating wave loads will be discussed.

When considering the load features where the load is transmitted gradually and continuously from a local structural member to an adjacent bigger supporting member, the best way to categorize loads on the hull structure is as follows:

- Longitudinal strength loads
- Transverse strength loads
- Local strength loads

(1) *Longitudinal Strength Load*: Longitudinal strength load means the load concerning the overall strength of the ship's hull, such as the bending moment, shear force and torsional moment acting on a hull girder. Since a ship has a slender shape, it will behave like a beam from the viewpoint of global deformation. Now let us assume a ship is moving diagonally across a regular wave as shown in Fig. 2.1. The wave generates not only a bending moment deforming the vessel in a longitudinal vertical plane but also a bending moment working in the horizontal plane, because of the horizontal forces acting on side shell. In addition, the wave causes a torsional moment due to the variation of the

Y. Okumoto et al., *Design of Ship Hull Structures*,
https://doi.org/10.1007/978-3-032-12500-2_2

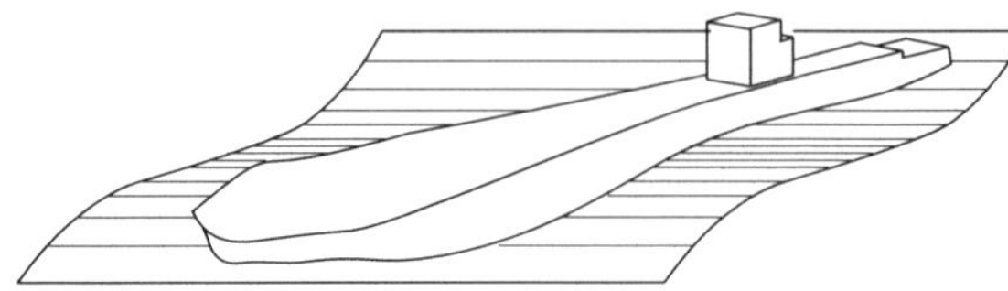

Fig. 2.1 Ship in oblique waves

wave surface at different sections along the ship's length. If the above longitudinal strength loads exceed the ultimate strength for the hull, the hull will suffer significant damages including hull girder collapse or sinking. Therefore, the longitudinal strength load is one of the most important loads when calculating the overall strength of a hull structure.

(2) *Transverse Strength Loads*: The transverse strength loads represent the loads which act on transverse members and cause structural distortion of a cross section. Transverse strength loads include hydrostatic pressure on the outer shell, weight of cargo load working on the bottom structure, ballast water pressure inducing the deformation of the ballast tank, etc. For instance, let us imagine a transverse section of a ship floating in still water as illustrated in Fig. 2.2. This section is subjected to: (a) hydrostatic pressure due to surrounding water, (b) internal loading due to self-weight and cargo weight. These loads are not always equal to each other at every point, consequently loads working on transverse members will produce transverse distortion as shown by the broken line in Fig. 2.2. When we consider transverse loads and longitudinal loads, the following characteristic is significant from the strength analysis point of view: The distortion due to longitudinal loads does not affect the deformation of the transverse section. For example, the longitudinal bending moment or shear force does not have an influence on the distortion of the cross section. It is therefore necessary to recognize the transverse deformation of the ship structure due to the transverse load, independently from the deformation induced by a longitudinal load. Transverse strength loads are commonly used in cases where we investigate the strength of primary members, such as transverse rings, transverse web frames, etc.

(3) *Local Strength Loads*: The local strength loads include loads which affect the local strength members such as shell panels, stiffeners and connecting constructions between stiffeners.

The above load categories are so convenient that they are extensively used for practical design purposes. A load acting on the structure can be treated independently by considering the load transferring from a local structure to a bigger structure. For example, let us consider the case where the designer commences the design of a bottom structure as shown in Fig. 2.3. Firstly, the strength of bottom shell panels must be determined regarding lateral water pressure, secondly, the strength of longitudinal stiffeners, which support the subject panels, must be evaluated, thirdly, the strength of transverse webs holding stiffeners at their ends must be estimated and finally, the global strength of the bottom structure must be discussed. Investigations can be done

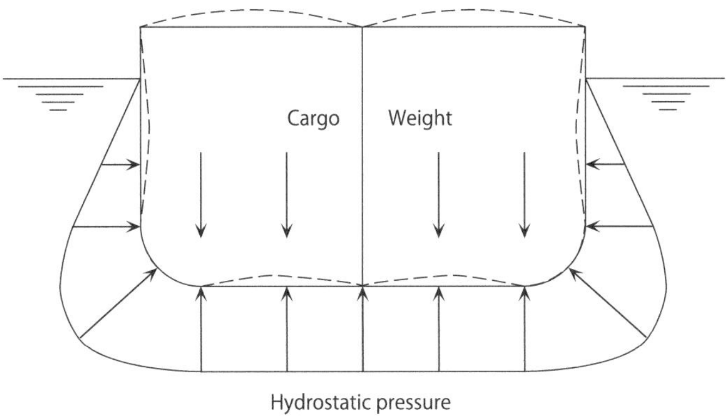

Fig. 2.2 Example of deformation due to transverse strength load

separately for each member by only considering the magnitudes of the loads which are transmitted by each member.

This is a convenient concept, however, time-dependent relations between those simplified loads have been intentionally omitted. Therefore, careful attention must be paid, particularly when analyzing the simultaneous response of the entire structure to combined loads, because the phase of the loads plays an important role when calculating the overall response.

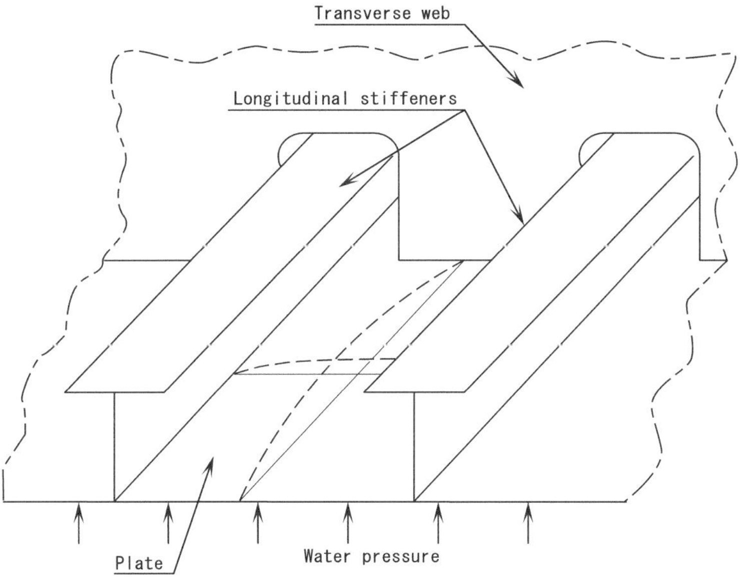

Fig. 2.3 Bottom structure under water pressure

2.2 Longitudinal Strength Load

Longitudinal strength loads are loads which affect the overall strength of a ship's hull girder, in which the ship is regarded as a beam or girder because of its slender profile. They are represented in terms of longitudinal bending moment, shear force and torsional moment. The longitudinal strength loads may be divided into two categories: static longitudinal loads and dynamic longitudinal loads.

Static longitudinal loads are induced by the local inequalities of weight and buoyancy in the still water condition. For instance, differences between weight and buoyancy along longitudinal direction cause a static bending moment and a static shear force, and asymmetrical cargo loading causes a static torsional moment.

Dynamic longitudinal loads are induced by waves. When the ship is on top of a wave crest in head sea condition, it causes a "hogging" bending moment and a shear force. When in a wave trough a "sagging" bending moment and shear force are experienced, as indicated in Fig. 2.4. These loads act alternately on the hull girder as the wave progresses along the ship. In cases where the ship encounters oblique seas a dynamic torsional moment is produced.

The magnitude of the dynamic longitudinal load used in strength calculations of wave bending moment and wave shear force used to vary with each Classification Society. However, the rules were standardized by the IACS (International Association of Classification Societies) in the Unified Rule Requirement in 1989 and were accepted by all Classification Societies. For example, IACS specifies the wave bending moments with the following equations, which are common equations used by the major Classification Societies that belong to IACS:

$$M_w(+) = +0.19 C_1 C_2 L_1^2 B C_b' \qquad (\mathrm{kN}-\mathrm{m}) \tag{2.2.1}$$

$$M_w(-) = -0.11 C_1 C_2 L_1^2 B \left(C_b' + 0.7\right) \qquad (\mathrm{kN}-\mathrm{m}) \tag{2.2.2}$$

where $M_w(+)$: the wave bending moment of hogging
$M_w(-)$: the wave bending moment of sagging
C_1: the parameter determined by ship length.

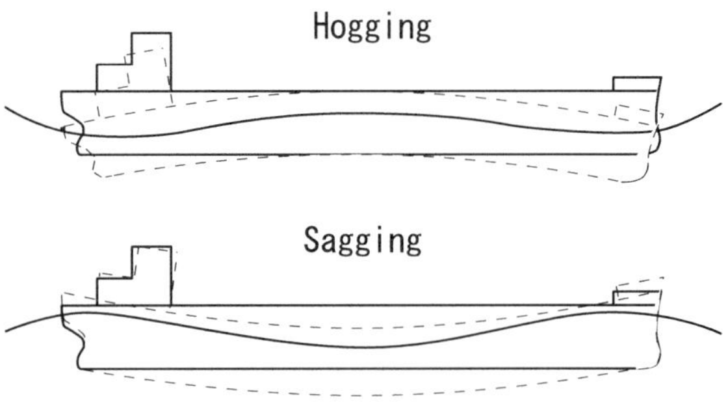

Fig. 2.4 Vertical bending due to waves

Fig. 2.5 Coefficient C_2: Distribution factor

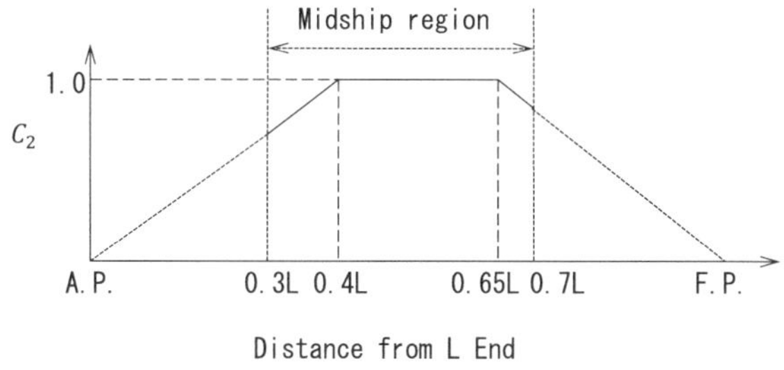

$$C_1 = \begin{cases} 10.75 - \left(\frac{300-L_1}{100}\right)^{1.5} & L_1 \le 300\,\text{m} \\ 10.75 & 300\ \text{m} \le L_1 \le 350\,\text{m} \\ 10.75 - \left(\frac{L_1-350}{100}\right)^{1.5} & 350\ \text{m} \le L_1 \end{cases} \quad (2.2.3)$$

C_2: distribution factor along ship length as specified in Fig. 2.5
L_1: ship length (m)
B: ship breadth(m)
C'_b: block coefficient.

The above dynamic longitudinal loads can be obtained by the long-term prediction method based on ship motion calculations by strip theory. The details will be discussed in Sect. 2.4. The long-term prediction method means a method to predict a response statistically over a rather long period, e.g. the 25 years life cycle of a ship. So, instead of using the Classification Societies' Rules formulae to calculate the dynamic loads in the conventional way, the direct calculation method for practical design purpose has become more popular for its convenience and accuracy. The wave induced bending moment of IACS is determined in such a way that the magnitude of the bending moment is expected to be approximately equal to the maximum value once in twenty-five years, i.e. the expected probability of occurrence is: $Q = 1 \times 10^{-8}$. The IACS formulae were found to be reasonable with the aid of the above-mentioned long-term prediction method carried out by several classification societies, with an appropriate selection of a wave scatter diagram and relevant considerations on ordinary ship operation to avoid harsh weather.

In the case of moderate seas the above prediction method, based on linear strip theory, is reliable enough to estimate the wave loads accurately. However, with rough sea conditions this linear theory cannot be applied, since the calculated results of ship motion and dynamic loads are influenced by the wave impact force and/or by the non-linear effect of the vertical change of ship breadth at each cross section. These impact loads can contribute to the increase of longitudinal loads, for they may generate both an additional bending moment and shear force in the hull girder.

Wave impact loads can be divided into two loads: (a) impact force induced by slamming and (b) impact force of green seas on the deck.

Slamming occurs when the hull hits the water surface hard, when ship movement is severe, particularly in heave and pitching conditions. When the bottom emerges from the wave surface, "bottom slamming" is induced on the ship's bottom plating,

Fig. 2.6 Types of slamming impact of a ship

as shown in Fig. 2.6. In the case of slamming on bow flare, it is called "bow flare slamming". Figure 2.7 shows also a container vessel subjected to wave impact in rough seas at the bow area.

If a ship's bow is pushed into the water in a severe downward pitch, the wave crest may come down on the forecastle deck, which may cause damage to the ship structure and deck machinery. This is called "deck wetness by green water."

Fig. 2.7 Wave impact at sea

2.3 Transverse Strength Load

Transverse strength loads denote the loads which cause distortion of transverse members due to unbalance of external and internal loads, including structural and cargo weights. These loads can be regarded as being independent of longitudinal strength loads, for the longitudinal loads only cause a ship to behave as a beam and they do not cause distortion of the transverse section. These loads are categorized as follows:

- Structural weight, ballast water weight and cargo weight
- Hydrostatic and hydrodynamic loads
- Inertia force of cargo or ballast due to ship motion
- Impact loads.

(1) *Structural weight, ballast water weight and cargo weight*: These loads are dead loads, which mean constant loads that are time independent, induced by gravity at the centers of gravity of the members.
(2) *Hydrostatic and hydrodynamic loads*: The hydrostatic load is the static pressure from the water surrounding a transverse section, which acts on the hull structure as an external load. Another external load is the hydrodynamic load induced by the interaction between waves and the ship motion and subjects the outer shell of the ship to fluctuating water pressure. It is superimposed on the hydrostatic load and creates the total water pressure.

These fluctuating external pressure by waves can be calculated for each specific ship and condition using ship motion and wave load analysis software. On the basis of the calculation results for vast varieties of ships, each classification society gives empirical formulae for the hydrodynamic loads for design purpose. For illustrative purpose, Fig. 2.8 shows external water pressure distribution including hydrostatic and hydrodynamic loads given in old NK rules. In the figure, H_0, H_1, H_2 are the symbols representing the magnitude of fluctuating pressure distribution at the wave trough position below the static draft, at the wave crest position above the static draft and at the bottom base line respectively.

Calculation results of the wave fluctuating pressure using ship motion and wave load analysis software are also available directly to design each specific ship, for example, with the aid of the long-term prediction method which is also the case when calculating the longitudinal strength load which will be described hereafter. For practical purposes this method is accurate enough to be used in structural design.

(3) *Inertia force of cargo or ballast due to ship motion*: The inertia force is induced by the reaction force of self-weight, cargo weight or ballast weight due to the acceleration of the ship motion. Assume that a tanker is rolling among waves in a fully loaded condition, then the cargo oil in the hold has a cyclic movement in the transverse direction. This must result in a fluctuating pressure of the hull structure of the tank due to the inertia force of the cargo oil movement. In addition, internal pressure is introduced not only by rolling but also by the

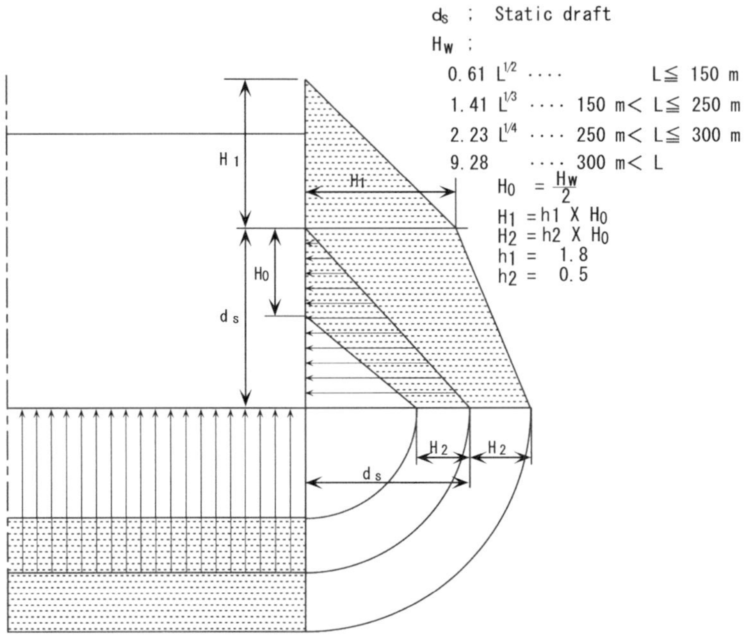

Fig. 2.8 Transverse strength calculation loads by NK

ship's other motions, such as heaving, pitching, etc. A similar phenomenon can be observed when a cargo oil tank is partially filled.

The same long-term prediction method is applicable in the prediction of the internal pressure, since it gives an accurate prediction of the acceleration.

(4) *Impact loads*: There are two impact loads classified as transverse strength loads: slamming and sloshing.

Slamming may be categorized as a transverse strength load, as well as a longitudinal strength load or a local strength load. It means the impact force as the shell plating hits the water surface severely. Therefore, it generates not only a longitudinal load but also a load affecting the transverse strength and local strength simultaneously. Many ships are damaged by slamming, resulting in denting of shell plating, in particular the bottom forward shell plating. Wave impact pressure is an item for which the pure theoretical approach is very difficult, so full-scale measurements and experiments are useful to estimate the impact pressure with reliable accuracy. The upper figure in Fig. 2.9 is an example of measured impact pressure on a bow flare shell plating of a container ship [10], and the lower figure shows one of the test results from a dropping test of an inclined 2-dimensional bow structure model.

Sloshing is a phenomenon where the fluid movement in the tank gets into resonance with the ship motion and creates an impact force between the moving free surface of the fluid and the tank structure. Sloshing is caused by the movement of

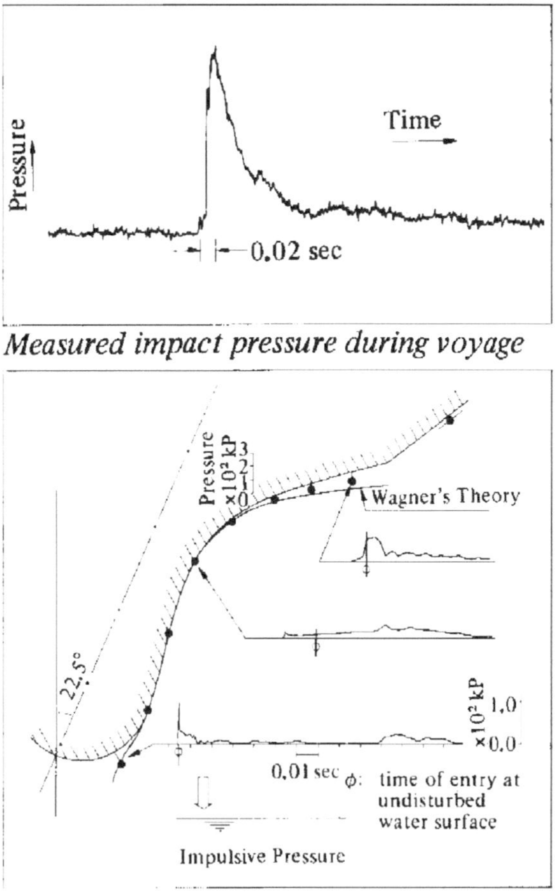

Fig. 2.9 Full-scale measurement and drop test results of bow flare impact pressure

the fluid's free surface, therefore, if the tank is fully filled with fluid, sloshing will never happen since the free movement of the liquid's surface is restricted. When the level of the liquid reaches to a certain portion of the tank, the liquid resonates with the movement of the tank and then sloshing occurs. The natural period of sloshing is determined by the tank dimensions and the level of the liquid as shown in Eq. (2.3.1).

$$T = 2\sqrt{\frac{\pi l}{g} \coth \frac{\pi d}{l}} \tag{2.3.1}$$

where l is the tank length, and d is the liquid depth.

Figure 2.10 shows one of the experimental results of a sloshing test of a rectangular tank model and Fig. 2.11 indicates the result of a computer simulation program utilizing the finite difference technique.

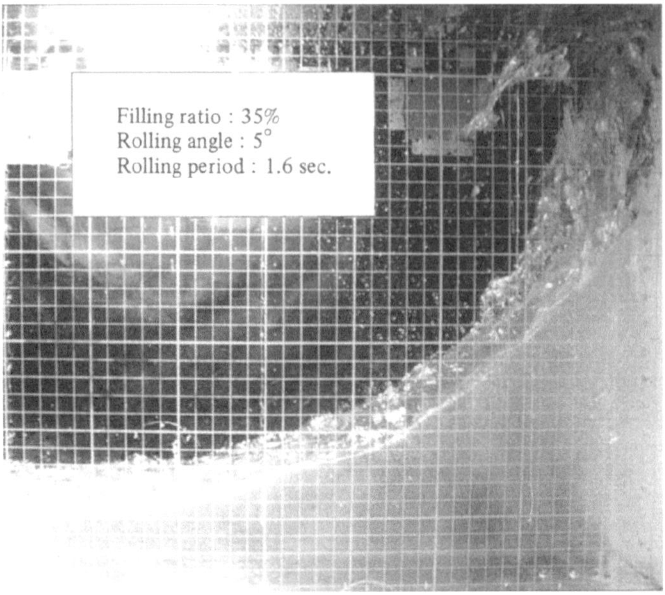

Fig. 2.10 Sloshing test of a rectangular tank

2.4 Ship Response Calculation in Waves

2.4.1 Introduction

In order to obtain an accurate estimation of wave induced loads, the designer has to know an exact wave load calculation method. In this article the evaluation procedure of wave loads is briefly introduced; for a more detailed discussion the reader is referred to references [11–13]. Varieties of numerical methods are available for ship motion and wave load calculation such as strip method, panel method, CFD and particle method, but we focus on the strip method here, which is theoretically simple but gives quite accurate and robust solutions. The calculated result can be used for the practical design of conventional vessels. The ordinary evaluation procedure follows the calculation steps as below:

- Calculation of ship responses in regular waves by the strip method
- Short-term prediction method for ship responses in irregular waves
- Long-term prediction method for ship responses in irregular waves

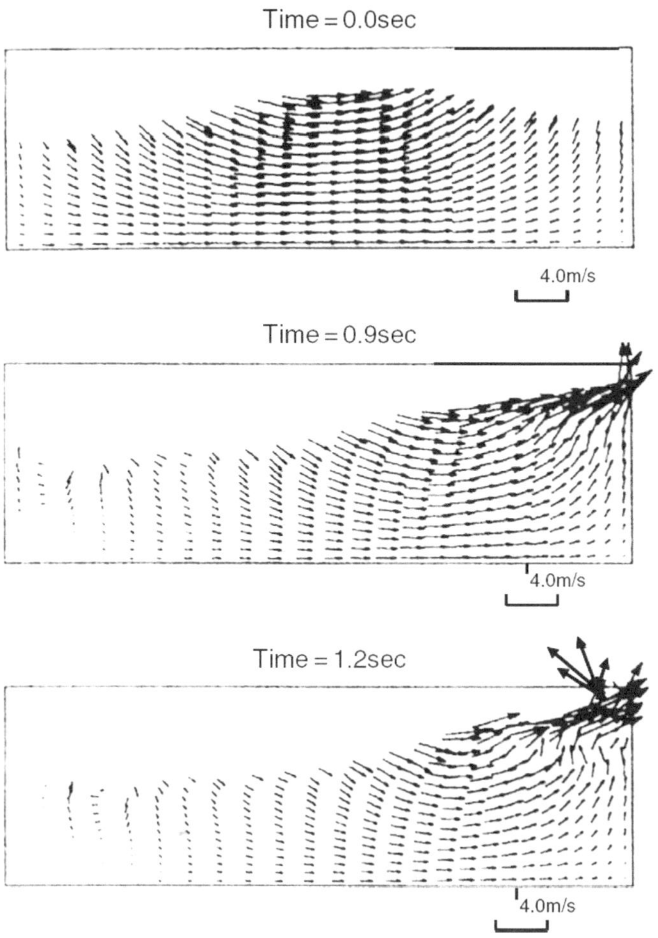

Fig. 2.11 Computer simulation of liquid sloshing in a tank

2.4.2 Strip Method

A ship has a long and slender hull shape in comparison with its breadth and depth. Taking advantage of the above assumption, fluid motion around hull surface due to ship motion can be regarded as if it moves in cross sectional plane. In other words, the fluid force can be obtained by integrating the force acting on transverse strips delivered as shown in Fig. 2.12; consequently the interaction force between each strip is neglected. The strip method is utilized to calculate ship motion, wave induced longitudinal strength loads and wave pressure distribution around the hull in regular waves. An example of the calculation results is presented in Fig. 2.13 and it shows the response functions due to a vertical bending moment at midship of double

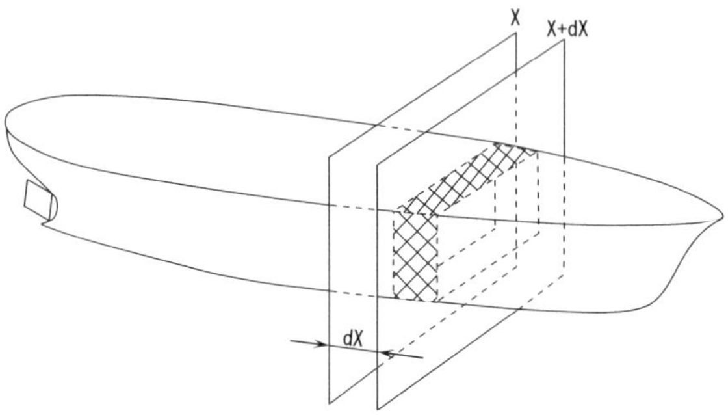

Fig. 2.12 Definition of ship's strip

hull tanker of 120,000 DWT (ship length = approximately 240 m) in regular waves [14]. According to the figure, it can be seen that the maximum bending moment occurs in the head sea condition (encounter angle of 180°) followed by the following sea condition (encounter angle of 0°), where the wave length is almost same as the ship length, i.e. wave length/ship length = 1.0.

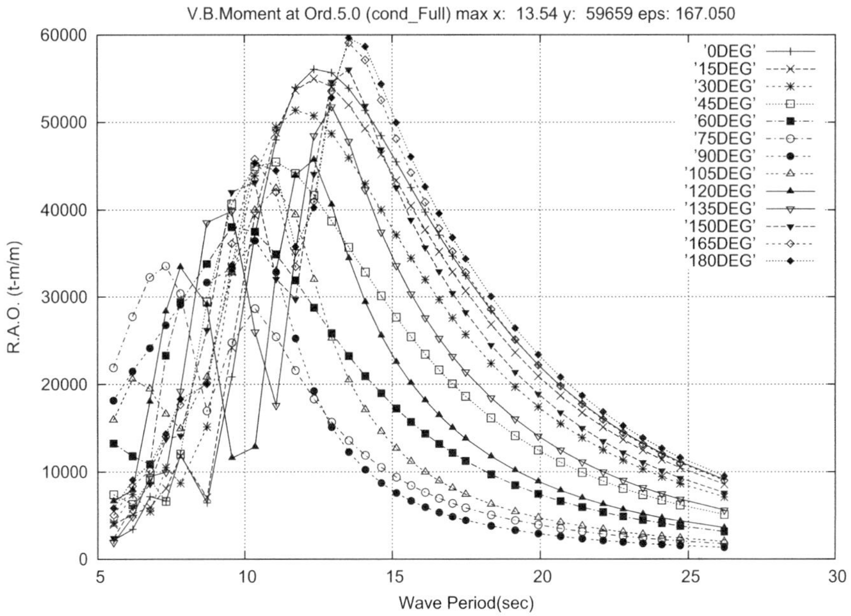

Fig. 2.13 Response fuction of vertical wave bending moment (150,000 DWT tanker)

2.4.3 Short-Term Prediction

A short-term prediction method is useful to evaluate the ship's response in irregular waves by the statistical prediction theory. Short-term means a short period, say, from 30 min to a few hours, where the significant wave height and the average wave period are considered to be stationary. This prediction method is used for calculating the ship's response over a relatively short period of time in terms of ship accelerations, relative wave height, deck wetness, slamming, etc.

Although the waves at sea are not regular waves, these waves can be dealt with by a combination of regular waves having various frequencies and various heights. If the relationship between wave height and frequency of irregular waves is given in spectrum form and the response function of the ship in regular waves is known, then the spectrum of the ship response against irregular waves can be predicted with the following equation:

$$S'(\omega, \gamma) = S'_w(\omega, \gamma) \cdot |A(\omega, \mu)|^2 \tag{2.4.1}$$

where

$S'(\omega, \gamma)$: power spectra of ship response,

$S'_w(\omega, \gamma)$: power spectra of component wave in the γ-direction,

$A(\omega, \mu)$: ship's response function in regular wave to component wave in the μ-direction,

δ: angle between ship's course direction and average wave direction of irregular waves as defined in Fig. 2.14,

γ: angle between average wave direction and component wave direction of irregular waves,

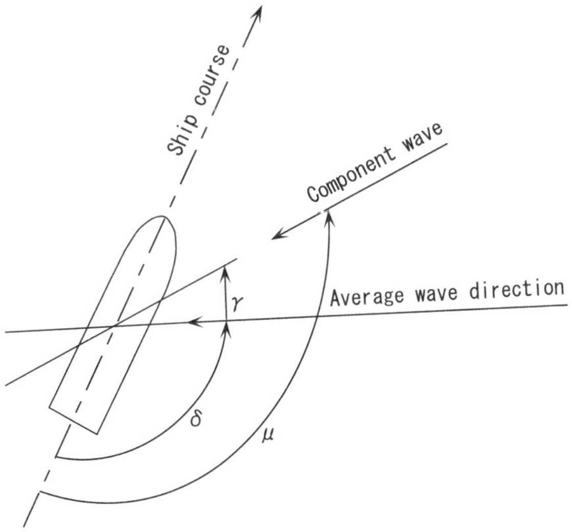

Fig. 2.14 Definition of ship's course snd wave incident direction

$\mu = \delta + \gamma$: angle between ship's course direction and component wave direction of irregular waves,

ω: circular frequency.

The standard deviation R of the response of a ship navigating with the relative direction of δ to the average wave direction can be obtained by integrating Eq. (2.4.1) for all frequencies ω and for all heading angles γ as in the form:

$$R(\delta)^2 = \int_{-\pi}^{\pi} \int_{0}^{\infty} S'_w(\omega, \gamma) \cdot |A(\omega, \delta + \gamma)|^2 d\omega \cdot d\gamma \tag{2.4.2}$$

Although several types of spectrum functions $S'_w(\omega)$ of irregular waves are proposed, the Modified Pierson-Moskowitz wave spectrum, which was proposed in the second ISSC (International Ship Structure Committee) held in 1964, is usually applied for practical design purposes to express the long crested wave. These are determined by only two parameters, the average wave period and the significant wave height. The ISSC's wave spectrum is in the following form:

$$\frac{S'_w(\omega)}{H^2} = 0.11\omega_1^{-1}\left(\frac{\omega}{\omega_1}\right)^{-5} \exp\left[-0.44\left(\frac{\omega}{\omega_1}\right)^{-4}\right] \tag{2.4.3}$$

where

H: significant wave height

$$\omega_1 = 2\pi/T$$

T: average wave period.

This spectrum looks as illustrated in Fig. 2.15. When the effect of short crested waves must be taken into account, we must assume a directional distribution of the density of wave power. Directional wave spectra $S'_w(\omega, \gamma)$ are expressed as $S'_w(\omega, \gamma) = S'_w(\omega)D(\gamma)$, where $D(\gamma)$ is the directional distribution function, and should be equal to unity when integrated all around.

$$\int_{-\pi}^{\pi} D(\gamma) d\gamma = 1 \tag{2.4.4}$$

One of representative directional distribution functions are the cos-2n distribution as shown in Eq. (2.4.5).

$$D(\gamma) = \frac{1}{\pi}\frac{(2n)!!}{(2n-1)!!}\cos^{2n}\gamma, \ |\gamma| \le \frac{\pi}{2} \tag{2.4.5}$$

When $n = 1, n = 1.5$ or $n = 2$, this equation reduces to

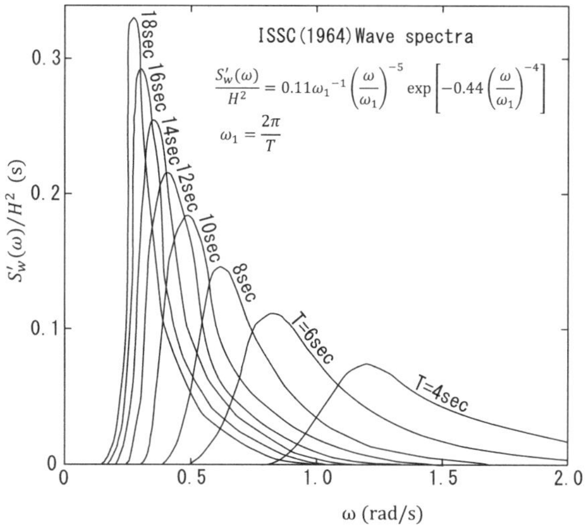

Fig. 2.15 Modified Pierson-Moskowitz wave spectra (ISSC spectra)

$$D(\gamma) = \begin{cases} \frac{2}{\pi}\cos^2\gamma & (n=1) \\ \frac{3}{4}\cos^3\gamma & (n=1.5) \\ \frac{8}{3\pi}\cos^4\gamma & (n=2) \end{cases} \tag{2.4.6}$$

On the other hand, the following Mitsuyasu-Goda's equation [15] contains spreading parameter of real number. $s = 5$ gives the closest distribution to the cos-2 distribution ($n = 1$) above.

$$D(\gamma) = \frac{2^{2s-1}}{\pi}\frac{\Gamma^2(s+1)}{\Gamma(2s+1)}\cos^{2s}\left(\frac{\gamma}{2}\right),\ |\gamma| \le \pi \tag{2.4.7}$$

Once the Standard Deviation R of the ship's response is obtained, a short-term prediction can be made by assuming that the probability distribution of the extreme response will follow Rayleigh's Distribution, that is:

$$\begin{aligned} \text{Average} &= 1.25R \\ \text{Average of } 1/3 \text{ max.} &= 2.00R \\ \text{Average of } 1/10 \text{ max.} &= 2.55R \\ \text{Expect of } 1/100 \text{ max.} &= 3.22R \\ \text{Expect of } 1/1000 \text{ max.} &= 3.87R \end{aligned} \tag{2.4.8}$$

The average value of the 1/3 largest response peak amplitudes ($2.00R$) is often referred to as the significant value, and is used as a representative magnitude of the irregular response.

2.4.4 Long-Term Prediction

The long-term prediction is a method to predict statistically the ship's response in irregular waves over a rather long period of time, such as 25 years of the life cycle of a ship. This method can be applied to cases such as the prediction of the maximum bending moment in 25 years, or of the fatigue damage life, etc. The prediction procedure is explained in Fig. 2.16. Once the Standard Deviation $R(\delta)$ of the ship response is obtained by the short-term prediction, it is possible to calculate $Q(r_1, \delta)$, probability of the response exceeding r_1, by providing the ship's wave data over the specified life time in the following equation:

$$Q(r_1, \delta) = \int_0^\infty \int_0^\infty \exp\left\{ -\frac{r_1^2}{2R^2(H_w, T_w, \delta)} \right\} \cdot p(H_w, T_w) dH_w \cdot dT_w \tag{2.4.9}$$

where

H_w: significant wave height

T_w: average wave period

$p(H_w, T_w)$: probability of wave occurrence in the specified sea area.

It is therefore possible to obtain the probability of the occurrence that an extreme response may exceed the specified value r_1:

$$Q(r_1) = \int_0^{2\pi} Q(r_1, \delta) \cdot p^*(\delta) d\delta \tag{2.4.10}$$

where

$Q(r_1)$: probability of occurrence

$p^*(\delta)$: probability density function of the ship's course direction.

Figure 2.17 shows an example of a long-term prediction of a midship vertical bending moment of a 120,000 DWT double hull tanker sailing in the North Atlantic.

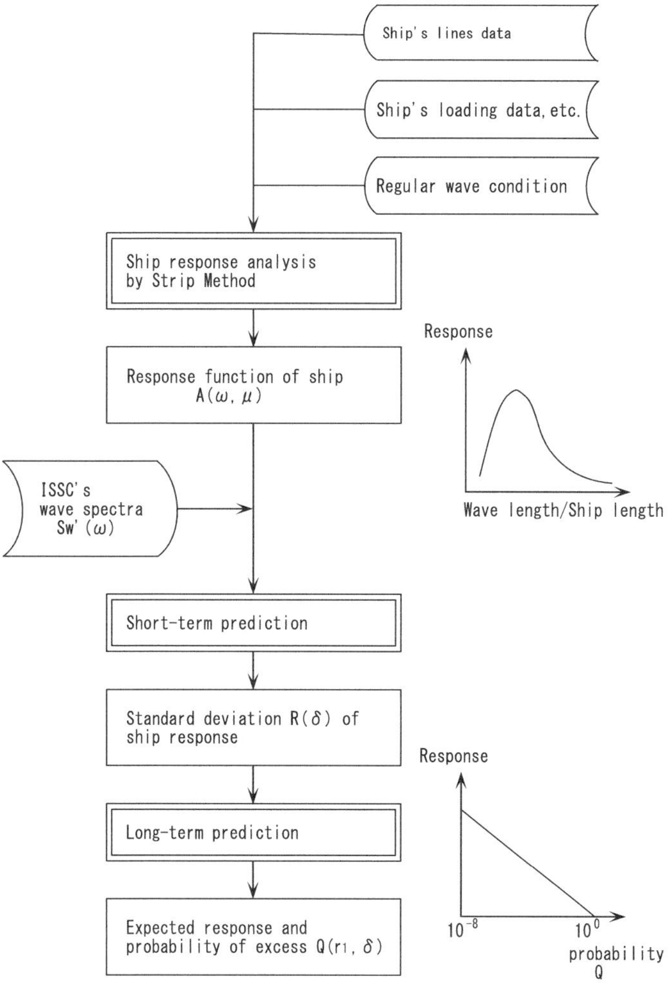

Fig. 2.16 Procedure of long-term prediction

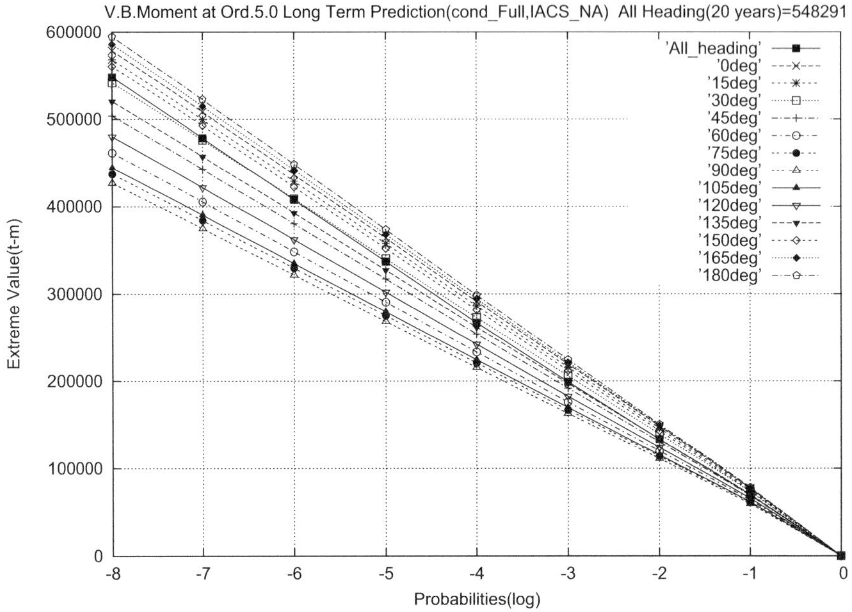

Fig. 2.17 Long-term prediction of vertical wave bending moment (150,000 DWT tanker)

Chapter 3
Strength Evaluation

3.1 General

3.1.1 Introduction

If structural damage to a ship occurs, it means that the load acting on the structure must have exceeded the maximum limit of structural strength. The hull structure has been planned originally by the designer in such a way that the structure can effectively resist the presumed maximum load, which is estimated from previous experience of failure modes of damage. Therefore, once damage occurs in the structure, it indicates the following facts:

- The magnitude of the load causing the failure is greater than that expected. and/or
- The actual failure takes place in a different mode from the presumed one.

By using that damage experience for planning the next structure, the designer is able to improve the accuracy of the load estimation and to accumulate knowledge relating to various modes of failure, and consequently the designer can achieve a design where there would be no future damage to the structure.

When the designer starts a hull structure design, the very first step is to establish the magnitude, direction, probability of occurrence of the load, and also the structural failure modes experienced in the past.

The behavior of the load was discussed in Chap. 2, and in this chapter several failure modes, such as yielding, buckling, fatigue and vibration are explained.

Y. Okumoto et al., *Design of Ship Hull Structures*,
https://doi.org/10.1007/978-3-032-12500-2_3

3.1.2 Procedure of Structural Strength Evaluation

We have already explained the summary of structural design procedure in Sect. 1. 2, but in order to consider particularly the evaluation of structural strength, we will repeat again the following design procedure:

(1) *Determine an initial system of structural members*
(2) *Presume a magnitude, direction, and probability of load*
(3) *Assume a failure modeof structure due to the load: Various kinds of loads act on a hull structure during a voyage and these may cause damage to the structure when the load reaches a certain critical level.*

Although there may be several failure modes, the following modes are significant for structural designers:

- yielding
- buckling
- fatigue.

Yielding: Now let us assume that a tensile load is gradually applied to a structure, then some elongation might be induced and be proportional to the load increment as long as the load is small. Once the load exceeds a certain critical value, then elongation would increase rapidly. That failure mode is called yielding. The designer usually takes care to maintain the strength of the structure so as not to exceed the yield point.

Buckling: In the case of a structure under compression load, the structure may suddenly be deflected when the load reaches a critical value. Such a failure mode is called buckling. Once a large deflection takes place, the structure may not recover its original shape even when the load is removed.

Fatigue: The structure may be fractured by small loads when the loads are provided repeatedly to the structure. That failure mode is called fatigue. Fatigue is very dangerous, because it may result even from substantially lower loads than yielding strength, especially where the number of cycles is very large. That type of fracture is sometimes caused by vibration, because its frequency is very high.

(4) *Select an appropriate analysis method*: Since a hull structure is mainly composed of several types of stiffened plate, the load on the hull is at first transmitted from panels to stiffeners, then from stiffeners to primary member such as transverse rings or girders, and finally from primary members to shell plating.

For instance, Fig. 3.1 illustrates the structure of the bottom of a cargo tank, which is composed of shell plating sustained by longitudinal stiffeners and bottom transverses. The working load may be induced by the difference of external pressure p_1 and internal pressure p_2, therefore, a panel is subjected to the lateral load of $(p_1 - p_2)sl$ where s, l are the spacing and span of longitudinal stiffeners, respectively. The panel is supported mainly by the adjacent two longitudinal stiffeners and the lateral load of $(p_1 - p_2)sl$ indicated by the hatched area of plate, is transmitted from the panel to each stiffener. Each stiffener is connected to transverse webs, then finally the lateral load of $(p_1 - p_2)sl$ is transmitted to these transverse webs.

As shown in the above example, we can deal with the strength of structural members individually according to their components. Thus we often investigate the

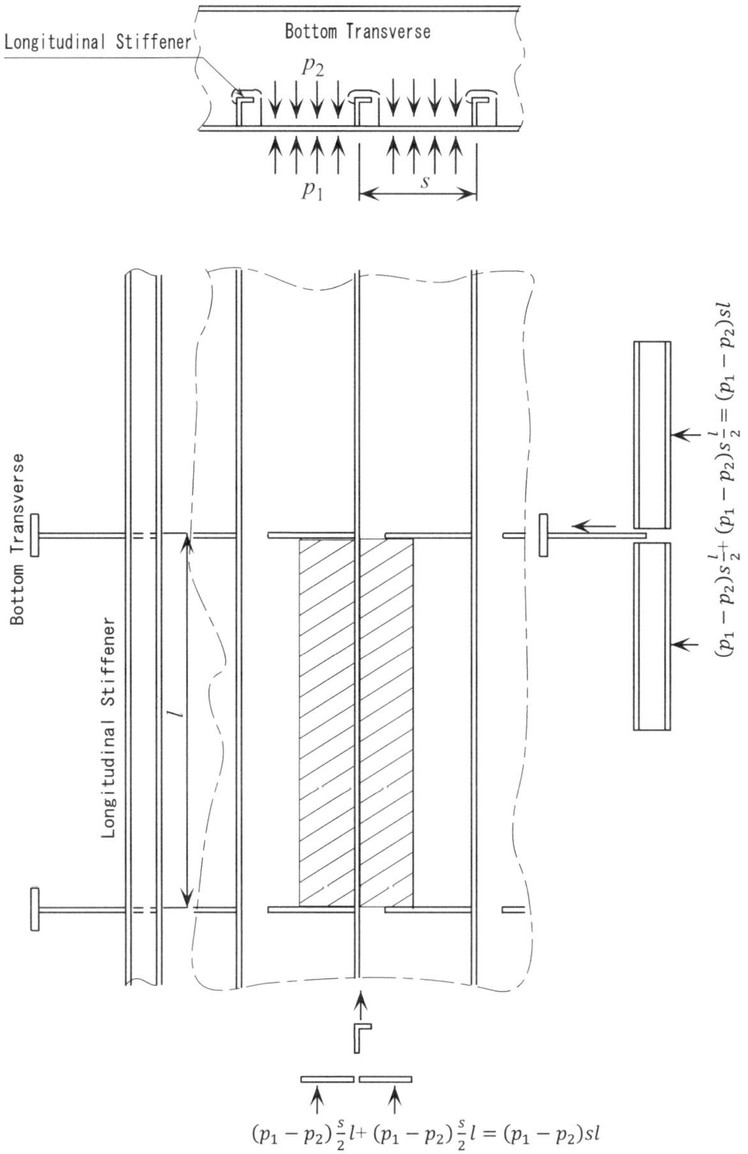

Fig. 3.1 Bottom structure under inner&outer pressure

strength of a panel by thin shell theory, and as for stiffeners and primary members, simple beam theory is usually applied. If we are obliged to study the stress concentration problem in association with fatigue strength, the above analysis method is no longer helpful, therefore, more sophisticated analysis such as finite element analysis must be adopted.

Anyhow, it is important to select an adequate analysis method to be able to evaluate the strength efficiently.

(5) *Choice of an acceptable strength criteria for a particular failure mode*: In starting a structural design, the designer is required to establish acceptable strength criteria of each member depending on the assumed failure mode. Although these strength criteria conform to the rules requirement for Classification Society, they are not always explicitly expressed in the rule formulae. Thus the designer should bear in mind the background of the rule on the basis of strength criteria which exist behind the rule formulae. Strength criteria are usually established in the following steps and they are included in the rule formulae in the form of safety factors:

Step 1: Long-term statistic analysis of waves is carried out by analyzing the waves in the whole life-cycle of the ship and the maximum load is settled as design load.

Step 2: The response of the structure against the design load is calculated to obtain the maximum response.

Step 3: Safety factors are established by considering the above maximum stress or force by considering the critical strength for each failure mode.

As mentioned in the above steps, the strength criteria are strongly associated with the load, so they cannot be dealt with separately. In other words, acceptable strength criteria are not always unchangeable i.e. they may vary where the prediction method of load is altered, or where the analysis method is improved.

(6) *Analysis of structural response*: In order to maintain the strength performance at a satisfactory level in the design of a ship structure, it is necessary to analyze the behavior of the structure under the subjected loads and to recognize the stress distribution over the structure. Therefore, we have to undertake some kind of experiments to simulate the behavior and the stress level. We used to carry out model tests, where a structural model of small size is utilized and the stress on the structural members is measured with the aid of strain gauges or photoelastic Moire experiment.

Nowadays, however, in parallel with the rapid progress of computers, we often execute numerical simulation successfully for the purpose of strength analysis by computer instead of the above mentioned structural model test.

(7) *Evaluation of the response for given criteria*: The calculated result is compared with the chosen strength criteria and the strength of the structure is evaluated. If it is acceptable, the design is completed but if not acceptable, the design must be revised and the calculation must be done again.

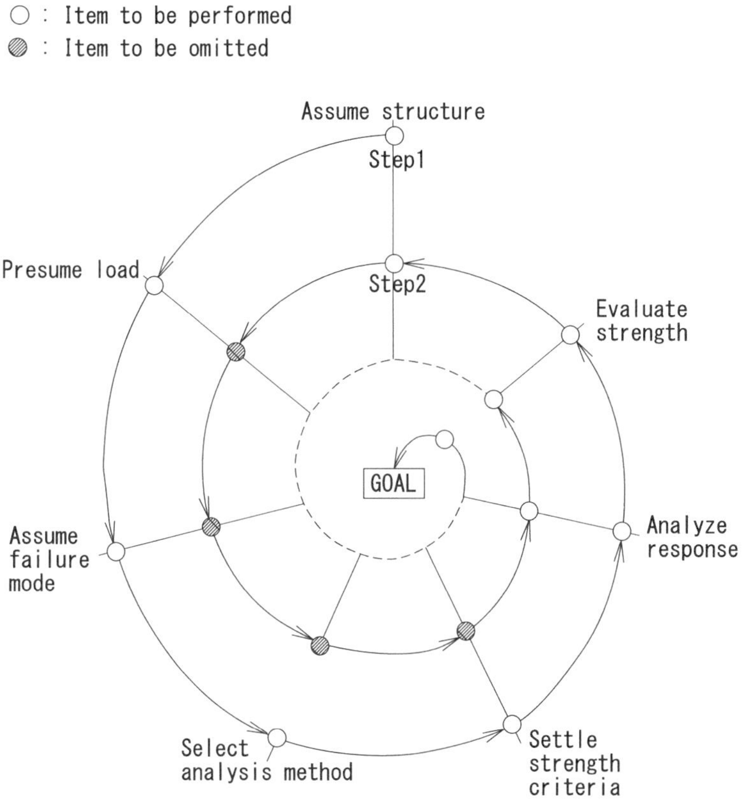

Fig. 3.2 Procedure of structural analysis and strength evaluation

By following the above design steps, the designer ensures the integrity of the structure and proceeds along a so-called "Design Spiral" as shown in Fig. 3.2 until the strength of the structure complies with the required strength. In the figure, a ○ mark shows that the process must be carried out but ● mark indicates the process is to be omitted usually in the second step or beyond.

3.2 Stress and Strain

3.2.1 Stress Pattern

Stresses acting on an element of a structure are thought to be a combination of stress patterns as shown in Fig. 3.3a–e [16]:

(a) simple stress condition

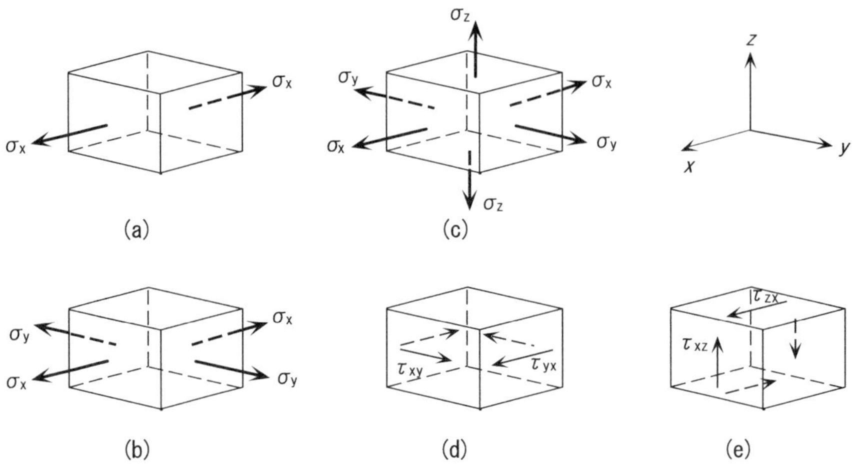

Fig. 3.3 Stress pattern

(b) biaxial stress condition (x-y plane)
(c) triaxial stress condition
(d) simple shearing stress condition (x-y plane)
(e) simple shearing stress condition (x-z plane).

In this coordinate system, when the stress in the z-direction is zero and the deformation in the z-direction is free, it is called a plane stress condition, and when the deformation in the z-direction is zero and the stress in the z-direction is acting, it is called a plane strain condition. Since a hull is mainly composed of thin plates, most of them are treated as being in the plane stress condition.

In the case where shearing stresses do not act - normal stress only acts - as shown in Fig. 3.3 (a)-(c), the normal stress at each condition is called the principal stress, and the corresponding plane is the principal plane, while the corresponding stress axis is the principal axis.

3.2.2 Biaxial Stress Condition [16]

Figure 3.4 shows a biaxial stress condition in the x–y plane, same as Fig. 3.3b; σ_x acts in the x-direction and σ_y acts in the y-direction. Considering a section m–n, where its normal is at φ degrees to the x-axis, the stress along this normal and the shearing stress along m–n line are given by the following equation, in case that only σ_x acts.

$$
\begin{aligned}
\sigma &= \sigma_x \cos^2 \varphi \\
\tau &= -(\sigma_x/2) \sin 2\varphi
\end{aligned}
\tag{3.2.1}
$$

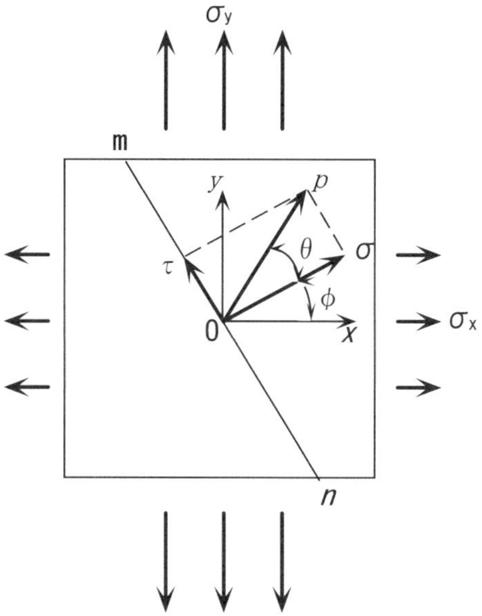

Fig. 3.4 Biaxial stress condition

where τ is defined as positive in the counter clockwise direction.

In the case that σ_y acts:

$$\begin{aligned} \sigma &= \sigma_y \sin^2 \varphi \\ \tau &= \left(\sigma_y/2\right) \sin 2\varphi \end{aligned} \tag{3.2.2}$$

Therefore, if both σ_x and σ_y act,

$$\begin{aligned} \sigma &= \sigma_x \cos^2 \varphi + \sigma_y \sin^2 \varphi \\ &= \frac{1}{2}\left(\sigma_x + \sigma_y\right) + \frac{1}{2}\left(\sigma_x - \sigma_y\right) \cos 2\varphi \end{aligned} \tag{3.2.3}$$

$$\tau = \frac{1}{2}\left(\sigma_y - \sigma_x\right) \sin 2\varphi \tag{3.2.4}$$

If $\sigma_x = \sigma_y$, τ is zero irrespective of the angle φ.

3.2.3 Combination of Normal Stress and Shearing Stress [16]

Figure 3.5 shows stress distribution acting on a small element $dx \times dy$ (thickness = 1) in the case of a combination of normal stress and shearing stress. Considering an equilibrium in triangle m–k-n along the x-axis:

Fig. 3.5 Biaxial and shearing stress

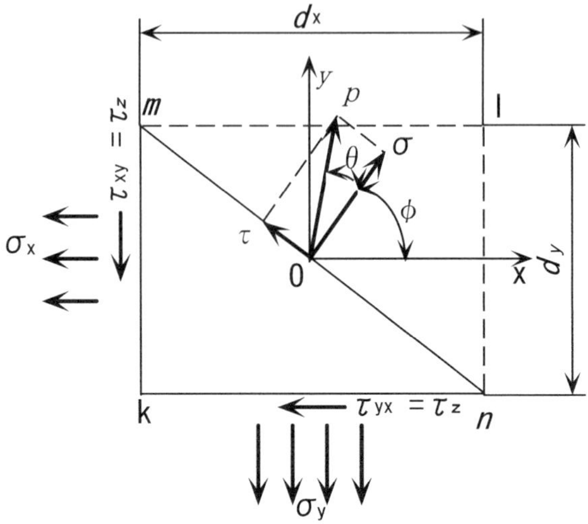

$$\sigma ds \cos\varphi - \tau ds \sin\varphi - \sigma_x dy - \tau_z dx = 0$$

where ds is the distance of m–n.

Being divided by ds,

$$\sigma \cos\varphi - \tau \sin\varphi - \sigma_x \cos\varphi - \tau_z \sin\varphi = 0 \tag{3.2.5}$$

For the y-axis

$$\sigma ds \sin\varphi + \tau ds \cos\varphi - \tau_z dy - \sigma_y dx = 0$$

Being divided by ds,

$$\sigma \sin\varphi + \tau \cos\varphi - \tau_z \cos\varphi - \sigma_y \sin\varphi = 0 \tag{3.2.6}$$

From (3.2.5) $\times \cos\varphi +$ (3.2.6) $\times \sin\varphi$,

$$\begin{aligned}\sigma &= \sigma_x \cos^2\varphi + \sigma_y \sin^2\varphi + \tau_z \sin 2\varphi \\ &= \frac{1}{2}(\sigma_x + \sigma_y) + \frac{1}{2}(\sigma_x - \sigma_y)\cos 2\varphi + \tau_z \sin 2\varphi\end{aligned} \tag{3.2.7}$$

From (3.2.6) $\times \cos\varphi -$ (3.2.5) $\times \sin\varphi$,

$$\tau = \frac{1}{2}(\sigma_y - \sigma_x)\sin 2\varphi + \tau_z \cos 2\varphi \tag{3.2.8}$$

The combined stress and angle to the normal line are

$$p = \sqrt{\sigma^2 + \tau^2} \tag{3.2.9}$$

$$\tan \theta = \tau / \sigma \tag{3.2.10}$$

3.2.4 Principal Stress and Principal Shearing Stress [16]

Principal stress and principal shearing stress are used for evaluation of hull strength. Principal stress is the normal stress to a plane, on which the stress becomes maximum or minimum. It can be derived by applying $d\sigma/d\varphi = 0$ at Eq. (3.2.7).

$$\left(\sigma_x - \sigma_y\right) \sin 2\varphi + 2\tau_z \cos 2\varphi = 0 \tag{3.2.11}$$

We represent φ_n for φ obtained from Eq. (3.2.11)

$$\therefore \tan 2\varphi_n = \frac{2\tau_z}{\sigma_x - \sigma_y} \tag{3.2.12}$$

If we substitute φ_n^m which satisfies Eq. (3.2.12), φ_n^m is

$$\varphi_n^m = \frac{1}{2} \tan^{-1} \frac{2\tau_z}{\sigma_x - \sigma_y} \pm \frac{\pi}{2} m$$

$$m = 1, 2, 3, \cdots$$

Hence the number of φ_n satisfying above equation is infinite, however a pair of φ_n^m and φ_n^{m+1} is always orthogonal. This means that there are two principal stresses in 2π and both (maximum principal stress and minimum principal stress) appear orthogonally. The shearing stress τ is zero on the principal plane from Eqs. (3.2.11) and (3.2.8). The principal stress can be obtained from Eqs. (3.2.12) and (3.2.7):

$$\cos 2\varphi_n = \frac{1}{\pm\sqrt{1 + \tan^2 2\varphi_n}} = \frac{\sigma_y - \sigma_x}{\pm\sqrt{\left(\sigma_x - \sigma_y\right)^2 + 4\tau_z^2}}$$

$$\sin 2\varphi_n = \frac{\tan 2\varphi_n}{\pm\sqrt{1 + \tan^2 2\varphi_n}} = \frac{2\tau_z}{\pm\sqrt{\left(\sigma_x - \sigma_y\right)^2 + 4\tau_z^2}}$$

Representing maximum principal stress as σ_1, and minimum principal stress as σ_2,

$$\sigma_1 = \frac{1}{2}\left(\sigma_x + \sigma_y\right) + \frac{1}{2}\sqrt{\left(\sigma_x - \sigma_y\right)^2 + 4\tau_z^2}$$

$$\sigma_2 = \frac{1}{2}(\sigma_x + \sigma_y) - \frac{1}{2}\sqrt{(\sigma_x - \sigma_y)^2 + 4\tau_z^2} \tag{3.2.13}$$

On the other hand, if we consider shearing stress, from $d\tau/d\varphi = 0$ in Eq. (3.2.8)

$$(\sigma_y - \sigma_x)\cos 2\varphi - 2\tau_z \sin 2\varphi = 0$$

$$\therefore \tan 2\varphi_t = \frac{\sigma_y - \sigma_x}{2\tau_z} \tag{3.2.14}$$

Since $\varphi_t \pm \pi/2$ also satisfies Eq. (3.2.14), the maximum shearing stress and minimum shearing stresses are also orthogonal with each other.

From

$$\tan 2\varphi_n \tan 2\varphi_t = \frac{2\tau_z}{\sigma_x - \sigma_y} \cdot \frac{\sigma_y - \sigma_x}{2\tau_z} = -1 \tag{3.2.15}$$

$\varphi_t = \varphi_n \pm \pi/4$ is obtained, i.e. the plane, at which maximum or minimum shearing stress acts, is at 45° to the principal plane. If we get $\cos 2\varphi_t$ and $\sin 2\varphi_t$ from Eq. (3.2.14), and substitute it into Eq. (3.2.8),

$$\tau_1, \tau_2 = \pm\frac{1}{2}\sqrt{(\sigma_x - \sigma_y)^2 + 4\tau_z^2} = \pm\frac{1}{2}(\sigma_1 - \sigma_2) \tag{3.2.16}$$

These values are called principal shearing stress.

3.2.5 Equivalent Stress

We denote principal stresses by $\sigma_1, \sigma_2, \sigma_3$ and principal shearing stresses by τ_1, τ_2, τ_3. Strain energy per unit volume is

$$\overline{U} = \frac{1}{2}(\sigma_1\varepsilon_1 + \sigma_2\varepsilon_2 + \sigma_3\varepsilon_3) \tag{3.2.17}$$

Eliminating $\varepsilon_1, \varepsilon_2, \varepsilon_3$ in Eq. (3.2.17) applying the relationship between stress and strain,

$$\begin{aligned}\overline{U} &= \frac{1}{2E}[\sigma_1\{\sigma_1 - \nu(\sigma_2 + \sigma_3)\} + \sigma_2\{\sigma_2 - \nu(\sigma_3 + \sigma_1)\} + \sigma_3\{\sigma_3 - \nu(\sigma_1 + \sigma_2)\}] \\ &= \frac{1 - 2\nu}{6E}(\sigma_1 + \sigma_2 + \sigma_3)^2 + \frac{1 + \nu}{6E}\left\{(\sigma_1 - \sigma_2)^2 + (\sigma_2 - \sigma_3)^2 + (\sigma_3 - \sigma_1)^2\right\} \\ &\equiv \overline{U}_1 + \overline{U}_2\end{aligned} \tag{3.2.18}$$

where

$$\overline{U}_1 = \frac{1-2\nu}{6E}(\sigma_1 + \sigma_2 + \sigma_3)^2 \tag{3.2.19}$$

$$\overline{U}_2 = \frac{1+\nu}{6E}\left\{(\sigma_1 - \sigma_2)^2 + (\sigma_2 - \sigma_3)^2 + (\sigma_3 - \sigma_1)^2\right\} \tag{3.2.20}$$

E: modulus of elasticity.

ν: Poisson's ratio.

$\overline{U}_1$ is dilatational strain energy by uniform tensile stress σ_m;

$$\sigma_m = \frac{1}{3}(\sigma_1 + \sigma_2 + \sigma_3) \tag{3.2.21}$$

and $\overline{U}_2$ is shearing strain energy without volume change.

In the case of simple tension, $\sigma_1 = \sigma_e$, $\sigma_2 = \sigma_3 = 0$, shearing strain energy is

$$\overline{U}_{2e} = \frac{1+\nu}{6E}\left(2\sigma_e^2\right) \tag{3.2.22}$$

In accordance with Mises yield stress criterion, the failure will occur when the shearing strain energy reaches that corresponding to the yield stress under simple tension, that is $\overline{U}_2 = \overline{U}_{2e}$. From Eqs. (3.2.20) and (3.2.22),

$$\sigma_e^2 = \frac{1}{2}\{(\sigma_1 - \sigma_2)^2 + (\sigma_2 - \sigma_3)^2 + (\sigma_3 - \sigma_1)^2\} \tag{3.2.23}$$

For arbitrary stresses $\sigma_1, \sigma_2, \sigma_3$, if taking $\sigma_0 = \sigma_e$

$$\sigma_0 = \sqrt{\frac{1}{2}\{(\sigma_1 - \sigma_2)^2 + (\sigma_2 - \sigma_3)^2 + (\sigma_3 - \sigma_1)^2\}} \tag{3.2.24}$$

σ_0 is termed equivalent stress or von Mises stress. In case the plane stress condition, $\sigma_3 = 0$.

$$\sigma_0 = \sqrt{\sigma_1^2 + \sigma_2^2 - \sigma_1\sigma_2} \tag{3.2.25}$$

$$\text{or } \sigma_0 = \sqrt{\sigma_x^2 + \sigma_y^2 - \sigma_x\sigma_y + 3\tau_{xy}^2} \tag{3.2.26}$$

In the case of simple tension and shear combination,

$$\sigma_0 = \sqrt{\sigma_x^2 + 3\tau_{xy}^2} \tag{3.2.27}$$

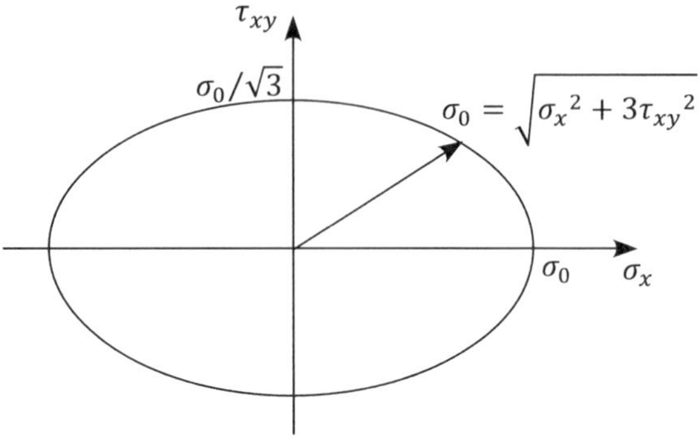

Fig. 3.6 Equivalent stress

Figure 3.6 shows the relation, where the elliptical curve represents the equivalent stress; if only the normal stress acts the failure will occur at $\sigma_x = \sigma_Y$ (yield stress), and if only the shearing stress acts the failure occurs at $\tau_{xy} = \sigma_Y/\sqrt{3}$.

3.2.6 Evaluation of Stress Calculated by FEM

In general, the FEM calculation produces several kinds of stresses. In the case of a 2-dimensional shell model, the following stresses are calculated:

- Stresses at nodes or at the center of an element
- Stresses on upper surfaces, lower surfaces, or in the middle
- Stress components (normal stress and shearing stress) in global or local (element) coordinate
- Maximum principal stress, minimum principal stress, or equivalent stress

Structure designers should recognize what kind of stress is output as a result of the FEM calculation, and should select them in accordance with each strength evaluation.

In general,

(1) Principal stress is used as failure criterion for brittle materials, and equivalent stress for ductile materials such as mild steel.
(2) Maximum principal stress is used for crack propagation analysis (Fig. 3.7).
(3) The membrane stress (mean stress between upper and lower surfaces) is used for plate buckling analysis, which is adjusted for the boundary stress of a panel. Figure 3.8 shows the procedure of panel buckling analysis of a bottom transverse web in a cargo tank [17].

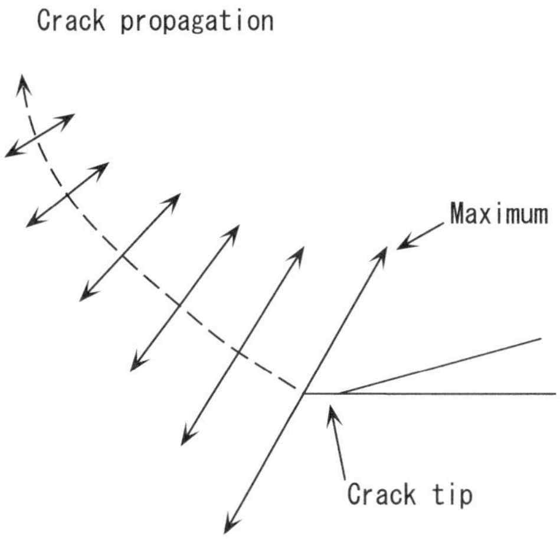

Fig. 3.7 Crack propagation

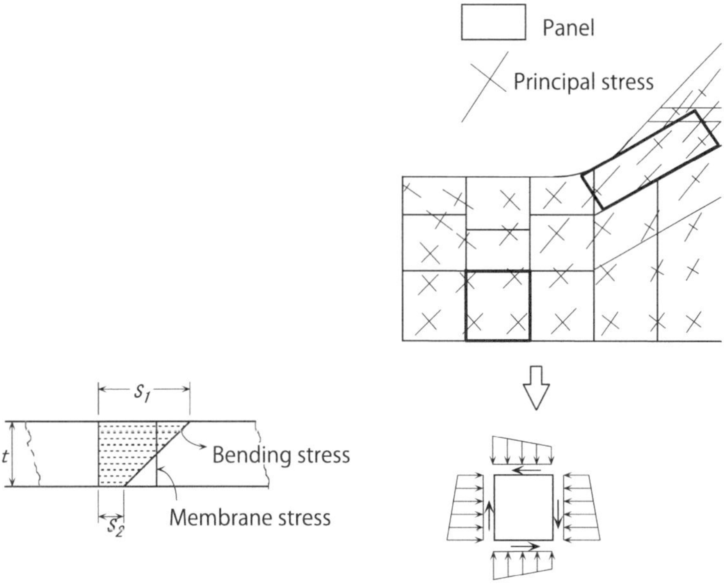

Fig. 3.8 Buckling check of a panel

3.3 Evaluation of Stress

3.3.1 Criteria of Failure

Ductile materials, such as mild steel and higher-strength steel used in ship hull structure, deform when the load exceeds the yield stress. Under such plastic deformation, materials reach finally failure. On the other hand, the failure of brittle materials such as cast iron may happen without such plastic deformation. Hence the strength of materials is based on the yield stress or the tensile stress for ductile materials, and the fracture stress for brittle materials.

The stress distribution in an actual structure is very complex because both the loading and the structure are complex. Although the strength criteria in such conditions are not known, some criteria have been proposed. Typical theories used for hull structure are introduced later, where principal stresses are denoted by $\sigma_1, \sigma_2, \sigma_3(\sigma_1 \geq \sigma_2 \geq \sigma_3)$, and stresses corresponding to failure in simple stress condition are $\sigma_t, \sigma_c, \tau_s$ for tension, compression, and shear respectively.

(1) *Maximum principal stress theory*: Rankine proposed this theory, which shows that materials fail when the maximum value of $\sigma_1, \sigma_2, \sigma_3$ reaches σ_t or σ_c.

$$\begin{aligned} \sigma_1 &= \sigma_t \quad \sigma_1 \geq |\sigma_3| \geq 0 \\ \sigma_3 &= \sigma_c \quad -\sigma_3 \geq |\sigma_1| \geq 0 \end{aligned} \tag{3.3.1}$$

According to this theory, the following premise must hold:

$$\tau_s/\sigma_t = 1$$

However, since measurement results for ductile materials show

$$\tau_s/\sigma_t = 0.5 \sim 0.6$$

This theory cannot be applied for ductile materials, but is applied widely for brittle materials.

(2) *Maximum shearing stress theory*: Guest or Tresca proposed this theory, which shows that materials fail when maximum shearing stress τ_{max} reaches maximum shearing stress τ_s corresponding to the yield stress of simple tension.

$$\tau_{max} = \tau_s$$

$$\tau_{max} = \frac{|\sigma_1 - \sigma_3|}{2}$$

$$\tau_s = \frac{\sigma_t}{2} \tag{3.3.2}$$

This theory means $\tau_s/\sigma_t = 0.5$, and this agrees with experiments for ductile materials. In addition, this shows that failure occurs along the plane of maximum shearing stress (45° to maximum principal stress). This also agrees with experiments on the material. Hence this theory is widely used for ductile materials.

(3) *Shearing strain energy theory*: This theory is called Mises-Hencky theory. As shown in Eq. (3.2.23), this theory denotes that failure occurs when the shearing strain energy in the material reaches that corresponding to the yield stress from simple tension. In this case $\tau_s/\sigma_t = 1/\sqrt{3} = 0.577$, which agrees with experiments of ductile materials such as mild steel. Accordingly, this theory is widely used to evaluate ship hull strength. According to this theory, the material yields when the von Mises stress as expressed in Eqs. (3.2.24)–(3.2.26) reaches the yield stress.

Figure 3.9 compares the above-mentioned theories for a 2-dimensional model [18]. Large differences between these theories appear especially in pure shearing conditions, that is

$$\sigma_1 = -\sigma_2 \geq 0 \,\text{or}\, \sigma_2 = -\sigma_1 \geq 0$$

For example, when $\sigma_1 = -\sigma_2 \geq 0$
$\sigma_1 = \sigma_t$ for maximum principal stress theory.
$\sigma_1 = \sigma_t/2$ for maximum shearing stress theory.
$\sigma_1 = \sigma_t/\sqrt{3}$ for shearing strain energy theory

3.3.2 Allowable Stress

As emphasized in Sect. 3.1.2, the strength criteria are strongly associated with the design load and strength assessment method and cannot be discussed separately. In this section, we briefly review the assumed allowable stresses based on the required scantlings in IACS CSR [19] for various levels of members (plate, stiffeners, primary supporting members and hull girder) under the design load and strength assessment method stipulated in CSR.

(1) *Plate*: Plate members are usually designed under plastic design scheme. The fully plastic moment of a plate of thickness t is:

$$M_p = \frac{\sigma'_Y t^2}{4} \tag{3.3.3}$$

where

σ'_Y: equivalent yield stress considering the Poisson's effect

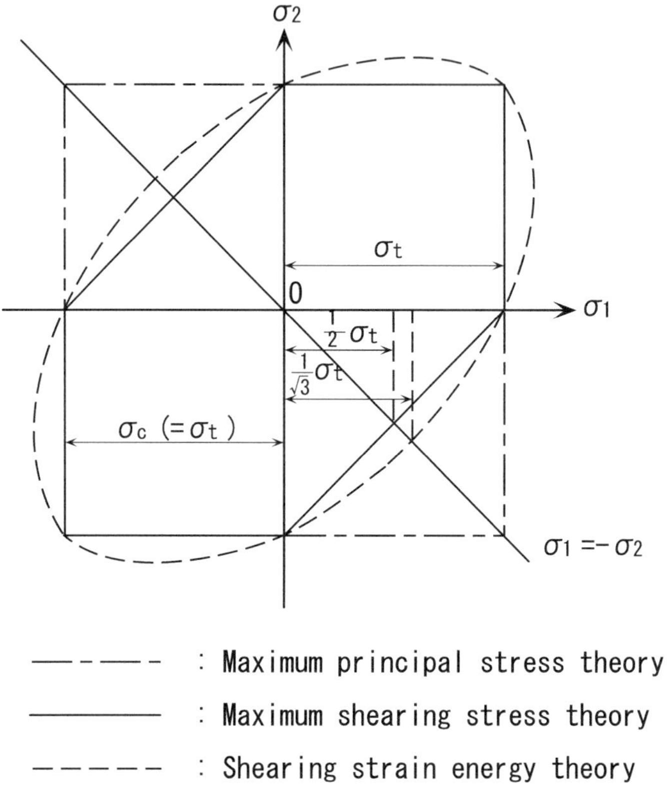

Fig. 3.9 Criteria of failure

$$\sigma'_Y = \frac{\sigma_Y}{\sqrt{1 - \nu_p + \nu_p^2}} = \frac{2\sqrt{3}}{3}\sigma_Y \tag{3.3.4}$$

σ_Y: yield stress

ν_p: Poisson's ratio in the plastic state (=0.5).

When the plate aspect ratio is sufficiently large, the collapse pressure P_c forming three plastic hinges is expressed in Eq. (3.3.5) for a plate strip of unit width with the span of the breadth of the plate panel.

$$P_c = \frac{16M_p}{s^2} \tag{3.3.5}$$

where

s: breadth of plate panel, i.e., stiffener spacing.

Substituting Eqs. (3.3.3) and (3.3.4) into Eq. (3.3.5), relationship between t and P_c is obtained:

$$P_c = \frac{8\sqrt{3}\sigma_Y t^2}{3s^2} \tag{3.3.6}$$

On the other hand, the net required thickness t is given as follows in the CSR. Net required thickness means the required thickness solely from viewpoint of strength without corrosion addition.

$$t = 0.0158\alpha_p s\sqrt{\frac{|P|}{\chi C_a \sigma_Y}} \tag{3.3.7}$$

where

α_p: correction factor for the panel aspect ratio. $\alpha_p = 1.0$ for sufficiently long plate.

χ: coefficient to take care of exceptional cases, and usually $\chi = 1.0$

C_a: permissible bending stress coefficient. $C_a = 1.0$ for static plus dynamic load cases without the action of axial stress for the members other than longitudinal strength members.

P: design pressure (kN/m^2).

Noting that t and s are given in units of mm and σ_Y is given in units of N/mm^2 in the rule formula, and substituting all the coefficients above into Eq. (3.3.7), the relationship between t and P after unit unification is obtained as follows:

$$t = 0.500 \times s\sqrt{\frac{|P|}{\sigma_Y}} \tag{3.3.8}$$

From Eqs. (3.3.6) and (3.3.8), the safety factor to the plate collapse (three plastic hinge formation) can be derived as follows:

$$\frac{P_c}{|P|} = 1.155 \tag{3.3.9}$$

Roughly speaking, plate collapse pressure is twice as large as the pressure that produces initial yielding of the plate surface, i.e., 1.5 (shape factor) × 1.33 (ratio of collapse load to two hinge formation load) = 2.0. Therefore, Eq. (3.3.9) means that the allowable stress for plate bending is much larger than the yield stress in the elastic design scheme.

(2) *Stiffener*: The required net section modulus, Z in cm^3, is given in CSR as follows:

$$Z = \frac{|P| s l_{bdg}^2}{f_{bdg}\chi C_s \sigma_Y} \tag{3.3.10}$$

where

l_{bdg}: effective bending span in m

f_{bdg}: bending moment factor taken as $f_{bdg} = 12$ for horizontal stiffeners

χ: coefficient to take care of exceptional cases, and usually $\chi = 1.0$

C_s: permissible bending stress coefficient. $C_s = 0.9$ for static plus dynamic load cases without the action of axial stress.

Substituting the coefficients above and unifying the units, Eq. (3.3.10) reduces to:

$$Z = \frac{|P|sl_{bdg}^2}{0.9 \times 12\sigma_Y} = \frac{M}{0.9\sigma_Y} \tag{3.3.11}$$

where $M = |P|sl_{bdg}^2/12$ is the bending moment at the span ends under the design pressure. From Eq. (3.3.11) we can find that the allowable stress in this case is $0.9\sigma_Y$.

CSR also provides required net thickness of the web plate of a stiffener, t_w in mm, as follows:

$$t_w = \frac{f_{shr}|P|sl_{shr}}{d_{shr}\chi C_t \tau_Y} \tag{3.3.12}$$

where

f_{shr}: shear force distribution factor taken as $f_{shr} = 0.5$ for horizontal stiffeners

l_{shr}: effective shear span in m

d_{shr}: effective shear depth in mm

C_t: permissible shear stress coefficient. $C_t = 0.9$ for static plus dynamic load cases.

τ_Y: shear yield stress, $\tau_Y = \sigma_Y/\sqrt{3}$.

Substituting the coefficients above and unifying the units, Eq. (3.3.12) reduces to:

$$t_w = \frac{|P|sl_{shr}}{2d_{shr} \times 0.9\tau_Y} = \frac{F}{d_{shr} \times 0.9\tau_Y} \tag{3.3.13}$$

where $F = |P|sl_{shr}/2$ is the shearing force at the span ends under the design pressure. From Eq. (3.3.13) we can find that the allowable shearing stress on the stiffener web plate is $0.9\tau_Y$.

(3) *Primary supporting member (Transverse strength)*: Strength of primary supporting members such as girders and transverse web frames is usually assessed using FEM. The assessment is based on the von Mises stress as defined in Eqs. (3.2.25) and (3.2.26). The allowable von Mises stress is stipulated to be σ_Y in general for static plus dynamic load cases in combination with the net scantling approach, specified loading conditions and specified modeling and analysis methodology of FEM.

(4) *Hull girder (Longitudinal strength)*: The most important hull girder strength is the vertical bending strength and the vertical shear strength. Hull girder bending stress and shearing stress are usually calculated using beam theory under the specified loading conditions and stipulated formulae of wave induced vertical bending moment and shear force, and they are compared with allowable stresses. CSR requires that the thickness used to calculate the stresses should be the gross offered thickness reduced by half the corrosion addition to take care of the practically severest case after thickness diminution for global strength.

The allowable bending stress is stipulated to be $190/k$MPa for the midship region in the seagoing condition where k is the material factor (1.0 for mild steel, 0.78, 0.72, 0.68 for yield point of 315, 355 and 390 MPa, respectively). This corresponds to the allowable stress of about $0.81\sigma_Y$ in case of mild steel. The allowable hull girder shearing stress is given as $120/k$MPa in the seagoing condition, i.e., about $0.88\tau_Y$ in case of mild steel.

3.4 Fatigue Strength

3.4.1 Introduction

Ship structures are subjected to cyclically fluctuating loads as well as to static loads. Figure 3.10 shows the time history of the longitudinal stress on the upper deck of a container ship during a voyage [10]. In such cases, they may sometimes fracture due to a large number of repeating loads, even when the magnitude of the cyclic load is much smaller than the strength against static load. It is a phenomenon called "fatigue failure" and the number of cycles causing fatigue failure is called "fatigue life". Fatigue is an important problem in the design of structural members exposed to cyclic loads, because cyclic loads due to wave are always present in almost all structural members of the ship during the voyage. The vibration is also regarded as a repeated load.

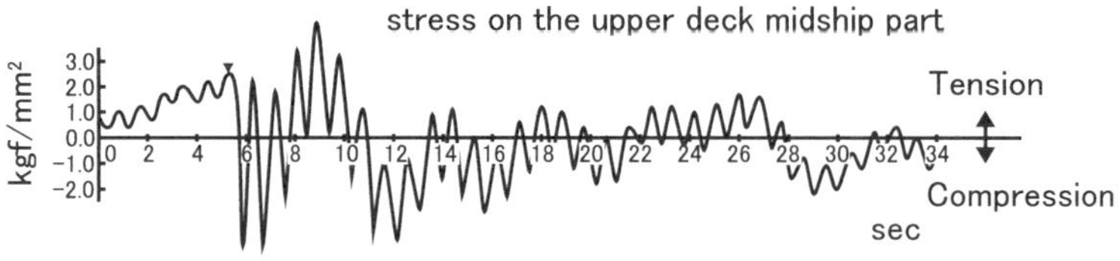

Fig. 3.10 Typical time history of stress

3.4.2 S–N Curve

Though the time history of stress is very complex as shown in Fig. 3.10, the pattern induced by cyclically repeating loads is simply illustrated in Fig. 3.11, where.

σ_{max}: maximum stress

σ_{min}: minimum stress

$\sigma_m = (\sigma_{max} + \sigma_{min})/2$: mean stress

$\sigma_a = (\sigma_{max} - \sigma_{min})/2$: stress amplitude.

Even in the case of the same material, the cycle number for fatigue life may differ depending on the magnitude or type of applied loads as shown in Fig. 3.11. Figure 3.11a–c show stress behavior corresponding to the test cases of repeated tension ($\sigma_{max} = 2\sigma_a$, $\sigma_{min} = 0$), reversed tension and compression ($\sigma_m = 0$), and fluctuating tension, respectively. The relation between stress amplitude *S* and fatigue life *N* can be obtained by a fatigue test as shown in Fig. 3.12, where the axis of abscissas is represented by a logarithmic scale and the axis of ordinates is a normal or logarithmic coordinate. The figure is simply called an "S–N curve".

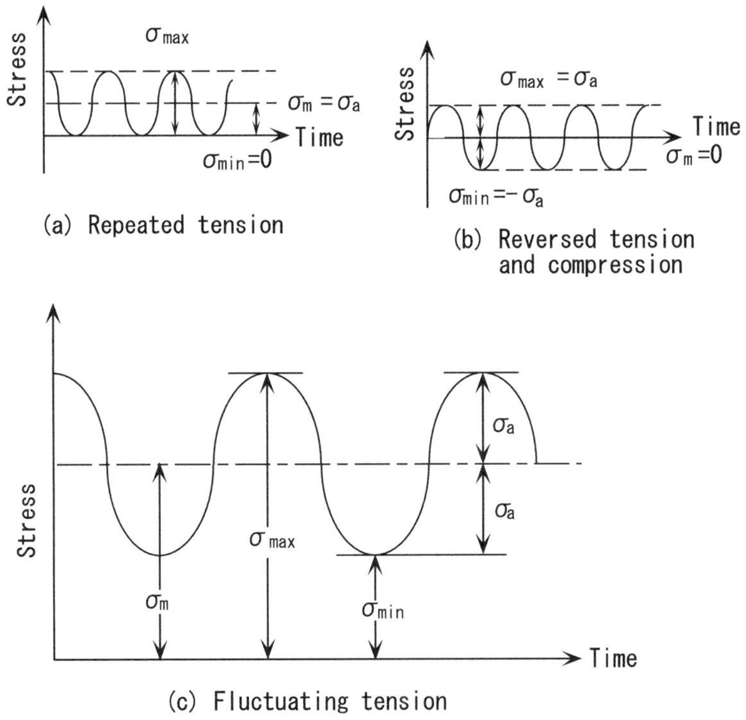

Fig. 3.11 Simplified time history of stress

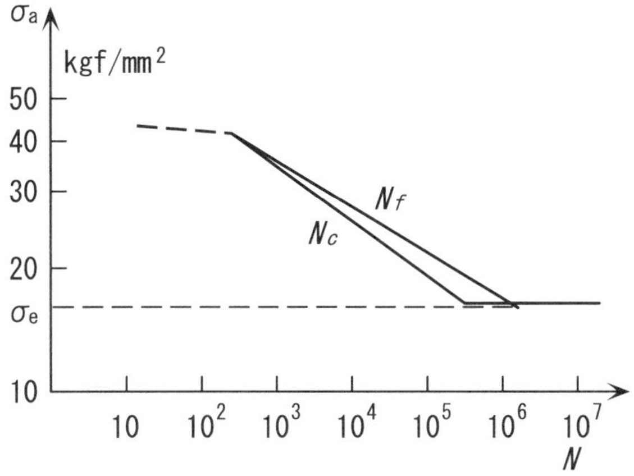

Fig. 3.12 S–N curve

In the figure, N_c represents the number of cycles in which a crack was initiated in the test specimen, i.e. crack initiation life, and N_f means the number of cycles when the specimen fractures, i.e. failure life.

If the stress amplitude becomes small, the fatigue life will increase. However, failure never occurs where the stress amplitude is lower than a certain stress level as indicated by a flat line in the figure. That critical stress which never induces fatigue failure is called the "fatigue limit" or "endurance limit". From a practical point of view, the fatigue limit can often be specified as the stress at $N = 10^6$, when the test specimen does not cause failure. The fatigue limit, in general, is much lower than the ultimate strength of the material due to static stress.

When the fatigue life is lower than $N = 10^5$, it is conveniently called "low cycle fatigue" and when higher than $N = 10^5$, it is "high cycle fatigue". In the case of low cycle fatigue, the stress amplitude may reach yield stress, because low cycle fatigue tends to take place in the area of high stress concentration. Therefore, in the low cycle fatigue test, the strain amplitude should be taken instead of the stress amplitude, then the fatigue strength is formulated as follows;

$$S = CN^{-k}, \varepsilon = CN^{-k} \tag{3.4.1}$$

where

S: stress amplitude

ε: strain amplitude

N: repetition number for crack initiation or failure

C, k: coefficient.

$$\text{or } S = a - b \log N, \varepsilon = a - b \log N \tag{3.4.2}$$

Table 3.1 shows the fatigue life for the base metal and welding joints of mild steel [20], which were tested under repeated tension. The fatigue life of welding joints is generally reduced because of the stress concentration around the beads, compared with the base metal.

In high cycle fatigue, the fatigue limit must be discussed. The fatigue limit depends not only on the stress amplitude but also on the mean stress. If we plot the mean stress and stress amplitude corresponding to the fatigue limit, we can obtain the line ABD in Fig. 3.13. Point A represents the fatigue limit of $\sigma_m = 0$, i.e. reversed tension and compression, and point B means the case of $\sigma_m = \sigma_a$, i.e. repeated tension ($\sigma_{max} = 2\sigma_a$, $\sigma_{min} = 0$). Fatigue failure never occurs in the case where the stress amplitude and mean stress are within the area below the line ABD. In addition, if we add another line YY' representing critical yield stress ($\sigma_m + \sigma_a = \sigma_Y$, σ_Y: yield

Table 3.1 Fatigue strength of welding joints

(kgf/mm^2)

Joints	Mean line		90% reliability	
	Formula	$N = 2\times10^4$	$N = 10^4$	$N = 2\times10^4$
Mild steel,as rolled	$\log S = \log 40.15 - 0.075 \log(N/10^4)$	26.8	29.0~40.0	23.0~32.0
,grinding	$\log S = \log 36.02 - 0.0611 \log(N/10^4)$	26.1	28.2~34.8	23.4~28.9
,scouring	$\log S = \log 38.18 - 0.0493 \log(N/10^4)$	29.7	30.6~38.2	25.2~32.6
Butt joint, transverse	$\log S = \log 39.66 - 0.184 \log(N/10^4)$	14.9	20.5~33.0	11.7~18.8
〃 , 〃 ,finishing	$\log S = \log 32.12 - 0.054 \log(N/10^4)$	24.0	21.2~38.0	17.9~32.8
Butt joint, longitudinal	$\log S = \log 38.22 - 0.1408 \log(N/10^4)$	18.2	25.0~30.8	16.4~20.2
〃 , 〃 ,finishing	$\log S = \log 42.59 - 0.1148 \log(N/10^4)$	23.3	30.7~35.2	21.6~24.7
Front fillet weld	$\log S = \log 23.41 - 0.1557 \log(N/10^4)$	10.3	11.0~25.3	6.6~15.9
〃 , compression	$\log S = \log 30.81 - 0.1053 \log(N/10^4)$	17.5	19.4~30.8	14.0~22.3
Side fillet weld,throat	$\log S = \log 24.47 - 0.1218 \log(N/10^4)$	13.5	16.7~22.5	11.6~15.7
〃 , base metal	$\log S = \log 12.53 - 0.0503 \log(N/10^4)$	9.6	7.0~17.7	6.0~15.3
Front and side fillet,throat	$\log S = \log 10.02 - 0.0751 \log(N/10^4)$	6.7	6.0~11.7	4.8~9.4
〃 , base metal	$\log S = \log 15.60 - 0.3387 \log(N/10^4)$	4.6	7.0~12.3	3.5~6.1
T-fillet,throat	$\log S = \log 25.93 - 0.2011 \log(N/10^4)$	13.6	–	11.4~16.6
Bead welding, longitudinal	$\log S = \log 42.54 - 0.1651 \log(N/10^4)$	17.7	24.4~34.5	14.7~21.0
〃 , transverse	$\log S = \log 33.34 - 0.1713 \log(N/10^4)$	13.4	–	12.2~14.4

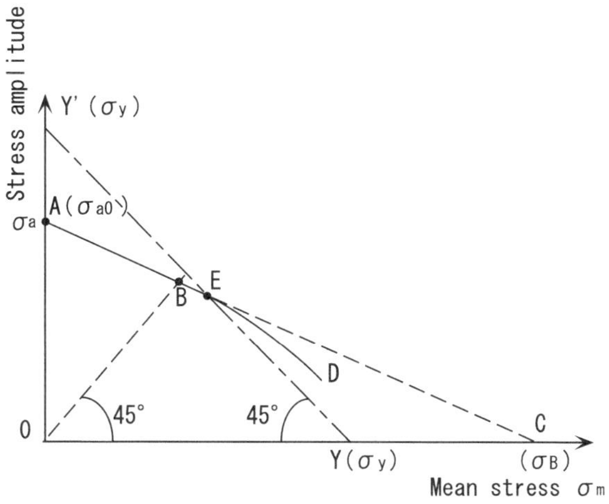

Fig. 3.13 Mean stress versus stress amplitude

stress), the area below the line ABEY indicates the safe area where failure will never be induced by fatigue or by yielding.

Cyclic loads in a marine environment result in the reduction of fatigue life due to corrosive conditions rather than in air. Corrosive conditions reduce fatigue life to a much greater extent in high cycle fatigue than in low cycle fatigue. In addition, fatigue tests in sea water do not show a clear fatigue limit.

3.4.3 Fatigue Damage

Loads acting on a ship's structure are not constant loads but random loads. On the other hand, S–N curves are generally obtained by fatigue tests with constant loads, therefore fatigue strength life is normally calculated based on the following linear assumption, which is called as Miner-Palmgren approach.

Assume that all stress cycles can be divided into the several stress blocks, i.e. the structure is subjected to a stress cycle n_1 with constant stress amplitude σ_1, n_2 with σ_2 and so on as shown in Fig. 3.14. The damage factor of the i-th stress block is defined by the ratio of number of encountered stress cycles n_i and failure life N_i at the stress amplitude of σ_i, and the total damage factor can be obtained by summing each damage factor by the following equation:

$$C_w = \sum_{i=1}^{k} \frac{n_i}{N_i} = \frac{n_1}{N_1} + \frac{n_2}{N_2} + \cdots + \frac{n_k}{N_k} \quad (3.4.3)$$

where

k: number of stress blocks

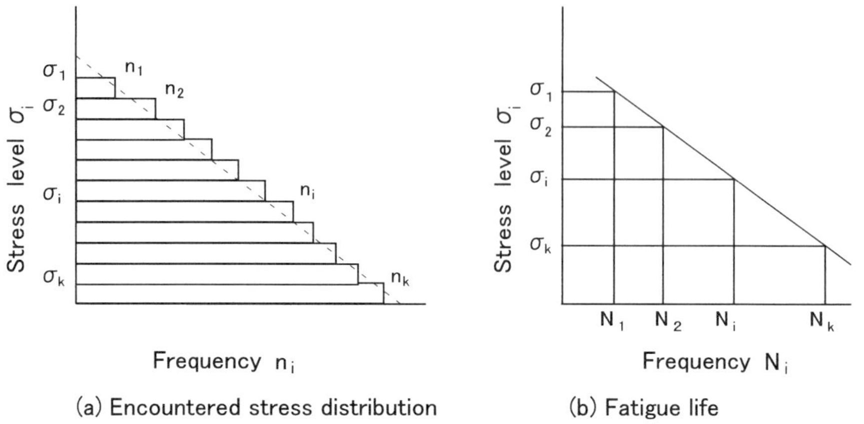

Fig. 3.14 Fatigue life in random fluctuation

n_i: number of stress cycles in i-th stress block with constant stress range σ_i

N_i: number of cycles to failure at constant stress range σ_i.

C_w is called the "cumulative fatigue damage factor" and fatigue failure would happen if C_w exceeded a certain critical value. According to the experiments, the critical value which gives fatigue failure varies from 0.1 to 10. In spite of that, $C_w = 1.0$ is often applied as a guideline for fatigue life assessment.

3.5 Buckling of Ship Structure

3.5.1 Introduction

Now let us consider the behavior of structures subjected to compressive loads. Under such circumstances the structure behaves in quite a different way to that under tensile loads. The structure under tensile stress may cause cracks, while compressive stress will result in buckling. Nearly all the parts of a structure of a ship consisting of plates and beams are sometimes loaded with compressive loads, and it is therefore important to consider the buckling problems.

Before starting the derivation of the buckling equation based on the references [21, 22], it is necessary to clarify the notation of buckling behavior. Figure 3.15 shows a simple compression rod subjected to an axial load P. The straight form of the rod would be maintained under small loads (stable condition). If the axial load increases and reaches to a certain amount, lateral deflection would suddenly take place (neutral condition) and the rod could no longer sustain loads exceeding P_{cr} (unstable condition). Lateral deflection would occur even where the load is applied in a uniform manner to the cross section of the column. This is due to imperfections

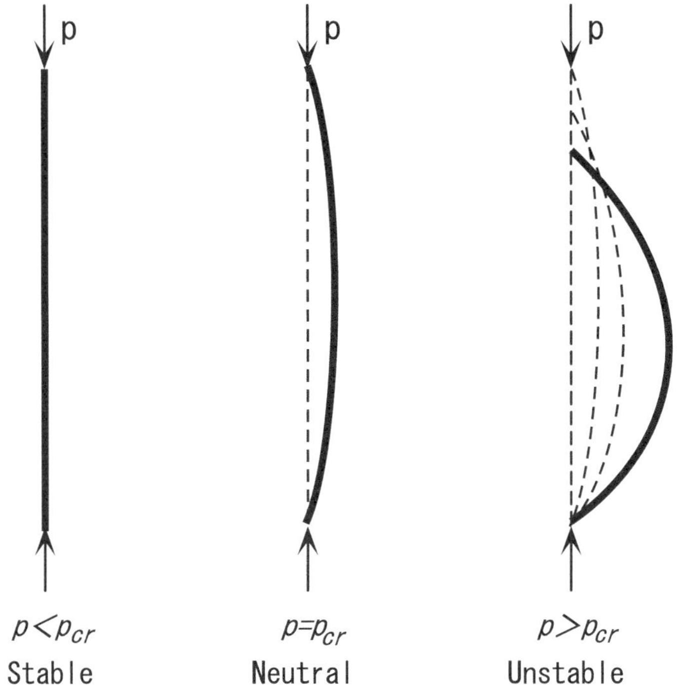

Fig. 3.15 Behavior of a rod subjected to an axial compression load

in the rod's straightness, non-homogeneous material, or non-uniformity on the cross-section of the rod.

The phenomenon where lateral deflection may arise in the athwart direction against the axial working load is called "buckling". In addition, the load P_{cr} which buckles the member is called the "buckling load" or "critical load".

3.5.2 *Column Buckling*

Figure 3.16 shows a hinged-ended column to which the critical load P_{cr} is applied. The column is slightly curved by that axial load. The bending moment created by the axial load due to the horizontal distance y between the load and the deflection of the column must be sustained by the bending moment of the column. It follows that

$$-EI\frac{d^2y}{dx^2} = P_{cr}y$$

$$\frac{d^2y}{dx^2} + \frac{P_{cr}}{EI}y = 0 \tag{3.5.1}$$

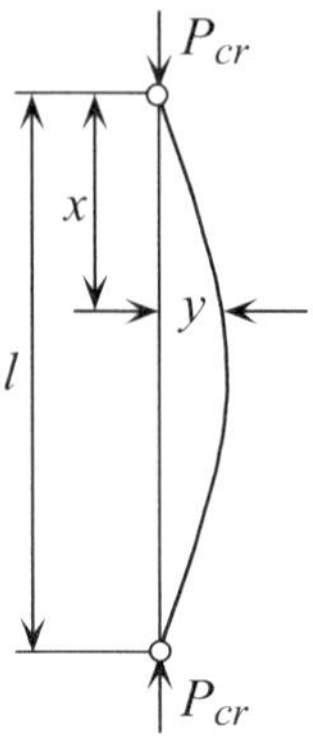

Fig. 3.16 A column under critical load with hinged supports

where I is the moment of inertia of the column section.

By assuming the following harmonic function, the solution of Eq. (3.5.1) can be obtained:

$$y = A\cos\sqrt{\frac{P_{cr}}{EI}}x + B\sin\sqrt{\frac{P_{cr}}{EI}}x \tag{3.5.2}$$

where A, B are arbitrary constants determined by boundary conditions.

For the hinged-edge boundary condition, we substitute $y = 0$ at $x = 0$ and $y = 0$ at $x = l$ into Eq. (3.5.3) and the following equations can be derived:

$$A = 0$$

$$B\sin\sqrt{\frac{P_{cr}}{EI}}l = 0 \tag{3.5.3}$$

In order to obtain a meaningful solution, the following relation must be satisfied from Eq. (3.5.3):

$$\sqrt{\frac{P_{cr}}{EI}}l = n\pi \tag{3.5.4}$$

where n is an arbitrary integer number. Equation (3.5.4) can be formed into

$$P_{cr} = \frac{n^2\pi^2 EI}{l^2} \tag{3.5.5}$$

The above equation means that the stable condition can be achieved only in the case where the load is less than the amount expressed by Eq. (3.5.5). For design purposes, the smallest value of the critical load is significant, therefore, we put $n = 1$ and the result is

$$P_{cr} = \frac{\pi^2 EI}{l^2} \tag{3.5.6}$$

Thus, the above formula provides the critical load for a uniform hinged-ended column.

Corresponding buckling form for $n = 1$ can be expressed as in the form

$$y = B \sin \frac{\pi x}{l} \tag{3.5.7}$$

The formula (3.5.6) was originally developed by L. Euler, in the eighteenth century, and is usually called "Euler's buckling". It gives the load at which a compression member will transform from a stable form of straight shape and becomes another stable form in curved shape.

A critical buckling stress is often used instead of a buckling load and it can be derived by dividing P_{cr} by A, the cross sectional area of the column

$$\sigma_{cr} = \pi^2 E \frac{I}{A} \frac{1}{l^2} = \pi^2 E \left(\frac{k}{l}\right)^2 \tag{3.5.8}$$

where $k\left(k^2 = I/A\right)$ is the radius of gyration of the section of the column. The ratio l/k, often called the slenderness ratio, is the main factor which governs the critical stress. When we plot σ_{cr} against l/k, the relation between them is as shown in Fig. 3.17 as Euler denoted by σ_E. For large value of l/k the critical stress tends toward zero, and at small values of l/k it tends to infinity. In Euler's formula, the buckling stress may become infinite for a small value of l/k, however, buckling stress never goes up above the yield stress of the material in actual conditions, because the material would fail if the stress exceeded the yield stress σ_Y. Therefore, by theoretical consideration, a horizontal line of yield stress connected to Euler buckling stress is specified as an upper limit of Euler's buckling curve.

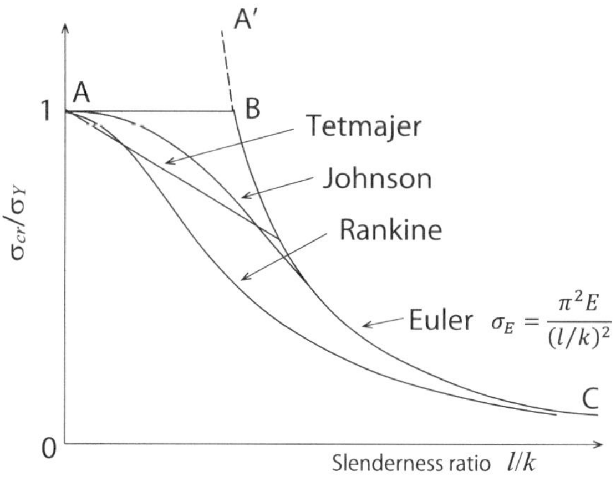

Fig. 3.17 Curves of buckling stress

The above-mentioned theory is obtained by analytical approaches, on the other hand, several buckling curves are proposed on the basis of experiments as shown in Eqs. (3.5.9)–(3.5.11) and plotted in Fig. 3.17.

$$\sigma_{cr} = a - b\left(\frac{l}{k}\right) \quad \text{Tetmajer's formula} \tag{3.5.9}$$

$$\sigma_{cr} = a - b\left(\frac{l}{k}\right)^2 \quad \text{Johnson's formula} \tag{3.5.10}$$

$$\sigma_{cr} = \frac{a}{1 + b(l/k)^2} \quad \text{Rankine's formula} \tag{3.5.11}$$

where a, b are empirical material coefficients.

Johnson's formula is widely used in the rules of classification societies in the form:

$$\sigma_{cr} = \sigma_Y - b\left(\frac{l}{k}\right)^2 \tag{3.5.12}$$

where b is decided so that this curve of the quadratic polynomial is tangent to the curve of Euler's formula at $\sigma_E = \sigma_Y/2$ as shown in Fig. 3.17, leading to:

$$\begin{aligned} \sigma_{cr} &= \sigma_Y - \frac{\sigma_Y^2}{4\pi^2 E}\left(\frac{l}{k}\right)^2 = \sigma_Y - \frac{\sigma_Y^2}{4\sigma_E} \quad \text{when } \sigma_E \geq \frac{\sigma_Y}{2} \\ \sigma_{cr} &= \sigma_E \quad \text{when } \sigma_E \leq \frac{\sigma_Y}{2} \end{aligned} \tag{3.5.13}$$

3.5.3 Plate Buckling

A ship hull is a stiffened-plate structure, the plating supported by a system of transverse or longitudinal stiffeners. The boundary conditions at the panel edges for buckling assessment depend upon the rigidity of the supporting members. For practical design purposes, it is often assumed that the panel is simply supported at all the four edges, since it gives the least critical stress and is on the safe side.

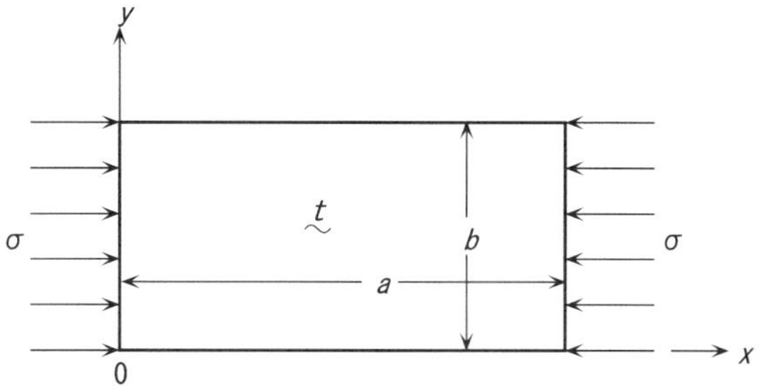

Fig. 3.18 Rectangular plate under uni-axial compression

Now let us consider the rectangular plate with simply supported edges as shown in Fig. 3.18. The equation of elastic buckling stress of the above panel under uni-axial compressive stress σ can be derived by referring to the textbooks [21, 22] as follows:

$$D\left(\frac{\partial^4 w}{\partial x^4}+2\frac{\partial^4 w}{\partial x^2\partial y^2}+\frac{\partial^4 w}{\partial y^4}\right)+\sigma t\frac{\partial^2 w}{\partial x^2}=0 \tag{3.5.14}$$

$$\text{where } D=\frac{Et^3}{12\left(1-\nu^2\right)} \tag{3.5.15}$$

The boundary condition where all four edges are simply supported, can be expressed in the form:

$$w=0,\ \frac{\partial^2 w}{\partial x^2}+\nu\frac{\partial^2 w}{\partial y^2}=0\ \text{ at } x=0, a \tag{3.5.16}$$

$$w=0,\ \frac{\partial^2 w}{\partial y^2}+\nu\frac{\partial^2 w}{\partial x^2}=0\ \text{ at } y=0, b \tag{3.5.17}$$

Let us assume that the buckling deflection of a panel is expressed by Eq. (3.5.18).

$$w=f\sin\frac{m\pi x}{a}\sin\frac{n\pi y}{b} \tag{3.5.18}$$

where m, n are integers representing the number of half waves of the buckling mode. Equation (3.5.18) automatically satisfies the boundary conditions Eqs. (3.5.16) and (3.5.17). Then, substituting Eq. (3.5.18) into Eq. (3.5.14), we can obtain the elastic buckling stress as follows:

$$\sigma=\frac{\pi^2 D}{b^2 t}\left(\frac{m}{\alpha}+n^2\frac{\alpha}{m}\right)^2 \tag{3.5.19}$$

where $\alpha=a/b$

Elastic buckling stress is a minimum critical stress, therefore, we put $n=1$ in Eq. (3.5.19) then

$$\sigma_e = \frac{\pi^2 E}{12(1 - \nu^2)} \left(\frac{t}{b}\right)^2 K \tag{3.5.20}$$

where K: minimum value of k

$$k = \left(\frac{m}{\alpha} + \frac{\alpha}{m}\right)^2 \tag{3.5.21}$$

If we plot the relation between α and K in Fig. 3.19, it can be seen that the least value of K is 4 at the point where α becomes m. An example is shown in Fig. 3.20a, where $a = 3b$, i.e., $\alpha = 3$. In this example, three half wave buckling mode may arise, one half wave in one divided square panel, as K values reach a minimum of 4 when the wave number m becomes 3. That is, in other words, a long panel of plating tends to buckle into the number of square panels. This will only be true if a is an exact multiple of b and, if it is not so, the panel will buckle into the nearest round number of half waves which will make the critical stress the minimum. In practical designing, $K = 4$ is usually applied for a panel of $\alpha \geq 1$, because $K = 4$ gives minimum buckling stress. When $\alpha < 1$, the number of half waves, m, is always equal to 1 as observed in Fig. 3.19, and its buckling mode is as shown in Fig. 3.20b.

For the small b/t, i.e. small b in comparison with t, the elastic buckling stress indicated by Eq. (3.5.20) becomes more than the yield stress of the plate material, however, this is obviously impractical. Therefore, a similar modification is used as in the case of column buckling. For this purpose, it is usual to use Johnson's modification factor η and the critical buckling stress σ_{cr} for the full range of value of t/b is as follows:

$$\sigma_{cr} = \frac{\pi^2 E}{12(1 - \nu^2)} \left(\frac{t}{b}\right)^2 K\eta \tag{3.5.22}$$

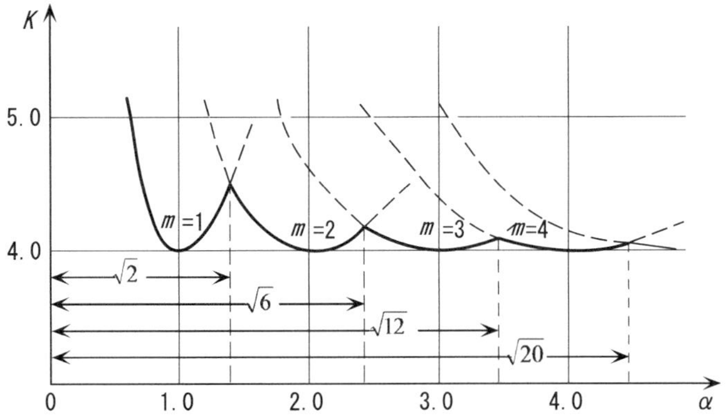

Fig. 3.19 Relation between aspect ratio α and K value

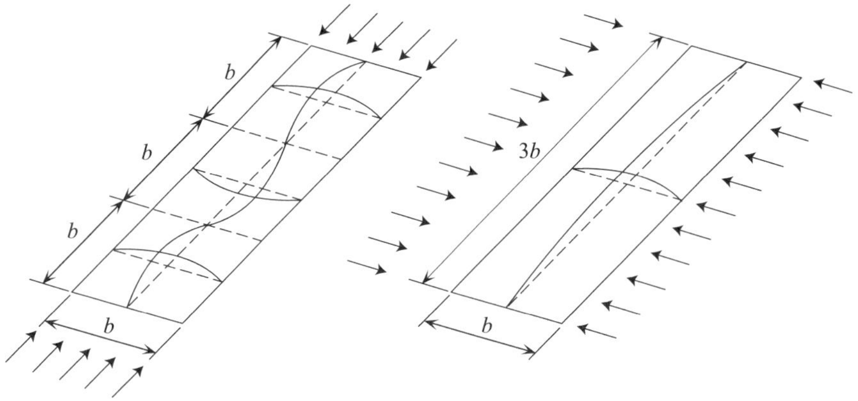

Fig. 3.20 Example of plate buckling mode

$$\text{where } \eta = \begin{cases} 1 & \sigma_e \le \sigma_Y/2 \\ \frac{\sigma_Y}{\sigma_e} - \frac{1}{4}\left(\frac{\sigma_Y}{\sigma_e}\right)^2 & \sigma_e > \sigma_Y/2 \end{cases} \quad (3.5.23)$$

3.6 Plastic Strength

3.6.1 Philosophy of Plastic Strength

In general, the strength criterion of a hull structure is where the elastic stress calculated for the design load does not exceed the allowable stress:

$$\sigma_{max} \le \sigma_{all} = \frac{\sigma_u}{\alpha} \text{or } \frac{\sigma_Y}{\alpha'} \quad (3.6.1)$$

where

σ_{max}: maximum design stress

σ_{all}: allowable stress

α, α': safety factor

σ_u: tensile strength

σ_Y: yield stress.

However, even if maximum design stress exceeds yield stress locally in some actual structure, fracture does not always occur because of redistribution of stress. Therefore, it is desirable to estimate the collapse load of a structure and to apply

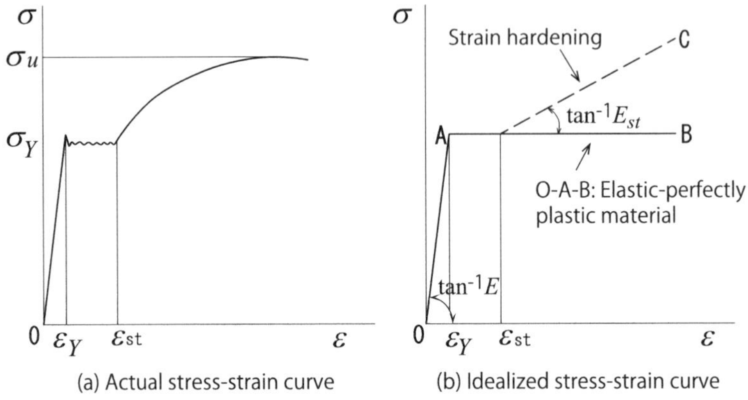

Fig. 3.21 Stress–strain curve

the safety factor for it. Plastic design is based on conditions where plastic regions develop in structural members leading to collapse, and defines the safety factor as the ultimate load divided by the design load.

Plastic design was introduced in ship structure design in the mid-1960s and now it is applied to the following design areas:

- bow structure loading by wave impact
- deck structure loading by local force such as wheel load
- bulkhead structure loading by water pressure
- tank boundary structure loaded by sloshing impact.

As a result of the tensile tests on mild steel used in ship structures, the stress–strain curve shown in Fig. 3.21a has been derived, and this curve is idealized as Fig. 3.21b, in which Young's modulus is $E = 2.1 \times 10^5$ MPa and the tangent modulus in the plastic zone is $E_{st} = 4000 \sim 7000$ MPa. If $E_{st} = 0$, the stress–strain curve is simply O-A-B, which is termed elastic-perfectly plastic body, neglecting strain-hardening.

Considering bending of a beam, the relation between the bending moment and curvature is shown in Fig. 3.22 for an elastic-perfectly plastic body, where the stress distributions in the rectangular sections are as illustrated in Fig. 3.23 in accordance with increasing bending moment. In Fig. 3.23a, the stress in the section is proportional to the distance from the neutral axis in O-A in Fig. 3.22, and the maximum stress reaches yield stress σ_Y at point A in Fig. 3.22. From this point, the stress does not exceed σ_Y and it spreads inside the section as shown in Fig. 3.23c. Finally the stress becomes $\pm\sigma_Y$ throughout the section as shown in Fig. 3.23d at point B in Fig. 3.22, at which the bending moment is called the fully plastic moment.

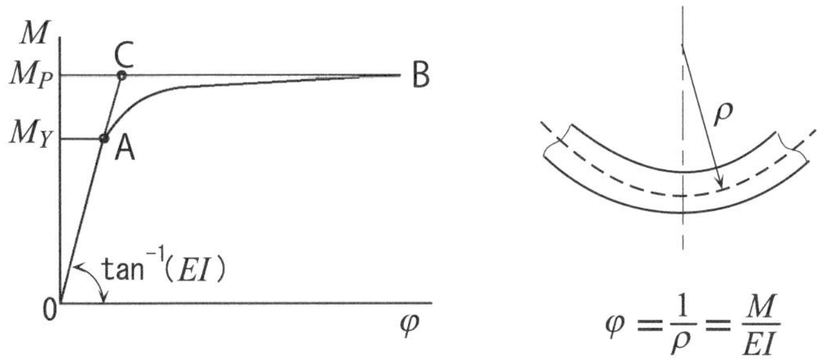

Fig. 3.22 Moment–curvature curve

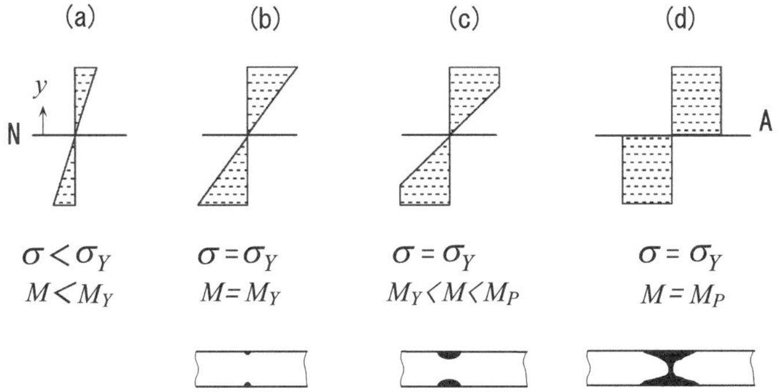

Fig. 3.23 Stress distribution along plate thickness

3.6.2 Plastic Bending and Plastic Section Modulus

Let us consider elastic bending of a beam as shown in Fig. 3.24. Since the total axial force at any section is zero,

$$0 = \int_A \sigma dA = \int_A E\varepsilon dA = \int_A E\frac{y}{\rho} dA = \frac{E}{\rho}\int_A y dA \tag{3.6.2}$$

$$\therefore \int_A y dA = 0 \tag{3.6.3}$$

Therefore, the neutral axis in the elastic state is located at the axis around which the 1st moment is zero (Fig. 3.25a), i.e., the centroid (center of gravity) of the section.

On the other hand, in the plastic design

$$0 = \int_A \sigma_Y dA = \int_{A_u} \sigma_Y dA_u - \int_{A_L} \sigma_Y dA_L \tag{3.6.4}$$

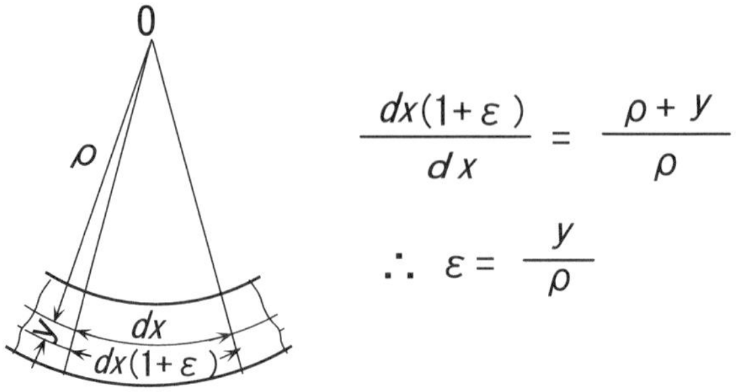

Fig. 3.24 Bending of beam

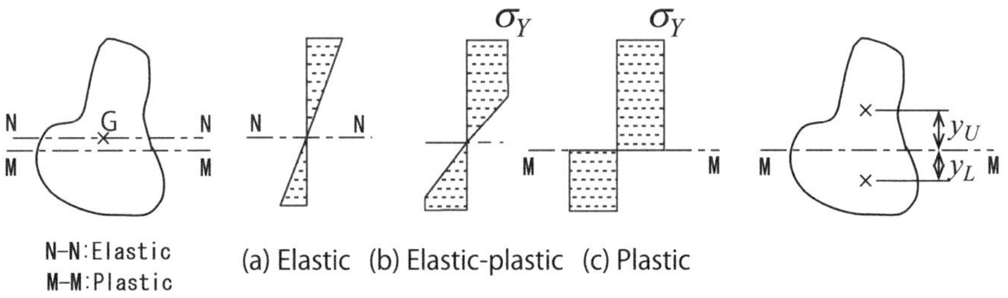

Fig. 3.25 Neutral axis and stress distribution in plastic condition

where

A_u: area of the upper part.

A_L: area of the lower part

$$\therefore A_u = A_L \tag{3.6.5}$$

That is, the neutral axis is located where the sectional area of the upper and lower parts is equal (Fig. 3.25c). Hence, if the cross section is symmetric vertically, the neutral axis in the plastic condition is the same as the elastic one, while it is not the same for non-symmetric sections.

As shown in Fig. 3.25c, the stress is $+\sigma_Y$ and $-\sigma_Y$ at the upper and lower areas of the axis M-M when the fully plastic moment M_P occurs. If the distance from the axis to the centroid of the upper area is denoted as y_U and of lower is y_L, then:

$$M_P = y_U \times \frac{A}{2}\sigma_Y + y_L \times \frac{A}{2}\sigma_Y = \frac{A}{2}\sigma_Y(y_L + y_U) \tag{3.6.6}$$

where

A: sectional area of the section

If we define plastic section modulus as $Z_P = A(y_L + y_U)/2$,

$$M_P = Z_P\sigma_Y \tag{3.6.7}$$

Since $M_Y = Z_e\sigma_Y$ at yield point (Z_e: section modulus),

$$f = \frac{M_P}{M_Y} = \frac{Z_P}{Z_e} \tag{3.6.8}$$

where

f: shape factor

The safety factor of plastic design is denoted as:

$$\text{S.F.} = \frac{M_P}{M} = \frac{M_P}{M_Y}\frac{M_Y}{M} = \frac{M_P}{M_Y}\frac{Z_e\sigma_Y}{Z_e\sigma_{all}} = f \cdot \frac{\sigma_Y}{\sigma_{all}} \tag{3.6.9}$$

where

M: design bending moment

σ_{all}: allowable stress of elastic design

i.e. multiplying shape factor by the safety factor of elastic design.

In the case of a rectangular section shown in Fig. 3.26,

$$Z_P = \frac{1}{2}A(y_U + y_L) = \frac{bh^2}{4} \tag{3.6.10}$$

On the other hand,

$$Z_e = \frac{bh^2}{6} \tag{3.6.11}$$

Therefore,

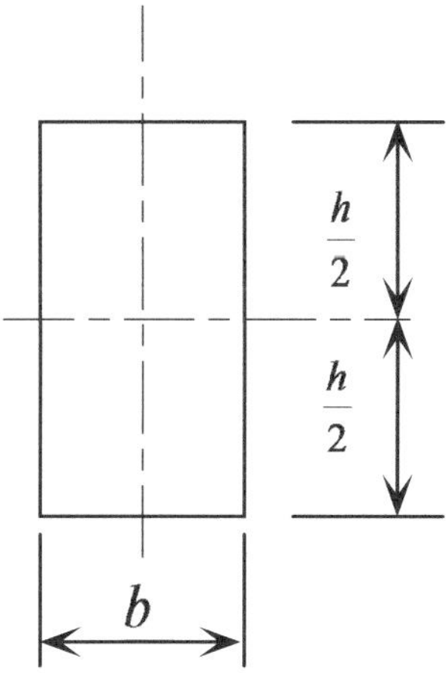

Fig. 3.26 Rectangular section

Table 3.2 Section properties

A: Sectional area I: Moment of inertia Z: Section modulus Z_P: Plastic section modulus

	A	I	Z	Z_P
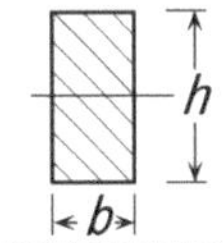	bh	$\frac{1}{12}bh^3$	$\frac{1}{6}bh^2$	$\frac{1}{4}bh^2$
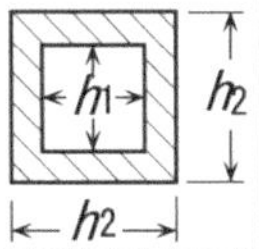	$h_2^2 - h_1^2$	$\frac{1}{12}(h_2^4 - h_1^4)$	$\frac{1}{6}\frac{(h_2^4 - h_1^4)}{h_2}$	$\frac{1}{4}(h_2^3 - h_1^3)$
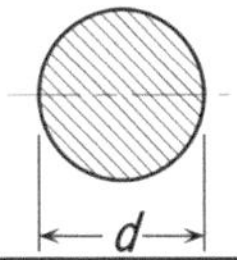	$\frac{1}{4}\pi d^2$	$\frac{1}{64}\pi d^4$	$\frac{1}{32}\pi d^3$	$\frac{1}{6}d^3$
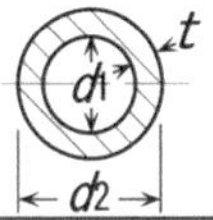	$\frac{1}{4}\pi(d_2^2 - d_1^2)$	$\frac{1}{64}\pi(d_2^4 - d_1^4)$	$\frac{\pi}{32}\frac{(d_2^4 - d_1^4)}{d_2}$	$\frac{1}{6}(d_2^3 - d_1^3)$

$$f = \frac{Z_P}{Z_e} = 1.5 \qquad (3.6.12)$$

In general, if the safety factor is the same as in the elastic design, the allowable stress of a plastic design may be greater than that of an elastic design.

Some examples of plastic section moduli and other sectional properties are shown in Table 3.2 [23] for simple sections, while Fig. 3.27 shows a calculation procedure for a complex section.

3.6.3 *Collapse of a Beam*

Let us consider a beam clamped at both the ends, and loaded by a lateral concentrated force P as shown in Fig. 3.28a. Within the elastic range the curvature of the beam is linearly proportional to the bending moment. If the load P is increased, the nearest support point (point A), will reach the yield stress, and then plasticity will spread throughout the depth of the beam until the section becomes fully plastic, known as the fully plastic moment M_P, as shown in Fig. 3.28b. Even if the load increases further, the beam can absorb no further bending moment in this section. It is as if a hinge has been inserted in the beam at this section, and hence this condition in a beam is termed plastic hinge. Thereafter with an increase of the load, the moment at

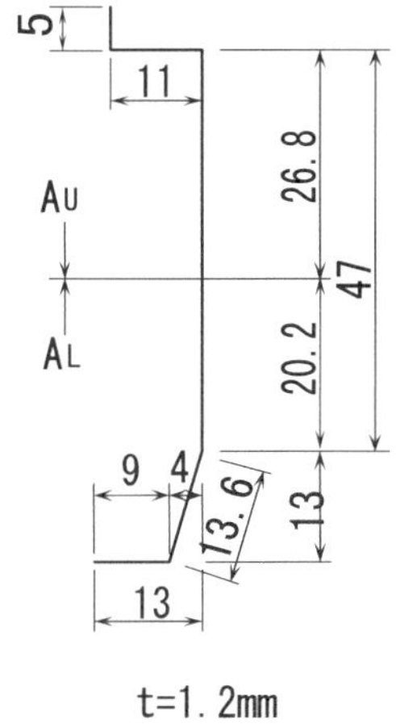

$A_U = A_L$

$Z_p = A_U \times y_U + A_L \times y_L = A_U(y_U + y_L)$

b	t	A_U	y	$A_U y$
5	x1.2	6.0	29.3	175.8
11	x1.2	13.2	26.8	353.76
26.8	x1.2	32.16	13.4	430.94
		51.36		960.50

b	t	A_L	y	$A_L y$
20.2	x1.2	24.24	10.1	244.82
13.6	x1.2	16.32	26.7	435.74
9	x1.2	10.8	33.2	358.56
		51.36		1039.12

$Z_p = \Sigma A_U y + \Sigma A_L y = 960.50 + 1039.12$
$= 2000\ \text{mm}^3$

Fig. 3.27 Calculation of plastic section modulus

point B becomes M_P and the plastic hinge occurs again, and finally at point C. The occurrence of such conditions makes the structure incapable of carrying any further load and it is called a collapse by mechanism, and the final load is called the collapse or ultimate load.

A typical method to estimate the ultimate load is the mechanism method, which applies the principle of virtual work, that is, the work done by an external load equals the work done at plastic hinges.

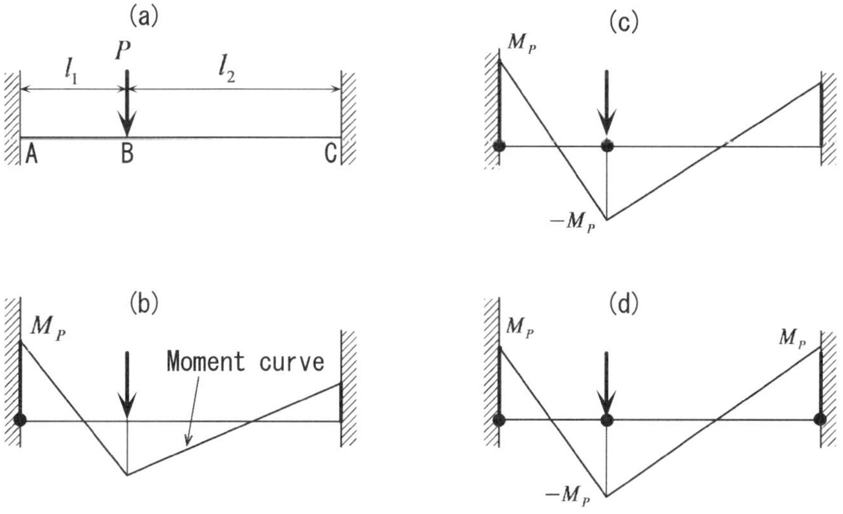

Fig. 3.28 Collapse of a beam

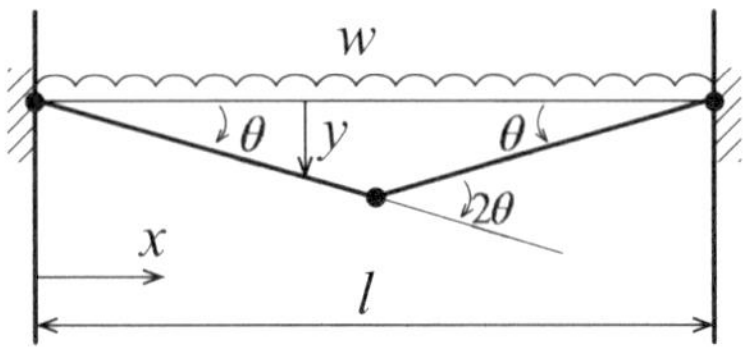

Fig. 3.29 Collapse of clamped beam

$$\sum Pu = \sum M_P\theta \tag{3.6.13}$$

where

P: external force

u: displacement

M_P: plastic moment

θ: rotation angle.

Let us take a simple example of a clamped beam which carries uniformly distributed load w, as shown in Fig. 3.29. The work done by the external load is

$$W_e = \int_0^l wydx = 2w\int_0^{l/2} \theta xdx = wl \cdot \frac{l\theta}{2} \cdot \frac{1}{2} = \frac{wl^2}{4}\theta \tag{3.6.14}$$

On the other hand, the work done by hinges is

$$W_i = M_P\theta(1 + 2 + 1) = 4M_P\theta \tag{3.6.15}$$

Therefore, as $W_e = W_i$,

$$M_P = \frac{wl^2}{16} \tag{3.6.16}$$

From $P = wl$,

$$P = \frac{16}{l}M_P \tag{3.6.17}$$

This P is the collapse load. Table 3.3 [23] shows some examples of the calculation formulae for collapse loads.

In actual design of a beam, the procedure is:

(1) calculate plastic section modulus, Z_P (see Table 3.2, Fig. 3.27)
(2) obtain fully plastic moment, by multiplying plastic section modulus by the yield stress,

Table 3.3 Ultimate load

P_C: Ultimate load M_P: plastic moment

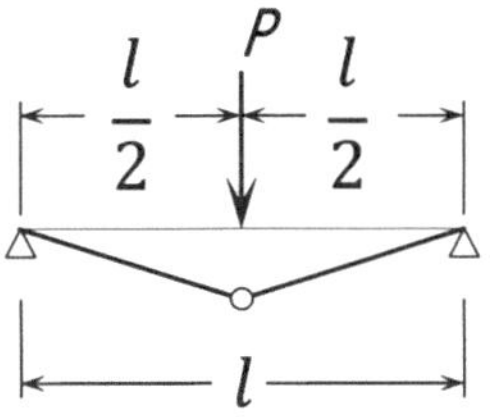 $P_C = \frac{4}{l} M_P$	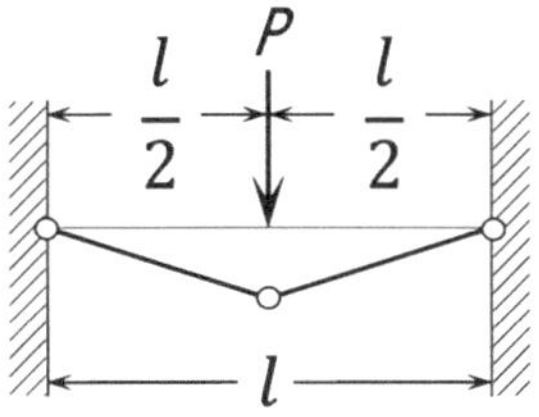 $P_C = \frac{8}{l} M_P$
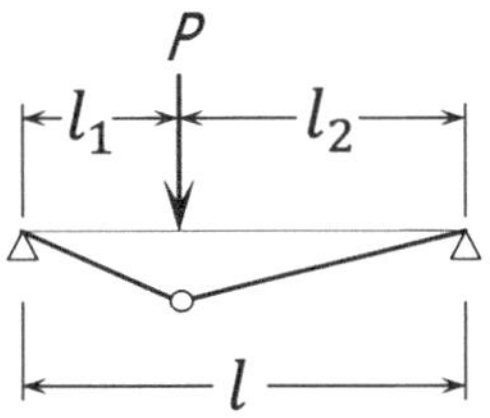 $P_C = \frac{l}{l_1 l_2} M_P$	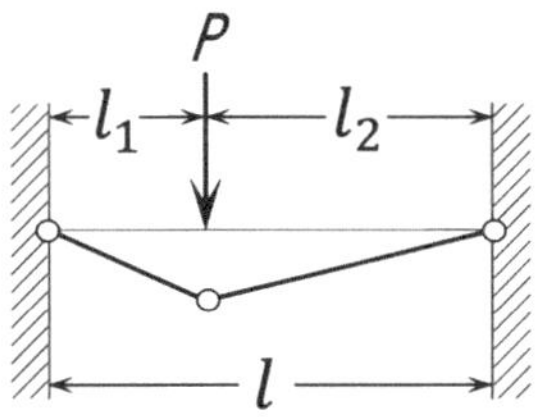 $P_C = \frac{2l}{l_1 l_2} M_P$
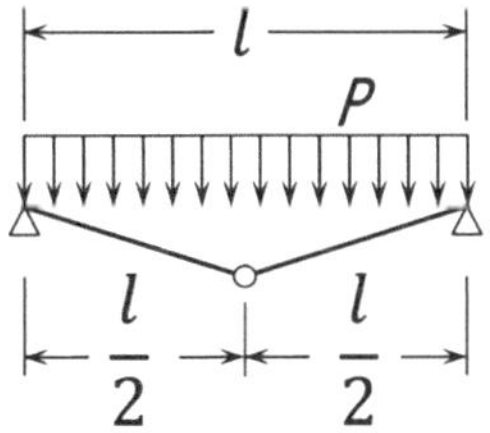 $P_C = \frac{8}{l} M_P$	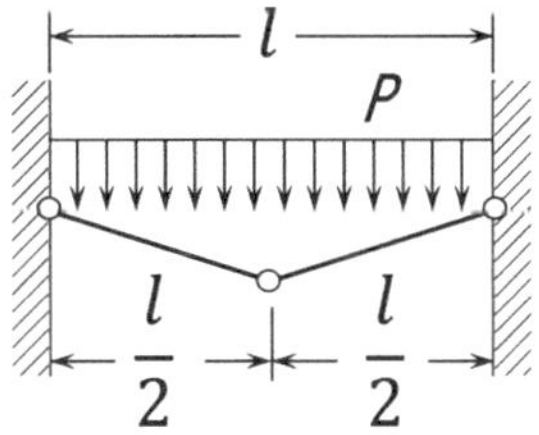 $P_C = \frac{16}{l} M_P$

$$M_P = Z_P \sigma_Y$$

(3) obtain the ultimate load, by multiplying M_P by a coefficient depending on the loading condition (see Table 3.3)
(4) calculate the safety factor, from ultimate load/working load.

3.6.4 Collapse of a Plate

Let us consider a square plate clamped on the four boundaries, which carries a uniform distributed lateral load, as shown in Fig. 3.30. If the load increases, plastic zones occur and spread in the mid-part of the four boundaries, as shown in (a) and (b) of the figure. Then these pass through the plate thickness and spread along the supporting lines (plastic hinge lines) and also plastic areas occur in the center as shown in (c) of the figure. Such ultimate load is also obtained using the principle of virtual work.

In the case of a rectangular plate simply supported around the entire boundary as shown in Fig. 3.31, the uniformly distributed collapse pressure is [24].

$$p_c = \frac{24M_P}{b^2\left(\sqrt{3+\beta^2}-\beta\right)^2} \tag{3.6.18}$$

where

M_P: fully plastic moment per unit length, $M_P = \sigma_Y h^2/4$.

h: plate thickness

$$\beta = b/a$$

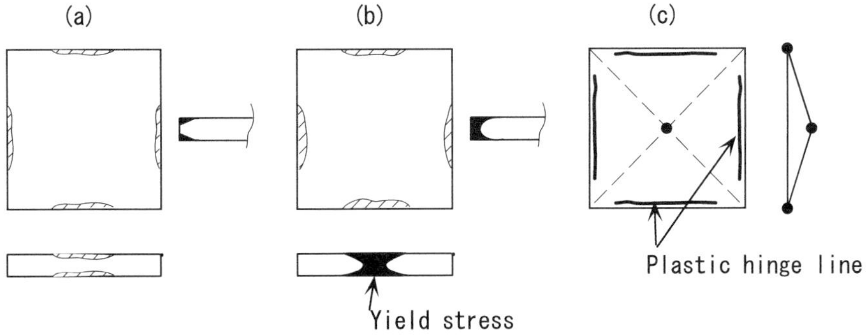

Fig. 3.30 Collapse of plate

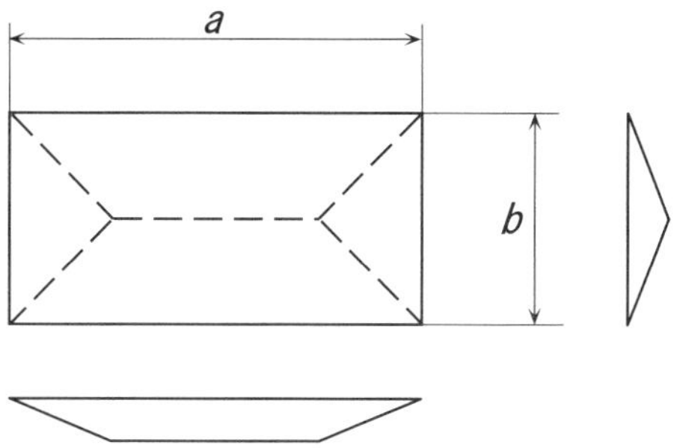

Fig. 3.31 Collapse of rectangular plate

In the case where the boundary around a rectangular plate is fully clamped, the uniformly distributed collapse pressure is [24]

$$p_c = \frac{48M_P}{b^2\left(\sqrt{3+\beta^2}-\beta\right)^2} \tag{3.6.19}$$

These formulas are used for the design of plates which are loaded by water pressure.

3.7 Vibration in Ship

3.7.1 Introduction

While buckling and yielding are mainly induced by static loading, in contrast, dynamic loading may cause vibrations in structures. It is important for the structural engineer to prevent vibration problems, because, cyclic forces due to the propeller or main engine may produce heavy vibrations in accommodation spaces, or they may result in failures of structural members due to vibration.

Thus, in order not to suffer from vibration problems, it is very important to obtain vibration characteristics of structures with high accuracy, especially in the initial stage of design.

3.7.2 Basic Theory of Single Degree of Freedom Vibration System

Regarding vibration phenomena, two important concepts "natural frequency" and "resonance" are explained hereafter. There are several well known textbooks [25,

26] on vibration and further details should be obtained by reference to these books, since the theory of vibration is a well established field.

(1) *Free Vibration*: Now let us consider a simple spring-mass system illustrated in Fig. 3.32, where m is the mass of weight and k is the spring stiffness. Suppose that the mass is forced downward an additional distance x and then suddenly released. The mass would vibrate vertically up and down around the equilibrium position. Let us assume that vertical displacement x is only allowed and horizontal movement is restricted; such systems are called one degree of freedom systems. The configuration will be determined completely by the vertical displacement x of the weight.

If the weight is pulled to the downward position x below the equilibrium state, an elastic force is produced on the weight by the spring. According to Newton's principle, the product of the mass and its acceleration is equal to the spring force acting in the reverse direction of displacement. Thus, a differential equation of motion is obtained in the following formula:

$$m\ddot{x} = -kx \tag{3.7.1}$$

where $\ddot{x} = d^2x/dt^2$

Introducing the notation

Fig. 3.32 Free vibration of a single degree of freedom

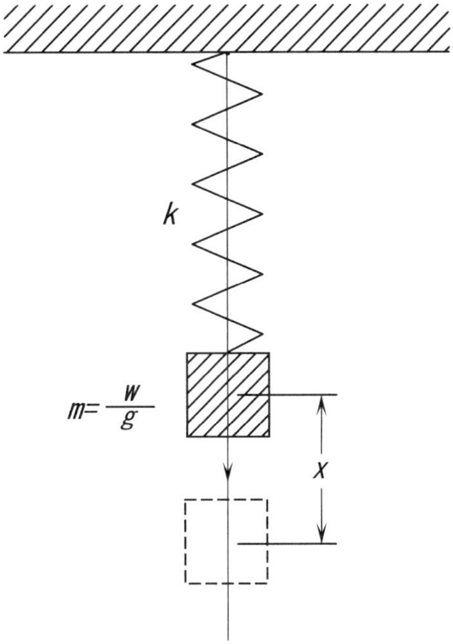

$$\frac{k}{m} = \nu^2 \tag{3.7.2}$$

Equation (3.7.1) can be represented in the following form:

$$\ddot{x} + \nu^2 x = 0 \tag{3.7.3}$$

This equation will be satisfied if we assume the following general solution:

$$x = A \cos \nu t + B \sin \nu t \tag{3.7.4}$$

where A and B are arbitrary constants, which are determined by initial conditions.

From the above equation, we find that the vertical motion of the weight has a vibrating characteristic, and such a vibration without external force or damping is called free vibration. From Eq. (3.7.2), the frequency of a vibrating system is:

$$\nu = \sqrt{\frac{k}{m}} \text{ (rad/s)}$$

$$f = \frac{1}{2\pi}\sqrt{\frac{k}{m}} \text{ (Hz)} \tag{3.7.5}$$

ν is often called an angular frequency representing an angle in radian per second, while f represents the number of cycles per second.

In the Eq. (3.7.5), the frequency of the system is determined only by the magnitude of mass and spring, and is independent of the magnitude of the oscillations, therefore, the above frequency is called the "natural frequency".

(2) *Forced Vibration*: Now let us consider a mass-spring system without damping subjected to periodical external force $P \cos pt$ as shown in Fig. 3.33. The equation of motion of this system is expressed in the form:

$$m\ddot{x} + kx = P \cos pt \tag{3.7.6}$$

Or it is transformed into

$$\ddot{x} + \nu^2 x = \frac{P}{m} \cos pt \tag{3.7.7}$$

The solution of Eq. (3.7.7) is a combination of (1) a general solution, which was already given in the form of free vibration, and (2) a particular solution to Eq. (3.7.7). To obtain the particular solution, substituting $x = C \cos pt$ into Eq. (3.7.7), we find:

$$C = \frac{P/m}{\nu^2 - p^2} \tag{3.7.8}$$

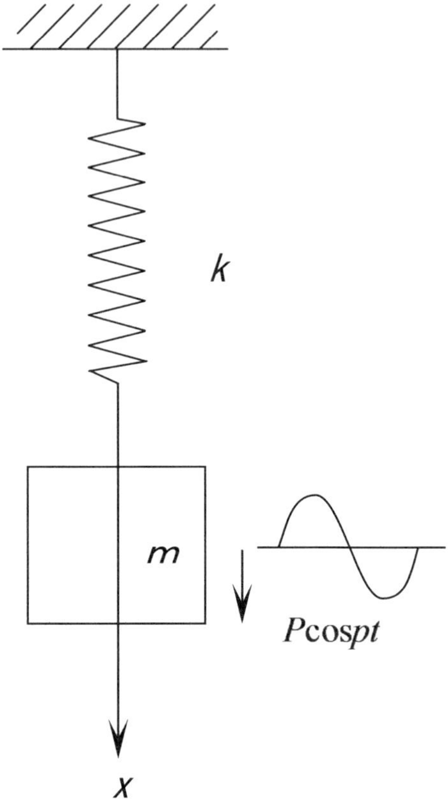

Fig. 3.33 Forced vibration of a single degree of freedom

Then the particular solution is given in the form:

$$x = \frac{P/m}{\nu^2 - p^2} \cos pt \tag{3.7.9}$$

Finally, the solution of Eq. (3.7.7) can be expressed as follows:

$$x = A \cos \nu t + B \sin \nu t + \frac{P/m}{\nu^2 - p^2} \cos pt \tag{3.7.10}$$

In Eq. (3.7.10), the first two terms indicate free vibration and the last one indicates forced vibration. Therefore, the motion of the mass is the combination of free vibration whose frequency is ν and forced vibration with frequency p of external force.

In practical cases, however, there may be some damping forces which decrease free vibration, the first two terms decrease with time, and the last forced vibration term only remains in a steady condition. Now let's consider the steady condition where free vibration has completely diminished, then the amplitude a of a mass in forced vibration is given by Eq. (3.7.10) and can be derived as follows:

$$a = \frac{P/m}{\nu^2 - p^2} = \frac{P/m\nu^2}{1 - p^2/\nu^2} = \frac{P/k}{1 - p^2/\nu^2} \tag{3.7.11}$$

where P/k means a static displacement of the mass and Eq. (3.7.11) can be modified in the following form by using the notation of static displacement x_{st}:

$$\frac{a}{x_{st}} = \frac{1}{1 - p^2/\nu^2} \tag{3.7.12}$$

The ratio a/x_{st} gives the magnification factor of dynamic amplitude in comparison with the static deformation. If we plot the magnification factor against p/ν, we obtain Fig. 3.34. This figure states that if p/ν is near to zero, i.e. exciting frequency is much lower than the natural frequency of the system, the amplitude of forced vibration is close to the static deflection. Conversely, as the exciting frequency increases, vibration amplitude becomes large and grows infinite when $p/\nu = 1$. When the exciting frequency exceeds $p/\nu = 1$, the amplitude is reduced and converges to zero eventually. In actual situations, we know that an infinite oscillation never happens because of the presence of damping which restricts movement of the mass.

Nevertheless, it is an important fact that the amplitude reaches a very large magnitude when the exciting frequency becomes very close to the natural frequency of the system, i.e. $p = \nu$. That phenomenon is called "resonance".

Fig. 3.34 Resonance curve of a single degree of freedom

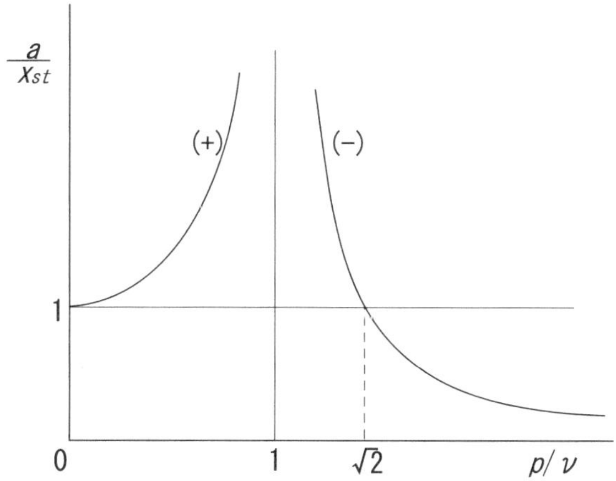

3.7.3 Vibration Problems in Ships

Although there are several exciting sources, such as propellers, main engines, auxiliary machines and wave force, which induce ship hull vibration and local structure vibration, serious vibration problems are caused mainly by propeller and by the main engine. As illustrated in Fig. 3.35, some part of the propeller exciting force occurs experienced as a fluctuating pressure acting on the outer shell plating of the afterbody above the propeller and is called the surface force. Another part is transmitted to the engine room double bottom structure through propeller shafting, and resulting in bearing forces and thrust forces. Those exciting forces induced by the propeller cause hull girder vibration, superstructure vibration, as well as local structure vibration. Hull girder vibration means a hull flexural vibration as shown in Fig. 3.36.

The exciting forces of the main engine, which take the form of unbalanced moments, guide forces, guide moments and thrust fluctuation of the line shafting, are transferred from the main engine bed or from the thrust block to the engine room double bottom and may finally induce hull girder vibration or superstructure vibration.

The above vibrations due to propeller or the main engine are steady vibrations, while another type of vibration, such as whipping and springing of hull girders, may be induced transiently by wave forces. A whipping vibration represents a transient hull girder response created by the impact force of slamming. A springing vibration is a constant hull girder vibration in a rather calm sea, which is induced by the resonance between hull girder natural frequency and higher-order frequencies of the ocean waves.

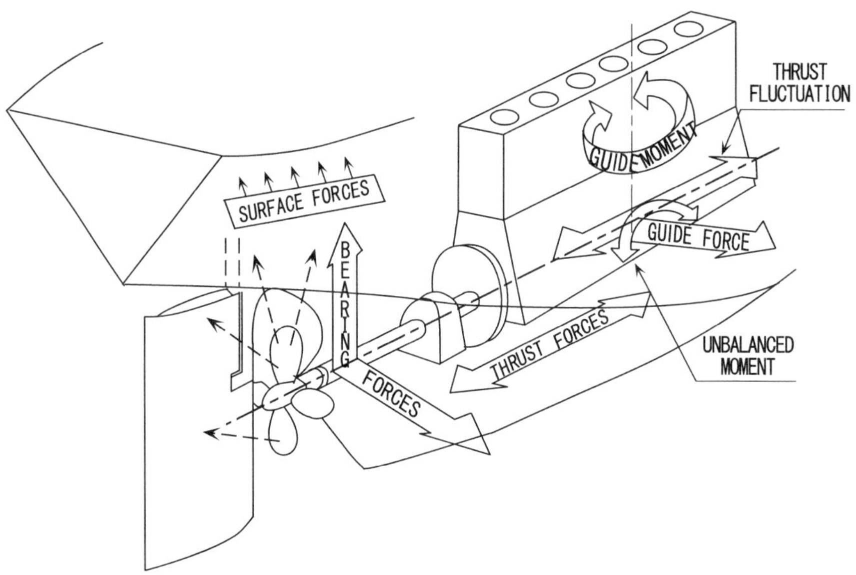

Fig. 3.35 Overview of ship vibration

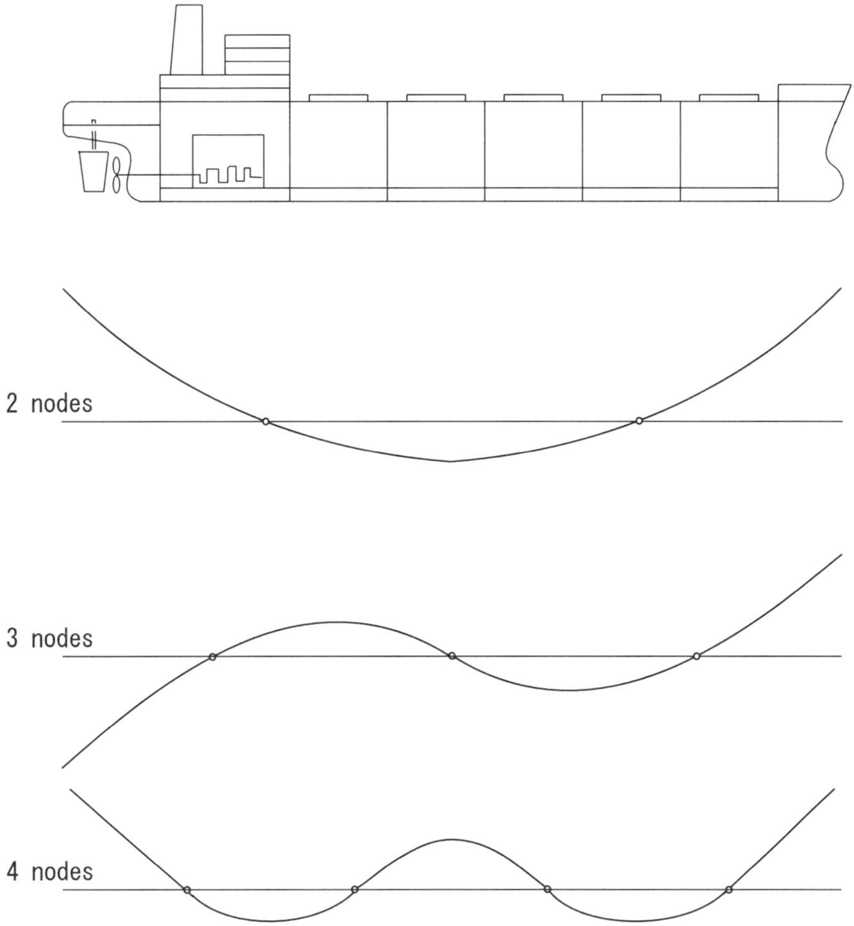

Fig. 3.36 Hull girder flexural vibration

3.7.4 Vibration Prevention Design

To prevent vibration in a ship, we have to design structures in such a way as to avoid resonance or to restrain the vibration amplitude of the structure below the allowable limit. This is done by estimating the vibration magnitude in the resonance condition in the case where resonance in the designed structure cannot be avoided.

(1) *Avoiding Resonance*: In the case where the natural frequency of a structure coincides with the exciting frequency of the external force, the structure will experience severe vibration in resonance. Therefore, to prevent a vibration problem, the structure should be designed so as to avoid resonance.

The design concept for avoiding resonance is achieved by the following procedure:

(1) **Establish the resonance frequency ranges for the exciting source**: For instance, with a propeller of 6 blades rotating at 100 rpm (revolution per minute), the principal order of resonance frequency is:

$$6 \times 100 = 600 \text{ cpm}$$

where cpm indicates a unit of cycle per minute.

Using a safety factor of 1.2, the target range of avoiding resonance is:

$$600/1.2 \sim 1.2 \times 600 = 500 \sim 720 \text{ cpm}$$

(2) **Design a structure whose natural frequency is less or greater than the resonance range**: Next, the natural frequency of the structures should be estimated accurately, and appropriate measure should be taken to make it out of the avoidance range of the exciting frequency. The natural frequency can be either greater than or lower than the exciting frequency, but in general greater natural frequency should be selected because:

- Lower natural frequency tends to be too flexible, causing large displacement.
- The structure may resonate in the lower shaft revolution than normal operation

The "avoiding resonance design" method is so convenient that it is often used in actual design stages, since we have only to examine the natural frequency of the structure once the exciting frequency is determined. This design method can be used under the following conditions:

- The exciting frequency or order of force is clearly known.
- The resonance peak of the structure is sharp and steep.
- The natural frequency of the structure can be accurately estimated.

(2) *Vibration Response Estimation*: If resonance between the structure and the exciting frequency cannot be avoided, a different design method is alternatively adopted, in which the structural response to the exciting force is estimated. In this design process, firstly the vibration amplitude or vibration stress of the structure resulting from the force is calculated by a suitable method; secondly this value is examined to check whether it satisfies the allowable limit or not. The design of the structure is then accepted when the estimated response is satisfactory.

Judgment on whether the design is valid or not is very dependent on the accuracy of vibration calculation, because the response estimation method is available only where:

- Mass, stiffness and damping of the structure are properly evaluated.
- Magnitude of the exciting force is accurately calculated.

Consequently, both highly advanced analysis techniques and much practical shipbuilding experience is necessary when the above-mentioned sophisticated estimation method is applied to actual designs.

3.8 Selection of Strength Analysis Method

3.8.1 Introduction

In this article, several analysis methods utilized in ship structure design are explained, because strength analysis is essential for the integrity of a designed structure. In order to maintain the strength performance to a satisfactory level, it is necessary to analyze the behavior of the structure under the subjected loads and to investigate the stress distribution over the structure. Therefore, we have to make some kind of experiment to simulate the behavior and the stress level. We used to carry out a model test, where a structural model of small size was used and the stress in the members was measured with the aid of strain gauges or photoelastic Moire experiment.

Nowadays, however, in parallel to the rapid progress of computers, we often execute numerical simulation successfully for the purpose of strength analysis by computer instead of the above structural model test.

3.8.2 Type of Analysis Method

A ship structure is, in general, so huge and statically indeterminate that 3-D analysis is not easy. For example, Fig. 3.37 indicates an example of a global 3-dimensional model of a container vessel. This model was specially utilized for the strength analysis of local structures around the hatch opening corners with regard to torsional strength analysis; however, such a huge structural model is not always essential in the design of every kind of vessel. In a practical design case, the designers strategically regard the whole structure as a collection of independent components, such as pillars, beams, plates, shell, etc., according to their mechanical characteristics and they select an appropriate analysis method for each component.

Those strength analysis methods are categorized as:

- Analysis using simple beam theory:
- Analysis of frame structure: Slope deflection method

  Transfer matrix method,
  Finite element method, etc.

- Analysis of 3 dimensional structure: Finite element method

  Boundary element method, etc.

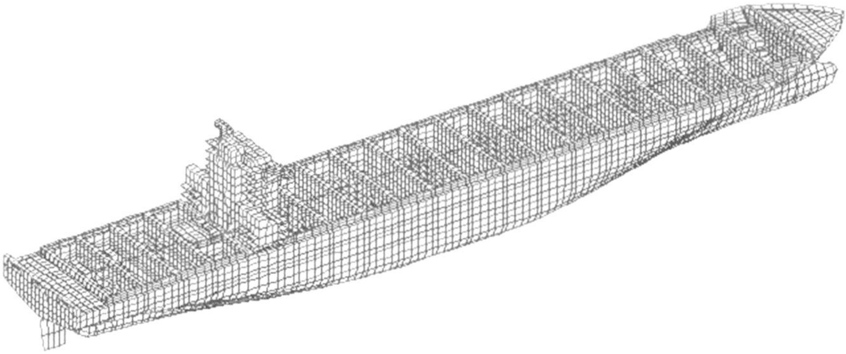

Fig. 3.37 Global FEM model of container vessel

Recently we have many commercially available numerical analysis programs for strength analysis. These programs are manly based on the finite element method, and have the following advantages:

(1) They can deal with any shape of structure.
(2) They can solve the structure with various types of boundary conditions.
(3) They can calculate the structural response for any arbitrary patterns of loads.
(4) They can take account of non-linear effects due to physical property changes in the materials, or geometric changes in the structure.

In choosing a suitable analysis program, the designers have to take into account the following:

(1) A program which has capable functions to solve the particular problem.
(2) A program which is able to calculate with sufficient accuracy.
(3) A program which performs the analysis with reasonable cost and time.

3.8.3 Analysis Procedure

Figure 3.38 illustrates the typical process of structural analysis or vibration analysis using the finite element method program. For the first step, an appropriate analysis program must be chosen for the specified problem, then the modeling is done by determining the appropriate size of the structure. It may be possible to reduce the size of the model by choosing suitable boundary conditions and loading conditions. The next step is to prepare the geometrical data of the finite elements with visual checking of the validity of the input data through the computer display. The loading data as well as the boundary conditions of the structure are then added.

After executing the analysis program, the calculated result must be assessed to check whether there could be some error in the input data in view of the calculated deformation, stress, etc. If there was a mistake in the input or misunderstanding of the problem, the procedure should be repeated from the beginning.

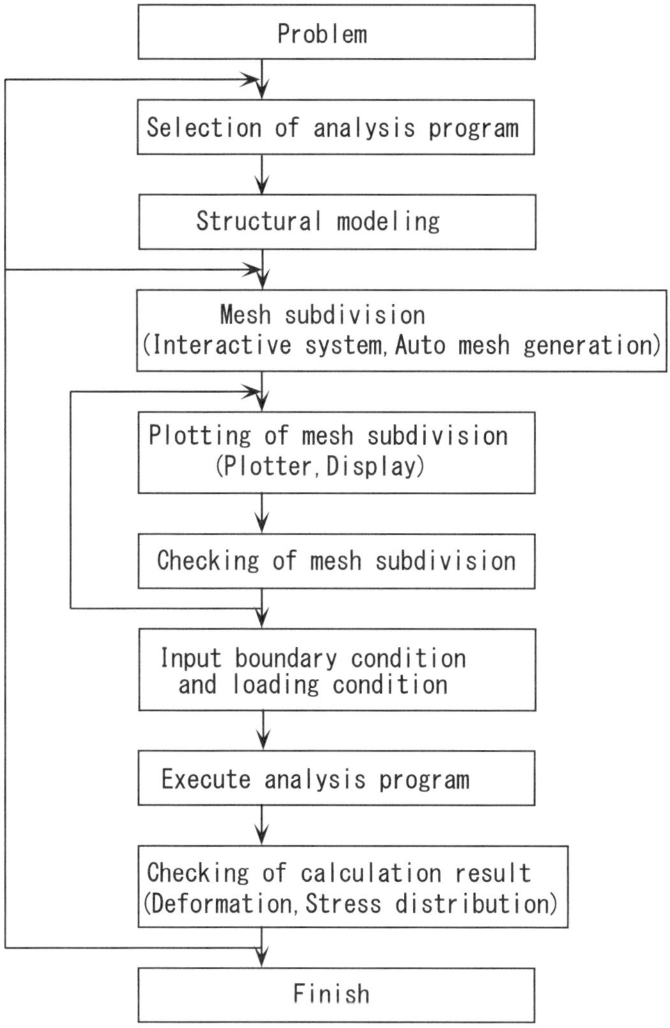

Fig. 3.38 Procedure of FEM structure analysis

3.8.4 Evaluation of Analysis Result

The most important point of strength analysis using a finite element program is to make a reasonable evaluation of the system. If you evaluate the output of the program in a wrong way, or if you fail to read the necessary information from the output, all the calculation efforts become useless for the structural design.

Since the finite element program is able to produce the analyzed result in numerical form, it is helpful to specify the stress or deformation at the point of concern. On the other hand, it is dangerous to believe in the output value without checking the total output data from the global point of view. Therefore, before making a final judgment, we must examine carefully the output data in total and obtain a better understanding of the behavior of the analyzed structure to the applied loads and to the boundary conditions.

It is recommended to use the following procedure in order that you can perform an exact calculation and obtain a reasonable evaluation efficiently:

(1) *Make preparations prior to calculation*: A prediction of the deformed shape of a structure and the stress distribution are needed to evaluate quickly the calculated output. Before starting a calculation, the magnitude of deformation and stress levels is roughly estimated. The analysis of a similar structure in the past is also to be studied.
(2) *Make an equivalent model of the behavior*: It is necessary to create an equivalent model which can give accurate structural response to the loads and which can also evaluate the strength of the designed structure. In addition, the cost and time of computer modeling encourage an economical approach.
(3) *Investigate the calculated result*: We should check whether the calculated results, such as the deformation mode, stress distribution, and eigen-value, resemble the predicted ones. If they differ largely, we had better be suspicious of the input data and consider the reasons why they are different. We should check whether the deformation mode is reasonable considering the direction of applied loads and the constrained degrees of freedom of the boundary condition. We should check whether the total value of input loads coincides with the total value of reaction forces in the output.

By checking the above items, we can judge both qualitatively and quantitatively the validity of the input data, i.e. the structural modeling, boundary conditions and loading conditions, consequently we can be assured of the reliability of the calculated analysis.

Chapter 4
Hull Structure Design System

4.1 Design Flow

The design of a hull structure is generally carried out in three stages: basic design, detail design, and production design.

(1) *Basic design*: Following the determination of a preliminary structural arrangement in the planning step, the midship section drawings are prepared, followed by the rule-based scantling calculations, strength and vibration calculations, and hull steel weight estimation.
(2) *Detail design*: Following the completion of the detailed midship section drawings, which also include the production method; the bow, stern, engine room, and superstructure are designed in detail. These designs take into account the fitting arrangement and hull block assembly process.
(3) *Production design*: To the above detailed design further information for structure manufacturing is added.

Finally the configuration of individual components is determined to provide supplementary information required for the fabrication process, and data is generated to feed the numerically-controlled (NC) machine tools.

Figure 4.1 illustrates the procedure of how to generate the components in each structural design stage, taking a floor plate in the double bottom as an example:

(1) *Basic design*: rough shape of boundaries, plate thickness, materials, outline of slots, arrangement of web stiffeners, etc.
(2) *Detail design*: determination of the 2-dimensional shape which is defined by the mould line, shapes of slots and scallops, as a result of consideration of the cutting, welding, and block assembly method.
(3) *Production design*: the 3-dimensional shape is developed taking into consideration of elongation to compensate for welding shrinkage, the preparation of plate edges. This is referred to the accuracy control and detail work practice.

Y. Okumoto et al., *Design of Ship Hull Structures*,
https://doi.org/10.1007/978-3-032-12500-2_4

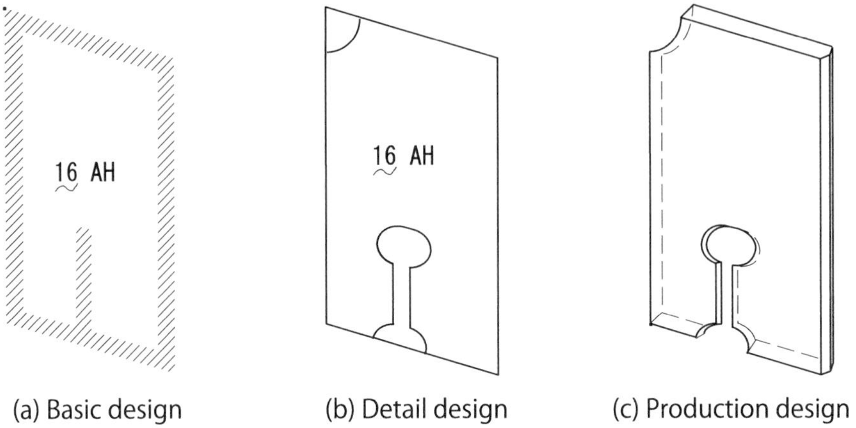

Fig. 4.1 Generation of parts in accordance with design stage

In recent design systems, design data is linked to a production system like CAD/CAM, as shown in Fig. 4.2 [27] and will be integrated in the near future by using a product model like a CIM system. The design data of the old systems was input manually for the production system. By using the product model, so-called concurrent engineering will be successfully applied as shown in Fig. 4.3, and the design time will be shortened.

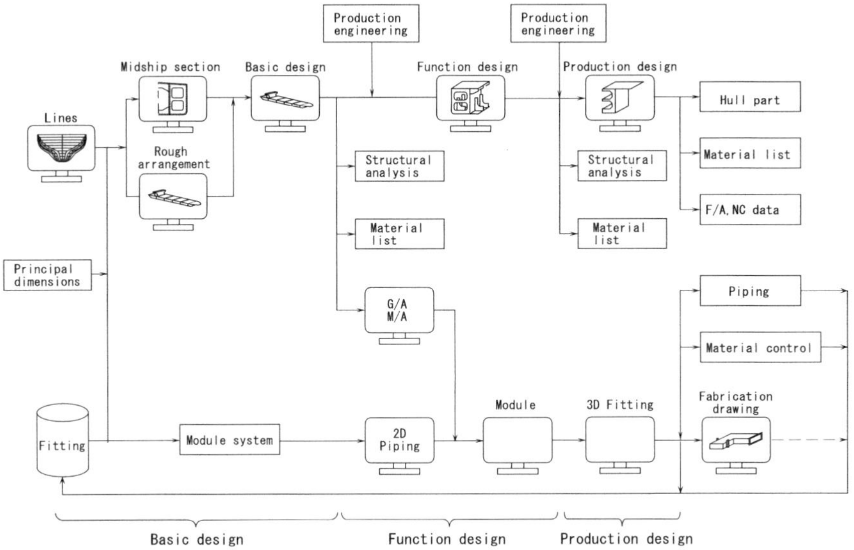

Fig. 4.2 CAD/CAM system

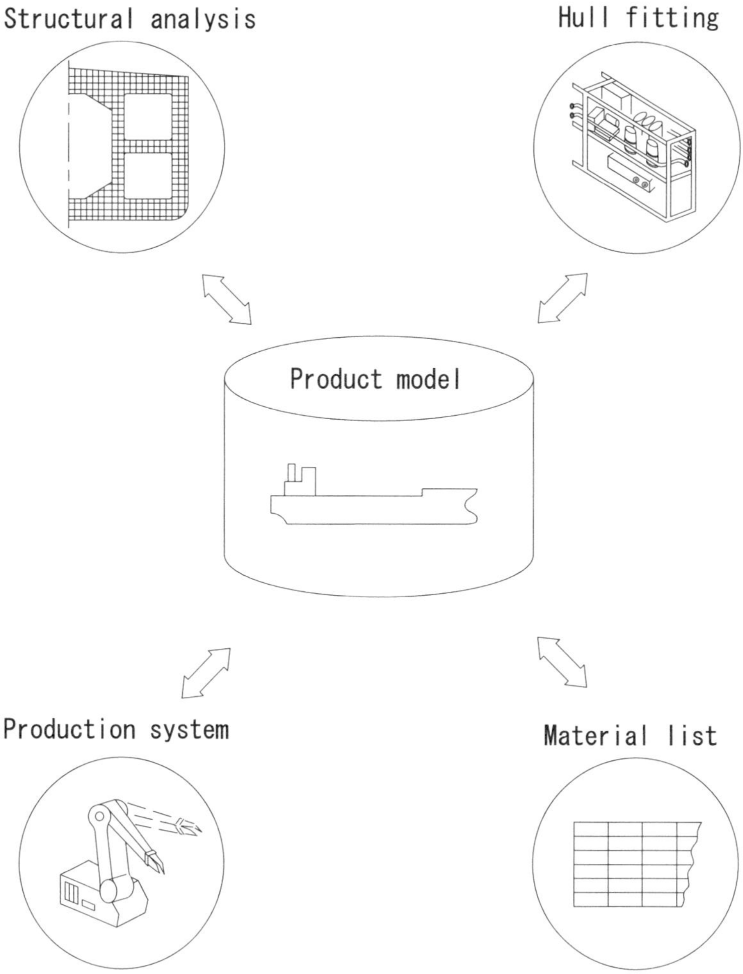

Fig. 4.3 Concurrent engineering using product model

The organization of the design department and the design schedule depends on the shipyard and the type of ship to be designed, however an example is shown in Fig. 4.4, in which hull structure design is carried out in four stages: (a) basic design, (b) function design, (c) detail design, and (d) part design.

Month (Before keel lay)	K-9 K-8	K-7 K-6	K-5 K-4	K-3	K-2 K-1
Stage	Basic design	Function design	Detail design / Parts design		
Drawings	Lines → General arrangement → Midship section → Machinery arrangement →	Shell landing → Shell expansion → Construction profile (panel) → Construction profile (section) →	Yard plan →	Piece table → Parts list →	
	∘ Negotiation to owner, classification society ∘ Functicnal design ∘ Strength analysis, vibration analysis	} do. ∘ Purchase of steel materials ∘ Study of production procedure	∘ Study of production procedure ∘ Definition of detail shape ∘ Naming of parts code ∘ Material list	∘ Definition of detail shape ∘ Welding data ∘ NC cutting data ∘ Nesting data	

Fig. 4.4 Design procedure of hull structure

4.2 Basic Design of Hull Structures

4.2.1 Role of Basic Design

Design work generally takes the form of the so-called design spiral—where a higher precision is progressively achieved at each cycle. The basic design of a ship is as shown in Fig. 4.5 [28]. This sequence starts with the determination of the principal dimensions, followed by the ship's speed and engine horsepower, as well as the type of engine, and continues with the determination of preliminary lines, arrangement of compartments, structural arrangement, fittings, engine room arrangement, etc. until a final setting of the list of materials, specifications, and ship price. Thus, starting from a blank sheet, basic ship design consists of making a series of decisions on alternative choices to obtain the outline of a ship possessing the required characteristics and performance in the shortest possible time.

The hull structure element in Step 1 consists of roughly determining the structural arrangement, while in Step 2 the same elements will be examined in more detail to consider strength and vibration. In Step 3, the scantlings are determined for the principal members forming the midship structure, and approval drawings are prepared, from which strength and vibration calculations are performed. The resulting data serve to estimate hull steel weight more accurately.

Figure 4.6 shows a hull structure basic design system COSMOS, which used to be applied to many new designs over the years in a shipbuilding company, as an example [28]. This system includes four modules as follows:

(1) Rule-based scantling calculations (COSMOS-R)
(2) Strength calculations (COSMOS-S)

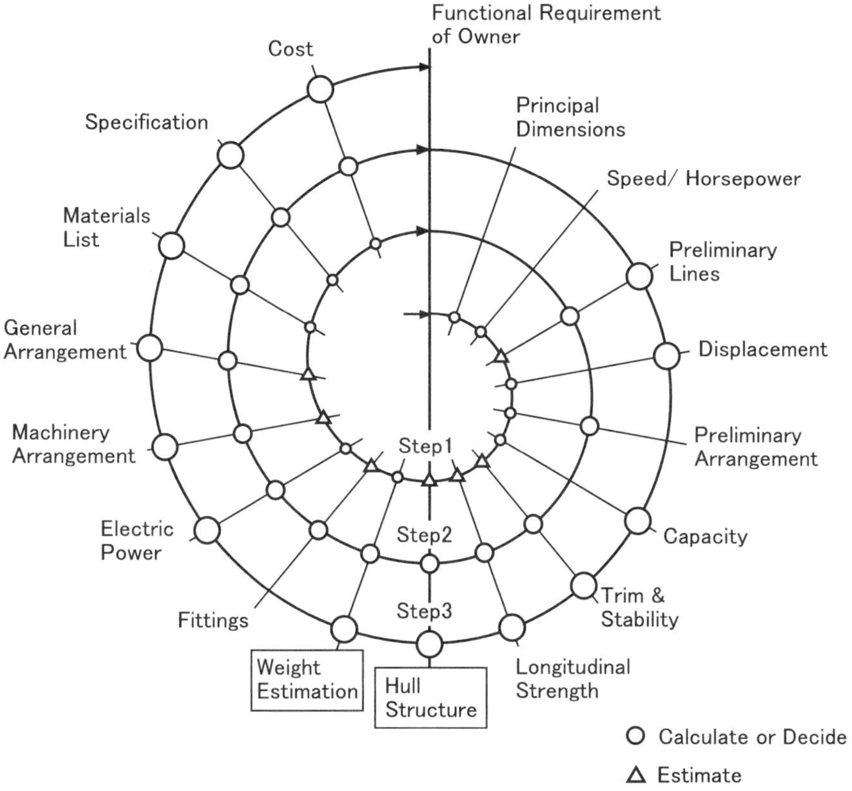

Fig. 4.5 Design spiral of basic design

- estimation of loads using hull motion calculations
- direct stress calculations using FEM analysis
- evaluation of stress, fatigue strength, and buckling strength.

(3) Vibration analysis (COSMOS-V)
(4) Hull steel weight estimation (COSMOS-W).

An important requirement of hull basic design is to respond quickly to inquiries from ship owners or other basic design sections, as well as to ensure safety and performance with respect to structural strength. To meet such demands, a procedure is needed for estimating the total hull steel weight quickly and accurately using simple formulas, charts from experience of sister ships, or trial-and-error method. A data base system, including a feedback system, is essential.

The drawing to be studied first is the "rough midship section", considering classification rules, design standards, production methods, etc. The important thing is that it satisfies not only the requirements of the classification society, but also experience from past ships including damage history of cracks and buckling. In addition, since the allowable value (or safety factor) depends on the method and accuracy of external

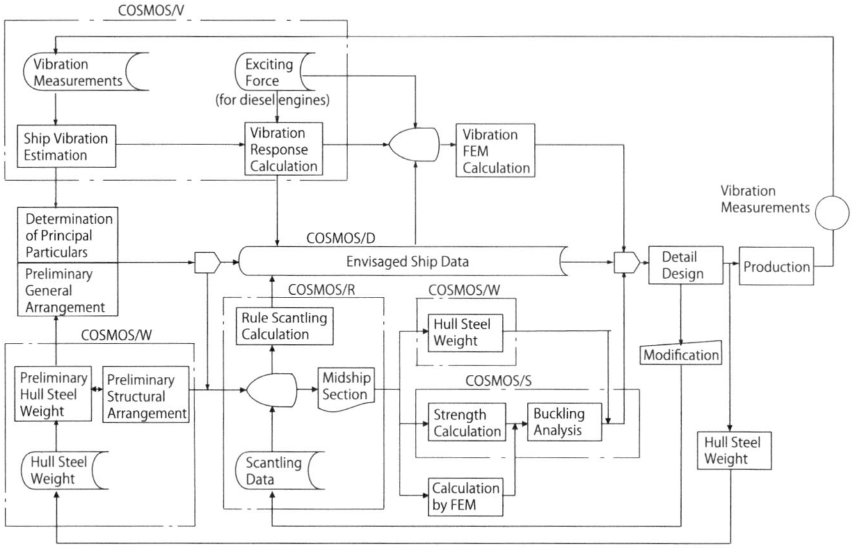

Fig. 4.6 System flow of COSMOS

load calculations and FEM structural analysis, the structural arrangement should be revised in accordance with any improvement in calculation methods.

In the arrangement of primary and secondary structural members and in their scantling calculations, optimum design technique should be applied to reduce hull steel weight and also working man-hours in cooperation with the design section in charge of the general arrangement. Hull steel weight directly influences ship cost and dead weight; therefore, it should be checked often during progress in the design.

Hull vibration affects not only comfort in living quarters on board, but also occurrence of cracks in hull structure and troubles with instruments. In hull girder vibration, possible resonance with unbalance forces or moments due to the diesel main engine, is checked. If resonance is expected, countermeasures to avoid resonance are considered, or ways of reducing the external force or moment are studied. An electric balancer is often provided in steering gear room to reduce the vibration amplitude.

In addition propeller exciting forces affect vibration in the superstructure; therefore, the number of blades and propeller aperture should be carefully studied. Local vibration should be checked in structural members in the engine room, in the aft construction, and in tanks in the hold part.

4.2.2 Check of General Arrangement

Since the general arrangement also affects strength of the ship's hull, the hull structure designer should watch it carefully and give suggestions from a strength point of view. The main items are as follows:

(1) Ratio of ship length and depth (L/D): If L/D is larger, deflection becomes larger, and hull weight becomes larger because of the difficulty to keep the midship section modulus.
(2) The arrangement of longitudinal and transverse bulkheads, double bottom, hatch openings in hold space—considerations for the structural continuity and global strength are needed
(3) Midship section shape—double bottom height, side tank shape including bilge and top side tank
(4) Type of bulkhead in cargo hold—plane, corrugated, or double plated
(5) Transverse (frame) space and longitudinal space, considering minimum steel weight and construction cost
(6) Position of superstructure, deck house, deck machinery
(7) Arrangement of steel walls in superstructure, considering vibration.

4.2.3 Check of Other Drawings

Engine particulars and machinery arrangement should be checked by the structure designer from strength and vibration points of view.

(1) Main engine particulars such as type and revolution number, regarding vibration
(2) Arrangement of the double bottom, flat, and bulkhead in the engine room, regarding engine room rigidity from engine vibration.

Propeller particulars and rough lines should be also checked by the structural designer from strength and vibration points of view.

(1) Propeller blade number, regarding vibration
(2) Bow shape, considering wave impact
(3) Propeller aperture, considering transmission of the propeller exciting force to the ship's hull
(4) Shape of body in way of propeller bossing, regarding vibration.

4.2.4 Optimization Technique in Basic Design Process [30]

As described in Sect. 4.2.1, the conventional practice in ship hull design is to follow what is known as the "design spiral", whereby the design parameters are first assumed on the basis of past experiences, and then modified through repeated cycles of trial and error across a variety of different design sections. This is quite time consuming, and as a result, varieties of simulated design parameters are limited to only a small number of cases. Therefore, the design remains a step-by-step satisfaction of design constraints rather than optimization.

From hull structure designers' point of view, for example, shorter ship in length will result in lighter hull steel weight, but it will be necessary to increase ship's depth and breadth to keep tank volume. This small L/B or L/D will make propulsive and maneuverability performance worse. To compensate this, if C_b is designed smaller, the extent of the curved shell will increase, resulting in increased construction cost. Regarding spacing of stiffeners and frames, smaller spacing will make the steel weight lighter, but also increase the number of pieces and welding length, leading to increased construction cost.

These are only a few examples. Actual investigation and decision making conducted in the basic design stage from various standpoint in each design section are very widely ranged. They are considered under various design conditions and constraints, such as international conventions, regulations, classification rules, internal design standards, constraints from operational aspects, constraints from construction aspects, etc.

Because of the complexity and variety of the problems, it has been difficult to apply a direct optimization approach to hull basic design. In the early times, the sequential unconstrained minimization technique was successfully applied to optimization problems of primary and secondary structural member arrangement [29]. However, to achieve true global optimum design, we must simultaneously deal with principal dimensions, hull form, and other general arrangement parameters. To deal with such complex problems, new optimization approaches such as the genetic algorithm have been developed, and are being successfully applied to each type of ship [30–32]. Figure 4.7 shows the conceptual flow chart of optimum structure design of a double hull tanker applying the genetic algorithm.

4.3 Design by Direct Load and Structural Analysis [33]

4.3.1 Approach to Strength Evaluation of Hull Structures

Historically, strength evaluations of hull structures have been conducted under limited load conditions due to the constraints imposed by the evaluation system and available time. Longitudinal strength was assessed using beam theory by evaluating the section modulus of longitudinal members against the class rule loads described in Sect. 2.2. Transverse strength, while directly calculated using the finite element method, was also evaluated under class rule loads, similar to longitudinal strength, and focused on a limited number of load conditions. The restriction to these specific load conditions was based on theoretical considerations of the relationship between

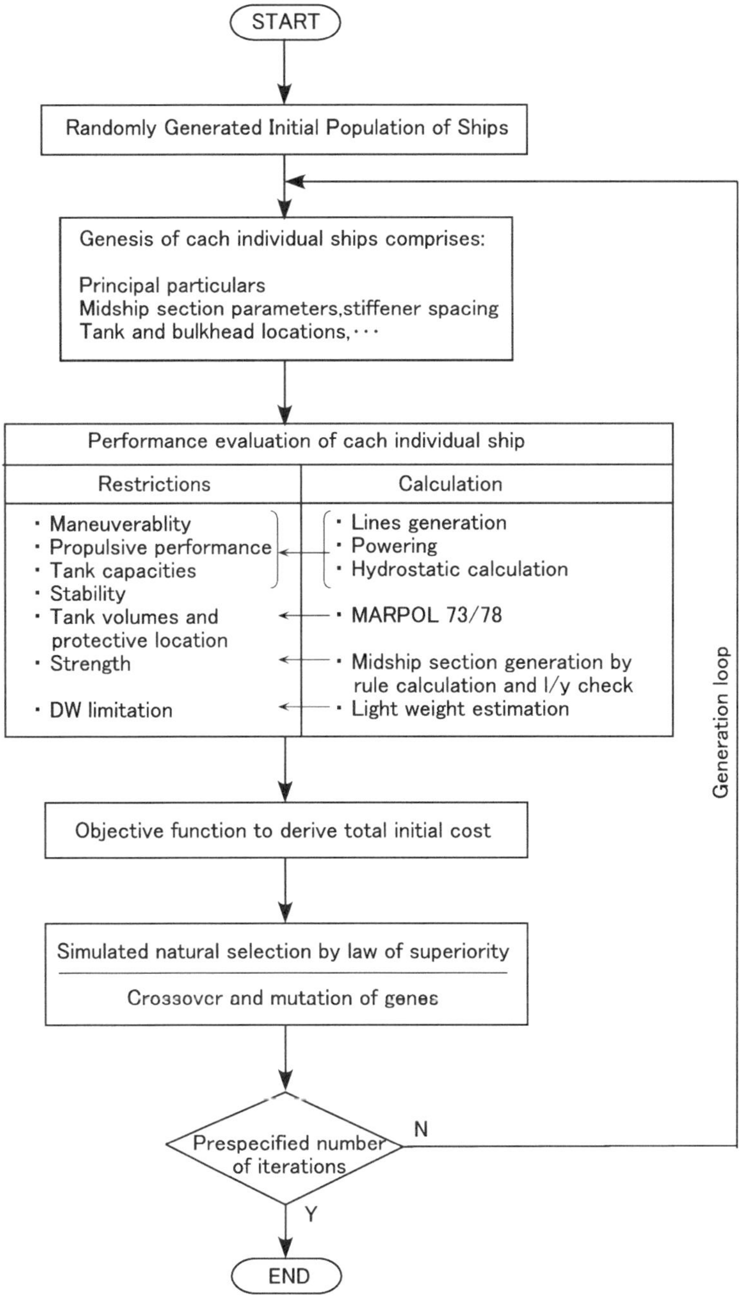

Fig. 4.7 Concept of optimum structure design of a double hull tanker applying genetic algorithm [30]

hull structure failure modes and loads, as well as accumulated experience from past damage incidents.

When making changes to hull size and structural configuration within the bounds of existing knowledge, strength evaluations under such limited load conditions were deemed sufficient. However, the recent rapid increase in hull size and complexity has raised concerns about damage occurring under stress conditions that cannot be reproduced by past experience-based limited wave conditions.

Therefore, the application of strength evaluation methods based on motion calculations, described in Sect. 2.4, has advanced. There are two main approaches:

(1) One method involves assuming wave conditions that generate the maximum load during the ship's lifetime and applying these waves or their equivalent moments as loads to a finite element model for strength evaluation. This wave is referred to as the "Equivalent Regular Wave."
(2) The other method involves assuming multiple waves of unit wave height, deriving stress response functions from the structural response to these waves, and calculating the expected value of the resulting stresses.

4.3.2 Strength Evaluation Using Equivalent Regular Waves

Using long-term prediction results based on the frequency response function (FRF) obtained from ship motion calculations, the expected value of the maximum load condition over a recurrence period of 20 or 25 years is determined. Next, an equivalent design wave is assumed, setting the wavelength and wave direction that reproduce these conditions. The maximum load condition refers to states where pitching/rolling angles, bending moments, and torsional moments are at their peak, chosen based on historical data. By applying this assumed equivalent design wave to a finite element model, stress analysis is conducted. The analysis flow is shown in Fig. 4.8.

This method is effective in the initial design stages as it requires evaluation of fewer load cases. However, there is a possibility of missing conditions where maximum stress occurs for certain structural members.

4.3.3 Strength Evaluation Using Stress Response Function

This method involves determining the extreme values of structural member stresses through long-term predictions based on stress response functions. Structural analysis is carried out for all combinations of wavelengths, wave directions, and phases. To cover all combinations, the number of load cases for the finite element analysis can be as high as about 2,000 cases, even if a minimum of 20–30 cases for wavelength, 12 cases for wave direction, and 6 cases for phase are selected.

The analysis flow is shown in Fig. 4.9. For each load condition, motion calculations are performed, and finite element calculations using the resulting hull motions

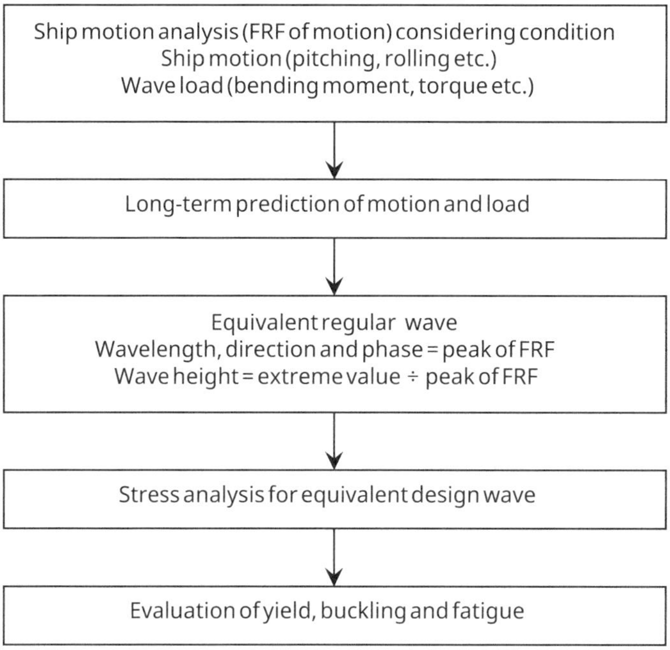

Fig. 4.8 Design flow of equivalent regular wave method

and loads are conducted to obtain the frequency response functions of the stresses of the elements included in the analysis model. Figure 4.10 is an example of applying loads to a finite element model, and Fig. 4.11 is an example of a stress response function. Using these stress response functions, long-term stress predictions are made, considering the ship's routes and the periods over which it navigates throughout its lifetime.

This method ensures that all members of the hull have sufficient strength. In shipyard hull structure design, this method is already in practical use to prevent unforeseen damage that could occur due to the enlargement of hulls, which cannot be predicted by conventional methods. In an example where this method was applied to the structural analysis of a large container ship of unprecedented size, it predicted the occurrence of high stress, which could not be anticipated by conventional methods, and measures were taken accordingly [34].

4.4 Structural Drawings

Each hull structural member is developed with regard to detail in accordance with each design stage, and is modeled on a computer recently. Therefore, drawings are now the only output of design by computer. The aim of a presentation by drawings

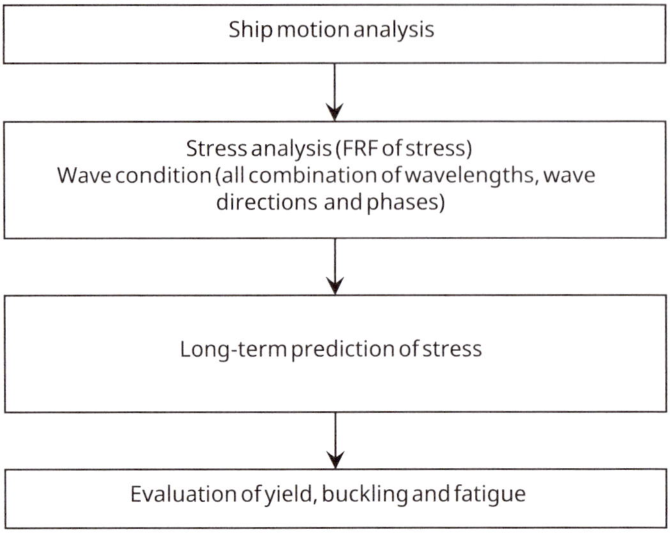

Fig. 4.9 Design flow of stress response function method

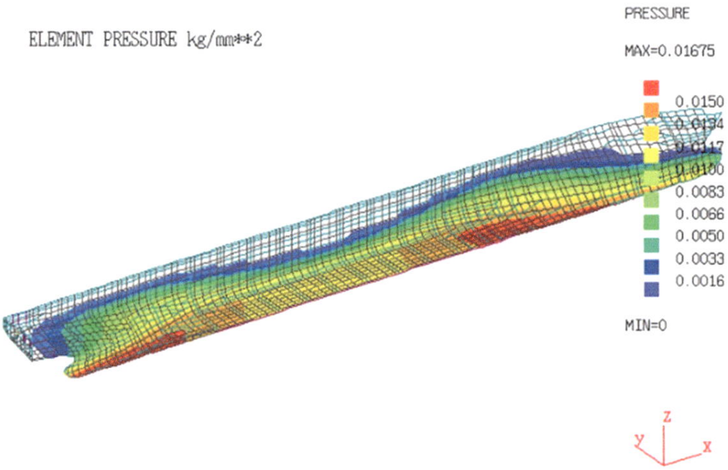

Fig. 4.10 Pressure distribution diagram due to wave under certain wave conditions

is to get approval from the ship owner and classification society, and to inform production engineering.

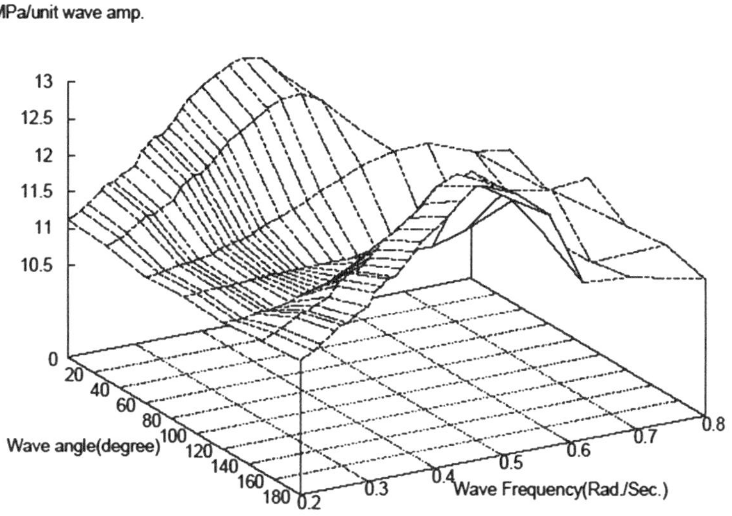

Fig. 4.11 An example of a stress response function diagram

4.4.1 Approval Drawings

Approval drawings are prepared to get approval from the ship owner and classification society, which are called the "key plan". The structural strength of each member and the ship as a whole, including vibration, should be included in the drawings. Since these drawings are also used to provide information to other design sections and the production engineering section, they should be submitted as soon as possible in order to shorten the design time.

After midship section drawings are completed, shell landing is firstly planned using lines, and panels representing decks, bulkheads, flats, etc. are successively designed. In these steps, the location of seams and butts of plates, and the position of longitudinals are decided, and these scantlings are calculated.

During the next step, transverse members are designed on cut sections of the above-mentioned panels. Both longitudinal and transverse members are now designed and calculated using computer.

The timing of the orders for the required materials is important in order to shorten the building period. Since steel sections such as angle and bulb plate need much time to procure, their scantlings and quantity should be decided as soon as possible, on discussion with the production engineering side.

4.4.2 Detail Drawings

Detail drawings are also called the "yard plan", which adds the following information in the key plan:

- detail information of the assembling method such as detail of joints
- welding scheme
- detail information of fitting design, such as piping holes, supports, reinforcements, etc.
- piece codes.

The drawing is usually developed by editing and adding to the key plan. These data give information to the next stages shown in the following:

- *piece drawing*: In general, a piece drawing is generated from CAD data representing contour shapes, markings, edge preparation data, etc. These data are called CL data (cutter location data). Figure 4.12 shows an example of nesting using CL data of each piece. On the other hand, long strip material such as shape steels, face plates, flat bar, built-up longitudinals, etc. are shown using table format which includes rough sketches, sizes, and piece codes. These data generate dimensional piece drawings of strip, and are called IL data (item list data), whose example is as shown in Fig. 4.13.
- *rough cutting plan and detailed cutting plan (nesting)*: Nesting is the process where many complicated hull structure shapes are cut from the rectangular steel

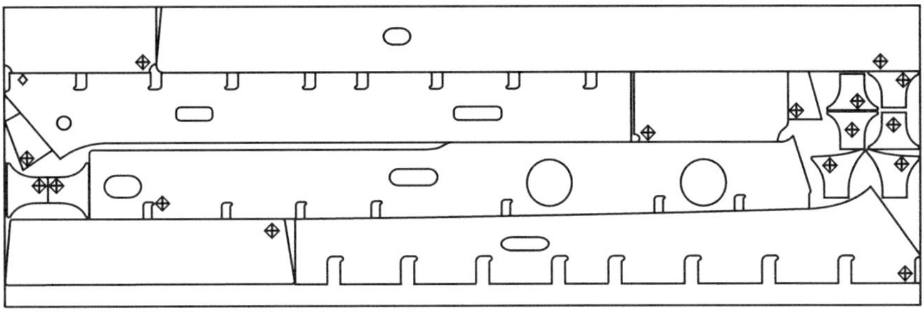

Fig. 4.12 Nesting of plates

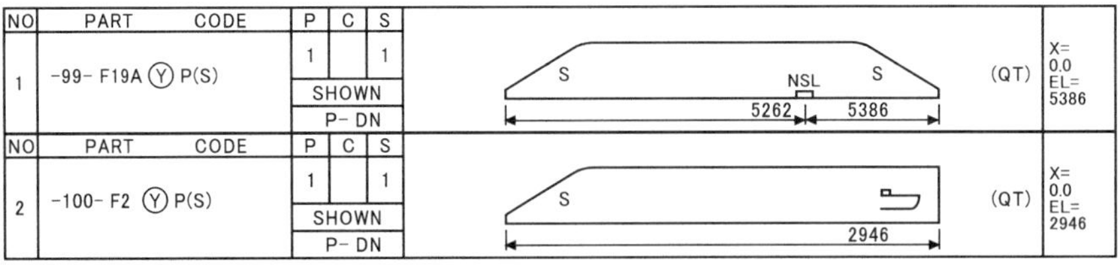

Fig. 4.13 Piece drawing (IL data)

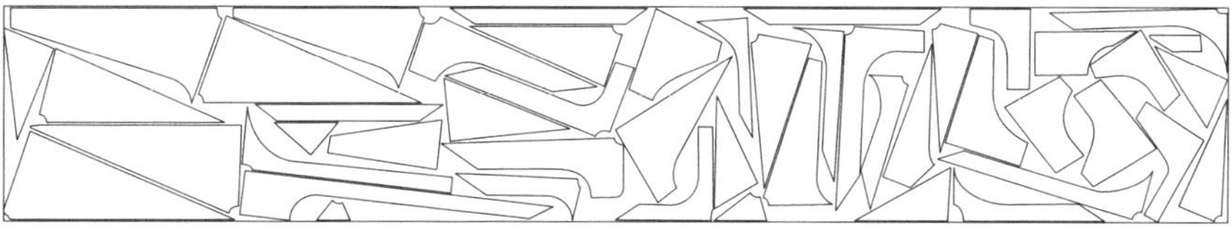

Fig. 4.14 Output from FINEST automatic nesting system

plate. There are two stages of nesting: one is nesting to produce a rough cutting plan, which is used to decide the dimensions and quantity of steel plates to be purchased. The other is nesting to produce a detailed cutting plan, where final piece data are arranged on rectangular plates and are directly transformed to NC cutting data.

Automation of the nesting process has been studied for many years, and some practical systems are in actual service. One example is the FINEST automatic nesting system [35, 36], which reduces the man-hours and the ratio of scrap. Figure 4.14 is an example of output from the FINEST system. It is observed that the pieces are not regularly arranged, yet the scrap ratio is excellent. In this system, narrow flat bars are required to be arranged in the longitudinal direction.

- the ordering of the steel
- welding length (for work management)
- fitting design
- working plan for each stage
- quality control.

4.4.3 Production Data

Production data includes cutter path for nesting data, and piece table data.

(1) *Cutter path data*: Cutter path data gives torch movement for NC cutting of the web plates, brackets, floor plates, girders, etc., which are nested on a steel plate. The geometric piece data includes:

- elongation to compensate welding shrinkage
- details of the groove to weld
- details of the shape of the slots, scallops, holes, etc.

(2) *Piece table*: Piece table includes:

- code name
- information for plate bending
- information for primer coating on steel surface
- place or stage to deliver
- number

- weight
- surface area etc.

4.5 Standardization

Standardization is essential in design, which contributes to design efficiency because the procedures are already defined, and it makes high quality design possible because it includes past experience. A feedback system of experiences in service is shown in Fig. 4.15.

On the other hand, standardization has some demerits:

- It is costly, because it is often necessary to review or renew.
- Creative attitude to new ideas may decrease gradually.

Standardization includes that of shape, arrangement, and design procedure. At early stages of design, a standard of arrangement is useful, while shape standards are applicable in the detail design stage. In these cases, a parametric design method may offer a more flexible approach. Standardization of the design procedure is needed for important structures from a strength point of view in order to improve safety and reliability.

Key points of standardization are as follows:

(1) Standardization of the superficial or external appearance is not effective, but a systematic combination of the various elements is useful by analysis of the function of each. Combination design or module design may be applicable. As shown in Fig. 4.16, standardization of each component is useful, but not for the shape as a whole.

(2) Having flexible standards is essential, to allow changes of ship type, design method, and production procedure. If the standard is too rigid, the applications become narrow. The economic factor and applicability related to flexibility are illustrated in Fig. 4.17, and the optimum zone should be investigated to ensure suitability of purpose. Thanks to factory automation and computer techniques, the shape of material or pieces can be cut and fabricated with a degree of flexibility.

(3) A global and long-term view is necessary for standardization, hence patience is required. Even if it is not profitable for one ship, an accumulation of applications yields a profit. A shortsighted view reduces the impact of standardization.

(4) *Top-down management*: In order to instigate the “standard based design system”, top-down management is necessary. A manager should enforce the application of standards by designers and also always make effort to improve them.

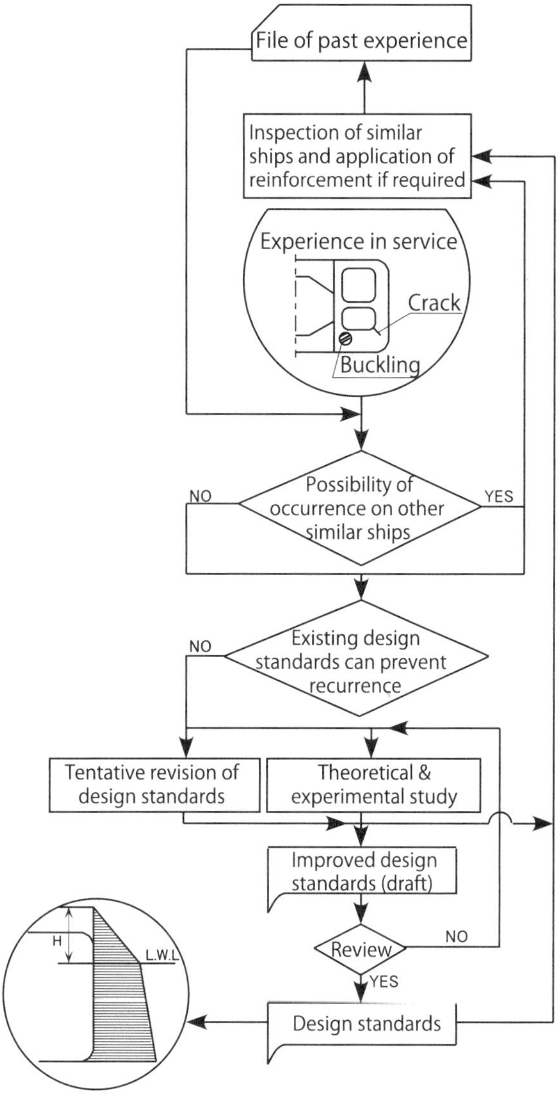

Fig. 4.15 Feedback system of experiences in service

4.6 Negotiations with Owner

Hull structure designers usually have little chance for discussion with the ship owner, compared with fitting designers. However it is very important to have good communication with the owner in order to avoid unnecessary troubles, and to get a satisfactory outcome.

Discussions with them are noted on "specifications" and approval drawings. In the specifications the following items should be included:

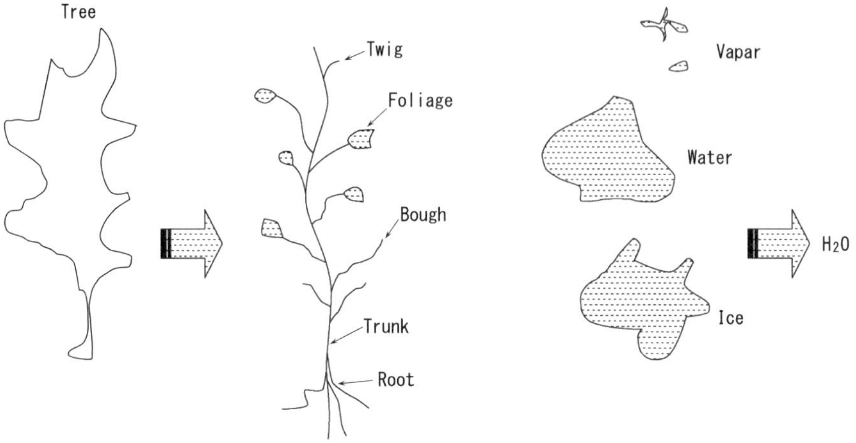

Fig. 4.16 Combination of functional elements

Fig. 4.17 Optimum degree of standard

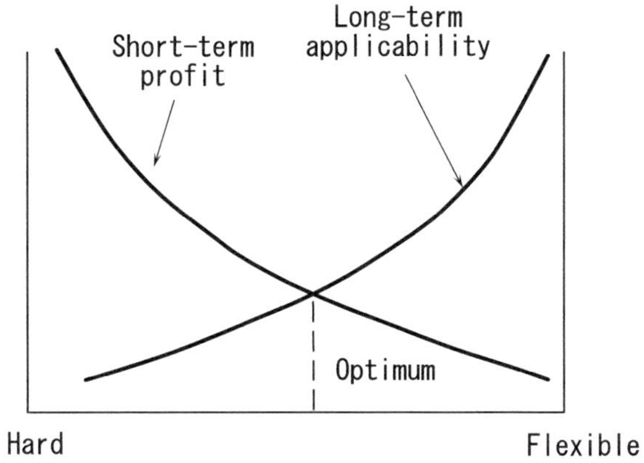

- application of hull steel materials, especially for higher-strength steel, or other special materials
- loading conditions including specific gravity of cargo load
- structural detail of stern frame, rudder, and rudder carrier.

The ship owner is usually interested in the rudder system because of the necessity of periodical inspection and maintenance.

- type of chain locker, sea chest, etc.
- vibration, especially allowable levels in living quarters, referring to ISO 6954 guidelines.

As to working practice,

- welding scheme

- hull construction standards
- hull construction and remedial quality standards, referring to JSQS (Japan Ship Quality Standard) or IACS Recommendation No. 47.

Chapter 5
Progress in Ship Design

5.1 Increase in Ship Dimensions of Tankers

The increase of ship dimensions (from VLCC to ULCC) started around 1950 and came to a stand-still in 1980 after the oil crisis. Here the dimension increase is reviewed from the hull structure design viewpoint.

From 1950 onwards it started with the standard tanker of 18,000 DWT or 20,000 DWT; the super tankers of 28,000 DWT (the *Yuhou-Maru* and *Kohou-Maru* of Iino Shipping Co., Ltd.), 35,000 DWT, and 45,000 DWT (the *Gohou-Maru* of Iino Shipping Co., Ltd.) were built one after the other. The mammoth tanker of 100,000 DWT (*Universe Apollo*) was also built at the end of the 1950's by NBC (National Bulk Carriers) Kure Shipyard.

These ships had common structural characteristics, i.e., transverse spacing of 3 m, tank length of 9 m for the standard tanker and 12 m for the super tanker and above. The tank arrangement was in general: three tanks next to each other, divided by 2 longitudinal bulkheads.

Even with the *Universe Apollo*, the length of the center tank was 12 m, divided by three longitudinal bulkheads; so there were 32 center tanks in 2 rows with the following dimensions: 12 m length, 12 m breadth and 21 m depth, just like a deep well. The length of the wing tanks was 24 m.

The main policy of the scaling up of ship size in the1950's was to apply already established technology. Then in the1960's this policy changed into rationalization of which the original idea was: "sardines and tuna have the same number of bones". It is natural that big fish have big bones which are arranged with more space in between and small fish have their smaller bones arranged more closely. Therefore, the number of bones is equal. Subsequently it would be natural to design a big ship with wider frame spacings and longer tank lengths.

Before the tank length could be increased, a lot of research was carried out on cargo oil movement in tanks. From this, the effectiveness of the swash bulkhead was confirmed. The 50 m long cargo oil tank with a swash bulkhead was designed, and

Y. Okumoto et al., *Design of Ship Hull Structures*,
https://doi.org/10.1007/978-3-032-12500-2_5

the transverse space was increased to about 6 m, thus breaking through the 3 m limit. In Fig. 5.1 tank arrangements are shown for 12 m tanks and longer tanks.

One important item which contributed to the scaling up of ships was the rationalization of longitudinal strength, shown in Fig. 1.5. A ship is subjected to hogging and

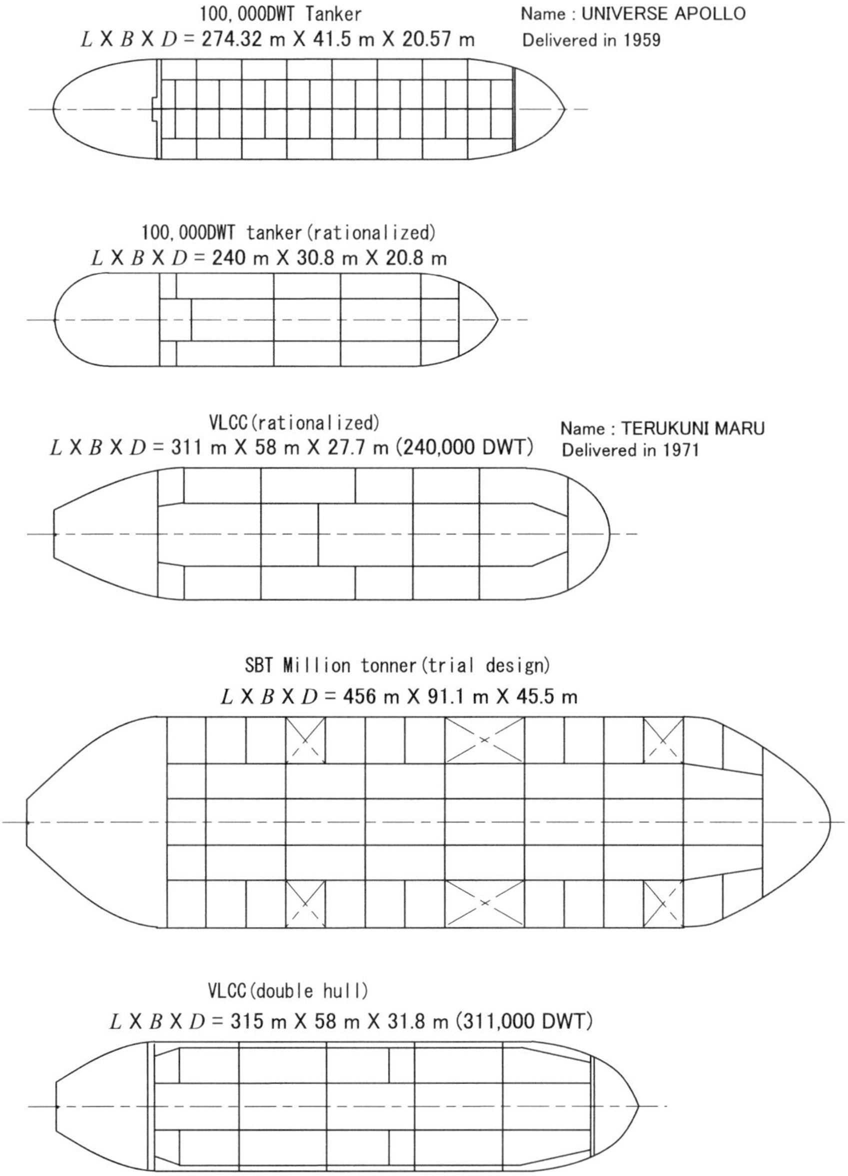

Fig. 5.1 Tank arrangement of each era

sagging moments in waves. The magnitudes of these moments had been calculated by assuming that the wave length is equal to the ship length and the wave height is 1/20 of the length. For a large ship this assumption cannot be applied. For example, the wave height of a 300 m long wave is 15 m, which is not realistic. The research results proved that the actual wave height for a long wave is smaller than expected, and therefore, the required longitudinal strength could be reduced. Consequently, a large ship can be constructed without riveted doubling plates on deck and bottom.

By these three technical innovations, longer tanks, wider spacing and reduced longitudinal strength, the scaling up of ships was achieved smoothly. Many large ships, designed and constructed with reduced longitudinal strength now have terminated their service lives without any serious failures. This means that the reduced longitudinal strength has still some margin.

On the other hand, many failures were caused by the longer tank and wider spacing in the transverse strength members of large ships built in 1960's. Cracking and buckling occurred at the cross points of longitudinals and transverse webs, the so called slot parts, where the load supported by the longitudinals is transmitted onto the web. Wrong design was the main reason for these failures, that is by not applying a thicker plate to the web which supports the increased load (about two times) from wider spacing. Moreover, as for the slot on bottom transverse, the longer tank had a relative displacement between the longitudinal bulkhead and the side shell which nullified the fixed condition on both ends (design condition) of the bottom transverse. This accelerated the failures. Examples of failures are shown in Fig. 5.2. Technical details of this kind of failures are explained in Sect. 10.5, Part II.

The main loads applied to a hull structure are weight and buoyancy, which act in the vertical direction. Against these vertical loads, vertical members, such as the side shell, longitudinal and transverse bulkheads, form the framework of the hull structure. For small ships the framework is rigid enough to keep both ends in the fixed boundary condition of the primary supporting members. For a large ship, which has longer tanks, the framework is not rigid enough and causes many failures.

In the hull structure this framework is most important and should be defined before longitudinal, transverse and local strength. The strength of this framework is called "basic strength" hereafter. In Table 5.1 the technical conditions, which led to the scaling up of ships, are shown.

In 1973 "The International Regulations for the Prevention of Pollution at Sea by Ships" was accepted, thus limiting tank size and tank arrangement. In addition to the minimally required structures for carrying cargo oil, a ship was to have extra structures to prevent sea pollution. This regulation constrained the scaling up of ships. The prevention of sea pollution regulation had a drastic effect on the rationalization which continued for a long time after World War II.

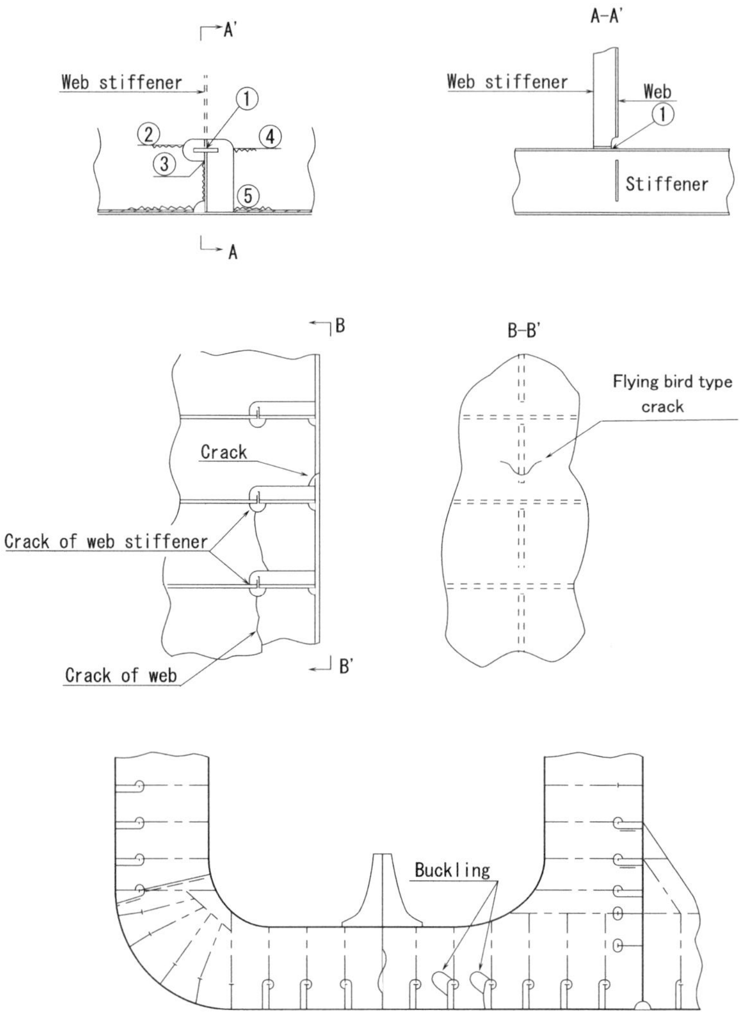

Fig. 5.2 Examples of damage around slot

5.2 Specialization of Ships

To achieve more efficiency in transportation, specialization of ships was promoted simultaneously with scaling up. In addition to tankers and general cargo ships previously operated, the ore carrier, bulk carrier, container ship, car carrier, liquefied gas carrier, car ferry, twin hull ship and the Techno-Super-Liner appeared as well as other special purpose ships. These specialized ships introduced new problems in structural design, which have been investigated and solved by hull structure designers. The

Table 5.1 Technical environments to realize ship's large sizing

Research items	Results	Problems	Countermeasures
Brittle fracture	Thicker plate & higher tensile steel can be applied	Higher material cost	Improved material (TMCP) (See Part I Chapter 6.3)
Welded construction	Lighter & bigger ship Block building method	Reduced buckling strength due to initial deformation	Improved structural system Allowable limit of deform
Sloshing pressure of liquid	Longer cargo oil tank	Relative displacement between side shell and longl. bhd.	Total strength design Rigid swash bhd
Estimation of wave load	Reduced longl. bending moment	Lack of trans. & local strength Unordinary casualty	Reinforcement Operation manual
Finite element analysis	Reliable design based on reduced piece number with wider spacing	Structural arrangement exceeding past experiences	Development of accurate direct calculation method
Shorter & wider hull form	Reduced longl bending moment	Higher resistance	Smaller Cb
		Poor maneuverability	Bigger rudder area
Plastic design	Rationalized trans. member	Lack of fatigue strength	Fracture control design
Unti-corrosion	Electrical & chemical methods	Anti-corrosion in ballast tank	Painting

in-service experiences with these special ships were fed back and applied in the next ship designs. Midship sections of such specialized ships are shown in Fig. 5.3.

The ore carrier has a long and big ore hold in the middle of ship which is sandwiched by ballast tanks. To achieve a more efficient handling of ore, less transverse bulkheads are provided in the cargo hold and the hatch opening is bigger than that of an ordinary cargo ship. The transverse bulkhead is an important member to maintain the strength of the transverse section of a ship, and special care must be taken when designing a hull structure with a long hold. In this case the double bottom and decks between hatches play a major role of maintaining the strength of the transverse section of a ship.

In the case of a bulk carrier, the hold volume has priority, for which a lower double bottom and holds without structures are a prerequisite. There are hopper tanks on both sides of the double bottom for good cargo handling and shoulder tanks below the upper deck, thus avoiding cargo movement and creating room for ballast water. The transverse strength of the double bottom and the side frame are retained by the torsional rigidity of hopper and shoulder tanks, supported by transverse bulkheads. The torsional rigidity depends on the distance between transverse bulkheads, namely the hold length. Then for the bulk carrier, hold length is a very important factor. For this reason, it once was common sense for ship designers that 7 holds for a 50,000 DWT, and 9 holds for a 100,000 DWT ship should be the norm. A cargo hold without a structure in it, means an extra difficulty to maintain the long spans of the upper deck and double bottom. However, the torsional rigidity of hopper and shoulder tanks makes it possible. It is appropriate to say that this was a rather clever design. The hopper and shoulder tanks are connected by an ordinary frame which is another clever design.

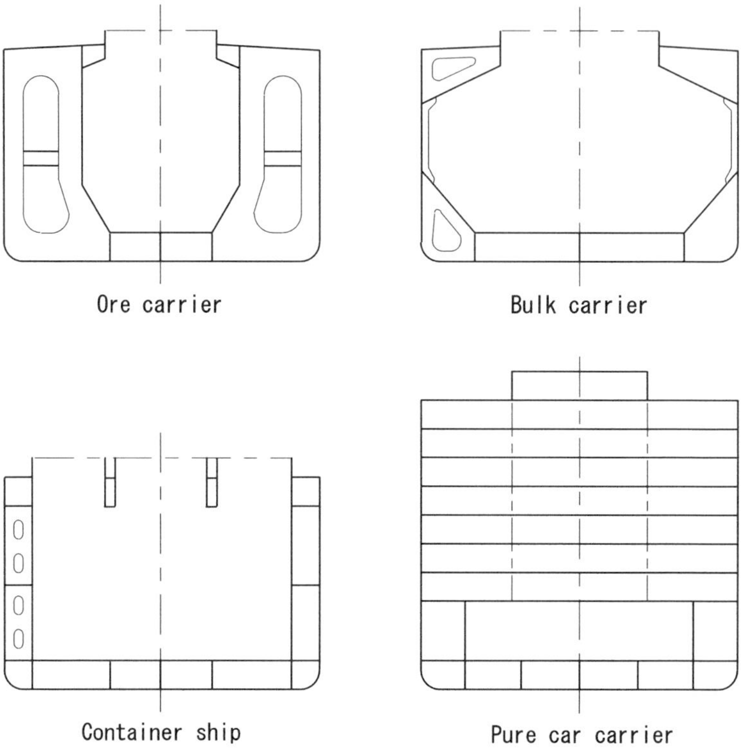

Fig. 5.3 Specialized ships

A container ship has very big hatch openings and a small upper deck area. This is a problem for the longitudinal and torsional strength, but transverse bulkheads are suitably arranged to maintain the strength in the transverse section.

One of the important characteristics of car carriers is that no transverse bulkheads are provided in the car hold, in order to maintain good accessibility. This causes difficult problems to be handled in the structural design such as racking deformation because a transverse bulkhead is like a node of bamboo and the total strength can be obtained from the transverse section of the node. Moreover, requirements are strict from the operational viewpoint; for example, the number of pillars should be minimum; girders and deck transverses should be as shallow as possible, and so on.

Liquefied gas carriers are dedicated to transport liquefied gases such as LPG and LNG. One of the important characteristics of this kind of vessels is low temperature of the cargo, typically −42 °C for LPG and −162 °C for LNG. To avoid brittle fracture due to this low temperature, special materials are employed for the cargo containment system depending on the design temperature. Specifically, LPG tanks are made of low temperature steel, and LNG tanks are made of aluminum alloy, stainless steel or nickel steel. Thermal contraction due to the low temperature and associated thermal stresses should also be taken care of in the design. Another important aspect of liquefied gas carriers is that their cargo is inherently explosive and dangerous. Therefore, accidental cargo leakage should be strictly controlled, for example, with a secondary barrier which can contain all the potential leakages safely for a specified period, specifically 15 days stipulated in the IGC Code [37]. The IGC Code requires provision of a full secondary barrier (type-A), while a reduced partial secondary barrier is permitted where an equivalent level of safety is demonstrated with respect to the tank failure and cargo leakage (type-B). No secondary barrier is required for the tanks based on pressure vessel criteria where structural failures and cargo leakages are unlikely to happen (type-C).

Typical midship sections with different kind of cargo containment systems are shown in Fig. 5.4. The world first LNG carrier, *Methane Pioneer*, delivered in 1958, employed independent prismatic tanks, called the *Conch* system, but the tank system was not successful commercially mainly due to their high cost associated with the full secondary barrier required as type-A tanks. After that, two major systems dominated the market; one is the *Moss* spherical tank system developed in Norway (Fig. 5.4a) and the other is the membrane tank system (Fig. 5.4b). The *Moss* system is made of an unstiffened spherical tank made of aluminum supported by a skirt at the equator. This system succeeded in demonstrating the equivalent safety thanks to its simple structure, and in reducing the cost by applying a partial secondary barrier as a type-B independent tank. In 1990's, IHI developed a stiffened prismatic tank system (*IHI-SPB* tanks; Fig. 5.4c) under the type-B concept, and several LNG carriers have been built under this concept. The *IHI-SPB* system succeeded in resolving many challenging problems such as weld bead shape control to ensure sufficient fatigue strength, high fidelity finite element analysis of the entire hull and tank structures to simulate the interaction between the hull and the tanks completely.

Sloshing of liquid cargo has been a vital problem for liquefied gas carriers. According to Eq. (2.3.1), the natural period of the internal liquid becomes larger when the tank is wider and the liquid level is lower. Therefore, for large tanks, the internal liquid motion tends to resonate with the ship motion, causing severe sloshing problems. The membrane system is inherently vulnerable to sloshing problems because of no internal structures to dampen the liquid motion, and has been practically operated under the filling restrictions prohibiting intermediate filling levels. The *IHI-SPB*

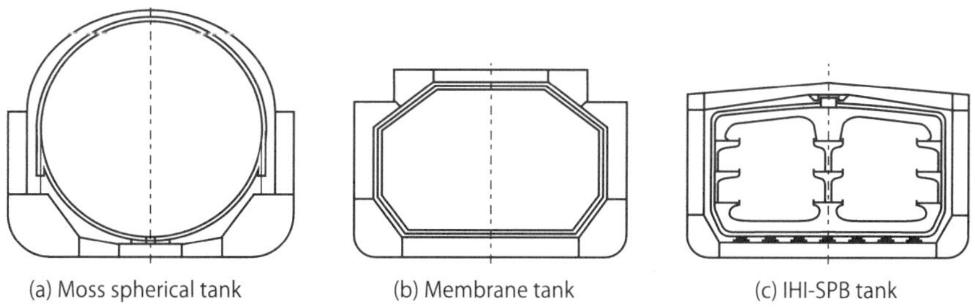

Fig. 5.4 Typical midship section of liquefied gas carriers

tanks have internal structures and swash bulkheads can be arranged arbitrarily, mitigating the risk of sloshing problems. Intermediate filling levels are inevitable in LNG fuel tanks and floating LNG units such as FLNG and FSRU; therefore, application of prismatic type-B tanks is expanding in these fields.

5.3 Increase in Container Ship Size

The maritime transport of containers, initiated by Malcom McLean in 1956, gradually expanded, and since the 1960s, the size of container ships has consistently increased. As shown in Fig. 1.2, until the mid-1990s, the size was capped at around 4,700 TEU due to the width restrictions of the then Panama Canal. However, with the rise of emerging industrial nations such as China, the demand for container logistics increased, and with the advent of Post-Panamax ship types, the barriers of 10,000 TEU and 20,000 TEU were quickly surpassed.

The main technical challenges in the enlargement of container ships were issues related to longitudinal strength and torsional strength due to large hatch openings, and impact loads due to large bow flares. One of the technological innovations that resolved the former issue of overall ship strength was the advancement in analytical technologies, enabling large-scale finite element analysis on the entire ship. Additionally, another technology was the development of higher strength, ultra-thick, high-toughness high-tensile steel as a material for large container ships, along with the establishment of welding techniques for these ultra-thick plates [38]. This ultra-thick high-tensile steel for container ships has been put into practical use up to the YP460MPa class [39]. The latter issue of impact loads on bow flares was also progressively resolved through numerical simulations of impact loads.

During the process of enlargement, the efficiency of container loading was pursued, leading to the elimination of longitudinal girder between hatches that existed initially, and efforts were made to minimize the spacing between containers. The width of double hull and the depth of double bottom were also made narrower and shallower, contrary to the increase in ship width and depth due to the enlargement of the hull, to increase the number of containers loaded in the hold. This was also made possible by the ability to perform large-scale three-dimensional finite element analysis on the entire ship. Following the process of this enlargement of container ships, typical midship sections over three generations are shown in Fig. 5.5.

In addition to advancements in analytical techniques, shipyards' proactive efforts in actual ship measurements have also contributed to solving various issues associated with the enlargement of container ships. With the aforementioned enlargement of the hull and the elimination of longitudinal girder between hatches, deformation of hatch openings was anticipated to be a significant issue. Measurements of the deformation shapes of hatch openings on container ships navigating in actual sea conditions were taken and fed back into the design process [34, 40]. In recent years, the increase in the number of hatch covers due to the widening of the ship and the application of ultra-high-strength steel to longitudinal strength members have made the deformation of

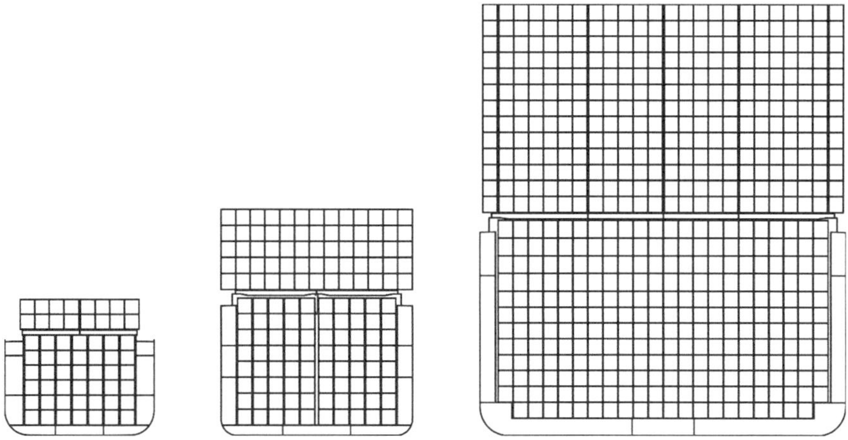

Fig. 5.5 Typical midship section of container ship

hatch openings and the hull more complex and significant. Therefore, it is crucial to continue analyzing hull deformations and making assumptions and evaluations based on actual measurement data.

However, damage due to longitudinal strength issues associated with the enlargement of ships still occurred. In 2007, the *MSC Napoli* suffered a fracture near the front of the engine room in the English Channel, and in 2013, the *MOL COMFORT* broke in half in the Indian Ocean. These incidents will be discussed in Part III, Chap. 20.

5.4 Change of Hull Form

In the1960's there appeared new hull forms with the scaling up of ships, thus obtaining higher speeds. They had a wider and shortened hull form with a bulbous bow. The so-called "Shinto hull form" was developed to reduce a ship's cost with greater breadth, limited length and smaller block-coefficient. The resistance increase caused by greater breadth and limited length was avoided by applying a smaller block coefficient. Limited length gave a smaller longitudinal bending moment. With a smaller bending moment and greater breadth, the longitudinal strength can be maintained without doubling plates on deck and bottom.

With the bulbous bow many structural problems were encountered. With high speed ships it is not a bulb any more, but a cylinder of which the top part is connected to the main hull structure by a thin plate, as shown in Fig. 5.6.

Container ships and ferry boats have big flares to improve sea worthiness and protect containers on deck from green sea. Very big wave impact forces are exerted on flare part, exceeding 45°.

The traditional hull form had a small radius, say 50R or 70R, in front of the stem. It was said that: the smaller the radius, the better the performance would be, and

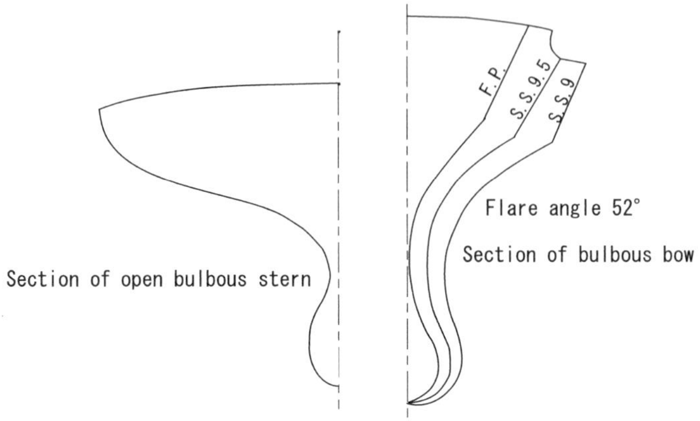

Fig. 5.6 Special hull form

therefore the smallest possible radius was applied. It is an old story that 100R could not guarantee the speed but 50R was OK. And the radius of curvature, in the vicinity of the fashion plate above the water line, is not so big that countermeasures against wave impact are necessary.

With the bulbous bow and the large radius of curvature on the stem, large flare was applied to protect the bulbous bow against damage from the anchors and cables. Due to the changed circumstances and hull forms, much bow flare damage occurred because the traditional design was still applied despite the increased wave impact pressure. The cylindrical bow was developed instead of the bulbous bow, which reduced bow damage because it had a big radius of curvature at the stem water line but no flare.

High-speed ships with a smaller block coefficient have no parallel part and the longitudinal strength cannot be maintained in the region of 0.4L midship with midship scantling of deck and bottom, because the deck breadth reduces rapidly away from the midship.

Actually around 1970 the forward part of a 10,000 DWT cargo ship had broken downwards at the section 1/4L from F.P., as if a jack knife was used. On this occasion, because of the damage, the traditional design standard of maintaining midship scantlings was improved in order to maintain the midship section modulus in the vicinity of 0.4L midship.

The open bulbous stern, which appeared in the energy saving era, had a very narrow stern construction and buckling occurred in the double bottom top plate in the aft part of the engine room because of a reduced section modulus in this section.

The hull structure designer did not anticipate this phenomenon well enough, which was caused by a change in hull form, but suitable countermeasures were designed and applied in time, based on lessons learned from the damage.

5.5 Ship Vibration Caused by Socio-Economical Change

It is said that ship vibrations are a problem that will never be solved by the ship designer because, as soon as one ship vibration problem is solved, a new one will appear. This is not due to the inability of the ship designer, but by the change of the boundary conditions around ship vibrations.

Before the 1950's a three islander, with the engine amidships, was the typical arrangement and ship vibrations were often caused by lateral vibrations of the long propelling shaft. In the 1960's, with the arrangement of the engine and the bridge on the aft end of the ship, vibrations appeared in the fore and aft direction of the aft bridge. Higher aft bridges vibrate in the fore and aft direction corresponding to vertical hull girder vibrations.

A high-speed turbine container ship, which has a 6-blade propeller with 130 rpm shaft revolutions, was investigated. The propeller exciting blade frequency is $130 \times 6 = 780$ cpm instead of the traditional $90 \times 5 = 450$ cpm for a turbine ship or $110 \times 5 = 550$ cpm for a diesel ship. By changing the propeller exciting frequency from 450 cpm ~ 550 cpm to 780 cpm (40% ~ 70% increase) the fittings in the accommodation quarters started to vibrate heavily. The lesson learned from this case was: it is most important to consider the exciting frequency closely.

The bigger structures of VLCC or ULCC tankers have lower natural frequencies than smaller tankers and there is possibility that the natural frequency will coincide with the exciting frequency of the propeller or main engine. This will cause resonance and heavy vibrations. Consequently, it became necessary to estimate the natural frequency of the structures of the tanks of VLCC and ULCC tankers. Generally, the natural frequency of a structure is calculated by the formula below, and in principle it can be estimated accurately.

$$f = c\sqrt{\frac{k}{m}} \tag{5.5.1}$$

where c is a constant, k is rigidity and m is mass.

Thus, an error in k or m of 1% causes a 0.5% error in f.

After the oil crises in 1973 and 1979, energy saving was of most importance. All efforts were made to reduce the ship's resistance and to increase the propulsion and machinery plant efficiency. Great success was achieved.

Power reduction was achieved by uniform wake thanks to hull form improvement and reduced propeller revolutions to achieve higher propulsion efficiency. This resulted in smaller propeller excitation forces. On the other hand, 4- and 5-cylinder diesel main engines were developed to reduce fuel consumption while requiring less maintenance compared with main engines with more cylinders. The 4- and 5-cylinder engines with a longer stroke have bigger exciting forces of the 4th or 5th order. In the energy saving era, propeller excitation was not the main concern but engine excitation affected the ship vibrations very much.

The change in boundary conditions of ship vibrations is shown in Table 5.2, and the existing boundary conditions caused by energy saving are shown in Fig. 5.7 [42].

Table 5.2 Environment and ship vibration problem

	Socioeconomic conditions	Technical conditions	Phenomena	Solution
1950-		Midship engine Midship bridge	Shaft lateral vib.	Rational alignment
1960-	Rationalization High power for economy	Aft engine Aft bridge 6RD90 (unbalanced MT) 10RD90 (crank shaft axial vib.)	Hull girder vibration Super structure fore & aft vib.	Balancer Damper
1970-	Specialize for economy	ULCC, VLCC High speed container ship (50,000HP 25kt) Bigger ship with high power engine	Vibration of structure in tank Higher blade freq (130rpm x 6=780cpm) Stern and local vibration	Avoid resonance Reinforcement Integral design system
1980-	Energy saving due to oil crisis	Less-cylinder engine Hull form improvement	Vibration caused by engine	engine
1990 | 2000	Leisure era Mega competition Less impact to environment	Cruise ship Large containership Double hull tanker Low revolution engine Increased engine power	Local vibration Long. vibration of superstructure Tank structure vibration in engine room	Elastic mounting of main engine Active mass damper Prediction method of tank vibration

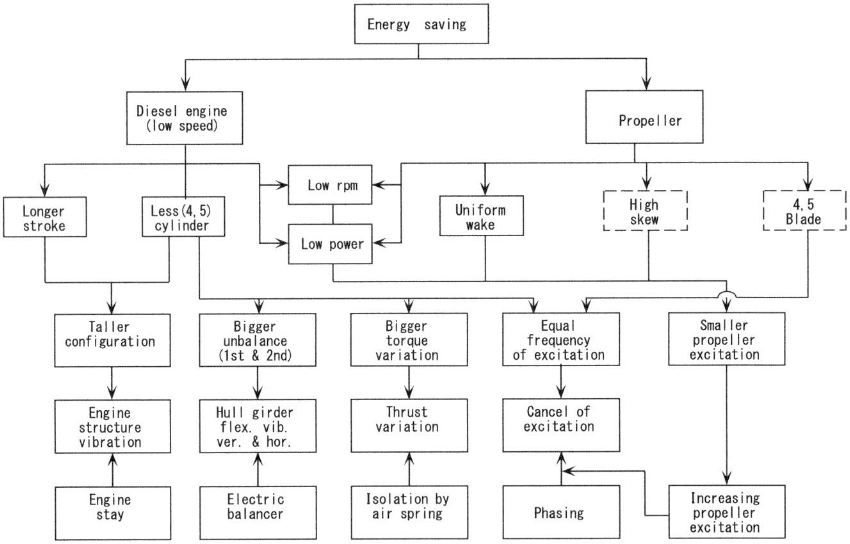

Fig. 5.7 Energy saving and ship vibration

5.6 Regulations for Environmental Conservation

After 1970, from the viewpoint of the prevention of pollution at sea, tank volumes in tankers were limited. The IMO (International Maritime Organization) issued the SBT (Segregated Ballast Tank) rule, which restricts the volume of oil outflow when damage occurs, and regulates the volume of exclusive ballast tanks to maintain the necessary draft, by estimating the damaged area by grounding.

By applying this SBT rule, 700,000 and 1,000,000 DWT tankers were designed. For a 700,000 DWT tanker, trial designs were carried out with combinations of the ratio $L/B = 5.0$ & 5.5 and the arrangement of three and four longitudinal bulkheads. Also 1,000,000 DWT tankers were designed, with a ratio $L/B = 5.0$ and four longitudinal bulkheads with horizontal and vertical main structures. It was possible to build these ships by stress, deflection and vibration analyses [43]. The tank arrangement of a 1,000,000 DWT tanker is shown in Fig. 5.1 together with other ships. She has many wing tanks with a breadth of 18.55 m, a length of 23.2 m and a depth of 45.5 m and she also has three transverses in a tank just like *Universe Apollo*, built before the rationalization using computer technology.

The trial designs of these big ships were carried out, but in practice such big ships were never built due to economic considerations and significant damage risk.

After the SBT rule was issued, which limits tank volume, VLCC *Exxon Valdez* grounded in May 1989 off the coast of Alaska and the out-flow of oil caused great damage to the sea and the seashore. After that the double hull arrangement was required for tankers, which increased the hull weight by approximately 20% in comparison with the single hull construction. The best efforts of hull structure designers to reduce steel weight by 1% or 0.5% through rationalization were wasted.

In the 1960's, the collision proof double hull construction for nuclear ships was investigated. Based on that knowledge modern research has been carried out by means of newly developed computer and measuring techniques.

The conventional VLCC with single hull construction has cross ties in the wing tanks, which connect vertical webs onto the side shell and longitudinal bulkhead. As the double hull side shell construction has enough strength by itself without cross ties, a new type of construction, which has cross ties in the center tank, was designed.

Double hull construction has been applied in the double bottom of general cargo ships and bulk carriers, and the bottom and side construction of container ships. However, in these cases the load on the double hull is not as high as for tankers. The rigidity of the double hull is proportional to the depth squared, when it works as a beam, but when the two panels work separately, the rigidity is just twice the amount of one panel, and independent of the depth. A suitable girder arrangement for double hull construction, to work as a beam, was investigated and the optimum design method was carried-out [43].

5.7 Technical Innovation

Technical innovation is a field that has a big effect on ship design. From the construction viewpoint, welding was the most important technical innovation in this century, and recently steel mill technology has also developed remarkably. One example is the wider plate. The width of steel plate for shipbuilding used to be maximum 2,400 mm (8'), but the development of high-power rolling mills enabled production of wider plates of 5,000 mm which reduced welding seams in the deck, shell and bulkheads.

Another example is TMCP high tensile steel which can be welded in the same way as mild steel and the price is not much higher; therefore, it has been widely applied in hull structures. TMCP steel is explained in Sect. 6.3 in detail.

Referring to advanced steel mill technology, tapered thickness plate and changing thickness plate can be rolled as shown in Fig. 5.8 [44]. Tapered thickness plate is used in bulkhead plate; changing thickness plate is used on decks and for the shell. These special plate are effective in reducing the hull steel weight. More varieties of thickness change as shown in Fig. 5.9 have also been made available as longitudinally profiled steel plates.

Moreover, SUF steel (Surface layer of Ultra Fine grain microstructure) has been developed, which has advantages in preventing buckling and plastic collapse [45].

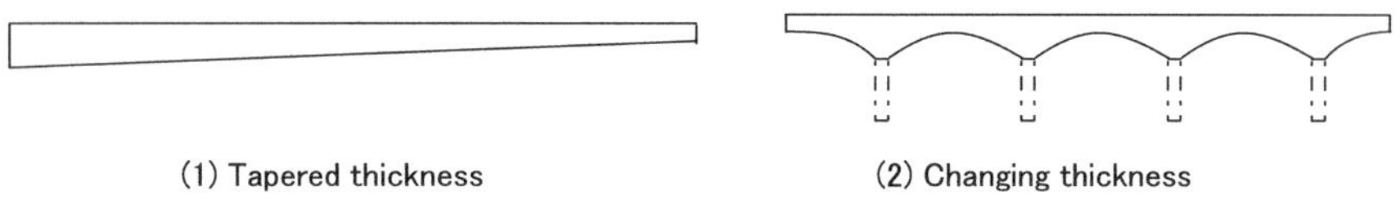

Fig. 5.8 Special section of plate

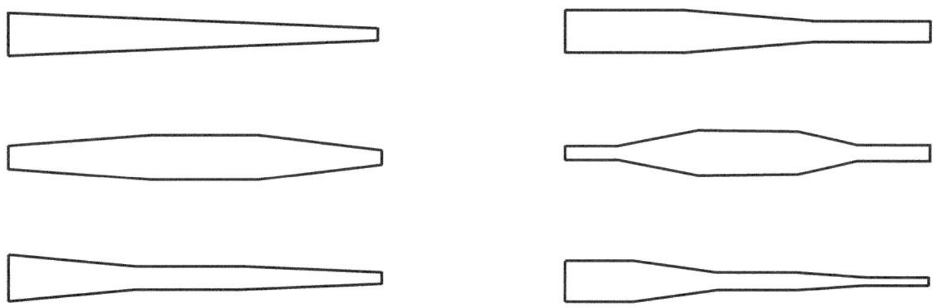

Fig. 5.9 Longitudinally profiled steel plate

Chapter 6
Materials

Selection of materials is very important in structure design, because it directly affects ship cost and strength. Hence, the structure designer should know well the characteristics of materials, especially newly developed ones.

6.1 Hull Steel

Mild steel and higher-strength steel are used for hull structure because of their excellence such as high strength, sufficient ductility, and low price.

Fundamental strength evaluations of the steel are done by a tensile test, which is illustrated in Fig. 6.1 as a stress–strain curve. When a steel specimen is loaded gradually, the stress–strain relationship shows a linear elastic behavior as shown in line O–A. The upper limit of this linear relationship is called the "proportional limit" and the maximum stress at which the strain returns to zero after unloading is called the "elastic limit". The proportional limit and the elastic limit are just below and very close to point A. Point A and B are called the upper yield point and lower yield point respectively, and the latter is the nominal yield stress, which is 235–280 MPa for mild steel in shipbuilding structures. Some materials have no dominant peak on yield point; in this case the stress corresponding to 0.2% permanent strain is assumed as the yield stress. As shown in the figure, the elongation becomes much larger after yield, and the stress reaches a maximum value at point C, which is termed the tensile strength, of 400–500 MPa for mild steel. After that, the strain becomes large and finally the steel fractures at point D. The solid line shows the nominal stress–strain relationship, where the nominal stress is calculated dividing the load by the original cross-sectional area of the specimen. The dashed line is the true stress–strain relationship, where the true stress is calculated using the actual cross-sectional area considering its reduction. Between the points B′ and C, the plastic strain increases and the sectional area reduces uniformly, but after point C, necking begins to be observed at the weakest section of the specimen. Therefore, the smallest

Y. Okumoto et al., *Design of Ship Hull Structures*,
https://doi.org/10.1007/978-3-032-12500-2_6

Fig. 6.1 Stress–strain curve

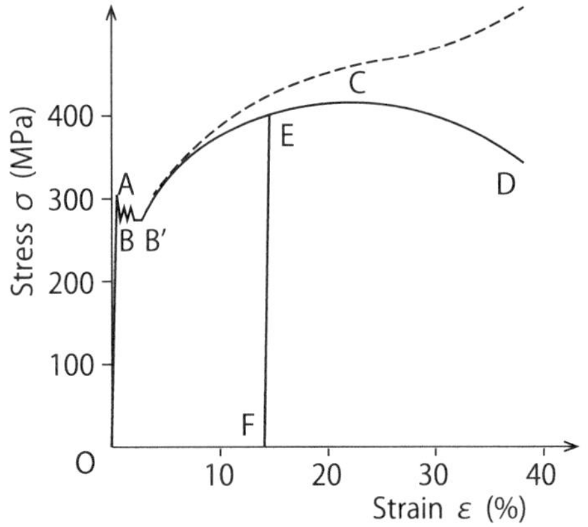

sectional area reduces rapidly causing large differences between the nominal stress and the true stress.

If it is unloaded in the plastic zone, it goes from point E to point F in the figure, parallel to O–A, and residual strain O–F remains, while in the elastic condition this value is zero. If it is loaded again after the plastic deformation, the stress increases again along line F–E, hence, the yield stress is greater than point A. This is termed work hardening.

When a compressive load is applied, the stress–strain curve is symmetric around point O of Fig. 6.1.

In the case of mild steel, strain at the beginning of yielding is in the order of 10^{-3}, and at the failure is around 0.3–0.4. Therefore, it retains some more strength after yielding until failure, and in addition, thanks to the large elongation after yielding, the applied loads will be redistributed to adjacent structural members resulting in a redundancy until collapse in statically indeterminate structures like ships. In this sense, ductile materials such as mild steel are superior and advantageous.

Many structural engineers often make the mistake that structural steel always start with a zero-stress condition at the beginning of a life of structures and the stress goes along line O–A–B in Fig. 6.1 during external loading. However, actual structure has residual stress initially as described in Sect. 6.8. Hence, we have to know stress and strain conditions at the initial state.

6.2 Grades of Steel

Since rivets were used to join plates and stiffeners in old steel ships before World War II, cracks often stopped at the structural discontinuity of rivet joints. However, if a crack occurs in a welded ship, it can propagate a long distance, and it may cause disaster. Hence, quality to arrest the crack propagation as well as weldability are required for hull steel. Welded structure ships newly built in USA during World War

II sometimes broke in two under sail in winter with a loud bang. Almost all the failure sections of the steel were crystalloid on the surface; therefore, it was concluded after the investigations that the failure was brittle fracture. Hence, IACS regulated the application of ship hull steel, categorized as A, B, C, D, and E in accordance with the notch toughness.

Grade A steel is widely used. There is no requirement for the impact test of a specimen. Impact test is additionally required for Grade B steel. Grade C is no longer used. Grade D steel is killed or killed and fine grain treated, and has much toughness as defined by impact tests. The value of absorbed energy is specified by whether cracks were arrested or not in previously built ships. Grade E steel is used as a crack arrester, hence it is highest grade. It is killed and fine grain treated steel.

Examples of mechanical properties and chemical composition from the NK rules are shown in Tables 6.1 and 6.2, respectively. Impact tests are defined for both the roll direction and the normal direction at each test temperature, because absorbed energy by the impact test depends on the direction of the specimen and temperature as shown in Fig. 6.2 [46]. Chemical compositions are defined by the percentage of C, Si, Mn, P, S, etc. The larger the percent of carbon, the stronger it becomes, but the ductility, toughness, and weldability become worse. Hence, the percentage of carbon is restricted to within about 0.2%, and Si and Mn are added in lieu of it. P and S are also limited within 0.035% each. In order to prevent welding cracks, the carbon equivalent as defined by Eq. (6.2.1) is required to be below a specified criterion for each material.

$$C_{eq} = C + \frac{Mn}{6} + \frac{Cr + Mo + V}{5} + \frac{Ni + Cu}{15} (\%) \tag{6.2.1}$$

KE47 is an extra-thick higher tensile steel specially developed for longitudinal members around upper deck and hatch side coaming of large container ships as explained in Sect. 5.3.

Application of the steel grade depends on the region of the hull and plate thickness. An example of a 60,000 DWT bulk carrier is shown in Fig. 6.3 [47], in which high grade steels are used for the bilge strake and hatch side plate on the upper deck.

6.3 Higher-Strength Steel

Higher-strength steel is used widely in main hull structures, because of smaller material costs and hull steel weight, resulting in increased dead weight and ship speed. At first, it was expensive and the welding procedure for it was difficult and complicated; however, it is now not so expensive and is easy to weld thanks to improvements in the steel making process. Hence, its application has expanded the extent that 60–70% of total hull steel uses it in recent merchant ships.

Around the 1980s, a novel higher-strength steel was developed and was termed CR (Controlled Rolling) or TMCP (Thermo Mechanical Controlled Process) steel,

Table 6.1 Mechanical properties of hull steel

<table>
<tr><th rowspan="3">Grade</th><th rowspan="3">Heat treatment</th><th colspan="3">Tensile strength</th><th colspan="3">Impact test</th></tr>
<tr><th rowspan="2">Yield stress (N/mm²)</th><th rowspan="2">Tensile strength (N/mm²)</th><th rowspan="2">Elongation (%)</th><th rowspan="2">Temperature (℃)</th><th colspan="2">Absorbed energy (J)</th></tr>
<tr><th>L</th><th>T</th></tr>
<tr><td>KA</td><td rowspan="2">AR</td><td rowspan="4">235~</td><td rowspan="4">400~520</td><td rowspan="4">22~</td><td>—</td><td>—</td><td>—</td></tr>
<tr><td>KB</td><td>0</td><td rowspan="3">27~</td><td rowspan="3">20~</td></tr>
<tr><td>KD</td><td>AR*, TMCP, N, CR</td><td>-20</td></tr>
<tr><td>KE</td><td>TMCP, N</td><td>-40</td></tr>
<tr><td>KA32</td><td rowspan="2">AR*, TMCP, N, CR</td><td rowspan="4">315~</td><td rowspan="4">440~590</td><td rowspan="4">22~</td><td>0</td><td rowspan="4">31~</td><td rowspan="4">22~</td></tr>
<tr><td>KD32</td><td>-20</td></tr>
<tr><td>KE32</td><td>TMCP, N</td><td>-40</td></tr>
<tr><td>KF32</td><td>TMCP, N, QT</td><td>-60</td></tr>
<tr><td>KA36</td><td rowspan="2">AR*, TMCP, N, CR</td><td rowspan="4">355~</td><td rowspan="4">490~620</td><td rowspan="4">21~</td><td>0</td><td rowspan="4">34~</td><td rowspan="4">24~</td></tr>
<tr><td>KD36</td><td>-20</td></tr>
<tr><td>KE36</td><td>TMCP, N</td><td>-40</td></tr>
<tr><td>KF36</td><td>TMCP, N, QT</td><td>-60</td></tr>
<tr><td>KA40</td><td>AR*, TMCP, N, CR</td><td rowspan="4">390~</td><td rowspan="4">510~650</td><td rowspan="4">20~</td><td>0</td><td rowspan="4">39~</td><td rowspan="4">26~</td></tr>
<tr><td>KD40</td><td>TMCP, N, CR</td><td>-20</td></tr>
<tr><td>KE40</td><td rowspan="2">TMCP, N, QT</td><td>-40</td></tr>
<tr><td>KF40</td><td>-60</td></tr>
<tr><td>KE47</td><td>TMCP</td><td>460~</td><td>570~720</td><td>17~</td><td>-40</td><td>53~75 depending on t</td><td>Appro-priate standards</td></tr>
<tr><td colspan="8">Notes: AR: As rolled
TMCP: Thermo-mechanical controlled processing
N: Normalizing
CR: Controlled rolling
QT: Quenching and tempering
*) AR permitted depending on the plate thickness</td></tr>
</table>

Table 6.2 Chemical properties of hull steel

<table>
<tr><th>Kind</th><th>Grade</th><th>Deoxidation</th><th>C</th><th>Si</th><th>Mn</th><th>P</th><th>S</th><th>Cu</th><th>Cr</th><th>Ni</th><th>Mo</th><th>Al</th><th>Nb</th><th>V</th><th>Ti</th><th>N</th></tr>
<tr><td rowspan="4">Mild steel</td><td>KA</td><td rowspan="2">Any method except rimmed</td><td rowspan="3">~0.21</td><td>~0.50</td><td>2.5xC~</td><td rowspan="4">~0.035</td><td rowspan="4">~0.035</td><td colspan="9" rowspan="2">$C + \frac{Mn}{6} \leq 0.4$</td></tr>
<tr><td>KB</td><td rowspan="3">~0.35</td><td>~0.80</td></tr>
<tr><td>KD</td><td>Killed or killed and fine grain treated</td><td>~0.60</td><td colspan="4" rowspan="2">—</td><td rowspan="2">0.015~</td><td colspan="4" rowspan="2">—</td></tr>
<tr><td>KE</td><td>Killed and fine grain treated</td><td>~0.18</td><td>~0.70</td></tr>
<tr><td rowspan="13">Higher strength steel</td><td>KA32</td><td rowspan="13">Killed and fine grain treated</td><td rowspan="9">~0.18</td><td rowspan="12">~0.50</td><td rowspan="12">0.90~ 1.60</td><td rowspan="9">~0.035</td><td rowspan="9">~0.035</td><td rowspan="13">~0.35</td><td rowspan="12">~0.20</td><td rowspan="9">~0.40</td><td rowspan="13">~0.08</td><td rowspan="13">0.015~</td><td rowspan="13">0.02~ 0.05</td><td rowspan="13">0.05~ 0.10</td><td rowspan="13">~0.02</td><td rowspan="9">—</td></tr>
<tr><td>KD32</td></tr>
<tr><td>KE32</td></tr>
<tr><td>KA36</td></tr>
<tr><td>KD36</td></tr>
<tr><td>KE36</td></tr>
<tr><td>KA40</td></tr>
<tr><td>KD40</td></tr>
<tr><td>KE40</td></tr>
<tr><td>KF32</td><td rowspan="3">~0.16</td><td rowspan="3">~0.025</td><td rowspan="3">~0.025</td><td rowspan="3">~0.80</td><td rowspan="3">~0.009</td></tr>
<tr><td>KF36</td></tr>
<tr><td>KF40</td></tr>
<tr><td>KE47</td><td>~0.18</td><td>~0.55</td><td>0.90~ 2.00</td><td>~0.020</td><td>~0.020</td><td>~0.25</td><td>~1.0</td><td>—</td></tr>
</table>

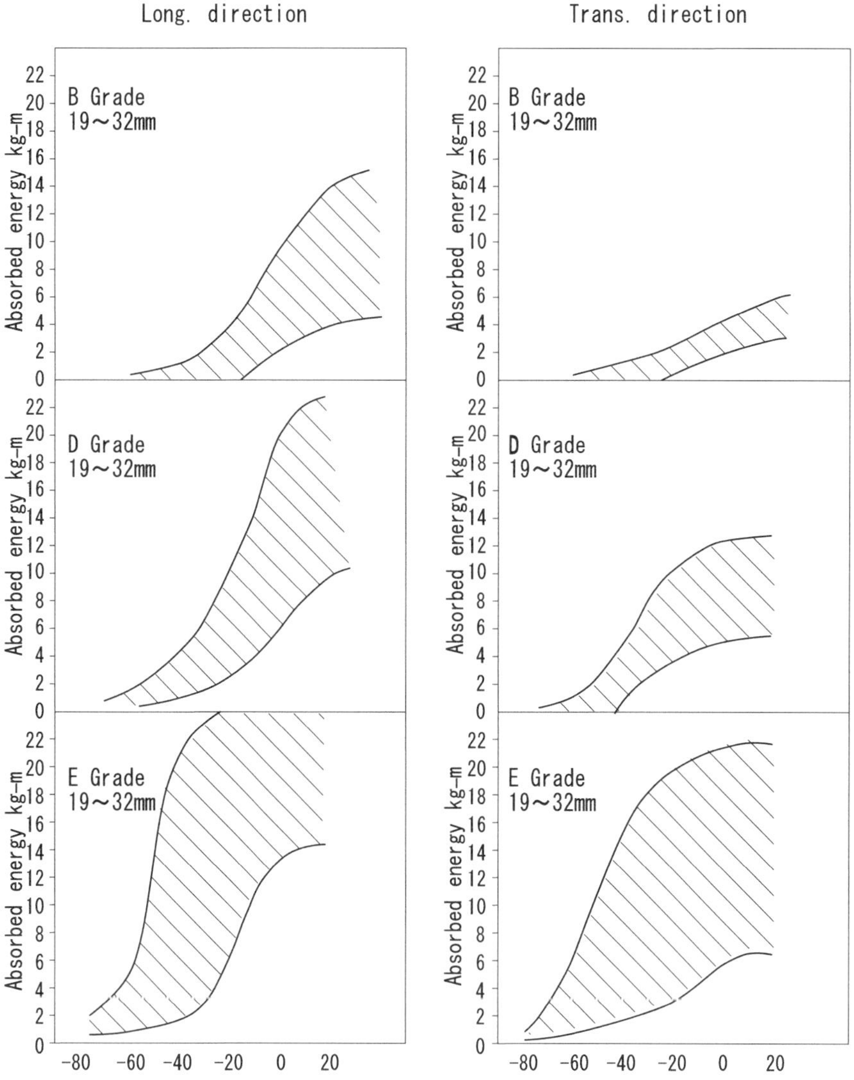

Fig. 6.2 Absorbed energy of mild steel

which improved notch toughness and weldability by the treatment of rapid water cooling using control cooling technology during steel making. The plate has a finer grain size than that of conventional higher-strength steel resulting in higher strength.

Classification societies regulate scantling reduction formulae for higher-strength steel by using a material coefficient k based on yield stress criterion, so that the allowable stress for higher-tensile steel is increased by dividing that for mild steel by k. Major classification society rules give the values of k as shown in Table 6.3. From Table 6.3, we can see that k is larger than the ratio of yield stress of the mild steel

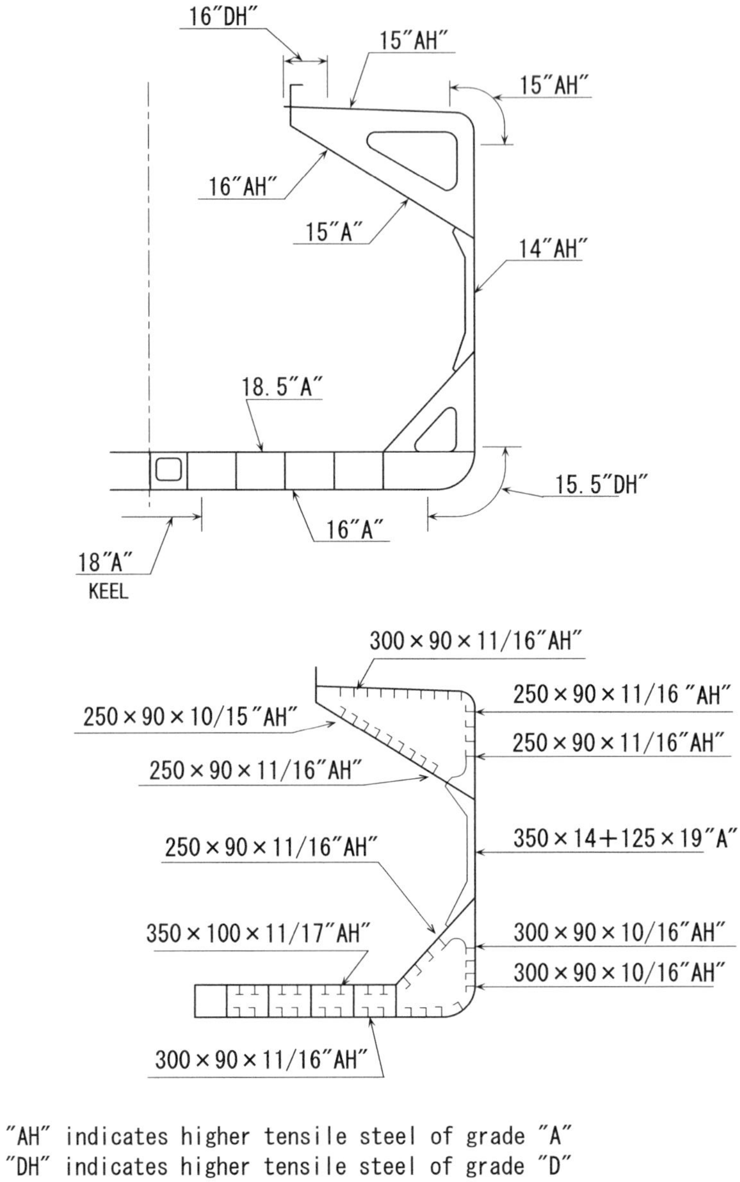

Fig. 6.3 Application of steel grade

Table 6.3 Material factor k of higher-strength steels

Material	σ_Y, yield stress (MPa)	$\sigma_Y(MS)/\sigma_Y$	k
Mild steel (MS)	235	1.0	1.0
HT32	315	0.75	0.78
HT36	355	0.66	0.72
HT40	390	0.60	0.68

and the higher-strength steel, $\sigma_Y(MS)/\sigma_Y$. There are various reasons for this, but the major reason is that the corrosion rate for steels is constant regardless of their yield stress, and thus, the strength of thinner higher-strength steel decreases more rapidly due to corrosion requiring greater safety margin.

The material factor k is applied for scantling reduction with respect to the strength of hull girder and primary supporting members. On the other hand, for local strength members such as plate and stiffeners, the yield stress ratio $\sigma_Y(MS)/\sigma_Y$ is usually applied to increase the allowable stress.

One important point to be mentioned is that the fatigue strength of higher-strength steel does not increase proportionally with yield strength. Therefore, it is necessary to take care of design details and working procedures, especially stress concentrations around the welding. In addition, consideration of deflection and buckling strength are necessary, because plate scantlings decrease according to the increased yield stress.

6.4 Steel Sections

There are steel sections as well as steel plates in hull steel materials, having several shapes as shown in Fig. 6.4; angle, bulb plate, and flat bar. Angle steel is widely used for longitudinals of 100–350 mm depth. Sectional geometry of the angle is standardized in steps. Welded built-up sections, using combinations of web plate and face plate, are applied for large sized sections.

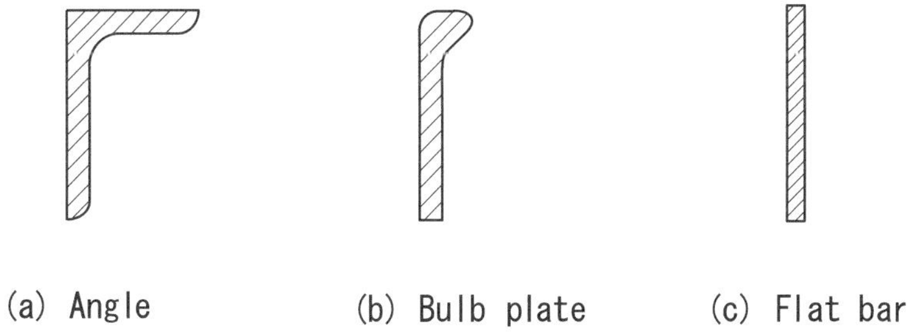

Fig. 6.4 Steel sections of stiffener

L-shaped angles used as stiffeners are prone to additional stress due to lateral displacement under bending as explained in detail in Sects. 8.8 and 10.5.4 of Part II. Therefore, symmetrically shaped built-up stiffeners are often used where fatigue strength is critical.

6.5 Other Materials

Cast steel is used in a part of the stern frame, rudder, propeller bossing, hawse pipe, etc., where the shape is complex and strength is needed. Compared with cast iron, its ductility, strength, and weldability are excellent. It is necessary to pay attention to avoiding high residual stress due to the welding of connecting parts to the steel plate. If needed, annealing or other stress relief process will be carried out. The stern frame was once made by the cast steel wholly, but now a combination of built-up steel is used.

Forged steel is used for rudder stock, rudder pintle, etc., where the shape is simple and strength is needed.

Aluminum alloy is used in high-speed boats and tank structures for liquefied gas, because of the advantage of the light weight, excellent corrosion resistance, and its toughness in a cryogenic environment against brittle fracture. The specific gravity of the aluminum alloy is one third of mild steel, and the yield strength is two thirds. Therefore, it is favorable in saving ship weight. However, special tools or facilities are necessary for its fabrication and assembly.

6.6 Scattering of Material Properties

It is very important to know well the characteristics of hull steel in order to evaluate the structural strength. Even for mild steels the values of yield stress and tensile strength vary widely in accordance with their grade, specification, size, temperature, steel making company, rolling conditions, and even part in one plate. Higher-strength steel also has a scattering in its strength. Table 6.1 only shows the lower limit or range of the strength and the actual value is not always the same. Figure 6.5 is an example of a mill sheet, which is the inspection certificate of each steel plate, describing actual mechanical properties and chemical compositions. The specified yield point of this plate is 235 MPa, but it is observed that actual yield point of this plate is 309 MPa.

Figure 6.6 shows the scattering of the yield stress and the tensile strength of mild steel, obtained from the mill sheet on delivery to the shipyard, in which the stress varies over a range of 100 MPa. These histograms seem approximately to have a normal distribution, shown as the curve in the figure. The standard deviation is 15–20 MPa.

Strength differs also with the roll direction (L) and the normal direction (T); usually that of the L direction is higher than that of the T direction. The difference

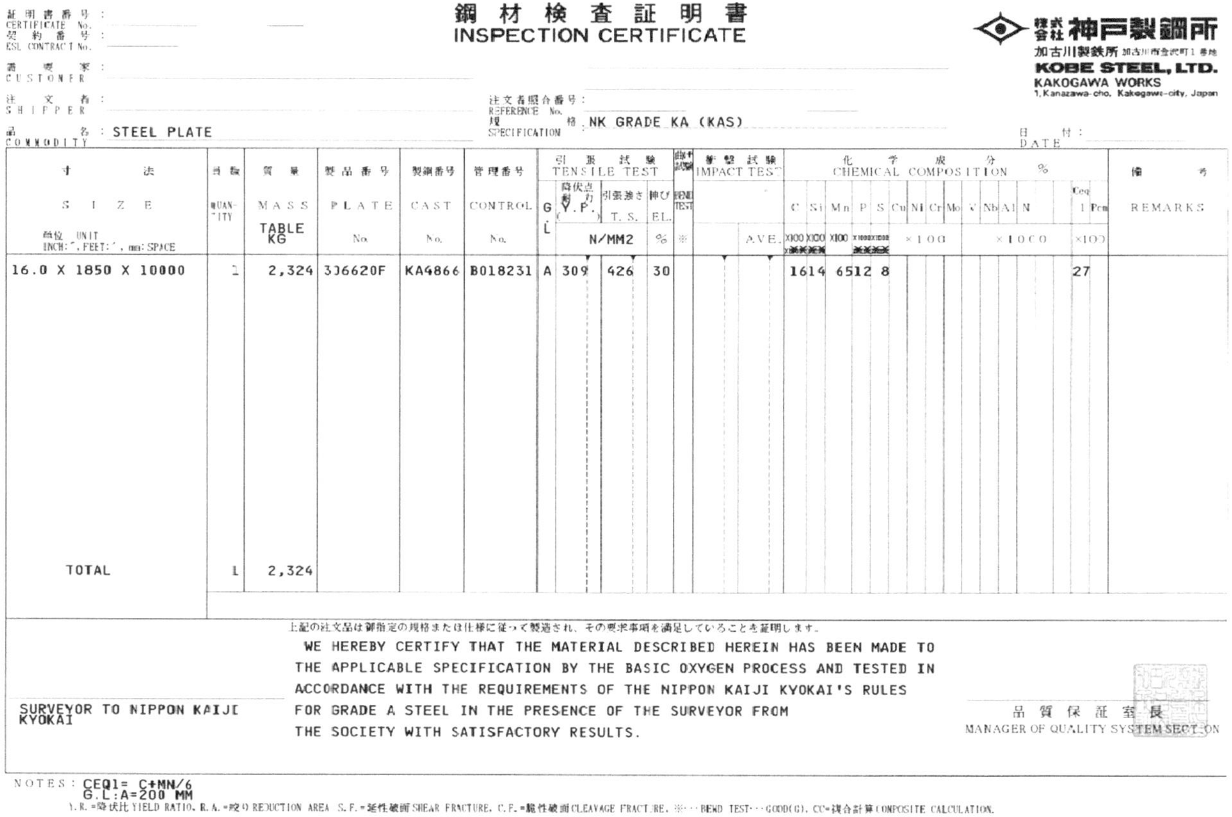

鋼材検査証明書
INSPECTION CERTIFICATE

株式会社 神戸製鋼所
加古川製鉄所
KOBE STEEL, LTD.
KAKOGAWA WORKS
1, Kanazawa-cho, Kakogawa-city, Japan

証明書番号 CERTIFICATE No.:
契約番号 KSL CONTRACT No.:
需要家 CUSTOMER:
注文者 SHIPPER:
品名 COMMODITY: STEEL PLATE
注文者照合番号 REFERENCE No.:
規格 SPECIFICATION: NK GRADE KA (KAS)
日付 DATE:

寸法 SIZE 単位 UNIT INCH:", FEET:', mm: SPACE	員数 QUANTITY	質量 MASS TABLE KG	製品番号 PLATE No.	製鋼番号 CAST No.	管理番号 CONTROL No.	GL	引張試験 TENSILE TEST 降伏点 Y.P. N/MM2	引張強さ T.S. N/MM2	伸び EL. %	曲げ試験 BEND TEST ※	衝撃試験 IMPACT TEST AVE.	化学成分 CHEMICAL COMPOSITION % C ×100	Si ×100	Mn ×100	P ×1000	S ×1000	Cu, Ni, Cr, Mo ×100	V, Nb, Al, N ×1000	Ceq ×100	備考 REMARKS
16.0 X 1850 X 10000	1	2,324	336620F	KA4866	B018231	A	309	426	30			16	14	65	12	8			27	
TOTAL	1	2,324																		

上記の注文品は御指定の規格または仕様に従って製造され、その要求事項を満足していることを証明します。

WE HEREBY CERTIFY THAT THE MATERIAL DESCRIBED HEREIN HAS BEEN MADE TO THE APPLICABLE SPECIFICATION BY THE BASIC OXYGEN PROCESS AND TESTED IN ACCORDANCE WITH THE REQUIREMENTS OF THE NIPPON KAIJI KYOKAI'S RULES FOR GRADE A STEEL IN THE PRESENCE OF THE SURVEYOR FROM THE SOCIETY WITH SATISFACTORY RESULTS.

SURVEYOR TO NIPPON KAIJI KYOKAI

品質保証室長
MANAGER OF QUALITY SYSTEM SECTION

NOTES: CEQ1= C+MN/6
G.L:A=200 MM
Y.R.=降伏比 YIELD RATIO, R.A.=絞り REDUCTION AREA, S.F.=延性破面 SHEAR FRACTURE, C.F.=脆性破面 CLEAVAGE FRACTURE, ※…BEND TEST…GOOD(G), CC=複合計算 COMPOSITE CALCULATION.

Fig. 6.5 Example of mill sheet

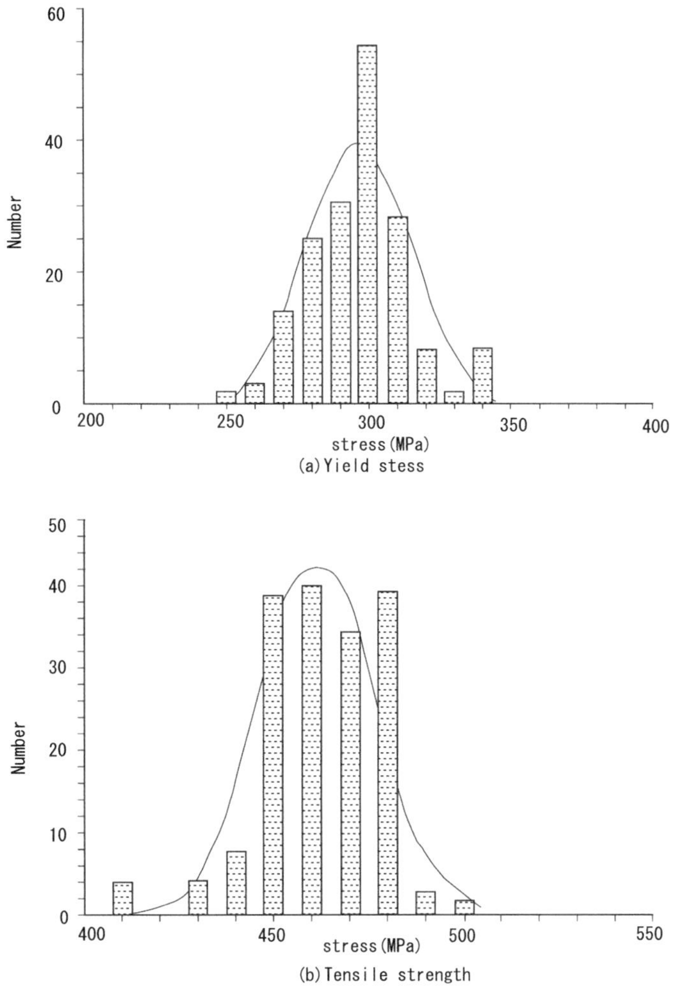

Fig. 6.6 Scattering of strength of mild steel

in the tensile strengths of L and T is now not so large, but that of toughness is larger. Hence, plates of longitudinal strength members must be arranged longitudinally along the L direction. The strength of a plate in the thickness direction (Z) is less than that in the in-plane direction. Therefore, if the tension force along the Z-direction is loaded through a rib or stiffener, tearing inside the plate may occur, which is termed lamellar tearing (Fig. 6.7). Classification societies recommend applying special steel plates with enhanced through thickness properties, known as "Z" quality steel, to such locations, e.g., the inner bottom plating just beneath the lower stools of a bulk carrier. Sulphur content deteriorates the through thickness properties, and it is limited under 0.008% for "Z" quality steel [48].

Fig. 6.7 Example of lamellar tears

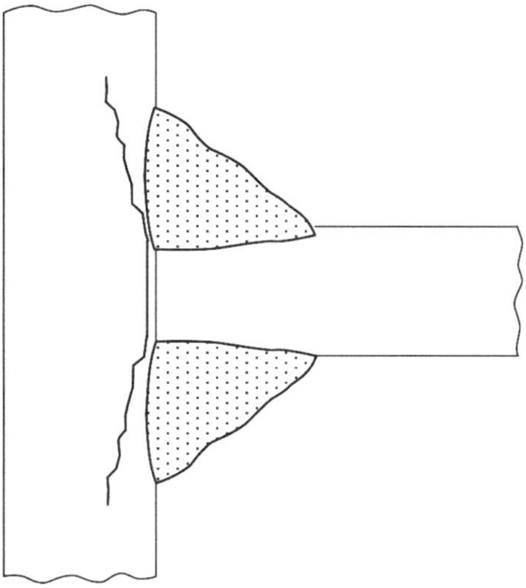

Material properties depend on temperature, and Fig. 6.8 shows temperature-dependent characteristics of Young's modulus, yield stress, and coefficient of linear expansion [49]. Steel reduces in strength beyond 400 °C and loses it at about 700 °C, and the coefficient of linear expansion increases gradually in accordance with the temperature. On the other hand, for temperatures below 0 °C the strength increases with temperature decrease; however, it decreases sharply when materials have defects or residual stresses caused by welding as shown in Fig. 6.9 [50]. These cause brittle fracture at low temperature.

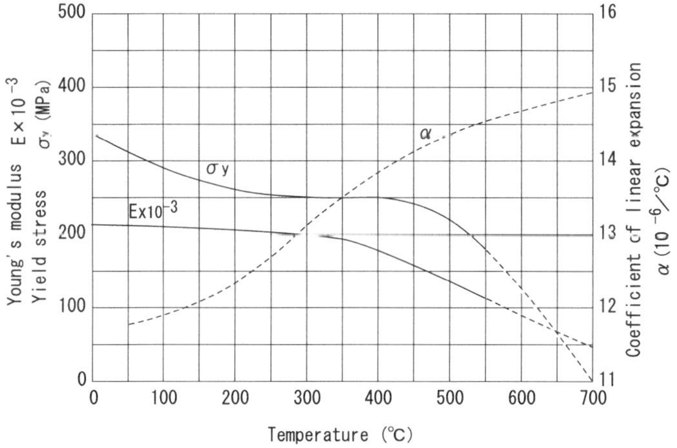

Fig. 6.8 Temperature dependency of material properties

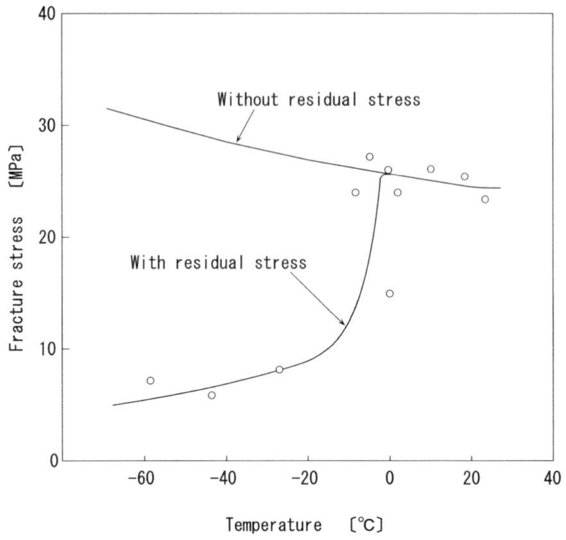

Fig. 6.9 Fracture stress in lower temperature

6.7 Scattering of Physical Properties

Plate thickness also varies depending on the roll condition in accordance with the steel maker, material properties, nominal thickness, etc. The classification societies used to allow the minus tolerance on nominal thickness being 0.3 mm anywhere on the plate. This regulation was revised in 2012 not to allow minus tolerance with respect to the average thickness of each plate, maintaining a local minus tolerance of 0.3 mm [51]. The plus error will be no problem, but we should pay attention to the minus error, because it causes a reduction in strength. Figure 6.10 shows histograms of actual thickness before fabrication in a shipyard, measured before the rule revision in 2012 [52]. It is known that the error spreads over a wide range. The plate thickness may be reduced due to the press bending with sharp angle.

Reduction of the plate thickness by corrosion is well known; it also causes cracks or buckling. Figure 6.11 shows a statistical analysis of the plate thickness reduction for both deep tank and water tight bulkheads as an example [53]. It is necessary to know the actual plate thickness in the case of damage analysis.

6.8 Residual Stress [54]

In accordance with the recent prevalence of computer hardware and software due to lower price and easy operation, hull structure design has become more reliable and accurate by applying direct calculation methods. This is especially true when structural analysis becomes more precise by using FEM analysis. If the structural FEM model is larger and finer, the stress will be calculated more precisely.

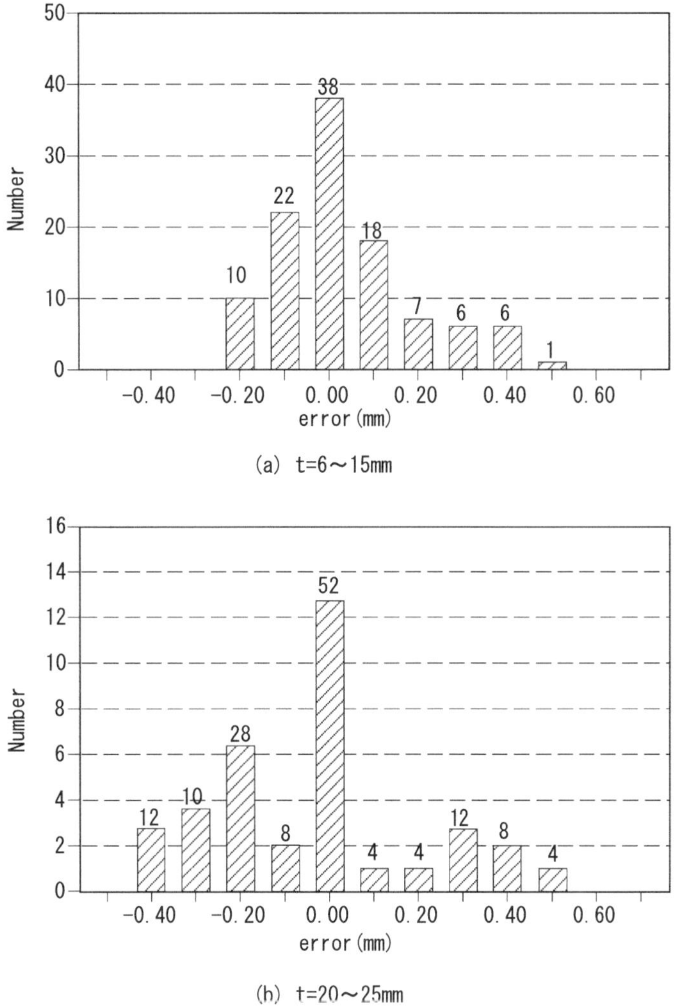

Fig. 6.10 Scattering of plate thickness

On the other hand, it is well known that structural materials have some residual stresses from the heat process of steel making and assembly processes (cutting, bending, welding, straightening, etc.), and these affect the ship hull strength. It is necessary to consider the residual stresses in structure design in order to evaluate the strength more precisely, while these are usually not considered in FEM calculations so far because the stresses are not exactly known.

Higher-strength steel such as TMCP (Thermo-Mechanical Control Process) steel, in conjunction with controlled rolling and controlled cooling, has been widely adopted recently in hull structure. It is said that the TMCP steel plate as rolled

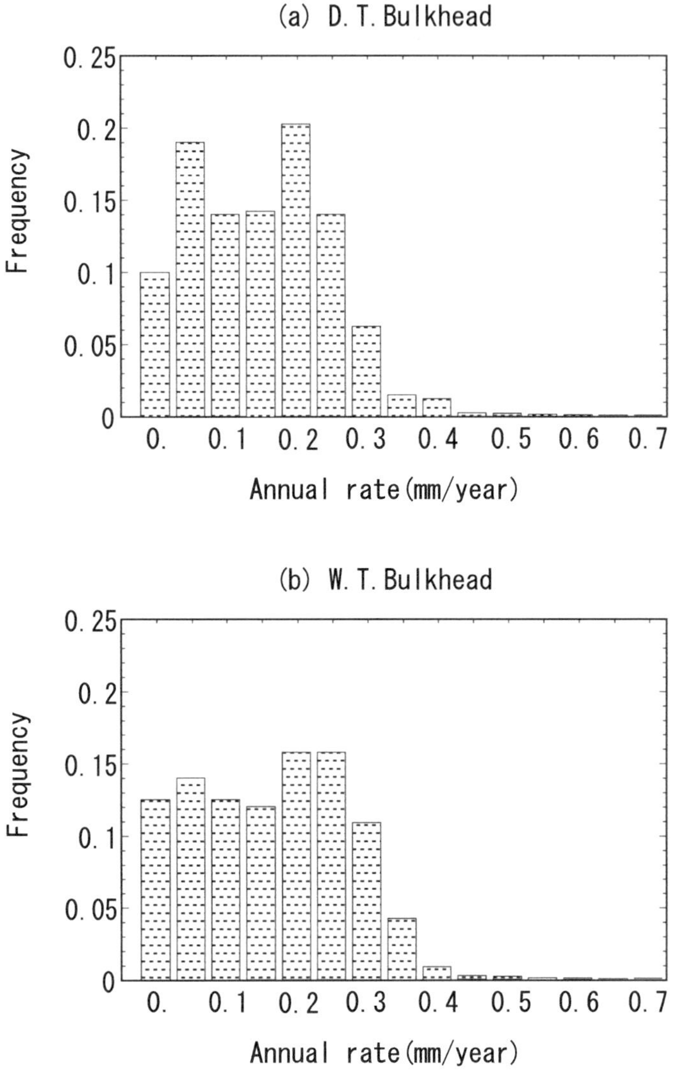

Fig. 6.11 Frequency distributions of annual corrosion rates

has greater residual stress around the plate edges, and also the stresses is distributed intricately inside the plate.

When constructing a ship's hull, residual stresses occur during surface preparation such as shot blasting, and also occur at various stages. The plastic deformation results after the heat treatment such as welding, cutting, straightening, and line heating. Residual stress occurs due to mechanical processing such as press bending, and the self weight and outside forces.

The residual stresses are thought to affect the hull strength as fol1ows;

- buckling in the skin plate
- fatigue cracks from welding beads
- brittle fractures in the vicinity of butt welding.

Chapter 7
Finite Element Method

7.1 Characteristics of FEM

Nowadays the finite element method (FEM) is an essential and powerful tool for solving structural problems not only in the field of shipbuilding but also in the design of most industrial products and even in non-structural fields. FEM can be used for a wide variety of problems in linear and nonlinear solid mechanics, dynamics, and ships' structural stability problems, in accordance with the development of computer technology and its popularization.

The conventional method in solving stress and deformation problems is an analytical one using theories of beams, columns and plates, etc. Hence its application is restricted to most simple structures and loads. On the other hand, FEM:

(1) divides a structure into small elements
(2) assumes each element to be a mathematical model
(3) assembles the elements and solves the overall as shown in Fig. 7.1 [55].

The element shown in Fig. 7.2 as a typical example is termed a finite element.

Characteristics of FEM are as follows:

It does not give an exact solution but solves approximately, because structures are modeled as a combination of simple elements and/or loads.

- It is a kind of numerical experiment without experimental devices, models, or instruments. Hence it is economical and time-saving.
- It can solve actual structural problems by using some models, although their shapes and loads are complex. It is even used for non-structural problems.
- It is used for a wide variety of steel, nonferrous materials and complex materials.
- It relies on computer technology for both hardware and software.
- It is applicable as "a black box tool", even engineers do not have to know the theory, because there are many general purpose FEM programs which are easy to operate.

Y. Okumoto et al., *Design of Ship Hull Structures*,
https://doi.org/10.1007/978-3-032-12500-2_7

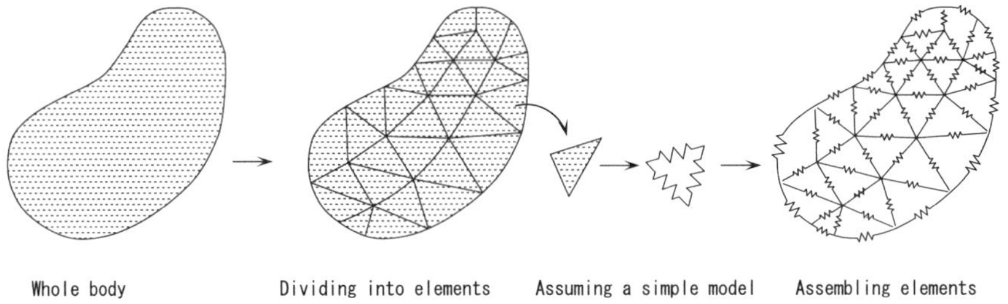

Fig. 7.1 Basic concept of FEM

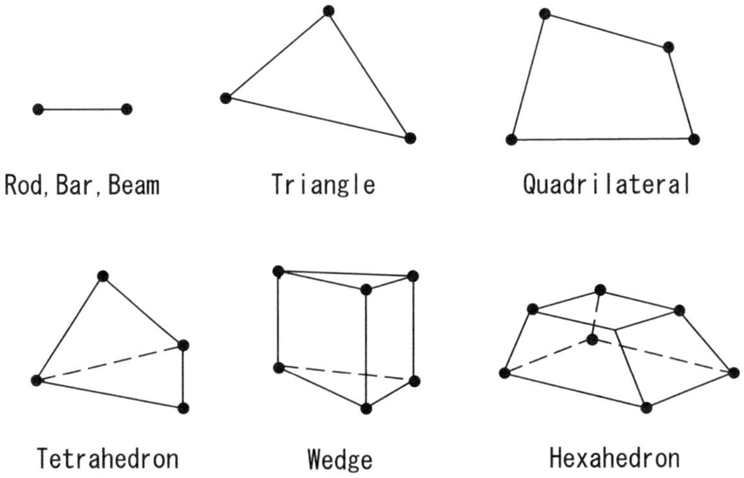

Fig. 7.2 Elements of FEM

Hence, FEM has merits and demerits.

7.2 Fundamentals of FEM

7.2.1 Stiffness Matrix

As a simple example, let us take a spring element and a combination of it, as shown in Fig. 7.3. When displacements of node 1 and 2 are u_1 and u_2 respectively and the spring constant is k_A as shown in Fig. 7.3a, the elongation of the spring is $u_2 - u_1$. Hence

$$\begin{aligned} f_1 &= -k_A(u_2 - u_1) \\ f_2 &= k_A(u_2 - u_1) \end{aligned} \tag{7.2.1}$$

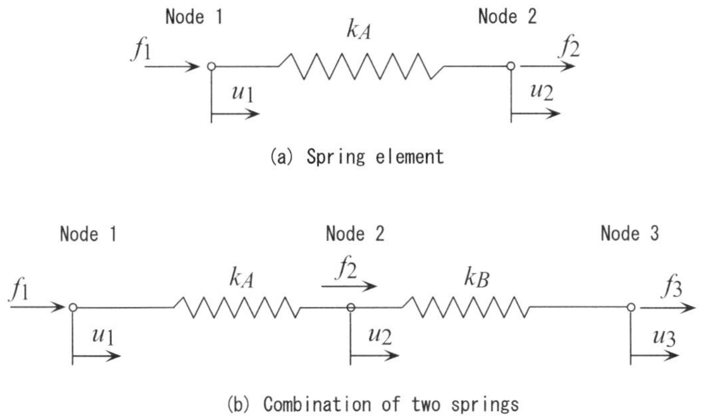

Fig. 7.3 Spring structure

Denoted in matrix form,

$$\begin{Bmatrix} f_1 \\ f_2 \end{Bmatrix} = \begin{bmatrix} k_A & -k_A \\ -k_A & k_A \end{bmatrix} \begin{Bmatrix} u_1 \\ u_2 \end{Bmatrix} \tag{7.2.2}$$

If two springs are connected as shown in Fig. 7.3b, the elongations of spring A and B are $u_2 - u_1$ and $u_3 - u_2$ respectively. Hence,

$$\begin{aligned} f_1 &= -k_A(u_2 - u_1) \\ f_2 &= k_A(u_2 - u_1) - k_B(u_3 - u_2) \\ f_3 &= k_B(u_3 - u_2) \end{aligned} \tag{7.2.3}$$

Then,

$$\begin{Bmatrix} f_1 \\ f_2 \\ f_3 \end{Bmatrix} = \begin{bmatrix} k_A & -k_A & 0 \\ -k_A & k_A + k_B & -k_B \\ 0 & -k_B & k_B \end{bmatrix} \begin{Bmatrix} u_1 \\ u_2 \\ u_3 \end{Bmatrix} \tag{7.2.4}$$

This equation, the load–displacement relation, can be expressed as follows:

$$\{f\} = [K]\{u\} \tag{7.2.5}$$

in which $\{f\}$ and $\{u\}$ are vectors of nodal loads and nodal displacements respectively and $[K]$ is termed the "stiffness matrix" of the structure. Equation (7.2.5) is also termed "stiffness equation".

When comparing Eqs. (7.2.2) and (7.2.4), it is clear that the combined stiffness matrix is the sum of each stiffness matrix, considering the order of the nodal points.

This is the same principle as with other elements, i.e., the overall stiffness matrix can be obtained only by the sum of each element.

When the structure is supported at node 1 and a force P acts at node 3 as in Fig. 7.3b, $u_1 = 0$, $f_3 = P$, $f_2 = 0$. The unknown values are then f_1, u_2, u_3. Hence, they can be solved by Eq. (7.2.4). Therefore, FEM will first assemble the whole stiffness matrix from the elements and then solve the simultaneous equations with respect to the given loads and displacements at the nodes.

7.2.2 Plane Stress [55]

A ship's structure consists mainly of thin plate structures and it is assumed to be flat locally, although there are curved shell plates. Therefore, we now examine the in-plane deformation of a plate element in the case of linear stress analysis in plane stress condition. A triangular element of constant thickness t is used for its versatility or simplicity, as shown in Fig. 7.4. The forces and displacements are represented by three nodes, as shown in the figure:

$$\{f\} = \begin{Bmatrix} f_{x1} \\ f_{y1} \\ f_{x2} \\ f_{y2} \\ f_{x3} \\ f_{y3} \end{Bmatrix}, \{u\} = \begin{Bmatrix} u_1 \\ v_1 \\ u_2 \\ v_2 \\ u_3 \\ v_3 \end{Bmatrix} \tag{7.2.6}$$

We choose a polynomial to describe the internal displacements in the triangle, and represent each of these displacement components by a linear polynomial in the x- and y-direction. This means that the strain in the element is constant in x and y.

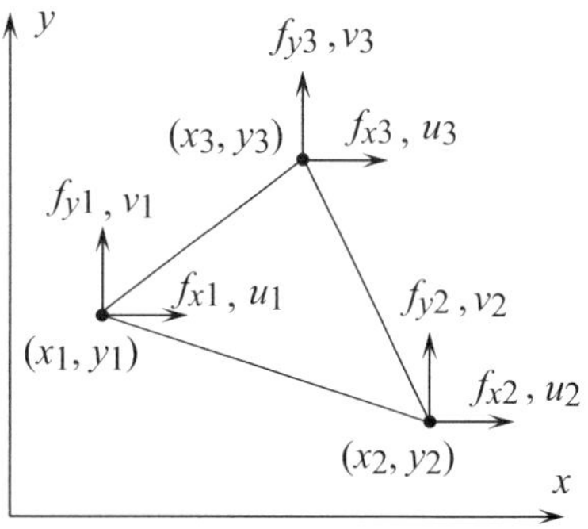

Fig. 7.4 Linear triangle element

$$u = c_1 + c_2 x + c_3 y$$
$$v = c_4 + c_5 x + c_6 y \tag{7.2.7}$$

Since there are six degrees of freedom as shown in Fig. 7.4, the coefficients c_1–c_6 can be represented by $u_1, v_1, u_2, v_2, u_3, v_3$.

$$\{u\} = \begin{bmatrix} 1 & x_1 & y_1 & 0 & 0 & 0 \\ 0 & 0 & 0 & 1 & x_1 & y_1 \\ 1 & x_2 & y_2 & 0 & 0 & 0 \\ 0 & 0 & 0 & 1 & x_2 & y_2 \\ 1 & x_3 & y_3 & 0 & 0 & 0 \\ 0 & 0 & 0 & 1 & x_3 & y_3 \end{bmatrix} \begin{Bmatrix} c_1 \\ c_2 \\ c_3 \\ c_4 \\ c_5 \\ c_6 \end{Bmatrix} \equiv [F]\{c\} \tag{7.2.8}$$

On the other hand

$$\varepsilon_x = \frac{\partial u}{\partial x},\ \varepsilon_y = \frac{\partial v}{\partial y},\ \gamma_{xy} = \frac{\partial u}{\partial y} + \frac{\partial v}{\partial x} \tag{7.2.9}$$

Then, strain vector $\{\varepsilon\}$ is

$$\begin{Bmatrix} \varepsilon_x \\ \varepsilon_y \\ \gamma_{xy} \end{Bmatrix} = \begin{bmatrix} 0 & 1 & 0 & 0 & 0 & 0 \\ 0 & 0 & 0 & 0 & 0 & 1 \\ 0 & 0 & 1 & 0 & 1 & 0 \end{bmatrix} \begin{Bmatrix} c_1 \\ c_2 \\ c_3 \\ c_4 \\ c_5 \\ c_6 \end{Bmatrix} \equiv [S]\{c\} \tag{7.2.10}$$

Therefore,

$$\{\varepsilon\} = [S]\{c\} = [S][F^{-1}]\{u\} \equiv [B]\{u\} \tag{7.2.11}$$

Matrix $[B]$ can be calculated in the following equation:

$$[B] = \frac{1}{2A} \begin{bmatrix} y_2 - y_3 & 0 & y_3 - y_1 & 0 & y_1 - y_2 & 0 \\ 0 & x_3 - x_2 & 0 & x_1 - x_3 & 0 & x_2 - x_1 \\ x_3 - x_2 & y_2 - y_3 & x_1 - x_3 & y_3 - y_1 & x_2 - x_1 & y_1 - y_2 \end{bmatrix} \tag{7.2.12}$$

$[B]$: strain coefficient matrix.
A: area of a triangular element.
Matrix $[B]$ is defined only by geometries of the element.
On the other hand, the stress–strain relationship is shown in the following equation for the plane stress condition.

$$\{\sigma\} = [D]\{\varepsilon\}$$

$$[D] = \frac{E}{1-\nu^2}\begin{bmatrix} 1 & \nu & 0 \\ \nu & 1 & 0 \\ 0 & 0 & \frac{1-\nu}{2} \end{bmatrix} \tag{7.2.13}$$

$[D]$: elasticity matrix.
E: Young's modulus.
ν: Poisson's ratio.
$\{\sigma\}$: stress vector.
Matrix $[D]$ is defined only by material properties of the element.

Next, let us consider the equilibrium of energy in an element. The external work for an element is the sum of the virtual displacement multiplied by the force at each node.

$$U_e = \delta\{u\}^T\{f\} \tag{7.2.14}$$

where δ denotes virtual infinitesimal variation and $\{\cdot\}^T$ denotes the transpose of a vector or a matrix. The internal work for an element is the sum of the virtual strain multiplied by the stress in an element. Using Eqs. (7.2.11) and (7.2.13),

$$U_i = \int \delta\{\varepsilon\}^T\{\sigma\}dV = \int\int \delta\{\varepsilon\}^T\{\sigma\}tdxdy = \int\int ([B]\delta\{u\})^T[D][B]\{u\}tdxdy \tag{7.2.15}$$

V: Volume of an element.
t: plate thickness of an element.

Using the principle of the virtual work, i.e. the external work equals the internal work in the element, i.e. $U_e = U_i$,

$$\delta\{u\}^T\{f\} = \int\int ([B]\delta\{u\})^T[D][B]\{u\}tdxdy \tag{7.2.16}$$

Then, the stiffness equation for the triangular element is

$$\{f\} = t[B]^T[D][B]\{u\}\int\int dxdy = At[B]^T[D][B]\{u\} \equiv [K]\{u\} \tag{7.2.17}$$

$[K]$: stiffness matrix

$$[K] = At[B]^T[D][B] \tag{7.2.18}$$

When combining many elements, we assemble the overall stiffness matrix from each element and solve the stiffness equation with the given loads and displacements at the nodes. We then can get the displacement of each node, the strain of each grid from Eq. (7.2.11), and the stress from Eq. (7.2.13), one by one.

7.3 Procedure of FEM

The procedure of displacement and stress calculations by computer, in the case of requiring linear solutions, is generally as follows:

(1) calculate [B] matrix using the geometry of an element, and calculate [D] matrix using material properties.
(2) calculate matrix [K] of an element. (1) and (2) are repeated for all elements.
(3) assemble the overall stiffness matrix
(4) calculate displacements of each node from load and support conditions by solving the stiffness Eq. (7.2.17).
(5) calculate the strains of each element by Eq. (7.2.11)
(6) calculate the stresses of each element by Eq. (7.2.13)
(7) calculate principal stresses, equivalent stresses, etc.

Figure 7.5 shows the process from (1)–(4) and Fig. 7.6 shows the program flow of FEM [55].

Figure 7.7 also shows the procedure of the FEM analysis. General purpose FEM program MSC/NASTRAN is widely used for linear structural problems and dynamic response. MARC, ANSYS, ABAQUS, etc. are also used, especially for nonlinear problems including geometric and material nonlinearity. User interfaces like pre-processor or post-processor have become more "user friendly and many commercial software packages are now available to create and evaluate large-scale 3-dimensional models.

7.4 Application of FEM

As mentioned earlier, FEM analysis is a numerical approximation method, hence some errors will inevitably be included. We have to reduce these to an acceptable level by making the model intrinsically adequate, in accordance with the theory of material strength.

7.4.1 Mesh Division

As shown by the basic theory, the stress and strain in an element are constant, hence the mesh size has to be sufficiently small, where the stress distribution is increasing sharply. Figure 7.8a shows an example of stress concentration around a hole for which the exact solution is given by the theory of elasticity, and this will allow us to assess the accuracy. Figure 7.8b is an FEM model using the MSC/NASTRAN linear program, and 7.8c is the calculation result [56]. The maximum theoretical nominal stress in this case is $\sigma_{yA} = 378$ MPa; however, the stress of the element at the corner

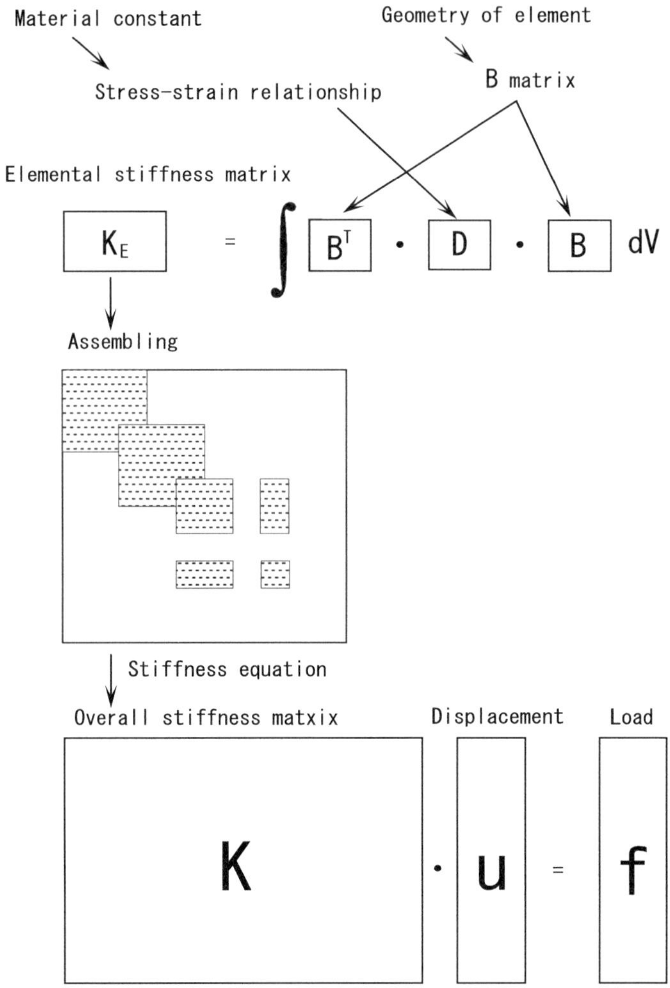

Fig. 7.5 Procedure to get stiffness equation

C obtained by FEM was $\sigma_{yE} = 294$ MPa, which is 22% smaller than the theoretical result. When the maximum stress is obtained by the linear extrapolation from the two elements as shown in Fig. 7.8c, it reaches $\sigma_{yC} = 336$ MPa but is still 11% smaller than the theory. Provision of bar elements along the circular edge as dummy elements to evaluate the stress is another method to obtain better results of boundary stresses around the edge. But anyway, mesh division with six elements for a quarter circle looks insufficient, and a finer mesh is desirable near the hole.

The aspect ratio of a rectangular element and the shape of a triangular element are important. An unusual slender form is not desirable.

In addition, it is important to avoid.

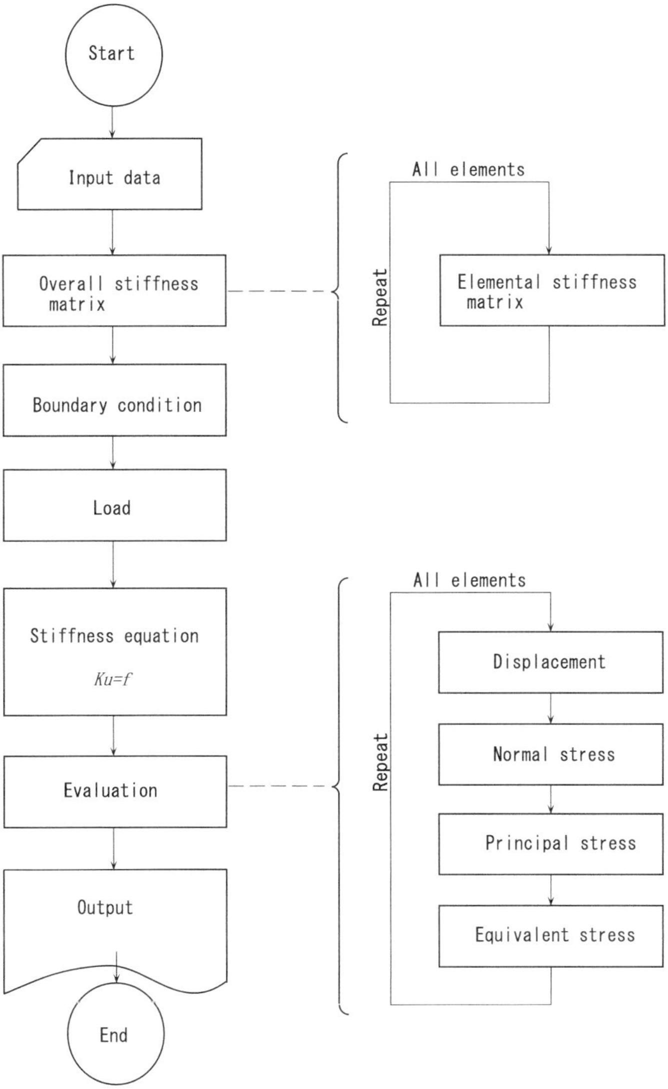

Fig. 7.6 Sequency of FEM calculation

- a connection with great differences in rigidity.
- a connection of a large element and a very small element.

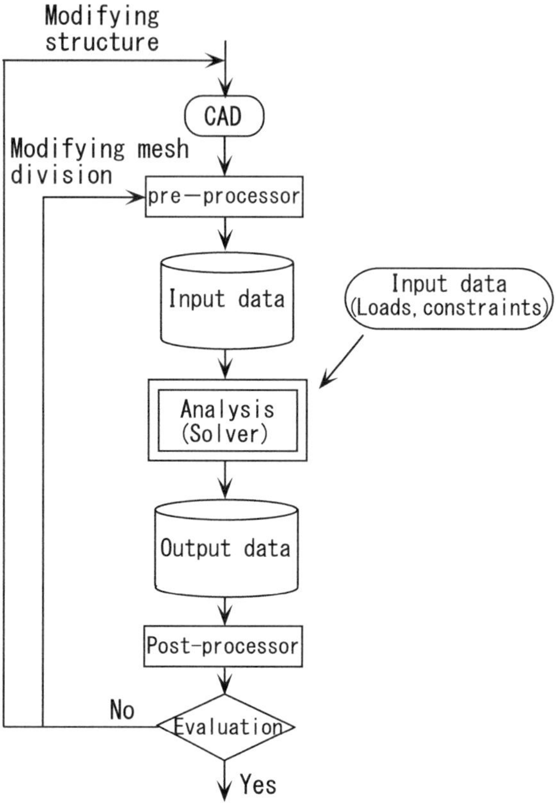

Fig. 7.7 Procedure of FEM analysis

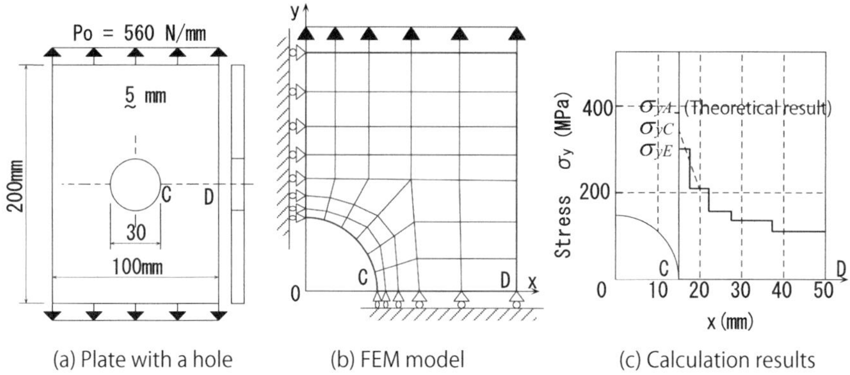

Fig. 7.8 Stress distribution around a hole

7.4.2 Loading and Supporting Condition

The loading condition should be carefully determined to correspond with the aim of the analysis, FEM model, and evaluation method. The balance of the accuracy between input load and output stress should be considered; usually the load is quite rough and the mesh division is more detailed. It is also necessary to know the characteristics of the applied load—static, dynamic, repeated, or random.

The use of symmetrical conditions allows us to reduce computer calculation time, as shown in Fig. 7.8b, which models only 1/4 of the whole model, considering appropriate support conditions.

In the case where we apply symmetrical boundary conditions, it is very important to keep in mind that boundary conditions should be determined based on the assumed deformation mode. Static stress analysis is quite simple, but we sometimes encounter careless mistakes in eigenvalue analysis such as buckling and vibration analysis.

Figure 7.9 is an example of buckling analysis of a simply supported panel with aspect ratio of 1:2. The resulting buckling mode must be of two half waves as shown in the figure. In the figure, the solid line circle means buckling deformation toward the fore side perpendicular to the panel, and the dotted line circle means buckling deformation toward the opposite (aft) side.

If we model this problem with a 1/4 model taking out only the hatched area in Fig. 7.9 taking advantage of the symmetrical geometry, and apply symmetrical conditions at the boundary, the resulting buckling mode becomes three half waves as shown in Fig. 7.10. That is, the calculated buckling stress is mistakenly higher than the actual buckling stress. This example tells us that we must set boundary conditions taking into account the deformation mode carefully.

In zooming calculations, the range of the FEM model and the loading method should be carefully considered.

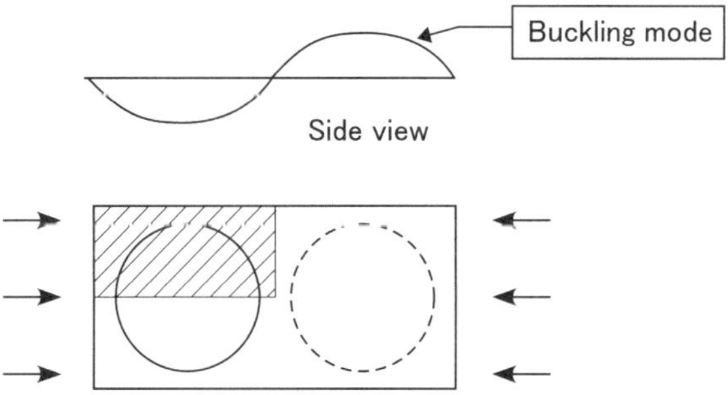

Fig. 7.9 Buckling mode of a panel of aspect ratio 1:2

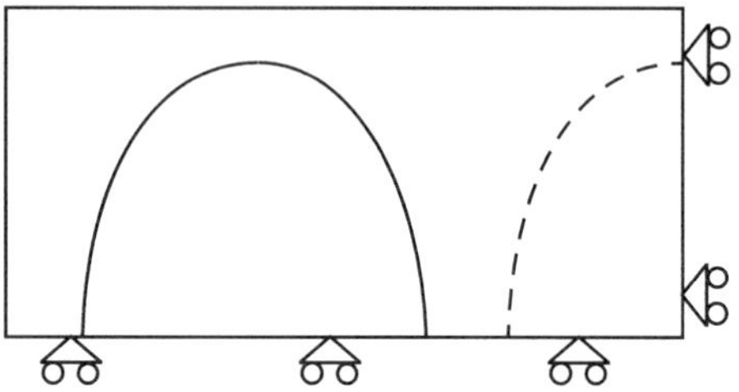

Fig. 7.10 Mistaken boundary conditions in buckling analysis

7.4.3 Degrees of Freedom

In actual application of finite element analysis, it is important to be conscious of degrees of freedom of each element.

a. plane stress (membrane) element, plane strain element 2: u, v
b. plate bending element 3: w, θ_x, θ_y
c. shell element (combination of a and b) 5: u, v, w, θ_x, θ_y
d. rod element (bar element) 1: u
e. beam element 6: u, v, w, θ_x, θ_y, θ_z
f. solid element 3: u, v, w

where u, v and w are the degrees of freedom in the translational direction of the x, y and z axes, while θ_x, θ_y and θ_z are the degrees of freedom in the rotational direction around the respective axes.

When membrane elements are assembled into three dimensional objects, each joint acquires three degrees of freedom (u, v, w). In this case, there is no rigidity in the out-of-plane direction, and the calculation becomes unstable. Therefore, the analysis system usually adds temporary support to the unstable direction automatically. Similarly, shell elements have six degrees of freedom when three-dimensionally assembled.

This knowledge of degrees of freedom helps us to avoid mistakes in connections between different kinds of elements. Figure 7.11a is a mistake. Since membrane or shell elements do not have degrees of freedom of rotation around the z-axis, the torsional moment of the beam element cannot be transferred to the plate elements. Figure 7.11b is also a mistake. In this case, if we want to transfer the load to the plate elements, we should insert the beam elements into the plate elements as shown in Fig. 7.11c.

It is appropriate to model typical pillars in engine room construction by bar elements. In this case, to avoid extraordinarily high stress at the connection between the pillar and the base structure, we should insert the pillar bar elements into the base structure as shown in Fig. 7.12. Such insertion of beam or bar elements is quite a simplified method in the computer model, but as long as we do not evaluate the stress just at this location, it is a useful solution to transfer load correctly.

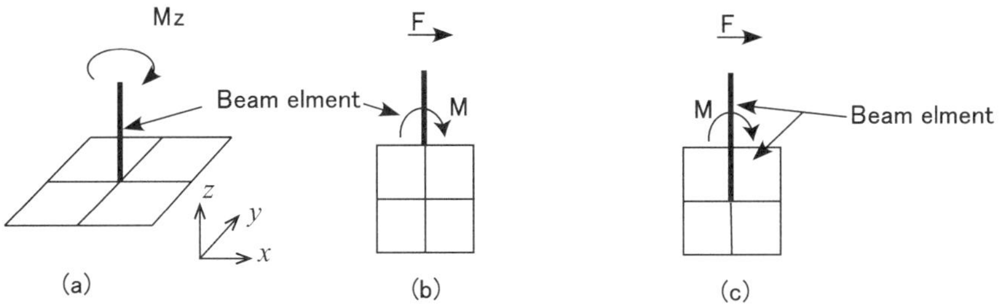

Fig. 7.11 Load transfer between beam and membrane elements

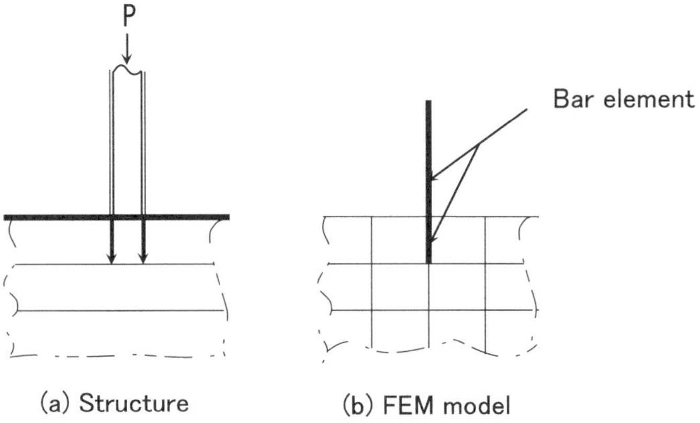

Fig. 7.12 FEM modeling of base of pillar

Part II
Theory

Chapter 8
Design of Beams

8.1 Effective Breadth of Attached Plates

Main hull structures such as decks, shells, and bulkheads are composed of stiffened panels, which include plates, secondary members and primary members. As explained in Sect. 3.1.2, Part I, lateral pressure acting on the plates is first transmitted to secondary members such as beams, stiffeners, frames and longitudinals, and then, the load is transmitted from the secondary members to primary members such as girders, web frames, transverse webs. In Part II, secondary members are collectively referred to as beams and primary members as girders.

These structures endure the lateral forces such as water pressure as well as in-plane loads. When the forces are imposed, the stress distribution in the plates and the stiffeners is complex as shown in Fig. 8.1. The longitudinal stress in the attached plate will be a maximum at the connection line to the stiffener and become smaller gradually beyond this line. Considering the strength, the stiffened panel will be assumed to be a collection of beams which include some parts of the attached plate. The breadth of this plate is called "effective breadth" or "effective width". The former term "effective breadth" relates to the so-called shear lag effect. When lateral pressure produces bending of a stiffener, the attached plate behaves as a flange of the stiffener. The coplanar stresses in the plate is loaded from the intersection with the stiffener, and diminishes as the distance from the stiffener increases. On the other hand, the latter term "effective width" is related to buckling of the plate. Compressive coplanar loading produces buckling of the plate between stiffeners causing reduction in the membrane rigidity of the plate.

Y. Okumoto et al., *Design of Ship Hull Structures*,
https://doi.org/10.1007/978-3-032-12500-2_8

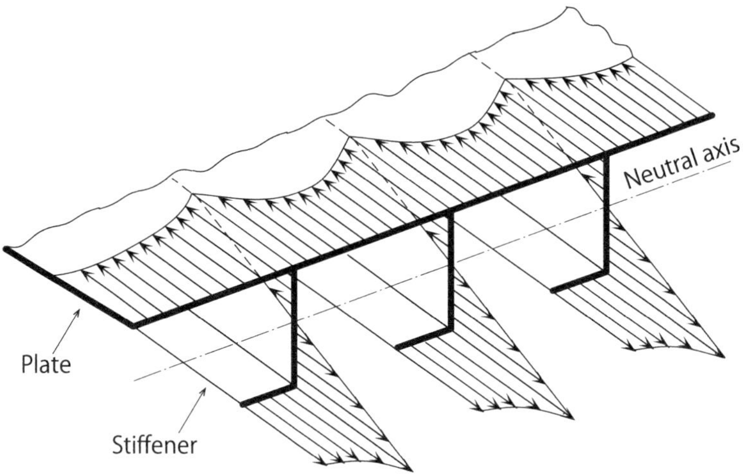

Fig. 8.1 Bending stress distribution of stiffened panel

8.1.1 Bending in Elastic Conditions

The effective breadth is represented as follows:

$$B_e = \frac{2}{\sigma_{\max}} \int_0^{B/2} \sigma_x dy \tag{8.1.1}$$

B_e: effective breadth
$\sigma_{\max}$: maximum stress at connection line of a stiffener
B: distance between stiffeners
σ_x: stress of longitudinal direction along with B

Though this value depends on loading conditions, structural dimensions, and end connections, it is related mainly to the ratio l (length between two points where the bending moment is zero) over B. If it is uniformly loaded, B_e/B is given in Fig. 8.2 [23]. If l/B becomes large, B_e/B becomes 1, i.e. full width, and if l/B becomes small, B_e/B becomes smaller than 1. NK rules simply state the effective breadth as follow:

$$\begin{aligned} B_e &= 0.2 \times (\text{stiffener span}) \\ B_e &= B \end{aligned} \tag{8.1.2}$$

whichever is lesser. This rule equation is drawn in Fig. 8.2 with a dashed line, considering that l in the horizontal axis is the distance between the two points with zero bending moment, meaning $l = (\text{stiffener span})/\sqrt{3}$. We can see that the rule equation fairly approximates the obtained B_e/B curves.

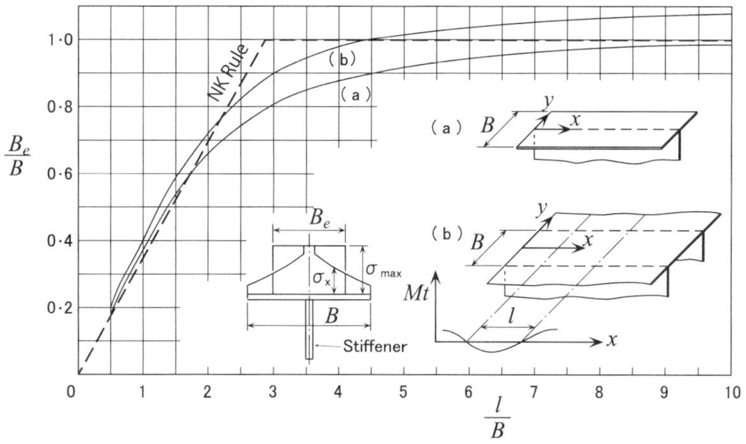

Fig. 8.2 Effective breadth of stiffened plate

8.1.2 Effective Width After Plate Buckling

If the stiffened panel buckles by axial forces, the middle part between the stiffeners will deform and the resistance to lateral loads will decrease. Considering the ultimate strength of the stiffener with effective width of the attached plate, the decrease of effective width should be investigated.

After the buckling, the plate near the stiffeners still offers some resistance as shown in Fig. 8.3 and the effective width after buckling b_e is defined in the figure. Faulkner shows the ultimate strength of a simply supported panel as follows [57]:

$$\frac{\sigma_u}{\sigma_y} = \frac{2}{\beta} - \frac{1}{\beta^2} \tag{8.1.3}$$

σ_u: ultimate strength
σ_y: yield stress

$$\beta = \frac{b}{t}\sqrt{\frac{\sigma_y}{E}}$$

t: plate thickness
b: plate breadth
E: Young's modulus

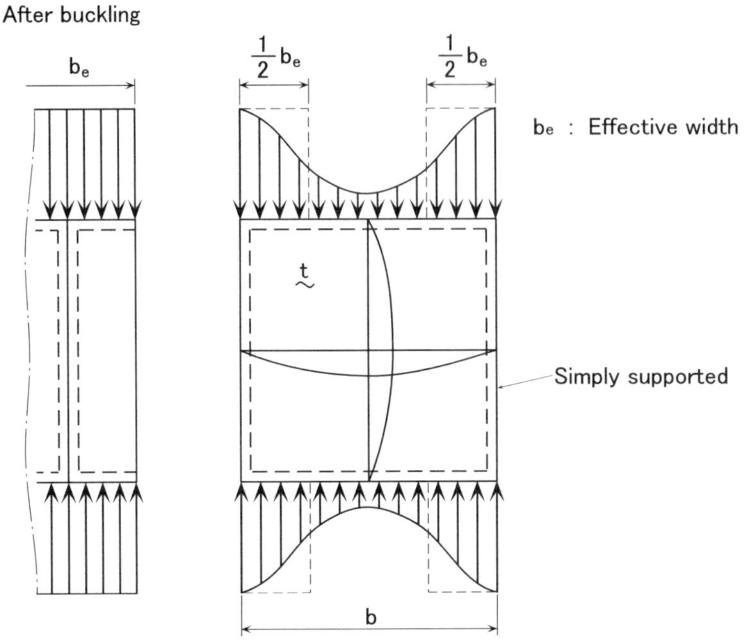

Fig. 8.3 Effective width after buckling

Then,

$$\frac{b_e}{b} = \frac{\sigma_u}{\sigma_y} \tag{8.1.4}$$

8.2 Span Point of Beams

Both ends of a beam are usually connected with brackets or other structures. Hence beam theory is not always applicable even if the cross section of middle part has uniform shape. It is convenient to assume a member to be a uniform section beam, having an equivalent length between two span points, and to assume the outside structures of the span points to be rigid bodies as illustrated in Fig. 8.4. The span point depends on structural details and loading conditions.

LR defines simply as follows [58]:

(1) For secondary stiffening members such as frames, beams, and stiffeners:

The span point is taken as the point where the depth of the end bracket, measured from the face of the secondary stiffening member, is equal to the depth of the member.

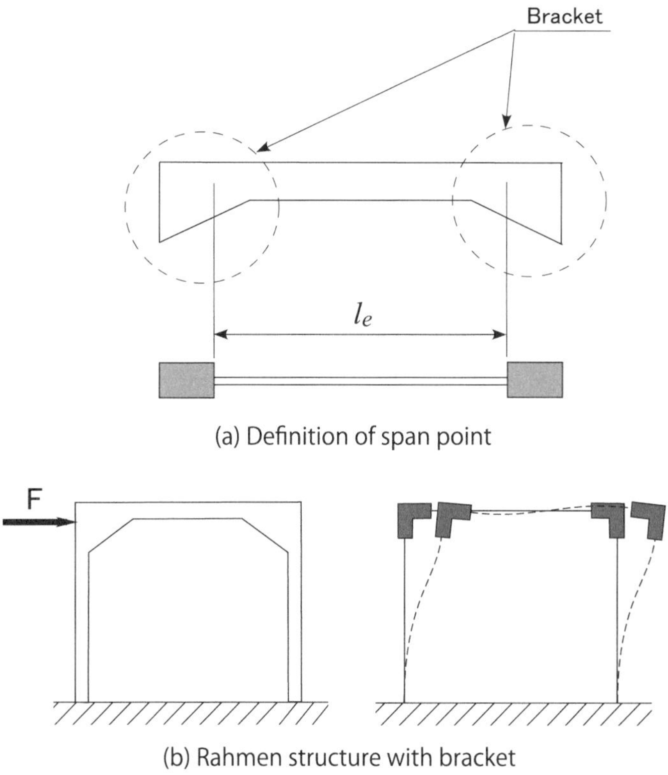

Fig. 8.4 Span point of beam

(2) For primary supporting members such as web frames, girders, and stringers:

The distance of span point from the end of the member is taken as,

$$b_e = b_b \left(1 - \frac{d_w}{d_b} \right) \tag{8.2.1}$$

The upper two figures of Fig. 8.5 illustrate the span points of secondary stiffening members, and the lower two figures illustrate those of primary supporting members.

On the other hand, Yamaguchi proposed a formula for bending which was obtained theoretically and experimentally [59] as shown in Fig. 8.6.

For straight type end connections

$$\frac{S^b}{b} = \frac{h_0}{h_A} \tag{8.2.2}$$

This equation is the same as Eq. (8.2.1) of LR rule.

For arc type end connections

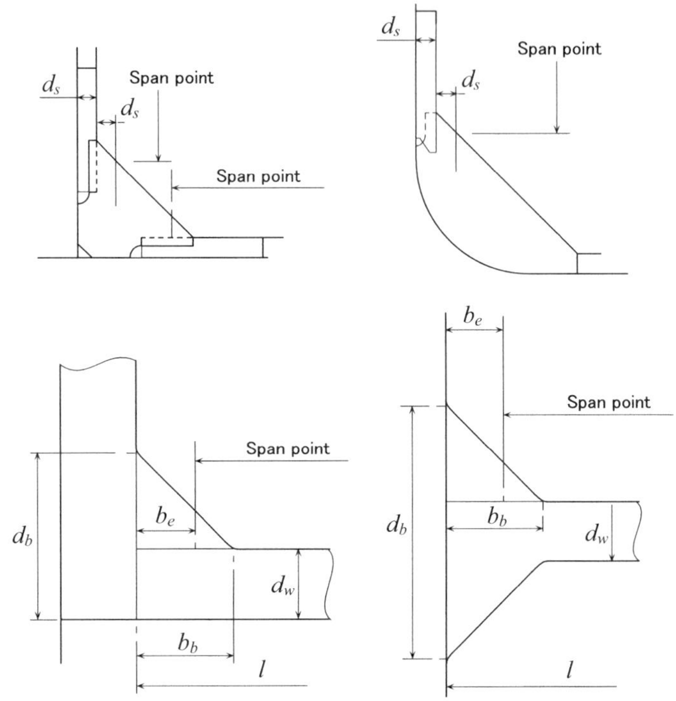

Fig. 8.5 Span points

$$\frac{S^b}{b} = 0.276 + \frac{0.724h_0}{0.724h_A + 0.276h_0} \tag{8.2.3}$$

Figure 8.7 shows the span point for the load of shearing forces [59].
For straight type end connections

$$\frac{S^b}{b} = \frac{h_0}{h_A} \log(1 + \frac{h_a}{h_0}) \tag{8.2.4}$$

For arc type end connections

$$\frac{S^b}{b} = \frac{2h_0/h_a(h_0/h_a + 1)}{\sqrt{(h_0/h_a)^2 + 2h_0/h_a}} \tan^{-1} \sqrt{1 + 2h_0/h_a} - \frac{\pi}{2} \frac{h_0}{h_a} \tag{8.2.5}$$

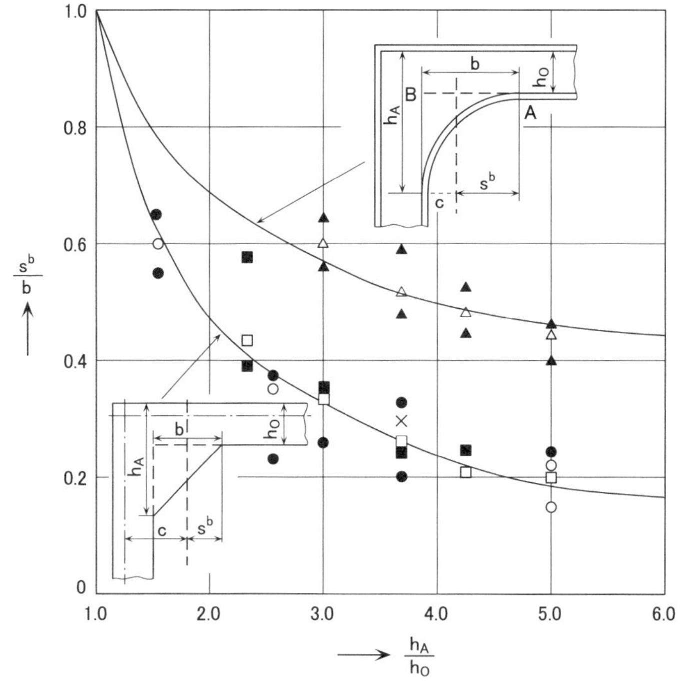

Fig. 8.6 Comparison of span point experiment versus calculation

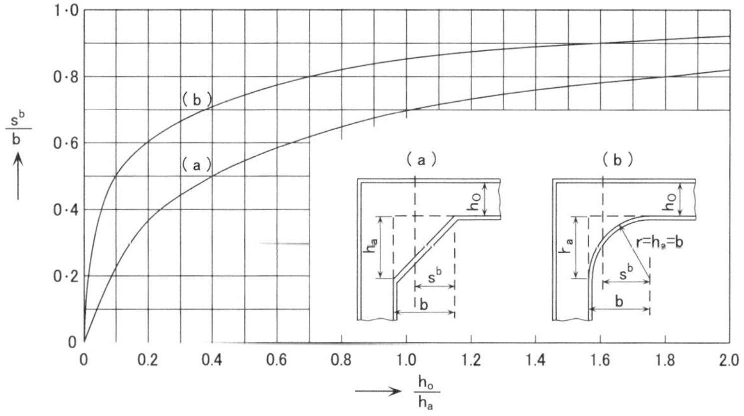

Fig. 8.7 Span points for shearing deformation

8.3 Design of Cross Section

The shape of cross section of a beam is generally designed as follows:

(1) decide the beam span between span points, derive the bending moment curve from the loading condition, and decide maximum bending moment $M_{\max}$
(2) estimate cross sectional shape, and calculate the section modulus Z

$$Z = I/y \tag{8.3.1}$$

I: sectional moment of inertia
y distance from neutral axis to the extreme tip of a beam
Z has two values of top and bottom.

(3) calculate the maximum stress

$$\sigma_{\max} = \frac{M_{\max}}{Z} \tag{8.3.2}$$

(4) compare $\sigma_{\max}$ with allowable stress σ_{all} or compare required section modulus Z_{req} to actual section modulus Z

$$Z_{req} = \frac{M_{\max}}{\sigma_{all}} \tag{8.3.3}$$

(5) adjust the sectional shape so as to be $\sigma_{\max} < \sigma_{all}$ or $Z_{req} < Z$.

8.3.1 Calculation of Section Modulus

In this section, we consider a cross section of a stiffener with an attached plate as shown in Fig. 8.8. Considering that thicknesses are much thinner than the overall dimension and neglecting the small terms of second order or more with respect to the thickness, the stiffener section can be modeled simply as shown in Fig. 8.8(2). Then, the sectional moment of inertia around the attached plate I_m is

$$I_m = A_w\left(\frac{d}{2}\right)^2 + A_f d^2 + \frac{A_w d^2}{12} = \frac{d^2(A_w + 3A_f)}{3} \tag{8.3.4}$$

where A_f, A_w and A_p are the sectional area of the face plate, web and attached plate, respectively. The distance between the neutral axis and the attached plate e_p is

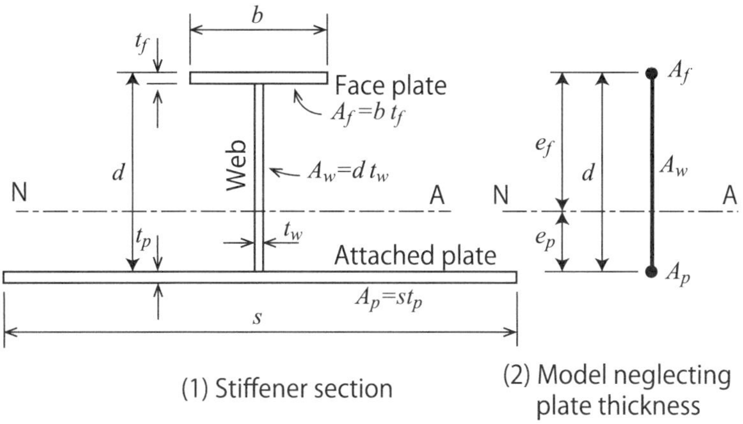

Fig. 8.8 Stiffener section with an attached plate

$$e_p = \frac{A_w \cdot \frac{d}{2} + A_f d}{A} = \frac{d(A_w + 2A_f)}{2A} \tag{8.3.5}$$

where the total area $A = A_f + A_w + A_p$.

Then, using the parallel axis theorem, the sectional moment of inertia around the neutral axis I is

$$\begin{aligned} I &= I_m - e_p^2 A \\ &= \frac{d^2(A_w + 3A_f)}{3} - \left\{ \frac{d(A_w + 2A_f)}{2A} \right\}^2 A \\ &= \frac{d^2(4A_pA_w + 4A_wA_f + 12A_pA_f + A_w^2)}{12A} \end{aligned} \tag{8.3.6}$$

The distance between the neutral axis and the face plate e_f is

$$e_f = d - e_p = \frac{d(2A_p + A_w)}{2A} \tag{8.3.7}$$

Therefore, the section modulus at the face plate is

$$\begin{aligned} Z &= \frac{I}{e_f} = \frac{d}{6} \cdot \frac{4A_pA_w + 4A_wA_f + 12A_pA_f + A_w^2}{2A_p + A_w} \\ &= \frac{d}{6} \cdot \frac{(12A_pA_f + 6A_wA_f) + (4A_pA_w - 2A_wA_f + A_w^2)}{2A_p + A_w} \\ &= d\left\{ A_f + \frac{4A_pA_w - 2A_wA_f + A_w^2}{6(2A_p + A_w)} \right\} \end{aligned}$$

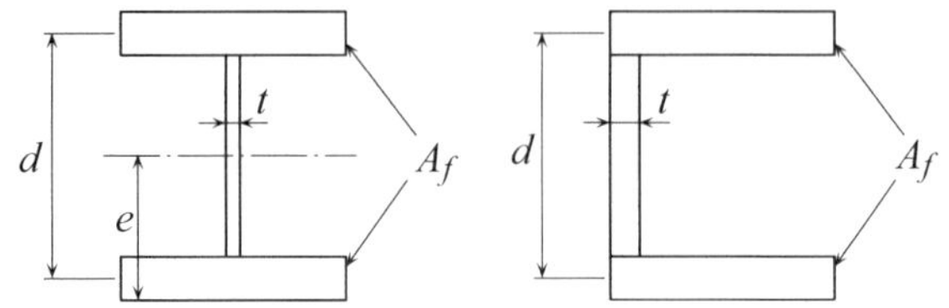

Fig. 8.9 Balanced section

$$= d\left(A_f + \frac{A_w}{6} \cdot \frac{4A_p - 2A_f + A_w}{2A_p + A_w}\right) = d\left\{A_f + \frac{A_w}{6} \cdot \frac{(2A_p + A_w) + 2A_p - 2A_f}{2A_p + A_w}\right\}$$

$$= d\left\{A_f + \frac{A_w}{6}\left(1 + \frac{1 - \frac{A_f}{A_p}}{1 + \frac{A_w}{2A_p}}\right)\right\} \tag{8.3.8}$$

This is the most general expression of the section modulus of a stiffener when the higher order terms of thickness are neglected.

(1) *Balanced section*: A beam having two flanges, which have the same sectional area, is called a balanced girder (Fig. 8.9). In this case, substituting $A_p = A_f$ into Eq. (8.3.6), the sectional moment of inertia I is obtained as follows:

$$I = \frac{d^2}{2}\left(A_f + \frac{A_w}{6}\right) \tag{8.3.9}$$

Then, the section modulus is

$$Z = \frac{I}{d/2} = d\left(A_f + \frac{A_w}{6}\right) \tag{8.3.10}$$

Hence

$$A_f = \frac{Z}{d} - \frac{A_w}{6} \tag{8.3.11}$$

Equation (8.3.11) is often represented in classification societies' rules.

(2) *Beam with large attached plate (when $A_p \gg A_f$)*: When the sectional area of the attached plate is much larger than that of the face plate, the neutral axis can be assumed to reside within the attached plate, and thus, the sectional moment of inertia I and the section modulus Z become

$$I = d^2A_f + \frac{A_w d^2}{12} + A_w\left(\frac{d}{2}\right)^2 = d^2\left(A_f + \frac{A_w}{3}\right) \tag{8.3.12}$$

$$Z = \frac{I}{d} = d\left(A_f + \frac{A_w}{3}\right) \tag{8.3.13}$$

Equations (8.3.12) and (8.3.13) can also be obtained by taking the limit of $A_p \to \infty$ for Eqs. (8.3.6) and (8.3.8), respectively.

(3) *Intermediate section*: Actual stiffeners of ship structures have intermediate properties between (1) balanced section and (2) beam with large attached plate. They can take various values depending on the structural arrangement and scantling, but if we boldly take the average between Eqs. (8.3.8) and (8.3.13), the section modulus can be expressed as follows:

$$Z = d\left(A_f + \frac{A_w}{4}\right) \tag{8.3.14}$$

This equation is useful for rough estimation of the section modulus without the information of the attached plate.

(4) *Built-up section with multiple members*: The section modulus of a beam built up by multiple plates is calculated as in Fig. 8.10.

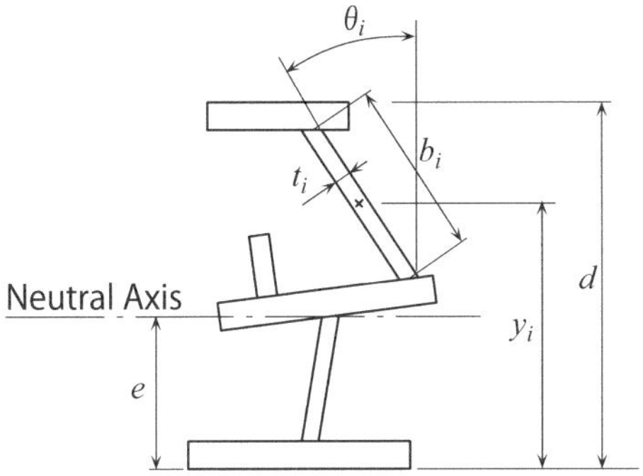

Member	Thickness	Breadth	Area	Lever	1st moment	$I\,(e{=}0)$	I_o
Plate i	t_i	b_i	$A_i{=}t_i b_i$	y_i	$A_i y_i$	$I_i{=}A_i y_i^2$	$I_{oi} = \frac{t_i b_i^3}{12}\cos^2\theta_i$
			$\sum A_i$		$\sum A_i y_i$	$\sum I_i$	$\sum I_{oi}$

$$e = \frac{\sum A_i y_i}{\sum A_i}$$

$$I = \sum I_i + \sum I_{0i} - e^2 \sum A_i$$

$$Z_1 = I/e$$

$$Z_2 = I/(d-e)$$

Fig. 8.10 Built-up section with multiple members

8.3.2 Simple Method to Adjust Member Thickness to Meet the Required Sectional Properties

Sizes of web plate and face plate of a beam are to be calculated so as to have the given section modulus I/y. In this section, we explain a simple calculation method of the required scantling adjustment to meet the required section modulus [60].

As an example, let us take up a hull girder section, where appropriate scantlings of longitudinal members should be designed to satisfy the required section modulus. For the hull girder midship section modulus to meet the required value, it is usual to adjust upper deck plate thickness after deciding the sizes of other members to have enough local strength. It can be achieved usually by a trial and error method because in this case by adjusting the upper deck plate thickness the position of the neutral axis moves accordingly. The method explained here is not a trial and error method but a direct method obtained by a simple approach.

Figure 8.11 shows a midship section of a tanker. For this midship section, section modulus at deck Z_{DK}, sectional area A, sectional moment of inertia I, and the distance between neutral axis and upper deck at side y_d are calculated based on the original scantlings of the longitudinal members with the minimum scantlings to satisfy local strength.

Let us assume that a sectional area of A_0 is added at the distance η above the neutral axis as shown in Fig. 8.11. Then, the neutral axis moves upward by a distance of d in the following equation.

$$d = \frac{A_0 \eta}{A + A_0} \tag{8.3.15}$$

The increment of the sectional moment of inertia ΔI is

$$\begin{aligned}\Delta I &= Ad^2 + A_0(\eta - d)^2 \\ &= A\left(\frac{A_0 \eta}{A + A_0}\right)^2 + A_0\left(\frac{A\eta}{A + A_0}\right)^2\end{aligned}$$

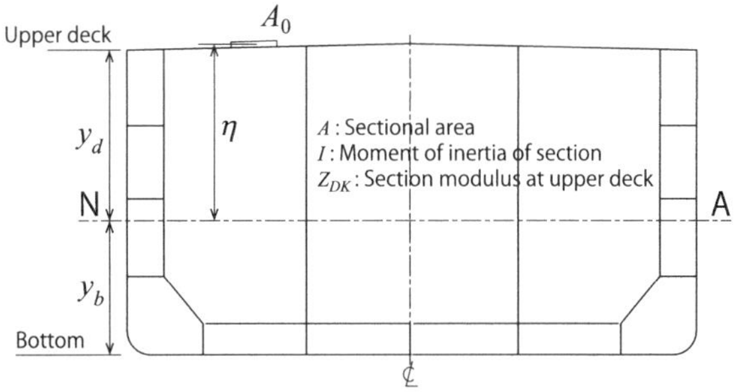

Fig. 8.11 Addition of longitudinal member at midship section of tanker

$$= \frac{AA_0(A + A_0)\eta^2}{(A + A_0)^2}$$
$$= \frac{AA_0\eta^2}{A + A_0} \tag{8.3.16}$$

The section modulus Z_0 on the upper deck at side after adding the additional area A_0 is calculated as follows:

$$Z_0 = \frac{I + \Delta I}{y_d - d} = \frac{I}{y_d} + \frac{\frac{\Delta I y_d + Id}{y_d^2}}{\frac{y_d - d}{y_d}} = \frac{I}{y_d} + \frac{\Delta I y_d + Id}{y_d(y_d - d)} = Z_{DK} + \frac{\Delta I y_d + Id}{y_d(y_d - d)} \tag{8.3.17}$$

Finally, substituting Eqs. (8.3.15) and (8.3.16) into Eq. (8.3.17), the increase in the section modulus ΔZ_{DK} is obtained as follows:

$$\Delta Z_{DK} = Z_0 - Z_{DK} = \frac{\Delta I y_d + Id}{y_d(y_d - d)} = \frac{\frac{AA_0\eta^2}{A+A_0} \cdot y_d + I \cdot \frac{A_0\eta}{A+A_0}}{y_d\left(y_d - \frac{A_0\eta}{A+A_0}\right)} = \frac{A_0\eta(A\eta + Z_{DK})}{y_d(A + A_0) - A_0\eta} \tag{8.3.18}$$

When additional area is put on the upper deck, η is nearly equal to y_d, thus the following simple relation is obtained:

$$\Delta Z_{DK} = \frac{A_0 y_d(A y_d + Z_{DK})}{A y_d} = \frac{A_0(A y_d + Z_{DK})}{A} \tag{8.3.19}$$

Therefore, the required sectional area to increase the section modulus by ΔZ_{DK} is

$$A_0 = \frac{A \Delta Z_{DK}}{A y_d + Z_{DK}} \tag{8.3.20}$$

Applying Eq. (8.3.20), necessary additional area on the upper deck can be calculated directly instead of by the traditional trial and error method.

Observing Eq. (8.3.18), we can find that ΔZ_{DK} is always positive when $0 < \eta < y_d$. That is, when the structural members are added above the neutral axis, the section modulus at deck always increases.

Attention should be paid to the case where η is negative, i.e., adding structural members below the neutral axis. In this case, by adding A_0, the section modulus may not always be increased. From the observation of Eq. (8.3.18), when $A\eta + Z_{DK}$ is positive, ΔZ_{DK} turns negative, i.e., the section modulus at deck reduces despite the addition of sectional area. The range of η which causes this reduction in the section modulus at deck can be expressed as follows:

$$-\frac{Z_{DK}}{A} < \eta < 0 \tag{8.3.21}$$

Conversely, the section modulus can be increased by taking off A_0 in this region. This phenomenon is caused by the bigger effect of moving the neutral axis than the increase in sectional moment of inertia.

8.4 Optimum Design of Beam Section

In the design of a beam, the designer should firstly pay attention to having least weight for the required strength, because a light and strong ship can carry more cargo more safely. By deciding the boundary conditions, loading conditions, and allowable stress of the beam, the required section modulus is given. Here some considerations for lighter section of beams for a required section modulus are explained.

Usually a beam section consists of three parts, attached plate, web and face plate, as shown in Fig. 8.12. The attached plate supports the lateral force such as water pressure, and the face plate supports the bending moment because of its greater distance from the neutral axis where bending stress is the maximum. The web supports the shearing force keeping the distance between the attached plate and the face plate.

In the case where the area of the attached plate is equal to that of the face plate, the neutral axis comes at the mid depth of the web as shown in Fig. 8.12a which is called "balanced girder". In the case where the area of the plate is far bigger than that of the face plate, the neutral axis lies almost on the plate as shown in Fig. 8.12c which is called "girder with large attached plate". Intermediate proportion is shown in Fig. 8.12b.

While a single beam or girder is often designed as a "balanced girder", a big girder supporting stiffeners such as a bottom transverse or a transverse ring of a tanker has a big attached plate compared with the face plate and this is treated as a "girder with large attached plate" which has the neutral axis on the plate. Many small stiffeners have an intermediate proportion whose neutral axis resides in the lower part of the web.

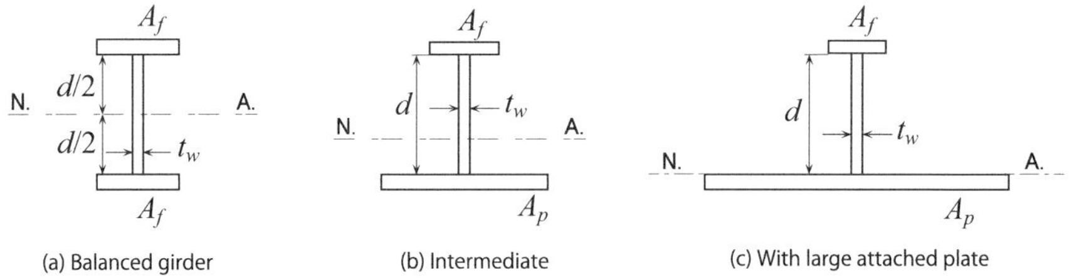

Fig. 8.12 Section of beam and girder

A different design philosophy will bring a different result. For the optimum section of a beam, an elastic design and a plastic design will also have different results. The optimum sections by an elastic design and by a plastic design are explained hereunder.

8.4.1 Elastic Design

(1) *Balanced girder*: In the case of a balanced girder, as explained in Sect. 8.3.1(1), the area of the face plate A_f to meet the required section modulus Z is obtained by following equation.

$$A_f = \frac{Z}{d} - \frac{t_w d}{6} \tag{8.4.1}$$

The sectional area A of the beam is the sum of the web area and face plate area.

$$A = A_f + t_w d = \frac{Z}{d} + \frac{5t_w d}{6} \tag{8.4.2}$$

Keeping the web thickness constant, the optimum depth of the web is obtained by differentiating Eq. (8.4.2) with respect to d and equating it with zero.

$$\frac{dA}{dd} = -\frac{Z}{d^2} + \frac{5}{6}t_w = 0 \tag{8.4.3}$$

$$\therefore d = \sqrt{\frac{6Z}{5t_w}} = 1.10\sqrt{\frac{Z}{t_w}} \tag{8.4.4}$$

In the above case the area of the face plate becomes

$$A_f = \left(\sqrt{\frac{5}{6}} - \frac{1}{6}\sqrt{\frac{6}{5}}\right)\sqrt{Zt_w} = 0.73\sqrt{Zt_w} \tag{8.4.5}$$

and the ratio between the web area and the face plate area is

$$\frac{A_w}{A_f} = 1.5 \tag{8.4.6}$$

It should be noted that Eq. (8.4.2) includes only one face plate, and therefore, this minimizes the weight of the sum of one face plate and the web. To minimize the total weight including both the face plate,

$$A = 2A_f + t_w d = \frac{2Z}{d} + \frac{2t_w d}{3} \tag{8.4.7}$$

$$\frac{dA}{dd} = -\frac{2Z}{d^2} + \frac{2}{3}t_w = 0 \tag{8.4.8}$$

$$\therefore d = \sqrt{\frac{3Z}{t_w}} = 1.73\sqrt{\frac{Z}{t_w}} \tag{8.4.9}$$

resulting in greater depth with smaller face plate.

(2) *Girder with large attached plate*: As explained in Sect. 8.3.1(2), a big girder such as a bottom transverse or transverse ring of a tanker has a big attached plate compared with the face plate and it is treated as a "girder with large attached plate" which has the neutral axis on the attached plate. In this case the sectional area of the face plate is shown by the following equation.

$$A_f = \frac{Z}{d} - \frac{t_w d}{3} \tag{8.4.10}$$

Keeping the web thickness constant, the optimum depth of the web is obtained.

$$\frac{dA}{dd} = -\frac{Z}{d^2} + \frac{2}{3}t_w = 0 \tag{8.4.11}$$

$$\therefore d = \sqrt{\frac{3Z}{2t_w}} = 1.22\sqrt{\frac{Z}{t_w}} \tag{8.4.12}$$

In the above case the area of the face plate becomes

$$A_f = \left(\sqrt{\frac{2}{3}} - \frac{1}{3}\sqrt{\frac{3}{2}}\right)\sqrt{Zt_w} = 0.41\sqrt{Zt_w} \tag{8.4.13}$$

and the ratio between the web area and the face plate area is

$$\frac{A_w}{A_f} = 3.0 \tag{8.4.14}$$

Comparing Eq. (8.4.4) with (8.4.12) and (8.4.5) with (8.4.13), it is observed that the "girder with large attached plate" is optimal with a deeper web and smaller face plate. On the other hand, the "balanced girder" is optimal with a shallow web and bigger face plate.

(3) *Intermediate section*: According to Sect. 8.3.1(3), the sectional area of the face plate of an intermediate section is expressed in the following equation.

$$A_f = \frac{Z}{d} - \frac{t_w d}{4} \tag{8.4.15}$$

Then, differentiating the total area with respect to d,

$$\frac{dA}{dd} = -\frac{Z}{d^2} + \frac{3}{4}t_w = 0 \tag{8.4.16}$$

$$\therefore d = \sqrt{\frac{4Z}{3t_w}} = 1.15\sqrt{\frac{Z}{t_w}} \tag{8.4.17}$$

In the above case the area of the face plate becomes

$$A_f = \left(\sqrt{\frac{3}{4}} - \frac{1}{4}\sqrt{\frac{4}{3}}\right)\sqrt{Zt_w} = 0.58\sqrt{Zt_w} \tag{8.4.18}$$

and the ratio between the web area and the face plate area is

$$\frac{A_w}{A_f} = 2.0 \tag{8.4.19}$$

Observing Eqs. (8.4.6), (8.4.14) and (8.4.19), optimum ratio of A_w/A_f is 1.5 for a balanced girder, 3.0 for a girder with large attached plate, and 2.0 for intermediate section. That is, when we design girders/stiffeners, A_w/A_f should be within the range 1.5 –3.0, and the optimal proportion for ordinary beams is such that the web area is approximately twice the face plate area. This is a very simple and useful design tip and guidance. Designers should take note if they find built-up beam proportions very different from the above value.

When we substitute the optimum proportion for the intermediate section $A_w = 2A_f$ into Eq. (8.3.8),

$$Z = \frac{dA_f(5A_p + 3A_f)}{3(A_p + A_f)} \tag{8.4.20}$$

When this agrees with Eq. (8.3.14), the following relationship is obtained.

$$A_p = 3A_f \tag{8.4.21}$$

This means that Eq. (8.3.14) assumes the sectional area of the attached plate three times larger than that of the face plate as a typical proportion between the stiffener and the attached plate. In this case, the position of the neutral axis e_p is calculated as follows using $A_p = 3A_f$ and $A_w = 2A_f$.

Table 8.1 Optimum section of beam

(a) Elastic design

		Balanced girder	Intermediate	With large attached plate
With face plate	Shape	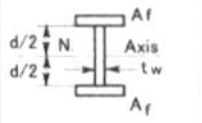		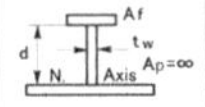
	d (cm)	$1.10\sqrt{Z/t_w}$	$1.15\sqrt{Z/t_w}$	$1.22\sqrt{Z/t_w}$
	Af (cm²)	$0.73\sqrt{Zt_w}$	$0.58\sqrt{Zt_w}$	$0.41\sqrt{Zt_w}$
	I (cm⁴)	$0.55Z\sqrt{Z/t_w}$	$0.77Z\sqrt{Z/t_w}$	$1.22Z\sqrt{Z/t_w}$
	w (kg/m)	$1.43\sqrt{Zt_w}$	$1.36\sqrt{Zt_w}$	$1.28\sqrt{Zt_w}$
Without face plate		Balanced girder	Intermediate	With large attached plate
	Shape	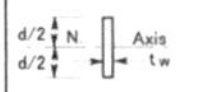		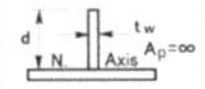
	d (cm)	$2.45\sqrt{Z/t_w}$	$2.00\sqrt{Z/t_w}$	$1.73\sqrt{Z/t_w}$
	I (cm⁴)	$1.22Z\sqrt{Z/t_w}$	$1.33Z\sqrt{Z/t_w}$	$1.73Z\sqrt{Z/t_w}$
	w (kg/m)	$1.92\sqrt{Zt_w}$	$1.57\sqrt{Zt_w}$	$1.36\sqrt{Zt_w}$

(b) Plastic design

		Balanced girder	Intermediate	With large attached plate
With face plate	Shape			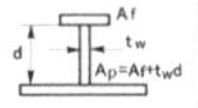
	d (cm)	$1.15\sqrt{Z_P/t_w}$	$1.28\sqrt{Z_P/t_w}$	$1.41\sqrt{Z_P/t_w}$
	Af (cm²)	$0.58\sqrt{Z_Pt_w}$	$0.29\sqrt{Z_Pt_w}$	0
	I (cm⁴)	$0.51Z_P\sqrt{Z_P/t_w}$	$0.72Z_P\sqrt{Z_P/t_w}$	$0.93Z_P\sqrt{Z_P/t_w}$
	w (kg/m)	$1.36\sqrt{Z_Pt_w}$	$1.23\sqrt{Z_Pt_w}$	$1.11\sqrt{Z_Pt_w}$
Without face plate		Balanced girder	Intermediate	With large attached plate
	Shape		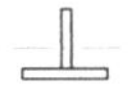	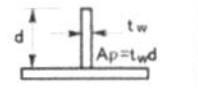
	d (cm)	$2.00\sqrt{Z_P/t_w}$	$1.71\sqrt{Z_P/t_w}$	$1.41\sqrt{Z_P/t_w}$
	I (cm⁴)	$0.67Z_P\sqrt{Z_P/t_w}$	$0.80Z_P\sqrt{Z_P/t_w}$	$0.93Z_P\sqrt{Z_P/t_w}$
	w (kg/m)	$1.57\sqrt{Z_Pt_w}$	$1.34\sqrt{Z_Pt_w}$	$1.11\sqrt{Z_Pt_w}$

$$e_p = \frac{A_w \cdot \frac{d}{2} + A_f d}{A} = \frac{d\left(A_f + A_f\right)}{A_f + 2A_f + 3A_f} = \frac{d}{3} \tag{8.4.22}$$

In Table 8.1a, the optimum sections of the elastic beams are shown for a "balanced girder" including "flat bar" which has no face plate, "girder with large attached plate", and the intermediate section between the two. The intermediate section is obtained to have an average section modulus between the balanced girder and the girder with large attached plate. In addition to the web depth and the face plate area, sectional moment of inertia I and weight per unit length w are shown in the table.

In designing a beam with the required section modulus Z, the derived section, sectional moment of inertia and weight per unit length depend on the designer's intention and are not unified. As shown in Table 8.1 under the optimal conditions, these values are standardized independent of the designer's intention. By optimization, the differences in design by each designer's intention are unified, and accordingly the reliability and the accuracy of the design can be improved.

8.4.2 Plastic Design

In a plastic beam the neutral axis divides the sectional area equally into two, above and below the neutral axis. Also, the neutral axis lies nearer the plate than the center of gravity for an elastic beam.

(1) *Balanced girder*: Expressing the plastic section modulus by Z_P, the area of the face plate A_f will be given by the following.

$$A_f = \frac{Z_P}{d} - \frac{t_w d}{4} \tag{8.4.23}$$

$$\frac{dA}{dd} = -\frac{Z_P}{d^2} + \frac{3}{4}t_w = 0 \tag{8.4.24}$$

$$\therefore d = \sqrt{\frac{4Z_P}{3t_w}} = 1.15\sqrt{\frac{Z_P}{t_w}} \tag{8.4.25}$$

$$A_f = \sqrt{\frac{Z_P t_w}{3}} = 0.58\sqrt{Z_P t_w} \tag{8.4.26}$$

(2) *Girder with large attached plate*: The plastic neutral axis is assumed to be in the attached plate.

$$A_f = \frac{Z_P}{d} - \frac{t_w d}{2} \tag{8.4.27}$$

$$\frac{dA}{dd} = -\frac{Z_P}{d^2} + \frac{1}{2}t_w = 0 \tag{8.4.28}$$

$$\therefore d = \sqrt{\frac{2Z_P}{t_w}} = 1.41\sqrt{\frac{Z_P}{t_w}} \tag{8.4.29}$$

$$A_f = \sqrt{\frac{Z_P t_w}{2}} - \sqrt{\frac{Z_P t_w}{2}} = 0 \tag{8.4.30}$$

In Table 8.1b the optimum sections of the plastic beams for plastic section modulus Z_P are shown. It can be found that in the plastic beam the web is more effective than the face plate while on the other hand in the elastic beam the face plate is more effective than the web.

8.4.3 Optimal Proportion for Beams

In Table 8.1, an intermediate shape is shown between a balanced girder and a girder with plate. This shape was determined so that the section modulus has the average value between that of the balanced girder and that of the girder with plate. This is a very rough and bold assumption but practically it can be accepted. In hull structure design such approximations bring very good and fruitful results.

Figure 8.13 shows the optimum depth and weight of an elastic beam for a required section modulus compared with those values for a standard angle. It can be said that the standard angle is very close to that for the optimum intermediate shape, i.e.,

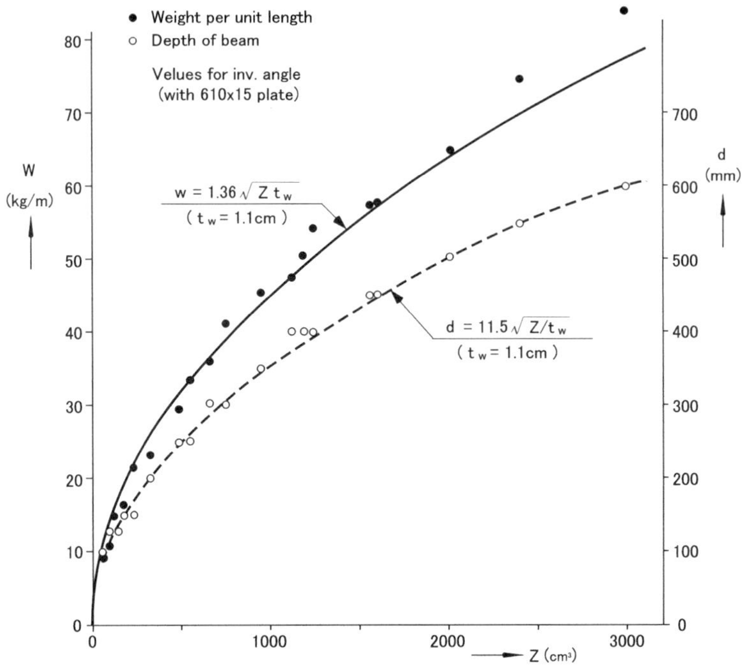

Fig. 8.13 Optimum section and standard section

the intermediate shape can be regarded to represent typical shipbuilding stiffeners. Therefore, Eqs. (8.4.17) and (8.4.19) can be used as a useful guide for efficient design.

However, everything has exceptions. Designers must not apply such guidance blindly. From the derivation described through Sects. 8.4.1 and 8.4.2, we find that the optimum proportion was obtained postulating that the web thickness is constant. Some longitudinal stiffeners such as bottom longitudinals are under large axial compression, and thus, web thickness has to be increased as the beam depth increases due to the web buckling criterion. Under these conditions, deep beams become inefficient, and the optimum proportion shifts to the side of smaller A_w/A_f ratio. In the case where a very large axial compression acts, such as in higher tensile steel application, this ratio may become down to around 1.0.

Similarly, under an action of forced displacement, deeper beams are also inefficient and optimum A_w/A_f ratio becomes smaller. This will be detailed in Sect. 8.6.6.

In actual design situations, various design constraints must be considered besides structural design issues. In many cases the beam depth is constrained due to accessibility, easy fabrication, fitting arrangement, etc. One typical example is a shallow deck beam with heavy face plate to ensure the required deck clear height for PCTC (Pure Car and Truck Carrier) and ferry boats.

8.5 Application of Beam Theory

8.5.1 Bending Moment

In designing a hull structure, many structural members can be regarded as beams, which are mainly subjected to lateral load. For instance, an entire hull structure is often considered as a beam, since it is deflected by the action of the vertical forces due to buoyancy and gravity. Another example of a beam is the shell plate. This may seem curious, but when a shell plate is sliced into plates of unit width, then these narrow band plates can be regarded as a group of beams under lateral pressure.

Most structural members in ship hulls are regularly arranged and subjected to nearly uniformly distributed loading as shown in Fig. 8.14. Therefore, the strength of a structural member is usually investigated by modeling one span as a beam with fixed ends, subjected to a uniform load as shown in the figure. Consequently, the maximum bending moment of the beam occurs at both ends and its magnitude can be calculated as follows:

$$M = \frac{wl^2}{12} = \frac{Wl}{12} \tag{8.5.1}$$

where

M: maximum bending moment
w: uniform load per unit length
W: total load, $W = wl$
l: beam span

On the other hand, if a concentrated load W is applied to a beam having both ends fixed as illustrated in Fig. 8.15, the maximum bending moment is generated at a fixed end and it is

$$M = -\frac{Wl_1^2(l - l_1)}{l^2} \tag{8.5.2}$$

Fig. 8.14 Fixed beam subjected to uniform load

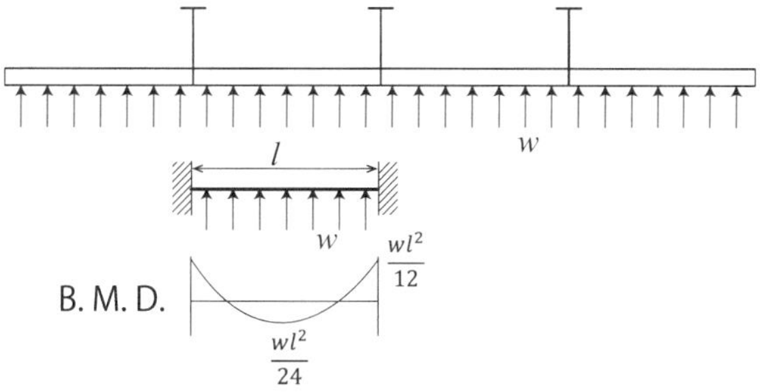

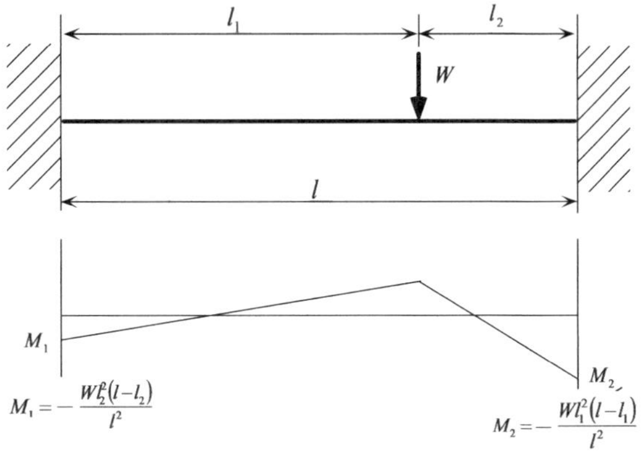

Fig. 8.15 Fixed beam subjected to concentrated load

where l_1: greater distance between loaded position and beam end

Since the maximum bending moment varies according to the position of the concentrated load applied, we have to fully understand the position where it creates the ultimate maximum bending moment. In order to investigate this position of load giving the maximum value, the first derivative of Eq. (8.5.2) is taken and then the following equation is obtained:

$$\frac{dM}{dl_1} = -\frac{W}{l^2}\left(2ll_1 - 3l_1^2\right) = 0$$

$$\therefore l_1 = 0 \text{ or } \frac{2}{3}l \tag{8.5.3}$$

$$M_{ultimate} = -\frac{4}{27}Wl \tag{8.5.4}$$

Therefore, the maximum moment is produced when the load is applied to the position of $l_1 = 2l/3$. This load creates a maximum bending moment of $4Wl/27$, which is about $4/27 \times 8 = 1.18$ times larger than when the load is applied to the center of the span. We are likely to misunderstand that the maximum bending moment be induced by a load applied to the mid-point of the beam span. But the above calculation shows this understanding is not correct. In addition, comparing this value with Eq. (8.5.1), the maximum bending moment of a concentrated load is about $4/27 \times 12 = 1.8$ times greater than that of uniform load.

In contrast to the above discussion, the tendency is sometimes opposite when the strength of beam is considered from the view point of plastic design. In the case of a beam rigidly fixed at both ends subjected to a concentrated load at the center of span, the load causing the collapse is calculated as $8M_P/l$, where M_P is the plastic moment of the cross section of a beam. On the other hand, the beam will collapse from a load of $9M_P/l$, when the load acts at a position $l/3$ from an end (see Table 3.6 2, Part I). That means the beam can carry greater load in the latter case. This example shows that the bending moment will differ due to a difference in the design philosophy applied.

8.5.2 *Easy Solution of Statically Indeterminate Beams*

A straight beam structure, which consists of at least 2 beam elements supported by 3 points or more, is usually called as "continuous beam". A structure as shown in Fig. 8.16 is an example of a continuous beam. The bending moment and shearing force distributions of continuous beams cannot be determined generally only by the equation of static equilibrium.

For example, in the case of Fig. 8.16, the number of equations at equilibrium is two, although the number of unknown reaction forces is four (R_1, R_2, R_3, M) as shown below. Therefore, the reaction forces cannot be determined, since the number of reactions exceeds the number of equations.

$$R_1 + R_2 + R_3 = P \tag{8.5.5}$$

$$R_2 \cdot l - P \cdot \frac{3}{2}l + R_3 \cdot 2l = M \tag{8.5.6}$$

If the unknown reactions cannot be determined simply by the equations of static equilibrium, the reactions of the beam are said to be statically indeterminate. In order to determine the unknown reaction forces, not only the static equilibrium but also the continuity of deflection and slope of the beam must be taken into account. Several kinds of useful methods to solve a continuous beam are proposed, e.g., the theorem of three moments. In this section, we introduce a simple method to obtain bending moments at each support point based on the concept of the theorem of three moments. In this method, a continuous beam is regarded as a set of simply supported beam elements. Let us make use of the example in Fig. 8.17 and divide the continuous beam into two simply supported beams (Beam I and Beam II) as shown in the upper figure in Fig. 8.17. The solution procedure follows the step (1)–(7):

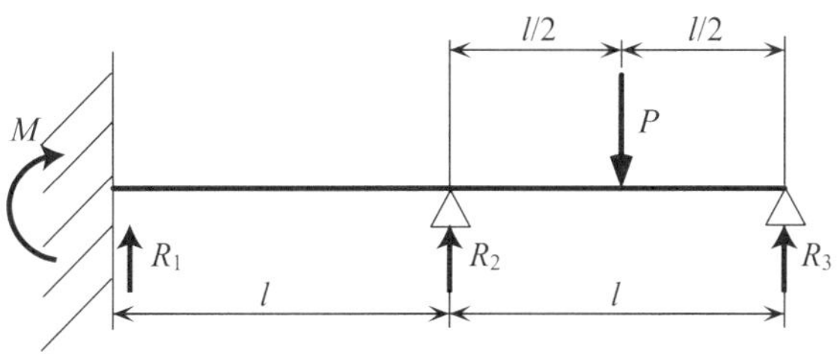

Fig. 8.16 Sample of continuous beam

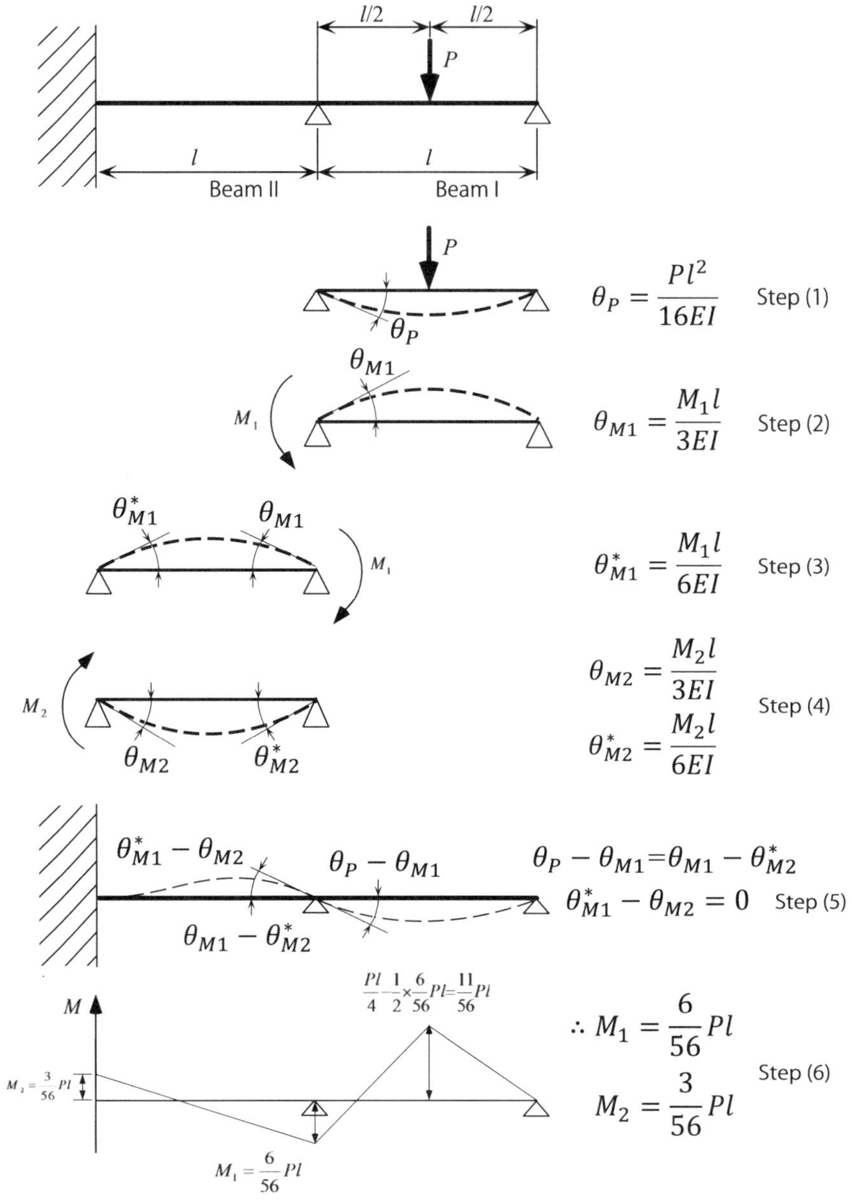

Fig. 8.17 Simplified method to solve continuous beam

(1) Force P causes a deflection of its slope θ_p in Beam I in the place of its connection to Beam II.
(2) The bending moment M_1 at the connection between Beam I and Beam II creates slope θ_{M1} at the left end of Beam I.
(3) The same bending moment M_1 creates deflection in Beam II and its slope is θ_{M1} at the right end and θ^*_{M1} at the left end.
(4) The bending moment M_2 at the left end causes deflection with a slope θ_{M2} at the left end and slope θ^*_{M2} at right end.
(5) Considering that the slope must be continuous at the connection of Beam I and Beam II, and that the slope must be zero at the left end of Beam II, then we obtain the equations of Step (5) as shown in the figure.
(6) By solving the above equations, the unknown moments M_1, M_2 can be expressed in the form of Step (6).
(7) Consequently, the distribution of bending moment in Beam I and II can be illustrated in the lowest figure in Fig. 8.17.

8.6 Irregularity in Regular Arrangement

As discussed in Sect. 8.5.1, most structural members in ship hulls are regularly arranged and subjected to nearly uniformly distributed loading, especially around midship region. Therefore, the strength of a structural member is usually investigated by modeling one span as a beam with fixed ends, subjected to a uniform load. This is practical and correct in many cases, but it should be noted that this assumption is not always valid when irregularity as enumerated below is included in the structure.

- Irregular span
- Simply supported end by a header of a stiffener
- Different load direction across a tank boundary
- Stiffeners penetrating a tank boundary under staggered loading
- Partially distributed or concentrated load on a continuous beam
- Forced displacement or forced rotation

The effect of each irregularity is discussed in the following sections.

8.6.1 Simply Supported End of a Continuous Beam [61]

As a prerequisite to discuss the effect of irregular arrangement of structures, we first derive rotational spring constants of basic beam elements. As shown in Fig. 8.18, the beam is supported by a rotational spring K_A at the left end and simply supported at the right end. The moment of force M_B is given at the right end, and the reaction moment at the left end is denoted by M_A. Then,

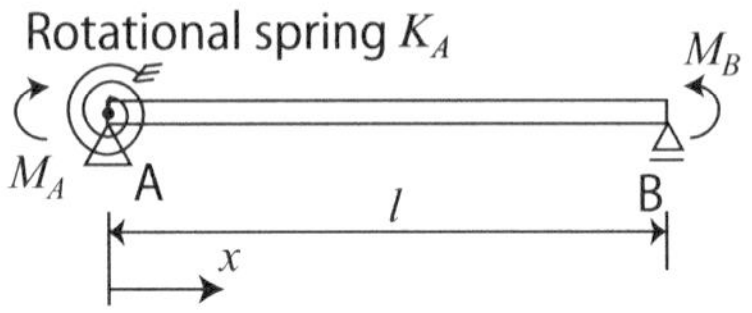

Fig. 8.18 Simply supported beam with a rotational spring at one end

$$K_A = -\frac{M_A}{\frac{dv(0)}{dx}} \tag{8.6.1}$$

$$k_A = \frac{l}{EI}K_A = -\frac{lM_A}{EI\frac{dv(0)}{dx}} \tag{8.6.2}$$

where $v(x)$ is deflection of the beam, and k_A is the non-dimensional spring constant. Now let us derive the spring constant at the right end of this beam. The bending moment along the beam $M(x)$ is

$$M(x) = \frac{M_B x}{l} + \frac{M_A(l-x)}{l} = \frac{(M_B - M_A)x}{l} + M_A \tag{8.6.3}$$

Thus, the differential equation of the elastic curve is obtained as follows:

$$-EI\frac{d^2v(x)}{dx^2} = \frac{(M_B - M_A)x}{l} + M_A \tag{8.6.4}$$

Integrating this,

$$-EI\frac{dv(x)}{dx} = \frac{(M_B - M_A)x^2}{2l} + M_A x + C_1 \tag{8.6.5}$$

Now, using Eq. (8.6.2) as the boundary condition at $x = 0$, C_1 is obtained. Then,

$$-EI\frac{dv(x)}{dx} = \frac{(M_B - M_A)x^2}{2l} + M_A x + \frac{lM_A}{k_A} \tag{8.6.6}$$

Integrating Eq. (8.6.6) again, and using the boundary condition $v(0) = 0$,

$$-EIv(x) = \frac{(M_B - M_A)x^3}{6l} + \frac{M_A x^2}{2} + \frac{lM_A x}{k_A} \tag{8.6.7}$$

Applying another boundary condition $v(l) = 0$,

$$\frac{(M_B - M_A)l^2}{6} + \frac{M_A l^2}{2} + \frac{M_A l^2}{k_A} = 0 \tag{8.6.8}$$

$$\therefore M_A = -\frac{k_A}{2(k_A + 3)}M_B \tag{8.6.9}$$

Now the reaction moment at the left end M_A is obtained. It depends on the magnitude of the spring constant. When $k_A = 0$, the left end is simply supported and therefore, M_A reduces to zero. When $k_A = \infty$, the left end is fixed, and therefore, $M_A = -M_B/2$.

Then, substituting Eq. (8.6.9) into Eq. (8.6.6),

$$-EI\frac{dv(x)}{dx} = M_B\left\{\frac{3(k_A+2)}{4(k_A+3)l}x^2 - \frac{k_A}{2(k_A+3)}x - \frac{l}{2(k_A+3)}\right\} \tag{8.6.10}$$

Therefore, at the right end of the beam,

$$\begin{aligned}-EI\frac{dv(l)}{dx} &= M_B\left\{\frac{3(k_A+2)}{4(k_A+3)l}l^2 - \frac{k_A}{2(k_A+3)}l - \frac{l}{2(k_A+3)}\right\}\\ &= M_B l\cdot\frac{k_A+4}{4(k_A+3)}\end{aligned} \tag{8.6.11}$$

This equation governs the relationship between the bending moment and the slope at the right end. Therefore, non-dimensional spring constant at the right end, k_B, is finally obtained as follows:

$$k_B = \frac{4(k_A+3)}{k_A+4} \tag{8.6.12}$$

Next, let us consider various cases with different situations at the left end.

- When the left end is simply supported, substituting $k_A = 0$ into Eq. (8.6.12), the rotational spring constant at the right end becomes $k_B = 3$.
- In case of two beams with simply supported left end, $k_A = 3$ and therefore, $k_B = 24/7 \approx 3.429$.
- When the left end is fixed, $k_A = \infty$ and therefore, $k_B = 4$.
- In case of two beams with fixed left end, $k_A = 4$ and therefore, $k_B = 7/2 = 3.5$.
- In case of infinite continuous beams, putting $k_A = k_B$ in the recursive Eq. (8.6.12), $k_B = 2\sqrt{3} \approx 3.464$.
- In case of one beam symmetric at the mid span, i.e., when the bending moment with the same magnitude and reverse sign is acting at the left end, putting $k_A = -k_B$ into Eq. (8.6.12),

$$(k_B - 2)(k_B - 6) = 0 \qquad \therefore k_B = 2, 6 \tag{8.6.13}$$

And the smaller $k_B = 2$ is the non-dimensional spring constant at the right end.

These results are summarized in Table 8.2. We can observe that the rotational spring of the two beams is already very close to that of the infinite continuous beams.

In case of the infinite continuous beam, putting $k_A = 2\sqrt{3}$ into Eq. (8.6.9),

Table 8.2 Non-dimensional rotational spring constant at the end of various continuous beams

Number of spans and boundary condition at left end	Non-dimensional spring constant at right end
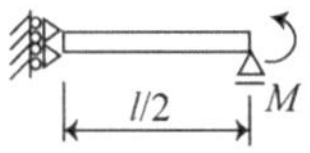	2
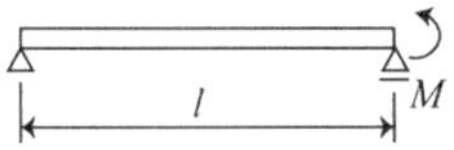	3
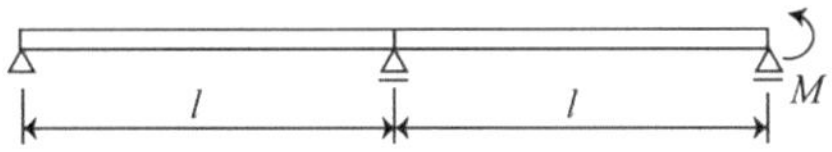	24 / 7 = 3.429
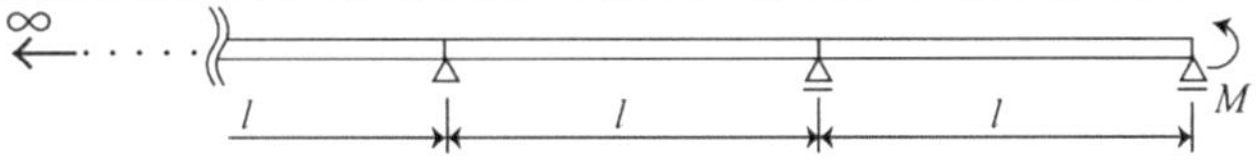	$2\sqrt{3}$ = 3.464
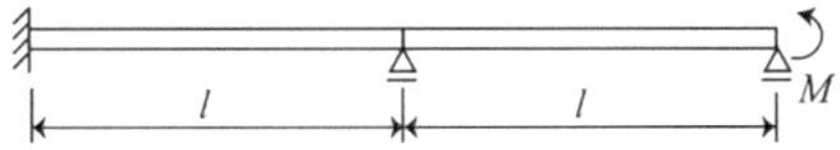	3.5
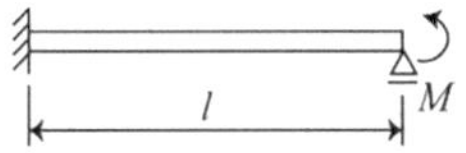	4

$$M_A = -\frac{\sqrt{3}}{3 + 2\sqrt{3}} M_B = \left(-2 + \sqrt{3}\right) M_B \approx -0.268 M_B \tag{8.6.14}$$

This means that the bending moment at each support reduces rapidly as it is apart from the right end with the n-th power to -0.268, as shown in Fig. 8.19.

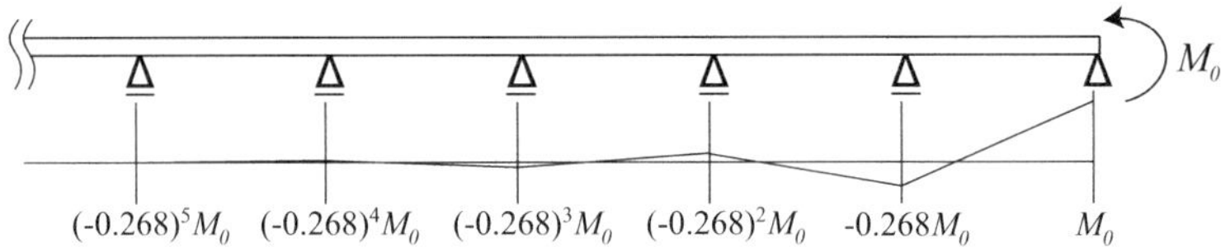

Fig. 8.19 B.M.D. of an infinitely continuous beam under a moment at the end

8.6.2 Irregular Span in a Uniformly Loaded Continuous Beam

First example of irregularities is an irregular span in a uniformly loaded continuous beam. Let us consider a continuous beam, where length of one span is shorter than the other spans as shown in Fig. 8.20. Such an arrangement is often seen as irregular transverse frame spacings in bottom structures of container ships and bulk carriers because the longitudinal width of transverse bulkheads is not necessarily consistent with the typical transverse spacing. Because stiffeners with shorter span is stronger, the stiffener scantling is usually determined based on the longer typical transverse spacing. Let us see whether such a practice is actually reasonable or not.

Thanks to the symmetry, only the left half is considered as shown in the lower figure of Fig. 8.20. Let us apply the virtual fixation method [61] and assume first that all the support point is fixed (rotation restricted), then the difference of moment at point B (ΔM_{eq}) from the left (M_L) and from the right (M_R) is

$$\Delta M_{eq} = M_L - M_R = \frac{wl^2}{12} - \frac{w\eta^2 l^2}{12} = \frac{wl^2\left(1-\eta^2\right)}{12} \tag{8.6.15}$$

Then, this difference is compensated by applying the equivalent moment ΔM_{eq} at point B, releasing the virtual rotational fixation. This equivalent moment is shared by the left side and the right side in proportion to their respective spring constants.

When the non-dimensional spring constant of the left side at point A is denoted by k_A, the non-dimensional spring constant of the left side at point B is obtained using Eq. (8.6.12). Then, the spring constant of each side, K_L and K_R, is determined as follows using Table 8.2.

$$K_L = \frac{4(k_A+3)EI}{(k_A+4)l},\ K_R = \frac{2EI}{\eta l} \tag{8.6.16}$$

Dividing ΔM_{eq} according to the ratio of K_L and K_R, the moment acting on the left side denoted by M_{BL} and that on the right side denoted by M_{BR} are expressed as follows:

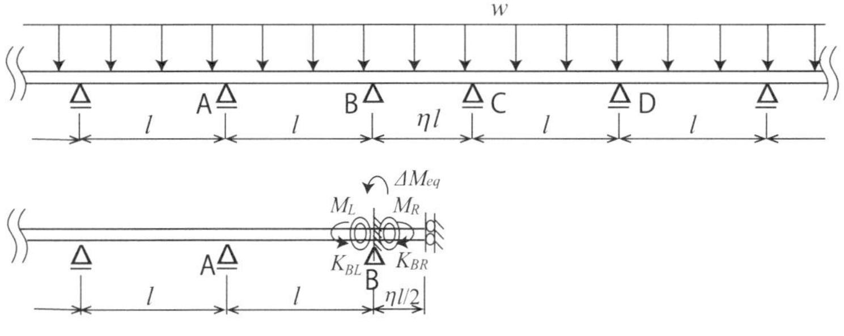

Fig. 8.20 Irregular span in a uniformly loaded continuous beam

$$M_{BL} = \Delta M_{eq} \cdot \frac{K_L}{K_L + K_R} = \frac{2(k_A+3)\eta(1-\eta^2)}{2(k_A+3)\eta + k_A + 4} \cdot \frac{wl^2}{12} \tag{8.6.17}$$

$$M_{BR} = \Delta M_{eq} \cdot \frac{K_R}{K_L + K_R} = \frac{(k_A+4)(1-\eta^2)}{2(k_A+3)\eta + k_A + 4} \cdot \frac{wl^2}{12} \tag{8.6.18}$$

The solution of the actual bending moment is the sum of that under the virtual fixation (M_L and M_R) and that under the action of ΔM_{eq} (M_L and M_R). Therefore, the bending moment at the support point B is obtained as follows:

$$M_B = -M_L + M_{BL} = -\frac{wl^2}{12}\left\{1 - \frac{2(k_A+3)\eta(1-\eta^2)}{2(k_A+3)\eta + k_A + 4}\right\} \tag{8.6.19}$$

Then, the bending moment at the support point A is obtained as follows using Eq. (8.6.9).

$$M_A = -\frac{wl^2}{12} - \frac{k_A}{2(k_A+3)}M_{BL} = -\frac{wl^2}{12}\left\{1 + \frac{k_A\eta(1-\eta^2)}{2(k_A+3)\eta + k_A + 4}\right\} \tag{8.6.20}$$

From Eqs. (8.6.19) and (8.6.20), we can find that when the inserted span is shorter than other typical spans (i.e., when $0 < \eta < 1$), $|M_B|$ becomes smaller than $wl^2/12$, but $|M_A|$ becomes greater than $wl^2/12$ conversely. The magnitude of M_A depends on the spring constant k_A. When point A is fixed (i.e., $k_A = \infty$), the bending moment at point A becomes

$$\lim_{k_A \to \infty} M_A = -\frac{wl^2}{12} \cdot \frac{1 + 3\eta - \eta^3}{1 + 2\eta} \tag{8.6.21}$$

Figure 8.21 shows the magnitude of M_A taking η in the horizontal axis for various values of k_A. We can observe approximately 20% increase of the bending moment when point A is fixed, and approximately 10% increase in case of an infinitely continuous beam when the shorter span length is about half or less of the typical span. This increase may be covered by the safety margin in many cases, but attention should be paid when the stiffener fatigue strength is critical at this location.

8.6.3 End of Longitudinal Stiffeners in Fore and Aft Part

The number of longitudinal stiffeners in fore and aft part reduces in accordance with the hull shape by ending a stiffener at a header as shown in Fig. 8.22. Let us examine the bending moments at support points A and B, which changes according to the length between the header and the transverse frame ηl.

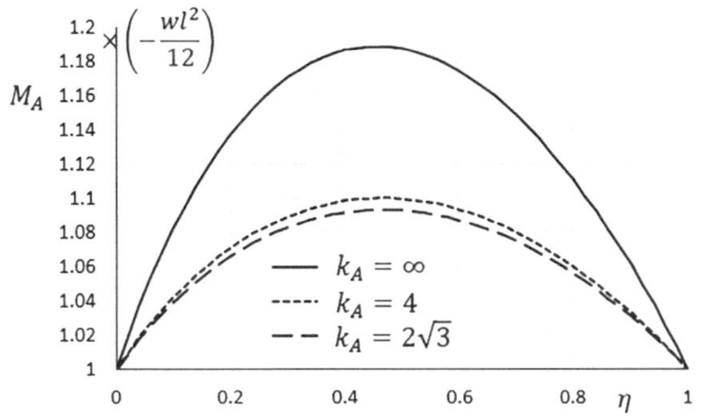

Fig. 8.21 Maximum bending moment versus the length of the inserted span

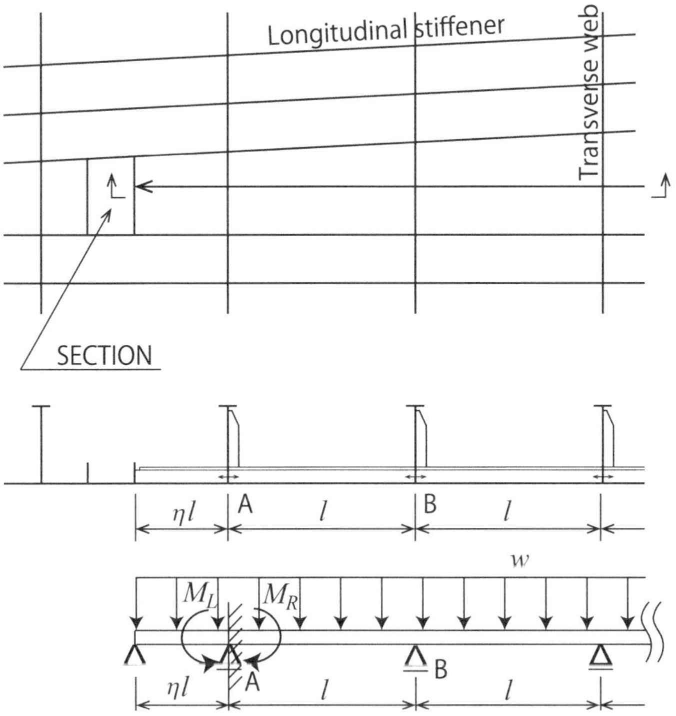

Fig. 8.22 Simply supported end of a longitudinal stiffener

The stiffener is modeled as shown in the lower figure in Fig. 8.22, and we first assume that point A is fixed in the same way as Sect. 8.6.2. Then, the difference of moment at point A (ΔM_{eq}) from the left (M_L) and from the right (M_R) is

$$\Delta M_{eq} = M_L - M_R = \frac{w\eta^2 l^2}{8} - \frac{wl^2}{12} = \frac{wl^2(3\eta^2 - 2)}{24} \tag{8.6.22}$$

Then, this difference is compensated by applying the equivalent moment ΔM_{eq} at point A, releasing the virtual rotational fixation. This equivalent moment is shared by the left side and the right side in proportion to their respective spring constants.

When the non-dimensional spring constant of the right side at point B is denoted by k_B, the non-dimensional spring constant of the right side at point A is obtained using Eq. (8.6.12). Then, the spring constant of each side, K_L and K_R, is determined as follows using Table 8.2.

$$K_L = \frac{3EI}{\eta l},\ K_R = \frac{4(k_B + 3)EI}{(k_B + 4)l} \tag{8.6.23}$$

Dividing ΔM_{eq} according to the ratio of K_L and K_R, the moment acting on the left side denoted by M_{BL} and that on the right side denoted by M_{BR} are expressed as follows:

$$M_{BL} = \Delta M_{eq} \cdot \frac{K_L}{K_L + K_R} = \frac{3(k_B + 4)(3\eta^2 - 2)}{6(k_B + 4) + 8(k_B + 3)\eta} \cdot \frac{wl^2}{12} \tag{8.6.24}$$

$$M_{BR} = \Delta M_{eq} \cdot \frac{K_R}{K_L + K_R} = \frac{4(k_B + 3)\eta(3\eta^2 - 2)}{6(k_B + 4) + 8(k_B + 3)\eta} \cdot \frac{wl^2}{12} \tag{8.6.25}$$

The solution of the actual bending moment is the sum of that under the virtual fixation (M_L and M_R) and that under the action of ΔM_{eq} (M_L and M_R). Therefore, the bending moment at the support point A is obtained as follows:

$$M_A = -M_R - M_{BR} = -\frac{wl^2}{12}\left\{1 + \frac{4(k_B + 3)\eta(3\eta^2 - 2)}{6(k_B + 4) + 8(k_B + 3)\eta}\right\} \tag{8.6.26}$$

Then, the bending moment at the support point B is obtained as follows using Eq. (8.6.9).

$$M_B = -\frac{wl^2}{12} + \frac{k_B}{2(k_B + 3)}M_{BR} = -\frac{wl^2}{12}\left\{1 - \frac{k_B\eta(3\eta^2 - 2)}{3(k_B + 4) + 4(k_B + 3)\eta}\right\} \tag{8.6.27}$$

When point B is fixed (i.e., $k_B = \infty$), the bending moments at point A and point B become

$$\lim_{k_B \to \infty} M_A = -\frac{wl^2}{12}\left\{1 + \frac{2\eta(3\eta^2 - 2)}{3 + 4\eta}\right\} \tag{8.6.28}$$

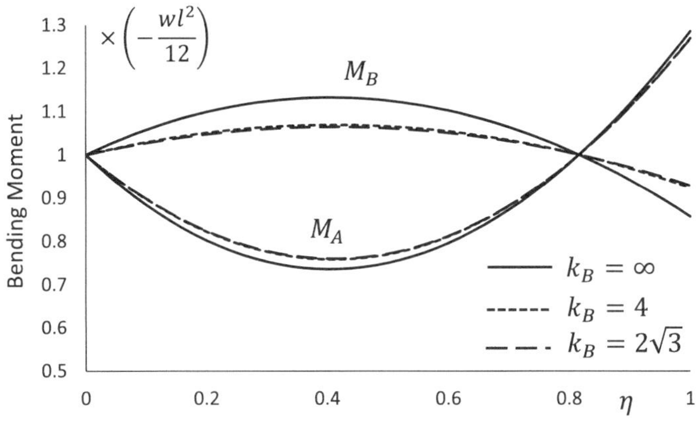

Fig. 8.23 Bending moment at support points versus the length between header and transverse frame

$$\lim_{k_B \to \infty} M_B = -\frac{wl^2}{12}\left\{1 - \frac{\eta(3\eta^2 - 2)}{3 + 4\eta}\right\}$$

Figure 8.23 shows M_A and M_B taking η in the horizontal axis for various values of k_A. When $\eta = 0$, all the support points are in fixed condition. When $0 < \eta < \sqrt{2/3}$, M_A is smaller than, and M_B is larger than the typical bending moment at the support. When $\eta = \sqrt{2/3}$, the bending moment from the both sides of point A is balanced and all the support points are again in fixed condition. When $\eta > \sqrt{2/3}$, M_A increases rapidly, and finally reaches 1.268 in case that the right side beam is infinitely continuous ($k_B = 2\sqrt{3}$). When point B is fixed ($k_B = \infty$), the increase in the bending moment is more prominent, while when the right side beam is infinitely continuous, the M_B increases by about 6.4% when $\eta \approx 0.4$. From this, we can say that it is best to fit a header at a distance of about $0.8l$ from the transverse frame to avoid increase in bending moment due to the stiffener end arrangement.

8.6.4 Longitudinal Stiffeners on Longitudinal Bulkhead Penetrating Transverse Bulkhead

Next example is a longitudinal stiffener on longitudinal bulkheads [62]. It experiences not only the bending moment caused by lateral water pressure but also axial tensile or compressive forces due to hull girder bending.

However, except in case which are near the top and bottom of the longitudinal bulkhead where the axial force is large, the axial force can be neglected and the longitudinal stiffener on longitudinal bulkhead can be designed considering only the bending moment caused by lateral water pressure.

The longitudinal stiffener on a longitudinal bulkhead is supported by transverse webs of equal span and can be assumed to be a continuous beam with equal spans and uniform load. In this case, it can be replaced by one of beam member with uniform load and both ends fixed as explained in Sect. 8.5.1. However, at the transverse bulkhead, the water pressure comes not always from one side so the assumption of one span with uniform load and of both-ends fixed cannot be applied. In this case it is reasonable to assume two continuous beams with both ends fixed or elastically supported in rotation with a support at the center. The load depends on the tank condition on both sides of the transverse bulkhead as shown in Fig. 8.24. This problem is solved in the same way as described in the previous sections with a model illustrated in Fig. 8.25, and the bending moments at the supporting points are given by the following equations.

$$M_A = -\frac{(5k + 12)w_1 - kw_2}{48(k + 3)} l^2 \tag{8.6.29}$$

$$M_B = -\frac{w_1 + w_2}{24} l^2 \tag{8.6.30}$$

$$M_C = \frac{kw_1 - (5k + 12)w_2}{48(k + 3)} l^2 \tag{8.6.31}$$

where k is the non-dimensional rotational spring constant at the transverse frame adjacent to the transverse bulkhead. $k = \infty$ means fixed condition and $k = 2\sqrt{3}$ means that the beam is continuous infinitely.

When the stiffener is fixed at the adjacent transverse frame (points A and C), M_A and M_C reduce as follows taking the limit of $k \to \infty$ for Eqs. (8.6.29) and (8.6.31).

$$M_A = -\frac{5w_1 - w_2}{48} l^2 \tag{8.6.32}$$

$$M_C = \frac{w_1 - 5w_2}{48} l^2 \tag{8.6.33}$$

When the stiffener is continuous infinitely, substituting $k = 2\sqrt{3}$ into Eqs. (8.6.29) and (8.6.31),

$$M_A = -\frac{\left(4 - \sqrt{3}\right)w_1 - \left(2 - \sqrt{3}\right)w_2}{24} l^2 \tag{8.6.34}$$

$$M_C = \frac{\left(2 - \sqrt{3}\right)w_1 - \left(4 - \sqrt{3}\right)w_2}{24} l^2 \tag{8.6.35}$$

Reaction forces at the supporting points are as follows:

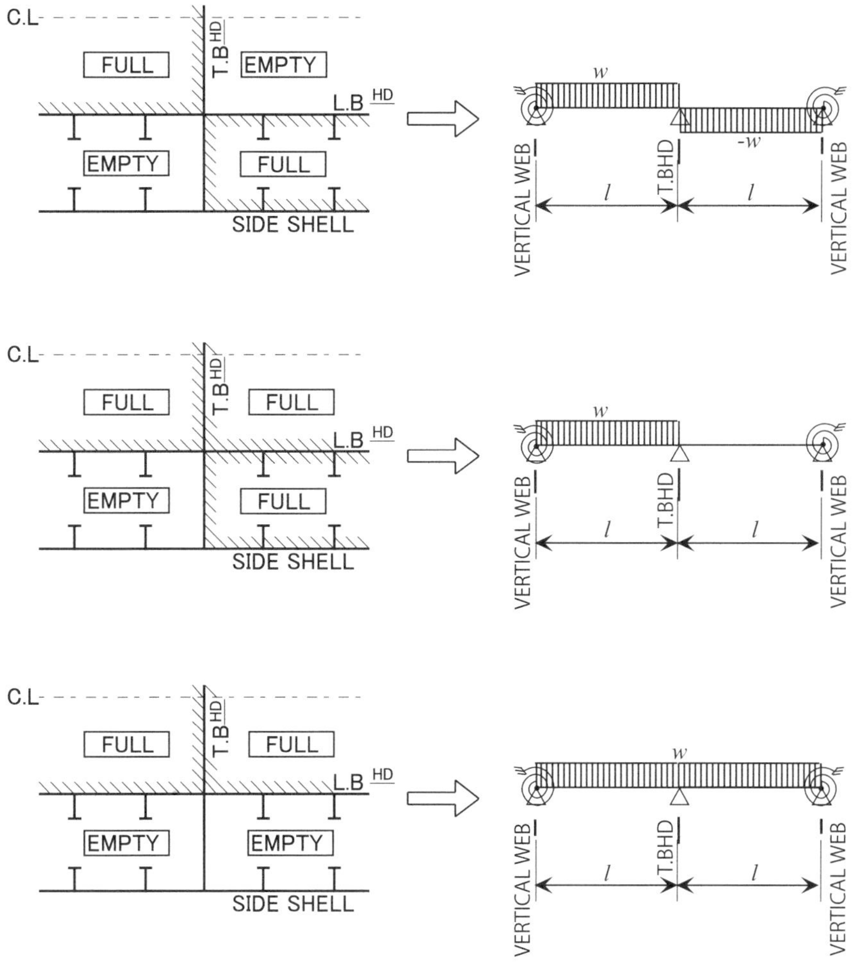

Fig. 8.24 Change of load by loading in tanks

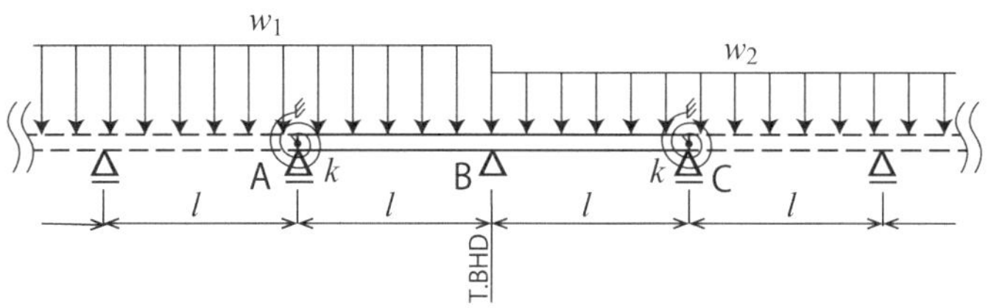

Fig. 8.25 Model of a stiffener penetrating a transverse bulkhead

$$R_B = \frac{w_1 + w_2}{2} l \tag{8.6.36}$$

When the stiffener is fixed at the adjacent transverse frame,

$$R_A = w_1 l + \frac{w_1 - w_2}{16} l \tag{8.6.37}$$

$$R_C = w_2 l + \frac{w_2 - w_1}{16} l \tag{8.6.38}$$

The reaction force becomes maximum when $w_2 = -w_1$, reaching $R_A = 1.125 w_1 l$. When the stiffener is continuous infinitely,

$$R_A = w_1 l + \frac{2 - \sqrt{3}}{4}(w_1 - w_2) l \tag{8.6.39}$$

$$R_C = w_2 l + \frac{2 - \sqrt{3}}{4}(w_2 - w_1) l \tag{8.6.40}$$

The reaction force becomes maximum when $w_2 = -w_1$, reaching $R_A = 1.134 w_1 l$.

With the change of w_2/w_1 from -1 to $+1$, these bending moments and reactions will change as shown in Fig. 8.26. The bending moment at the transverse bulkhead ($|M_B|$) is always less than or equal to $w_1 l^2/12$. The bending moment at the adjacent transverse frames ($|M_A|$ and $|M_C|$) are bigger than $w_1 l^2/12$ in some cases. In case of $w_2/w_1 = -1$, they reach $1.5 w_1 l^2/12$ when the stiffener is fixed at the adjacent transverse frames and $1.268 w_1 l^2/12$ when the stiffener is continuous infinitely.

In addition to the bending moment, the reaction at the transverse frames adjacent to the transverse bulkhead reaches values greater than expected. $|R_B|$ is always less than $w_1 l$, but in case of $w_2/w_1 = -1$, $|R_A|$ and $|R_C|$ reach $1.125 w_1 l$ when the stiffener is fixed at the adjacent transverse frames and $1.134 w_1 l$ when the stiffener is continuous infinitely. The load, bending moment and shearing force in the above case are shown in Fig. 8.27.

From the above consideration we can learn that in the case of staggered loading the force on the transverse web adjacent to the transverse bulkhead is 1.125–1.134 times that of other transverse web and the bending moment on the longitudinal stiffener on longitudinal bulkhead is 1.268–1.5 times of point at the ordinary transverse web.

This is an example that hull structure designers should carefully check, not only structural arrangement but also the loading pattern to assume appropriate boundary conditions. For design improvement such practical and detailed boundary conditions are to be considered.

However, we can design the structure to reasonably reduce such additional bending moments and reaction forces. In the case of staggered loading, the transverse bulkhead also supports staggered loading from the cargo oil. If the longitudinal stiffener on the longitudinal bulkhead is designed so as to be rigidly connected to the adjacent vertical stiffener on the transverse bulkhead as shown in Fig. 8.28, rotation is

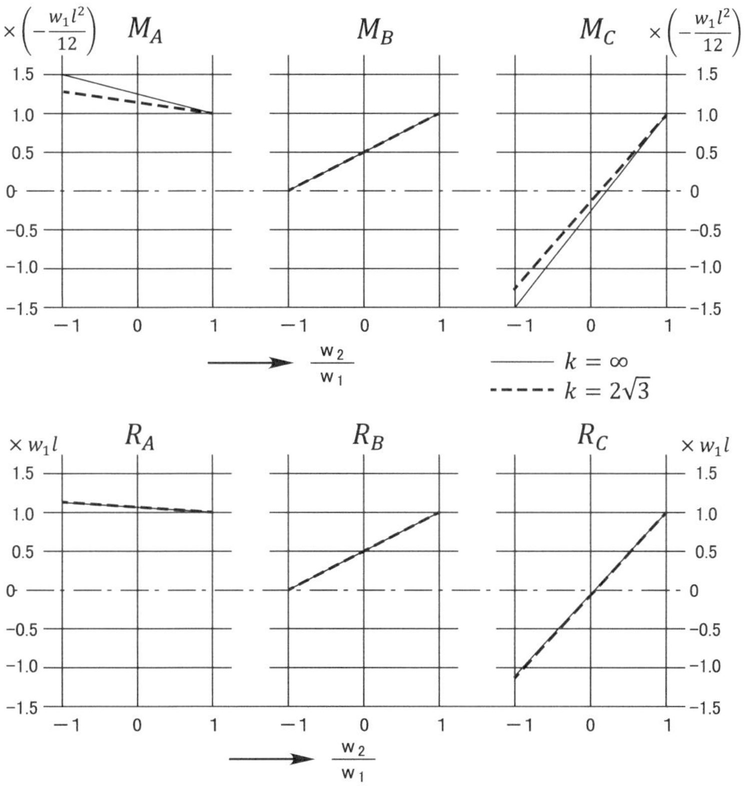

Fig. 8.26 Change of bending moment and shearing force by loading

restricted and boundary condition at the transverse bulkhead can again be assumed to be fixed depending on the rigidity of the adjacent members.

8.6.5 Partially Distributed or Concentrated Load on a Continuous Beam

In the case where a partial load is applied to a beam as shown in Fig. 8.29, the beam is to be designed as a continuous beam which takes more time, labour and expense. Approximately the beam may be designed as a simply supported beam. Here the differences due to such simple modeling will be examined.

In Fig. 8.29a, assuming both the ends of beam A are elastically supported in rotation with a non-dimensional rotational spring constant of k, the maximum bending moment M at the midspan is obtained by the following equation.

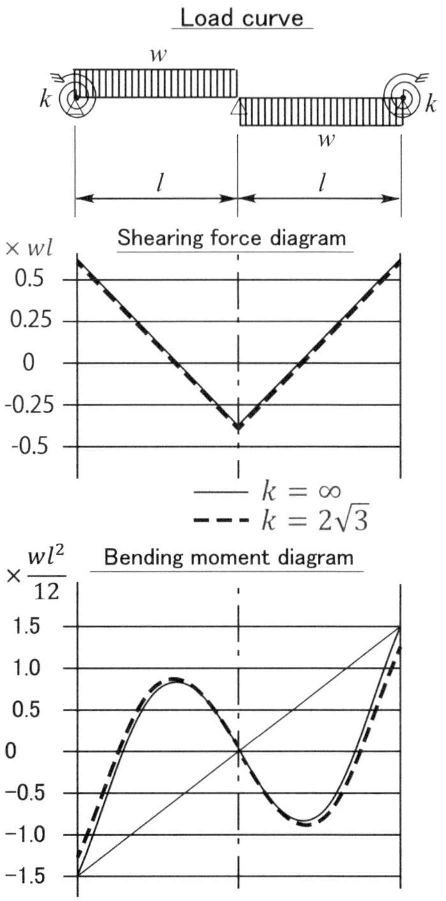

Fig. 8.27 B.M.D. and S.F.D. on longitudinal under staggered load

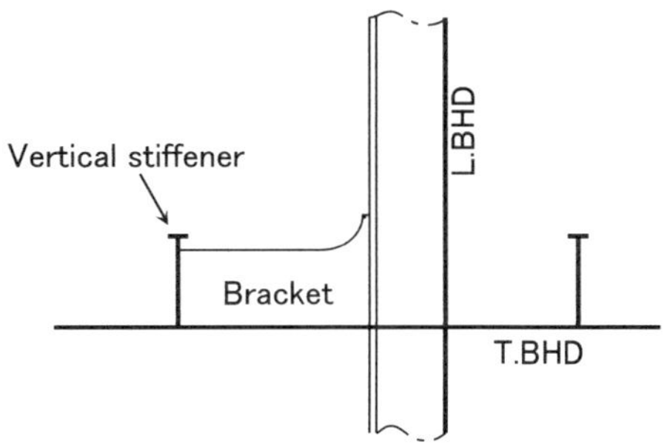

Fig. 8.28 Bracket connection between longitudinal and vertical stiffeners

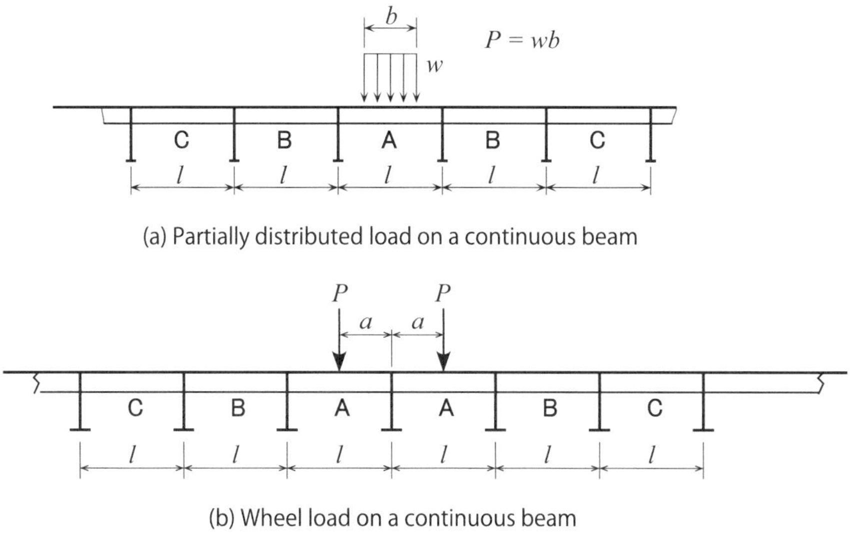

Fig. 8.29 Partially distributed or concentrated load on a continuous beam

$$M = \frac{Pl}{4} \cdot \frac{3(k+4) - 3(k+2)\frac{b}{l} + k\left(\frac{b}{l}\right)^2}{6(k+2)} \tag{8.6.41}$$

where $P(=wb)$ denotes the total load. Then, let us consider the effect of considering various number of spans, substituting corresponding values of k according to Table 8.2.

Then, when beam A is assumed to be simply supported at both the ends, substituting $k = 0$ into Eq. (8.6.41),

$$M = \frac{Pl}{4}\left(1 - \frac{b}{2l}\right) \tag{8.6.42}$$

Including adjacent spans B, substituting $k = 3$ into Eq. (8.6.41),

$$M = \frac{Pl}{4}\left\{\frac{7}{10} - \frac{1}{2}\left(\frac{b}{l}\right) + \frac{1}{10}\left(\frac{b}{l}\right)^2\right\} \tag{8.6.43}$$

Including adjacent spans B and C, substituting $k = 24/7$,

$$M = \frac{Pl}{4}\left\{\frac{13}{19} - \frac{1}{2}\left(\frac{b}{l}\right) + \frac{2}{19}\left(\frac{b}{l}\right)^2\right\} \tag{8.6.44}$$

When the beam is infinitely continuous beyond beam C, substituting $k = 2\sqrt{3}$,

$$M = \frac{Pl}{4}\left\{\frac{1+\sqrt{3}}{4} - \frac{1}{2}\left(\frac{b}{l}\right) + \frac{3-\sqrt{3}}{12}\left(\frac{b}{l}\right)^2\right\} \tag{8.6.45}$$

The above relations are shown in Fig. 8.30. In case of a concentrated load at the midspan (left end of Fig. 8.30), the result of the simply supported assumption gives about 46% higher value than that of the infinitely continuous beam, but when we consider the adjacent spans B, the result becomes very close to that of the infinitely continuous beam. When we further consider the spans C, the difference from the infinitely continuous beam is almost invisible. Designers often assume a single simply supported beam to evaluate strength of a continuous beam subjected to a concentrated load, but this has been found to be quite conservative. It should be noted, however, that if the loaded span is reinforced against the concentrated load with greater bending rigidity than the adjacent spans, the boundary conditions for the reinforced span will be closer to the simply supported condition, and some of the conservatism mentioned above will be lost. In Fig. 8.30, the effect of the width of the load is also shown. We can see that the wider load reduces the bending moment very much.

In the case where the load is applied separately on the adjacent two spans of beam as shown in Fig. 8.29b, assuming the connections between beams A and beams B are elastically supported in rotation with a non-dimensional rotational spring constant of k, the bending moment at the mid support point M_0 and the bending moment at the loaded point M_P are obtained by the following equations.

$$M_0 = Pl \cdot \frac{-(k+4)\frac{a}{l} + 2(k+3)\left(\frac{a}{l}\right)^2 - (k+2)\left(\frac{a}{l}\right)^3}{k+4} \tag{8.6.46}$$

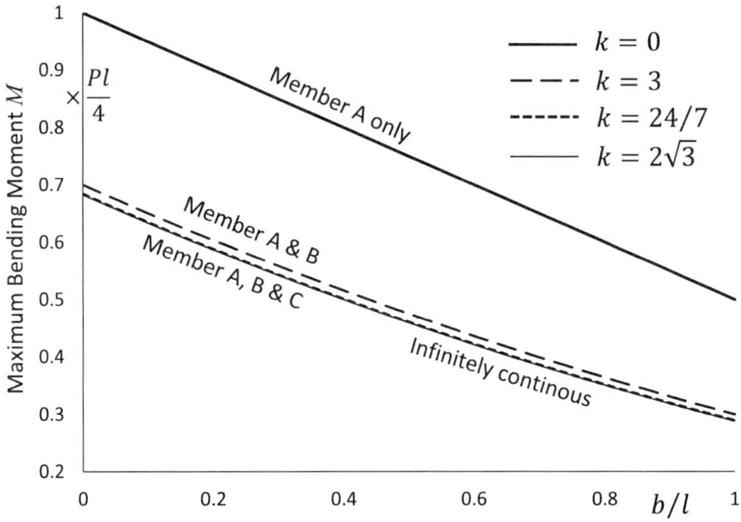

Fig. 8.30 Maximum bending moment versus loading width under partially distributed load on a continuous beam

$$M_P = Pl \cdot \frac{2(k+3)\left(\frac{a}{l}\right)^2 - 4(k+2)\left(\frac{a}{l}\right)^3 + 2(k+1)\left(\frac{a}{l}\right)^4}{k+4} \tag{8.6.47}$$

where a is the distance between the load and the mid support point. Substituting various values of k according to Table 8.2, we can evaluate the effect of the considered number of spans.

Considering only the spans A in Fig. 8.29b, substituting $k = 0$ into Eqs. (8.6.46) and (8.6.47),

$$M_0 = Pl\left\{-\frac{a}{l} + \frac{3}{2}\left(\frac{a}{l}\right)^2 - \frac{1}{2}\left(\frac{a}{l}\right)^3\right\} \tag{8.6.48}$$

$$M_P = Pl\left\{\frac{3}{2}\left(\frac{a}{l}\right)^2 - 2\left(\frac{a}{l}\right)^3 + \frac{1}{2}\left(\frac{a}{l}\right)^4\right\} \tag{8.6.49}$$

Considering the spans A and B, substituting $k = 3$,

$$M_0 = Pl\left\{-\frac{a}{l} + \frac{12}{7}\left(\frac{a}{l}\right)^2 - \frac{5}{7}\left(\frac{a}{l}\right)^3\right\} \tag{8.6.50}$$

$$M_P = Pl\left\{\frac{12}{7}\left(\frac{a}{l}\right)^2 - \frac{20}{7}\left(\frac{a}{l}\right)^3 + \frac{8}{7}\left(\frac{a}{l}\right)^4\right\} \tag{8.6.51}$$

Then considering the spans A, B and C, substituting $k = 24/7$,

$$M_0 = Pl\left\{-\frac{a}{l} + \frac{45}{26}\left(\frac{a}{l}\right)^2 - \frac{19}{26}\left(\frac{a}{l}\right)^3\right\} \tag{8.6.52}$$

$$M_P = Pl\left\{\frac{45}{26}\left(\frac{a}{l}\right)^2 - \frac{38}{13}\left(\frac{a}{l}\right)^3 + \frac{31}{26}\left(\frac{a}{l}\right)^4\right\} \tag{8.6.53}$$

When the beam is infinitely continuous beyond beam C, substituting $k = 2\sqrt{3}$,

$$M_0 = Pl\left\{-\frac{a}{l} + \sqrt{3}\left(\frac{a}{l}\right)^2 - \left(\sqrt{3}-1\right)\left(\frac{a}{l}\right)^3\right\} \tag{8.6.54}$$

$$M_P = Pl\left\{\sqrt{3}\left(\frac{a}{l}\right)^2 - 4\left(\sqrt{3}-1\right)\left(\frac{a}{l}\right)^3 + \left(3\sqrt{3}-4\right)\left(\frac{a}{l}\right)^4\right\} \tag{8.6.55}$$

The above relations are shown in Fig. 8.31. In case of an infinitely continuous beam, M_0 takes the maximum value of $-0.170Pl$ when $a/l = 0.380$. In case only the spans A are considered, M_0 takes the maximum value of $-0.192Pl$ when $a/l = 0.423$, and this is only 13% higher than that for the infinitely continuous beam.

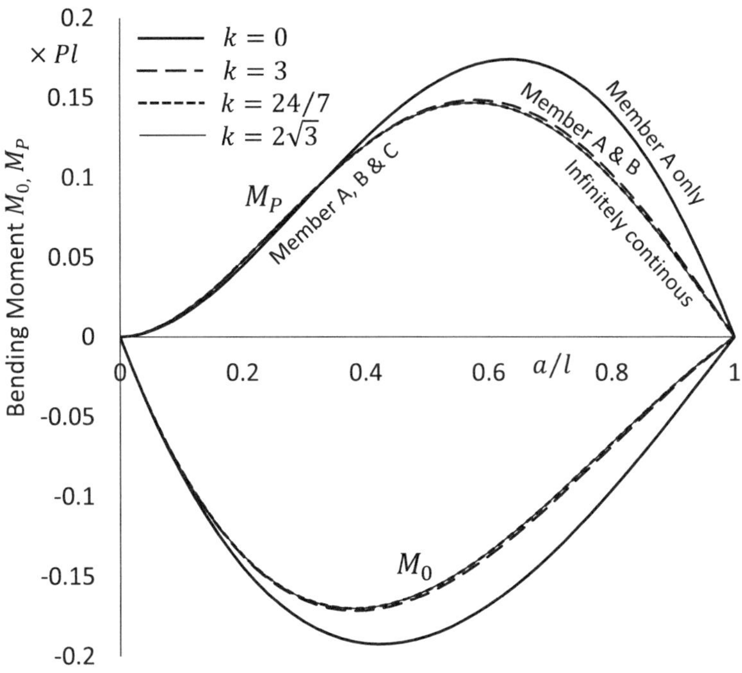

Fig. 8.31 Bending moment versus concentrated load position

8.6.6 Forced Displacement and Forced Rotation

A secondary member such as a beam supported by a primary member as a girder, is usually designed assuming that the primary member has no deflection. However primary members are not 100% rigid and an additional bending moment is generated in the secondary member by the deflection of the primary members. One example is the additional bending moment at the bottom bracket of the vertical stiffener on a transverse bulkhead which is caused by deflection of the horizontal girders on the transverse bulkhead as shown in Fig. 24.10, Part III. Another example is a double bottom structure of a bulk carrier as shown in Fig. 8.32, where the total bending stress distribution including the effect of the forced displacement is shown in dashed lines.

To overcome this additional bending moment due to forced displacement, increasing only the sectional moment of inertia and section modulus of the beam is not sufficient.

Let us consider a continuous beam where a forced displacement or a forced rotation is given at support point A as shown in Fig. 8.33. Assuming the beam is elastically supported in rotation at point B with a non-dimensional rotational spring constant of k, the bending moment at point A is obtained by the following equations where M_δ denotes the bending moment due to the forced displacement δ and M_θ denotes the bending moment due to the forced rotation θ.

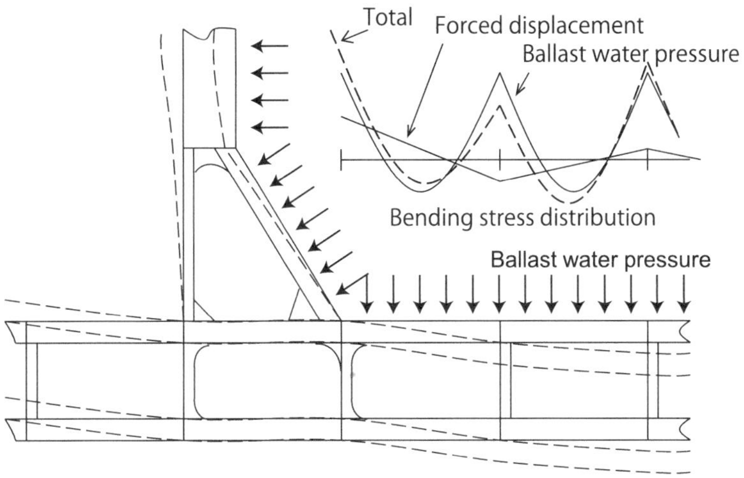

Fig. 8.32 Forced displacement of bulk carrier double bottom and bending stress increase on a longitudinal stiffener

$$M_\delta = \frac{6(k+2)EI\delta}{(k+4)l^2} \tag{8.6.56}$$

$$M_\theta = -\frac{4(k+3)EI\theta}{(k+4)l} \tag{8.6.57}$$

When points B is fixed, M_δ and M_θ become maximum and can be expressed as follows, substituting $k = \infty$ into Eqs. (8.6.56) and (8.6.57),

$$M_\delta = \frac{6EI\delta}{l^2} \tag{8.6.58}$$

$$M_\theta = -\frac{4EI\theta}{l} \tag{8.6.59}$$

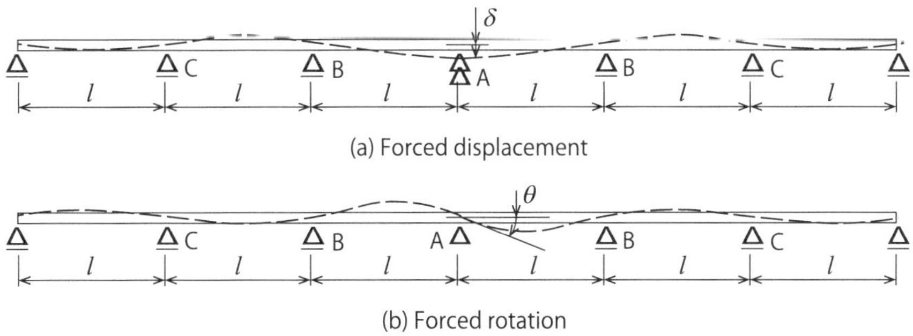

Fig. 8.33 Forced displacement and forced rotation on a continuous beam

When points C is fixed, substituting $k = 4$,

$$M_\delta = \frac{3}{4} \times \frac{6EI\delta}{l^2} \tag{8.6.60}$$

$$M_\theta = -\frac{7}{8} \times \frac{4EI\theta}{l} \tag{8.6.61}$$

When the beam is infinitely continuous, substituting $k = 2\sqrt{3}$,

$$M_\delta = \left(\sqrt{3} - 1\right) \times \frac{6EI\delta}{l^2} \tag{8.6.62}$$

$$M_\theta = -\frac{\sqrt{3}}{2} \times \frac{4EI\theta}{l} \tag{8.6.63}$$

When points B is simply supported, substituting $k = 0$,

$$M_\delta = \frac{1}{2} \times \frac{6EI\delta}{l^2} \tag{8.6.64}$$

$$M_\theta = -\frac{3}{4} \times \frac{4EI\theta}{l} \tag{8.6.65}$$

When the allowable stress for this additional moment is σ_a, the following equation must be satisfied.

$$\sigma_a \geq \frac{M_\delta}{Z} \text{ or } \frac{M_\theta}{Z} \tag{8.6.66}$$

In case points B are fixed, substituting Eqs. (8.6.58) and (8.6.59) into Eq. (8.6.66),

$$\frac{I}{Z} \leq \frac{\sigma_a l^2}{6E\delta} \text{ or } \frac{\sigma_a l}{4E\theta} \tag{8.6.67}$$

$Z = I/y$, then $I/Z = y$ which is the distance between the neutral axis and the extreme fiber. In the case of a balanced girder $y = d/2$, then the following equation is obtained for the forced displacement.

$$\frac{I}{Z} = \frac{d}{2} \leq \frac{\sigma_a l^2}{6E\delta} \tag{8.6.68}$$

$$\therefore \frac{l}{d} \geq \frac{3E}{\sigma_a} \cdot \frac{\delta}{l} \tag{8.6.69}$$

From the above result the slenderness ratio of the beam should be increased proportionally to the amount of the forced displacement to keep the additional stress within the allowable limit. From these discussions, it is important for a designer to find and feel in his bones the fact that the bending stress due to forced displacement depends on (or is proportional to) the distance from the neutral axis only, and is independent from other sectional properties such as I or Z. Based on this fact, in case of an ordinary beam with skin plate under forced displacement, we can easily understand intuitively and quantitatively that to make the beam shallower is extremely effective, on the other hand to make the face plate thicker is less effective because it makes the distance from the neutral axis closer only slightly.

A flexible (smaller Young's modulus) and slender beam is good for the forced displacement. "A willow twig will never be broken by strong wind." However, a weak and slender beam will not support ordinary lateral loads. The correct solution in this case is to reduce the deflection of the primary member itself or the relative displacement of the adjacent primary members. However, if such relative displacement is inevitable, it is necessary to make the combined stress due to lateral loads and forced displacement less than allowable value. As discussed above, the stress due to forced displacement depends only on y (the distance from the neutral axis), and the stress due to lateral loads depends only on Z (section modulus). Therefore, when we design beam under lateral loads and forced displacement, we must first decide the appropriate depth of the beam, and then control the section modulus by adjusting the scantling of the face plate, roughly imaging the amount of each stress component.

In Sect. 8.4, we discussed optimal design of beam section and found that the best proportion of ordinary beams is that the area of web is about 2 times of the area of face plate. In case of beams under simultaneous forced displacement, the optimal proportion shift to the side of shallower beam and bigger face plate, sometimes resulting in the ratio of web area to face plate area to be about 1.0.

Damages caused by forced displacements are often observed where the balance between primary supporting members and stiffeners is somewhat awkward. Figure 8.34a shows a horizontal plan of side structure of a container ship around a transverse bulkhead. To facilitate the erection joint welding work in the building dock, a construction engineer proposed to fit built-in stages for every three longitudinal spacings as shown by hatching in the figure. However, the hull structural designer in charge rejected this proposal at a glance.

Primary supporting members (transverse web frames) support the longitudinal stiffeners, but the longitudinal stiffeners do not support the primary supporting members. The deflection of the primary supporting members will be as shown in Fig. 8.34b. Because rigid stages are fitted, the span length of the longitudinal stiffeners under forced displacement becomes very short. The bending moment due to the forced displacement is inversely proportional to the span length squared. Therefore, excessive bending moment on the longitudinal stiffener is anticipated. The structural designer made a counter-proposal to fit full rigid flat which can effectively support the primary supporting member with an access hole for the erection workers as shown in Fig. 8.34c. In this case, length of the butt joint increases, but the adoption of the stages might be decided on trade-offs among various factors.

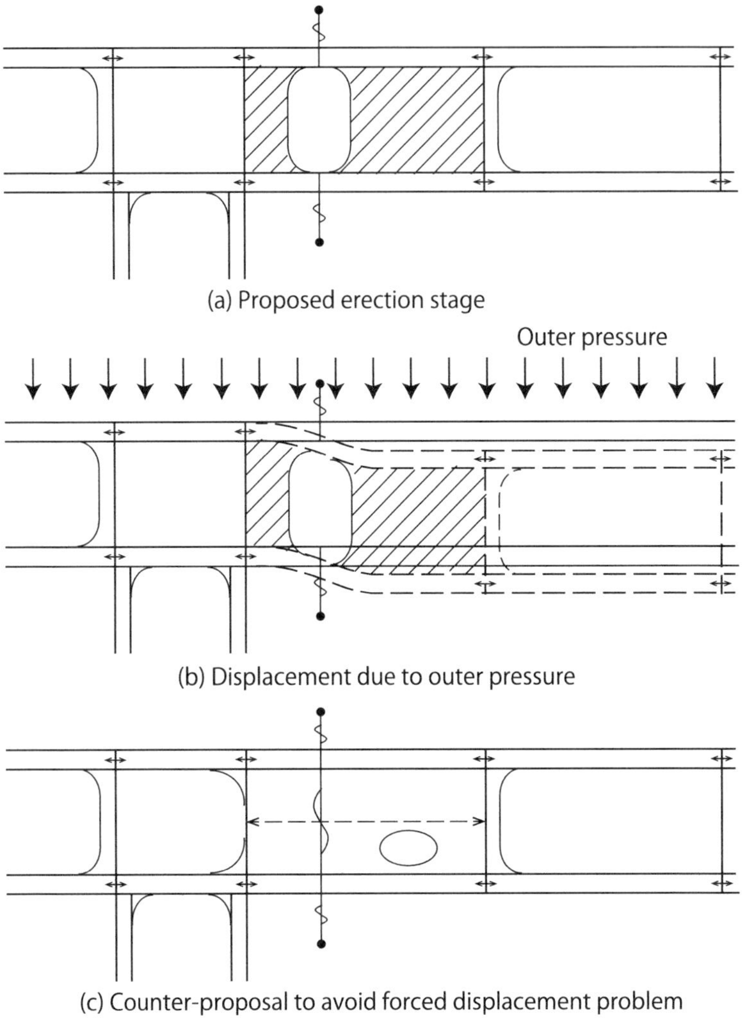

Fig. 8.34 Forced displacement of container ship side structure

8.7 Effect of Struts

In the case that two parallel beams are connected by a strut at their midspan as shown in Fig. 8.35a, the applied force on the beam deforms the whole structure and causes a reaction distribution which depends on the rigidity ratio of each member. One example is the longitudinal framing of a cargo ship double bottom where the inner bottom and the bottom longitudinals are connected by a strut.

In this case the bending moments on the beams and the axial force in the strut are expressed by the following equations:

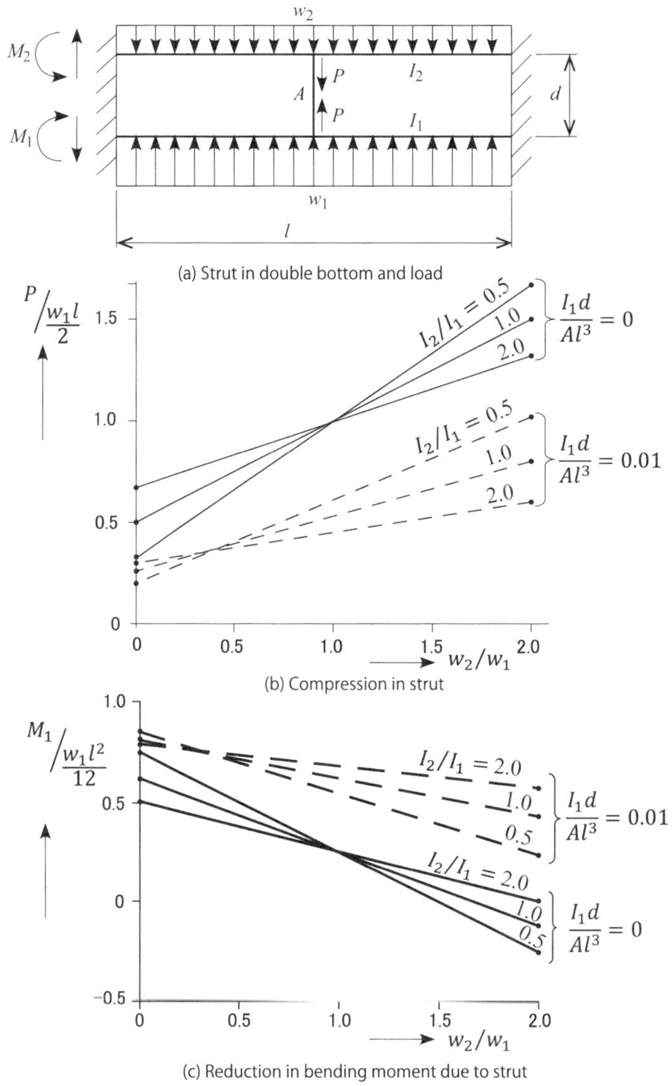

Fig. 8.35 Effect of strut in double beams

$$P = K\frac{w_1 l}{2} \tag{8.7.1}$$

$$M_1 = \left(1 - \frac{3}{4}K\right)\frac{w_1 l^2}{12} \tag{8.7.2}$$

where $K = \frac{1+\frac{w_2}{w_1}\cdot\frac{I_1}{I_2}}{1+\frac{I_1}{I_2}+\frac{192 I_1 d}{A l^3}}$

w_2, w_1: loads on top and bottom beams respectively
I_2, I_1: sectional moment of inertia of top and bottom beams respectively.
A, d: sectional area and the length of the strut
P: axial compressive force in the strut
M_1: bending moment at the fixed end of the bottom beam.

Figure 8.35b shows that the axial compressive force in the strut is the mean value of the loads applied on the top and the bottom beams where the strut is rigid and the rigidities of the top and the bottom beams are equal. In the case of an equal load on the top and the bottom beams, the rigidity of the beam has no effect at all. The axial compressive force in the strut is increased by increasing the rigidity of the beam on which the smaller load is applied. Assuming elasticity in the strut (dashed lines in Fig. 8.35), the axial compressive force in the strut is reduced. The bigger axial compressive force in the strut causes a smaller bending moment at the fixed ends. A bigger load on the top beam reduces the bending moment of the bottom beam at its fixed ends. Where the load on the bottom beam is bigger than that on the top beam, increasing the rigidity of the top beam is effective. Having a more rigid strut gives a better design.

8.8 Lateral Bending of Asymmetrical Stiffener

The face plate of a beam lies at a certain distance from the neutral axis and the assumption that the bending stress in the face plate is uniform is usually applied in beam design. However as shown in Fig. 8.36, it is not uniform but is maximum at the attachment point to the web and minimum at the free edge. The concept of the effective breadth b_e is such that the force in the face plate is equal to the product of the maximum bending stress and the face plate area which assumes an effective breadth b_e.

In the case where a beam is subjected to bending, the beam deforms in a vertical direction and it may move in lateral direction simultaneously according to the shape of its cross-section. For example, an asymmetrical section (L section) beam will move in a lateral direction due to bending, while a beam with a symmetric cross-section does not cause lateral movement. This lateral movement causes lateral bending of the face plate and the bending stress linearly changing along the width of the face plate.

As shown in Fig. 8.37, in the case of a tensile stress in the face plate due to outer pressure, the beam wants to move in the left direction, whereas for a compressive stress in the face plate due to inner pressure, the beam wants to move in the right direction. This phenomenon is explained by noting that the beam has a torsional axis called a "shear center" as described in Sect. 14.5. When a torsional moment is applied to the beam, the beam twists around the shear center. The position of this shear center depends only on the shape of the cross section of the beam and it lies on the opposite side of the flange in the case of an L section beam, as illustrated

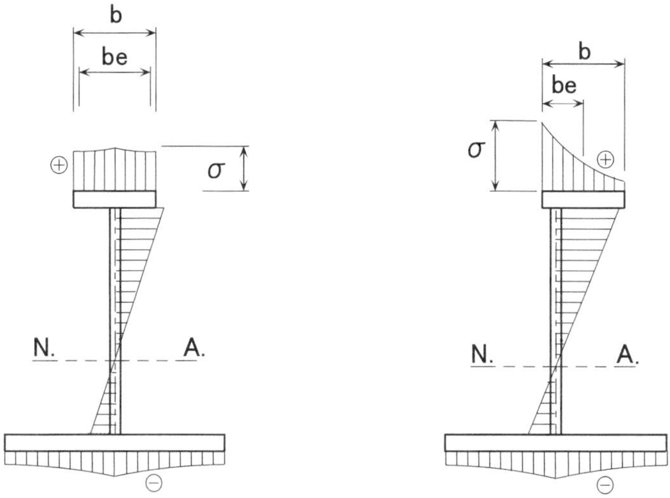

Fig. 8.36 Bending stress distribution in beam section

in Fig. 8.37. Hence, as presented in Fig. 8.37a, a beam subjected to inner pressure will move to the right side due to a clockwise torsional moment around the shear center induced by the pressure. Assuming a both-end-fixed beam, the stress on the face plate at the fixed ends is tensile under this inner pressure. Because the face plate moves to the right direction, the left edge of the face plate is tensile and the right edge is compressive with respect to the face plate bending. Therefore, the stiffener bending stress and the lateral bending stress of the face plate superimpose each other at the face plate edge close to the web, and cancel each other at the face plate edge far from the web as shown in Fig. 8.37. This additional stress often causes fatigue problems because fatigue strength is sensitive to the maximum stress. Actually, one of the root causes of the fatigue cracks propagating into side longitudinals explained in Sect. 10.5.4 was wide application of L section for side longitudinals. In contrast, in the case of a beam of a symmetric cross-section, lateral movement of the beam never occurs, because the shear center of this beam is in the middle of the web plate and no torsional moment is created.

The stress distribution in a section considering this lateral movement can be obtained analytically using a beam theory. Let us assume a both end fixed beam with a section as shown in Fig. 8.38 under a uniform load per length w. The offset of the face plate is expressed by e. When $e = 0$, the section is completely symmetric (T section), while when $e = b/2$, the section is a complete L section. In this case, the longitudinal stress at both the face plate edges σ_{fe} is expressed as follows:

$$\sigma_{fe} = wl^2 \frac{e_f}{I_{ow}\left(1 - \frac{I_{fw}^2}{I_{of} I_{ow}}\right)} \left\{ \left(I_f \pm \frac{A_f eb}{2} \right) \frac{1}{I_{of}} f_1(\xi) - \left(1 - \frac{I_{fw}^2}{I_{of} I_{ow}} \right) (f_1(\xi) - f_0(\xi)) \right\} \tag{8.8.1}$$

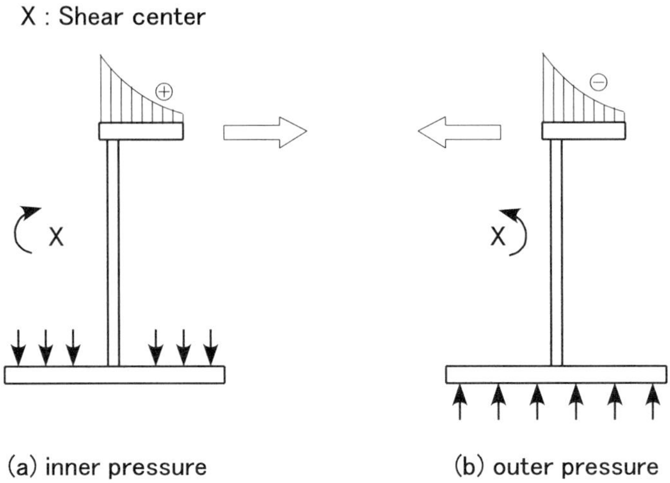

Fig. 8.37 Lateral movement of beam due to torsional moment induced by pressure

where

e_f: the distance to face plate from the neutral axis of the integrated beam,

$$e_f = \frac{\left(A_w + 2A_p\right)}{A_o}\frac{d}{2} \tag{8.8.2}$$

A_o: total sectional area of the integrated beam, $A_o = A_f + A_w + A_p$
I_{ow}: total moment of inertia of the integrated beam about the horizontal neutral axis,

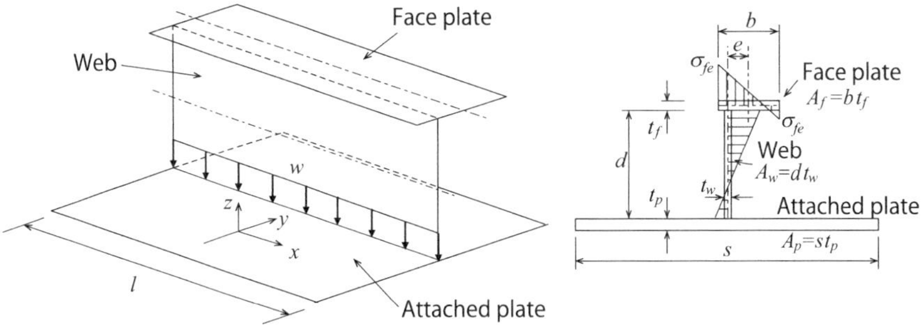

Fig. 8.38 Analytical model of asymmetrical stiffener and stress distribution

$$I_{ow} = I_w + \frac{A_w(A_f + A_p) + 4A_f A_p}{A_o}\left(\frac{d}{2}\right)^2 \tag{8.8.3}$$

I_w: sectional moment of inertia of the web, $I_w = d^3 t_w/12$
I_{of}: total moment of inertia of the integrated beam about the vertical neutral axis,

$$I_{of} = I_f + \frac{A_f(A_w + A_p)}{A_o} e^2 \tag{8.8.4}$$

I_f: sectional moment of inertia of the face plate, $I_f = b^3 t_f/12$.

$$I_{fw} = A_f e e_f \tag{8.8.5}$$

$$f_1(\xi) = \frac{1}{\lambda^2(\sinh\lambda + \sin\lambda)}\left\{\left(\cosh\frac{\lambda}{2}\sin\frac{\lambda}{2} + \sinh\frac{\lambda}{2}\cos\frac{\lambda}{2}\right)\sinh\lambda\xi\sin\lambda\xi - \left(\cosh\frac{\lambda}{2}\sin\frac{\lambda}{2} - \sinh\frac{\lambda}{2}\cos\frac{\lambda}{2}\right)\cosh\lambda\xi\cos\lambda\xi\right\} \tag{8.8.6}$$

ξ: non-dimensional coordinate in longitudinal (x) direction. $\xi = 0$ at the midspan, and $\xi = \pm 1/2$ at the span ends.

At the fixed span ends, Eq. (8.8.6) reduces to:

$$f_1\left(\frac{1}{2}\right) = \frac{\sinh\lambda - \sin\lambda}{2\lambda^2(\sinh\lambda + \sin\lambda)} \tag{8.8.7}$$

$$\lambda = l \cdot \sqrt[4]{\frac{\kappa_n}{4I_{of}\left(1 - \frac{I_{fw}^2}{I_{of} I_{ow}}\right)}} \tag{8.8.8}$$

$f_0(\xi)$: non-dimensional bending moment distribution function along the stiffener length,

$$f_0(\xi) = \frac{1}{2}\xi^2 - \frac{1}{24} \tag{8.8.9}$$

κ_n: non-dimensional spring constant of lateral displacement per unit length,

$$\kappa_n = \frac{1}{\frac{1}{\kappa_{n1}} + \frac{1}{\kappa_{n2}}} \tag{8.8.10}$$

κ_{n1}: non-dimensional spring constant of lateral displacement due to the bending rigidity of the attached plate, corresponding to the relationship $p = E\kappa_{n1}v_1$ in Fig. 8.39.

$$\kappa_{n1} = \frac{t_p^3}{sd^2} \tag{8.8.11}$$

κ_{n2}: non-dimensional spring constant of lateral displacement due to the bending rigidity of the web, corresponding to the relationship $p = E\kappa_{n2}v_2$ in Fig. 8.39.

$$\kappa_{n2} = \frac{t_w^3}{4d^3} \tag{8.8.12}$$

In case of a T section, the stress on the face plate σ_T is uniform if the shear lag effect is neglected, and is expressed as follows:

$$\sigma_T = f_0(\xi)wl^2 \frac{e_f}{I_{ow}} \tag{8.8.13}$$

Then, dividing Eq. (8.8.1) by Eq. (8.8.13), the rate of stress change due to the asymmetry γ_{fe} is expressed as follows:

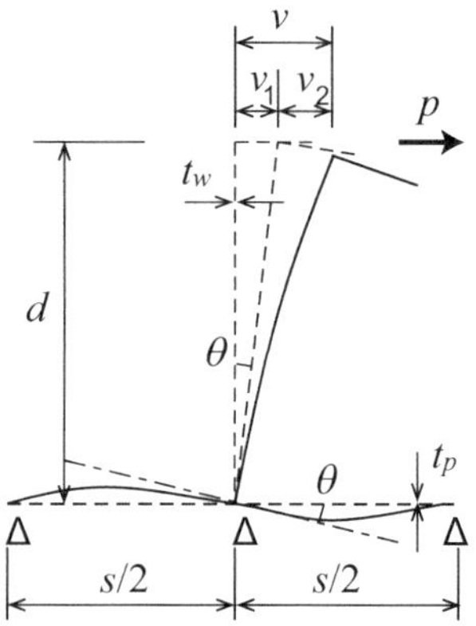

Fig. 8.39 Spring for lateral movement of stiffener

$$
\begin{aligned}
\gamma_{fe} = \frac{\sigma_{fe}}{\sigma_T} &= \frac{\left(I_f \pm \frac{A_f eb}{2}\right)\frac{1}{I_{of}} f_1(\xi) - \left(1 - \frac{I_{fw}^2}{I_{ow} I_{of}}\right)(f_1(\xi) - f_0(\xi))}{\left(1 - \frac{I_{fw}^2}{I_{of} I_{ow}}\right) f_0(\xi)} \\
&= \left\{ \frac{\left(I_f \pm \frac{A_f eb}{2}\right)\frac{1}{I_{of}}}{\left(1 - \frac{I_{fw}^2}{I_{of} I_{ow}}\right)} - 1 \right\} \frac{f_1(\xi)}{f_0(\xi)} + 1
\end{aligned} \tag{8.8.14}
$$

The first term in the curly brackets is the rate of stress change due to asymmetry in case of no spring for the lateral movement of the stiffener ($\kappa_n \to 0$). Therefore, $f_1(\xi)/f_0(\xi)$ means the reduction ratio of the stress change due to the effect of the spring. At the fixed span ends, Eq. (8.8.14) reduces to:

$$
\gamma_{fe} = \left\{ \frac{\left(I_f \pm \frac{A_f eb}{2}\right)\frac{1}{I_{of}}}{\left(1 - \frac{I_{fw}^2}{I_{of} I_{ow}}\right)} - 1 \right\} \cdot \frac{6}{\lambda^2} \cdot \frac{\sinh\lambda - \sin\lambda}{\sinh\lambda + \sin\lambda} + 1 \tag{8.8.15}
$$

Figure 8.40 shows γ_{fe} at the left and right edges of the face plate in the span end section with and without the spring, taking $2e/b$ in the horizontal axis. The dimensions of the stiffener are $d \times b \times t_w/t_f + s \times t_p$: $250 \times 90 \times 10/15 + 800 \times 15$ with the span length of 2400 in the unit mm. The left end of the graph represents a complete T section and the right end represents a complete L section. We can observe that the spring of the web and the attached plate works well to reduce the additional stress. The maximum stress approaching $\gamma_{fe} = 1.7$ occurs at the left edge of the face plate when $2e/b > 0.6$ (i.e. $e > 0.3b$). The stress at the right edge of the face plate almost reduces to zero when $2e/b \approx 0.6$, and as the section approaches to a complete L shape ($2e/b \to 1$), the face plate bending stress exceeds the uniform axial stress on the face plate due to the stiffener bending, and the total stress turns to negative.

8.9 Effect of Axial Stress and Shear Stress

8.9.1 *Effect of Axial Stress*

Longitudinal stiffeners support lateral load due to water or cargo pressure and axial load due to hull girder bending simultaneously, especially near the upper deck or the bottom. When the axial stress has the same sign with the stiffener bending stress on the face plate at the span ends as shown in Fig. 8.41a, it is known that the stiffener bending strength is deteriorated by the superposition of the axial stress. Conversely, when the axial stress has the opposite sign with stiffener bending stress as shown in Fig. 8.41b, the bending strength is not reduced (but rather enhanced) when the axial stress is small. However, when the axial stress is quite significant, it superimposes on

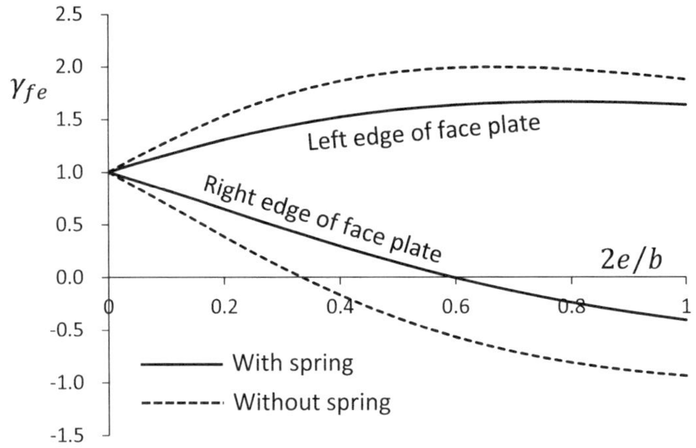

Fig. 8.40 Stress change rate due to lateral bending of asymmetrical stiffener

the stiffener bending stress at the midspan, accelerates the plastic hinge formation at the midspan and deteriorates the stiffener bending strength.

This bending strength reduction is mainly caused by the reduction in the fully plastic moment due to the simultaneous action of the axial stress. Nitta [63] derived formulae of the fully plastic moment of stiffener section under the action of axial stress rigorously using plastic design theory. When the thickness of each member is sufficiently smaller than the overall dimension, we can neglect the geometrical thickness of the face plate and the attached plate and consider only the concentrated sectional area at the nominal position. Then, the formulation can be simplified and is summarized as follows [64].

Let us assume a stiffener section as shown in Fig. 8.8 under an axial stress of σ_a. In case of $\sigma_a = 0$, the plastic neutral axis (PNA) resides within the attached plate because $A_p > A_w + A_f$ usually holds. When axial stress is working, the position of the PNA changes, and three different equations for the fully plastic moment M_P can be derived depending on the member where the PNA resides.

When

$$-\frac{\sigma_Y\left(\gamma_p s t_p + \gamma_w d t_w + b t_f\right)}{A_o} < \sigma_a < \frac{\sigma_Y\left(\gamma_p s t_p - \gamma_w d t_w - b t_f\right)}{A_o} \tag{8.9.1}$$

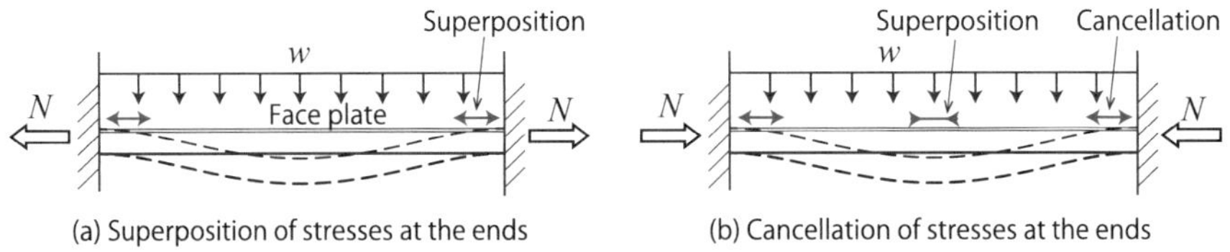

Fig. 8.41 Superposition and cancellation of bending stress and axial stress of a stiffener

holds, the PNA resides within the attached plate, where

σ_Y: Yield stress
γ_w: Material coefficient of web (ratio of the yield stress of web to that of face plate).
γ_p: Material coefficient of attached plate (ratio of yield stress of attached plate to that of face plate).
A_o: Total sectional area ($= A_f + A_w + A_p$).

In this case, the fully plastic moment becomes

$$M_P = (\sigma_Y + \sigma_a)dbt_f + (\gamma_w\sigma_Y + \sigma_a)\frac{d^2t_w}{2} \tag{8.9.2}$$

where M_P is taken positive when the face plate is compressed, and σ_a is taken positive in tension.

When σ_a is greater and enter within

$$\frac{\sigma_Y\left(\gamma_p st_p - \gamma_w dt_w - bt_f\right)}{A_o} < \sigma_a < \frac{\sigma_Y\left(\gamma_p st_p + \gamma_w dt_w - bt_f\right)}{A_o} \tag{8.9.3}$$

the PNA moves into the web, and the fully plastic moment becomes

$$\begin{aligned} M_P &= (\sigma_Y + \sigma_a)bt_f(d - \eta) + (\gamma_w\sigma_Y + \sigma_a)\frac{(d-\eta)^2t_w}{2} \\ &\quad + (\gamma_w\sigma_Y - \sigma_a)\frac{\eta^2t_w}{2} + \left(\gamma_p\sigma_Y - \sigma_a\right)st_p\eta \end{aligned} \tag{8.9.4}$$

where

$$\eta = \frac{bt_f + \gamma_w dt_w - \gamma_p st_p + \left(\frac{\sigma_a}{\sigma_Y}\right)A_o}{2\gamma_w t_w} \tag{8.9.5}$$

Finally, when

$$\frac{\sigma_Y\left(\gamma_p st_p + \gamma_w dt_w - bt_f\right)}{A_o} < \sigma_a < \frac{\sigma_Y\left(\gamma_p st_p + \gamma_w dt_w + bt_f\right)}{A_o} \tag{8.9.6}$$

the PNA moves into the face plate, and the fully plastic moment becomes

$$M_P = (\gamma_w\sigma_Y - \sigma_a)\frac{d^2t_w}{2} + \left(\gamma_p\sigma_Y - \sigma_a\right)h_w st_p \tag{8.9.7}$$

Figure 8.42 shows an example of M_P curve taking σ_a/σ_Y in the horizontal axis. Assumed scantlings and material properties are as show in the figure. In the left half of the figure, the bending stress and the axial stress superimpose each other at the face plate, and in the right half, they cancel each other. When $\sigma_a/\sigma_Y = 0$, the

PNA is located within the attached plate and Eq. (8.9.2) applies. Compressive axial stress moves the PNA downward and tensile axial stress moves it upward when the bending moment is positive. Therefore, in the cancellation case of the right-hand side of Fig. 8.42, M_P is first governed by Eq. (8.9.2) when the axial stress is small. Then, when the PNA moves into the web as the axial stress increases, it changes to be governed by Eq. (8.9.4), and finally when the PNA moves into the face plate, it is governed by Eq. (8.9.7) as shown in Fig. 8.42.

Next, let us consider the bending strength of a stiffener with respect to the plastic hinge formation. The lateral pressure corresponding to the two hinge formation at the span ends (denoted by P_{2H}), the one hinge formation at the midspan (denoted by P_{1H}), and the collapse, i.e., three hinge formation (denoted by P_C) are expressed in the following equations.

$$P_{2H} = \frac{12M_{PE}}{sl^2} \tag{8.9.8}$$

$$P_{1H} = \frac{24M_{PM}}{sl^2} \tag{8.9.9}$$

$$P_C = \frac{8(M_{PE} + M_{PM})}{sl^2} \tag{8.9.10}$$

where

M_{PE}: fully plastic moment at the span ends
M_{PM}: fully plastic moment at the midspan.
l: span of the stiffener.

These pressures are plotted in Fig. 8.43 for a stiffener under the lateral pressure from the plate side (the opposite side of the stiffener attachment). They are non-dimensionalized divided by the pressure P_{2H0}, forming two plastic hinges at the

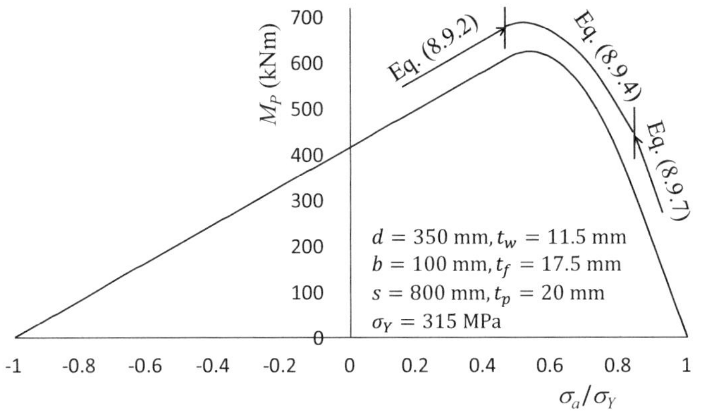

Fig. 8.42 Correlation between fully plastic moment and axial stress

span ends when no axial stress is working. The two hinge formation (P_{2H}) and the one hinge formation (P_{1H}) are not independent. When the axial stress is negative (left half of the figure), the bending stress and the axial stress are superimposed on the face plate at the span ends and two hinges are formed first at the span ends. In this case the P_{1H} curve has no meaning. Conversely when $\sigma_a > \sigma_Y/3$, a plastic hinge is formed at the midspan before the plastic hinge formation at the span ends. In this case the P_{1H} curve governs and the P_{2H} curve has no meaning. Thus, the initial plastic hinge formation phenomenon is governed by the lower envelope of the P_{1H} curve and the P_{2H} curve, which is shown by gray dashed lines in the figure. When the pressure is further increased after the initial hinge formation in accordance with the lower envelope curves, the stiffener collapses with three hinge formation in accordance with the P_C curve, which is shown by a chained-dotted line in the figure. We can observe from the figure that when σ_a is around $\sigma_Y/3$, the plastic hinges at the span ends and at the midspan occur almost simultaneously, and the stiffener collapses immediately.

Now, let us discuss the required plastic section modulus. If we employ the initial hinge formation as the criterion, we can obtain the relationship between the required plastic section modulus and the axial stress using the $M_P - \sigma_a$ curves for various stiffener section with different plastic section moduli. Figure 8.44 shows the required plastic section modulus Z_{Preq}, given as a ratio to the required plastic section modulus without the axial stress Z_{Preq0}.

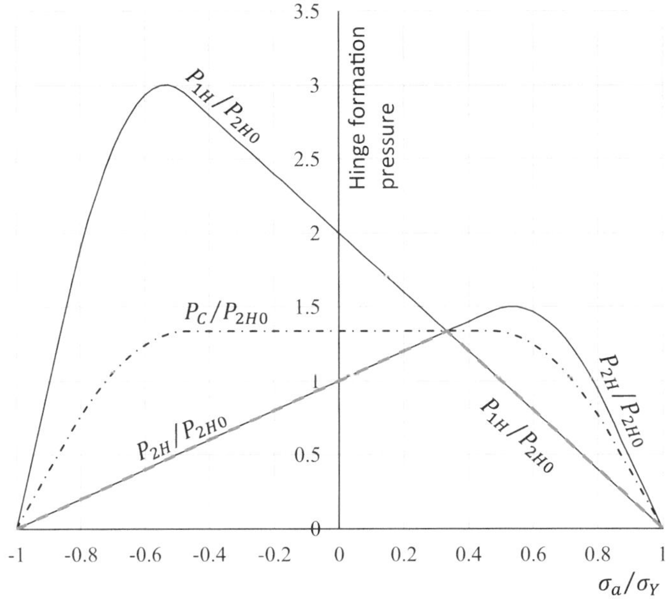

Fig. 8.43 Correlation between critical pressure for hinge formation and axial stress (plate side pressure)

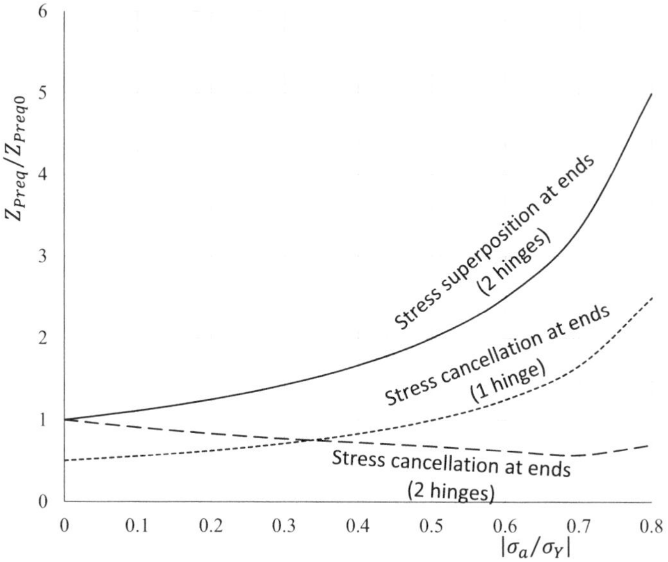

Fig. 8.44 Correlation between required plastic section modulus and axial stress

When the bending stress and the axial stress superimpose each other at the span ends (i.e., lateral pressure is from the plate side under axial compression, or lateral pressure is from the stiffener side under axial tension), the required plastic section modulus increases monotonically according to the axial stress as shown by the solid line in Fig. 8.44. On the other hand, when the bending stress and the axial stress cancel each other at the span ends (i.e., lateral pressure is from the plate side under axial tension, or lateral pressure is from the stiffener side under axial compression), the required plastic section modulus first decreases while the axial stress is small as shown by the dashed lines in Fig. 8.44, but it turns to increase when $|\sigma_a| > \sigma_Y/3$ due to the initial hinge formation at the midspan as shown by the dotted line in Fig. 8.44.

Based on the observations of Fig. 8.44, Zhang et al. [64] proposed axial stress coefficients as follows.

(1) In case of the stress superposition at the span ends:

$$\frac{Z_{Preq}}{Z_{Preq0}} = \frac{\sigma_Y}{\sigma_Y - |\sigma_a|} \tag{8.9.11}$$

(2) In case of the stress cancellation at the span ends:

$$\frac{Z_{Preq}}{Z_{Preq0}} = \max\left(1.0, \frac{\sigma_Y}{2(\sigma_Y - |\sigma_a|)}\right) \tag{8.9.12}$$

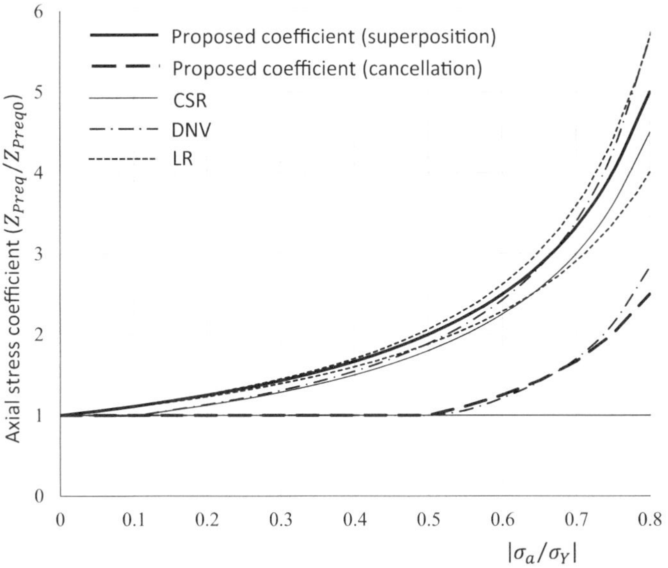

Fig. 8.45 Axial stress coefficient and comparison with class rules

These proposed coefficients are shown in the thick solid and dashed lines in Fig. 8.45. Axial stress coefficients in some classification society rules are also plotted in the figure. The rules are different class by class, but we can observe that the proposed coefficients are close to those by DNV including the stress cancellation case among various rules.

8.9.2 *Effect of Shear Stress [65]*

In Sect. 8.9.1, the fully plastic moment was calculated based on solely bending stresses, and the effect of shear stress was neglected. However, it is well known that stiffener bending strength is adversely affected by shear stresses on the web plate, especially in case of a deep and short stiffener or girder under large lateral pressure while it can be neglected in a shallow and long stiffener. This effect of the shear stress on the fully plastic moment can be evaluated by assuming that the effective yield stress of the web σ_{Yw} is reduced by the acting shear stress τ_{xy} according to the following equation based on von Mises yield criterion.

$$\sigma_{Yw} = \sqrt{\sigma_Y^2 - 3\tau_{xy}^2} \tag{8.9.13}$$

where τ_{xy} is the shear stress on the web. In this case, the web material coefficient γ_w in Eqs. (8.9.1)–(8.9.7) can be effectively expressed as follows:

$$\gamma_w = \frac{\sigma_{Yw}}{\sigma_Y} = \frac{\sqrt{\sigma_Y^2 - 3\tau_{xy}^2}}{\sigma_Y} \tag{8.9.14}$$

The relevant fully plastic moment M_P can be calculated considering the effect of the shear stress using this γ_w as shown in Fig. 8.46, where M_P reduction due to smaller γ_w is illustrated. However, this does not work as a design formula because σ_{Yw} changes according to the structural arrangement and scantling. Therefore, in the classification rules, shear stress is not directly considered in the bending strength formula, but considered separately as a web thickness criterion as explained in Sect. 3.3.2(2) of Part I. In this section, we discuss how we can consider the shear stress effect directly in the bending strength formula. In addition, the conditions to neglect the shear stress effect are discussed.

Assuming that lateral pressure P induces the bending moment M_P at the span ends, the following two equations hold.

$$M_P = \frac{Psl^2}{12} \tag{8.9.15}$$

$$\tau_{xy} = \frac{Psl}{2A_w} \tag{8.9.16}$$

Eliminating P from these two equations,

$$\tau_{xy} = \frac{6M_P}{LA_w} \tag{8.9.17}$$

Substituting this into Eq. (8.9.14),

$$\gamma_w = \frac{\sqrt{L^2A_w^2\sigma_Y^2 - 108M_P^2}}{LA_w\sigma_Y} \tag{8.9.18}$$

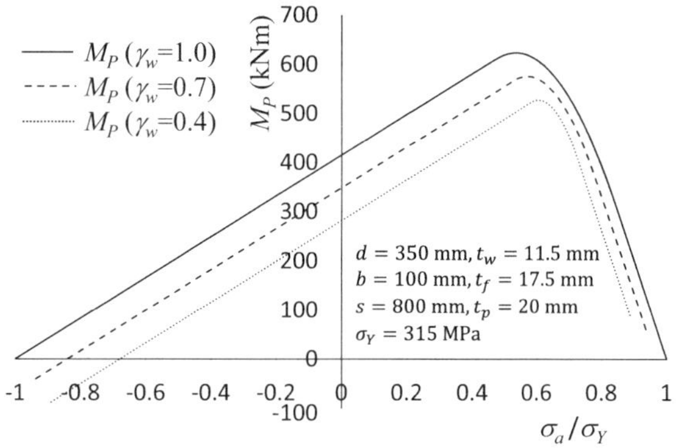

Fig. 8.46 Correlation between fully plastic moment and axial stress under reduced web strength

Then, substituting Eq. (8.9.18) into Eq. (8.9.2), we can obtain

$$M_P = (\sigma_Y + \sigma_a) d A_f + \left(\frac{\sqrt{l^2 A_w^2 \sigma_Y^2 - 108 M_P^2}}{l A_w} + \sigma_a \right) \frac{d A_w}{2} \tag{8.9.19}$$

Solving Eq. (8.9.19) with respect to M_P and taking the feasible one among the two real solutions,

$$M_P = \frac{\sigma_Y d A_w}{1 + 27\left(\frac{d}{l}\right)^2} \left[\left(1 + \frac{\sigma_a}{\sigma_Y}\right) \frac{A_f}{A_w} + \frac{1}{2} \cdot \frac{\sigma_a}{\sigma_Y} \right.$$
$$\left. + \frac{1}{2} \sqrt{1 + 27\left(\frac{d}{l}\right)^2 \left(1 + \frac{\sigma_a}{\sigma_Y}\right) \left(1 + 2 \cdot \frac{A_f}{A_w}\right) \left\{ 1 - 2 \cdot \frac{A_f}{A_w} - \frac{\sigma_a}{\sigma_Y} \left(1 + 2 \cdot \frac{A_f}{A_w}\right) \right\}} \right] \tag{8.9.20}$$

Considering $d \ll l$, the second term in the square root can be assumed to be sufficiently less than one. Then, expanding the square root and neglecting the small terms of the second order or more,

$$M_P = \frac{\sigma_Y d A_w \left(1 + \frac{\sigma_a}{\sigma_Y}\right)\left(1 + 2 \cdot \frac{A_f}{A_w}\right)}{2\left\{1 + 27\left(\frac{d}{l}\right)^2\right\}} \left[1 + \frac{27}{2} \left(\frac{d}{l}\right)^2 \left\{ 1 - 2 \cdot \frac{A_f}{A_w} - \frac{\sigma_a}{\sigma_Y} \left(1 + 2 \cdot \frac{A_f}{A_w}\right) \right\} \right] \tag{8.9.21}$$

Dividing Eq. (8.9.21) by the fully plastic moment without the shear stress effect M_{P0}, we can finally obtain the equation representing M_P reduction due to the shear stress effect as follows:

$$\frac{M_P}{M_{P0}} = C_\tau C_{pa} \tag{8.9.22}$$

where

$$C_\tau = \frac{1}{1 + 27\left(\frac{d}{l}\right)^2} \tag{8.9.23}$$

$$C_{pa} = 1 + \frac{27}{2} \left(\frac{d}{l}\right)^2 \left\{ 1 - 2 \cdot \frac{A_f}{A_w} - \frac{\sigma_a}{\sigma_Y} \left(1 + 2 \cdot \frac{A_f}{A_w}\right) \right\} \tag{8.9.24}$$

Now, sectional proportion of ordinary stiffeners follows the relationship $1 - 2A_f/A_w \approx 0$ when the stiffener is designed to minimize weight under a given section modulus as we discussed in Sect. 8.4.1. Therefore, when $\sigma_a = 0$ at the same time, the term denoted by C_{pa} reduces to unity. Then, we can regard C_τ as a fundamental coefficient representing the fully plastic moment reduction for ordinary stiffeners without the action of the axial stress, and C_{pa} as a correction coefficient to take account of the deviation of the stiffener sectional proportion and the action of the axial stress. We can see that the fully plastic moment reduction is dependent on three non-dimensional parameters, i.e., depth to span length ratio (d/l), face plate area to web area ratio (A_f/A_w), and the axial stress to yield stress ratio (σ_a/σ_Y).

Next, let us examine the conditions to neglect the shear stress effect. With respect to a stiffener complying with the standard proportion, the depth d can be expressed approximately using the required section modulus Z as follows according to Sect. 8.4.1.

$$d = \sqrt{\frac{4Z}{3t_w}} \tag{8.9.25}$$

When we take σ_Y as the allowable stress for the stiffener, the required section modulus Z is expressed as follows according to the proposed axial stress effect discussed in Sect. 8.9.1.

$$Z = \frac{Psl^2}{12\left(1 + \frac{\sigma_a}{\sigma_Y}\right)\sigma_Y} \tag{8.9.26}$$

Substituting Eq. (8.9.26) into Eq. (8.9.25),

$$\frac{d}{l} = \frac{1}{3}\sqrt{\frac{Ps}{t_w\left(1 + \frac{\sigma_a}{\sigma_Y}\right)\sigma_Y}} \tag{8.9.27}$$

This means that the non-dimensional parameter d/l depends only on P and σ_a, assuming that t_w and s are constant. Substituting Eq. (8.9.27) into Eq. (8.9.22) bearing in mind that $A_f/A_w \approx 1/2$ for the stiffeners with ordinary proportion, M_P/M_{P0} reduces to

$$\frac{M_P}{M_{P0}} = \frac{t_w\left(1 + \frac{\sigma_a}{\sigma_Y}\right)\sigma_Y - 3Ps \cdot \frac{\sigma_a}{\sigma_Y}}{t_w\left(1 + \frac{\sigma_a}{\sigma_Y}\right)\sigma_Y + 3Ps} \tag{8.9.28}$$

Now, ordinary small stiffeners have plentiful track records without considering the shear stress effect. For example, 350 mm depth stiffener with 4000 mm span length employed in Fig. 8.42 has been successfully applied without shear stress consideration, and in this case M_P reduces by a factor of 0.84 due to the action of the shear stress. Therefore, when the allowable limit of M_P/M_{P0} is denoted by $\gamma(\approx 0.84)$, the condition to neglect shear stress effect is

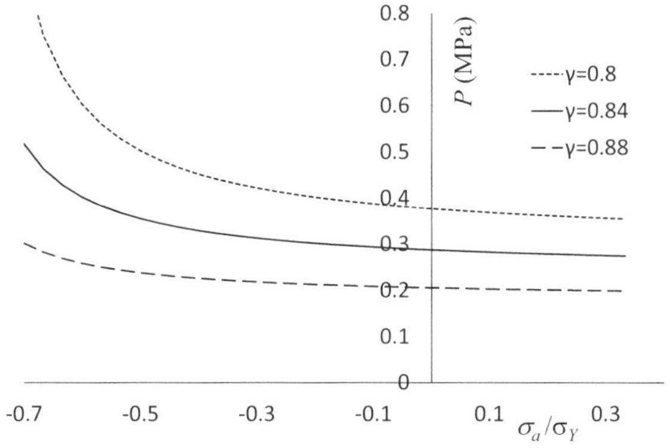

Fig. 8.47 Design pressure curve which gives constant fully plastic moment reduction

$$\frac{t_w\left(1+\frac{\sigma_a}{\sigma_Y}\right)\sigma_Y - 3Ps\cdot\frac{\sigma_a}{\sigma_Y}}{t_w\left(1+\frac{\sigma_a}{\sigma_Y}\right)\sigma_Y + 3Ps} > \gamma \tag{8.9.29}$$

Solving this with respect to P,

$$P < \frac{(1-\gamma)\left(1+\frac{\sigma_a}{\sigma_Y}\right)t_w\sigma_Y}{3\left(\gamma+\frac{\sigma_a}{\sigma_Y}\right)s} \tag{8.9.30}$$

This criterion is plotted in Fig. 8.47 for three different γ values, assuming $t_w = 11.5$mm and $s = 800$mm. When the design pressure P is less than these curves, it is not necessary to consider the shear stress effect for the stiffeners properly designed under this condition. If $\gamma = 0.84$ is employed as the criterion, we can observe from the figure that as long as the design pressure is less than about 0.27 MPa, the shear stress effect can effectively be neglected. This covers almost all the stiffeners in ship hull structures.

Finally, if M_P reduction by γ is allowable, the formula for the required section modulus can be proposed as follows:

$$Z = \frac{\gamma Z_0}{C_\tau C_{pa}} \tag{8.9.31}$$

where Z_0 is the required section modulus without considering the shear stress effect. This formulation will result in a stricter requirement for stiffeners whose M_P/M_{P0} is less than γ, but will allow smaller section modulus than the current practice for most of the shipbuilding stiffeners, whose M_P/M_{P0} is greater than γ.

Chapter 9
Design of Girders

In Chap. 8, it was explained that beams resist the bending moment caused by lateral forces and that the section modulus is important in resisting this bending moment. In the design of mammoth tankers, 40,000–50,000 DWT, built around 1960, girders were designed based only on bending theory. For the bigger ships built after that era, shearing theory has been also applied.

A girder is a structural member which supports lateral forces imposed by beams. Beams, stiffeners, frames, longitudinals, etc. are called secondary members, and girders, web frames, transverse webs, etc. are called primary members. The former is called beam and the latter is called girder collectively in Part II. The shearing force has to be taken into consideration during design for girders. Simply speaking, a beam is designed against bending and a girder is designed against bending and shearing.

9.1 Shearing Force

In a book of strength of materials, shearing force and shearing stress are explained in detail based on the equilibrium of forces. However, for the design of a girder, the information shown in Fig. 9.1 will simply illustrate. Complicated theory will not help to understand this phenomenon.

In Fig. 9.1 a cantilever girder with a concentrated load F at the free end is shown. Taking out a small part Δx from the girder, the mean shearing stress τ in this part is F/A, where F is the shearing force and A is the sectional area of the web. The shearing strain γ is shown by the following equation:

$$\gamma = \frac{F}{AG} \tag{9.1.1}$$

where G is the shear modulus.

Y. Okumoto et al., *Design of Ship Hull Structures*,
https://doi.org/10.1007/978-3-032-12500-2_9

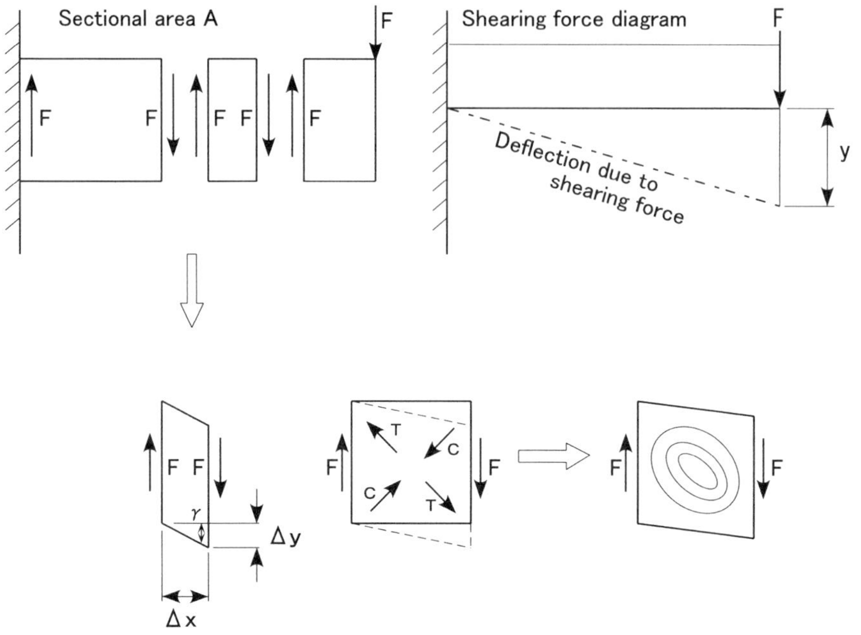

Fig. 9.1 Shearing force

Then the shearing deflection Δy in the vicinity of Δx is expressed as follows when we assume an approximately uniform distribution of the shear stress within the section.

$$\begin{aligned}\Delta y &= \gamma \Delta x \\ &= \frac{F \Delta x}{AG}\end{aligned} \tag{9.1.2}$$

$$\therefore y = \int_0^x \frac{F}{AG} dx \tag{9.1.3}$$

Equation (9.1.3) shows that the shearing deflection y is proportional to x where F is constant, therefore the shearing deflection is a straight line as shown in Fig. 9.1.

Under the action of the shearing force, the part of the girder, which has a rectangular shape, initially deforms just like a squashed match box, and consequently a tensile force T in a diagonal direction and a compressive force C in the other diagonal direction, are generated. The tensile force T sometimes causes cracks and the compressive force C causes buckling in the web plate. This is the reason why shearing buckling occurs in a diagonal direction.

A girder consists of a web plate, a face plate, and an attached plate similar to a beam, and the web plate supports shearing force. The behaviors can be easily understood from Fig. 9.1. The face plate and the attached plate has only their thickness in the shearing force direction and can be easily deformed without resisting the shearing force applied.

The distributions of the shearing stresses are shown in Fig. 9.2a, b in the cases of web plates and built-up sections respectively. The shearing stress is 0 at the free edge, and in the case of a web plate, the maximum shearing stress is 1.5 times of the mean value. In the case of a built-up section the shearing stress distribution in the web plate is more even and its maximum stress is considered to be nearly equal to the mean value, which is obtained by dividing the shearing force by the sectional area of the web plate.

Plates connected at the top and bottom of the web plate do not support the shearing force very much, but they even out the distribution of shearing stress in the web plate.

The shearing stress τ_1 in the section parallel to and at the distance y_1 from the neutral axis is obtained as shown in Fig. 9.3.

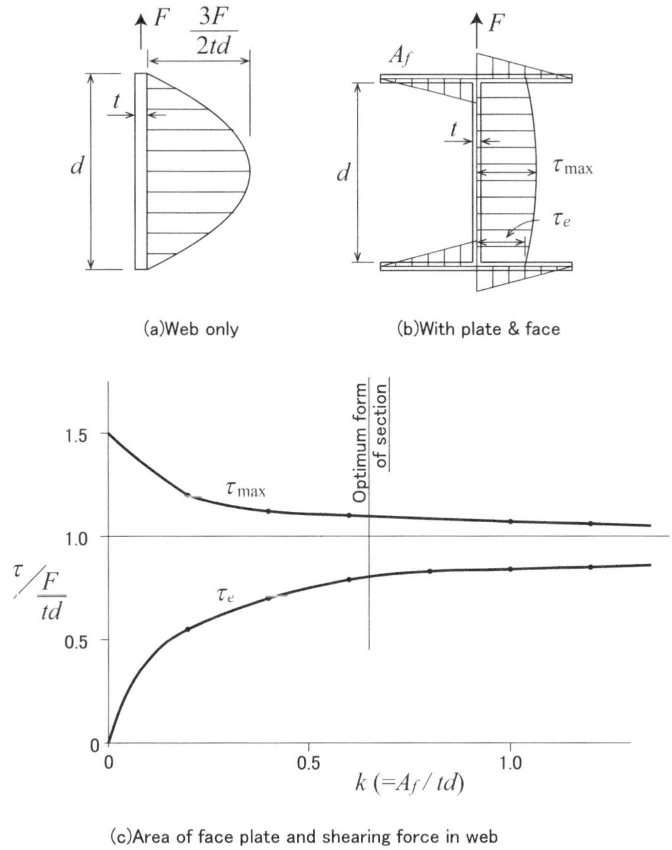

Fig. 9.2 Shearing stress distribution

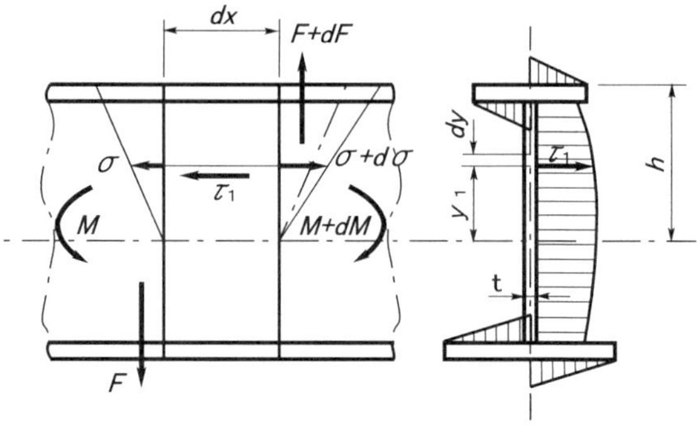

Fig. 9.3 How to obtain shearing stress distribution

$$\tau_1 t dx = \int_{y_1}^{h} d\sigma t dy \tag{9.1.4}$$

$$\sigma = \frac{M}{I} y \tag{9.1.5}$$

$$\therefore d\sigma = \frac{y}{I} F dx \tag{9.1.6}$$

From Eqs. (9.1.4) and (9.1.6) the following is derived.

$$\begin{aligned} \tau_1 &= \frac{F}{It} \int_{y_1}^{h} y t dy \\ &= \frac{F m_1}{It} \end{aligned} \tag{9.1.7}$$

where

M: bending moment

F: shearing force

t: thickness of web plate

I: sectional moment of inertia

m_1: first moment of area of the part farther than distance y_1 from the neutral axis

In the case of a balanced girder, the shearing stress in the web plate is calculated as below. The area of the face plate is A_f and the depth and the thickness of the web plate are d and t respectively. The maximum shearing stress τ_{max} at the neutral axis can be calculated as follows:

$$I = \frac{A_f d^2}{2} + \frac{t d^3}{12} \tag{9.1.8}$$

$$m = \frac{A_f d}{2} + \frac{t d^2}{8} \tag{9.1.9}$$

$$\therefore \frac{m}{I} = \frac{\frac{A_f}{2} + \frac{td}{8}}{d\left(\frac{A_f}{2} + \frac{td}{12}\right)} \tag{9.1.10}$$

here putting $A_f = ktd$,

$$\frac{m}{I} = \frac{12k + 3}{d(12k + 2)} \tag{9.1.11}$$

$$\therefore \tau_{\max} = \frac{Fm}{It} = \frac{F(12k + 3)}{td(12k + 2)} \tag{9.1.12}$$

The shearing stress of web at top and bottom τ_e is obtained as follows:

$$m = \frac{A_f d}{2} \tag{9.1.13}$$

$$\therefore \frac{m}{I} = \frac{A_f}{2} \cdot \frac{1}{d\left(\frac{A_f}{2} + \frac{td}{12}\right)} = \frac{6k}{d(6k + 1)} \tag{9.1.14}$$

$$\tau_e = \frac{6Fk}{td(6k + 1)} \tag{9.1.15}$$

The relationship of Eq. (9.1.12) with (9.1.15) is shown in Fig. 9.2c. The origin, $k = A_f/td = 0$, corresponds to the section without face plates.

F/td, the value of the shearing force divided by the sectional area of web plate (assuming the attached plate and the face plate do not support the shearing force), is the mean shearing stress. It can be said that the shearing stress distribution in a web plate is nearly uniform and its value is nearly equal to the mean shearing stress expressed by F/td where the sectional area of the face plate is sufficient.

9.2 Rational Design of Girders

The rational design of girders is explained below for the bottom transverse in a center tank of a tanker under the assumption that the web plate supports the shearing force and the face plate supports the bending moment.

As shown in Fig. 9.4 the bottom transverse in the center tank of a tanker can be assumed to be a girder with both ends fixed and with uniform load. Accordingly, the shearing force diagram is a straight line passing through the 0 point at the midspan. The thickness of the web plate t and the depth of the girder at the ends h are to be expressed as follows:

$$h = \frac{wl}{2t\tau_a} \tag{9.2.1}$$

where τ_a: allowable shearing stress

The bending moment at midpoint in the span is $M = wl^2/24$, and for an allowable bending stress given by σ_a the required section modulus Z is as follows:

$$Z = \frac{wl^2}{24\sigma_a}$$

Using an optimal section shown in Sect. 8.4, the depth d and the sectional area of the face plate A_f of the girder are obtained as follows.

$$d = 1.23\sqrt{\frac{wl^2}{24t\sigma_a}} \tag{9.2.2}$$

$$A_f = 0.41\sqrt{\frac{wl^2 t}{24\sigma_a}} \tag{9.2.3}$$

From Eqs. (9.2.1) and (9.2.3) the ratio of h and d can be obtained.

$$\frac{h}{d} = \frac{wl\sqrt{24t\sigma_a}}{2t\tau_a \times 1.23\sqrt{wl^2}} = \frac{\sqrt{6\sigma_a w}}{1.23\tau_a\sqrt{t}} \tag{9.2.4}$$

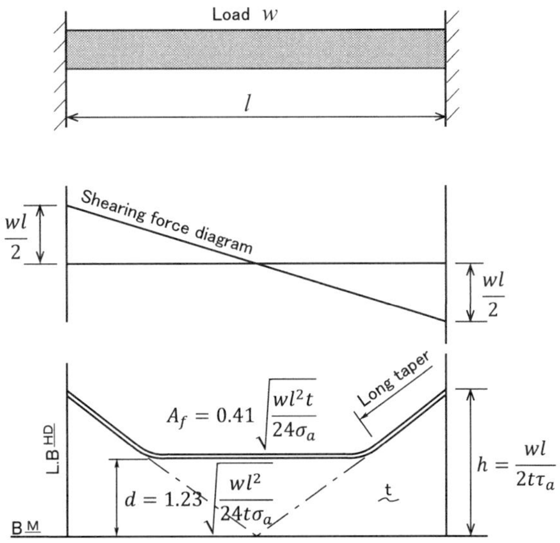

Fig. 9.4 Rational girder design

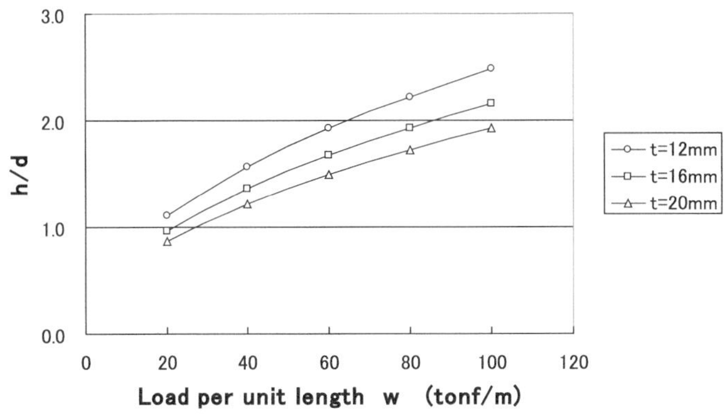

Fig. 9.5 Ratio of depth at end and center of rationally designed girder

Assuming that the allowable bending stress and the shearing stress are $\sigma_a = 120\,\text{MPa}$ and $\tau_a = 80\,\text{MPa}$ respectively, h/d is obtained as shown in Fig. 9.5.

Judging from the appearance of the girder, h/d will be around 1.5–2.5. By using Fig. 9.5 the thickness of the web plate t for a given load w can be derived for the design h/d. After that the area of the face plate A_f is derived using Eq. (9.2.3).

The sectional area of the face plate A_f is to be maintained in the vicinity of the parallel part of the girder. However, in the vicinity of the end bracket where the depth of the girder is deeper than the parallel part, the required section modulus can be obtained even with a reduced sectional area of the face plate. In this consideration the long taper of the face plate is developed as shown in Fig. 9.6 [62]. The long tapered face plate brings not only a weight saving but also a decrease in the scrap percentage thus killing two birds with one stone.

9.3 Bottom Transverses Supported by Centerline Girder [62]

In Sect. 8.6.6, the increase of the bending moment due to forced displacement is explained. In this section the additional bending moment and shearing force in the bottom transverse due to the displacement of the bottom centerline girder are explained.

The bottom transverse shown in Fig. 9.7 used to be designed assuming a fixed boundary condition both at the longitudinal bulkhead and at the centerline girder. However, the centerline girder has less rigidity than the longitudinal bulkhead and a relative displacement between the longitudinal bulkhead and the centerline girder will be generated. Therefore, it will be more practical to assume that the bottom transverse is fixed at the longitudinal bulkheads and is elastically supported by the centerline girder. With the reaction force at the centerline girder W and the deflection

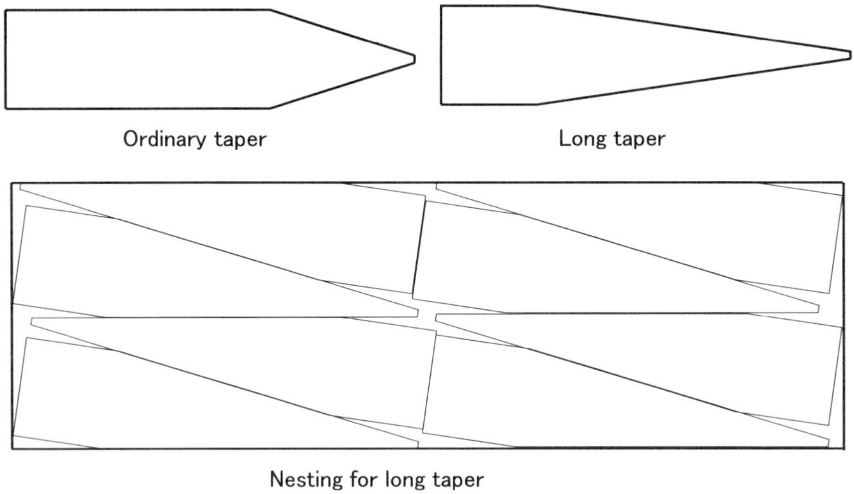

Fig. 9.6 Long tapered face plate end

at the centerline δ, the following equation is obtained:

$$\delta = \frac{wl^4}{384EI} - \frac{Wl^3}{192EI} \tag{9.3.1}$$

where

l: span of bottom transverse (distance between longitudinal bulkheads).
E: Young's modulus.
I: sectional moment of inertia of bottom transverse.
w: uniform load per unit length.

Expressing the shearing force and the bending moment at fixed ends by F and M respectively, following equation can be obtained:

$$F = \frac{wl}{2} - \frac{W}{2} \tag{9.3.2}$$

$$M = \frac{wl^2}{12} - \frac{Wl}{8} \tag{9.3.3}$$

Eliminating W from Eqs. (9.3.1)–(9.3.3), the relation between δ, F and M can be derived as follows:

$$F = \frac{96EI}{l^3}\delta + \frac{wl}{4} \tag{9.3.4}$$

$$M = \frac{24EI}{l^2}\delta + \frac{wl^2}{48} \tag{9.3.5}$$

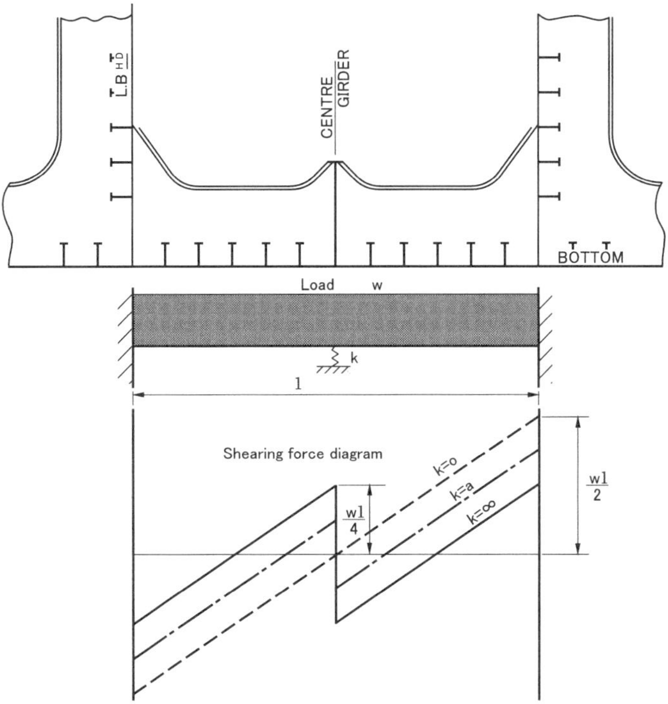

Fig. 9.7 Bottom transverse supported by center line girder

F and M where $\delta = 0$ are expressed by F_0 and M_0 respectively, then F_0 and M_0 will be as follows:

$$F_0 = \frac{wl}{4} \tag{9.3.6}$$

$$M_0 = \frac{wl^2}{48} \tag{9.3.7}$$

F_0 and M_0 are the shearing force and the bending moment at the fixed ends when the bottom transverse is designed as a girder fixed at the longitudinal bulkhead and the centerline girder. Accordingly the rate of increase in the shearing force F/F_0 and the bending moment M/M_0 of the bottom transverses, where the displacement at the centerline girder occurs, is as follows:

$$\frac{F}{F_0} = \frac{384EI}{wl^4}\delta + 1 \tag{9.3.8}$$

$$\frac{M}{M_0} = \frac{1152EI}{wl^4}\delta + 1 \tag{9.3.9}$$

From the above equations it can be seen that the rate of increase of the bending moment is bigger than that of the shearing force.

In reality the rate of increase can be calculated, assuming that the allowable bending stress σ_a in the case of no displacement at the centerline is 118 MPa, and the ratio between span and depth of the bottom transverse is 4.0.

$$M_0 = \frac{I}{y}\sigma_a \tag{9.3.10}$$

$$\therefore \frac{wl^2}{48} = 118\frac{I}{y} \tag{9.3.11}$$

$$\frac{l}{y} = 4 \tag{9.3.12}$$

where y is the distance between the neutral axis and face plate of bottom transverse (assuming neutral axis lies in the attached plate, and y equals to the depth of bottom transverse)

Using Eqs. (9.3.11) and (9.3.12) into Eqs. (9.3.8) and (9.3.9), the following equations are obtained assuming $E = 206{,}000$ MPa.

$$\frac{F}{F_0} = 3.5 \times 10^3\frac{\delta}{l} + 1 \tag{9.3.13}$$

$$\frac{M}{M_0} = 1.05 \times 10^4\frac{\delta}{l} + 1 \tag{9.3.14}$$

From the above results, we can see that with 1 mm displacement in the 10 m span of a bottom transverse at the mid-span ($\delta/l = 10^{-4}$), the shearing force and the bending moment increase by 1.35 times and 2.05 times respectively.

As shown above the shearing force at both ends increases, but the shearing force at the centerline reduces by the same value as the increase at the fixed ends. In the case where the centerline girder is not effective, the shearing force at the center is zero.

9.4 Deflection of Girders

In Fig. 9.1 the shearing deflection mode of a cantilever beam with a concentrated load at the free end is shown. The mode seems strange for those who are familiar with the bending deflection mode.

In Fig. 9.8 several shearing deflection modes are shown. Figure 9.8a shows the deflection mode of a cantilever beam with a concentrated load at the mid-point. The end part far from the load has no shearing force, consequently no shearing deflection appears. Figure 9.8b shows the deflection mode of a cantilever beam with a uniform load. The part near the fixed end has a higher shearing force then bigger shearing deformation with bigger curvature. At the free end the shearing force is zero and the deflection curve is tangential to the horizontal.

Figure 9.8c, d show beams with a concentrated load at the midspan, with Fig. 9.8c simply supported at both ends and Fig. 9.8d fixed at both ends. In these cases, the end conditions do not make any difference.

In a deep girder the shearing deflection is dominant and with a shallow girder the bending deflection is dominant. Below, the relationship between the slenderness of a girder and the ratio of the bending and the shearing deflections is investigated. The maximum bending deflection δ_b of a girder with both ends fixed for uniform load is expressed by the following equation:

$$\delta_b = \frac{wl^4}{384EI} \tag{9.4.1}$$

The maximum shearing deflection δ_s is obtained as follows:

$$\delta_s = \int_0^{\frac{l}{2}} \frac{1}{AG}\left(\frac{wl}{2} - wx\right)dx = \frac{wl^2}{8AG} \tag{9.4.2}$$

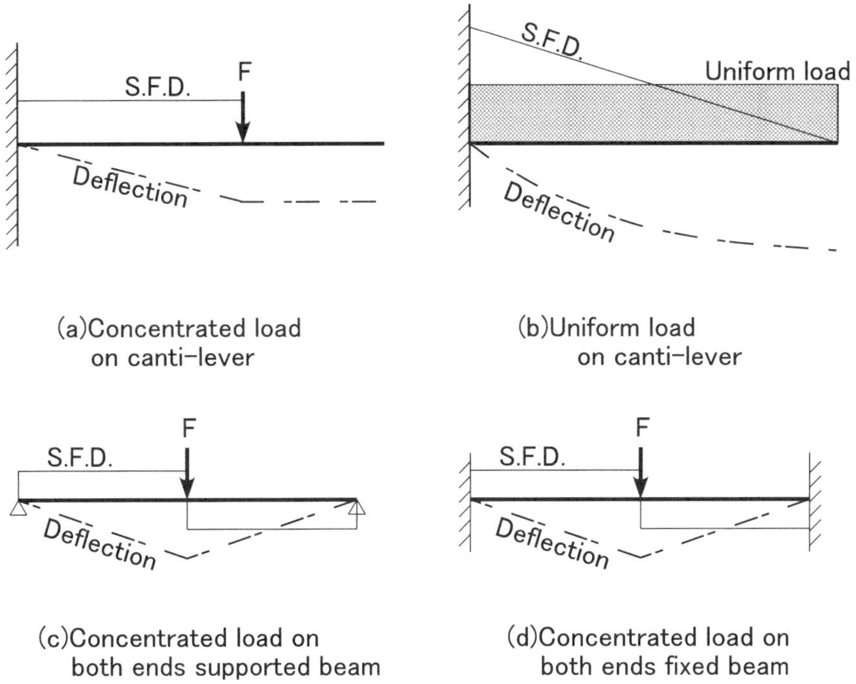

Fig. 9.8 Shearing deflection mode of girder

From Eqs. (9.4.1) and (9.4.2) the following equation is delivered:

$$\frac{\delta_s}{\delta_b} = \frac{48EI}{AGl^2} \tag{9.4.3}$$

I and A can be expressed by allowable bending and shearing stresses respectively.

$$I = \frac{wl^2}{12\sigma_a}y \tag{9.4.4}$$

$$A = \frac{wl}{2\tau_a} \tag{9.4.5}$$

$$\therefore \frac{I}{A} = \frac{ly\tau_a}{6\sigma_a}$$

Putting Eqs. (9.4.4) and (9.4.5) into Eq. (9.4.3), the following equation can be obtained.

$$\frac{\delta_s}{\delta_b} = \frac{8E\tau_a}{G\sigma_a} \times \frac{y}{l} \tag{9.4.6}$$

where

E: Young's modulus (206,000 MPa)

G: shear modulus (78,000 MPa)

σ_a: allowable bending stress (assumed to be 120 MPa)

τ_a: allowable shearing stress (assumed to be 80 MPa)

y: distance between neutral axis and face plate (in the case of a section with a plate, equals the depth of girder)

l: span of girder

Putting the above figures into Eq. (9.4.6), the following simple relation is delivered.

$$\frac{\delta_s}{\delta_b} = 14\frac{d}{l} \tag{9.4.7}$$

Equation (9.4.7) means that in the case of an ordinary girder design (σ_a is 120 MPa and τ_a is 80 MPa) the bending and shearing deflections are comparable at a slenderness ratio of 14. Usually the span of a girder is 4–8 times the depth and therefore the shearing deflection will be 2–3 times that of the bending deflection.

9.5 Large Opening on Girder Web Plate

When holes for piping, access, etc. are opened on the web plate of primary supporting members such as girders and transverse web frames, the most important effect on the strength is the reduced shear area. The remaining part of the web plate should have a sufficient shear area to withstand the shear force acting on that section. However, when the openings are much larger, this is not enough.

Figure 9.9 shows a vertical web on a transverse bulkhead of a bulk carrier. Such vertical webs are sometimes employed in the part adjacent to the side shell instead of corrugation to ensure in-plane rigidity of the transverse bulkhead (see Sect. 24.10, Part III). A large access hole is opened at the upper part of the vertical web so that an inclined ladder can pass through it. When an opening is such large, we must be careful about secondary bending stresses exerted at both the sides of the opening as shown in the right figure of Fig. 9.9, especially when the opening is located where large shear force is acting.

This problem can be modeled using two parallel local beams (Beam 1 and Beam 2) at both the sides of the opening as shown in Fig. 9.10. At the section of the opening center, the bending moment and the shearing force on the web are denoted by M_{0m} and F_{0m}, respectively. On the local beams, linearly distributed local loads are acting, and w_{i0} and w_{i1} denote the local loading per unit length on Beam i at the left end and the right end, respectively. Then the shearing force, bending moment and axial force on Beam i are obtained as follows:

$$F_{i0} = \frac{I_i}{I_1 + I_2}\left(F_{0m} - \frac{1}{20}\Delta w_0 l_e\right) + \frac{1}{2}w_{im}l_e\left(1 - \frac{2}{5}\cdot\frac{\Delta w_i}{w_{im}}\right) \tag{9.5.1}$$

$$F_{i1} = \frac{I_i}{I_1 + I_2}\left(F_{0m} - \frac{1}{20}\Delta w_0 l_e\right) - \frac{1}{2}w_{im}l_e\left(1 + \frac{2}{5}\cdot\frac{\Delta w_i}{w_{im}}\right) \tag{9.5.2}$$

$$M_{i0} = -\frac{I_i}{2(I_1 + I_2)}\left(F_{0m}l_e - \frac{1}{20}\Delta w_0 {l_e}^2\right) + \frac{I_i}{I_0}\left(M_{0m} - \frac{1}{24}w_{0m}{l_e}^2\right) - \frac{1}{12}w_{im}{l_e}^2\left(1 - \frac{1}{5}\cdot\frac{\Delta w_i}{w_{im}}\right) \tag{9.5.3}$$

$$M_{i1} = \frac{I_i}{2(I_1 + I_2)}\left(F_{0m}l_e - \frac{1}{20}\Delta w_0 {l_e}^2\right) + \frac{I_i}{I_0}\left(M_{0m} - \frac{1}{24}w_{0m}{l_e}^2\right) - \frac{1}{12}w_{im}{l_e}^2\left(1 + \frac{1}{5}\cdot\frac{\Delta w_i}{w_{im}}\right) \tag{9.5.4}$$

$$P_1 = -P_2 = -\left(1 - \frac{I_1 + I_2}{I_0}\right)\frac{1}{e}\left(M_{0m} - \frac{1}{24}w_{0m}{l_e}^2\right) \tag{9.5.5}$$

where

F_{i0}: shearing force on Beam i at the left end

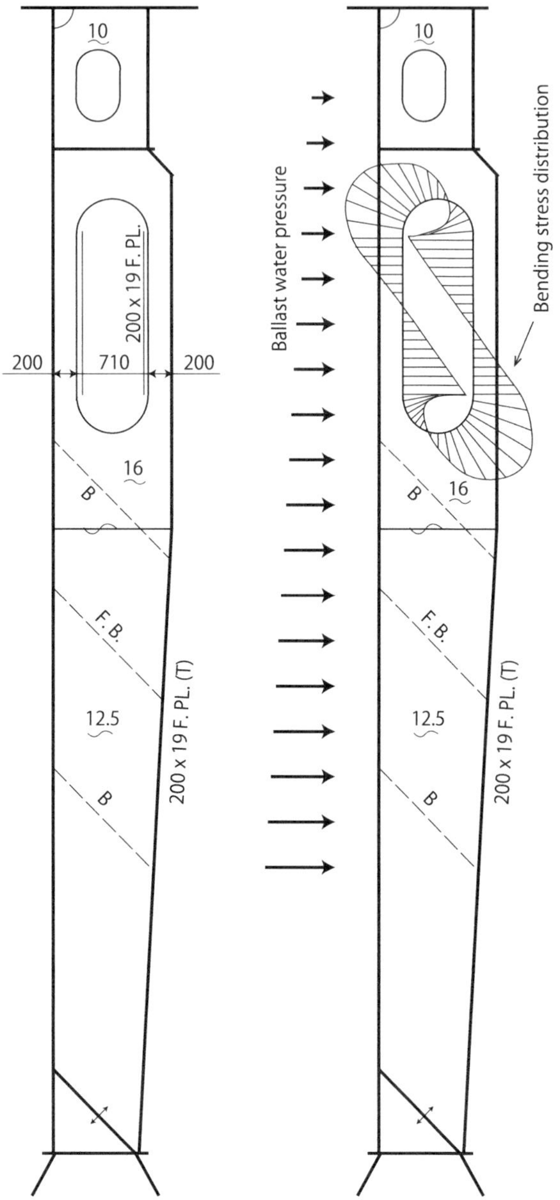

Fig. 9.9 Large opening on a vertical web plate

F_{i1}: shearing force on Beam i at the right end.
M_{i0}: bending moment on Beam i at the left end
M_{i1}: bending moment on Beam i at the right end
P_i: axial force on Beam i
I_i: sectional moment of inertia of Beam i

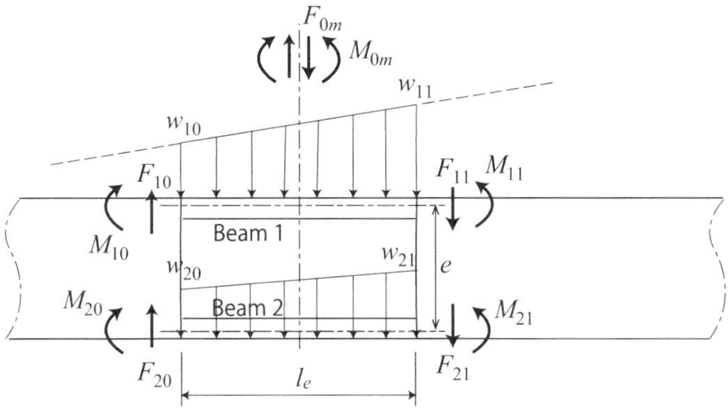

Fig. 9.10 Modeling of large opening by two parallel local beams

I_0: sectional moment of inertia of the integrated beams of Beam 1 and Beam 2

$$I_0 = I_1 + I_2 + \frac{A_1 A_2 e^2}{A_1 + A_2} \tag{9.5.6}$$

A_i: sectional area of Beam i
e: distance between the neutral axis of Beam 1 and that of Beam 2
l_e: effective span of the local beams (Beam 1 and Beam 2)
w_{im}: local load per unit length at the midspan of Beam i

$$w_{im} = \frac{w_{i0} + w_{i1}}{2} \tag{9.5.7}$$

Δw_i: half the difference between the left and right unit load on Beam i

$$\Delta w_i = \frac{w_{i1} - w_{i0}}{2} \tag{9.5.8}$$

w_{0m}: total local load at the center of the opening

$$w_{0m} = w_{1m} + w_{2m} \tag{9.5.9}$$

Δw_0: half the difference between the left and right unit load on both the local beams

$$\Delta w_0 = \Delta w_1 + \Delta w_2 \tag{9.5.10}$$

The equations of the bending moments on the local beams (Eqs. (9.5.3) and (9.5.4)) are composed of three terms, each of which has the meaning as follows:

- first term: Secondary bending moment on Beam i due to shearing force on the web
- second term: Bending moment on Beam i due to bending moment on the web
- third term: Bending moment on Beam i due to the effect of local pressure.

From the observation of the second term, the exerted local stress on Beam i due to bending moment on the girder will not be significant as long as $I_0 \gg I_i$. On the other hand, attention should be paid to the first term and the third term. The first term can be significant when the magnitude of the shearing force on the web is large, and the third term can be significant when the local pressure is large.

When only the shearing force acts on the web, Eqs. (9.5.1)–(9.5.4) reduce to:

$$F_{i0} = F_{i1} = \frac{I_i}{I_1 + I_2} F_{0m} \tag{9.5.11}$$

$$M_{i0} = -M_{i1} = -\frac{I_i}{2(I_1 + I_2)} F_{0m} l_e \tag{9.5.12}$$

From Eqs. (9.5.11) and (9.5.12), we can find that Beam 1 and Beam 2 support the shearing force and bending moment to the amount proportionate to the respective sectional moment of inertia. The bending moment on each local beam due to the shearing force on the web has opposite sign at each end of the local beam, and distributes linearly between both the ends, whose typical stress distribution is as shown in the right figure of Fig. 9.9.

When Beam 1 and Beam 2 have the same sectional moment of inertia, Eqs. (9.5.11) and (9.5.12) further reduce to the following simple formulae, which are useful when we roughly evaluate the strength of the local beams beside a large opening on the web plate of a girder.

$$F_{i0} = F_{i1} = \frac{F_{0m}}{2} \tag{9.5.13}$$

$$M_{i0} = -M_{i1} = -\frac{F_{0m} l_e}{4} \tag{9.5.14}$$

Chapter 10
Damage of Girders

With the large increase in the size of ships since 1965, many kinds of damage on the transverse strength members, especially on the transverse rings of tankers, appeared. Initially the web plate of the transverse ring buckled under water pressure tests. Then cracks appeared at slots where the load is transmitted to the transverse web from the longitudinal stiffeners.

The main reason for these damages was the lack of a thicker web plate in spite of the increased load which was brought on by the deeper ship depth and the wider longitudinal and transverse spaces. The local buckling or the cracking developed sometimes up to the extensive damage which is shown in Fig. 10.1 [66].

For a reliable design it is important that the hull structure design is based on a design standard which is revised according to feedback from ships in service. Damage is the most important feedback obtained from ships in service.

For the large-increase in the size of ships which started around 1955 the classification society's rules were not suitably revised until 1970, and the hull structure designer had to design bigger ships based on his own experience as well as the classification society's rules.

This chapter explains how the hull structure designer established his/her own design standard for new designs based on the damage experienced by girders in previous ships in service. In Fig. 4.13 of Part I, a feedback system of damage is shown.

10.1 Buckling Caused by Compression

Figure 10.2 shows buckling damage in a wide panel (4,000 mm × 800 mm) at the connection between a horizontal stringer and a strut in a fore peak tank. Here the buckling happened first which then caused the crack later.

In this case the load was the water pressure from the outside. Assuming the water head of 1.35d (d is full load draft) above base line, the actual water head at the

Y. Okumoto et al., *Design of Ship Hull Structures*,
https://doi.org/10.1007/978-3-032-12500-2_10

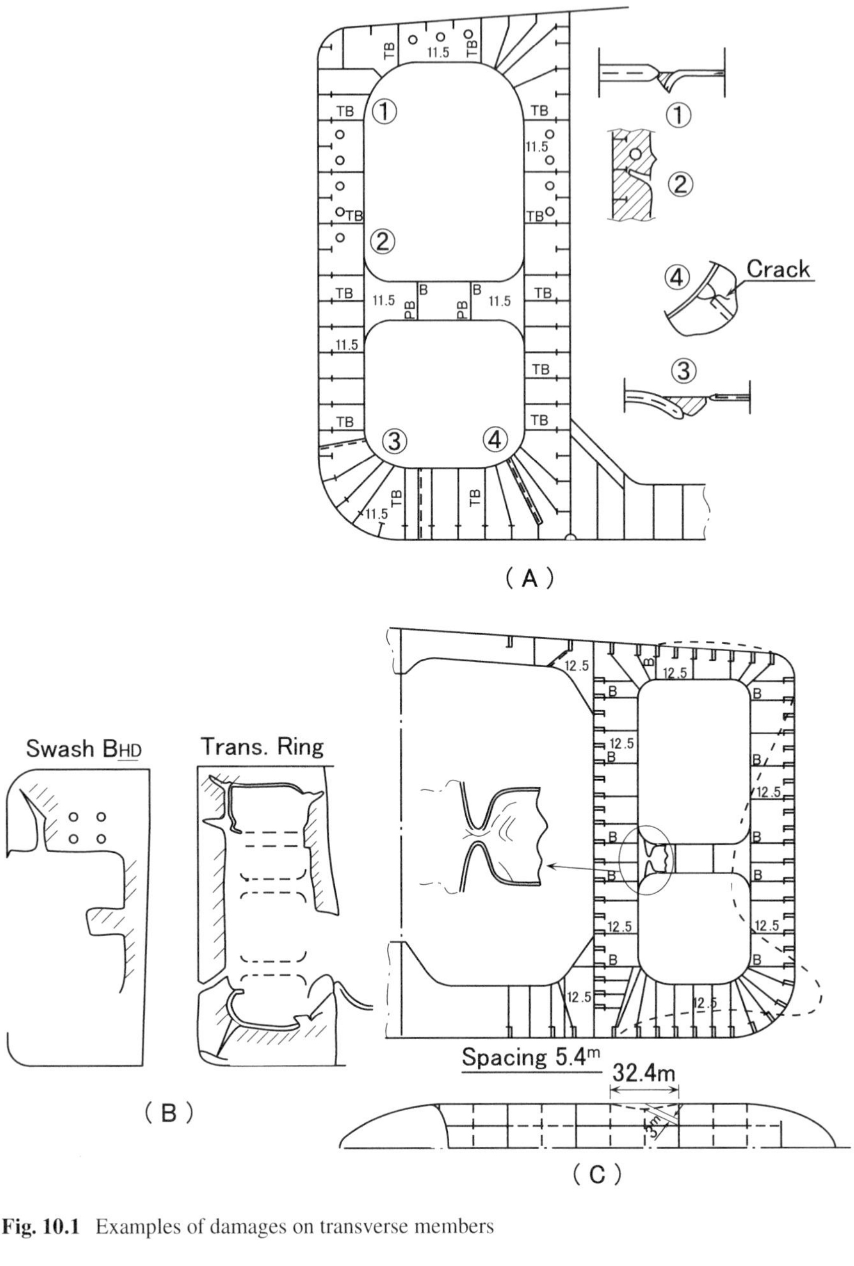

Fig. 10.1 Examples of damages on transverse members

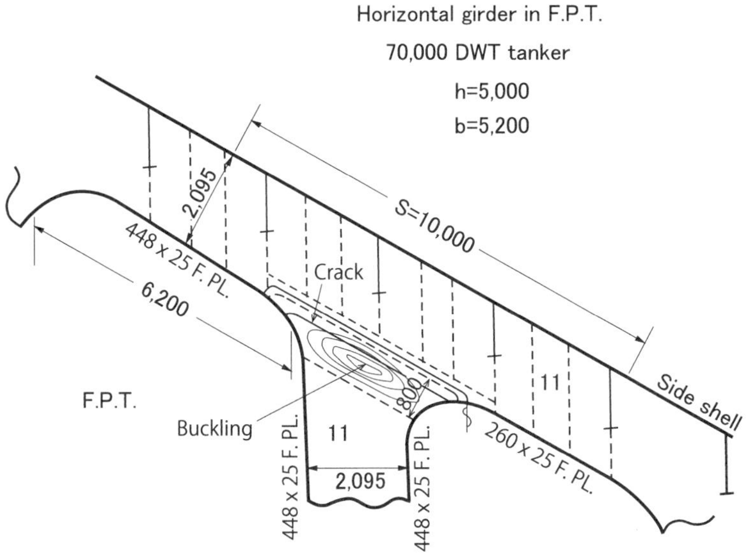

Fig. 10.2 Buckling due to compression

horizontal stringer level is 5 m. If the side shell area supported by the strut is 10 m (breadth) × 5.2 m (depth) = 52 m^2, then the load supported by the buckled part is 52 m^2 × 5 tf/m^2 = 260 tf = 2.55 × 10^6 N. And the sectional area at the buckled part is 224 cm^2 (face plates) + 230 cm^2 (web plate) = 454 cm^2. Accordingly, the compressive stress at the buckled section σ_w is 2.55 × 10^6 N/454 cm^2 = 56 MPa.

On the other hand, the critical buckling stress of this panel σ_{cr} is calculated as follows with the assumption that the rectangular plate is simply supported at all edges.

$$\sigma_{cr} = K\frac{\pi^2 E}{12(1-\nu^2)}\left(\frac{t}{b}\right)^2 = 39$$
$$\therefore \sigma_{cr} < \sigma_w \tag{10.1.1}$$

It is clear that the possibility of buckling existed.

Based on this buckling damage the design standard of $\sigma_{cr}/\sigma_w \geq 2.0$ was established. Similar parts to which attention should be paid are the "ends of cross tie", "the lower parts of vertical web on longitudinal bulkheads in wing tanks" and "the upper deck between the hatches of ore carriers".

In applying this design standard special attention should be paid to the following two points:

(1) The external load is to be calculated with a water head of 1.35d.
(2) The critical buckling stress is to be calculated assuming that the rectangular plate is simply supported at all edges.

The design standard of $\sigma_{cr}/\sigma_w \geq 2.0$ is only valid with a unified load and unified calculation method.

When similar buckling happens after applying this design standard, the safety factor of 2.0 is to be increased.

10.2 Buckling Caused by Bending

The buckling shown in Fig. 10.3 is caused by the compression due to the bending of the horizontal girder on the transverse bulkhead.

Assuming that the water head at the water pressure test is 2.45 m above the upper deck, the bending stress of this part is about 100 MPa. On the other hand, σ_{cr} is 75 MPa. In this case also $\sigma_{cr} < \sigma_w$ and there exists the possibility of buckling.

The horizontal girder on the transverse bulkhead is designed to sustain a bending stress of about 100 MPa, and the design standard of $\sigma_{cr} \geq 150$ MPa was established.

The safety factor is 1.5 which is smaller than 2.0 mentioned in Sect. 10.1 because the water pressure on the transverse bulkhead can be estimated more accurately than the water pressure on the side shell plate.

Parts in which similar attention should be paid are "bracket parts of horizontal girders in the fore peak tank", "the corner of the bottom transverse at the lower part of the vertical web of longitudinal bulkhead in the wing tank", etc.

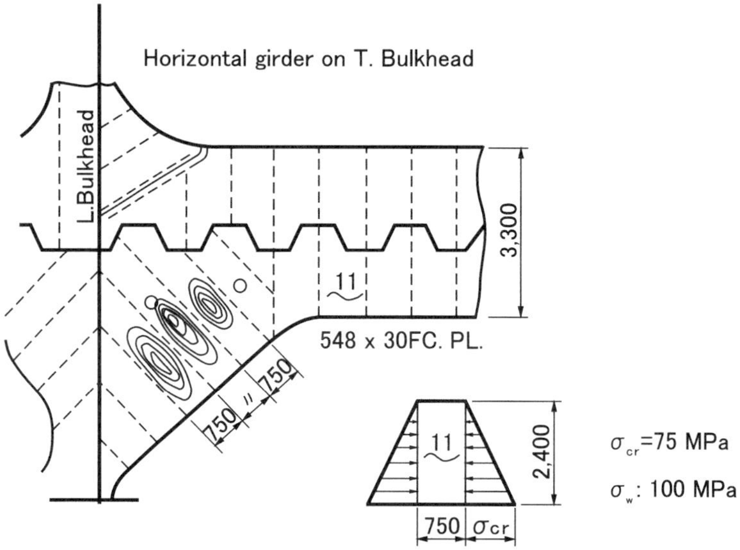

Fig. 10.3 Buckling due to bending

10.3 Buckling Caused by Shearing

Figure 10.4 shows buckling on the bottom transverse of a tanker. The buckling pattern takes a diagonal direction in a rectangular plate and indicates that the buckling is caused by the shearing force as explained in Sect. 9.1. In this case the free edge of a slot buckled first.

The critical buckling stress of a plate with a slot is not clear theoretically and an experiment was carried out on a 3/10 scale model of the bottom transverse of 100,000 DWT tanker as shown in Fig. 10.5 [62]. Two models were tested, one had the original depth of the slot and the other had a 1.5 times deeper slot than that of the original.

The buckling started at the corner of the free edge of the slot in both cases. Actually, the buckling was generated not in the vicinity of the slot where the shear area is reduced due to the slot cutout, but between the vertical web stiffeners where the web plate extends from the bottom shell plate to the upper face plate. Therefore, the shearing stress for buckling evaluation was calculated by dividing the shear force by the shear area corresponding to the full depth of the web.

The buckling stresses obtained by the experiment are plotted in Fig. 10.6, where the horizontal axis is the ratio of the slot depth to the web depth. The left graph shows the buckling stress and the right graph shows its ratio to the theoretical shearing buckling stress without the slot cutout under the assumption of simply supported boundaries.

Figure 10.6 shows that in the range where the ratio of slot depth and web depth is 0.3–0.5, the value of the shearing critical buckling stress decreases in a straight line with an increase in this ratio.

In the model shown in Fig. 10.5 the orientation of the slot is not symmetrical for the center bar and the buckling happened only at right side slots. This is a simple but very important finding. The buckling is caused only by compression not by tension. The free edge of the slot in the left side of the center bar has tensile stress and has no buckling in this case. In designing a slot, the designer should pay attention to the direction of the shearing force and decide the direction of the slot which does not cause any compression at the free edge. The direction of the slot does not affect the fabrication cost and a clever designer can design a sound slot without any extra expense.

By means of Fig. 10.6 the shearing critical buckling stress τ_{cr} can be obtained for a rectangular plate with a slot. The actual working shearing stress τ_w is calculated with the following assumptions:

(1) Water pressure is supported by the bottom transverse.
(2) The bottom transverse in a wing tank is fixed at both ends, the longitudinal bulkhead and the side shell.
(3) The relative displacement between the longitudinal bulkhead and the side shell is the same value as the shearing deflection of the side shell.

Applying the above method, the calculations for several ships were carried out and the results are shown in Fig. 10.7 in which a proposed design standard is shown

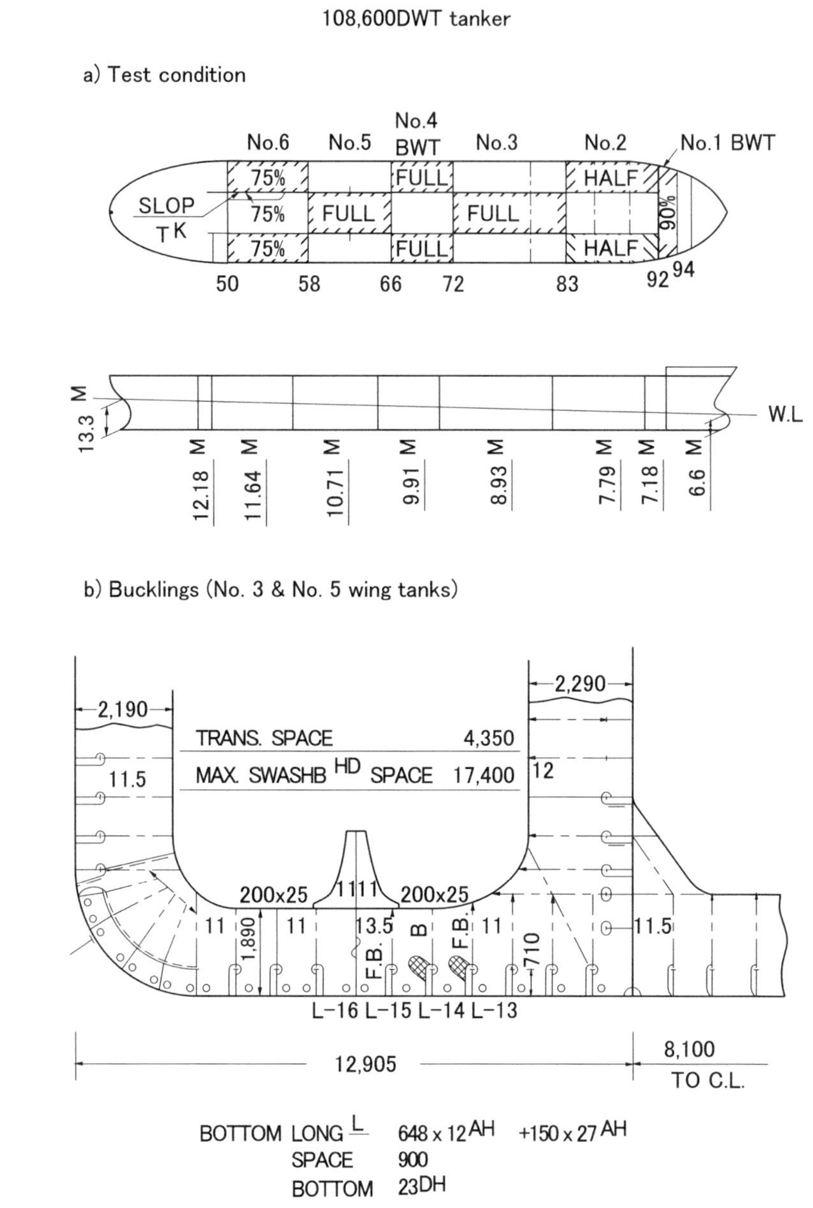

Fig. 10.4 Example of buckling on bottom transverse

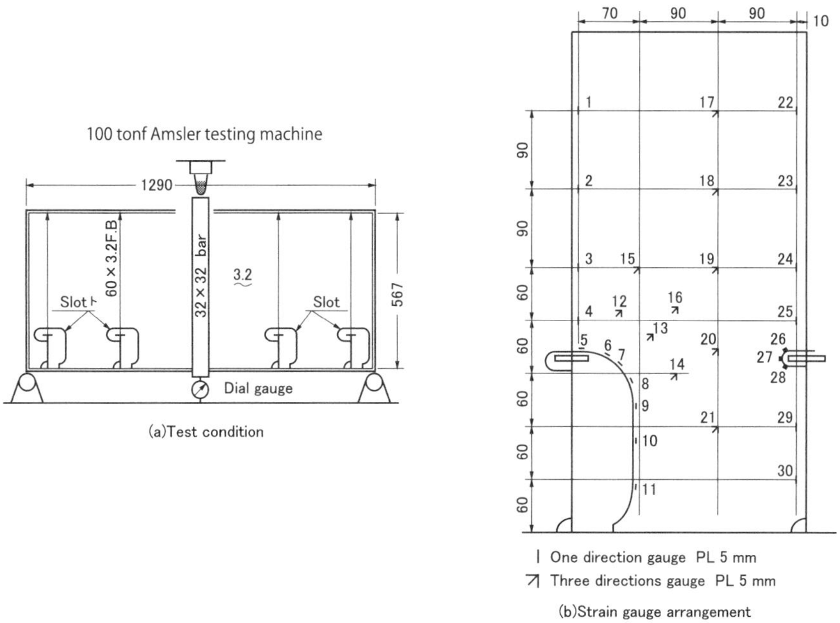

Fig. 10.5 Buckling tests of panels with slots

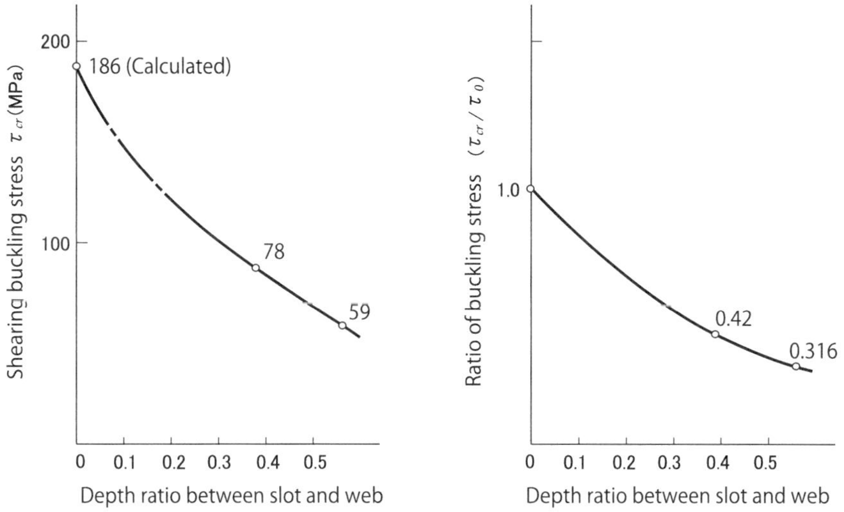

Fig. 10.6 Buckling stress of web plate and size of slot

by the full line. The design standard without a slot is as follows:

$$\frac{\tau_{cr}}{\tau_w} \geq 2.0 \qquad (10.3.1)$$

When there is a slot, the standard is divided into two parts as follows:

$$\frac{\tau_{cr}}{\tau_w} \geq \frac{20}{9}\lambda + \frac{14}{9} \quad \text{for } 0.2 \leq \lambda \leq 0.3 \qquad (10.3.2)$$

$$\frac{\tau_{cr}}{\tau_w} \geq \frac{400}{63}\lambda + \frac{20}{63} \quad \text{for } 0.3 \leq \lambda \leq 0.4 \qquad (10.3.3)$$

where λ is the ratio of slot depth to web depth.

It is clear from Fig. 10.7 that the design standards shown by Eqs. (10.3.1)–(10.3.3) can avoid the buckling due to the shear forces, shear buckling, in the web plate.

Nowadays with computers well developed, buckling on the transverse web is checked by computer. However, a paper model is also useful to assess buckling conveniently. To establish suitable countermeasures for the many buckling damages on transverse webs as well as cracks which appeared around the end of 1960s, after finishing the detailed design of the midship section of a 150,000 DWT tanker, a paper model (1/50) of the midship section was prepared. Buckling was investigated by applying possible loads by finger pressure and applying reinforcement against the

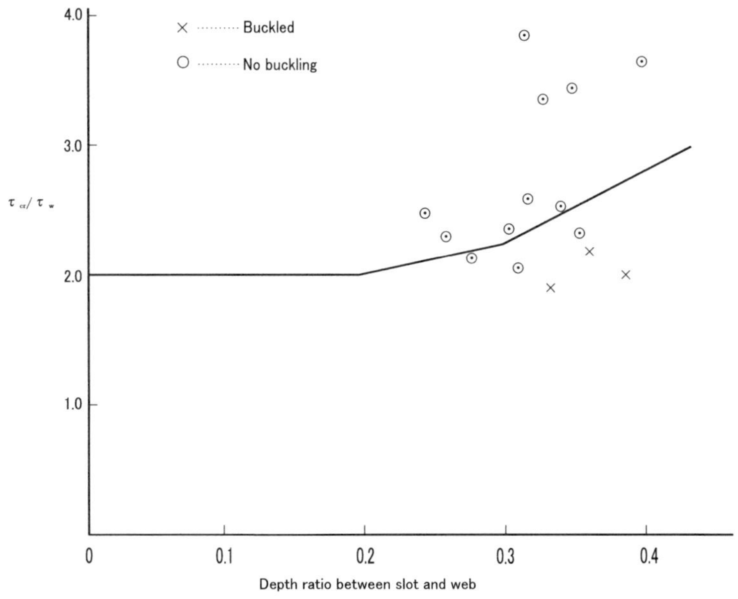

Fig. 10.7 τ_{cr}/τ_w on bottom transverse web

observed buckling pattern. In this way the designer can apply any load he/she wants and observe the buckling pattern for that load.

Paper models were prepared as shown in Fig. 10.8 for the study of buckling of the bottom transverse shown in Fig. 10.4. Corrosion control was applied in this ship and structures in the tank were coated by epoxy paint. In this case it was important to make clear whether any buckling was caused by shearing force or bottom pressure. In the former case reinforcement was necessary only to the high shearing stress area, but in the latter case it had to be applied to all of the bottom area. Reinforcement by means of welding will cause damage to the epoxy paint and it will take much work and expense to recover the paint damage.

A shearing force and bottom pressure were applied by finger pressure to the paper models. The buckling patterns were clearly different. Figure 10.8a shows the buckling pattern by a shearing force which runs in a diagonal direction. Figure 10.8b shows the buckling pattern caused by bottom pressure which happened locally close to the bottom. By this confirmation, the actual buckling pattern is quite similar to that caused by shearing force, and reinforcement was applied only to the high shearing stress area, saving much labor and much expense.

Figure 10.8c showing no buckling was caused by a shearing force which was opposite to the shearing force shown in Fig. 10.8a. In this case a tensile stress was generated at the free edge of the slot which cannot cause any buckling, similarly to the model test results shown in Fig. 10.5.

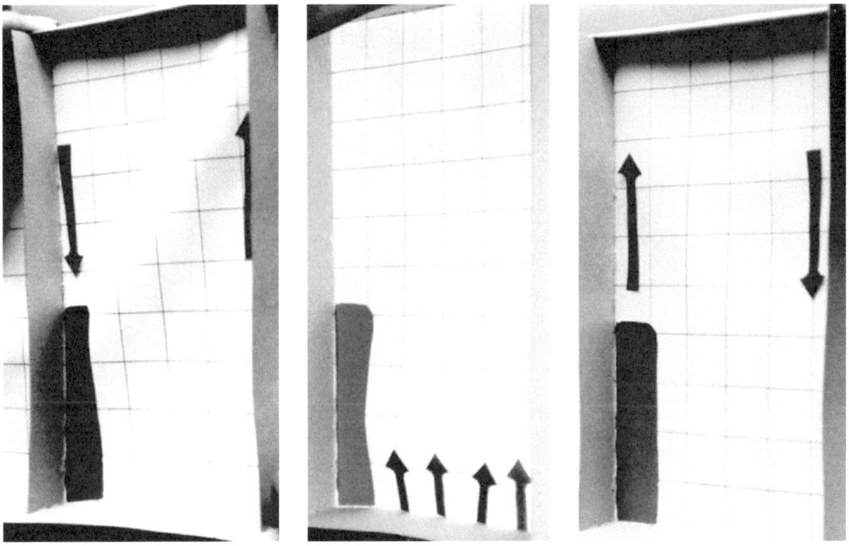

(a) Buckling due to shearing force

(b) Buckling due to vertical force of bottom pressure

(c) No buckling by shearing force opposite to (a)

Fig. 10.8 Paper craft model of transverse web buckling

10.4 Buckling Caused by Concentrated Loads

Figure 10.9 shows buckling of a web plate of a vertical web on a longitudinal bulkhead. In this case no web stiffener was fitted where the buckling occurred.

In smaller ships, a web stiffener is provided alternately or on every third longitudinal stiffener. Smaller ship practices are no longer effective for larger ships. The

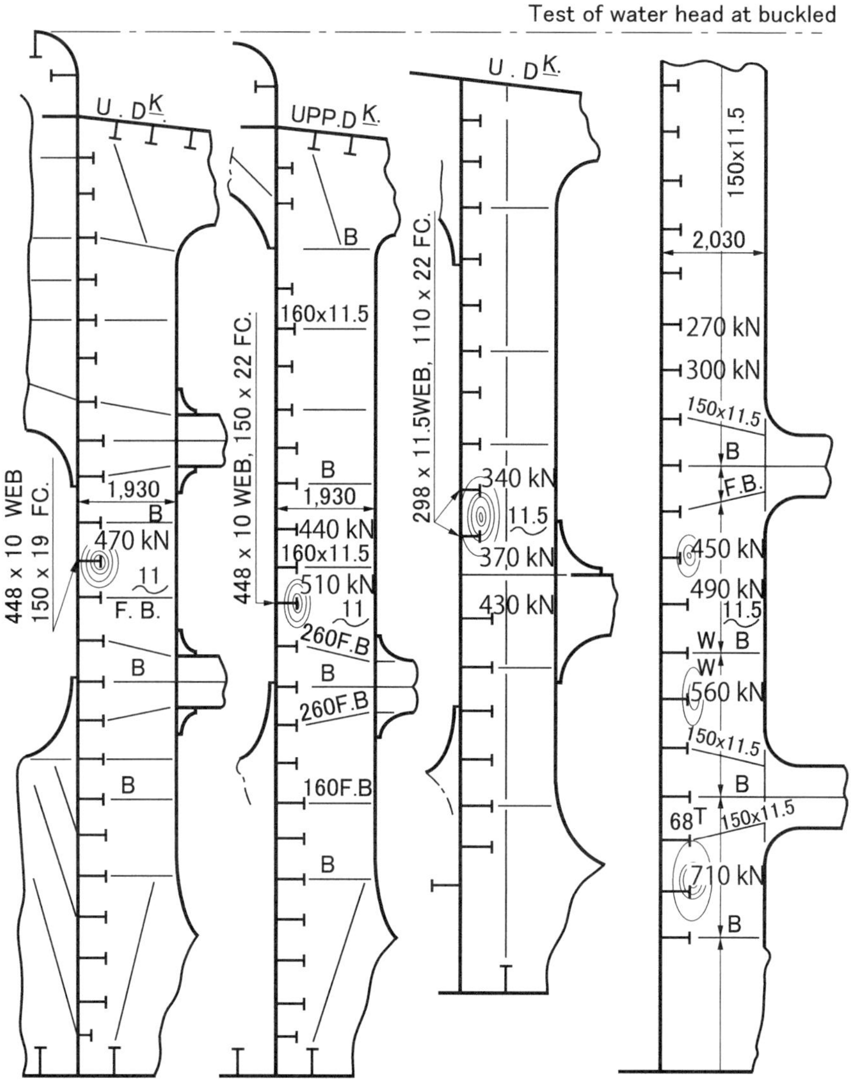

Fig. 10.9 Buckling of web due to concentrated load

load transmitted from the longitudinal stiffener is big enough to cause buckling of the web where no web stiffener is fitted.

With the water head at the water pressure test (2.45 m above upper deck) the load transmitted from the longitudinal stiffener to the web plate is calculated and shown in Fig. 10.9. From this, it can be said that in cases of 11 mm or 11.5 mm thickness of the web plate the following values of the loads can cause buckling:

(1) alternately arranged web stiffener—440 kN.
(2) every third arranged web stiffener—300 kN.

The critical buckling stress by compression σ_{cr} is expressed as follows.

$$\sigma_{cr} = K\left(\frac{t}{b}\right)^2$$

$$\therefore K = \sigma_{cr}\left(\frac{b}{t}\right)^2 = \frac{F}{bt}\left(\frac{b}{t}\right)^2 = \frac{Fb}{t^3} \tag{10.4.1}$$

where

t: thickness of web plate

b: breadth of edge on which load is applied (i.e., space of longitudinal stiffener)

In case (1) for the alternately arranged web stiffener, $F = 440$ kN, $t = 11$ mm and assuming $b = 800$ mm, K is calculated as follows.

$$K = \frac{440\,\text{kN} \times 800\,\text{mm}}{(11\,\text{mm})^3} = 264\,\text{kN/mm}^2$$

$$\therefore \sigma_{cr} = 264\left(\frac{t}{b}\right)^2\ \text{kN/mm}^2 \tag{10.4.2}$$

In case (2) for web stiffeners on every third longitudinals, $F = 300\,\text{kN}$, $t = 11.5\,\text{mm}$ and assuming $b = 800\,\text{mm}$, K is calculated as follows:

$$K = \frac{300\,\text{kN} \times 800\,\text{mm}}{(11.5\,\text{mm})^3} = 158\,\text{kN/mm}^2$$

$$\therefore \sigma_{cr} = 158\left(\frac{t}{b}\right)^2\ \text{kN/mm}^2 \tag{10.4.3}$$

$\sigma_{cr}/\sigma_w \geq 1.5$ will be the design criteria then the necessary web plate thickness is obtained from Eqs. (10.4.2) and (10.4.3).

Hull structure designers often apply the equation of critical buckling stress of uniform loading to the buckling caused by a concentrated load. This is not a correct way because the buckling pattern caused by a uniform load and a concentrated load is quite different. Only in cases where the buckling pattern is similar is it correct to apply this equation. It is very important for hull structure designers to keep in mind that a critical buckling stress corresponds to each buckling mode in the same way as a natural frequency corresponds to each vibration mode.

Bigger loads from a longitudinal stiffener sometimes cause buckling at the root of web stiffener. To establish the maximum load which can be supported by a web stiffener, an experimental test was carried out using a 1/4 scale model of the structure as shown in Fig. 10.10. In the model test the web stiffener started to buckle at point C at the load of 98 kN which corresponds to 1570 kN (98 kN × 4^2) in the actual structure.

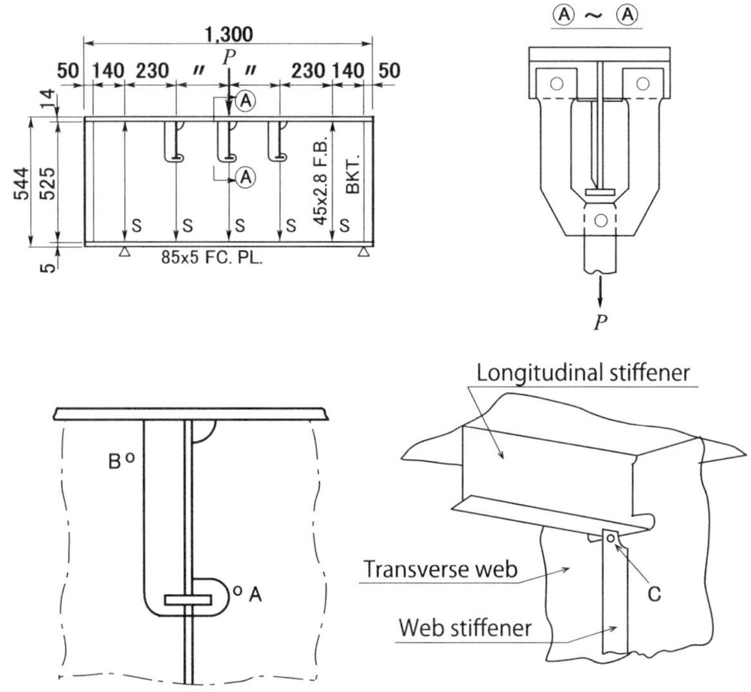

Buckling load			Ultimate load
A	B	C	
15^T.000	16^T.600	10^T.000	18^T.500

Fig. 10.10 Buckling test of web stiffener root

10.5 Cracks Around Slot

10.5.1 Cracks of First Generation

Buckling is caused by compression and a crack is caused by tension. In a discussion of the strength of a structure it is very important to make clear the magnitude and the direction of the force to be applied on the structure. Most of the cracks which occurred in a web plate were around a slot, which is a cutout in the web plate to allow the passage of a longitudinal stiffener at its intersection with the web plate, as shown in Fig. 10.11. The load, from the water pressure, is supported firstly by the longitudinal stiffener then transmitted to the web plate through the welded connection between the longitudinal stiffener and the web plate. Many cracks happened around the welded connection between the longitudinal stiffener and the web plate where there is a neck in the load transmission. In addition to the load transmission explained above, the shearing force on the transverse web plate causes the deformation of the slot cutout, generating additional stress at the root of the web stiffener.

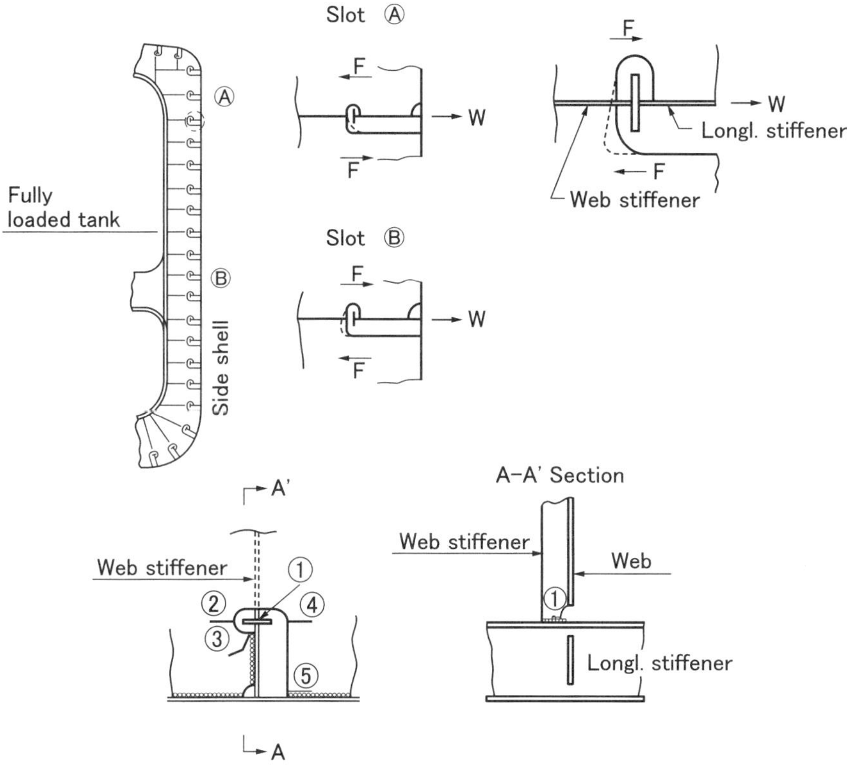

Fig. 10.11 Detail of slot and example of crack around slot

The points where cracks occur are shown in Fig. 10.11 as No. 1, No. 2, No. 3, No. 4 and No. 5. No. 1 crack, which happened at the root of the web stiffener, is the most common and sometimes is accompanied by other cracks. It is considered that the No. 1 crack happens first then other cracks follow.

As a result of leaving these cracks they grow and become extensive, for example No. 2 and No. 4 cracks cause the loss of the connection between web plate and the side shell plate as shown in Fig. 10.12. No. 5 crack goes into the side shell plate which causes oil leakage in the vicinity of a tank. The pattern of this crack on side shell plate looks like a flying bird spreading her wings which is why it is called the flying bird crack as shown in Fig. 10.12b.

As it will take a while for these cracks to grow to dangerous ones, sound inspection and early repair can prevent major damage. However from the maintenance viewpoint it is important to prevent such cracks.

To prevent cracking around a slot it is most important to prevent crack No. 1 which happens at the connection between a web stiffener and a longitudinal frame. For the prevention of crack No. 1, empirical formulae including corresponding allowable stresses were proposed just after the damage experiences in 1960s, which is explained in Sect. 10.5.2 [67]. Then in Sect. 10.5.3, we explain more comprehensive formulation of the exerted stresses based on the load share among the web stiffener and the transverse web attachment [68].

10.5.2 Empirical Formulae Established After the First Generation Cracks [67]

The water head of the inside pressure is $D + 2.45$ m where D is the ship's depth, and the outside pressure is $0.32d_f$ where d_f is full draft. For the longitudinal and the transverse bulkheads the water head is $D + 2.45$ m assuming the adjacent tank is empty.

Then the load W transmitted from the longitudinal stiffener to the web plate is expressed by the following equation:

$$W = \rho \times S \times l \times h_i \tag{10.5.1}$$

where

ρ: specific gravity of the liquid.

S, l: spaces of longitudinal and web frame.

h_i: water head at the longitudinal stiffener.

The shearing force is to be calculated as a both ends fixed girder. The stress σ_{AP} at the bottom of the web stiffener caused by the load W is given by the following equation:

$$\sigma_{AP} = \frac{W}{A_1} \tag{10.5.2}$$

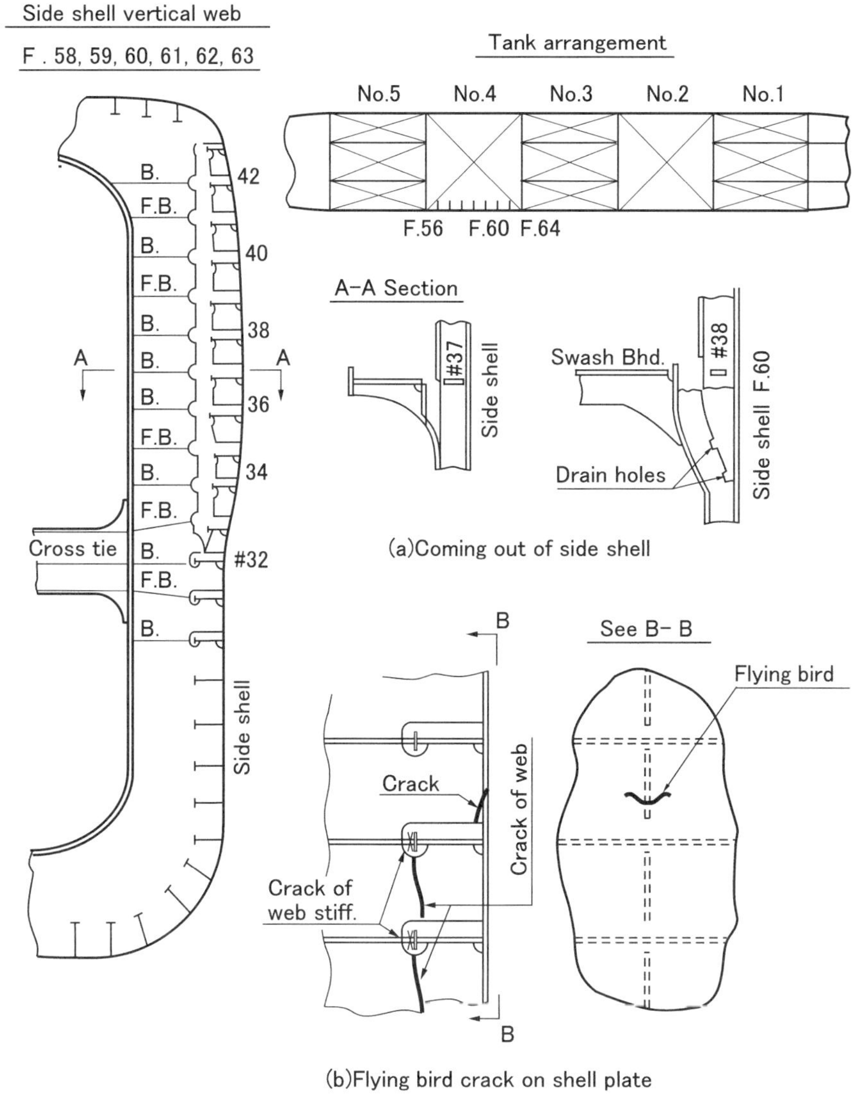

Fig. 10.12 Example of cracks developed around slot

where

$$A_1 = Ct_s b_s + \alpha t_w h_w + \beta t_c h_c$$

$$\alpha = \frac{Gh_e}{El_w}, \beta = \frac{Gh_e}{El_c}$$

$t_s, b_s, t_w, h_w, t_c, h_c, h_e, l_w$ and l_c are shown in Fig. 10.13.

$C = 2$ where back bracket is fitted.
$C = 1$ where no back bracket is fitted.

The stress σ_{AF} at the bottom of the web stiffener caused by the shearing force F is given by the following equation:

$$\sigma_{AF} = \frac{F}{A_2} \cdot \frac{t_w h_w - t_c h_c}{A_1} \tag{10.5.3}$$

where

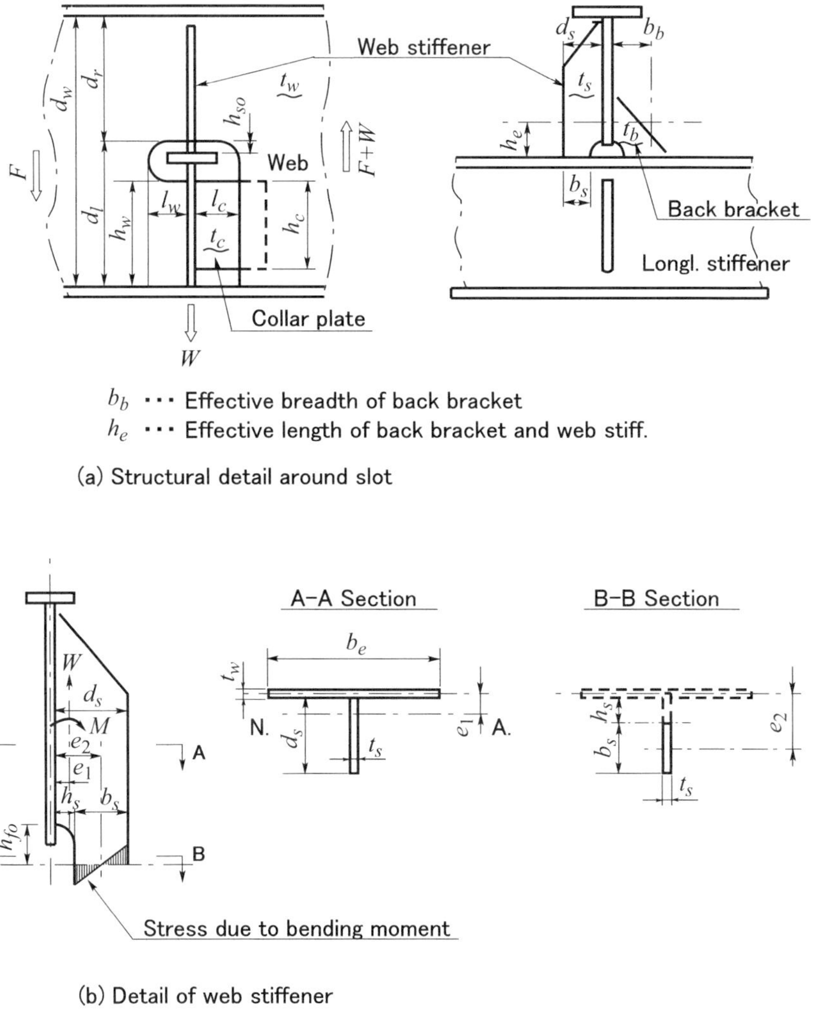

Fig. 10.13 Detail of structure around slot

$$A_2 = (d_w - d_l)t_w + t_c h_c$$

d_w and d_l are shown in Fig. 10.13.

By putting W and F into Eqs. (10.5.2) and (10.5.3), σ_{AP} and σ_{AF} can be obtained simplifying α and β as follows.

h_e is the effective height where the axial force is supported by the web stiffener and is assumed to be at least equal to the breadth b_s of the web stiffener at the bottom. Then $\alpha = \frac{Gb_s}{El_w}$ is obtained. For actual stiffeners α is calculated and the results are as follows:

For a T section stiffener α is 0.39–0.7, say 0.5.

For a L section stiffener α is 0.77–1.9, say 1.0 because of smaller l_w than that of the T section. In the case of a slot closed by a collar plate α is to be 1.0 because of continuous web plate and added rigidity of the shell plate, and β is also simplified to be 1.0 for the same considerations.

Applying the above simplified method on the slots of ships in service, σ_{AP} and σ_{AF} are plotted in Fig. 10.14 from which the following results are observed.

In the case of a web stiffener only, the border line of no crack is $\sigma_{AP} + \sigma_{AF} \leq$ 70 MPa, which is the mean stress at the bottom of the web stiffener, and the maximum stress including bending effect and stress concentration will be 210 MPa.

In the case of a web stiffener and back bracket, the border line of no crack is $\sigma_{AP} + \sigma_{AF} \leq 185$ MPa.

In the case where a collar plate is fitted σ_{AF} is small, less than 10 MPa, which means the shearing force can have very little effect. This makes the markers with a collar plate invisible in the graph. To make them visible, σ_{AF} of such slots is plotted by multiplying a factor of 10 in the graph.

First the calculation method of the stress of the web stiffener at the bottom is established and by applying this method to the slots of many ships in service, some of which cracked and others did not, Fig. 10.14 was prepared. The empirical reliable design is established as the above, that is, firstly an analytical method is decided then by analyzing many data an allowable limit is decided. By applying the empirical reliable design method, a similar level of reliability can be kept for similar conditions.

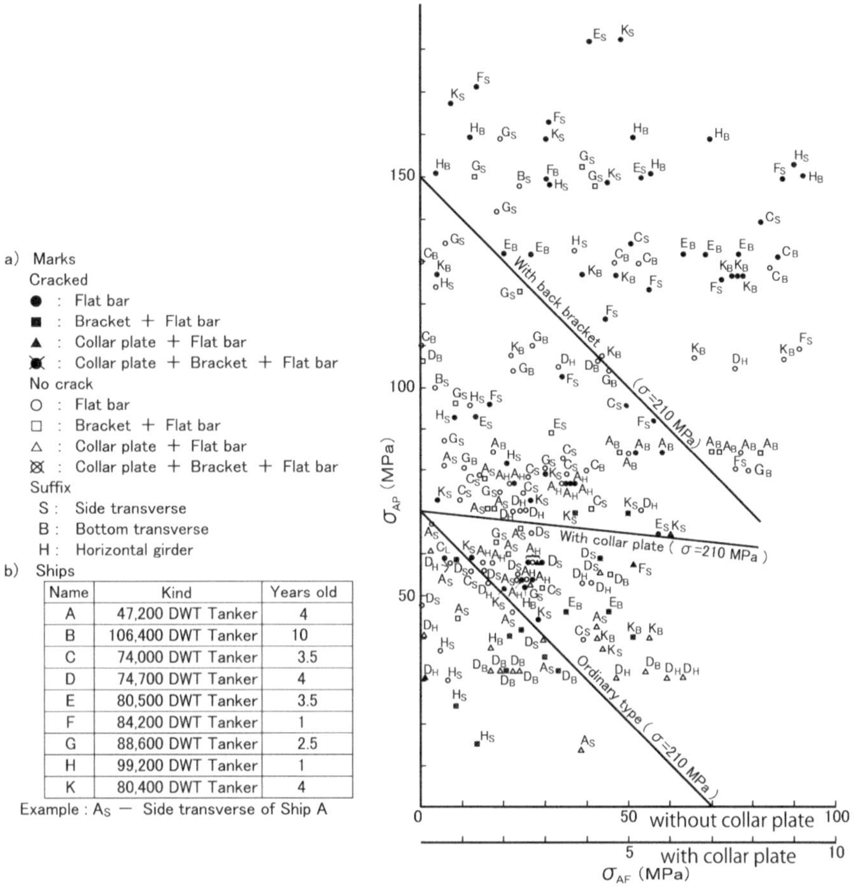

Name	Kind	Years old
A	47,200 DWT Tanker	4
B	106,400 DWT Tanker	10
C	74,000 DWT Tanker	3.5
D	74,700 DWT Tanker	4
E	80,500 DWT Tanker	3.5
F	84,200 DWT Tanker	1
G	88,600 DWT Tanker	2.5
H	99,200 DWT Tanker	1
K	80,400 DWT Tanker	4

Fig. 10.14 How to decide allowable stress for web stiffener root

From Fig. 10.14 the allowable limit, in this case it is the design standard, is as follows.

(a) web stiffener only:

$$\sigma_{AP} + \sigma_{AF} \leq 70\,\text{MPa},\ \text{maximum stress} = 210\ \text{MPa}$$
$$\text{(stress concentration factor 3.0)}$$

(b) web stiffener and back bracket:

$$\sigma_{AP} + \sigma_{AF} \leq\ 150\text{MPa},\ \text{maximum stress} \approx 210\ \text{MPa}$$
$$\text{(stress concentration factor 1.5)}$$

Figure 10.14 shows the allowable limit will be 185 MPa, but to have similar maximum stress as in case of (a), 150 MPa is adopted as a design standard.

(c) web stiffener and collar plate:

$$\sigma_{AP} + \sigma_{AF} \leq 70\text{MPa}, \text{ maximum stress} = 210 \text{ MPa}$$
$$\text{(stress concentration factor 3.0)}$$

In this case $\sigma_{AF} \approx 0$ and σ_{AP} plays a main role.

In the design standard above, when σ_{AP} and σ_{AF} have the same sign, the effect is superposed, however as shown in Fig. 10.11 there is a possibility for σ_{AP} and σ_{AF} to have opposite signs. In this case σ_{AP} and σ_{AF} will cancel each other and the resultant will be smaller than σ_{AP}. This is a simple but important phenomenon, and by deciding the slot direction that σ_{AP} and σ_{AF} have opposite signs, the stress of the web stiffener at the bottom can be reduced.

However the direction of the shearing force is changed by loading conditions, external wave load etc. and the accuracy of the shearing force calculation is poorer than that for a direct load W, then the design standard is decided as follows.

$$\sigma_{AP} + \sigma_{AF} \leq 30 \text{ MPa where } \sigma_{AF} \leq -40 \text{ MPa}$$

$$\sigma_{AP} \leq 70 \text{ MPa where } -40 \text{ MPa} \leq \sigma_{AF} < 0$$

In the actual design, first $\sigma_{AP} + \sigma_{AF}$ is calculated for the case of a web stiffener only for a given load condition. If the value of $\sigma_{AP} + \sigma_{AF}$ exceeds the allowable value a collar plate is to be fitted. Then $\sigma_{AP} + \sigma_{AF}$ is calculated again with a collar plate and if $\sigma_{AP} + \sigma_{AF}$ still exceeds the allowable value a back bracket is fitted. Where $\sigma_{AP} + \sigma_{AF}$ exceeds the allowable value for the case where a web stiffener only is fitted, then to fit a back bracket or to increase the scantling of the web stiffener is also a correct alternative, but to provide a collar plate is better because it is effective not only in reducing $\sigma_{AP} + \sigma_{AF}$ but also in reinforcing the free edge of the slot.

The mechanism of crack generation in Fig. 10.11 shows that where σ_{AF} is negative the stress on the free edge of the slot is positive, that is tension and good for reducing buckling as shown in Fig. 10.8. The direction of the slot which reduces the stress of the web stiffener at the bottom is best in avoiding buckling, killing two birds with one stone. The designer should apply the best direction of the slot to prevent cracking and buckling.

The crack at the bottom of the web stiffener is caused by fatigue. A longer service period causes more fatigue cracks. In Fig. 10.15 the relationship between the service period and the allowable stress is shown which is obtained from ships of different service periods. Figure 10.15 shows that a ship with a service period 3.5–4.0 years has no crack where $\sigma_{AP} + \sigma_{AF} \leq 70$ MPa (maximum stress 210 MPa) and the ship with a service period of 1.0 year has no crack where $\sigma_{AP} + \sigma_{AF} \leq 100$ MPa (maximum stress 300 MPa).

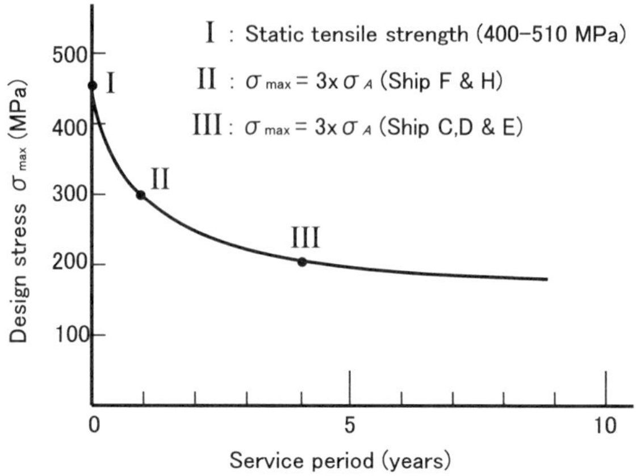

Fig. 10.15 Relation between service period and allowable stress

The crack at the bottom of a web stiffener is generated generally in base metal not in deposited metal.

10.5.3 Comprehensive Formulation Considering All the Load Paths

(1) *Share of load supported by three load paths*: Here we explain more comprehensive formulation to determine the stress at the root of the web stiffener through detailed modeling of the rigidity of all the load paths around the slot cutout structure [68].

The total load W transmitted from the longitudinal stiffener to the transverse web is supported by three load paths: the web stiffener, one side attachment of the transverse web to the web of the longitudinal, and the other side attachment of the transverse web to the web of the longitudinal. When the load shares of each load path are denoted by P_f, S_w, and S_c, respectively as shown in Fig. 10.16, they are obtained as follows:

$$\begin{pmatrix} P_f \\ S_w \\ S_c \end{pmatrix} = \begin{pmatrix} \beta_1 - \beta_3 \\ \alpha_w \beta_1 \\ \alpha_c(\beta_1 - \beta_2) \end{pmatrix} W \tag{10.5.4}$$

The coefficients α_w and α_c are the ratio of the shear spring constants of each side attachment of the transverse web to the normal spring constant k_p of the web stiffener as shown in Fig. 10.16. They are obtained as follows:

$$\alpha_w = \frac{1}{2(1+\nu)} \cdot \frac{h_w t_w}{b_s t_s} \cdot \frac{h_e}{l_w} \tag{10.5.5}$$

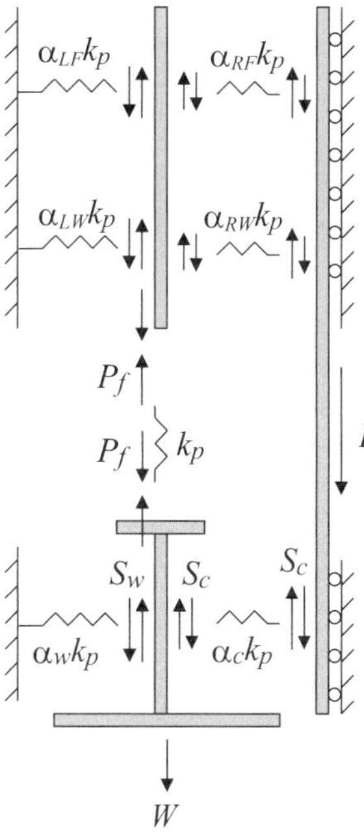

Fig. 10.16 Model of load paths and spring constants

$$\alpha_c = \frac{1}{2(1+\nu)} \cdot \frac{h_c t_c}{b_s t_s} \cdot \frac{h_e}{l_c} \tag{10.5.6}$$

where all the parameters are defined in Fig. 10.13. The effective length of the web stiffener h_e is determined more rigorously here in the following equation based on the beam theory because this parameter is very important to estimate the share of load supported by each load path accurately.

$$h_e = h_{fo} + 2.58\sqrt{d_s t_s} \tag{10.5.7}$$

The coefficients β_1, β_2, and β_3 are as follows:

$$\begin{pmatrix} \beta_1 \\ \beta_2 \\ \beta_3 \end{pmatrix} = \frac{1}{|A|} \begin{pmatrix} (\alpha_c + \alpha_{RF})\alpha_1 - \alpha_{RF}{}^2 & \alpha_c\alpha_1 + \alpha_{RF} & \alpha_c\alpha_{RF} + \alpha_c + \alpha_{RF} \\ \alpha_c\alpha_1 + \alpha_{RF} & \alpha_2\alpha_1 - 1 & \alpha_{RF}\alpha_2 + \alpha_c \\ \alpha_c\alpha_{RF} + \alpha_c + \alpha_{RF} & \alpha_{RF}\alpha_2 + \alpha_c & \alpha_2(\alpha_c + \alpha_{RF}) - \alpha_c{}^2 \end{pmatrix} \cdot \begin{pmatrix} 1 \\ n - \frac{1}{2} + \frac{\alpha_{RW} - \alpha_{RF}}{\alpha_2(1+\alpha_{LW}+\alpha_{RW})-1} \\ \frac{-\alpha_{LW}+\alpha_{LF}-\alpha_{RW}+\alpha_{RF}}{\alpha_2(1+\alpha_{LW}+\alpha_{RW})-1} \end{pmatrix} \tag{10.5.8}$$

where

$$|A| = \alpha_1\alpha_2(\alpha_c + \alpha_{RF}) - 2\alpha_c\alpha_{RF} - \alpha_c - \alpha_{RF} - \alpha_1{\alpha_c}^2 - \alpha_2{\alpha_{RF}}^2$$

$$\alpha_1 = 1 + \alpha_{LF} + \alpha_{RF}$$

$$\alpha_2 = 1 + \alpha_w + \alpha_c.$$

The coefficients α_{LF}, α_{RF}, α_{LW}, and α_{RW} are the ratio of the shear spring constants of each side transverse web in the upper part of the slot cutout to the normal spring constant k_p of the web stiffener as shown in Fig. 10.16. The load transmitted through this portion of the transverse web is divided into two components; one is the load uniformly distributed along the length d_r which corresponds to the transmission of the shear force acting on the transverse web, and the other is the load transmitted from the web stiffener to the transverse web, which flows into the transverse web through the limited area of the web stiffener in the vicinity of the longitudinal stiffener. The former is denoted by the subscript "F" and the latter is denoted by "W". These coefficients are obtained as follows:

$$\alpha_{LF} = \frac{1}{2(1+\nu)} \cdot \frac{d_r t_w}{b_s t_s} \cdot \frac{h_e}{l_w} \tag{10.5.9}$$

$$\alpha_{RF} = \frac{1}{2(1+\nu)} \cdot \frac{d_r t_w}{b_s t_s} \cdot \frac{h_e}{l_c} \tag{10.5.10}$$

$$\alpha_{LW} = \frac{h_e t_w}{b_s t_s}\left[\frac{h_e - h_{so}}{2(1+\nu)l_w} + \frac{l_w}{2(d_r + h_{so} - h_e)}\right] \tag{10.5.11}$$

$$\alpha_{RW} = \frac{h_e t_w}{b_s t_s}\left[\frac{h_e - h_{so}}{2(1+\nu)l_c} + \frac{b_c}{2(d_r + h_{so} - h_e)}\right] \tag{10.5.12}$$

The parameter n in Eq. (10.5.8) is the number of longitudinal stiffeners between the subject slot position and the location where the shear stress on the transverse web is equal to zero. This parameter represents the magnitude of the shearing force F on the transverse web at the subject slot position through the following relationship:

$$F = \left(n - \frac{1}{2}\right)W \tag{10.5.13}$$

The load share ratio of the web stiffener P_f/W according to Eq. (10.5.4) was compared with the results of an experiment of a scaled model [67] and the results of finite element analysis, whose results are plotted in Fig. 10.17 taking the position of the slot (n) in the horizontal axis. The scantling and dimensions of the scaled model is shown in Fig. 10.18. From Fig. 10.17, we can find that the results of Eq. (10.5.4) agree well with the results of finite element analysis and that the experimental results also follow this tendency. Observing the overall tendency, we can find that the load due to the shearing force on the transverse web superimposes on the direct load

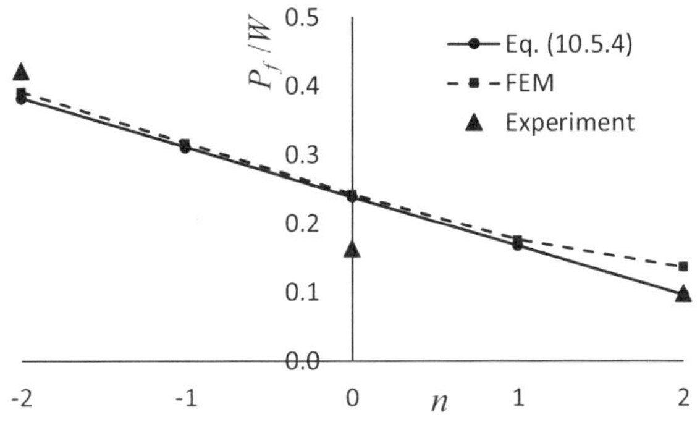

Fig. 10.17 Comparison of load share ratio of the web stiffener

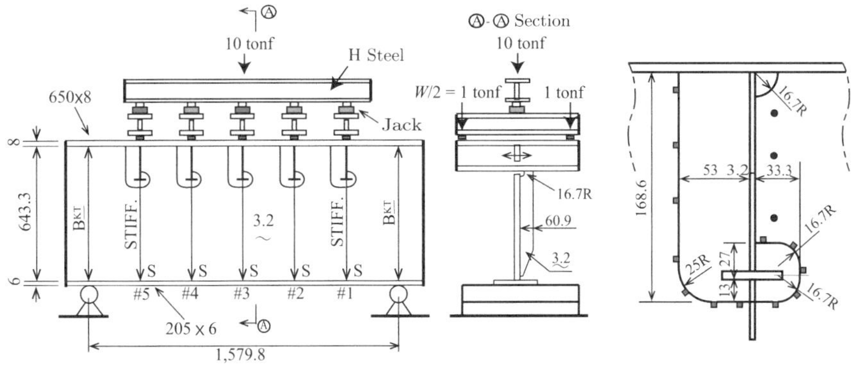

Fig. 10.18 Experimental set up and dimensions of scaled model [67]

from the longitudinal stiffener in the negative n side, and conversely cancels it in the positive n side.

Then, let us investigate the effect of various types of reinforcement using Eq. (10.5.4). The base case is the structural arrangement shown in Fig. 10.18, and the following three reinforcement cases are considered as illustrated in Fig. 10.19.

- Case 1: Increase the web stiffener scantling ($d_s \times t_s$ are made twice as large proportionately).
- Case 2: Fit a collar plate (collar plate height h_c is about 65% of h_w, $t_c = t_w$).
- Case 3: Change the slot type to the slit connection (the right-hand side attachment of the transverse web in Fig. 10.19 (Case 3) is completely copied to the left-hand side, i.e., l_c and h_c is changed to be equal to l_w and h_w).

The results are plotted in Fig. 10.20 taking the position of the slot in the horizontal axis. Figure 10.20a shows the load share ratio supported by the web stiffener P_f/W, and Fig. 10.20b shows the nominal stress $\sigma_f = P_f/b_s t_s$ of each case divided by the nominal stress $\sigma_{f0} = W/b_s t_s$ of the base case. Let us focus on the result at $n = -5$,

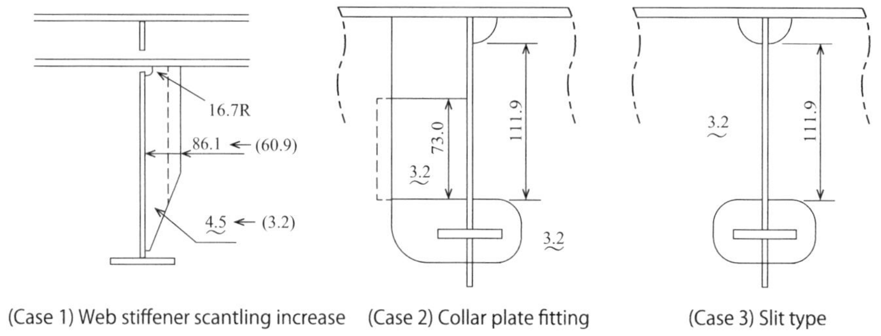

Fig. 10.19 Slot reinforcement cases

where P_f/W is large due to the superposition of the effects from the shearing force on the transverse web and the direct force from the longitudinal stiffener. Observing the result of Case 2, the scantling increase of the web stiffener actually works and the stress reduced to 70% of the base case as shown in Fig. 10.20b. However, this reduction is less than the value simply calculated from the doubled sectional area of the web stiffener. This is because the larger rigidity of the web stiffener has gathered more load and P_f/W has increased by 36% because of this reinforcement, as we can observe in Fig. 10.20a.

When a collar plate is fitted (Case 2), the load reduces remarkably to 41% of the base case thanks to the mitigation of the effect from the shearing force on the transverse web, because the shearing forces on both the sides of the slot cutout balances to some extent. Application of completely symmetric slit cutout structure (Case 3) can be found to be further effective. The effect of the shearing force on the transverse web disappears completely, reducing the load on the web stiffener to 21% of the base case.

Another effective countermeasure is to change the direction of the slot cutout so that the load due to the shearing force on the transverse web and the direct load from the longitudinal stiffener cancel each other. In this case, the value at $n = 5$ in Fig. 10.20 shows that the absolute value of the load reduces to 20% of the base case. This is the most cost effective way, needing no additional members nor scantling increase, but its feasibility may depend on the construction processes and facilities.

(2) *Bending of web stiffener due to asymmetric arrangement*: Web stiffeners are usually attached on one side of the transverse web, and due to this asymmetric arrangement, bending moment is exerted on the web stiffener causing non-uniform stress distribution along the root of the web stiffener attached to the face plate of the longitudinal as shown in Fig. 10.13b and as a result, much larger stress at the heel of the web stiffener. In Sect. 10.5.2, a stress concentration factor of 3.0 was assumed including this bending effect, but in this section this phenomenon is treated a little more consistently using beam theory.

We model the web stiffener as shown in Fig. 10.21. Average axial stress σ_f at the root of the web stiffener is:

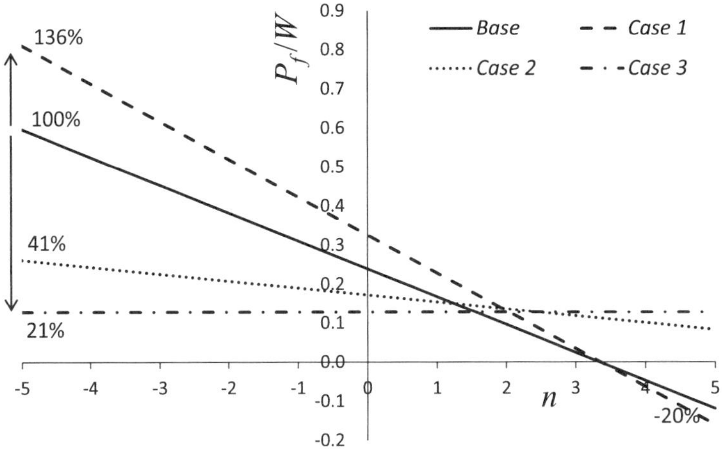

(a) Load share ratio of web stiffener

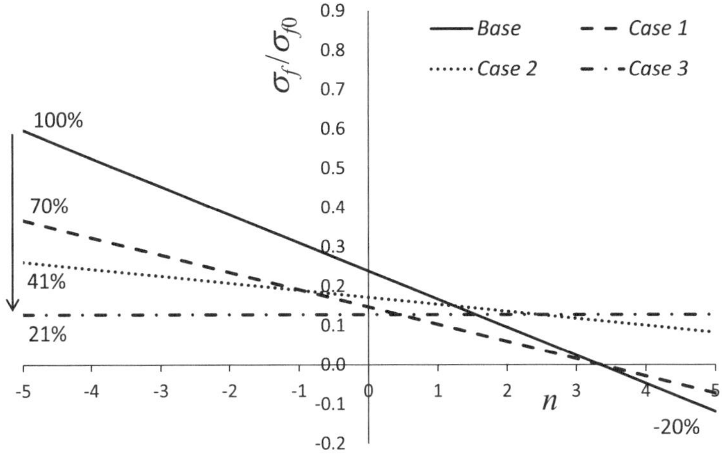

(b) Nominal stress ratio at the root of web stiffener

Fig. 10.20 Effect of reinforcement on the load share ratio of web stiffener

$$\sigma_f = \frac{P_f}{b_s t_s} \tag{10.5.14}$$

The bending moment M caused by the couple of forces of P_f and the reaction force from the transverse web is:

$$M = P_f \cdot \frac{2d_s - b_s}{2} \tag{10.5.15}$$

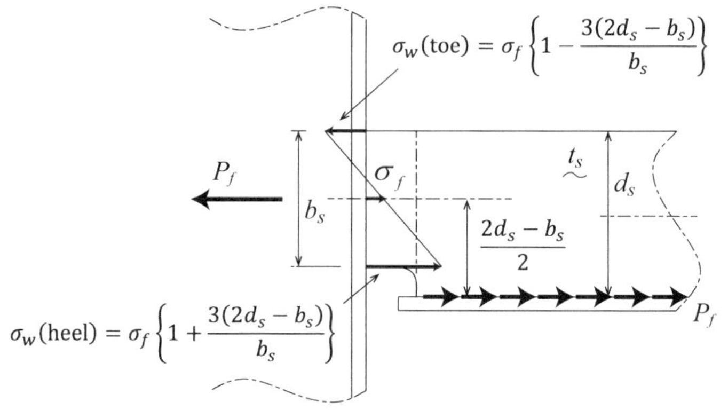

Fig. 10.21 Theoretical model of web stiffener bending

Then, the bending stress σ_b is obtained as follows dividing Eq. (10.5.15) by the section modulus at the root.

$$\sigma_b = P_f \cdot \frac{2d_s - b_s}{2} \cdot \frac{6}{t_s {b_s}^2} = \sigma_f \cdot \frac{3(2d_s - b_s)}{b_s} \tag{10.5.16}$$

Therefore, the total stress σ_w at the toe and heel and toe of the web stiffener is:

$$\sigma_w = \sigma_f \pm \sigma_b = \sigma_f \left\{ 1 \pm \frac{3(2d_s - b_s)}{b_s} \right\} \tag{10.5.17}$$

From Eq. (10.5.17), we can see that when there is no scallop at the heel of the web stiffener ($d_s = b_s$), $\sigma_w = 4\sigma_f$ and $-2\sigma_f$, that is, the total stress at the heel of the web stiffener is four times larger than the average axial stress. When there is a scallop, the total stress becomes even larger.

This result is compared with the results of finite element analysis and the experiment in Fig. 10.22. There are five strain gauges in the experiment and four elements in the finite element analysis along the attachment of the web stiffener to the longitudinal stiffener. Therefore, five or four markers are shown for each in Fig. 10.22. In general, we can observe fair agreement, but a little steeper inclination of Eq. (10.5.17) compared with the results of the finite element analysis and the experiment. This conservative deviation may be attributed to the assumption that the web stiffener is completely free to rotate. The peak stresses at the heel are observed to be large in the results of the finite element analysis and the experiment, which are considered to be caused by the stress concentration around the scallop.

In a double hull structure, the rotation of the web stiffener is effectively restricted and the effect of the bending of the web stiffener is mitigated as shown in Fig. 10.23. In the range of typical double hull structural arrangement, this reduction in the bending effect is about 20–30% compared with that of the single hull structure [69]. In fact,

Fig. 10.22 Comparison of stress distribution along the root of web stiffener

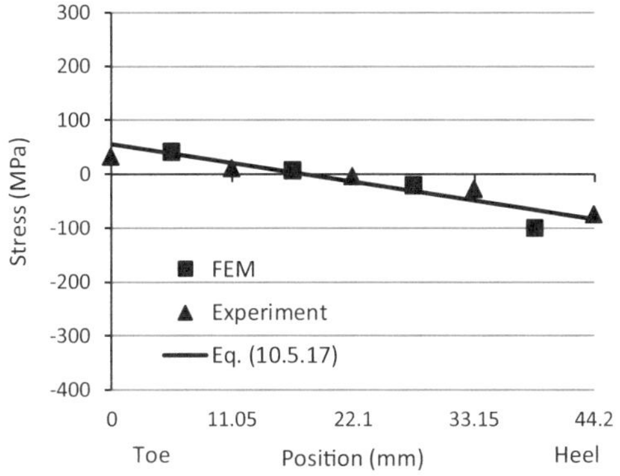

(a) Longitudinal #1

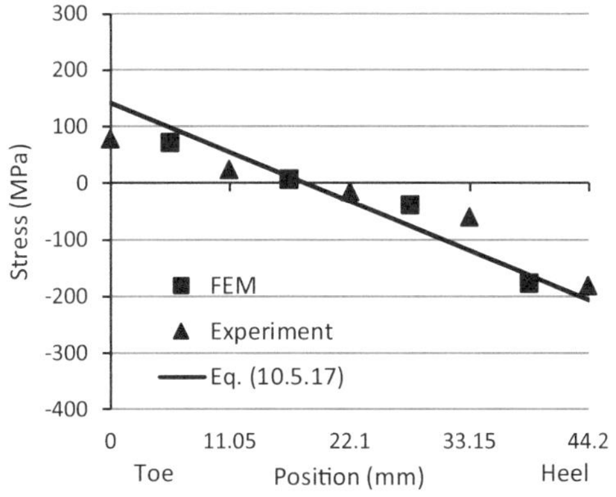

(b) Longitudinal #3

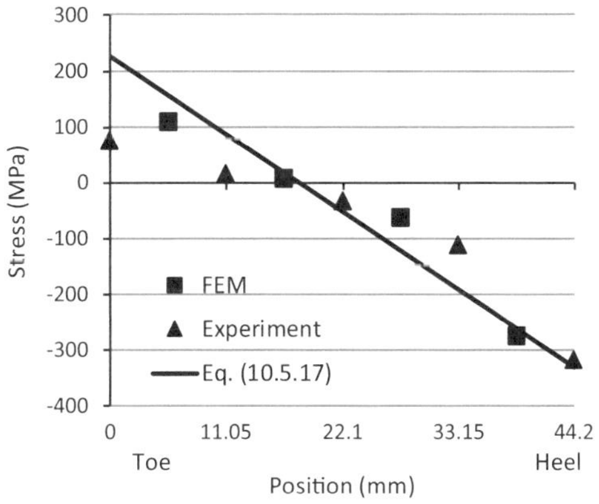

(c) Longitudinal #5

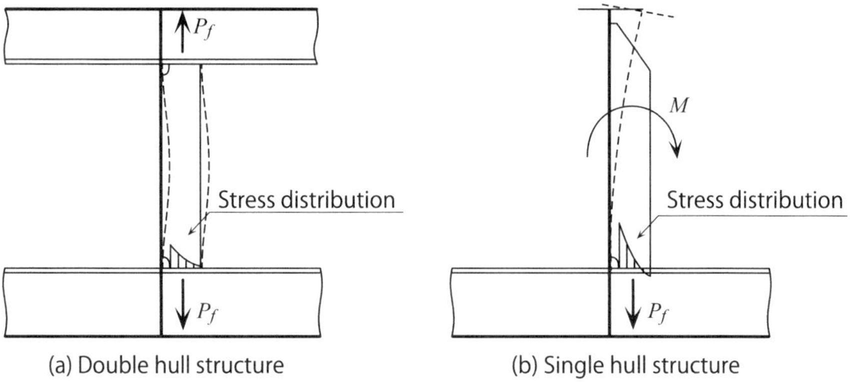

Fig. 10.23 Stress distribution along web stiffener root in double hull and single hull structure

cracks were seldom reported around the slot cutouts in the double bottom floors of bulk carriers, ore carriers and other cargo ships.

In the current major classification societies' rules on slot strength, the effect of the shearing force on the transverse web is not considered explicitly. Therefore, when the effect of the shearing force on the transverse web cancels the effect of the direct load from the longitudinal stiffener, it may result in too conservative design. Conversely, even if the rules are satisfied, the designer should pay special attention to the slot strength when the slot is located in a single hull structure where the shearing force on the transverse web is large and the slot cutout is directed so that the shearing force on the transverse web and the direct load from the longitudinal stiffener superimpose on each other. Fitting a collar plate or other countermeasures should be considered carefully, for example, at the circled location in Fig. 10.24.

10.5.4 Cracks of Second Generation Propagating into Longitudinals

After the 1990s, new type of cracks began to be observed around slot cutouts of side longitudinals. These new cracks are characterized by propagation into the longitudinals as shown in Fig. 10.25 [70], while the first-generation cracks propagated through the welding connection between the web stiffener and the longitudinal. Many of the new single-hull VLCC's in service for 2–5 years between the Persian Gulf and Japan suffered these cracks. Further characteristics of these cracks were observed as follows:

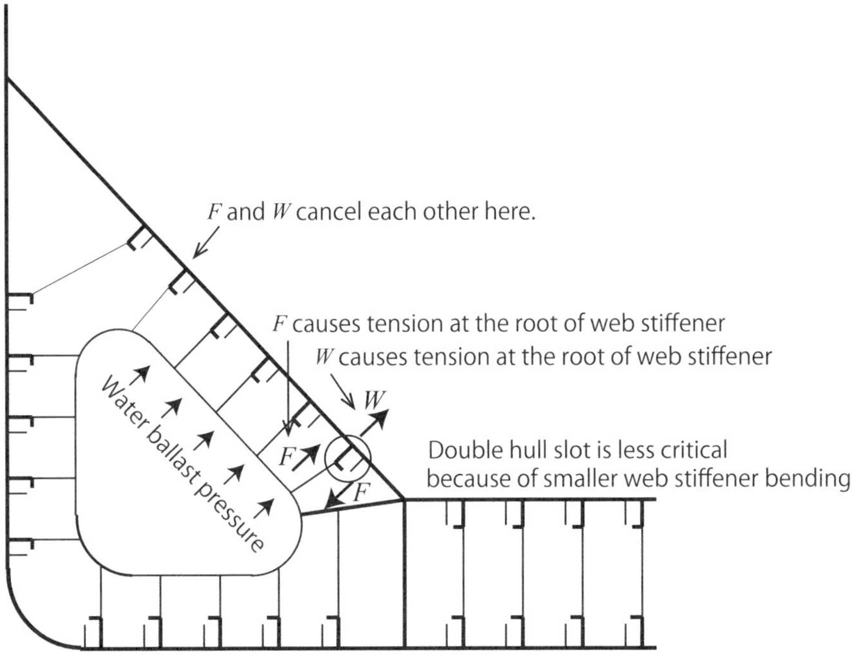

Fig. 10.24 Slot structure to be examined carefully

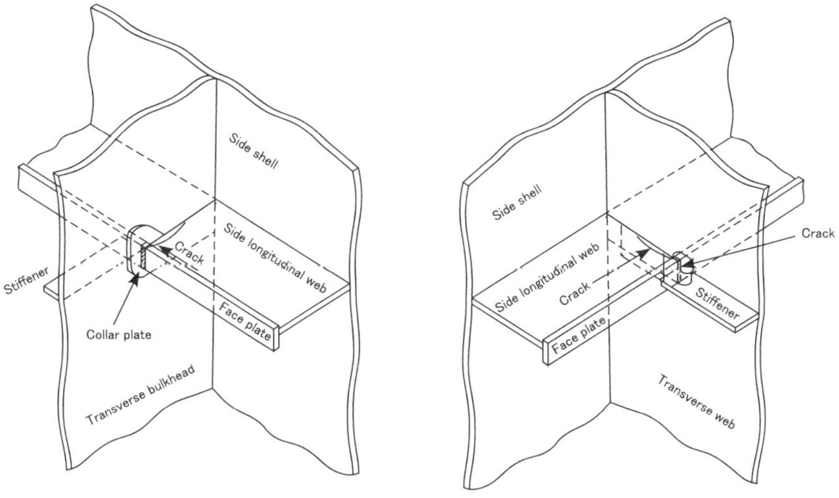

Fig. 10.25 Fatigue crack propagating into longitudinals

(a) Many damages were distributed between 2–5 m below the full load draft.
(b) More damages were found in way of transverse bulkheads than ordinary transverse web.
(c) Damages were more dominant in cargo oil tanks (No. 1, 3, 5 tanks) than in water ballast tanks (No. 2, 4 tanks).
(d) The number of damages were much larger in the starboard side than in the port side.
(e) Cracks occurred only in L-shaped longitudinals, not in T-shaped longitudinals.

Comprehensive study was carried out [71] and the following facts were found.

The most essential cause of the damages was wider application of higher tensile steel of higher specified yield stress such as 355–390 N/mm^2 to the side longitudinals. The scantling of the longitudinals was determined by the rule formulae of the classification societies, considering only yielding criteria. The application of higher tensile steel resulted in smaller scantling and increased stresses on the longitudinals with regard to both static and fluctuating stress. Wave variable pressure is relatively large near the waterline level, and become smaller toward the bottom. On the other hand, static pressure is larger at the bottom, leading to larger scantling toward the bottom. The result was the cracks of the side longitudinals near the waterline level, where the component of the fluctuating wave pressure is relatively large. This explains the observation (a) stated above.

Wider application of higher tensile steel even for transverse webs caused larger relative deflection between the transverse bulkhead and adjacent transverse webs. Additional bending moment on the side longitudinals due to this forced displacement accelerated the crack damage, explaining the reason why the cracks were dominant in way of transverse bulkheads (observation (b)).

Another factor, which caused these cracks, was the L type section usually applied to side longitudinals (observation (e)). In cargo oil tanks, accumulated sludge on the side longitudinals is a problem, and L-shaped side longitudinals were considered to be beneficial in this regard, and were widely used. However, additional longitudinal stress is exerted in the case of an L section, as explained in Sect. 8.8. This phenomenon is considered to have accelerated the crack damage.

In fully loaded voyages, the internal pressure from the tank is larger than the external pressure in fully loaded cargo oil tanks, and conversely, the external pressure is larger than the internal pressure in empty water ballast tanks. As a result, the mean bending stress on side longitudinals at the slot connection is tensile in cargo oil tanks and compressive in water ballast tanks. It is considered that the compressive mean stress in the water ballast tanks alleviated the fatigue damage, causing the observation (c).

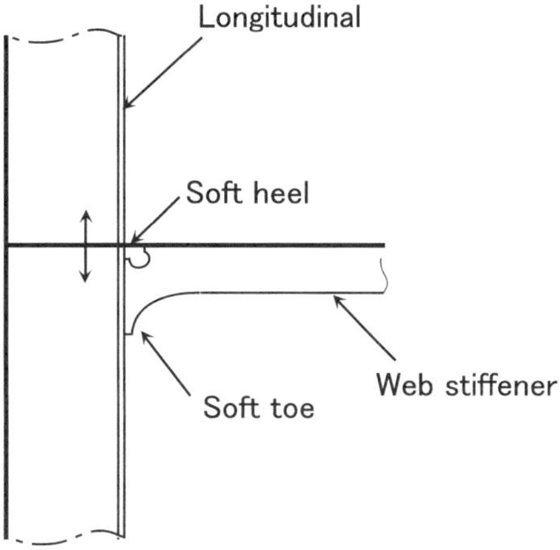

Fig. 10.26 Soft heel and toe of web stiffener

In the Indian Ocean, waves from the south are dominant, corresponding to the incident waves from port side in westbound ballast voyages, and from starboard side in eastbound fully loaded voyages. The side longitudinals near the full load draft suffer fatigue damage accumulation only in fully loaded voyages, and therefore, starboard side damages were dominant (observation (d)).

Countermeasures were soon taken. Each classification society also studied this problem, and established new rules, which took into account not only the yielding criteria but also the fatigue strength due to the fluctuating stress caused by the waves. Moreover, further countermeasures to enhance fatigue strength were applied, such as the application of a T section to side longitudinals, soft heel and toe of the web stiffeners at the connection to the side longitudinals as shown in Fig. 10.26, etc. Thanks to these countermeasures, cracks of this type have been seldom observed in recent years.

10.5.5 Cracks Around Slots Due to High Stress on Transverses

According to Sects. 10.5.2–10.5.3, a collar plate is a very effective member to enhance fatigue strength, because shearing force on the transverse web is canceled from both sides of the longitudinal, and does not cause additional stress in the web stiffener. Moreover, the collar plate supports some portion of the load transmitted and reduces stress on the web stiffener.

However, if the transverse web itself was designed using higher tensile steel, and the wave variable pressure was large, high stress is exerted at the connection between the collar plate and the transverse web, i.e. near the end of the transverse web where the shear stress is large.

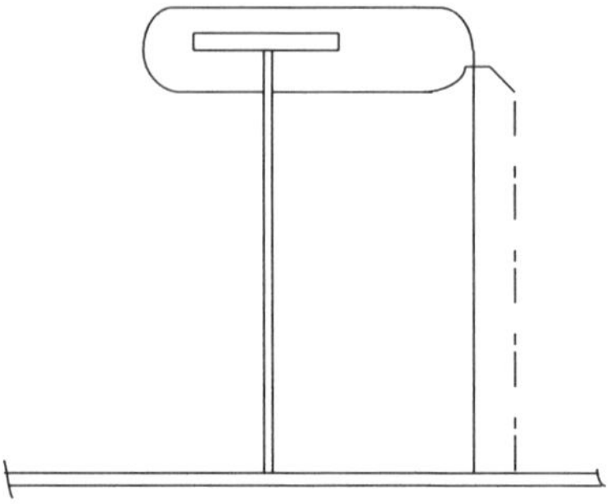

Fig. 10.27 Softly shaped collar plate attached to skin plate

As countermeasures for this type of cracks, it is effective to apply softly shaped collar plate attached to the skin plate as shown in Fig. 10.27.

Chapter 11
Design of Pillar

A pillar supports axial force; it generally supports axial compression but sometimes supports axial tension. For axial tension, pillars are very effective. For example, 40 cm^2 sectional area of mild steel can support 1,000 kN force ($\sigma = 250$ MPa). Considering that the pillar sometimes supports tension, attention should be paid to the connection at both ends of the pillar.

For axial compression, buckling phenomena are important. A slender pillar can buckle at a relatively low compressive stress. As the buckling happens in the weakest direction of the pillar, the section of the pillar should be designed so as to have homogeneous strength in all directions. From this view point square or circular pillars (solid or hollow) are desirable.

The end connections of pillars have a significant effect on buckling strength. Brackets are fitted usually at the end connection of a built-up pillar but rarely for a circular pillar.

The cross tie in a wing tank of a tanker is this kind of the pillar. Section of the cross ties are also studied in this chapter.

11.1 Slenderness Ratio of Pillar

A slender pillar under a compressive load will break by bending when the load exceeds some limit which is called buckling. The stress at buckling is decided by the slenderness ratio of the pillar. The slenderness ratio is given by l/k where l is the length of the pillar and k is the radius of gyration. The radius of gyration $k = \sqrt{I/A}$ where I is the sectional moment of inertia and A is the sectional area.

k values for several sections are shown in Table 11.1 [72]. It is important to note that comparing a solid pillar with a hollow pillar of the same diameter it has smaller k than that of the hollow pillar. However, the solid pillar has a bigger sectional area and can support a bigger load accordingly.

Comparing a solid circular pillar with a solid square pillar, k for the former is $0.25d$ (d is diameter of the pillar) and k for the latter is $0.289h$ (h is length of an

Y. Okumoto et al., *Design of Ship Hull Structures*,
https://doi.org/10.1007/978-3-032-12500-2_11

Table 11.1 I, A and k for each section

Sectional form	I	A, k
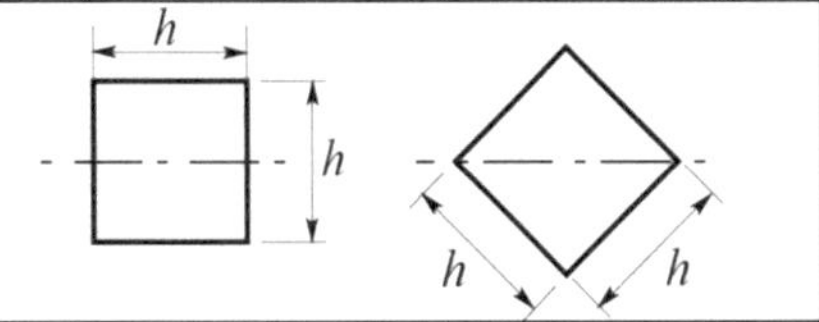	$\frac{1}{12}h^4$	$A = h^2$ $k = 0.289h$
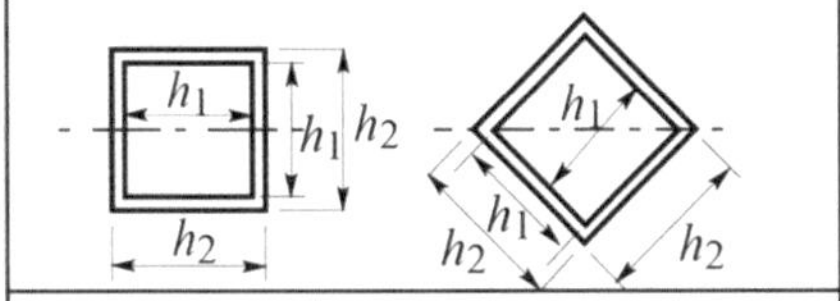	$\frac{1}{12}\left(h_2{}^4 - h_1{}^4\right)$	$A = h_2{}^2 - h_1{}^2$ $k^2 = \frac{1}{12}\left(h_2{}^2 + h_1{}^2\right)$
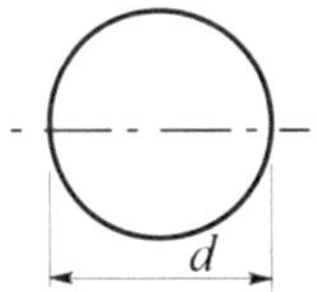	$\frac{\pi}{64}d^4$	$A = \frac{\pi}{4}d^2$ $k = 0.25d$
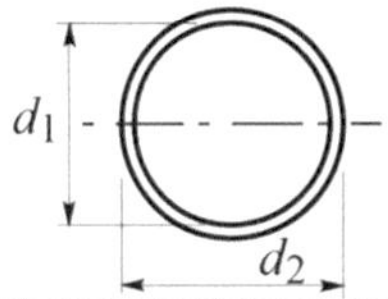	$\frac{\pi}{64}\left(d_2{}^4 - d_1{}^4\right)$	$A = \frac{\pi}{4}\left(d_2{}^2 - d_1{}^2\right)$ $k = 0.25\sqrt{d_2{}^2 + d_1{}^2}$
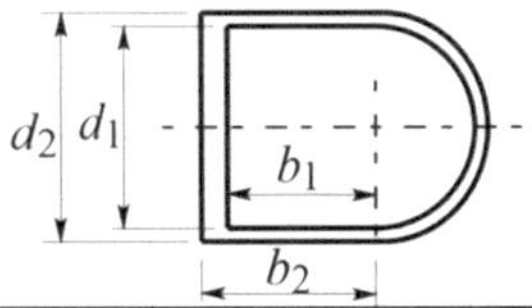	$\frac{\pi}{128}\left(d_2{}^4 - d_1{}^4\right)$ $+\frac{1}{12}\left(b_2d_2{}^3 - b_1d_1{}^3\right)$	$A = \frac{\pi}{8}\left(d_2{}^2 - d_1{}^2\right)$ $+(b_2d_2 - b_1d_1)$
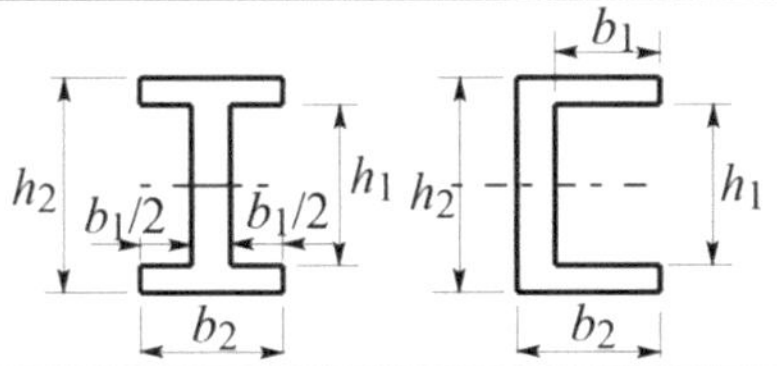	$\frac{1}{12}\left(b_2h_2{}^3 - b_1h_1{}^3\right)$	$A = b_2h_2 - b_1h_1$ $k^2 = \frac{1}{12}\cdot\frac{b_2h_2{}^3 - b_1h_1{}^3}{b_2h_2 - b_1h_1}$
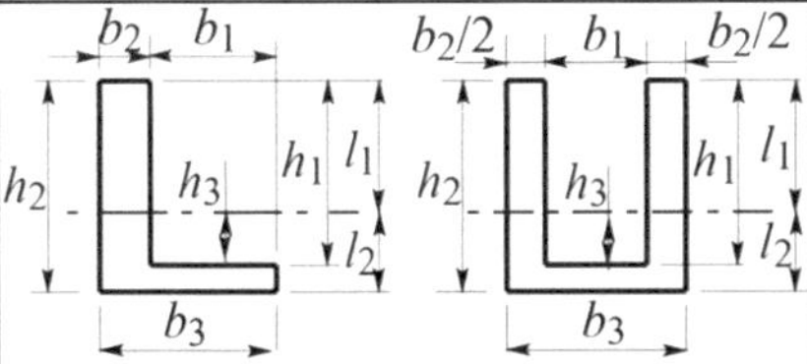	$\frac{b_3l_2{}^3 - b_1h_3{}^3 + b_2l_1{}^3}{3}$	$A = b_3h_2 - b_1h_1$

edge) which is bigger than k for the solid circular pillar. This is reasonable because the solid square pillar has a bigger sectional area than the solid circular pillar. Then with the condition that the sectional area is the same, both of them are compared as follows:

$$\frac{\pi}{4}d^2 = h^2$$

$$\therefore \frac{\sqrt{\pi}}{2}d = h$$

The radius of gyration of a solid circular pillar of diameter d is $0.25d$, whereas the radius of gyration of a solid square pillar with the same sectional area as the solid circular pillar is $0.289 \times \sqrt{\pi}d/2 = 0.256d$. From the above comparison it can be said that a solid square pillar is better than a solid circular pillar. The value of the radius of gyration for square sections is homogeneous for all axes passing through the center of gravity just as in circular sections (this is proved in Sect. 11.3).

In Fig. 11.1 the relation between critical buckling stress and the slenderness ratio $\lambda\,(=l/k)$ is shown [72]. Here, as for the critical buckling stress for slender pillars of mild steel with both ends hinged, Euler's theoretical formula, Rankine's and Tetmajer's experimental formula, and the Tetmajer's experimental results are summarized. The required sectional area A (cm^2) for an I-section pillar, according to the earlier rules by Nippon Kaiji Kyokai (ClassNK) was given by the following equation, and is also shown in Fig. 11.1.

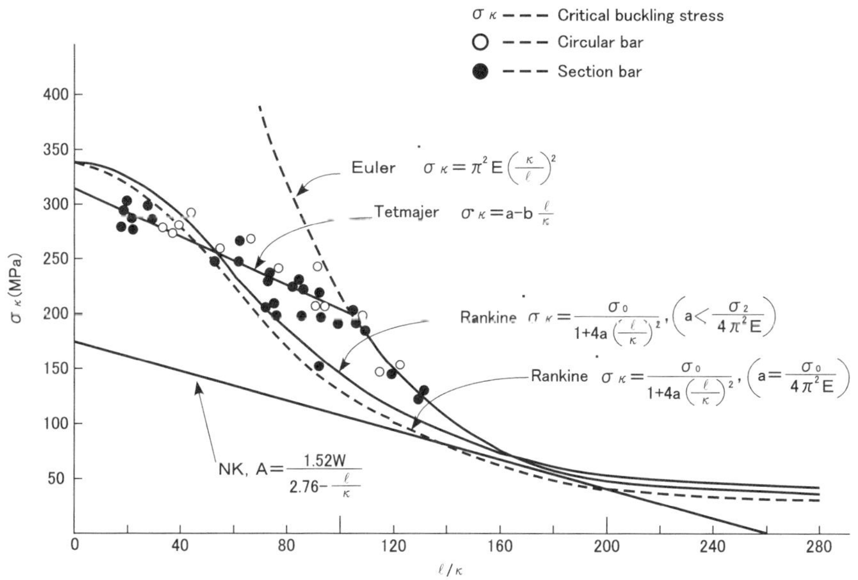

Fig. 11.1 Slenderness and buckling critical stress of pillar

$$A = \frac{1.52W}{2.76 - \frac{l}{k}} \tag{11.1.1}$$

where W is the axial load (tf), l is the length of the pillar (m) and k is the radius of gyration (cm).

If the size of a hollow circular pillar under the forecastle deck, for example, is $l =$ 3~4 m and the diameter $d = 200$~300 mm, the slenderness ratio λ is 33 ~ 67. If the size of a solid circular pillar in the accommodation space, for example, is $l = 2.5$~3 m and the diameter $d = 50$~70 mm, the slenderness ratio λ. is 140 ~ 240. The solid pillar is slenderer than the hollow one.

11.2 Sectional Shape of Pillars

Here characteristics of hollow and built up pillars are explained. The applicable sectional shapes are shown in Table 11.1; D shape pillars support hatch end coaming at the hold center of a cargo ship. The circular part faces the hatch opening and on the flat part a ladder is fitted.

A pillar will buckle in its weakest direction and therefore it should have a homogeneous strength in all directions (homogeneous k for each axis through the center of gravity). A circular and a square section satisfy this requirement.

The above condition for an I section is studied as follows. Assuming the thickness of web and flange plates is equal, I_x. and I_y. in Fig. 11.2a are given by the following equation:

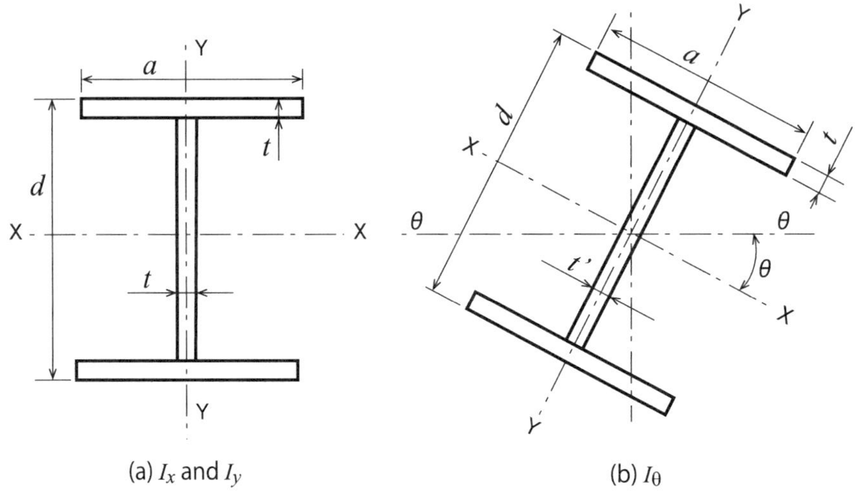

Fig. 11.2 Sectional moment of inertia of I shaped section

$$I_x = 2at\left(\frac{d}{2}\right)^2 + \frac{td^3}{12} \tag{11.2.1}$$

$$I_y = 2 \times \frac{ta^3}{12} \tag{11.2.2}$$

$$\text{Equating these two equations, } \frac{atd^2}{2} + \frac{td^3}{12} = \frac{ta^3}{6} \tag{11.2.3}$$

$$\therefore d^3 + 6ad^2 - 2a^3 = 0$$

Then $d \approx 0.55a$ is derived. This is an I section with wide flanges and narrow web. The sectional moment of inertia I_θ for the axis $\theta - \theta$ in Fig. 11.2b is calculated as follows. Here t and t' are the thickness of the flange and web plates, respectively.

$$\begin{aligned} I_\theta &= \frac{t'd(d\cos\theta)^2}{12} + 2\left\{\frac{at(a\sin\theta)^2}{12} + at\left(\frac{d}{2}\cos\theta\right)^2\right\} \\ &= \frac{ta^3}{6}\sin^2\theta + \left(\frac{tad^2}{2} + \frac{t'd^3}{12}\right)\cos^2\theta \end{aligned} \tag{11.2.4}$$

Putting $d \equiv \alpha a$ and $t' \equiv \beta t$, the following relation can be obtained.

$$I_\theta = \frac{ta^3}{12}\left\{2\sin^2\theta + \left(6\alpha^2 + \alpha^3\beta\right)\cos^2\theta\right\} \tag{11.2.5}$$

From Eq. (11.2.5) it can be clear that in the condition of $6\alpha^2 + \alpha^3\beta = 2$, I_θ has a constant value of $ta^3/6$ without regard to θ. When the thicknesses of the flange and web are equal ($\beta = 1$), this yields $\alpha = 0.55$. In general, when the sectional moments of inertia about two orthogonal axes are identical, I_θ is constant about any axes passing through the centroid of the section regardless of θ. In conditions other than the above, I_θ will change depending on θ as shown below. Taking the differential of I_θ with respect to θ, the following equation can be obtained.

$$\begin{aligned} \tfrac{dI_\theta}{d\theta} &= \tfrac{ta^3}{12}\left\{4\sin\theta\cos\theta - 2\left(6\alpha^2 + \alpha^3\beta\right)\cos\theta\sin\theta\right\} \\ &= \tfrac{ta^3}{12}\left\{2 - \left(6\alpha^2 + \alpha^3\beta\right)\right\}\sin 2\theta \end{aligned} \tag{11.2.6}$$

From Eq. (11.2.6) it can be said that where $2 - (6\alpha^2 + \alpha^3\beta) < 0$, that is, $I_x > I_y$, I_θ has a maximum value at $\theta = 0°$ and a minimum value at $\theta = 90°$. Conversely, where $2 - (6\alpha^2 + \alpha^3\beta) > 0$, that is $I_x < I_y$, I_θ has a minimum value at $\theta = 0°$ and a maximum value at $\theta = 90°$. As I_x is usually bigger than I_y, from Eq. (11.2.5) I_x/I_y is calculated as shown below.

$$\frac{I_x}{I_y} = \frac{2\sin^2 0° + \left(6\alpha^2 + \alpha^3\beta\right)\cos^2 0°}{2\sin^2 90° + \left(6\alpha^2 + \alpha^3\beta\right)\cos^2 90°} = \frac{6\alpha^2 + \alpha^3\beta}{2} \tag{11.2.7}$$

The relation obtained by Eq. (11.2.7) is shown in Fig. 11.3. I section pillar's I_x/I_y is little affected by β, which is the ratio of web and flange plate thickness, but mainly affected by α, which is the ratio of web depth and flange breadth. In an ordinary I section pillar, α will be nearly 1.0 and I_x/I_y will be more than 3.0.

For a Π shape section as shown in Fig. 11.4a, a similar equation is derived as follows:

$$I_\theta = \frac{ta^3}{12}\left\{\left(6\alpha^2 + \alpha^3\beta\right)\cos^2\theta + \frac{4(2\alpha\beta + 1)}{\alpha\beta + 2}\sin^2\theta\right\} \tag{11.2.8}$$

Furthermore, under the condition of $(6\alpha^2 + \alpha^3\beta)(\alpha\beta + 2) = 4(2\alpha\beta + 1)$, I_θ has a constant value of $\frac{ta^3}{12}\left(6\alpha^2 + \alpha^3\beta\right)$ without regard to θ. When the thicknesses of the flange and web are equal ($\beta = 1$), this condition yields $\alpha = 0.73$, which is bigger than $\alpha = 0.55$ for an I shape section. In the case of $\alpha > 0.73$, in a similar way to that of an I shape section, I_x will be the maximum and I_y will be the minimum. The ratio

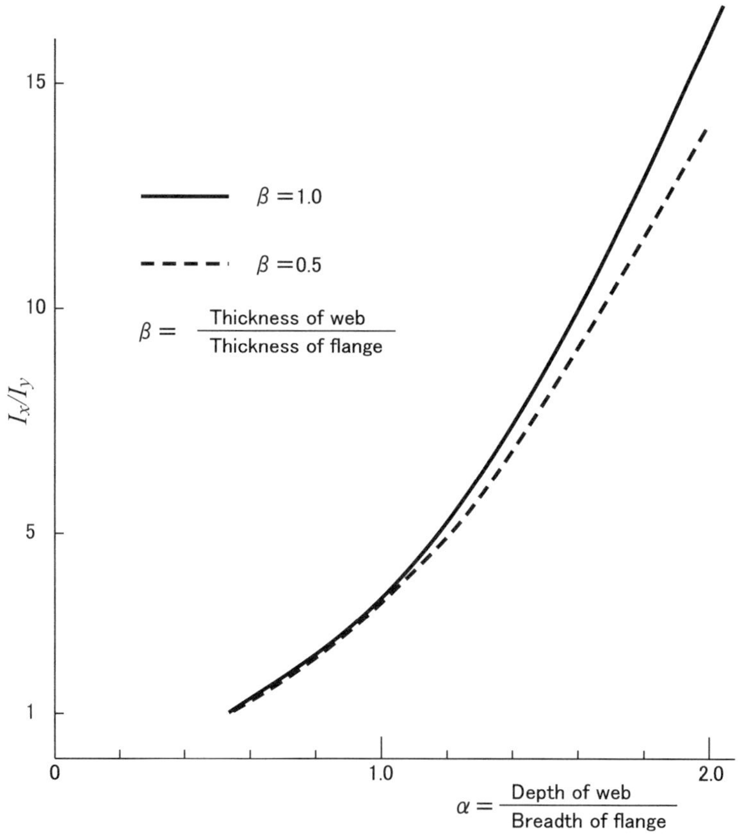

Fig. 11.3 α and I_x/I_y of I shaped section

I_x/I_y is expressed by the following equation:

$$\frac{I_x}{I_y} = \frac{\alpha^2(\alpha\beta + 6)(\alpha\beta + 2)}{4(2\alpha\beta + 1)} \tag{11.2.9}$$

The relation of Eq. (11.2.9) is shown in Fig. 11.5. Comparing Fig. 11.3. with Fig. 11.5 from the viewpoint of the I_x/I_y ratio, the Π shape section is better than the I shape section. For example, in the case of $\beta = 1$ and $\alpha = 1$, $I_x/I_y = 3.5$ for the I shape section and 1.75 for the Π shape section. In this case I_x is the same value for both sections; therefore, I_y of the Π shape section is twice that for the I shape section.

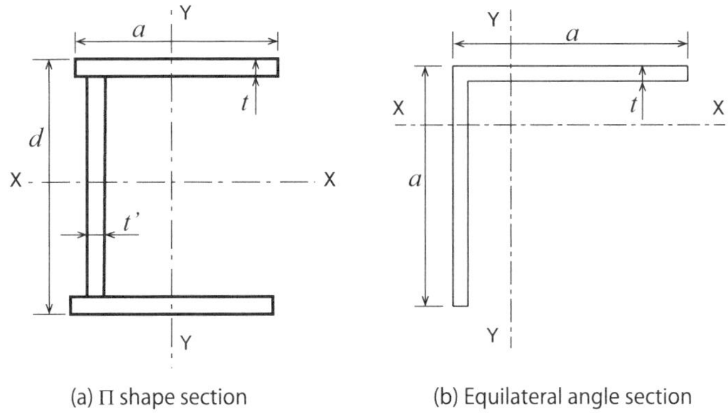

Fig. 11.4 Π shape section and angle section

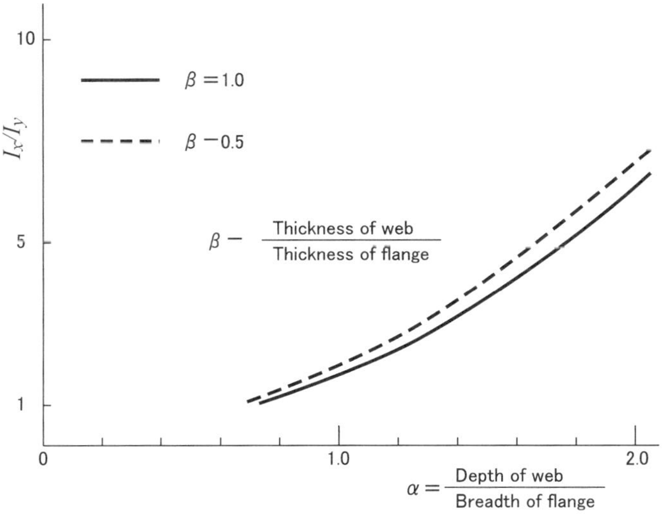

Fig. 11.5 α and I_x/I_y of Π shaped section

For equal flanges of an angle section as shown in Fig. 11.4b, it should be kept in the designer's mind that in some cases of θ, I_θ is smaller than I_x or I_y. Assuming the length of a flange is a and the thickness is t, I_θ is obtained as follows.

$$I_\theta = \frac{ta^3}{12} + \frac{ta^3}{8}(1 + \sin 2\theta) \tag{11.2.10}$$

In the case of $\theta = 0$, $I_\theta \equiv I_x$ or I_y and its value is as follows:

$$I_x = I_y = \frac{ta^3}{12} + \frac{ta^3}{8} = \frac{5}{24}ta^3 = 0.208ta^3$$

I_θ becomes maximum at $\theta = 45°$ and minimum at $\theta = -45°$ for which the values are as follows:

$$I_{max} = \frac{ta^3}{12} + \frac{ta^3}{4} = \frac{1}{3}ta^3 = 0.333ta^3$$

$$I_{min} = \frac{ta^3}{12} = 0.083ta^3$$

The minimum value is 40% of I_x and the maximum value is four times that of the minimum value.

11.3 Radius of Gyration of Square Section

In a square section the radius of gyration is homogeneous for each axis passing through the center of gravity. This is easily understood as being based on a very important principle on the sectional moment of inertia "In a plane, the sum of the sectional moment of inertia for the axes crossing perpendicularly to each other is equal to the polar moment of inertia (the sectional moment of inertia for the axis perpendicular to the plane and passing through the cross point of the axes in the plane)."

Keeping the sectional area of the square section constant, discussion on the radius of gyration is the same as that for the sectional moment of inertia. Assuming I_θ is the sectional moment of inertia for an axis passing through the center of gravity, $I_{\theta+90}$ is the sectional moment of inertia for the axis crossing perpendicularly the above axis at the center of gravity, and I_0 is the polar moment of inertia of the square section, the following relations are obtained.

$$I_\theta = I_{\theta+90}$$

$$I_\theta + I_{\theta+90} = I_0$$

$$\therefore I_\theta = \frac{I_0}{2}$$

From the above relations it can be said that I_θ is constant. In fact, it is known that the sectional moment of inertia of a regular polygon with any number of vertices is constant irrespective of the angle of the axis passing through its centroid. Proof can be found, for example, in Ref. [61].

11.4 Pillar Supporting Tensile Force

A pillar in a tank sometimes supports a tensile force. The tank top plate supports an upwards load when it is filled up to the overflow pipe. In this case a pillar connecting the top plate and the bottom of the tank has a tensile load.

A long superstructure of a ship deforms in the opposite direction to the main hull girder at hogging and sagging condition. In the case of hogging the upper deck has tensile strain longitudinally and the bottom of the long superstructure also has tensile strain longitudinally which means the long superstructure will bend in a sagging form. The hogging of the main hull girder and the sagging of a long superstructure causes tensile forces at the both ends of the long superstructure in the vertical direction. This mechanism is explained in Part III Chap. 31 in detail. In this case pillars below the upper deck at both ends of the long superstructure will suffer from tensile forces.

A pillar is stronger against tension than against compression because there is no buckling phenomenon. However, attention should be paid to the connection at the top and bottom. In the case of compression, the connection at the top and bottom is not so important. But in the case of tension the welded sectional area should be equivalent to the sectional area of the pillar. The doubling plate at the connection of the top and bottom of a pillar is effective in distributing the compressive load but it is weak for the tensile load. In the case of a tensile load the doubling plate should be omitted or a thick and narrow doubling plate should be fitted. The examples of cracks at the connections of the top and bottom of a pillar are shown in Fig. 11.6 [73].

11.5 Connection of Pillar at Top and Bottom

In Fig. 11.1. the critical buckling stress of a pillar is shown against the slenderness ratio $\lambda(=l/k)$ in the case of hinged ends. In the case of fixed ends, the critical buckling stress is higher than that of hinged ends.

In Fig. 11.7 the buckling modes of hinged and fixed-ended pillars are shown. The midpart $l/2$ (l is the length of the pillar) of a fixed-end pillar has the same buckling mode as that of a hinged-ended pillar. Accordingly, the critical buckling stress of the

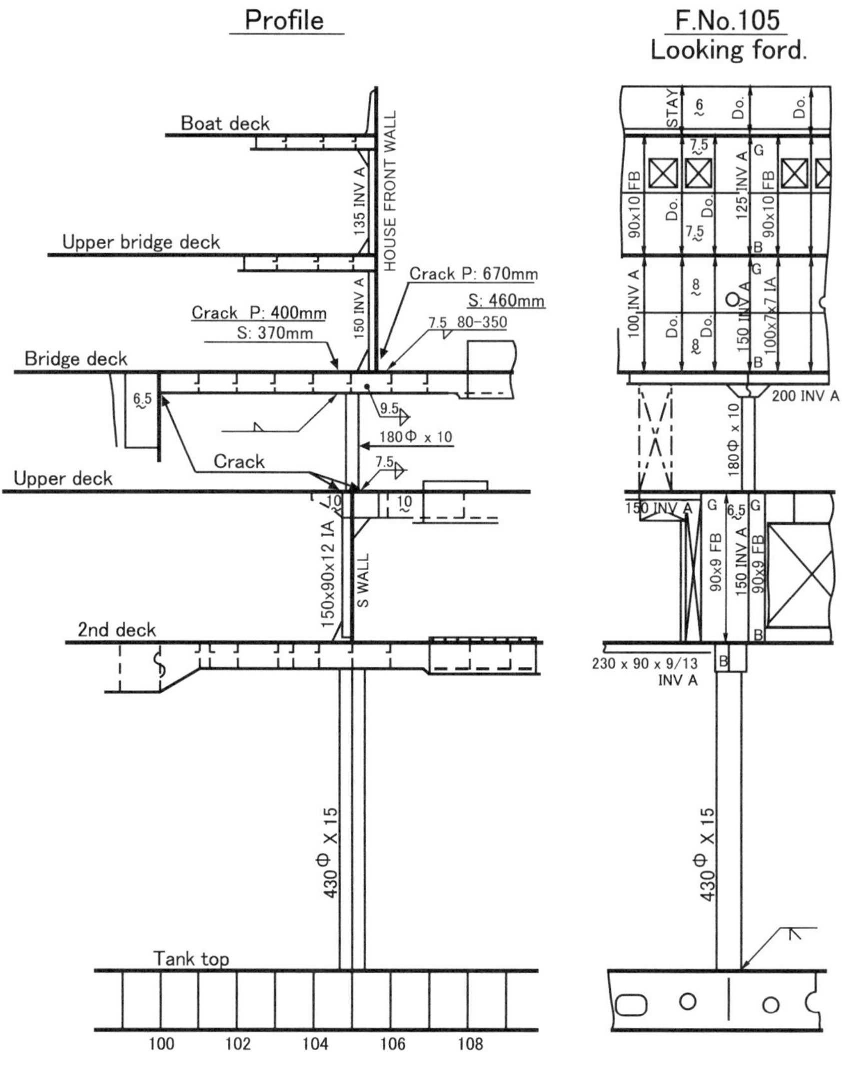

Fig. 11.6 Crack at fixed end of circular pillar due to tension

fixed-end pillar will be the same as that for the hinged-ended pillar of half length ($l/2$).

As far as Euler's theoretical formula $\sigma_k = \pi^2 E\left(\frac{k}{l}\right)^2$ can be applied, σ_k is inversely proportional to l^2, and the critical buckling stress of a fixed-ended pillar is four times of that for hinged-ended pillar.

The above consideration can be applied not only to a pillar but also to a beam. The deflection curve of a beam supported at both ends (length is $2l$) with a concentrated

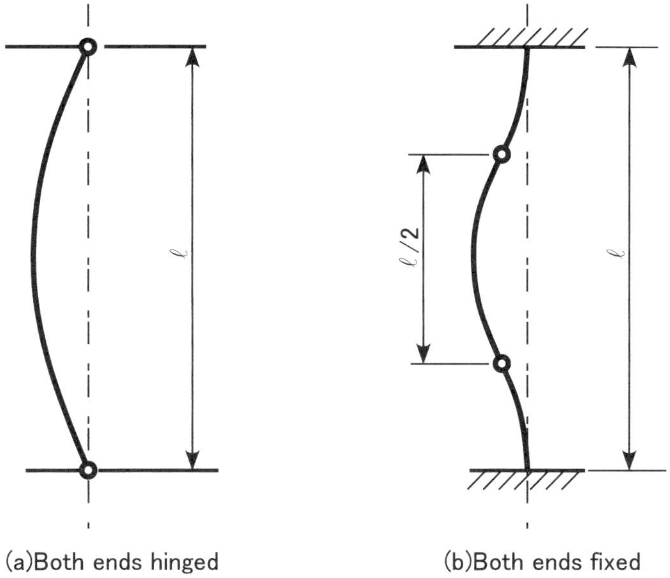

Fig. 11.7 Buckling mode of pillar

load W at midspan is equivalent to that of cantilever (length is l) with a concentrated load $W/2$ at the free end. The maximum bending moment in each case is $Wl/2$.

The critical buckling stress depends on the buckling mode, and the equation for critical buckling stress for one mode cannot be applied to a different buckling mode. On the other hand, as explained above the critical buckling stress can be easily found for each buckling mode by comparing with the other mode.

In Sect. 11.2 it is explained that the circular and square sections have homogeneous sectional moment of inertia in every axis through the center of gravity, whereas I, Π, and L sections are not homogeneous in all axes, and the ratio of maximum and minimum sectional moment of inertia is usually about 3. In practice the actual condition of the end connections of a pillar will be between hinged and fixed. By increasing the rigidity of the end connections in the weak direction, the above unbalance can be reduced.

Considering the above, bracket arrangements at the top and bottom of a pillar are proposed as shown in Fig. 11.8. The dotted lines in the figure mean snipped flanges because no structural member lies underneath.

It is reasonable not to fit brackets on circular and square pillars. This is especially the case of brackets on a hollow circular pillar which will not work well because they are connected only to the outside of the thin plate of the pillar. In Fig. 11.9 examples of buckled pillars are shown [73]. From the figure it can be said that without brackets the top and bottom connections of a circular pillar have a nearly fixed condition. Therefore, a damage plan or sketch is very useful to show the actual condition, and the buckling mode and crack pattern should be shown correctly on the drawings.

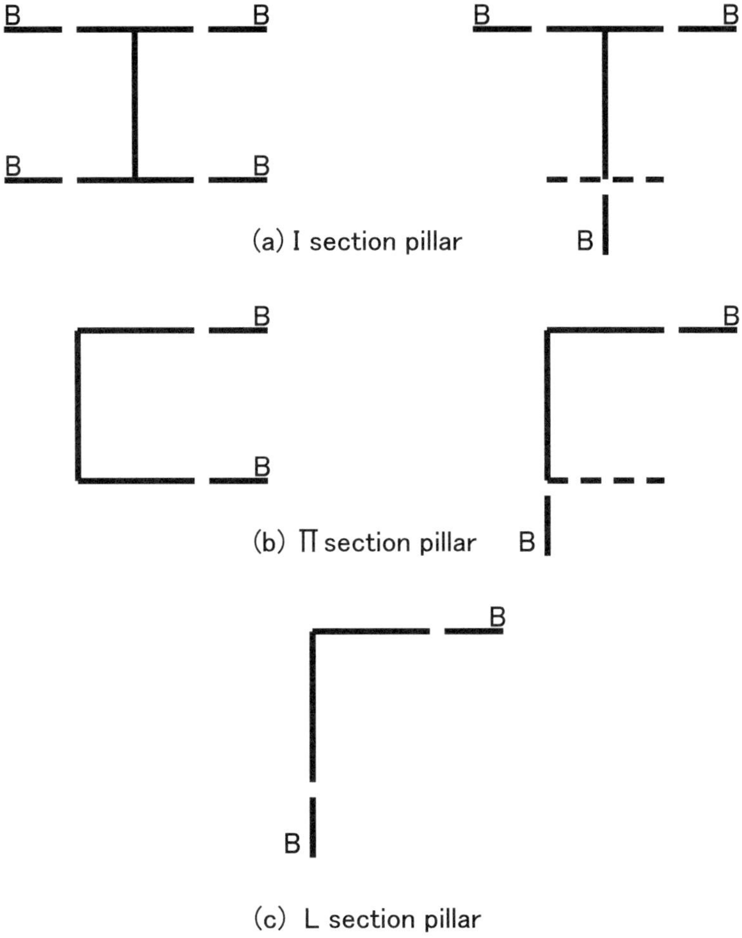

Fig. 11.8 Brackets at both ends of pillar

11.6 Cross Ties

The cross tie in a wing tank of a tanker is a kind of pillar. It supports not only compression and tension as a pillar but also as part of the transverse ring it supports bending and shearing. As explained in Sect. 11.2 a pillar should preferably have homogeneous strength for each axis through its center of gravity, however the cross tie has a deep web and narrow flanges and consequently I_x/I_y is very big as shown in Fig. 11.10a, because such cross ties used to be designed as part of a transverse ring to maintain transverse strength.

However, the cross ties are too weak to maintain the transverse strength and deeper cross ties will suffer from the effect of forced deformation caused by the relative displacement between the side shell and longitudinal bulkhead. To reduce

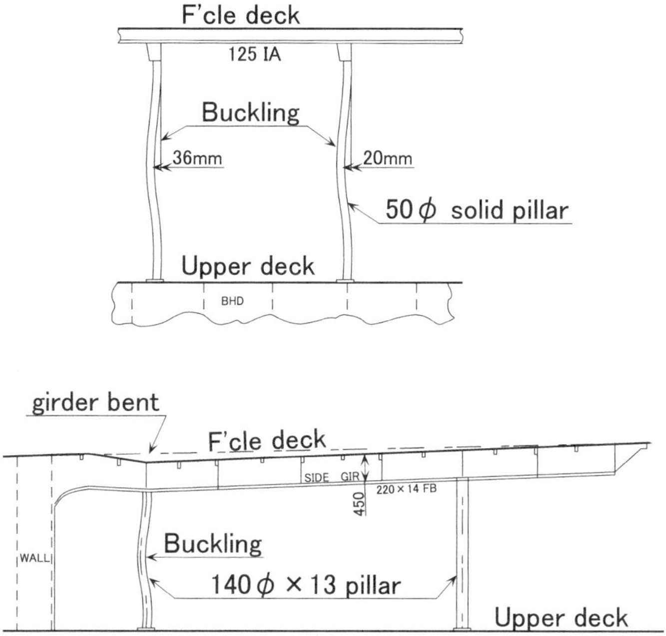

Fig. 11.9 Buckling of circular pillar

the additional bending stress due to the relative displacement between side shell and longitudinal bulkhead, a cross tie with shallow depth is proposed and applied as shown in Fig. 11.10b. To reduce the relative displacement between the side shell and the longitudinal bulkhead, a partial bulkhead at midpoint of the tank is the best way.

In the shallow web cross tie, the balance between I_x and I_y is improved. Furthermore, by providing a side stringer on the longitudinal bulkhead and extending the horizontal web plate of the cross tie, the connection at the end in the horizontal plane can be improved. These horizontal members are useful for access in a tank, killing two birds with one stone.

In Fig. 11.11 examples of damage to cross ties are shown, one caused by compression and the other by tension. There are two kinds of construction for the connection between a cross tie and the vertical web on the side shell or longitudinal bulkhead. One is a straight face plate type with brackets and the other is a curved face plate type (It is called R type.). The latter is usually applied because of good continuity of face plate and its light weight.

In the case of an R type, tension and compression are generated between the web and the face plate in the vicinity of the R part in the direction perpendicular to the face plate, by the tension and the compression in the cross tie as shown in Fig. 11.12. In the design of land structures attention is paid to this tension and compression between web and face plate in the vicinity of the R part, but hull structure designers pay very little attention to it.

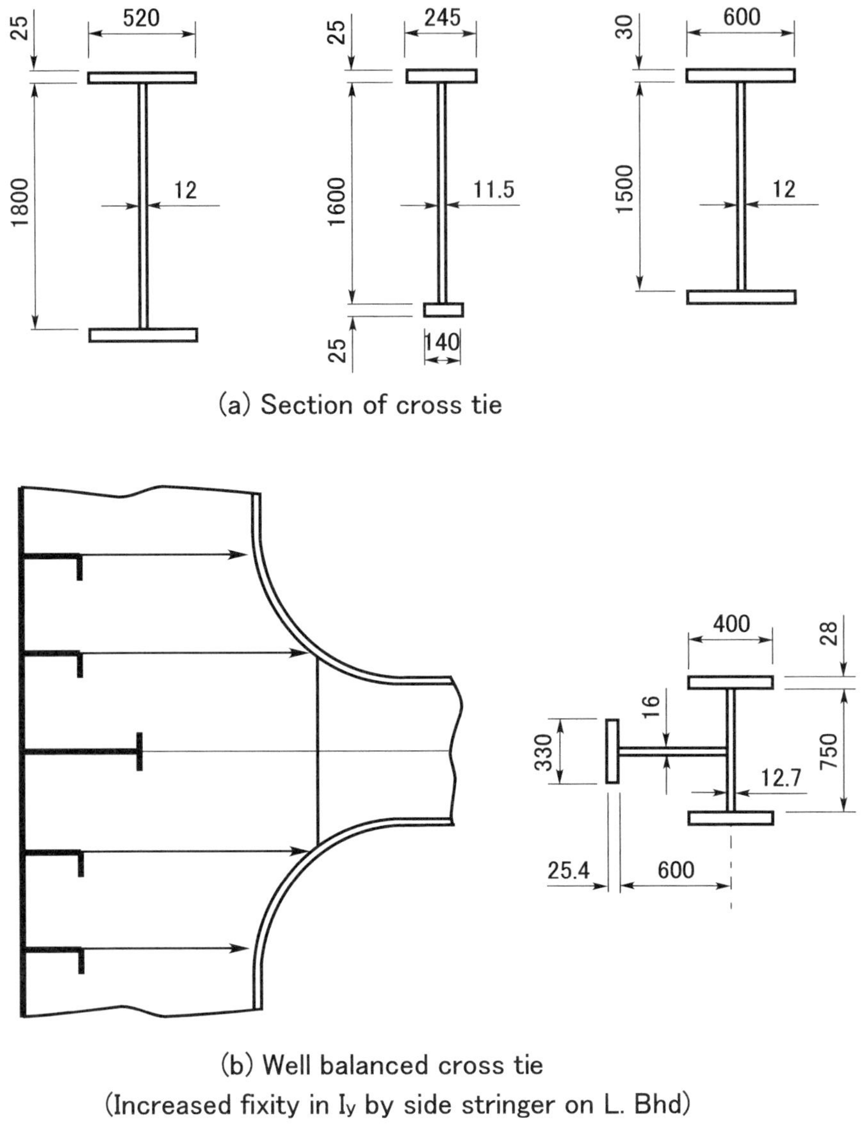

Fig. 11.10 Sectional shape of cross tie

Figure. 11.12 shows the mechanism of generating tension and compression. P is the axial force in the face plate and the normal component N to the face plate is $N = Pd\theta/2$. The normal component acting on the ds part is $2N = Pd\theta$. If the thickness of the web plate is t, the area to support the force of $2N$ is tds. Then the stress σ_w in the web perpendicular to the face plate is expressed by the following equation:

Fig. 11.11 Buckling and crack at the end of cross tie

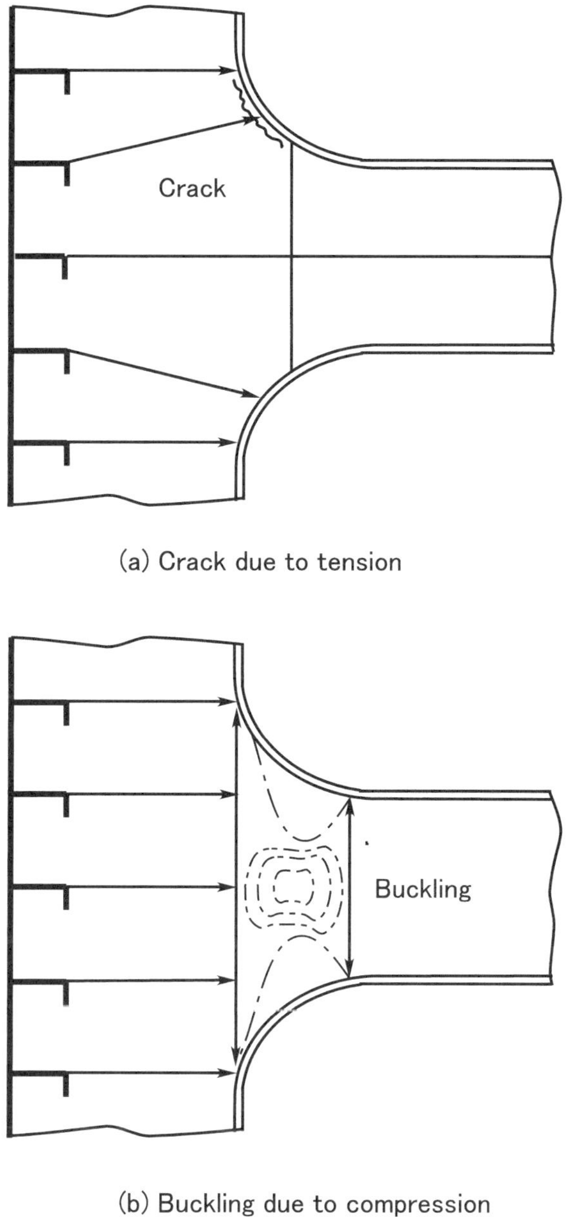

Fig. 11.12 Secondary stress at curved face plate

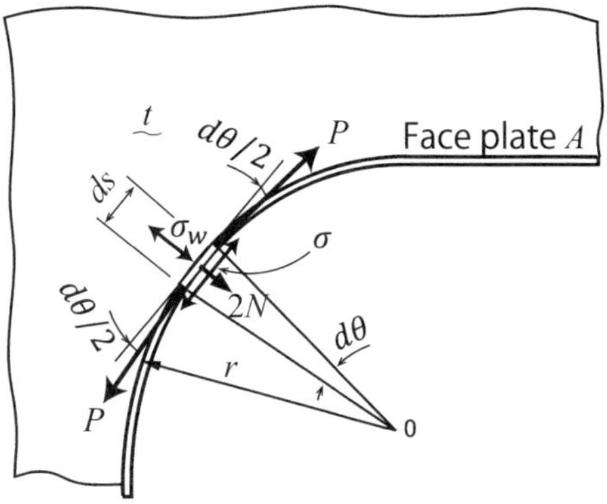

$$\sigma_w = \frac{2N}{tds} = \frac{Pd\theta}{tds} = \frac{Pd\theta}{trd\theta} = \frac{P}{tr} = \frac{A\sigma}{tr} \quad (11.6.1)$$

where A denotes the sectional area of the face plate and σ denotes the axial stress acting in the face plate.

From Eq. (11.6.1) it can be deduced that a bigger sectional area in the flange and a smaller radius results in a bigger σ_w. One example is calculated below.

$$A = 500 \text{ mm} \times 30 \text{ mm}$$

$$t = 12 \text{ mm}$$

$$r = 600 \text{ mm}$$

$$\sigma = 150 \text{ MPa}$$

$$\sigma_w = \frac{500 \text{ mm} \times 30 \text{ mm} \times 150 \text{ MPa}}{12 \text{ mm} \times 600 \text{ mm}} = 312 \text{ MPa}$$

In this case σ_w is about twice that of σ. In designing R-type cross tie ends attention should be paid to the above σ_w and the thickness of the web t and the corner radius r should be chosen to make σ_w equivalent to σ.

Chapter 12
Design of Plates

12.1 Boundary Conditions of Plates

Plates make up the main hull structure such as shells, decks and bulkheads, in conjunction with secondary supporting members such as stiffeners and primary supporting members such as girders. The strength of such plates is most important since many kinds of load act directly on the plates, especially water pressure and longitudinal bending. The strength of plates is usually taken as a plane rectangular plate having uniform thickness. The boundaries, i.e., the four sides of a plate, are considered as fixed, simply supported, or elastically supported, in accordance with the kind of loads. Decisions relating to the boundary supporting conditions are very important because the calculated deformation and stress depend on these conditions.

Usually it is assumed as fixed when uniformly distributed pressure is acting on a plate, while as simply supported when concentrated lateral loads and in-plane loads are acting; the latter is also assumed for vibration analysis of a plate. The reason can be understood if we consider their deformations.

Figure 12.1 illustrates this concept. Figure 12.1a shows uniformly distributed pressure on a continuous plate having equal stiffener spacing. In this case the plate is assumed to have fixed boundaries, the same as for a continuous beam, because slope of the stiffeners is zero due to symmetrical conditions around the stiffeners. Figure 12.1b shows a concentrated load in the mid-region of a plate. In this case simply supported boundaries are assumed, because the slope at the stiffeners is not zero. Figure 12.1c is a case of plate buckling by in-plane forces. Since the fundamental buckling mode of a continuous plate is convex-concave figure alternately, their boundaries are considered to be simply supported, as shown in case (c).

If the stiffeners, which support a plate, deform vertically, the boundary condition is considered to be elastically supported. When the deflection of the adjacent stiffeners is same, these effects are neglected and both ends are assumed as not to move. If the deflection of two adjacent stiffeners differs, additional stress occurs as shown in Fig. 12.2, which is an example in the vicinity of a bulkhead; the shell plate is fixed

Y. Okumoto et al., *Design of Ship Hull Structures*,
https://doi.org/10.1007/978-3-032-12500-2_12

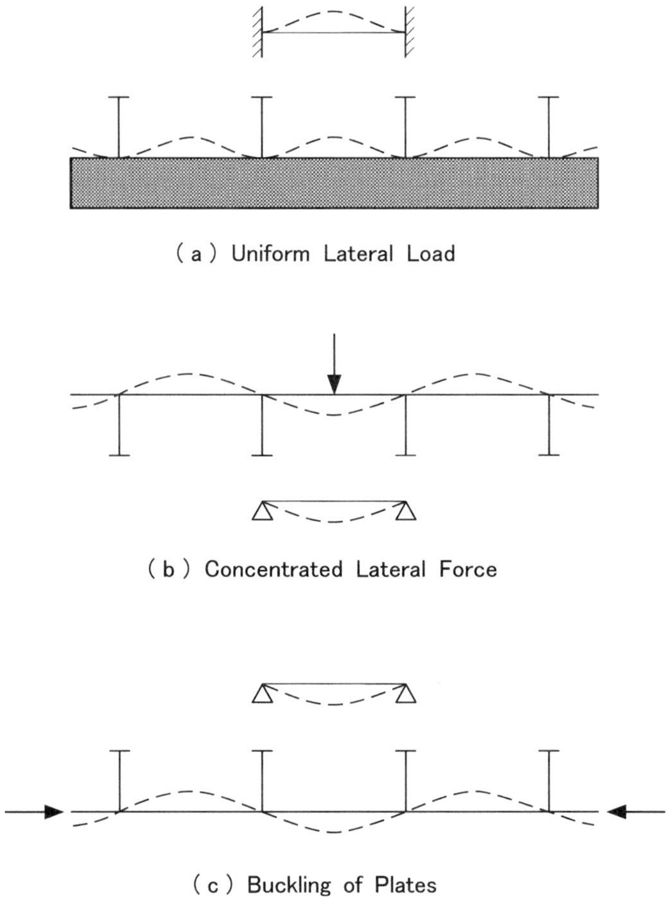

Fig. 12.1 Deformation of plates

along the longitudinal bulkhead and the bottom longitudinal is deformed by water pressure. Here the deformation is the sum of that due to water pressure as both ends fixed and of the forced displacement. This additional stress by forced displacement often causes some defects such as cracking. This additional stress can be calculated using beam theory in the same way as explained in Sect. 8.6.6 for a continuous beam.

12.2 Strength of Plates Under Lateral Loads

Plates such as shells, decks, and bulkheads are mostly loaded by uniform lateral pressure. The strength of the plate is based on a rectangular plate with fixed edges as shown in Fig. 12.3.

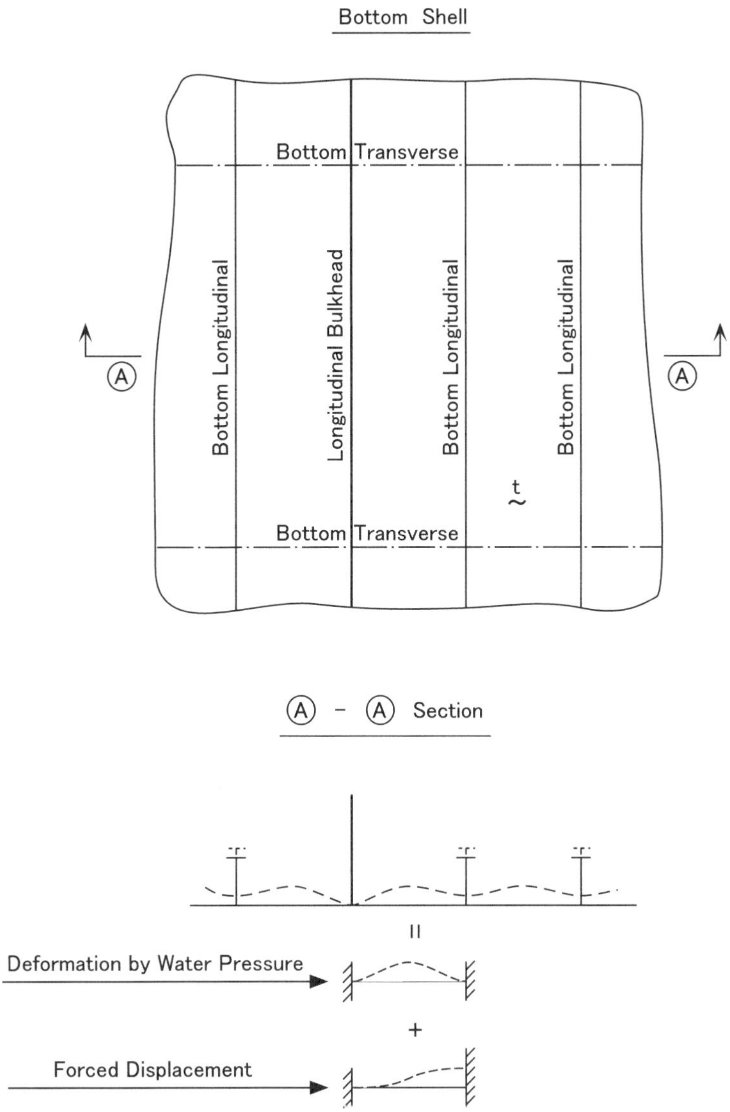

Fig. 12.2 Additional stress due to forced displacement

(1) *Elastic design*: The deflection and stress of a plate having constant thickness are obtained as follows [74]:

$$w = \alpha \frac{qa^4}{D} \tag{12.2.1}$$

where

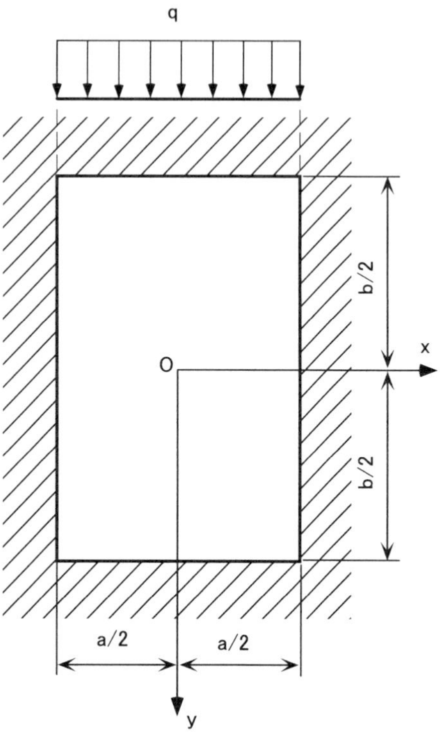

Fig. 12.3 Rectangular plate with all edges built in

w: deflection at plate center.

α: coefficient (see Fig. 12.4).

q: intensity of uniformly distributed load.

a: length of shorter edge of plate.

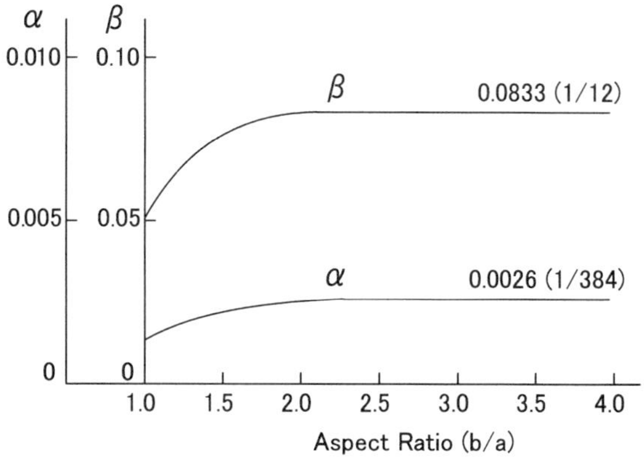

Fig. 12.4 Coefficient α, β

b: length of longer edge of plate.

D: flexural rigidity of a plate

$$D = \frac{Et^3}{12\left(1 - \nu^2\right)} \tag{12.2.2}$$

E: Young's modulus.

ν: Poisson's ratio.

t: plate thickness.

The bending moment at the midpoint of the longer edge is:

$$M = \beta q a^2 \tag{12.2.3}$$

where

M: bending moment at $x = a/2, y = 0$

β: coefficient (see Fig. 12.4).

Then, the bending stress σ is:

$$\sigma = \frac{M}{Z} = M / \frac{t^2}{6} \tag{12.2.4}$$

If the aspect ratio of a plate b/a exceeds 2.0, the maximum deflection w and maximum stress σ become constant as follows from Fig. 12.4.

$$w = \frac{qa^4}{384D} = \frac{qa^4}{384 \cdot \frac{E}{1-\nu^2} \cdot \frac{t^3}{12}} \tag{12.2.5}$$

$$\sigma = \frac{qa^2}{12 \cdot \frac{t^2}{6}} \tag{12.2.6}$$

Equations (12.2.5) and (12.2.6) show that the deflection and bending stress of a rectangular plate with a large aspect ratio agree with those of a both-end-fixed plate strip having unit breadth and a span of a, by replacing Young's modulus with $E/\left(1 - \nu^2\right)$. This apparent Young's modulus is due to the fact that the plate has the same displacement between the two adjacent plate strips, so that the bending displacement around the x-axis is constrained.

(2) *Plastic design:* There are two equations for the collapse load in accordance with the upper and lower bound theorem. When uniformly distributed pressure acts on a rectangular plate, the collapse load is given [24]:

$$p_u = \gamma_u \frac{M_p}{a^2}$$

$$p_l = \gamma_l \frac{M_p}{a^2} \quad (12.2.7)$$

where

p_u: upper collapse load per unit area.

p_l: lower collapse load per unit area.

γ_u, γ_l: coefficient.

M_p: fully plastic moment per unit length

$$M_p = \frac{t^2 \sigma_y}{4} \quad (12.2.8)$$

σ_y: yield stress

t: plate thickness.

a: length of shorter edge of plate.

b: length of longer edge of plate.

Coefficients γ_u, γ_l are given in Table 12.1 and Fig. 12.5. If the aspect ratio $\beta = b/a$ becomes large, the coefficients converge gradually to a constant value 8 for a simply supported plate and 16 for a plate with fixed boundaries, which corresponds exactly to the collapse load of a plate strip with unit breadth, similarly to the discussion in (1) "Elastic design". The collapse pressure q with all the four edges fixed condition is expressed as follows:

Table 12.1 Coefficient γ_u, γ_l

Boundary condition		Upper bound (γ_u)	Lower bound (γ_l)
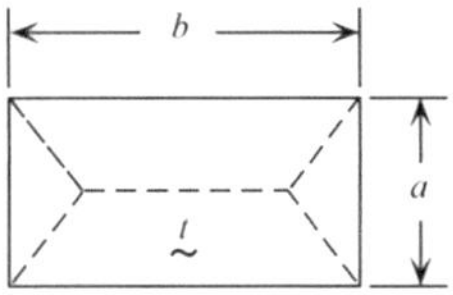 ---- Plastic hinge; $k = \frac{b}{a}$ (Aspect ratio)	Simply supported	$\frac{24k^2}{\left(1-\sqrt{1+3k^2}\right)^2}$	$8\left(1+\frac{1}{k}+\frac{1}{k^2}\right)$
	Fixed	$\frac{48k^2}{\left(1-\sqrt{1+3k^2}\right)^2}$	$16\left(1+\frac{1}{k^2}\right)$

Fig. 12.5 Coefficient γ_u, γ_l

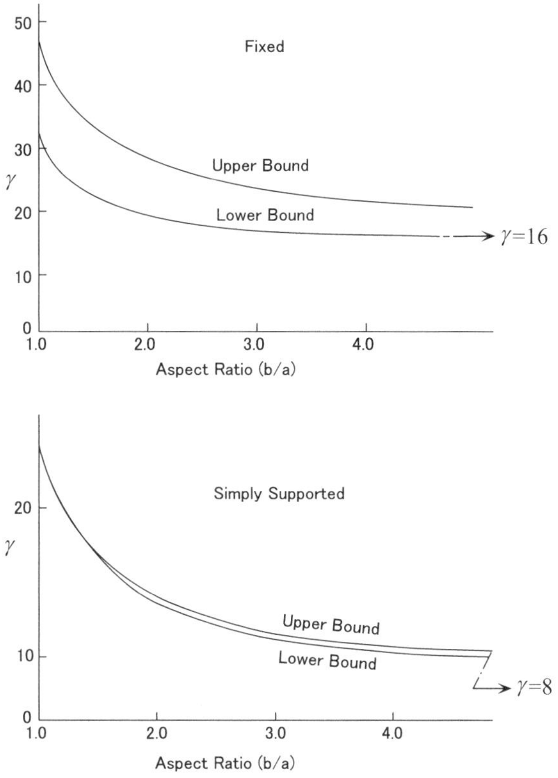

$$q = \frac{16}{a^2} M_p = \frac{4t^2\sigma_y}{a^2} \tag{12.2.9}$$

Comparing Fig. 12.5 with Fig. 12.4, we can find that the strength increase due to a small aspect ratio compared with a long plate is much more significant in plastic design than in elastic design.

Classification societies' scantling formulae of the required thickness t_{req} of plate members are usually in the form of:

$$t_{req} = Ca\sqrt{q} \tag{12.2.10}$$

where C is a constant representing safety factor and other coefficients to account for various conditions. When we solve Eq. (12.2.6) or Eq. (12.2.9) with respect to t, we can find that the resulting equations have the same form as Eq. (12.2.10), i.e., t is proportional to the product of shorter edge length a and the square root of uniform pressure q. This means that the scantling formulae are based on the initial yielding criteria (Eq. (12.2.6)) or plate collapse criteria (Eq. (12.2.9)) with a certain amount of safety factor. As we have already observed in Sect. 3.3.2 (1) in Part I, net plate

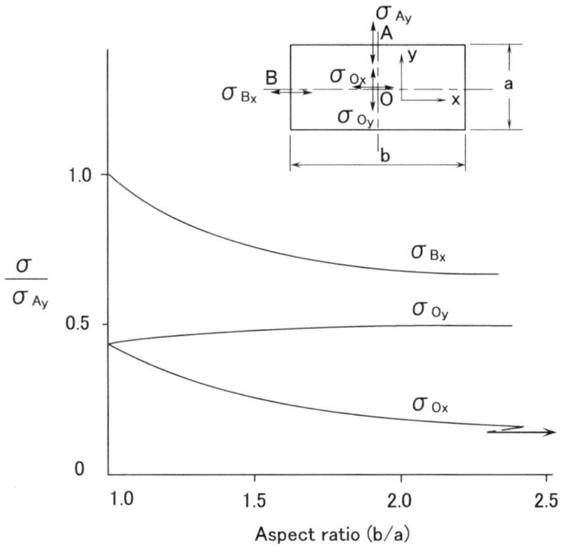

Fig. 12.6 Aspect ratio of plate versus stress at each point

thickness in the CSR has a safety factor of 1.155 to the plate collapse, and the bending stresses exerted along the longer edge of a panel is well beyond the yield stress.

(3) *Stress at various points in the panel:* In the previous discussion, panel strength is investigated in association with bending stress at the mid-point of the longer edge of a panel, because the maximum stress occurs at this point. Let us consider the stress level at other main points.

According to Fig. 12.6, we find that the stress at the middle of shorter edge is 60 to 70% of the maximum and that the stress at the panel center in both directions is lower than that of the shorter edge.

(4) *Panel strength considering membrane stress [75]:* In actual ships, the stress in a bulkhead panel, of which thickness is specified by the Classification Society's rule, exceeds the yield stress, and the safety factor for collapse load in plastic design is not so high. Nevertheless, strictly speaking, the real strength of a panel under water pressure is much higher than the calculated strength based on simple elastic design or plastic design, because the membrane stress increases the strength of the panel and it is not taken into account in the above.

Experimental results are shown in Fig. 12.7, which shows the effect of the membrane stress on increasing the collapse load of the panel. This experiment was carried out on a mild steel plate specimen, breadth $b = 30.05$ cm, length $l = 90$ cm, thickness $h = 0.634$ cm, which is loaded by gradually increasing the water pressure. The vertical and horizontal axes denote applied water pressure and deflection measured at the panel center, respectively. In Fig. 12.7, $q_p = 6.87$ kgf/cm^2 indicates the collapse load of a panel $4(h/b)^2\sigma_y$ without considering the effect of membrane stress. This figure represents the panel deflection against the change of water pressure through three inflection points. First, both ends of a panel become plastic hinges, then, a plastic hinge appears at the panel center. It must be noted that the panel neither collapses nor produces a rapid increase in deflection, even when the plastic hinges occur in the panel edges and panel center. This phenomenon is explained by the fact that the water pressure increase does not create an increase in bending stress but the increase in membrane stress. Finally, the area reaching the yield stress covers the whole plate and the plate becomes a plastic membrane. Then the deflection of the plate increases approximately proportionally to the increment of water pressure. Therefore, the plate can bear loads well exceeding the collapse load.

Once the whole plate becomes a plastic membrane, failure of the plate may be caused by some defect in welding or in the material. In addition, this membrane action depends on the boundary rigidity of the structure, and when the boundaries are allowed to move in the plane, full reinforcement effect due to the membrane action cannot be expected. Therefore, this additional redundancy due to the membrane action is not usually considered in the design.

Figure 12.8 shows the panel ratio b/h, in terms of panel breadth and thickness on the horizontal axis and water pressure q in the vertical axis. In this figure, collapse pressure is plotted according to various assumptions: i.e. earlier NK rule pressure on deep tank bulkheads, critical pressure without membrane stress, pressure causing

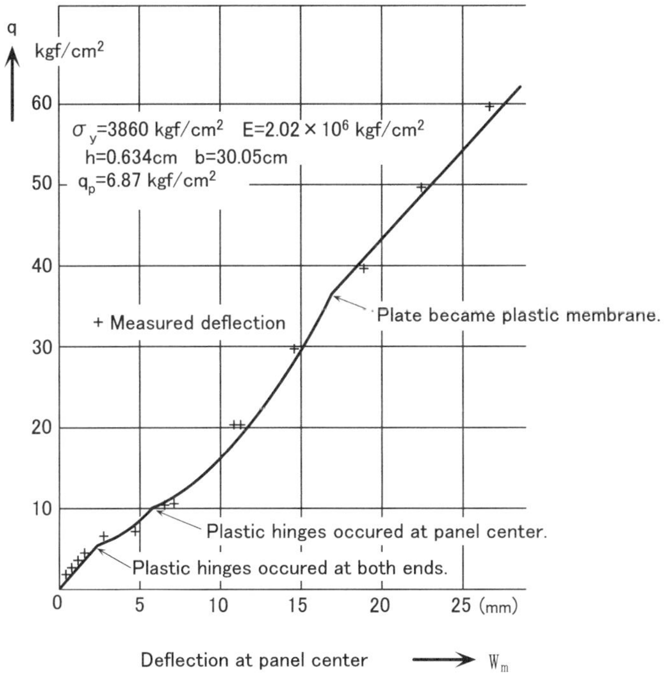

Fig. 12.7 Plastic design of plate affected by membrane force

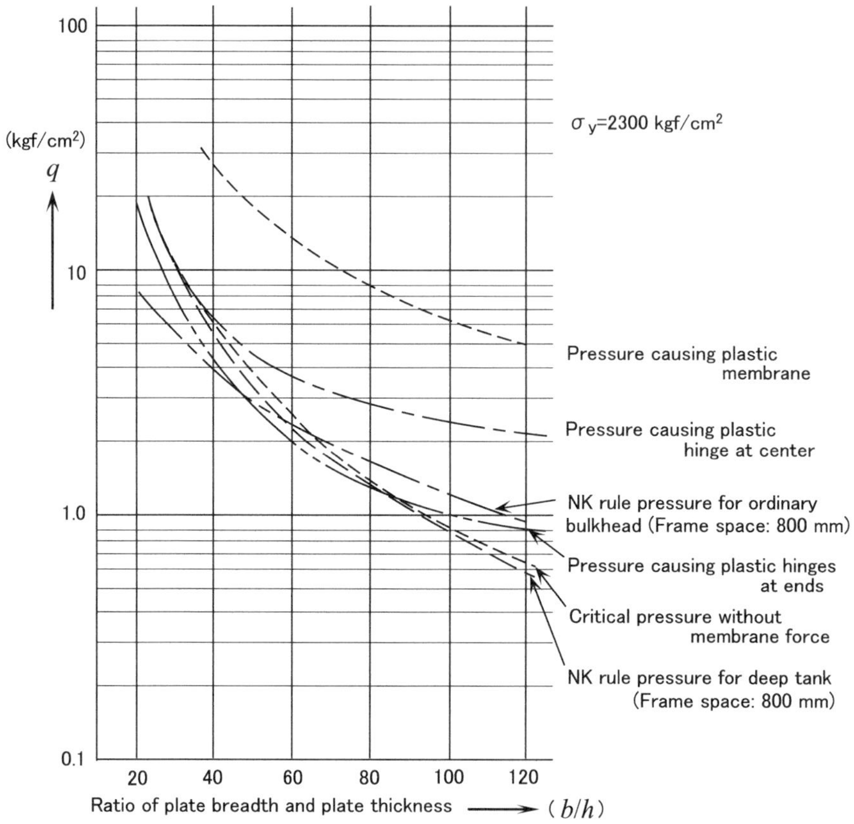

Fig. 12.8 Comparison of critical pressure based on different assumptions

plastic hinges at plate ends, earlier NK rule pressure for ordinary bulkheads, pressure causing the plastic hinge at the plate center, and pressure causing the plastic membrane. We can observe that the deep tank rule has a small amount of safety factor to the critical pressure without membrane force, but has a considerably large safety factor to the plastic hinge formation at center or the plastic membrane formation, especially when the plate is slenderer with large b/h.

12.3 Plates Supporting Bending and Axial Load Simultaneously [76]

Let us suppose that a rectangular panel is subjected to bending due to water pressure and axial load due to longitudinal bending of the hull girder simultaneously. In this case, the axial stress is assumed to be uni-axial because it is caused by hull girder bending. There are two configurations for a rectangular plate which supports

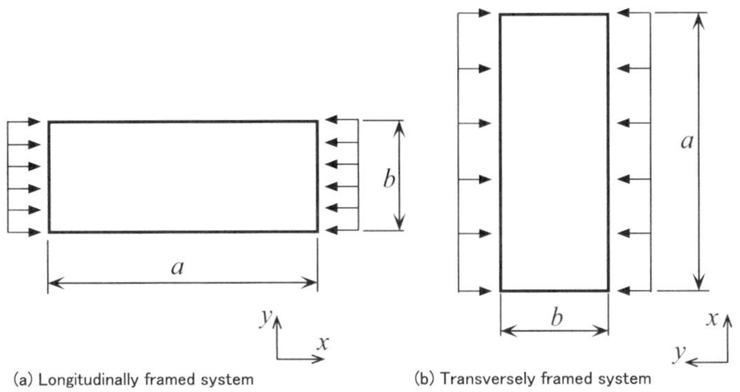

Fig. 12.9 Axial stress acting on a plate according to framing system

axial stress: axial stress is loaded on the longer edge of the plate (transversely framed system) and on the shorter edge (longitudinally framed system), as shown in Fig. 12.9. In this section, x-axis is taken along the longer edge and y-axis is taken along the shorter edge.

12.3.1 Influence of Axial Stress on Fully Plastic Moment

(1) *Transversely framed system:* Let us consider a situation where a plate is bent to form a plastic hinge along the longer edge due to the lateral pressure. When we assume that the bending stress to the y-direction is σ_0, Poisson's effect generates the stress of $\nu_p\sigma_0$ to the x-direction because the displacement to the x-direction due to the bending around the y-axis is constrained, where $\nu_p(= 0.5)$ is the Poisson's ratio in the plastic state. In this situation, the von Mises stress must be equal to the yield stress σ_Y, and the following equation holds.

$$\sigma_0^2 - \sigma_0 \cdot \nu_p\sigma_0 + \left(\nu_p\sigma_0\right)^2 = \sigma_Y^2 \tag{12.3.1}$$

$$\therefore \sigma_0 = \pm \frac{\sigma_Y}{\sqrt{1 - \nu_p + \nu_p^2}} = \pm\alpha\sigma_Y \tag{12.3.2}$$

$$\alpha = \frac{1}{\sqrt{1 - \nu_p + \nu_p^2}} = \frac{2\sqrt{3}}{3} \approx 1.15 \tag{12.3.3}$$

This means that in the fully plastic condition, the normal stress to the y-direction is effectively greater than σ_Y by the ratio of α.

When the axial stress σ_y is loaded on the longer edge, the axial stress acts in the same direction as the bending stress of the plate. Then, in the fully plastic condition, the stress distribution through the thickness is as shown in Fig. 12.10, where η is the ratio within the plate thickness where the stress other than the bending component is acting. Equating the total axial force $\sigma_y t$ to the integration of this stress distribution,

$$\sigma_y t = \eta t \alpha \sigma_Y \tag{12.3.4}$$

where t is the plate thickness.

The plastic moment M_P is given as the couple of the force $\{(1 - \eta)/2\} t \alpha \sigma_Y$ with the lever $\{(1 + \eta)/2\} t$ as follows:

$$M_P = \left(\frac{1-\eta}{2}\right)\left(\frac{1+\eta}{2}\right) t^2 \alpha \sigma_Y = \frac{t^2 \alpha \sigma_Y (1 - \eta^2)}{4} = \frac{\alpha \sigma_Y t^2}{4}\left\{1 - \left(\frac{\sigma_y}{\alpha \sigma_Y}\right)^2\right\} \tag{12.3.5}$$

However, for the axial load, the displacement to the orthogonl direction is not completely constrained, and full consideration of the coefficient α as shown in Eq. (12.3.15) is somewhat non-conservative. Therefore, M_P in the following form is usually considered.

$$M_P = \frac{\alpha \sigma_Y t^2}{4}\left\{1 - \left(\frac{\sigma_y}{\sigma_Y}\right)^2\right\} \tag{12.3.6}$$

From Eq. (12.3.16), the effect of the axial stress on the fully plastic moment is obtained as follows as the ratio of M_P to the fully plastic moment M_{P0} when no axial stress is working.

$$\frac{M_P}{M_{P0}} = 1 - \left(\frac{\sigma_y}{\sigma_Y}\right)^2 \tag{12.3.7}$$

Therefore, the required thickness increase due to the axial stress can be expressed as follows as the ratio to the required thickness t_0 when no axial stress is working.

$$\frac{t}{t_0} = \frac{1}{\sqrt{1 - \left(\frac{\sigma_y}{\sigma_Y}\right)^2}} \tag{12.3.8}$$

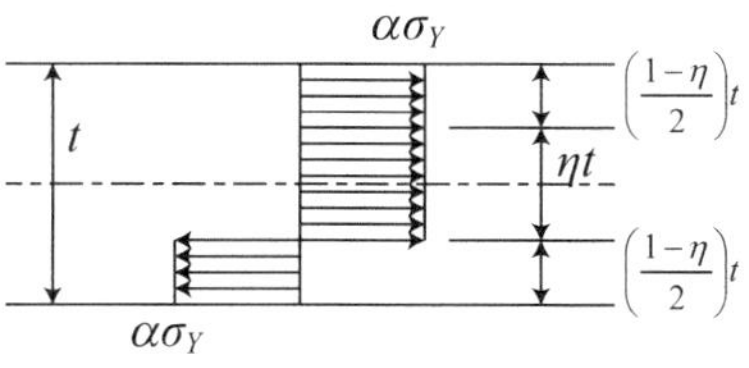

Fig. 12.10 Fully plastic condition under simultaneous action of bending and axial stress

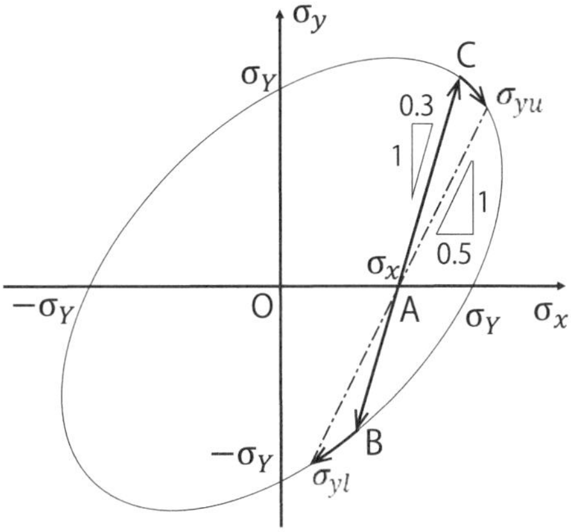

Fig. 12.11 Von Mises yield curve for plate bending in longitudinally framed system

(2) *Longitudinally framed system:* When the axial stress σ_x is loaded on the shorter edge, the axial stress acts in the orthogonal direction to the bending stress of the plate. In this case, the stress conditions can be modeled in a von Mises yield curve as shown in Fig. 12.11. Firstly, we assume the condition designated by point A, where only the uniform axial stress of σ_x is working on the plate. Subsequently, when lateral pressure is gradually loaded, bending stress σ_y increases and σ_x also changes according to the Poisson's effect with a ratio of $\nu = 0.3$. After yielding at point B and C, the stresses follow the von Mises yield curve. After yielding, Poisson's ratio changes to 0.5 and the stresses eventually settle at σ_{yu} and σ_{yl}. Then, using the equation of von Mises yield criterion, σ_{yu} and σ_{yl} are obtained as follows:

$$\sigma_{yu} = -\sigma_{yl} = \alpha\sqrt{\sigma_Y^2 - \sigma_x^2} \tag{12.3.9}$$

and because the total axial force to the y-direction must be zero, the plastic neutral axis resides at the mid-thickness.

Therefore, the fully plastic moment M_P is:

$$M_P = \sigma_{yu} \cdot \frac{t}{2} \cdot \frac{t}{2} = \frac{\alpha\sigma_Y t^2}{4}\sqrt{1 - \left(\frac{\sigma_x}{\sigma_Y}\right)^2} = M_{P0}\sqrt{1 - \left(\frac{\sigma_x}{\sigma_Y}\right)^2} \tag{12.3.10}$$

$$\therefore \frac{M_P}{M_{P0}} = \sqrt{1 - \left(\frac{\sigma_x}{\sigma_Y}\right)^2} \tag{12.3.11}$$

Finally, the required thickness increase ratio is:

$$\frac{t}{t_0} = \frac{1}{\left\{1 - \left(\frac{\sigma_x}{\sigma_Y}\right)^2\right\}^{\frac{1}{4}}} \tag{12.3.12}$$

12.3.2 Influence of Additional Lateral Force Due to Axial Load

When the axial stress is loaded on the longer edge, the effect of the additional lateral force caused by the axial load is significant. When we consider a plate strip under an axial tensile load N as shown in Fig. 12.12, the equation of the elastic curve can be expressed as shown in Eq. (12.3.13), where the last term is the additional lateral force due to the axial load.

$$EI\frac{d^4v}{dx^4} = w + N\frac{d^2v}{dx^2} \tag{12.3.13}$$

where E is Young's modulus and I is the sectional moment of inertia.

Solving this equation under relevant boundary conditions and equating the bending moment at the span end or at the midspan with the fully plastic moment M_P, we can obtain the load necessary to form 2-point plastic hinges at the span ends or to form 3-point plastic hinges at the span ends and the midspan as follows:

$$w_{2H-t} = \frac{2\beta^2 M_P\left(e^{\beta l} - 1\right)}{\beta l + \beta l e^{\beta l} - 2e^{\beta l} + 2} \tag{12.3.14}$$

$$w_{C-t} = \beta^2 M_P\left(\frac{1 + e^{\frac{\beta l}{2}}}{1 - e^{\frac{\beta l}{2}}}\right)^2 \tag{12.3.15}$$

$$w_{2H-c} = \frac{2\beta^2 M_P(-1 + \cos\beta l)}{\beta l \cdot \sin\beta l - 2 + 2\cos\beta l} \tag{12.3.16}$$

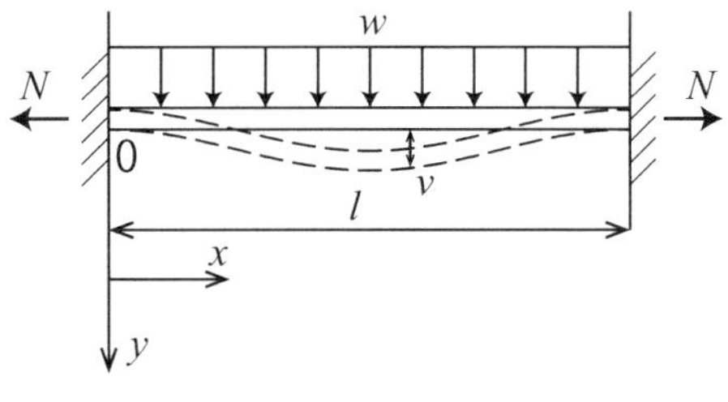

Fig. 12.12 Bending of plate under tensile axial load

$$w_{C-c} = \beta^2 M_P \cdot \frac{1 + \cos\frac{\beta l}{2}}{1 - \cos\frac{\beta l}{2}} \tag{12.3.17}$$

where the subscripts "$2H$" and "C" mean 2-point and 3-point plastic hinge formation, respectively; the subscripts "$-t$" and "$-c$" mean tensile and compressive axial load, respectively, and

$$\beta = \sqrt{\frac{|N|}{EI}} \tag{12.3.18}$$

The strength ratios with and without the action of the axial stress are summarized in Figs. 12.13, 12.14, 12.15 and 12.16. Figures 12.13 and 12.15 show the strength ratio with respect to the 2-point plastic hinge formation and Figs. 12.14 and 12.16 show the strength ratio with respect to the 3-point plastic hinge formation. The thick dashed lines are the strength reduction due to Eq. (12.3.17), i.e., the results without consideration of the additional lateral force due to the axial load.

From these figures, we can observe that the effect of the additional lateral force due to the axial load is prominent, especially for thinner plates. Tensile axial load enhances the strength while compressive axial load deteriorates the strength. The effect is more prominent in 3-point plastic hinge formation than in 2-point plastic hinge formation.

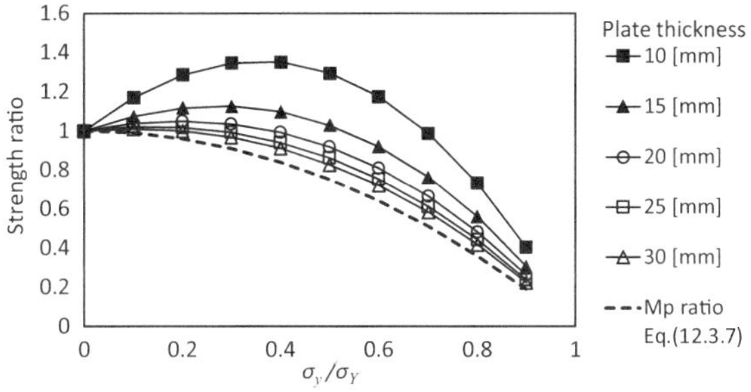

Fig. 12.13 Strength ratio of 2-point plastic hinge formation (tensile axial stress)

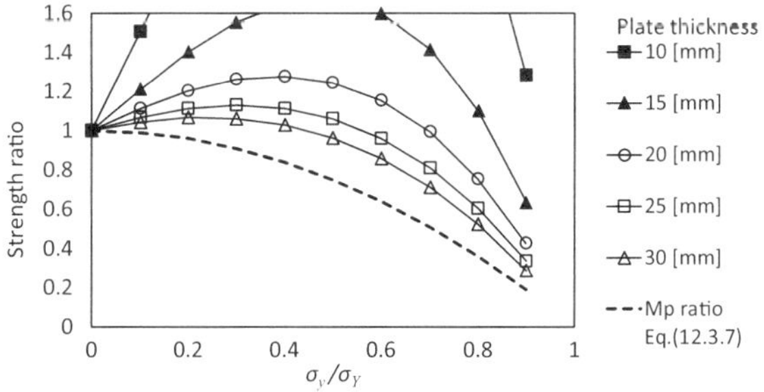

Fig. 12.14 Strength ratio of 3-point plastic hinge formation (tensile axial stress)

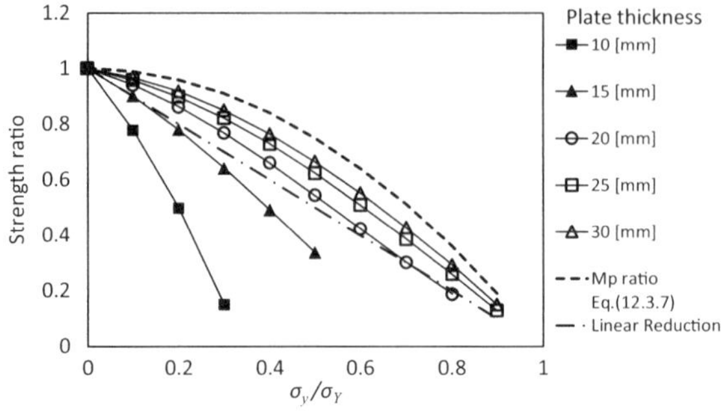

Fig. 12.15 Strength ratio of 2-point plastic hinge formation (compressive axial stress)

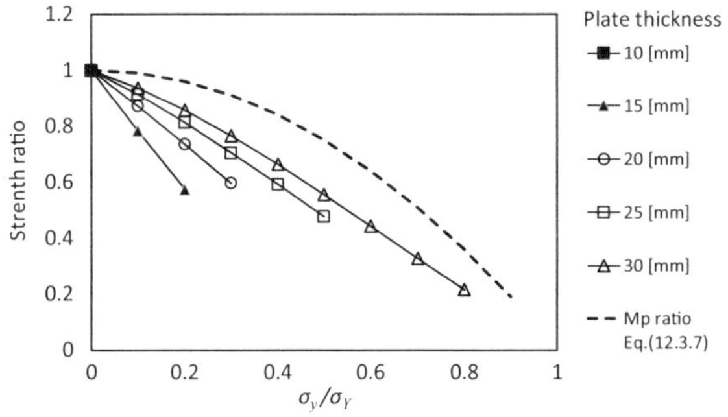

Fig. 12.16 Strength ratio of 3-point plastic hinge formation (compressive axial stress)

In the case of tensile axial load, Fig. 12.13 shows that the strength enhancement is not so significant when the plate thickness is 20 mm or more. Therefore, it may be prudent to employ Eq. (12.3.17) as the strength criteria and to consider the small strength enhancement due to the additional lateral force as a strength margin.

From Fig. 12.15, we can understand that Eq. (12.3.17) is non-conservative and stricter criteria are necessary to account for the adverse effect of the additional lateral force due to the compressive axial load. It is observed that the chained-dotted straight line can cover the results of the thickness greater than 20 mm in conservative side, and this can be employed as the strength criterion. The strength ratio for the thickness of 15 mm or less is lower than this criterion, but these thicknesses are not actually feasible because they cannot satisfy the buckling criteria.

In summary, strength ratio and required thickness ratio under the effect of axial stress are tabled in Table 12.2. For transversely framed system under compressive axial load, linear strength reduction model is employed, while for the other cases, strength reduction model with von Mises yield criterion (Eqs. (12.17) and (12.21)) is employed. These formulations are adopted in ClassNK rules [77] as coefficients of axial force effect, other than the case of transversely framed system under tensile axial load because practically, both tensile and compressive axial loads act on most plate members under alternating hogging and sagging conditions.

Figure 12.17 represents the required thickness ratio against the working axial stress. According to this figure, we find that the required thickness does not increase

Table 12.2 Strength ratio and required thickness ratio under the effect of axial stress

		Applied criterion	Strength ratio	Required thickness ratio t/t_0
Longitudinally framed system		Von Mises criterion with orthogonal axial and bending stress	$\sqrt{1-\left(\frac{\sigma_x}{\sigma_Y}\right)^2}$	$\left\{1-\left(\frac{\sigma_x}{\sigma_Y}\right)^2\right\}^{-\frac{1}{4}}$
Trans-versely framed system	Tension	Von Mises criterion with parallel axial and bending stress	$1-\left(\frac{\sigma_y}{\sigma_Y}\right)^2$	$\left\{1-\left(\frac{\sigma_x}{\sigma_Y}\right)^2\right\}^{-\frac{1}{2}}$
	Compression	Linear strength reduction	$1-\frac{\lvert\sigma_y\rvert}{\sigma_Y}$	$\left(1-\frac{\lvert\sigma_y\rvert}{\sigma_Y}\right)^{-\frac{1}{2}}$

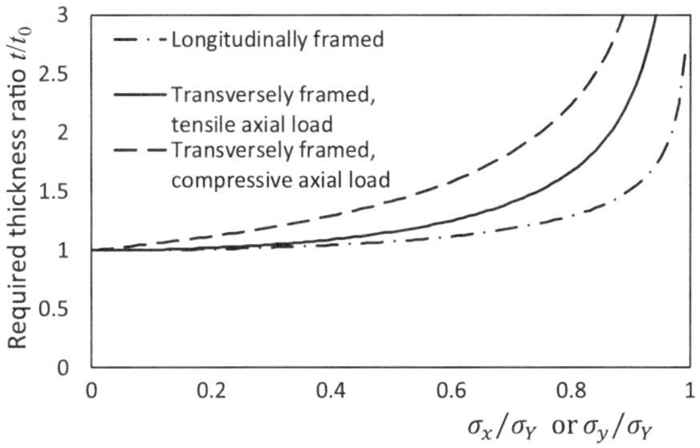

Fig. 12.17 Required thickness ratio due to axial load effect

significantly within the area where the axial stress is less than half of the yield stress in case of the transversely framed system under tension and the longitudinally framed system. For the transversely framed system under compression, the thickness increase is more significant, requiring more attention. Whereas, for greater axial stresses, the required thickness rapidly increases.

12.3.3 Axial Stress Effect for Plate with Small Aspect Ratio [78]

In Table 12.2, coefficients of axial stress effect are given separately for longitudinally framed system and transversely framed system. This is not rational when the plate aspect ratio is close to 1.0 because the required thickness becomes discontinuous at the aspect ratio of 1.0 between the longitudinally long plate and the transversely long plate. Therefore, it is necessary to apply intermediate coefficient of axial stress effect when the aspect ratio is close to 1.0.

From the observation of the formulations of the strength ratio in Table 12.2, they all can be generalized in the form of Eq. (12.3.19), where $\alpha = 2$ and $\beta = 0.5$ for

Table 12.3 Interpolation of parameters α and β

	Axial stress	s/l	α	β
Longitudinally framed system	Common to tensile and compressive	$\frac{s}{l} \leq 0.5$	2	0.5
		$0.5 \leq \frac{s}{l} \leq 1$	2	$\frac{s}{l}$
Transversely framed system	Tensile	$1 \leq \frac{s}{l}$	2	1
	Compressive	$1 \leq \frac{s}{l} \leq 2$	$2\frac{l}{s}$	1
		$2 \leq \frac{s}{l}$	1	1

Note
s: Length of the edge orthogonal to the axial load
l: Length of the edge parallel to the axial load

longitudinally framed system; $\alpha = 2$ and $\beta = 1$ for transversely framed system with tensile axial load; and $\alpha = 1$ and $\beta = 1$ for transversely framed system with compressive axial load.

$$(\text{Strength ratio}) = \left\{1 - \left(\frac{|\sigma_a|}{\sigma_Y}\right)^{\alpha}\right\}^{\beta} \tag{12.3.19}$$

where σ_a is σ_x or σ_y depending on whether the axial load is acting on the shorter edge or on the longer edge. ClassNK rules [77] have employed the interpolation of α and β between the aspect ratio of 0.5 and 2.0 as shown in Table 12.3, while applying the coefficients of axial stress effect of Table 12.2 when the aspect ratio is less than 0.5 or greater than 2.0 (other than the case of transversely framed system under tensile axial load).

When the aspect ratio is close to 1.0, the plate is stronger than the assumption of the rule scantling formula where only the length of the shorter edge is used as the span length of the plate strip bent by the lateral pressure, as we have already seen in Figs. 12.4 and 12.5. Classification society rules usually give a coefficient to account for this strengthening effect, and they are applied in the scantling formula in combination with the coefficients of axial stress effect.

12.4 Strength of Plates by In-Plane Loads

The upper deck plates and bottom plates are subjected to in-plane loads, for they act as face plates of the hull girder against longitudinal bending. The in-plane load has two nominal components; tension and compression. The former may cause cracking, which will be discussed in Sect. 12.7. Let us now consider the latter, which may cause buckling of a plate in the same manner as for a pillar.

Let us investigate the buckling stress according to stiffening arrangement: i.e. longitudinally stiffened system and transversely stiffened system. The critical buckling stress can be obtained as follows by referring to Eq. (3.5.22) in Part I.

$$\sigma_{cr} = \frac{\pi^2 E}{12(1-\nu^2)}\left(\frac{t}{b}\right)^2 K\eta \tag{12.4.1}$$

where

K: minimum value of k

$k = \left(\frac{m}{\alpha} + \frac{\alpha}{m}\right)^2$ (see Fig. 12.18)

$$\alpha = a/b$$

η: Johnson's yield modification factor.

Regarding the longitudinally stiffened system as illustrated in Fig. 12.9a, the aspect ratio α of the panel is much greater than 1.0, because the length of the unloaded edge a is larger than that of the load carrying edge b in this system. Thus buckling coefficient K can be taken approximately as 4.0 from Fig. 12.18. Hence the critical buckling stress of the longitudinally framed panel σ_{cr}^{l} is expressed in the following form:

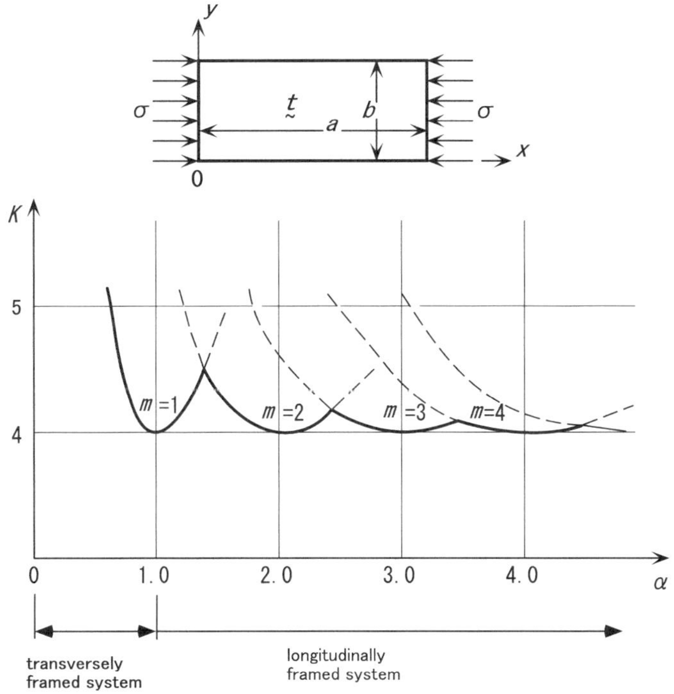

Fig. 12.18 Buckling coefficient K against aspect ratio *a*

$$\sigma_{cr}^{l} = \frac{4\pi^2 E}{12(1-\nu^2)}\left(\frac{t}{b}\right)^2 \eta \tag{12.4.2}$$

Next, let us imagine a transversely stiffened panel where the length of the unloaded edge is b and the length of the load carrying edge is a. In this case, the length of the load carrying edge is a instead of b in Eq. (12.4.2), and the critical buckling stress of transversely-framed panel σ_{cr}^{t} is given as in the form:

$$\begin{aligned} \sigma_{cr}^{t} &= \frac{\pi^2 E}{12(1-\nu^2)}\left(\frac{t}{a}\right)^2 K\eta \\ &= \frac{\pi^2 E}{12(1-\nu^2)}\left(\frac{t}{\alpha b}\right)^2 K\eta \\ &= \frac{K}{\alpha^2}\frac{\pi^2 E}{12(1-\nu^2)}\left(\frac{t}{b}\right)^2 \eta \end{aligned} \tag{12.4.3}$$

By taking the ratio σ_{cr}^{t} and σ_{cr}^{l}, we find the following relation:

$$\frac{\sigma_{cr}^{t}}{\sigma_{cr}^{l}} = \frac{K}{4\alpha^2} \tag{12.4.4}$$

In the case of a transversely framed system, K takes a minimum value by putting $m = 1$ within the range where the aspect ratio is less than 1.0, therefore, the following relationship can be derived:

$$K = \left(\alpha + \frac{1}{\alpha}\right)^2$$

$$\therefore \frac{\sigma_{cr}^{t}}{\sigma_{cr}^{l}} = \frac{1}{4}\left(1 + \frac{1}{\alpha^2}\right)^2 \tag{12.4.5}$$

When the aspect ratio $1/\alpha$ of this panel is close to zero, this buckling stress ratio converges to 1/4. Consequently, roughly speaking, the buckling strength of a transversely framed panel is one-fourth of that of a longitudinally framed panel.

Figure 12.19 illustrates one example of buckling damage observed in transversely framed upper deck plates of a cargo vessel. The measurement of the buckle was taken at 1,250 mm from the hatch side on the starboard side, and 1,500 mm on the port side. Unlike longitudinally framed vessels, transversely framed vessels sometime suffer from such buckling damage.

However, it is sometimes inevitable to arrange framing in the weaker direction. In such a case, we must be careful to maintain buckling strength by increasing the thickness of the plate or by fitting additional stiffeners.

Now, let us consider a quiz. Suppose you must reinforce a panel shown in Fig. 12.9b by adding only one piece of carling, how would you arrange the carling to

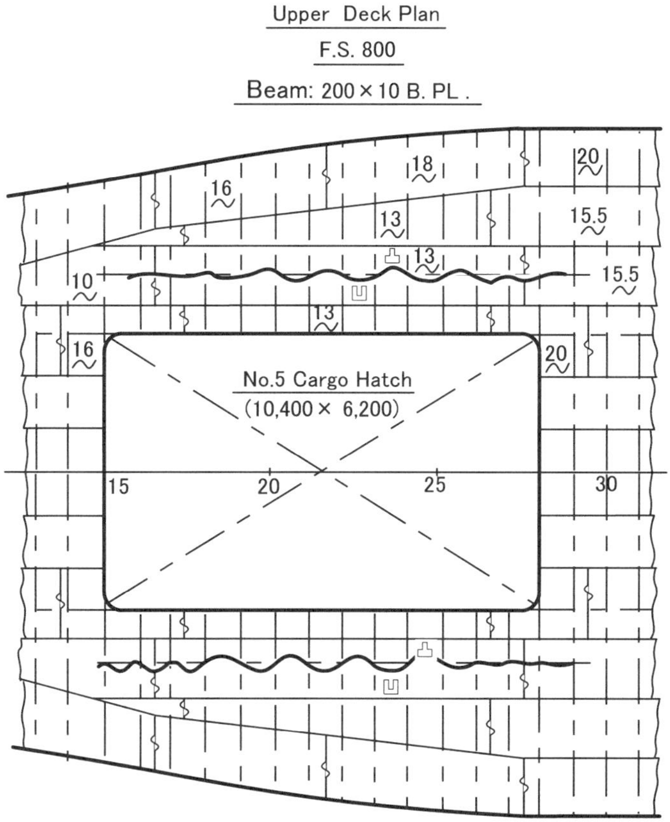

Fig. 12.19 Example of buckling on transversely framed upper deck (maximum deflection 17 cm)

achieve the maximum increase in buckling strength? You may have a fixed idea that it is better to arrange stiffeners parallel to the direction of force, and consider adding a small piece of carling parallel to the stress as shown in Fig. 12.20a. However, the actual carling arrangement shown in Fig. 12.20b is much stronger. It is very important and useful for designers to have a quantitative sense of the efficiency of each arrangement of additional carling, and the readers are expected to calculate buckling strength by themselves and feel it in his bones once in his technical career.

Regarding the web stiffeners and carlings fitted to divide panels to ensure sufficient buckling strength, how should we determine scantlings? There are many criteria to consider such as vibration, etc., but fundamental idea is that the overall buckling should not precede the buckling of each divided plate. To achieve this, the stiffeners should have minimum flexural rigidity, and usually designers apply an appropriate depth / length ratio (such as 0.08) to flat bar stiffeners to keep this minimum flexural rigidity. The thickness of a flat bar stiffener is determined from the viewpoint of torsional buckling. If the stiffener is under significant axial compression, the depth /

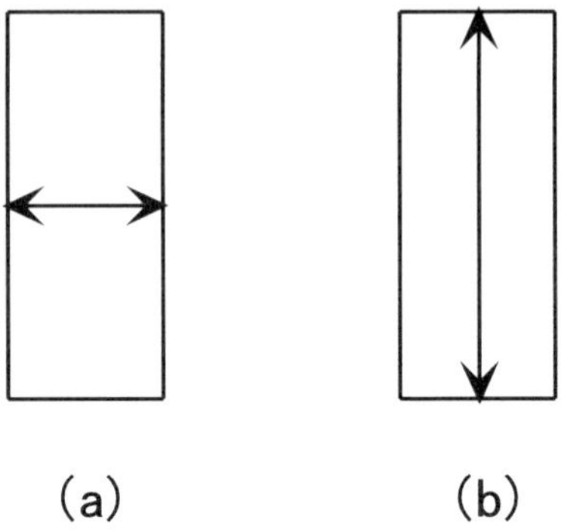

Fig. 12.20 Two ways of buckling reinforcement

thickness ratio should be less than 15, but otherwise it can be about 18 or 20. "Know-why" is more important than "know-how." The theoretical background of these values can be learned from many other textbooks available, and practical designers should realize why such values are required in order to correctly apply them to actual design.

12.5 Stress Concentration Around Openings

If an in-plane load acts on a plate having an opening such as the upper deck plate or bottom plate, the opening induces a stress concentration around its free edge. It is a well-known fact that the stress concentration factor around a circle in an infinitely wide plate subjected to a uniform tensile stress is 3.0 maximum. On the other hand, the stress concentration factor of an ellipse is 2.0, in the case where the elliptic opening, of which length ratio of the longer axis and shorter axis is 2:1, is subjected to uniform tension along the longer axis. The ellipse in Fig. 12.21 is normally applied in hull structure design. The maximum stress occurs at Point A of the ellipse. In providing this shape of ellipse to an upper deck opening, the free edge of the opening is not reinforced as long as the size of opening is not so big compared with the breadth of upper deck. We will discuss this point later.

Although the ellipse, with length ratio 2:1, has the advantage of reducing the stress concentration around the opening, it has an economic disadvantage because of the size of the associated manhole cover plate or hatch coaming. To solve this problem, a smart shape of opening, which has smaller length ratio between longer and shorter axis than the ordinary ellipse, was developed [79] as shown in Fig. 12.22. This shape of opening was discovered in 1965 from the results of photo-elastic experiments. The length ratio of the longer axis against the shorter axis of this opening is 1.4 and it is called a "semi-elliptic opening".

This semi-ellipse is developed on the basis of the following idea as shown in Fig. 12.23:

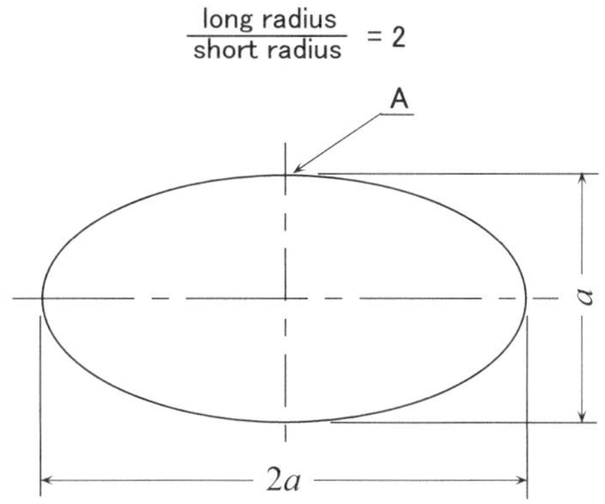

Fig. 12.21 Elliptic opening

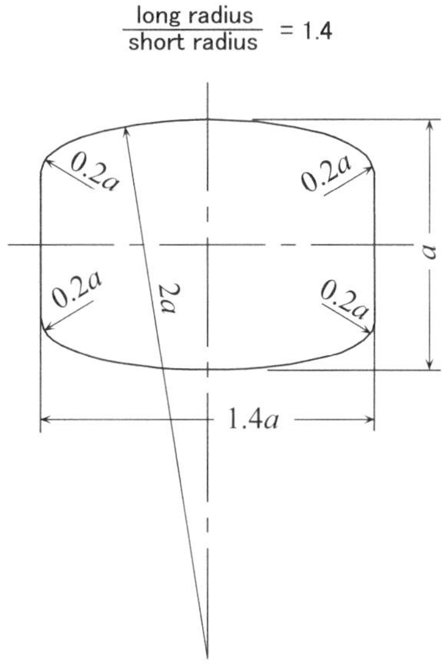

Fig. 12.22 Semi-elliptic opening

(1) A large arc C touching the ellipse B at point A can be drawn in such a way that the curvature of the arc ought to be about the same as that of the ellipse. The stress concentration is approximately proportional to the curvature, therefore the stress concentration factor of this large arc becomes around 2.0, similar to that of the ellipse at point A. In this case, we found the radius of the large arc should be $2a$.

(2) Both ends of a large arc are cut off by line D and line E, in order to decrease the length along the longer axis. Here we took the length of $1.4a$.

(3) To avoid a stress concentration due to abrupt cutting of the large arc by line D, arc C and line D should be connected with a small arc F. The stress concentration in arc F can be less than 2.0, if the radius of arc F is $0.2a$.

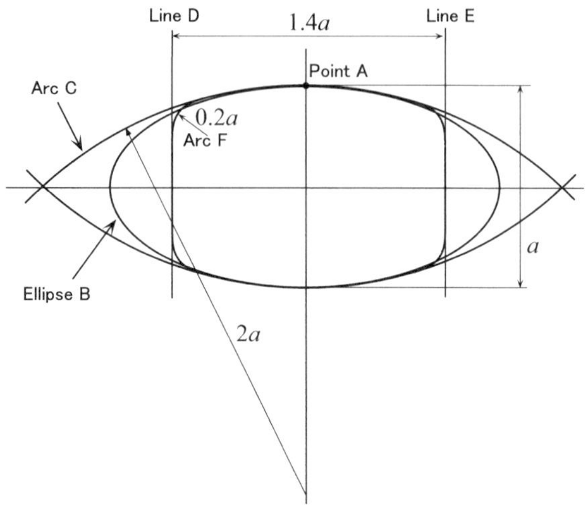

Fig. 12.23 Concept of semi-elliptic opening

In this manner, we obtained a suitable shape called a semi-elliptic opening in Fig. 12.22. Figure 12.24 represents the stress distribution around the semi-elliptic opening under uni-axial tension, which indicates that the stress in the small arc is lower than in the center of the large arc, thanks to the interchanging tension and compression within the small arc.

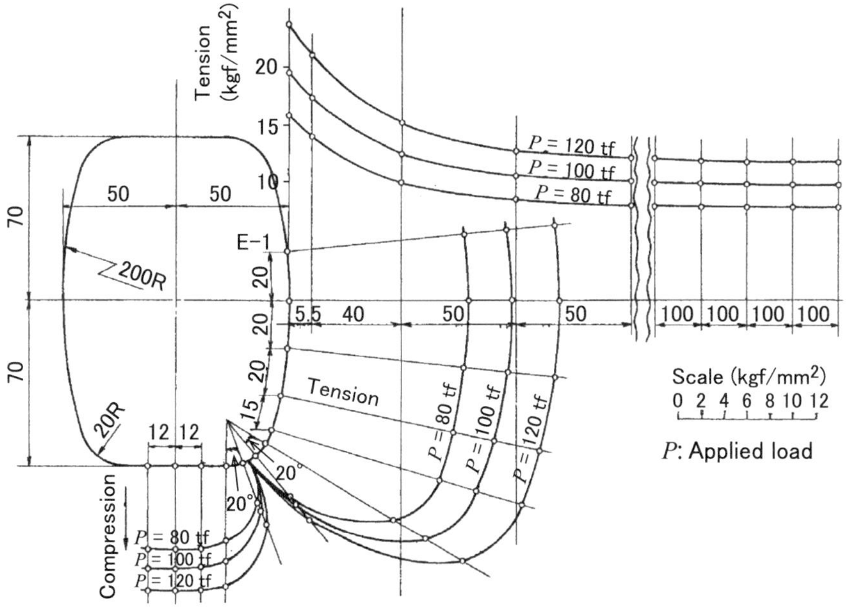

Fig. 12.24 Improved shape of opening and stress distribution

By further investigations, the stress concentration in the elliptic opening and semi-elliptic opening were measured for the case where the longer axis of both openings was slightly inclined against the tensile direction. This measurement was aimed at confirming the increase of the stress concentration around the openings due to a small inclination, because the actual opening in the vessel has a certain fabrication tolerance and because the direction of the longitudinal stress due to bending is distorted in oblique waves. From this additional measurement, the stress concentration around the opening increased by the increment of inclination angle both in the case of the elliptic opening and the semi-elliptic opening. Comparing these two openings, the stress concentration of a semi-elliptic opening is more affected by the inclination angle than the elliptic opening.

Indeed, the above discussion is true, as long as the opening exists in a plate with infinite breadth. But the situation is different if the opening is made in a plate of finite width such as an upper deck plate, since the stress around the opening varies according to the opening size. When the opening size is sufficiently large compared with the breadth of the upper deck plate, the remaining plate area around the opening becomes so small that the nominal stress increases at the opening section. In addition to this increased nominal stress, the stress around the opening becomes large due to the stress concentration of the opening itself. To prevent such large stresses, a reinforcement is necessary to compensate for the reduced plate area.

In order to clarify the upper limit of the opening size allowable without reinforcement, finite element analyses of the stress distribution around a semi-elliptic opening were performed for changes in the opening size in a plate with constant breadth. Figure 12.25 represents the results [80]. According to Fig. 12.25, we find the stress concentration at the free edge of the opening remains at about 2.0 where a/b is less than 10%. This concentration is almost the same as that of a semi-elliptic opening in a plate with infinite breadth.

Keeping such a situation in mind, let us consider the critical limit of opening without reinforcement specified by the Classification Societies' rules. NK allows an opening without reinforcement when its breadth is within 6% of the upper deck breadth, and LRS allows an opening within 10%. These requirements seem to be quite reasonable considering the above calculation results.

12.6 Material and Roll Direction

It is very important to design correct arrangement of steel plate materials. This is because if different grades of steel plate are incorrectly applied, there is nothing the designer can do other than replace the incorrect steel plate with another good plate. It is less serious to have a thinner thickness of plate or stiffener scantlings by mistake, because the former can be compensated by adding stiffeners to the plate and the latter can be solved by providing suitable reinforcement such as additional face plate. A mistake in the selection of grade of steel sometimes occurs in refrigerated cargo vessels constructed with low temperature steel plates. Such steel plates are

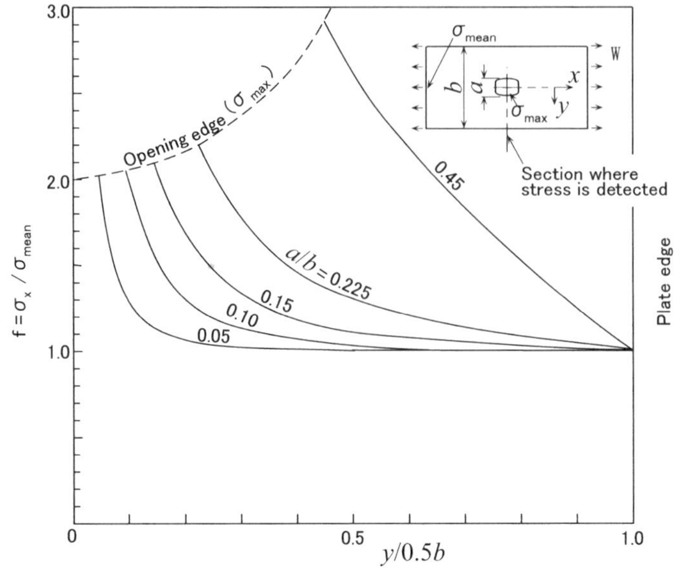

Fig. 12.25 Relation between opening size and stress concentration

usually covered by insulation material; therefore, replacing this steel owing to a design mistake causes serious problems at the ship construction stage. Thus, the designer must pay careful attention not to use improper steel material.

With regard to steel material, let us discuss the strength of steel plate which is different in different axes of the plate; i.e. roll direction (L), the athwart direction to the roll (T) and the plate thickness direction (Z). In general, the strength of normal steel plate in the L direction is higher than that of in T or in Z. Hence, plates of longitudinal strength members such as upper deck plates and bottom shell plates are usually arranged along the L direction, not along the T direction.

Nevertheless, these strengths have a scatter in the L, T, and Z direction depending on the method used to produce the steel plate. For instance, Table 12.4 [81] shows strength for each direction obtained from a steel plate produced by the controlled rolling method (CR steel), which shows a similar strength in all the directions. If such steel is adopted, the T direction can be applied even as the longitudinal direction of a longitudinal strength member. This type of steel is beneficial for easy construction.

Figure 12.26 [81] shows results of Charpy V-notch impact test of the CR plate given in Table 12.4. Although the yield stress and tensile strength are similar in all the directions, regarding absorption energy given by the Charpy V-notch test, the absorption energy in the T direction is less than that of the L direction. Furthermore, the energy in the Z direction is much lower.

A steel plate formed by the cold bending process such as bilge shell plate or gunwale can absorb less energy than a plate without form bending, and the reduction of absorption energy depends on the amount of bending strain. Figure 12.27

Table 12.4 Deviation of steel strength due to difference of direction

Steel	Direction	Yield stress (kgf/mm²)	Tensile strength (kgf/mm²)	Elongation (for L, T) or Area contraction (for Z) (%)
A	L	40.4	53.6	29.2
	T	41.8	54.3	25.8
	Z	39.3	54.7	68.2
B	L	36.9	50.3	29.9
	T	37.9	51.0	26.8
	Z	37.0	51.8	61.7

Note: A and B steel are higher tensile strength steel of 36kgf/mm^2 with grade E

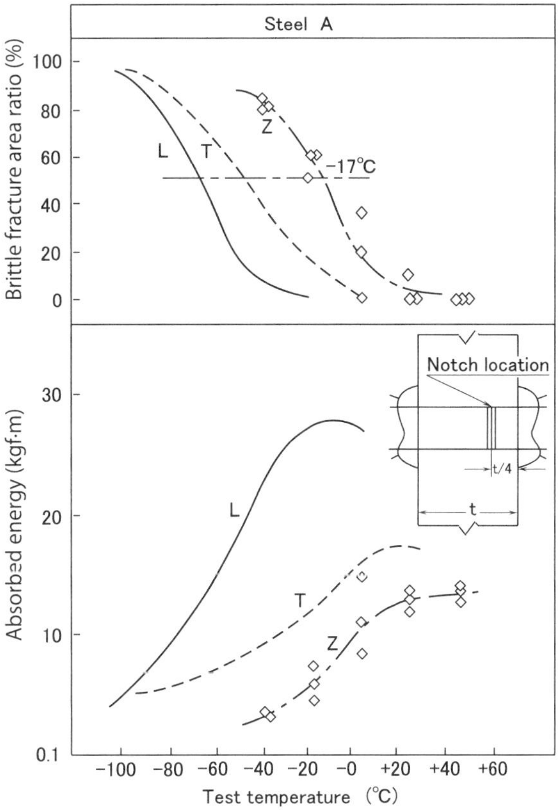

Fig. 12.26 Charpy V-notch impact test according to specimen's roll direction

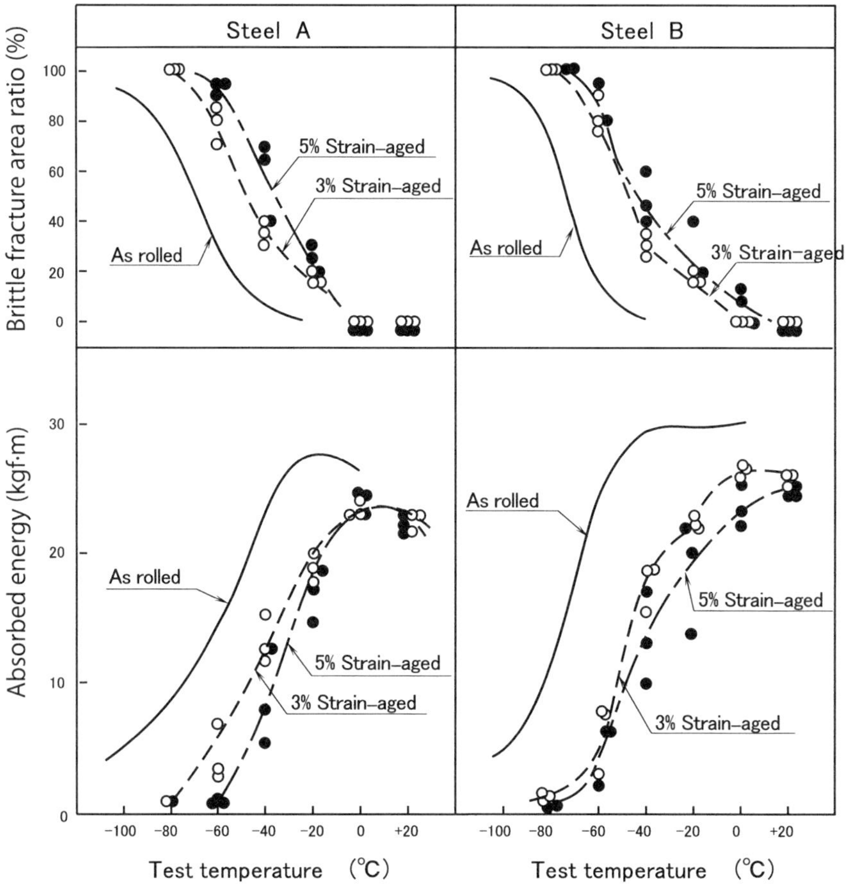

Fig. 12.27 Impact test result of strain-aged steel

represents the results of V-notch Charpy impact tests of strain-aging steel plate of two types of E grade steel; Steel A and Steel B. The absorption energy of strain aged steel is considerably less than that of as rolled steel. A bilge shell plate or gunwale, if constructed by cold bending, may have less absorption energy; however, we can say no problem would arise because the direction of bending strain of the plate is perpendicular to the longitudinal stress due to hull girder bending.

12.7 Damage of Plate

In the tensile test of a rod specimen, necking begins at some part of the specimen and finally it breaks at this section with reduced sectional area. In such tests the fractured surface is not normal to the tensile direction but is inclined at about 45°. This is a typical ductile fracture, whereas, a plate fracture observed in an actual vessel often shows a different way of breaking. The breaking rarely happens in association with the necking phenomena, or some significant sectional area reduction. In addition, the fractured surface is usually approximately normal to the plate surface. This fact suggests that the breaking damage in a plate is brought about by lower stress in the actual vessel compared to the breaking stress measured in experiments.

According to this actual damage of vessels, we consider it is not reasonable to apply the design concept based on tensile strength σ_B to practical design. Figure 12.28 illustrates examples of plate fractures in damaged vessels. Figure 12.28a indicates that the damage is caused by fatigue cracking. This crack firstly initiated from Point A, the highest stress part around the slot opening, and secondly it propagated to Point B tracing the high stress path in the web plate. There is neither reduction in plate thickness nor an inclination of the broken surface. Figure 12.28b shows a structural damage of the outer shell plating and transverse web plate. This damage was caused by wave impact force; i.e. the breaking of the shell plate was initiated by tensile stress due to wave impact acting on the curved shell plate at Point A. There was no plate thinning and the fractured surface was normal to the shell plate. On the other hand, the broken surface of the web plate at B-B section showed the same tendency as experimental results; namely the web plate was constricted at the broken line and the fractured surface was not normal to the plate surface. The crack in the above outer shell plate is called a rapid fracture due to wave impact force and shows the Chevron pattern. After the breaking of the shell plate, a very big force was applied on the web plate which showed a ductile fracture.

In the case of the damage on transverse members shown in Fig. 10.1b, the web frame was damaged by the breaking down of the cross tie, then the side shell was broken. On the other hand, in the case of the damage shown in Fig. 10.1c the side shell was dented in about 3 m and the bottom shell about 1 m. The bilge plate deformed very much. In such a situation the shell plate kept water tightness.

As for the stranding damage of the bottom shell plate, one example indicated that the bottom shell was dented about 30 cm, which created the failure of the internal member but did not cause the breaking of the shell plate.

From the above examples of plate damage, we can say that plate has sufficient toughness that in most situations it rarely fractures. However, the plate might fracture under an unlucky case as in Fig. 12.28b.

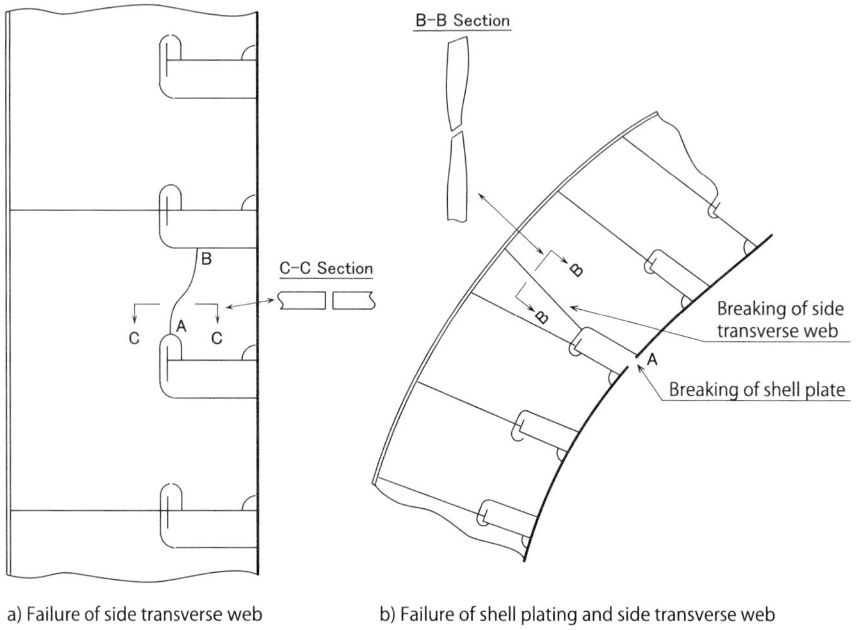

Fig. 12.28 Examples of plate failure

Chapter 13
Design of Stiffened Panel

The hull structure consists of stiffened panels; bottom construction, side shell construction, upper deck construction, bulkhead, etc. Usually stiffened panels consist of plates, beams (small member, secondary member) and girders (big member, primary member). The plate receives loads such as water pressure; the beam supports the loads from the plate and the girder supports the loads from the beam.

In Sect. 8.4 Optimum Design of Beam Section and Sect. 9.2 Rational Design of Girders, the investigation was carried out based on the concept that optimum means minimum weight. The concept stands on the understanding that for a single structural member, the lighter member maintaining the same strength is preferable.

With a stiffened panel many beams and girders are arranged in longitudinal and transverse directions. In designing a definite area of the stiffened panel, spacing between each member is very important not only for saving weight but also saving construction labour costs. The weight and the labour are important parameters to manage in a shipyard.

Under the above considerations "minimum building cost" depends on the design philosophy for the stiffened panels. However, construction methods are being improved year by year and the construction cost is getting smaller; therefore, mainly minimum weight viewpoint discussions will be presented below.

In this chapter firstly, a rational grillage structure which has girders crossing each other (mutually-supported structure) is explained; then secondly the optimum spacing of girders which are arranged only in one direction is discussed, and finally optimum beam space is introduced.

The girder arrangement of the grillage structure and the space of girders are studied from a strength viewpoint, and the beam arrangement is studied not only from a strength viewpoint but also from a vibration viewpoint especially to the stern and engine room construction where vibration problems are generated so often.

In the case of a simple lateral load, the results in this section can be applied as they are, but in the design of a stiffened panel for an upper deck, the longitudinal strength of the hull girder is to be considered where another criterion, the minimum cost of the hull girder including upper deck structure, will be applicable

Y. Okumoto et al., *Design of Ship Hull Structures*,
https://doi.org/10.1007/978-3-032-12500-2_13

13.1 Grillage Construction

Figure 13.1. shows a stiffened panel with length of longer edge a, shorter edge b and uniform load p. It is common sense for the girders to be arranged in the direction of the shorter span as the girders in the longer span are not so effective. Here common sense is proved quantitatively.

As the results for fixed-boundary conditions and for simply supported conditions are similar hereafter, the simply-supported boundary condition is: to be applied [23].

The maximum stress σ_y and the maximum deflection δ are generated at the midpoint O in Fig. 13.1, and they are obtained as follows:

$$\sigma_y = \frac{\pi^2 \delta E e_y}{b^2}$$

$$\delta = \frac{abp}{\frac{\pi^6 E}{16}\left\{\frac{I_x(m+1)}{a^3} + \frac{I_y(n+1)}{b^3}\right\}} \tag{13.1.1}$$

Applying the following notations, maximum stress σ_y is described below.

I_x and I_y are sectional moment of inertia with effective breadths.

m and n are numbers of girders.

e_x and e_y are distances from center of gravity of section to face plates.

Spaces of girders: $l_x = b/(m+1)$ and $l_y = a/(n+1.)$

Rigidity ratio per unit breadth: $i_x = I_x/l_x$ and $i_y = I_y/l_y$

Rigidity ratio in longer and shorter edge directions (mutually-supporting ratio)

$$\alpha = i_x/i_y$$

$$\sigma_y = \frac{16b^2 p l_y}{\pi^4 \frac{I_y}{e_y}\left(1 + \alpha \frac{b^4}{a^4}\right)} \tag{13.1.2}$$

Considering σ_y as the allowable stress, the required section modulus Z_y is given by the following equation.

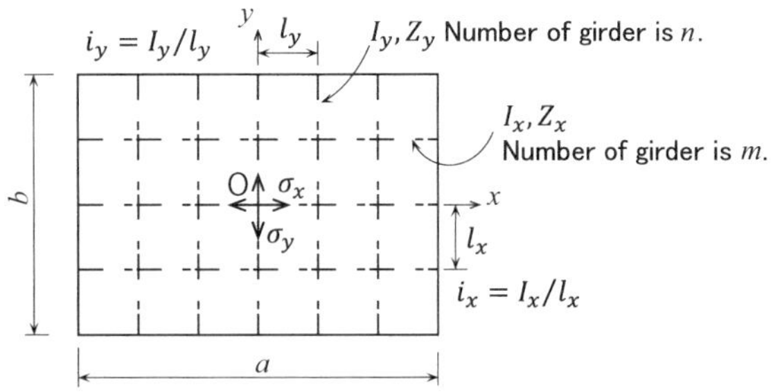

Fig. 13.1 Grillage structure

$$Z_y = \frac{I_y}{e_y} = \frac{16b^2 p l_y}{\pi^4 \sigma_y \left(1 + \alpha \frac{b^4}{a^4}\right)} \tag{13.1.3}$$

Putting weight per unit area of grillage structure as W_1, and weights per unit length of girders in longer and shorter directions as W_x and W_y respectively, the following relation is: obtained.

$$W_1 = \frac{W_x}{l_x} + \frac{W_y}{l_y} \tag{13.1.4}$$

Usually the same scantling girders are to be applied for x and y directions in a mutually-supported grillage structure. Applying this principle, the following results are obtained.

$$I_x = I_y, Z_x = Z_y, W_x = W_y \tag{13.1.5}$$

$$\alpha = \frac{i_x}{i_y} = \frac{I_x l_y}{l_x I_y} = \frac{l_y}{l_x} \tag{13.1.6}$$

In Sect. 8.4, Optimum Design of Beam Section, the relation between the section modulus of a beam and its weight per unit length is explained. This principle can be applied to girders assuming the web thickness is 12 mm. The result is shown in Eq. (13.1.7).

$$W_y = 1.5\sqrt{Z_y} \tag{13.1.7}$$

where

W_y: weight per unit length of girder in kgf/m

Z_y: section modulus of girder in cm^3

Putting Eqs. (13.1.7) and (13.1.3) into (13.1.4) and with the conditions described by (13.1.5) and (13.1.6), the weight per unit area W_1 of girders is given by the follong equation.

$$W_1 = \frac{1.5}{l_y} \sqrt{\frac{16b^2 p l_y}{\pi^4 \sigma_y}} \cdot \frac{1 + \alpha}{\sqrt{1 + \alpha \frac{b^4}{a^4}}} \tag{13.1.8}$$

Where a grillage (mutually-supported) arrangement is not applied, a girder arrangement in one direction is usually applied. In this case the weight per unit area W_0 is obtained by putting $\alpha = 0$ in Eq. (13.1.8). The weight ratio W_1/W_0 of the mutually-supported and the one-direction girder arrangement is given as follows:

$$\frac{W_1}{W_0} = \frac{1 + \alpha}{\sqrt{1 + \alpha \frac{b^4}{a^4}}} \tag{13.1.9}$$

Fig. 13.2 Weight increase due to mutually-supported arrangement

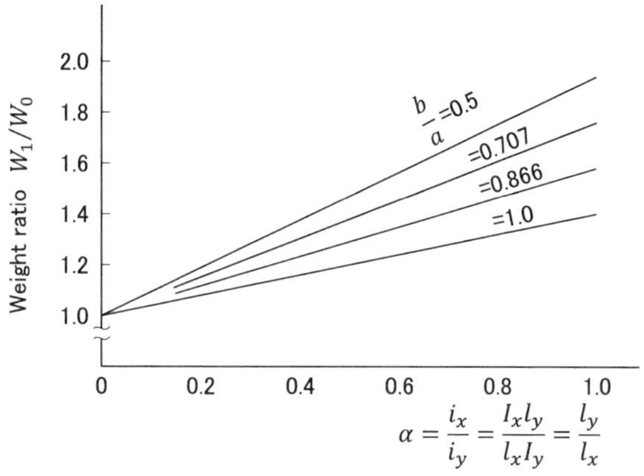

The relation of Eq. (13.1.9) is shown in Fig. 13.2 with α along the horizontal axis and b/a as a parameter. It can be seen that the smaller b/a, which means a slender rectangular, and bigger α, mutually-supporting ratio, will bring a bigger weight difference. It is important to note that the ratio W_1/W_0 is always bigger than 1.0 which means the mutually-supported arrangement is always heavier than the one-direction girder arrangement. This result agrees with the designer's common sense.

The stress σ_y at the point O in Fig. 13.1 in the shorter edge direction is expressed by Eq. (13.1.1) and the stress σ_x at the same point O in the longer edge direction is expressed by the following equations:

$$\sigma_x = \frac{\pi^2 \delta E e_x}{a^2}$$

$$\frac{\sigma_x}{\sigma_y} = \frac{b^2 e_x}{a^2 e_y}$$

σ_x/σ_y is proportional to the square of b/a which means for a slender panel, σ_x is very much smaller than σ_y and the mutually-supported condition will disappear. Even in the case of a square panel the one-direction arrangement is better than the mutually-supported arrangement because the weight ratio between two girders with section modulus Z_y and one girder with section modulus $2Z_y$ is:

$$\frac{2 \times 1.5\sqrt{Z_y}}{1.5\sqrt{2Z_y}} = 1.41$$

and in Fig. 13.2, $W_1/W_0 = 1.41$ for $b/a = 1.0$ and $\alpha = 1.0$ is the same story.

Regarding the minimum weight of grillage structure, Yagi and Yasukawa's study is famous [82], and Kitamura's study as a non-linear programming method is also useful [83].

13.2 Optimum Space of Girders

As shown in Fig. 13.3 it is usual for a panel to be stiffened by beams parallel to the longer edge which are supported by girders parallel to the shorter edge. Assuming that the girders are fixed at both ends, the maximum stress in the girder σ_y is expressed as follows.

$$\sigma_y = \frac{M}{z_y} = \frac{plb^2}{12Z_y}. \tag{13.2.1}$$

The maximum stress in the beam σ_s is as follows:

$$\sigma_s = \frac{M}{Z_s} = \frac{psl^2}{12Z_s} \tag{13.2.2}$$

where

Z_y and Z_s: section modulus with effective breadth

p: uniform load

Considering σ_y and σ_s as allowable stresses, the required section modulus Z_y and Z_s are given by the following equations:

$$Z_y = \frac{plb^2}{12\sigma_y} \tag{13.2.3}$$

$$Z_s = \frac{psl^2}{12\sigma_s} \tag{13.2.4}$$

Applying Eq. (13.1.7) to the girder and beam then the weight per unit area of beam and girder W_2 is derived as follows:

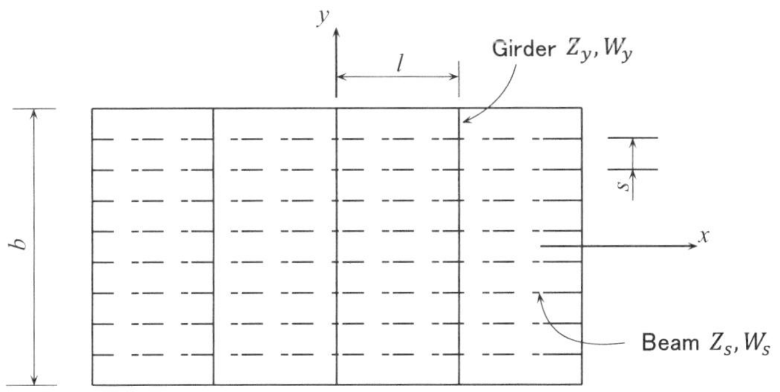

Fig. 13.3 One direction girder arrangement

$$W_2 = \frac{W_s}{s} + \frac{W_y}{l} = \frac{1.5}{s}\sqrt{\frac{psl^2}{12\sigma_s}} + \frac{1.5}{l}\sqrt{\frac{plb^2}{12\sigma_y}} \tag{13.2.5}$$

where W_s and W_y: weight per unit length of beam and girder respectively

Assuming the allowable stresses for the girder and beam are equal, the optimum girder space l is obtained as below putting $dW_2/dl = 0$.

$$l = 2^{-\frac{2}{3}} \cdot s^{\frac{1}{3}} \cdot b^{\frac{2}{3}} = 0.63s^{\frac{1}{3}} \cdot b^{\frac{2}{3}} \tag{13.2.6}$$

In the case of a girder with both ends supported, the following relation obtained because $\sigma_y = plb^2/8Z_y$.

$$l = \frac{3^{\frac{1}{3}}}{2} s^{\frac{1}{3}} \cdot b^{\frac{2}{3}} = 0.72s^{\frac{1}{3}} \cdot b^{\frac{2}{3}} \tag{13.2.7}$$

The optimum girder spaces obtained by Eqs. (13.2.6) and (13.2.7) are shown in Fig. 13.4. It shows that for a girder with a longer span, a wider space is preferable. This reminds us of the idea of designers in the early days: "sardines and tuna have the same number of bones" as mentioned in Sect. 5.1 in Part I, i.e., wider spacings are more beneficial for larger ships. It is recommended to study the weight changing ratio using Eq. (13.2.5) together with Fig. 13.4 because in the case where the weight changing ratio at the optimum point is small, a wider space than optimum point will be decided.

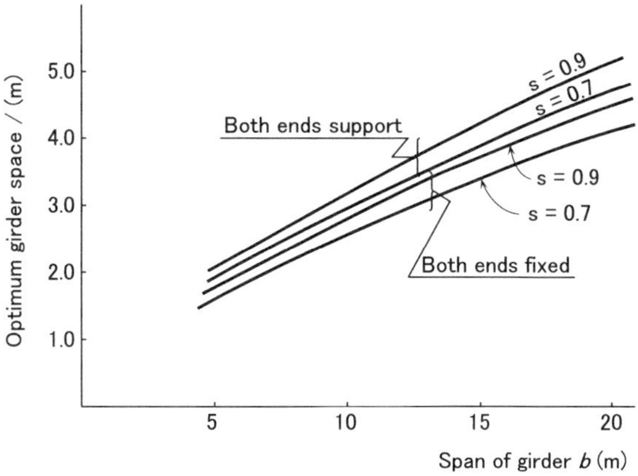

Fig. 13.4 Optimum girder space

13.3 Optimum Space of Beams

Stiffened plates around the stern and machinery room are to be designed not only from the strength viewpoint but also from the vibration viewpoint. Accordingly, the presentation below will be done from both viewpoints.

13.3.1 *Design Condition Against Lateral Load like Water Pressure*

As shown in Fig. 6.2.1 in a one-direction girder arrangement, the plate thickness and the section modulus of the beams are given by ClassNK's earlier rules for watertight bulkheads as follows:

$$t = 3.6s\sqrt{h} + 3.5 \text{ (mm)} \tag{13.3.1}$$

$$Z = 7shl^2 \left(\text{cm}^3\right) \tag{13.3.2}$$

where

s: beam space
l: span of beam
h: water pressure head on beam and plate

13.3.2 *Design Conditions from Vibration Viewpoint*

(1) Plate thickness: The stiffened plate vibrates heavily usually in the case of contact with water. The natural frequency of the panel of size $l \times s$ and thickness t, as shown in Fig. 13.3 with the boundary simply supported condition, is as follows:

$$N_P = \alpha \cdot 1.46 \times 10^2 \frac{t}{s^2}\left\{1 + \left(\frac{s}{l}\right)^2\right\} \text{ (cpm)}$$

where

t: plate thickness in mm
l: span of beam in m
s: space of beam in m

α is the coefficient of frequency reduction by contact with water and usually it is about 0.3—0.5 for hull structures. Here it is assumed that $\alpha = 0.3$ to be on the safety side, and $(s/l)^2$ is negligibly small in comparison with 1.0, because usually s/l is about 1/4. Then the following simplified relation can be obtained.

$$N_P = 0.3 \times 1.46 \times 10^2 \times \frac{t}{s^2} \text{ (cpm)}$$

On the other hand the frequency of the exciting force on the stiffened panel is usually lower than 1000 cpm; therefore, to avoid resonance with the exciting frequency $N_P \geq 1000$ condition will be applied. Accordingly, the following relation is: obtained:

$$t = \frac{1,000}{43.8} s^2 = 22.8s^2 \tag{13.3.3}$$

(2) Scantling of beams: Considering a stiffened plate $l \times b$ surrounded by girders and boundaries in a boundary supported condition, the natural frequency of the stiffened panel is nearly equal to the natural frequency N_s of a beam being cut out from the stiffened panel. This is expressed by the following equation:

$$N_s = \frac{4,830}{l^2} \sqrt{\frac{I_S}{A}} \text{ (cpm)}$$

$$A = A_S + A_P + A_W$$

where

A_S: sectional area of beam in cm^2

A_P: sectional area of plate; i.e., $10ts$ (t in mm and s in m)

A_W: equivalent sectional area due to added mass of water, usually taking the value of around $400ls$ (l and s in m)

I_S: sectional moment of inertia including effective breadth of attached plate in cm^4

Neglecting A_S and A_P, because in usual hull structure they are negligibly small in comparison with A_W, the following relation can be obtained:

$$N_s = \frac{4,830}{l^2} \sqrt{\frac{I_S}{4,000ls}} \text{(cpm)}$$

Similar to the case of plate thickness where $N_s \geq 1000$, the following simple relation is: obtained:

$$I_S = 17.1sl^5 \tag{13.3.4}$$

where s and l are in m, and I_S in cm^4.

13.3.3 Minimum Plate Thickness

In addition to the strength and vibration viewpoints explained in Sects. 13.3.1 and 13.3.2, consideration should be given also to construction methodology with regard to the minimum plate thickness. The classification societies give the minimum plate thickness in their rules. Here considering welding distortion during assembly, the following is proposed for strength members:

$$t_{min} \geq 8 \text{ (mm)} \tag{13.3.5}$$

13.3.4 Optimum Beam Space

The weight W_3 per unit area, which consists of plate and beams, is expressed by the following equations:

$$W_3 = W_P + \frac{W_S}{s} = \gamma t + \frac{\gamma A_S}{10s} \tag{13.3.6}$$

where

W_P: weight of plate per unit area in kgf/m^2

γ: weight of plate per unit volume, that is, 7.85 kgf/m^{2}mm

s: beam space in m

W_S: weight of stiffened panel per unit length in kgf/m

A_S: sectional area of beams in cm^2

In Fig. 13.5 the relation between sectional area A_S and sectional moment of inertia I is shown, from which the following formula derived:

$$A_S = \left(\frac{I}{2}\right)^{\frac{1}{2.35}} \tag{13.3.7}$$

(1) *Case 1: Plate and beams are designed from strength viewpoint:* Putting Eqs. (13.3.7), (13.3.1), and (13.3.2) into (13.3.6) the following relationship can be obtained.

$$W_3 = 7.85 \times \left(3.6s\sqrt{h} + 3.5\right) + \frac{1.5}{s}\sqrt{7shl^2} \tag{13.3.8}$$

By the condition of $dW_3/ds = 0$, the optimum beam space s is as follows:

$$s = 0.17l^{\frac{2}{3}} \tag{13.3.9}$$

In Table 13.1 the optimum beam spaces for different span l are shown. Here again, wider spacings are more beneficial for stiffeners with longer spans. The results in Table 13.1 coincide well with the results of Nagasawa and Honma's study [84] relating to minimum weight analysis, including plastic deformation. However, they are considerably smaller in comparison with 0.6 ~ 0.9 m which are usually applied in practice because the condition of the minimum plate thickness for fabrication is not included.

Figure 13.6 shows the weight difference caused by the change of beam space s based on Eq. (13.3.8) assuming that the stiffener span is 3 m. In Fig. 13.6 the

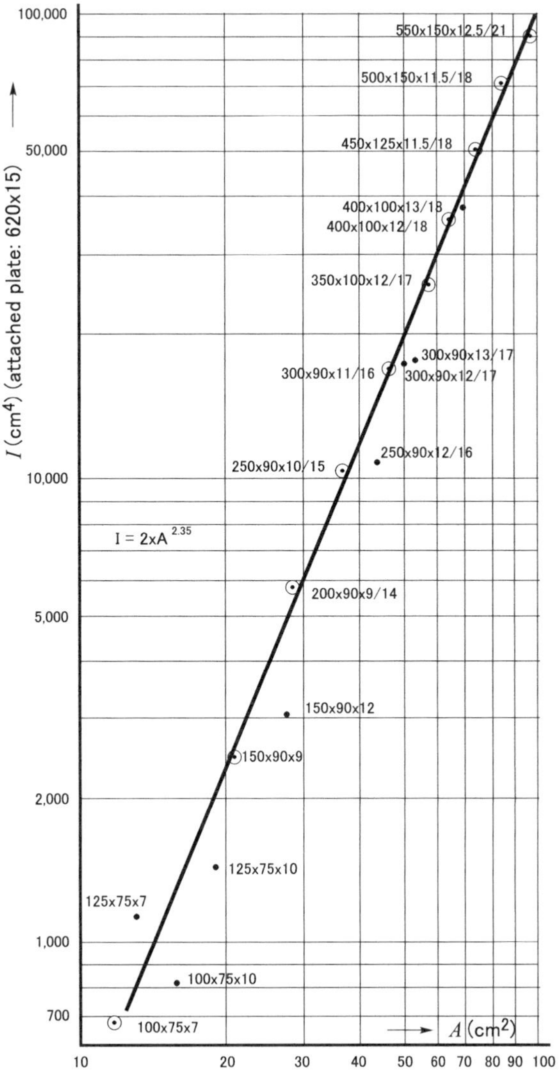

Fig. 13.5 Sectional area and sectional moment of inertia of beam

Table 13.1 Optimum beam space (without account of t_{min})

Span of stiff.	Case 1	Case 2	Case 3		Case 4	
			h = 9 m	h = 16 m	h = 9 m	h = 16 m
	Plate: strength Beam: strength	Plate: vibration Beam: vibration	Plate: vibration Beam: strength		Plate: strength Beam: vibration	
2 m	0.288 m	0.190	0.244	0.273	0.164	0.137
3	0.354	0.265	0.287	0.321	0.284	0.237
4	0.429	0.336	0.322	0.361	0.419	0.350
5	0.497	0.405	0.352	0.394	0.566	0.473

weight difference using the condition of the minimum thickness of Eq. (13.3.5) is also shown. With this condition the optimum beam space becomes bigger than that shown in Table 13.1. Figure 13.6 also shows that the weight is not sensitive to the change in beam space, and a wider space can be applied from the construction viewpoint with very small weight increase.

(2) *Case 2: Plate and beams are designed from vibration viewpoint:* Putting Eqs. (13.3.3), (13.3.4) and (13.3.7) into (13.3.6) the following relationship can be obtained:

$$W_3 = 7.85 \times 22.8s^2 + \frac{0.785}{s}\left(\frac{17.1}{2}sl^5\right)^{\frac{1}{2.35}} \tag{13.3.10}$$

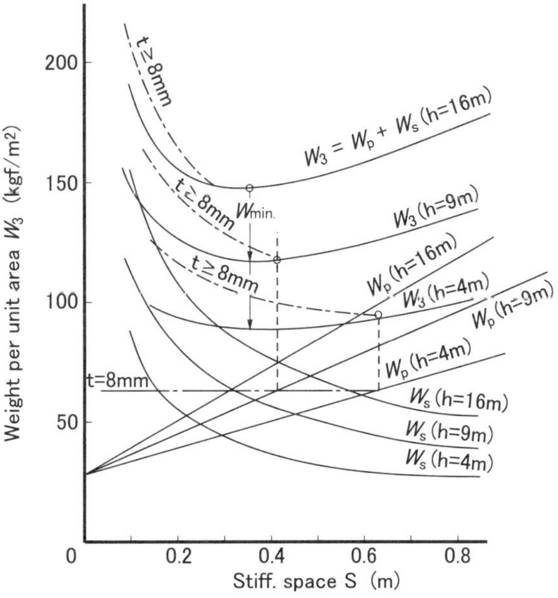

Fig. 13.6 Weight per unit area (strength design for plate and beam)

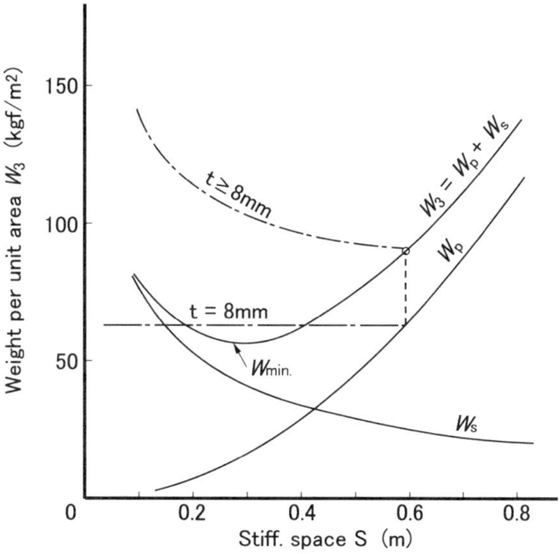

Fig. 13.7 Weight per unit area (vibration design for plate and beam)

With the condition of $dW_3/ds = 0$, the optimum beam space s is as follows:

$$s = 0.107l^{0.826} \tag{13.3.11}$$

In Table 13.1 Case 2, the optimum beam spaces for different spans l are shown. Assuming the beam span is $l = 3.0$ m, the weight difference due to the change of space is shown in Fig. 13.7 The optimum space shown in Table 13.1 is narrow; however, by considering the minimum thickness it comes up to the space usually applied. The weight difference due to a change of space is considerably bigger in the case of the design from a vibration viewpoint, relative to that in the case of the strength design viewpoint shown in (1) Case 1.

(3) *Case 3: Plate is designed from vibration viewpoint and beams are designed from strength viewpoint:* Putting Eqs. (13.1.7), (13.3.2) and (13.3.3) into (13.3.6) the following relation can be obtained:

$$W_3 = 7.85 \times 22.8s^2 + \frac{1.5}{s}\sqrt{7shl^2} \tag{13.3.12}$$

By the condition of $dW_3/ds = 0$, the optimum beam space s is as follows:

$$s = \left(4.9 \times 10^{-3}\sqrt{h}\right)^{\frac{2}{5}} \cdot l^{\frac{2}{5}} \tag{13.3.13}$$

In Table 13.1 Case 3, the optimal beam spaces for different spans of l and different water head h are shown. The weight difference due to the change of space is shown in Fig. 13.8

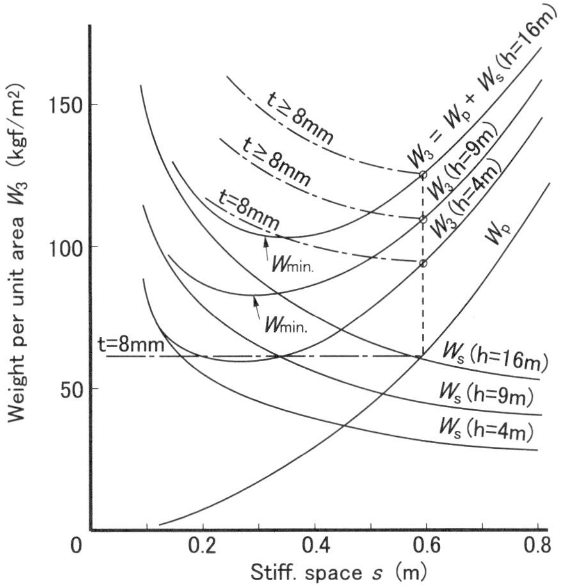

Fig. 13.8 Weight per unit area (vibration for plate and strength for beam)

(4) *Case 4: Plate is designed from strength viewpoint and beams are designed from vibration viewpoint:* Putting Eqs. (13.3.1), (13.3.4) and (13.3.7) into (13.3.6) the following relation can be obtained:

$$W_3 = 7.85 \times \left(3.6s\sqrt{h} + 3.5\right) + \frac{0.785}{s}\left(\frac{17.1}{2}sl^5\right)^{\frac{1}{2.35}} \tag{13.3.14}$$

With the condition of $dW_3/ds = 0$, the optimum beam space s is as follows:

$$s = \left(3.95 \times 10^{-2}\frac{1}{\sqrt{h}}\right)^{\frac{5}{3.7}} \cdot l^{\frac{5}{3.7}} \tag{13.3.15}$$

In Table 13.1 Case 4 the optimum beam space s for different spans of l is shown. Figure 13.9 shows the weight difference based on Eq. (13.3.15).

Table 13.1, Figs. 13.6, 13.7, 13.8 and 13.9 show considerably smaller optimum beam spaces than usually applied, they are, 0.6 ~ 0.9 m. However, it is recommended that by applying the minimum thickness condition they come up close to the practical values, and the effect of beam space on weight is not so large that a little wider space than the optimum can be applied.

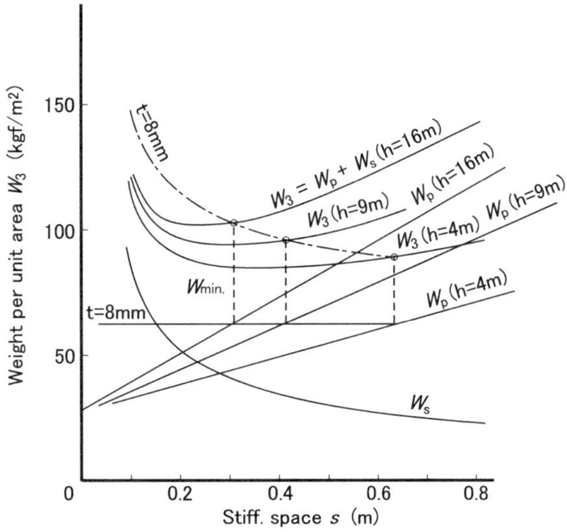

Fig. 13.9 Weight per unit area (strength for plate and vibration for beam)

Chapter 14
Torsion

14.1 Overview of the Theory

Let us discuss the torsion of a bar. The discussion will be proceeded on the basis of the following assumptions:

Assumption

(A) A bar is straight.
(B) A bar is twisted uniformly.
(C) A bar has the same cross sectional shape at any point.

The theory of a bar subjected to torsion can be roughly categorized by the following two groups according to the cross sectional shape of the bar as listed in Table 14.1.

- solid cylindrical section or closed section composed of thin plate elements
- open section made up of thin plate elements such as I-shape section, U channel section, etc.

In this Chapter, the following items are explained.

Section 14.2 will go on to explain the torsion theory of a cylindrical solid section bar. The simple case of a twisted bar, and then describes "pure torsion" theory established by Saint-Venant [85]. That is a general theory applicable to a twisted bar with a closed section. This theory is derived on the basis of the assumption that the plane of a cross section of a twisted bar still remains in the original plane. That is the reason why the theory is called pure torsion. And then, the analogy of a membrane under lateral pressure is introduced, since the analogy can explain the behavior of the twisted bar.

In Sect. 14.3, several examples are shown to calculate the torsional rigidity of various cross sections. The membrane analogy is fully utilized in calculating the torsional rigidity.

Y. Okumoto et al., *Design of Ship Hull Structures*,
https://doi.org/10.1007/978-3-032-12500-2_14

Table 14.1 Category of torsional theory

Category	Pure torsion	Torsion with warping
Typical section	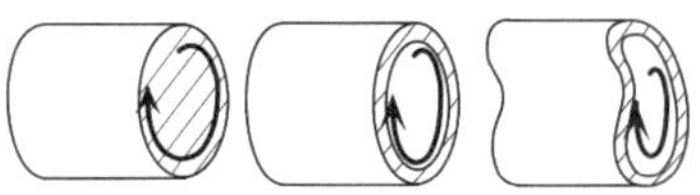	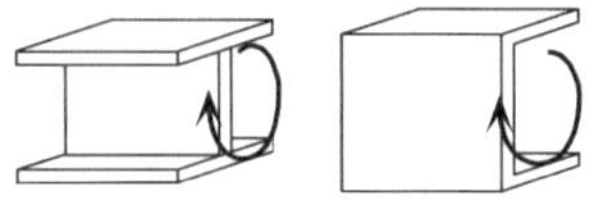
Feature	· Solid section · Closed section · No deformation of cross section due to torsion	· Open section · Deformation of cross section causing additional stress
Theory	Saint-Venant	Wagner
Equation	$T = C\theta$ or $T = C\dfrac{d\varphi}{dz}$	$T = C\dfrac{d\varphi}{dz} - \Gamma\dfrac{d^3\varphi}{dz^3}$

In Sect. 14.4, the torsion of an I-section is investigated, since this section behaves in a different way from the above closed section bar. An I-section bar is a simple case of a twisted bar which has an open cross section. In that case, the assumption of Saint-Venant's theory is not applicable, because the cross section is distorted and the plane of the cross section deforms. The flanges of an I-section generate additional shearing stress due to bending of the flange induced by twisting. The deformation of the plane is called "warping" [22].

In Sect. 14.5, the general theory of torsion with warping is presented. Torsion with warping occurs usually in the case of twisting of an open section bar; the problem of torsion with warping was theoretically analyzed by Wagner. The theory assumes that the cross section of a twisted bar causes deformation of the section and it results in a normal stress along the bar's length. Wagner's theory is usually called "torsion with warping", in contrast with "pure torsion" by Saint-Venant. Table 14.2 provides a list of torsional rigidity and warping rigidity for various sections.

14.2 Torsion Theory of Closed Section Bars

Before discussing the general case of a twisted bar with a closed section, let us consider a cylindrical solid bar subjected to twisting because this is the simplest and important case. Let us assume a circular bar fixed at one end is subjected to twisting at the other end. This creates shearing stress τ due to the twisting angle φ as shown in Fig. 14.1, and the shearing deformation γ occurs

$$\gamma dz = r d\varphi \tag{14.2.1}$$

Table 14.2 Shear center, torsional rigidity and warping rigidity

	Cross section	e J Γ
1	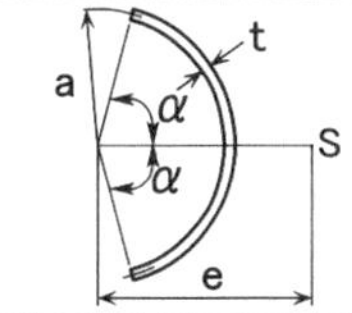	$e=\dfrac{2a(\sin\alpha-\alpha\cos\alpha)}{\alpha-\sin\alpha\cos\alpha}$ $J=\dfrac{2}{3}a\alpha t^3$ $\Gamma^*=\dfrac{2}{3}a^5t\left\{\alpha^3-\dfrac{6(\sin\alpha-\alpha\cos\alpha)^2}{\alpha-\sin\alpha\cos\alpha}\right\}$
2	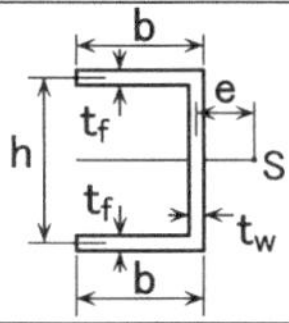	$e=\dfrac{3b^2t_f}{6bt_f+ht_w}$ $J=\dfrac{1}{3}(2bt_f^3+ht_w^3)$ $\Gamma^*=\dfrac{b^3h^2t_f}{12}\dfrac{3bt_f+2ht_w}{6bt_f+ht_w}$
3	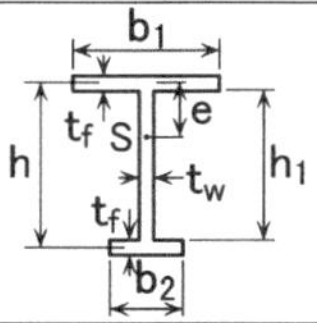	$e=\dfrac{I_2}{I_1+I_2}h,\quad I_1=\dfrac{1}{12}b_1^3t_1,\quad I_2=\dfrac{1}{12}b_2^3t_2$ $J=\dfrac{1}{3}(h_1t_w^3+b_1t_1^3+b_2t_2^3)$ $\Gamma^*=\dfrac{1}{12}\{b_1^3t_1e^2+b_2^3t_2(h-e)^2\}$
4	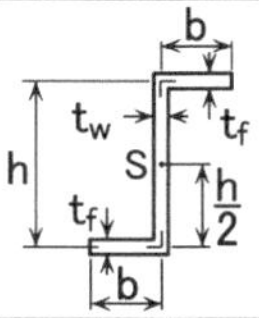	$J=\dfrac{1}{3}(2bt_f^3+ht_w^3)$ $\Gamma^*=\dfrac{b^3h^2}{12(2b+h)^2}\{2(b^2+bh+h^2)t_f+3bht_w\}$
5	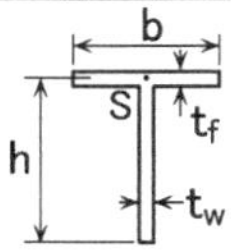	$J=\dfrac{1}{3}(bt_f^3+ht_w^3)$ $\Gamma^*=\dfrac{b^3t_f^3}{144}+\dfrac{h^3t_w^3}{36}$
6	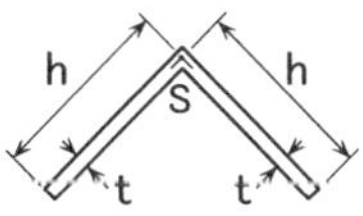	$J=\dfrac{2}{3}ht^3$ $\Gamma^*=\dfrac{1}{18}h^3t^3$

e : Location of shear center (S)

J : Torsional constant $C=GJ$

Γ^* : Warping rigidity $\Gamma^*=\dfrac{\Gamma}{E}$

Corresponding to this deformation, we have shearing stress as shown in Eq. (14.2.2) where G denotes the shear modulus of elasticity.

$$\tau = G\gamma \tag{14.2.2}$$

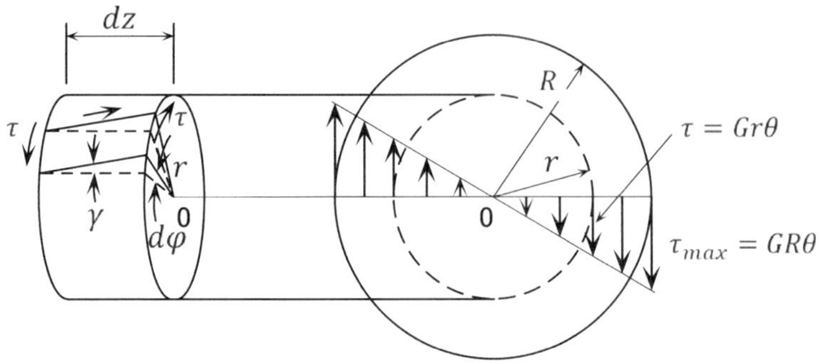

Fig. 14.1 Shearing stress distribution in cicular cylinder section

Substituting Eq. (14.2.2) in Eq. (14.2.1), we find

$$\tau = Gr\frac{d\varphi}{dz} \tag{14.2.3}$$

Let us introduce the twisting ratio θ as the twisting angle per unit length, i.e., $\theta = d\varphi/dz$, then Eq. (14.2.3) can be rewritten

$$\tau = Gr\theta \tag{14.2.4}$$

Since θ is a constant value at any point in a cross section because of the assumption (B), the shearing stress increases in proportion to the radius of the circular bar. Consequently, maximum shearing stress occurs at the outermost surface of the cylinder. If we sum the shearing stress around the z-axis in the area of a cross section, this balances the twisting moment T applied to the bar and T becomes:

$$T = \int_0^R \tau \cdot (2\pi r \cdot dr) \cdot r = C\theta \tag{14.2.5}$$

where $C = \frac{1}{2}G\pi R^4$

or

$$T = C\frac{d\varphi}{dz} \tag{14.2.6}$$

In Eqs. (14.2.5) and (14.2.6), the constant C is called "torsional rigidity".

The generalized theory of a closed section is established by Saint-Venant and that is usually called "pure torsion" as indicated in Table 14.1. The reason for this naming is because the theory is based on the following assumption:

Assumption

(D) The cross section of a twisted bar never distorts, it only rotates perpendicularly around the torsional axis. Even after twisting, the plane of a cross section still remains in a plane. This, in other words, means that there is no normal stress along the bar's length due to twisting.

Figure 14.2a illustrates a cylindrical bar twisted by a torsional moment T around the z-axis. The stress condition of a small portion of the bar is expressed as in Fig. 14.2d, where the normal stress in the z-direction is neglected on the above assumption (D) and only the shearing stresses τ_{xz}, τ_{yz} are taken into account. The equation for the shearing force equilibrium in the z-direction can be obtained as follows:

$$\frac{\partial \tau_{xz}}{\partial x} + \frac{\partial \tau_{yz}}{\partial y} = 0 \tag{14.2.7}$$

In Fig. 14.2c, when Point A of coordinate (x, y) moves to Point B due to a twisting by angle $\varphi(= \theta z)$, the displacement of Point A can be obtained by the following form:

$$\begin{aligned} u &= -y\varphi = -y\theta z \\ v &= x\varphi = x\theta z \end{aligned} \tag{14.2.8}$$

The shearing strain caused by deformation of u and w can be obtained in the following form as shown in Fig. 14.2b:

$$\gamma_{xz} = \frac{\partial u}{\partial z} + \frac{\partial w}{\partial x} = -y\theta + \frac{\partial w}{\partial x} \tag{14.2.9}$$

In the same manner, the following equation can be derived:

$$\gamma_{yz} = \frac{\partial v}{\partial z} + \frac{\partial w}{\partial y} = x\theta + \frac{\partial w}{\partial y} \tag{14.2.10}$$

Therefore, from Eqs. (14.2.8)–(14.2.10), the shearing stress is related to the displacement by the following equations:

$$\begin{aligned} \tau_{xz} &= G\gamma_{xz} = G\left(-y\theta + \frac{\partial w}{\partial x}\right) \\ \tau_{yz} &= G\gamma_{yz} = G\left(x\theta + \frac{\partial w}{\partial y}\right) \end{aligned} \tag{14.2.11}$$

Equations (14.7) and (14.11) are differential equations in terms of the two unknown parameters τ_{xz}, τ_{yz} and they are solved for specified boundary conditions and applied torsional moments.

In order to put the equations in a more simple form, let us introduce a stress function Ψ as in the following formula, which automatically satisfies Eq. (14.2.7):

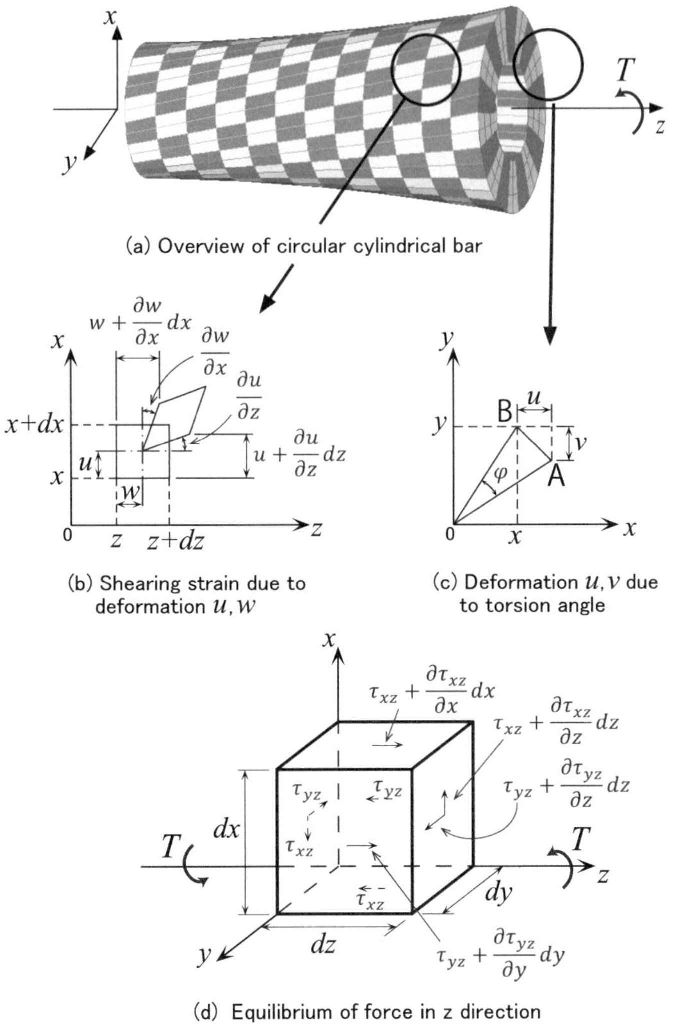

Fig. 14.2 Circular cylindrical bar under twisting

$$\tau_{xz} = \frac{\partial \Psi}{\partial y}, \tau_{yz} = -\frac{\partial \Psi}{\partial x} \tag{14.2.12}$$

From Eqs. (14.2.11) and (14.2.12), the fundamental equation of torsion can be expressed as follows:

$$\frac{\partial^2 \Psi}{\partial x^2} + \frac{\partial^2 \Psi}{\partial y^2} = -2G\theta \tag{14.2.13}$$

Considering the equilibrium of the torsional moment created by τ_{yz} and τ_{xz}, the torsional moment is obtained as follows using Gauss' divergence theorem.

$$
\begin{aligned}
T &= \int \left(\tau_{yz}x - \tau_{xz}y\right)dxdy \\
&= -\iint \left(\frac{\partial \Psi}{\partial x}x + \frac{\partial \Psi}{\partial y}y\right)dxdy \\
&= -\iint \left[\frac{\partial}{\partial x}(\Psi x) - \Psi + \frac{\partial}{\partial y}(\Psi y) - \Psi\right]dxdy \\
&= -\oint_C \left(\Psi x n_x + \Psi y n_y\right)ds + 2\iint \Psi dxdy \\
&= 2\iint \Psi dxdy
\end{aligned}
\tag{14.2.14}
$$

The integral over the boundary C vanishes because $\Psi = 0$ on the boundary.

This means that the torsional moment is equal to twice the volume inside the closed surface indicated by the stress function Ψ. This equation is the governing equation of torsion, however, mathematical knowledge is necessary to handle this equation head-on.

Instead of solving the above torsional equation by mathematical analysis, membrane analogy suggested by Plandtl can be applied, because this approach is easier to understand than dealing with the above equation.

Plandtl has indicated that the behavior of a twisted bar is equivalent to that of a membrane under pressure, which has the same outline as that of the cross section of the bar. Let us imagine the membrane is deformed in the form of ζ by the pressure p as shown in Fig. 14.3. Taking the tension of the membrane as F, we can obtain the equilibrium value for all normal forces:

$$
Fdy\frac{\partial^2 \zeta}{\partial x^2}dx + Fdx\frac{\partial^2 \zeta}{\partial y^2}dy + pdxdy = 0 \tag{14.2.15}
$$

$$
\therefore \frac{\partial^2 \zeta}{\partial x^2} + \frac{\partial^2 \zeta}{\partial y^2} = -\frac{p}{F} \tag{14.2.16}
$$

Comparing Eqs. (14.2.15) and (14.2.16) with Eqs. (14.2.13) and (14.2.14), we find the following relations between torsion and membrane analogy;

	Torsion	Membrane	
Function	Ψ	$: \zeta$	
	$2G\theta$	$: p/F$	(14.2.17)
Shearing stress	$\tau = \frac{\partial \Psi}{\partial x}, \frac{\partial \Psi}{\partial y}$	$: \tau = \frac{\partial \zeta}{\partial x}, \frac{\partial \zeta}{\partial y}$	
Torsional moment	$T = 2\iint \Psi dxdy$	$: T = 2\iint \zeta dxdy$	

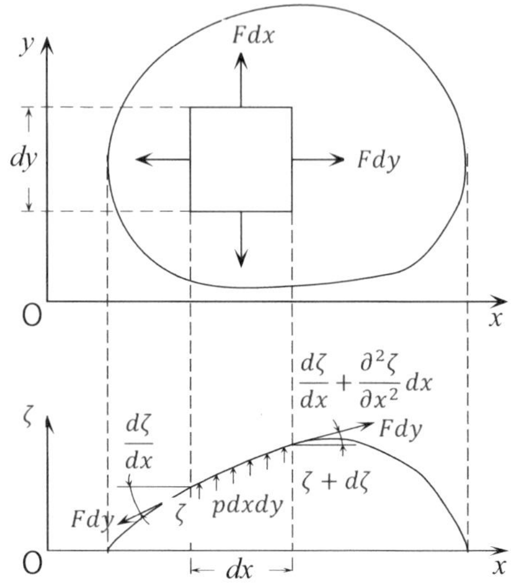

Fig. 14.3 Equilibrium of deflected membrane under pressure

Therefore, twice the volume bounded by the expanded membrane equals the external torque T, and the shearing stress τ is equal to the slope of the membrane's deflection. Consequently, instead of solving the torsional equation, the stress distribution by torsion can be obtained by investigating the deformation of a membrane.

For example, let us consider the case of a cylindrical section as we have already examined in Eqs. (14.2.5) and (14.2.6). Figure 14.4 shows the equilibrium of the upward force caused by the pressure and the downward membrane force. Hence we have the following differential equation:

$$p \cdot \pi r^2 = -F\frac{d\zeta}{dr} \cdot 2\pi r$$

$$\therefore d\zeta = -\frac{rp}{2F}dr = -G\theta r dr \tag{14.2.18}$$

Solving the above differential equation by considering $\zeta = 0$ at $r = R$, we find the deformation of the membrane:

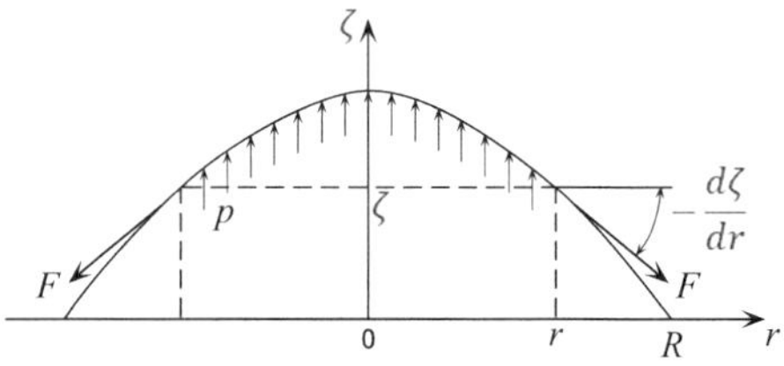

Fig. 14.4 Membrane analogy of cylindrical circular section

$$\therefore \zeta = \frac{1}{2}(R^2 - r^2)G\theta \tag{14.2.19}$$

Differentiating the above deformation, we obtain the shearing stress in the form

$$\tau = -\frac{d\zeta}{dr} = Gr\theta \tag{14.2.20}$$

As the torsional moment is twice the volume bounded by the deformed membrane, T can be obtained:

$$T = 2\int_0^R \zeta \cdot 2\pi r dr = \frac{1}{2}G\pi R^4\theta \tag{14.2.21}$$

The result of the above calculation agrees with Eqs. (14.2.4) and (14.2.5).

14.3 Torsional Rigidity of Various Sections

Let us calculate torsional rigidity of several cross sections by applying the membrane analogy.

Cylindrical thin plate section: Let us imagine a thin cylindrical shell, where the inner radius is R_1 and the outer radius R_2 as shown in Fig. 14.5. The plate thickness is $t = R_2 - R_1$. With the aid of membrane analogy, the deformation of the membrane increases in the thin plate area and the deformation of the inside area of the inner circle becomes constant, because there is no shearing stress in this area. Let us assume the height of this plateau to be h.. Then, according to the relation between torsion and membrane as explained in Eq. (14.2.17), we have the shearing stress expressed by the slope of the membrane in the same way as in Eq. (14.2.20). Hence,

$$\tau = \frac{h}{t} = Gr\theta \tag{14.3.1}$$

Since the torsional moment is twice that of the volume bounded by the deformed membrane, we have the following equation by putting $R = (R_1 + R_2)/2$.

$$T = 2\pi R^2 h = 2G\pi R^3 t\theta \tag{14.3.2}$$

Hence, the torsional rigidity will be

$$C = 2\pi GR^3 t \tag{14.3.3}$$

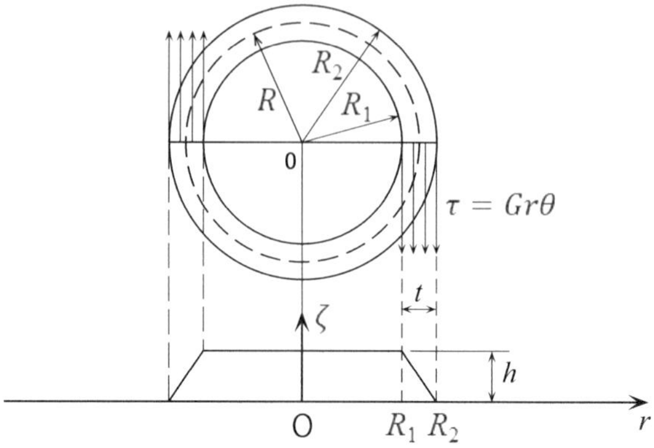

Fig. 14.5 Cylindrical thin plate section

Closed section composed of thin plate: Let us consider a bar having a closed cross section composed of thin plate as shown in Fig. 14.6. The membrane is deformed by pressure and it creates the height h of the membrane. A is the area surrounded by the centerline of plate thickness t and s is the local coordinate along that centerline. By considering the equilibrium of the upward force due to pressure p and the downward force induced by the membrane tension F, we have

$$pA = \oint F\frac{h}{t}ds \tag{14.3.4}$$

On the other hand, we have the following relation as expressed by Eq. (14.2.16):

$$\frac{p}{F} = 2G\theta \tag{14.3.5}$$

The torsional moment equals twice the volume of the membrane deformation, thus we find

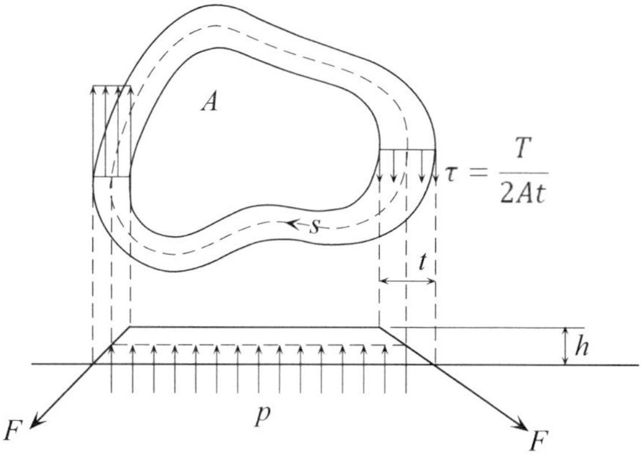

Fig. 14.6 Closed thin plate section

$$T = 2Ah = C\theta \tag{14.3.6}$$

By eliminating $h, p/F$ and θ from Eqs. (14.3.4)–(14.3.6), we find

$$C = \frac{4GA^2}{\oint \frac{ds}{t}} \tag{14.3.7}$$

The above equation is applicable to the general case of a closed section with one chamber inside the cross section. A typical application of Eq. (14.3.7) is indicated in Fig. 14.7. In this example, the area surrounded by the centerline of box plates is $A = ab$, therefore, we find the torsional rigidity:

$$C = \frac{4G(ab)^2}{2(\frac{a}{t} + \frac{b}{t})} = \frac{2Ga^2b^2t}{a+b} \tag{14.3.8}$$

Thin plate section: Let us imagine a thin plate section with the length of l and thickness of t varying along its length as shown in Fig. 14.8. Assuming a large aspect ratio ($l \gg t$), equilibrium in the $x =$ const section is considered. Then, we have:

$$p \cdot 2y = -2F \cdot \frac{d\zeta}{dy} \tag{14.3.9}$$

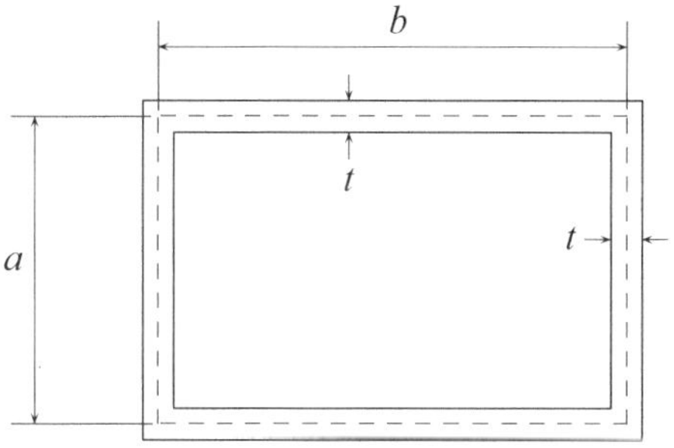

Fig. 14.7 Box shape section composed of thin plates

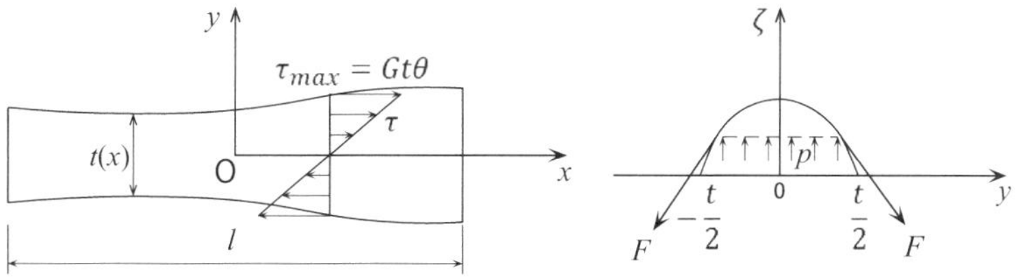

Fig. 14.8 Thin plate section

Hence, the following differential equation can be obtained

$$d\zeta = -\frac{p}{F} y dy = -2G\theta y dy \tag{14.3.10}$$

By integrating this differential equation using the boundary condition $\zeta = 0$ at $y = t/2$, we have the deformation of the membrane

$$\zeta = \left(\frac{t^2}{4} - y^2\right) G\theta \tag{14.3.11}$$

Using the above deformation, we have the torsional moment

$$\begin{aligned} T &= 2\int_0^l \int_{-\frac{t}{2}}^{\frac{t}{2}} \zeta \, dy dx \\ &= 2G\theta \int_0^l \int_{-\frac{t}{2}}^{\frac{t}{2}} \left(\frac{t^2}{4} - y^2\right) dy dx \\ &= \frac{G\theta}{3} \int_0^l t^3 dx \end{aligned} \tag{14.3.12}$$

Thus, the torsional rigidity of the plate section is

$$\therefore C = \frac{G}{3} \int_0^l t^3 dx \tag{14.3.13}$$

When the thickness t is constant, the torsional rigidity reduces to

$$C = \frac{1}{3} G l t^3 \tag{14.3.14}$$

It is important to note that the coefficient of 1/3 in Eq. (14.3.14) is valid only when $l \gg t$. It reduces to the following values when l/t is smaller [85].

0.312 when $l/t = 10$
0.281 when $l/t = 4$
0.141 when $l/t = 1$

Equations (14.3.13) and (14.3.14) can be applied to any combination of thin shell open sections without a closed cell. When a section is composed of several plate members, the total torsional rigidity can be calculated as the sum of torsional rigidity of each plate member. For example, the torsional rigidity of the sections shown in Fig. 14.9 can be expressed as follows.

I-shaped section

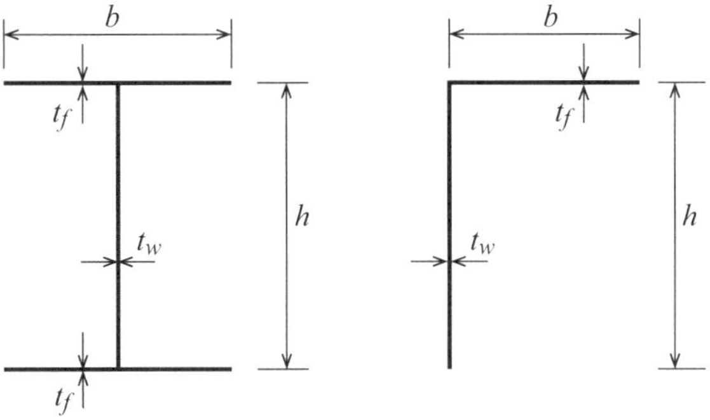

Fig. 14.9 Combination of thin plate sections

$$C = G\left(\frac{2bt_f^3}{3} + \frac{ht_w^3}{3}\right) \quad (14.3.15)$$

L-shaped section

$$C = G\left(\frac{bt_f^3}{3} + \frac{ht_w^3}{3}\right) \quad (14.3.16)$$

Thin cylindrical section with slit: Applying Eq. (14.3.16) to a thin cylindrical section with a slit as shown in Fig. 14.10a, we can obtain the torsional rigidity as follow using the circumference length of the cylinder of $2\pi R$.

$$C = \frac{1}{3}G(2\pi R)t^3 = \frac{2}{3}G\pi Rt^3 \quad (14.3.17)$$

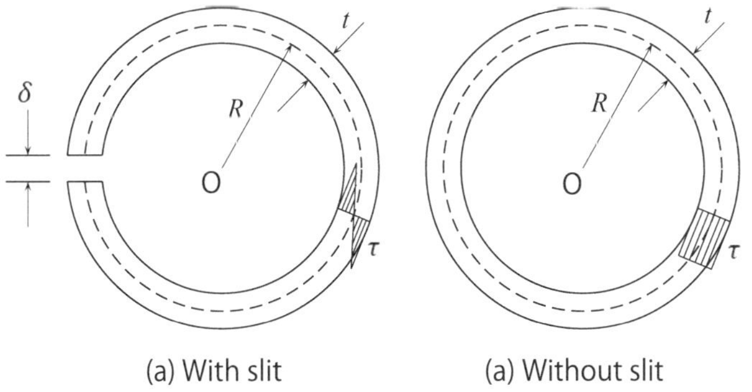

Fig. 14.10 Thin cylindrical section with and without slit

Comparing the above rigidity with Eq. (14.3.3), in which the cylindrical section has no slit, the ratio of the torsional rigidity is

$$\frac{C_{\text{with slit}}}{C_{\text{without slit}}} = \frac{\frac{2}{3}\pi GRt^3}{2\pi GR^3 t} = \frac{1}{3}\left(\frac{t}{R}\right)^2 \tag{14.3.18}$$

Considering that $R \gg t$, closed section is much stronger than open section against torsional moment. In the case of $t/R = 1/5$, the rigidity of a cylinder with a slit becomes 1/75 of that of the cylinder without a slit.

In addition, it may be worthwhile to note the difference in the shearing stress distribution as shown in Fig. 14.10. The uniform τ in the closed section as shown in Fig. 14.10b can produce much more torsional moment due to the larger lever.

14.4 Torsion Theory of I-Section

Let us consider an I-section bar which is fixed at one end and twisted as illustrated in Fig. 14.11. From Fig. 14.11b, the plane composed of flanges and webs is distorted at the free end of the bar. Hence, the assumption (D) is not satisfied in this I-section bar. This distortion of the plane is called "warping".

When the section rotates around the z-axis, the web of the section experiences only torsion, on the other hand, the two flanges will experience bending as well as torsion. If the section rotates by an angle φ, the rotation will cause a horizontal displacement of the flanges of $h\varphi/2$. These displacements induce shearing forces in the upper and lower flanges and each component is

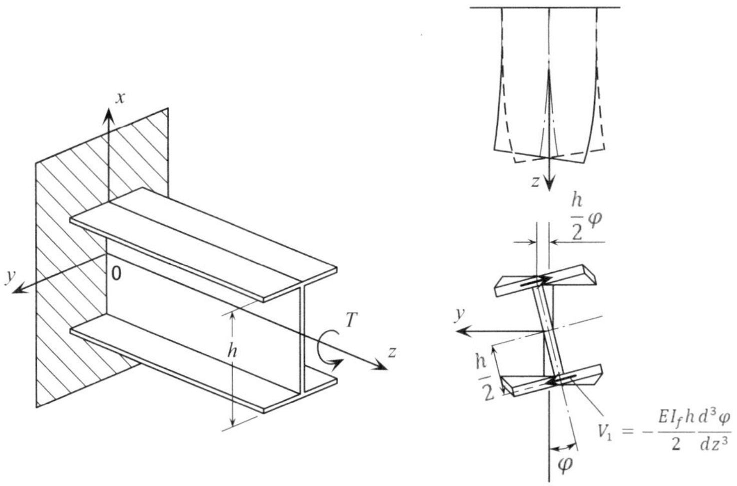

Fig. 14.11 Torsion with warping deformation of I section bar

$$V_1 = \frac{h}{2} EI_f \frac{d^3\varphi}{dz^3} \tag{14.4.1}$$

where h is the height of the web and I_f is the sectional moment of inertia of the flange around the x-axis. These shearing forces in a flange make a twisting moment T_2:

$$T_2 = -\frac{EI_f h^2}{2} \frac{d^3\varphi}{dz^3} \tag{14.4.2}$$

In addition, pure torsion is induced due to twisting as follows:

$$T_1 = C\frac{d\varphi}{dz} \tag{14.4.3}$$

Therefore, equilibrium of external torsional moment and internal torsional moment can be expressed as in the following form:

$$\begin{aligned} T &= T_1 + T_2 \\ &= C\frac{d\varphi}{dz} - \frac{EI_f h^2}{2} \frac{d^3\varphi}{dz^3} \end{aligned} \tag{14.4.4}$$

This is a fundamental equation of torsion with warping of an I-section bar. This equation is different from that of "pure torsion" in terms of the existence of $EI_f h^2/2$, which is called "warping rigidity" and expressed by Γ. Therefore, Eq. (14.3.4) can be rewritten in the form:

$$T = C\frac{d\varphi}{dz} - \Gamma\frac{d^3\varphi}{dz^3} \tag{14.4.5}$$

This example of an I-section bar is an easy case to understand in that the flange creates an additional moment due to warping.

14.5 Torsion Theory of Open Section Bars

In the case of a twisted bar with an open section, the bar generates not only torsional deformation but also warping deformation as explained in Sect. 14.4. Since that warping is not constant at any point in a cross section, it induces normal stress as a result. The general theory of torsion with warping was established by Wagner.

Before discussing the torsion with warping, let us investigate the shear center. When a shearing force is applied at an arbitrary point on a bar having an open section, the force creates a torsional moment around the center of twisting and consequently it induces torsional deformation and bending deformation simultaneously. However, in the case where the force acts on the center of twisting, torsional deformation will never occur and only bending deformation will be induced. That twisting center point is called "shear center".

In order to obtain the shear center, the shearing distribution of a cross section must first be considered. Figure 14.12 represents the mechanism where a change in normal stress induces shearing stress. Figure 14.12a illustrates the cross section. Let us consider the equilibrium force in the lower portion of the section; i.e. hatched area. In the longitudinal direction of this bar, the normal stress σ due to bending of the bar is distributed as shown in Fig. 14.12b. By cutting out the lower portion from this section at $y = y_1$ level, we can observe that the normal stress is acting on both sides of the portion and the shearing stress is working on the cut out surface in Fig. 14.12c. The difference in the normal stress $\Delta\sigma$ on both sides is caused by the change of moment $V \cdot \Delta x$, where V is a shearing force acting on the cross section, and the sum of the differences in the normal stresses must equal the shearing stress τ_{yx} as explained in Fig. 14.12d. Hence from the equilibrium of the horizontal forces acting on the lower portion in Fig. 14.12e, the following equation can be derived:

$$\tau_{yx} \cdot \Delta x \cdot b = \int_{y_1}^{C} V \cdot \Delta x / \frac{I}{y} \cdot dA$$

$$\therefore\ \tau_{xy} = \tau_{yx} = \frac{V}{Ib} \int_{y_1}^{C} y dA \tag{14.5.1}$$

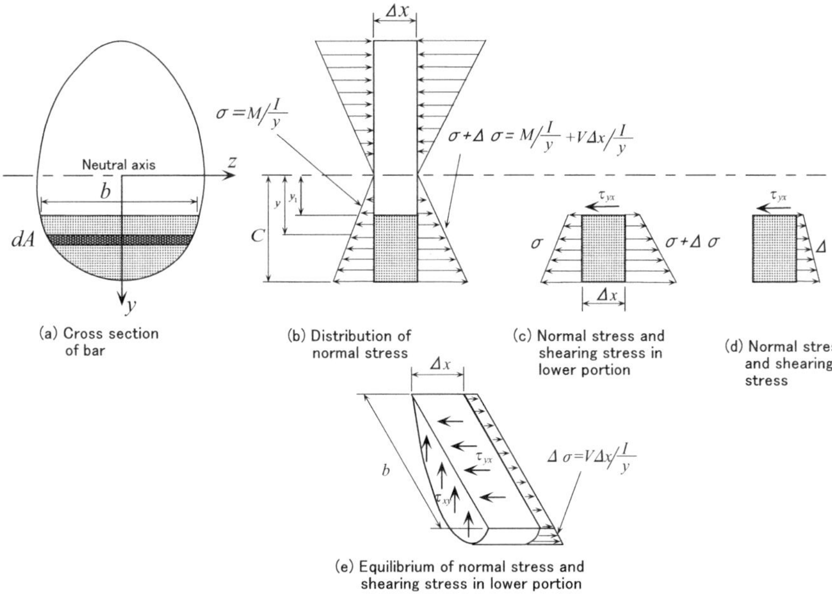

Fig. 14.12 Shearing stress distribution due to change of normal stress

The derivation of this equation is already explained by Eq. (9.1.7) in Sect. 9.1. Equation (9.1.7) deals with the shearing stress in the web plate, whereas, the above Eq. (14.5.1) is more generalized.

By using this relation between shearing stress and bending stress, let us calculate the shear center of a channel section subjected to a vertical force of 7500 kgf in Fig. 14.13a. The sectional moment of inertia I is obtained as follows:

$$I = 2bt\left(\frac{h}{2}\right)^2 + \frac{th^3}{12} = 2 \times 20 \times 1 \times 15^2 + \frac{1 \times 30^3}{12} = 11,250\ \left(\mathrm{cm}^4\right) \quad (14.5.2)$$

When a slice of length $\Delta x = 1(\mathrm{cm})$ is cut from the channel, the bending moment created by a shearing force V becomes $V\Delta x$, then the difference in the normal stress of the flange σ due to this bending moment is calculated in the following form as shown in Fig. 14.13b:

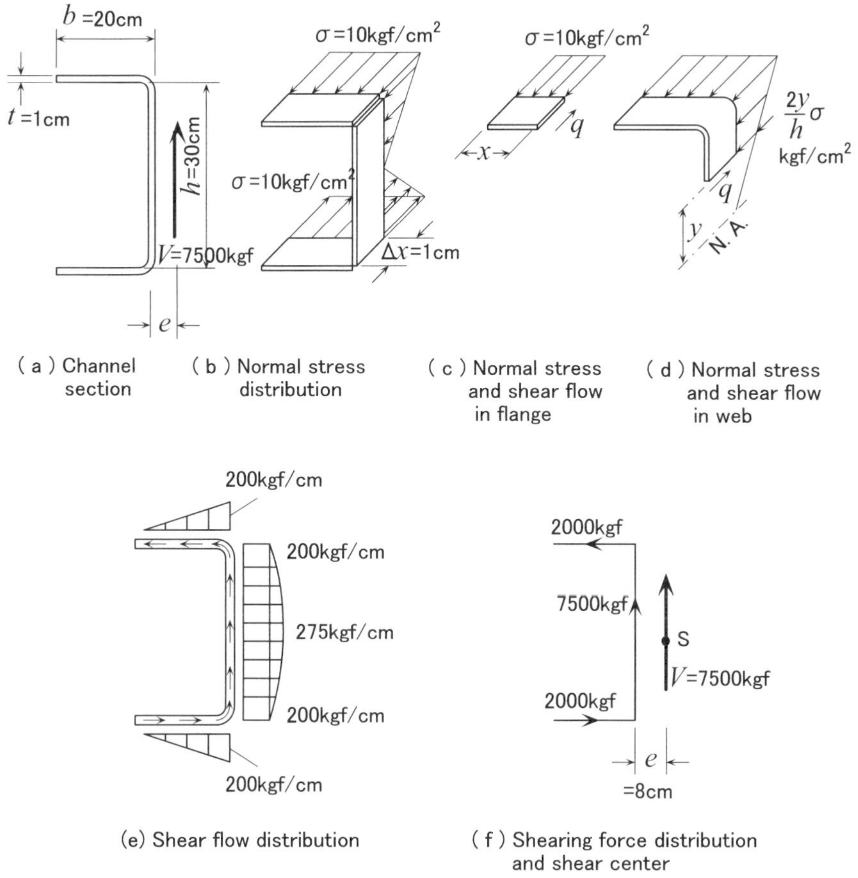

Fig. 14.13 Detail of shearing stress distribution

$$\sigma = V\Delta x/\frac{I}{y} = 7,500 \times 1/\frac{11,250}{15} = 10\ (\text{kgf/cm}^2) \tag{14.5.3}$$

If we define the shear flow q by multiplying shearing stress τ and plate thickness t, then the shear flow of the flange at the point x from the free edge is obtained as follows by considering that the shearing stress must equal the sum of the normal stress as shown in Fig. 14.13c:

$$q = \frac{\sigma x t}{t\Delta x} \cdot t = \frac{10x \times 1}{1} = 10x\ (\text{kgf/cm}) \tag{14.5.4}$$

In the vicinity of the web plate of the channel, the shear flow at the point y from the neutral axis must support the sum of the normal stress of the flange and the normal stress in the upper part of web as shown in Fig. 14.13d. Therefore, it becomes

$$q = \frac{t}{\Delta x}\left\{\sigma b + \frac{1}{2}\left(\sigma + y\frac{2}{h}\sigma\right) \times \left(\frac{h}{2} - y\right)\right\} = 275 - \frac{1}{3}y^2\ (\text{kgf/cm}) \tag{14.5.5}$$

In the lower part of the web or in the lower flange, the shear flow can be calculated in the same manner, thus the shear flow distribution in this channel section is obtained as illustrated in Fig. 14.13e. The total shearing force in the web plate is estimated by integrating the shear flow over the web plate height and this is

$$F = \int_{-h/2}^{h/2} q dy = 275h - \frac{h^3}{36} = 7,500\ (\text{kgf}) \tag{14.5.6}$$

This shearing force equals the applied shearing force V.

From Fig. 14.13e, the shearing force in the flanges and web is calculated by integrating the shear flow and these forces are plotted in Fig. 14.13f. The torsional moment around the shear center S in the flanges must be equivalent to that of the web plate, hence, the equilibrium of these torsional moments is expressed in the following form by taking from the distance of shear center to the web plate as e:

$$2000\ \text{h} = 7500\ \text{e}$$

$$\therefore \quad e = \frac{2,000}{7,500} \times 30 = 8\ (\text{cm})$$

In more general form, the shear center S of the open section as illustrated in Fig. 14.14 is proved by the following equation:

$$x_s = \frac{1}{I_x}\int_A^B r_s\left(\int_0^s ytds\right)ds$$

Fig. 14.14 Shear center of channel section

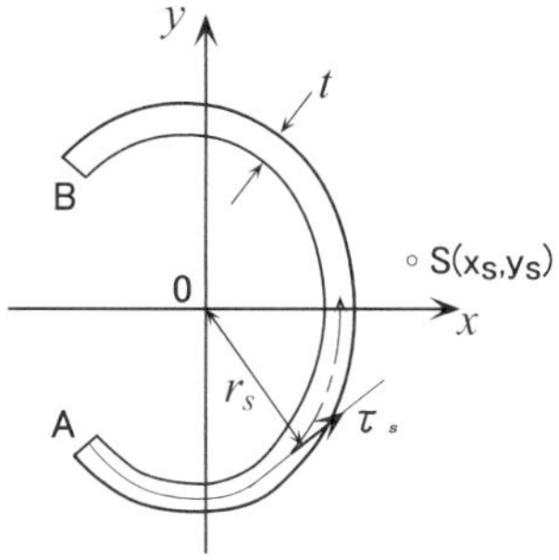

$$y_s = \frac{1}{I_y} \int_A^B r_s \left(\int_0^s xtds \right) ds \tag{14.5.7}$$

where r_s: projection length between the origin of coordinate and the subject point.

As explained in Sect. 14.4, the general form of the fundamental equation of torsion with warping is expressed by the following equation:

$$C\frac{d\varphi}{dz} - \Gamma\frac{d^3\varphi}{dz^3} = T \tag{14.5.8}$$

where the warping rigidity $\Gamma = E\int_A^B w^2 \cdot tds$ can be obtained by performing the following integration in terms of warping function w:

$$w = \frac{\int_A^B t\left(\int_0^s rds\right)ds}{\int_A^B tds} - \int_0^s rds \tag{14.5.9}$$

As shown in the above equation, torsional moment T includes the pure torsional moment as well as moment due to warping. Typical examples of torsional rigidity and warping rigidity are listed in Table 14.2.

Chapter 15
Deflection of Hull Structures

Generally, in designing a structure there are three important considerations: strength, deflection, and vibration. Strength is the most important item because any structures cannot hold its intended form without strength.

Deflection depends on the rigidity of the structure. Vibration causes trouble by resonance where the exciting frequency coincides with the natural frequency of the structure. From the failure viewpoint deflection and vibration are not so important as they cause only small additional stresses, but from the habitability and fitting viewpoints they are becoming more important.

For a bending member such as a beam, the strength is governed by I/y; the rigidity for deflection is governed by I and the vibration depends on I/A, where I is the sectional moment of inertia, y is the distance from the neutral axis, and A is the sectional area. It is important to note that in all cases I is involved, which is a fundamental property of the structure.

Up to here, from Chapters 8–14 in Part II, the following five items are explained about deflection.

1. Sect. 8.6.6 Forced Displacement and Forced Rotation: A beam supported by a girder will have additional bending moment which causes additional stress due to displacement of the girder
2. Sect. 9.3 Bottom Transverses Supported by Centerline Girder: In a bottom transverse, additional bending moments and shearing forces will be generated by deflection of the centerline longitudinal girder
3. Sect. 9.4 Deflection of Girders: The deflection pattern caused by shearing and the ratio of bending and shearing deflections are explained
4. Sect. 12.1 Boundary Conditions of Plates: Additional stresses will be induced in a plate by deformation of the boundary of the plate
5. Sect. 13.3 Optimum Space of Beams: The conditions by which stiffened panels will not be resonant with exciting frequencies lower than 1000 cpm are explained.

Y. Okumoto et al., *Design of Ship Hull Structures*,
https://doi.org/10.1007/978-3-032-12500-2_15

All items except Sect. 9.4 “Deflection of Girders” are concerned with the additional stress in a secondary member caused by the deflection of a primary supporting member.

In a design based on stress in the case of elastic design as well as plastic design, the deflection itself cannot be a main design condition but is a supplementary condition limiting additional stress due to deflection.

On the other hand, classification society’s rules, which have been established based on long years of experience, restrict the deflection to some extent. From the viewpoint of avoiding failure, the restrictions on deflection do not appear to be necessary. The fact that the classification societies have some restrictions on deflection means that some problems have been experienced due to excessive deflection in the past.

In this chapter the deflection of hull structure, hull girder deflection, deflection of beams, deflection of girders and webs, additional stress in secondary members caused by the deflection of a primary supporting member, and the limitation of deflection from viewpoints other than strength, are explained.

15.1 Deflection of Hull Girder

For the components in ships to limit the deflection of a hull girder, the following items are to be considered but no limitation exists from the strength viewpoint. A bigger hull girder deflection will be expected for a ship with bigger L/D, to which attention will be paid.

1. Expansion and contraction of pipes, rods, and handrails fitted in longitudinal direction on the upper deck or bottom
2. Increase of draft caused by deflection of hull girder
3. Generation of secondary stress by the deflection of hull girder
4. Flexural vibration of hull girder, “whipping”

For item (1), pipes are usually fitted with expansion joints where necessary. Generally, hull girder bending stress is estimated to be about 200 MPa which corresponds to a strain of $\varepsilon = \sigma/E = 200$ MPa/206, 000 MPa $\approx$ 1/1000. Accordingly, an expansion joint of ± 1 cm is enough for a 10 m long fitting. Actually, a fitting designer designs long pipes with expansion joint of ± 2 cm per 10 m in length.

Handrails along both the sides of the upper deck are also expanded or contracted by the hull girder bending. To avoid any adverse effect from this forced displacement, they are usually separated by a certain length. In this case, the stanchions located at the fore and aft ends of one continuous handrail are bent in the form of a cantilever beam, and bending stresses occur at their base as shown in Fig. 15.1. Assuming the deck strain of 1/1000 as stated above, the deflection of one stanchion δ can be expressed as:

$$\delta = \frac{l}{2000} \tag{15.1.1}$$

Fig. 15.1 Deformation of handrail stanchion due to hull girder bending

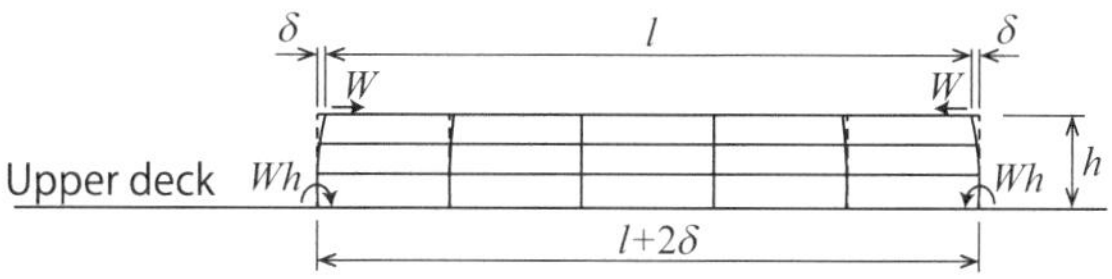

where l denotes the length of one continuous handrail.

Modeling this stanchion as a cantilever under a concentrated load W at the top handrail, the deflection δ can also be express using W as follows:

$$\delta = \frac{Wh^3}{3EI} \quad (15.1.2)$$

where h is the height of the handrail, E and I are the Young's modulus and the sectional moment of inertia of the stanchion.

From Eqs. (15.1.1) and (15.1.2),

$$W = \frac{3EIl}{2000\mathrm{h}^3} \quad (15.1.3)$$

Therefore, the bending stress σ_b at the base of the stanchion is:

$$\sigma_b = \frac{Wh}{Z} = \frac{3EIl}{2000\mathrm{h}^2 Z} = \frac{3Ely}{2000\mathrm{h}^2} \quad (15.1.4)$$

where Z is the section modulus of the stanchion and y is the distance between the outermost fiber and the neutral axis regarding the stanchion bending. When the stanchion is made of a plate whose thickness is t, $y = t/2$. Then, substituting $E = 206,000\ \ \mathrm{MPa}$, Eq. (15.1.4) finally reduces to:

$$\sigma_b = 154.5\frac{lt}{h^2} \quad (15.1.5)$$

When $t = 10\ \ \mathrm{mm}$, $h = 1100\ \ \mathrm{mm}$ and the allowable stress is 10 MPa,

$$l = \frac{\sigma_b h^2}{154.5t} = \frac{10 \times 1100^2}{154.5 \times 10} = 7832\,\mathrm{mm}$$

meaning that the handrail should be separated approximately every 8 meters.

Item (2), the increase of draft will be a problem when a ship with a deep draft passes through some shallow strait. Another problem is the decrease of deadweight in the case of sagging deflection because the Plimsol mark is at midships.

A welded ship usually has sagging deflection even with no load because the deck part is welded at the last stage and contracts because of shrinkage of welded metal. Because of this sagging there will be a loss of deadweight as mentioned above. To avoid this loss of deadweight a ship will be constructed originally with a hogging deflection which is called initial hogging.

The amount of the initial hogging is expressed by the following equation depending on the L/D of the ship and the number of erection butt joints on the upper deck n, which represents the welding volume in the upper deck [86].

$$\delta = 0.75\frac{L}{D}n - 77\,\text{mm}$$

where.

δ: amount of initial hogging.

L: length of ship.

D: depth of ship.

n: number of erection butt joints on upper deck.

This value of δ is aimed to be equal to the sum of the sagging deflection due to the welding shrinkage in the building dock and the deflection due to the still water bending moment just after launching. In this initial hog condition, there remains tensile residual stress in the upper deck and compressive residual stress in other parts of the ship such as bottom and side members.

The main concern regarding (3), i.e., secondary stress generated by the deflection of the hull girders, may be the compression in vertical members such as transverse bulkheads. In Fig. 15.2, the mechanism of how compressive stress is generated by hull girder deflection is shown. This is the same mechanism as explained in Fig. 11.6.3. Here, let us show that such compressive stress is very small and can be neglected.

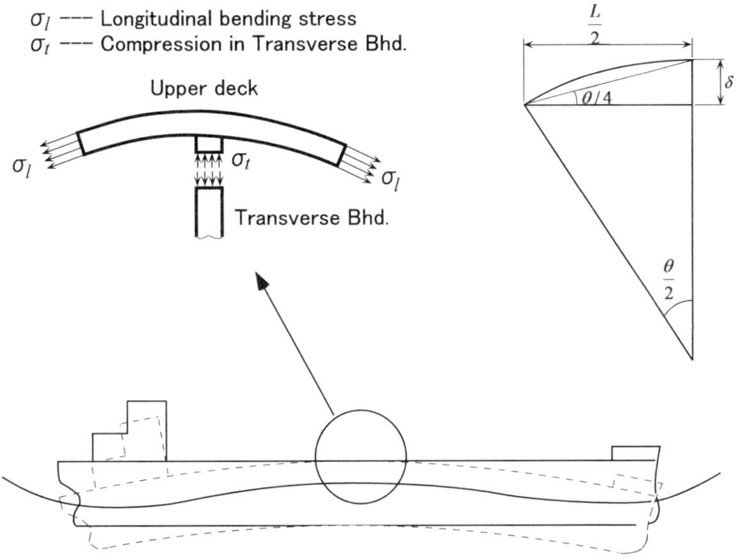

Fig. 15.2 Compression in vertical member due to hull girder bending

Assuming the sectional area of the upper deck is A_l; the longitudinal bending stress in the upper deck is σ_l; the sectional area of the transverse bulkhead is A_t and the compressive stress in the transverse bulkhead is σ_t, the relationship between σ_l and σ_t can be very roughly estimated as follows:

$$2A_t\sigma_t = A_l\sigma_l\theta \tag{15.1.6}$$

where θ is the included angle for the full length of hull girder bending as shown in Fig. 15.2.

θ is approximately expressed as follows where δ is the deflection of the hull girder and L is length of the ship.

$$\delta \approx \frac{L}{2} \cdot \frac{\theta}{4}$$

$$\therefore \frac{\delta}{L} \approx \frac{\theta}{8} \tag{15.1.7}$$

From Eqs. (15.1.6) and (15.1.7), the relation of Eq. (15.1.8) can be obtained.

$$\sigma_t = 4\frac{A_l}{A_t} \cdot \sigma_l \cdot \frac{\delta}{L} \tag{15.1.8}$$

Assuming $A_l/A_t = 2.0$ and $\delta/L = 1/1000$, $\sigma_t/\sigma_l = 1/125$ which is negligible.

The last issue is item (4), flexural vibration of hull girder. The amplitude of the flexural vibration of a hull girder is proportional to the static deflection of the hull girder and inversely proportional to the sectional moment of inertia of the hull girder. However it is very difficult to establish a limit for the hull girder rigidity from the vibration viewpoint because in the vibration problem, resonant phenomena are very important. Recently, the direct calculation methods of the vibration response caused by whipping have been developed and it may be possible to have an allowable limit for whipping stress [87–89].

From the longitudinal strength viewpoint, deflection of the hull girder cannot be limited. However the classification societies have a rule which limits the deflection of the hull girder based on their long experience. That is because hull girder deflection increases in accordance with the increase in L/D and the usage of higher strength steel such as HT32 and HT36.

The following relation is established where the longitudinal strength is evaluated keeping upper deck and bottom stresses lower than σ_a.

$$\delta = K\frac{WL^3}{EI} = \frac{K \cdot WL \cdot L^2}{E \cdot \frac{I}{y} \cdot y} = 2K\frac{\sigma_a}{E} \cdot \frac{L}{D} \cdot L \tag{15.1.9}$$

where.

δ: deflection of hull girder at midship.

W: load applied on hull girder.
E: Young's modulus.
I: sectional moment of inertia of hull girder.
L: ship length.
D: ship depth.
y: distance between neutral axis and upper deck or bottom ($y \approx D/2$).
K: constant.

From the above analysis it is found that the ratio of deflection to ship length δ/L is proportional to the ratio of ship length and ship depth L/D. Classification society rules provide for an upper limit of L/D. For example, the Lloyds Register of Shipping 1976 Rules give 17 as the maximum value of L/D. In cases where L/D is over 17, the allowable longitudinal stress has to be modified to $\sigma_a \times 17/(L/D)$ which makes the hull girder deflection equivalent to the case of $L/D = 17$. In the case where high tensile steel HT32 or HT36 is applied on the deck and bottom, the reduction coefficient of section modulus has a limitation of $0.059L/D$ where 0.059 corresponds to 1/17.

The later classification rules, for example, the Nippon Kaiji Kyokai (NK) 1996 rules have the following regulation which follows the IACS unified rule:

$$I \geq 3ZL \tag{15.1.10}$$

where.
I: sectional moment of inertia at midship section in cm^4
Z: section modulus at midship section in cm^3
L: length of ship in m.

Eq. (15.1.10) can be rewritten using cm units as follows:

$$I \geq \frac{3ZL}{100} \tag{15.1.11}$$

Putting $Z = I/y \approx 2I/D$ into Eq. (15.1.11), the following relationship is obtained:

$$I \geq \frac{6IL}{100D}$$

$$\therefore L/D \leq 16.7 \tag{15.1.12}$$

Equation (15.1.12) corresponds well to $L/D = 17$ of the Lloyds Register of Shipping 1976 Rules.

Figure 15.3 shows the actual data for the hull girder deflection δ in still water against ship length L. Recently, δ/L is getting bigger than 1/1000.

Hull girder deflection used to be calculated by integrating the M/I curve twice, where M is the bending moment acting on the hull girder. As an approximation it can be estimated using only the midship bending moment. The former requires time and labour, and the latter is preferred to the deflection in a wave where the bending

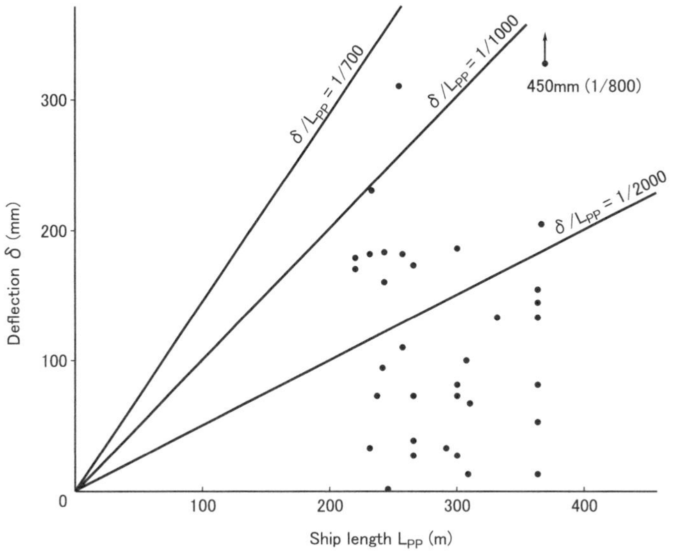

Fig. 15.3 Deflection of hull girder in still water

moment distribution is quite simple and similar to the assumption, but not good for still water because the bending moment distribution is changed by the loading condition.

Assuming the hull girder as a simply supported beam, the deflection for each load at midships can be calculated. By summing up these deflections the total deflection of the hull girder can be obtained easily and accurately [90].

Figure 15.4 shows a beam supported at both ends. The deflection δ at midship from a load W is expressed by the following equation.

$$\delta = \frac{W}{12EI}\left(2l^3 - 3lx^2 + x^3\right) \qquad (15.1.13)$$

where.

W: concentrated load.

x: distance between midship and loaded point.

Fig. 15.4 Deflection of both ends supported beam

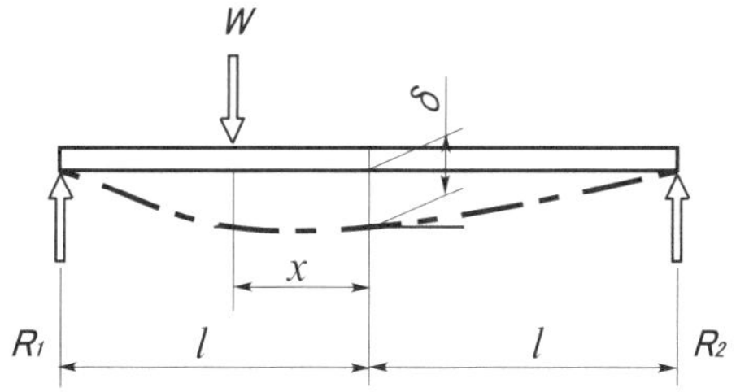

l: half length of ship.
E: Young's modulus.
I: sectional moment of inertia of midship section.

Applying the above formula, the deflections at midship from each load and buoyancy are calculated and added.

For each load a reaction is generated at the supporting points, i.e., the fore and aft ends of the ship, but the sum of the total reaction will be zero, which means that both ends are free, that is, a floating condition.

15.2 Deflection of Beams with Optimum Section

The deflection of the optimum beam explained in Sect. 8.4 is studied hereafter. Assuming the beam is fixed at both ends and has uniform load, the maximum bending moment M and the maximum deflection δ are expressed by the following equations:

$$M = \frac{Wl}{12} \tag{15.2.1}$$

$$\delta = \frac{Wl^3}{384EI} \tag{15.2.2}$$

where.
W: total weight applied to the beam.
l: span of the beam.
E: Young's modulus.
I: sectional moment of inertia.

Then, the allowable bending stress σ can be expressed as follows assuming that the section modulus Z of the beam is designed under this allowable stress.

$$\sigma = \frac{M}{Z} = \frac{Wl}{12Z} \tag{15.2.3}$$

$$\therefore Wl = 12Z\sigma$$

Putting the above into Eq. (15.2.2), the following equation is obtained:

$$\frac{\delta}{l} = \frac{12Z\sigma l}{384EI} \tag{15.2.4}$$

Now, the beam depth d and the sectional moment of inertia I are expressed as follows when the beam is designed to have the optimum section according to Table 8.4.1a.

$$d = \alpha\sqrt{\frac{Z}{t_w}} \tag{15.2.5}$$

$$I = \beta Z\sqrt{\frac{Z}{t_w}} \tag{15.2.6}$$

$$\therefore I = \frac{\beta}{\alpha} Zd \tag{15.2.7}$$

where α and β are the coefficients given in Table 8.4.1a.

Substituting Eq. (15.2.7) into Eq. (15.2.4),

$$\frac{\delta}{l} = \frac{12Z\sigma l}{384E\frac{\beta}{\alpha}Zd} = \frac{12\sigma}{384E} \cdot \frac{\alpha}{\beta} \cdot \frac{l}{d} \tag{15.2.8}$$

Equation (15.2.8) shows that the deflection per length of a beam δ/l is proportional to the allowable stress σ and the slenderness ratio l/d. α/β is 2.0 for a balanced girder, 1.0 for a girder with large attached plate, and 1.5 for an intermediate section. The deflection of a balanced girder is twice that of a girder with plate on the condition that σ and l/d are equal in both cases. For a given span and load, the bending moment is calculated irrespective of whether the balanced girder or not, and for an allowable stress σ, the necessary section modulus is derived. For a given section modulus, the optimum depth of the balanced girder is $1.10\sqrt{Z/t_w}$ and that of the girder with large attached plate is $1.22\sqrt{Z/t_w}$. Accordingly in the optimum condition, l/d for a balanced girder is bigger than that for a girder with large attached plate by 12%.

Comparing $(\alpha/\beta) \times (l/d)$ in Eq. (15.2.6) for the balanced girder and for the girder with large attached plate, the former is 2.24 times bigger than the latter. As seen from the above consideration, a balanced girder has a smaller sectional moment of inertia for a certain section modulus.

In Fig. 15.5 the relation between δ/l and l/d is shown for the optimum section of a balanced girder. For example, in the case that δ/l is 1/800 and σ is 200 MPa, $l/d = 20$ is optimum.

15.3 Deflection of Girders and Web Frames

As known from Eqs. (15.1.9) and (15.2.8), the ratio of deflection and span δ/l is proportional to the design stress σ and to the ratio between span and depth l/d. Hull structure designers have to pay particular attention so as to detect extraordinary cases where the deflection of a beam or a girder may cause troubles.

A hull girder acts like a free-free beam (both ends supported and no reaction at both ends), and the deflection is bigger than that for a both ends fixed beam. As shown in Fig. 15.3, the hull girder deflection in still water is in the order of $L/1{,}000$. In waves, assuming the longitudinal bending stress in waves is twice that in still

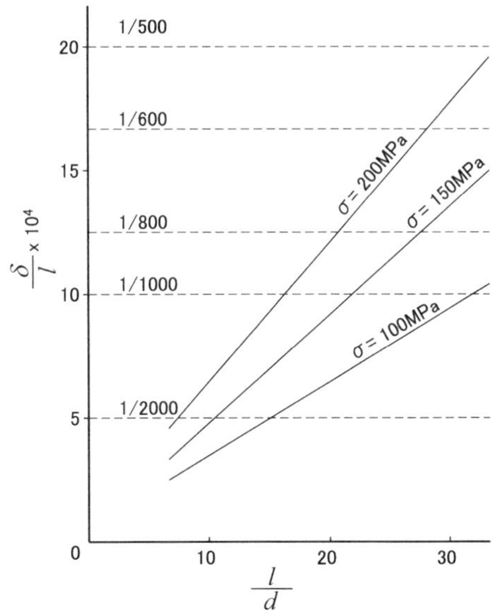

Fig. 15.5 Deflection of optimum beam (balanced girder)

water, the deflection will be *L*/500. These values are for ordinary ships with $L/D =$ 11~13, but for special ships with a *L*/*D* of over 14, special attention has to be paid to the deflection of the hull girder.

In the 1960's some troubles were experienced regarding the alignment of propeller shafts in ships with bigger *L*/*D* than ordinary ships. The deflection of the hull girder might have badly affected the shaft alignment.

For secondary members such as beams, frames, and stiffeners, *l*/*d* is usually 20~25, for deck girders 15 ~ 20, and for girders in a tank around 8.

Figure 15.5 shows that for a design stress of 150 MPa, deflection is 1/1000 of the span for the secondary member, 1/1200 for the deck girder, and 1/3, 000 for the girder in tank.

The hatch side girder and the hatch end strong beam of a cargo ship are designed to have a shallow depth to ease cargo handling. When a girder is designed under an allowable stress criterion, shallower depth causes greater deflection. A senior designer chose a value of $\delta/l \leq 1/1000$ based on his/her know-how with both ends supported condition, which seems to be quite severe.

As stated in Eq. (9.4.7) in Sect. 9.4. in the case of a girder with small *l*/*d*, i.e. less than 14, the shearing deflection is not negligible but sometimes bigger than the bending deflection. In this case both bending and shearing deflections are to be considered.

15.4 Additional Stress Caused by Deflection

In Sects. 8.6.6 and 9.3, the additional stresses in a beam and girder caused by the deflection are explained. The deflection at the supporting point causes an additional stress in the supported member.

In Sect. 8.6.6 "Forced Displacement and Forced Rotation", it is explained that the displacement δ at the supporting point will cause an additional bending moment M_δ in the supported member in both-end fixed condition as follows.

$$M_\delta = \frac{6EI}{l^2}\delta \quad (15.4.1)$$

where.

l: span of beam.

E: Young's modulus.

I: sectional moment of inertia of beam.

In this case the following reaction force R_δ is generated at the supporting points at the same time.

$$R_\delta = \frac{12EI}{l^3}\delta \quad (15.4.2)$$

As shown in Fig. 15.6, the reaction force is added at the rigid supporting point and deducted from the deflected supporting point. This is a natural philosophy, that is, the stronger supports a bigger load and the weaker a less load.

Supposing that on the longitudinal stiffener supported by a transverse bulkhead and the adjacent vertical web, the deflection of the vertical web is bigger than that of the transverse bulkhead, when water pressure is applied. Accordingly a longitudinal stiffener has an additional bending stress and simultaneously a reaction is added to the supporting point of the transverse bulkhead. The bending moment M_w and the reaction force R_w of the longitudinal stiffener due to the water pressure is expressed as follows:

$$M_w = \frac{Wl}{12} \quad (15.4.3)$$

$$R_w = \frac{W}{2} \quad (15.4.4)$$

Fig. 15.6 Reaction due to deflection at support

Therefore, from Eqs. (15.4.1)—(15.4.4), the ratios of the forced displacement component to the water pressure component are:

$$\frac{M_\delta}{M_w} = \frac{6EI}{l^2}\delta\frac{12}{Wl} = \frac{72EI\delta}{Wl^3} \tag{15.4.5}$$

$$\frac{R_\delta}{R_w} = \frac{12EI}{l^3}\delta\frac{2}{W} = \frac{24EI\delta}{Wl^3} \tag{15.4.6}$$

From Eqs. (15.4.5) and (15.4.6), we can find that $(M_\delta/M_w)/(R_\delta/R_w) = 3$, that is, the effect of the forced displacement relative to the effect of water pressure is three times severer for the bending moment than for the reaction force. In Sect. 10.5, we explained two generations of slot damages. The cracks occurred at the root of the web stiffeners, and propagated along the welding between the web stiffener and the face plate of the longitudinal stiffener in the first generation damages. On the other hand, in the second generation damages, the cracks occurred at the same location but propagated into the web plate of the longitudinal stiffener. The first generation damages were governed by the reaction force at the transverse bulkhead and the vertical web, while the second generation damages were governed by the bending moment on the longitudinal stiffener. Therefore, the effect of the relative displacement between the transverse bulkhead and the adjacent vertical web was much more significant for the second generation damages, and has been considered in the design to prevent the second generation damages.

Taking a side longitudinal as a specific example as shown in Fig. 15.7, the following calculation is carried out.

$$M_w = \frac{Wl}{12} = 52.3\,\text{kN}\cdot\text{m}$$

$$R_w = \frac{W}{2} = 78.5\,\text{kN}$$

When the side longitudinal is designed with allowable stress σ of 150 MPa,

$$Z = \frac{M_w}{\sigma} = 349\,\text{cm}^3$$

The thickness of the web plate is 10 mm and the optimum section is adopted.

$$I = 0.77Z\sqrt{\frac{Z}{t_w}} = 5020\,\text{cm}^4$$

Then,

$$M_\delta = \frac{6EI}{l^2}\delta = \frac{6\times 206\times 10^6\times 5020\times 10^{-8}}{4^2}\delta\times 10^{-3} = 3.88\delta\,\text{kN}\cdot\text{m}$$

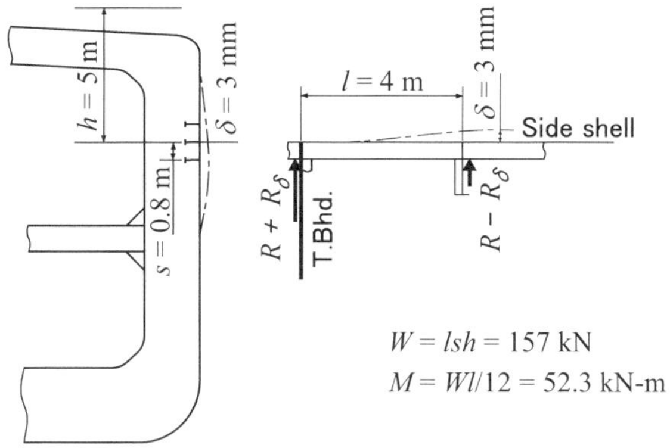

Fig. 15.7 Additional stress due to forced displacement on side longitudinal

$$R_\delta = \frac{12EI}{l^3}\delta = \frac{12 \times 206 \times 10^6 \times 5020 \times 10^{-8}}{4^3}\delta \times 10^{-3} = 1.94\delta \text{ kN}$$

Now, assuming $\delta = 3$ mm, the following results are obtained:

$$M_\delta = 11.6\,\text{kN} \cdot \text{m}$$

$$R_\delta = 5.82\,\text{kN}$$

$M_\delta/M_w = 0.222$, $R_\delta/R_w = 0.074$, then the additional bending stress is 22% and the additional reaction force is 7.4% of the component due to water pressure.

15.5 Shearing Deflection

When the hull girder is in a sagging condition, the aft part including the aft engine room deforms in a convex upwards mode. This phenomenon has been known for a long time but not clearly understood. It can be explained by the shearing deflection of a cantilever beam under uniformly distributed load stated in Fig. 9.4.1b and is simple but very important for design [91]. Common sense does not expect such deflection to be convex downwards as shown in Fig. 9.4.1b, because one is familiar with the bending deflection which is convex upwards.

When one of the authors calculated the hull girder shearing deflection in the vicinity of the aft part of a ship including the aft engine room, the result was different from the initial expectation. The aft part of the ship was considered as a deep girder with dominant shearing deflection. The load was the buoyancy difference between fully loaded and ballast conditions, that is, a downward load from the loss of the buoyancy. The results showed that the deflection mode was convex downwards. Repeated calculations showed the same results and finally it was understood that in this case the deflection by bending is convex upwards and the deflection by shearing is convex downwards.

These phenomena can be applied in the control of the deflection of a double bottom in an aft engine room in which shafting is installed. This problem is explained in detail in Chap. 28, Part III.

Recently with the use of computers, the deflection can be calculated quickly and easily but not by separating that by bending and that by shearing. The mode of the shearing deflection can be understood only by means of a manual calculation. Such considerations should be kept in a designer's mind even in the existing computer era.

Chapter 16
Welding

A hull structure is constructed by welding steel plates together. Before the use of welding methods in hull construction, the steel plates were joined by rivets. In the case of riveted joints, the many rivet drill holes in the plates reduce the strength of the plate compared to a plate without drill holes, i.e. the joint efficiency is reduced, and also drilling needs much manpower. In addition to that, from the viewpoint of the structural designer, the design of a rivet connection needs much work, since it is required to determine the location of every rivet hole. Conversely, in case of welding, the designer has only to draw a line to present a weld line, instead of every rivet hole, and to indicate the edge preparation and to decide the welding sequence, if necessary.

In applying a rivet joint to a connection, the adjacent connections were designed so as not to be in line with each other, because the efficiency of the rivet joint is reduced. Such a practice of joint arrangement remained long after welding was adopted.

In the case that rivets are applied for both seam and butt joints, the plates around the joint are intricately stacked in three layers. Such connections were inconvenient for both the construction people and the design people, because of difficulty to arrange the plates of three layers accurately. Then in the 1950's, welding was applied to the butts in the outer shell plate and deck plate connections, eliminating the three-layer structure.

Rivet joints were gradually replaced by welding one after another, and in the 1970's only the riveting of bilge keel to the adjacent shell plating was left. The purpose of this rivet joint was to prevent crack propagation from the bilge keel to the shell plating; however, nowadays this joint has been changed to welding due to the development of reliable welding processes, and rivet joints are utilized only for a special connection, such as between steel plate and aluminum plate.

In recent ship design, the designer does not pay so much attention to welding. However, it is recommended that designers consider the importance of welding connections because the vessel will break into pieces if welding is destroyed.

Y. Okumoto et al., *Design of Ship Hull Structures*,
https://doi.org/10.1007/978-3-032-12500-2_16

In designing for welding, it is important to maintain welding strength as well as to maintain productivity, since the welding process takes the most man-hours in the hull construction process.

Several unexpected happenings have taken place during the transition from riveted construction to welded construction. One example occurred in tank testing when a tank was filled with water to confirm the tightness of the tank boundary. There was no leakage of water around the tank, but the water flowed out at a point far away from the tank. The reason was that water went through a gap in the fillet welding of the tank and it flowed out from pin holes at the far point. We usually regard a welded structure as a water and gas tight structure; however, it is not correct to consider that a welded structure is 100% tight against water and gas because welding contains a certain percentage of defects.

When welding was first introduced to the production of vessels, strength and toughness of the welding joint was thought to be less than that of the base metal because of welding defects. This thought was reflected in the welding practice that welding lines should be staggered. Another example was seen in the practice of a welding line not being designed to interfere with another welding line. It was because the strength of the steel plate might be damaged by being heated twice.

Comparing butt welding and fillet welding, the latter is useful for the saving of manpower in some cases. In joining an upper plate and lower plate by horizontal welding, butt welding can be replaced by fillet welding.

This chapter explains butt welding and fillet welding, which are the basis of welding, and then gives examples of welding design to achieve high productivity. Finally, detail design for welding is described in association with methods to prevent water leakage, scallops to prevent interference of welding lines, replacement of butt welding by fillet welding.

16.1 Butt Welding

The strength of a butt welding joint is considered to be satisfactory for static loads. That is because

(1) The ultimate strength of deposited metal is said to be more than that of the base metal.

For example, in the case of mild steel, the ultimate strength σ_B of base metal is 41–50 kgf/mm^2 and σ_B of deposited metal is 41–57 kgf/mm^2.

(2) Reinforcement of the weld is provided.
(3) Small welding defects do not affect static strength.

On the other hand, fatigue strength of a butt-welded joint reduces from that of the base plate as shown in Fig. 16.1 [92, 93]. The reason for that fatigue strength reduction is explained as

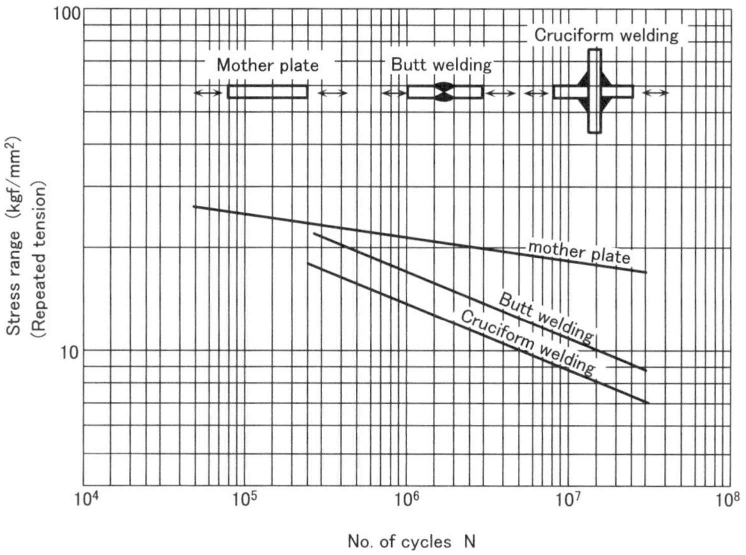

Fig. 16.1 Example of fatigue strengh of mild steel welding

(1) There is a discontinuity of shape between the deposited metal and base plate which causes stress concentration.
(2) Small defects inside welding affect fatigue strength.

The fatigue strength of a welded joint can be evaluated in terms of the cumulative fatigue damage factor calculated by (1) the S-N curve obtained by fatigue tests of welded joints under repeated tension and (2) the distribution of encountered stress cycles. The failure of a welded joint due to fatigue usually occurs in the base plate, not in the deposited metal.

Regarding the butt welding method for connecting flat plates such as outer shell platings, deck platings etc., the submerged arc welding process is applied for downward welding and the electrogas welding process for vertical welding. These welding processes are used in the workshop and in the building dock, which can contribute to efficient production. In contrast with this situation, manual welding or semi-automatic welding is normally applied to the connection of internal members.

Let us consider the relation between the edge preparation and the sectional area of deposited metal, as shown in Fig. 16.2. From this figure, we find that the sectional area of the X-shape groove is considerably smaller than that of the V-shape. For example, comparing the sectional area of the X-shape groove, which is vertically symmetric, with the area of the V-shape, the former is just half of the latter. Therefore, from the view point of productivity, a symmetric X-shape edge preparation is helpful to save welding man-power for vertical and horizontal butt welding. In spite of that, a V-shape edge preparation, or X-shape of which the upper groove is larger than the lower, may be applicable to downward butt welding.

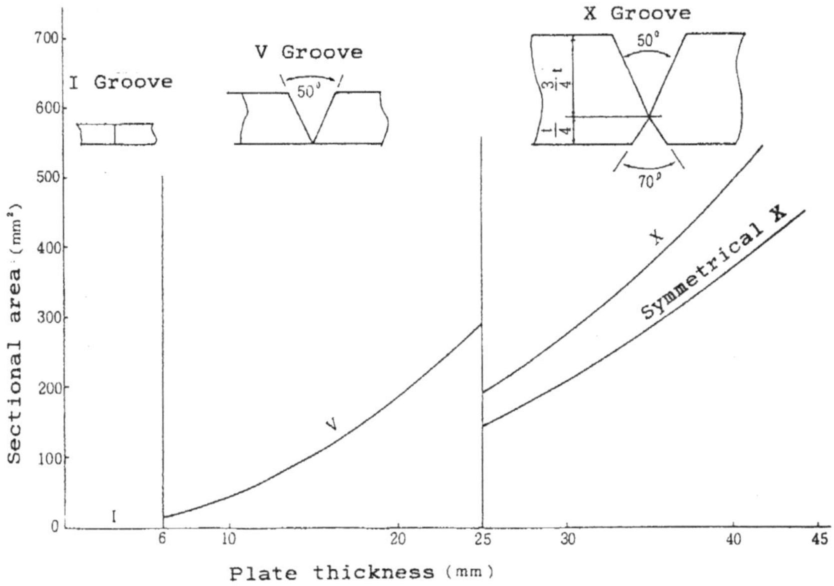

Fig. 16.2 Shape and sectional area of welding groove

Another effective way to reduce the manpower for welding is to decrease welding length. In order to do this, the application of wider plates is the simplest and most useful solution. If a large panel of uniform thickness is fabricated by several small panels, the use of wider plates does not cause an increase in steel weight. But in cases where the thickness of a large panel varies such as in a transverse bulkhead, the application of wider plates increases the steel weight.

As an example, let us calculate the increase of weight by applying wider plates based on the earlier NK deep tank rule. The plate thickness t of deep tank bulkheads is given by the following Equation:

$$t = 3.6s\sqrt{h} + 3.5 \text{ mm} \tag{16.1.1}$$

where

s: stiffener space.
h: water head at the lowest point of plate.

Assuming the transverse bulkhead of depth D is divided equally by n into narrow plates of width D/n, and the water head is taken as the distance from the top of the bulkhead to the bottom of the subject plate. Thus, the plate thickness at each height is obtained as follows neglecting the minimum thickness criterion:

$$\left.\begin{aligned} t_1 &= 3.6s\sqrt{\tfrac{D}{n}} + 3.5 \\ t_2 &= 3.6s\sqrt{\tfrac{2D}{n}} + 3.5 \\ &\cdot \\ &\cdot \\ t_n &= 3.6s\sqrt{\tfrac{nD}{n}} + 3.5 \end{aligned}\right\} \tag{16.1.2}$$

Hence the total area of the bulkhead plate A is calculated by summing up the multiplication of plate thickness $t_1, t_2, \cdots t_n$ and plate width D/n as follows:

$$\begin{aligned} A &= \frac{D}{n} \times 3.5 \times n + \frac{D}{n} \times 3.6s\sqrt{\frac{D}{n}}\left\{\sqrt{1} + \sqrt{2} + \cdots + \sqrt{n}\right\} \\ &= 3.5D + 3.6s\sqrt{\frac{D^3}{n^3}}\left\{\sqrt{1} + \sqrt{2} + \cdots + \sqrt{n}\right\} \end{aligned} \tag{16.1.3}$$

Introducing area A_1 is the total area of the bulkhead in the case of $n = 1$, i.e. bulkhead plate thickness is constant, the area ratio A/A_1 represents the weight ratio and it becomes

$$A_1 = 3.5D + 3.6s\sqrt{D^3} \tag{16.1.4}$$

$$\begin{aligned} \therefore \frac{A}{A_1} &= \frac{3.5D + 3.6s\sqrt{\frac{D^3}{n^3}}\left\{\sqrt{1} + \sqrt{2} + \cdots + \sqrt{n}\right\}}{3.5D + 3.6s\sqrt{D^3}} \\ &= \frac{1}{\sqrt{n^3}}\left\{\sqrt{1} + \sqrt{2} + \cdots + \sqrt{n}\right\} + \frac{3.5D\left[1 - \frac{1}{\sqrt{n^3}}\left\{\sqrt{1} + \sqrt{2} + \cdots + \sqrt{n}\right\}\right]}{3.5D + 3.6s\sqrt{D^3}} \end{aligned} \tag{16.1.5}$$

When the corrosion addition of 3.5 mm is neglected and only the net thickness is considered,

$$\frac{A}{A_1} = \frac{1}{\sqrt{n^3}}\left\{\sqrt{1} + \sqrt{2} + \cdots + \sqrt{n}\right\} \tag{16.1.6}$$

$$\lim_{n\to\infty} \frac{A}{A_1} = \frac{2}{3} \tag{16.1.7}$$

Figure 16.3 represents the calculated relation of area ratio A/A_1 according to Eq. (16.1.6) against the dividing number n. When n approaches infinity, A/A_1 approaches 2/3 theoretically according to Eq. (16.1.7). In Fig. 16.3, a realistic example is also shown in which $D = 14.4$ m is assumed, the top of water head

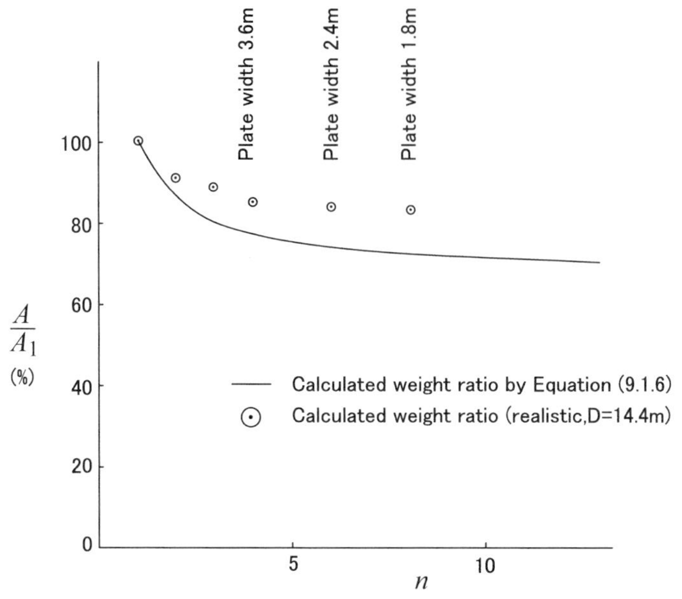

Fig. 16.3 Dividing number n of transverse bulkhead versus weight ratio

is taken from 2.44 m above D and the plate thickness is expressed in round figures of 0.5 mm interval. From the figure, we find that the total weight changes only slightly when the number of plates becomes more than 5 or 6 in both cases. In Fig. 16.3, the calculated weight ratio by Eq. (16.1.6) is always less than those of a realistic example. That is because Eq. (16.1.6) is derived by neglecting the second term of Eq. (16.1.5), the effect of the corrosion addition.

16.2 Fillet Welding

A fillet welded joint is generally subjected to the following forces, as illustrated in Fig. 16.4.

(1) tension, where the fillet welding acts as a cruciform joint against tension. (Fig. 16.4a)
(2) bending moment (Fig. 16.4b)
(3) shearing force, where the fillet welding transmits the shearing force between the web plate and the face plate of a beam (Fig. 16.4c)

The strength of a fillet welded joint was investigated in the report [94–96]. Referring to this report, let us explain its summary.

A fillet welded joint subjected to tension is generally called a cruciform joint in Fig. 16.4a. In this case, the summation of throat thickness w of both sides must

Fig. 16.4 Fillet weld joint

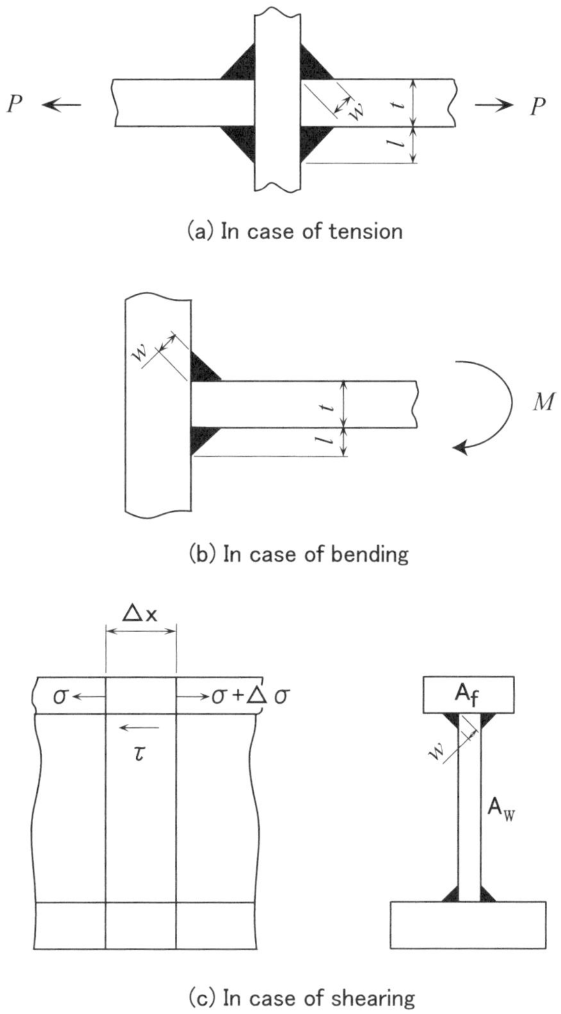

be equal to the plate thickness t to fully compensate for the plate thickness under tension; therefore, leg length l of the fillet welding is given by the following equation:

$$t = 2w \tag{16.2.1}$$

$$\therefore l = \frac{\sqrt{2}}{2}t = 0.7t \tag{16.2.2}$$

This kind of load carrying cruciform joints can be seen everywhere in ship structures. When the acting tensile force P is very large, such as in the case of a transverse web sandwiched between two abutting longitudinal members, it is recommended to apply non-load carrying cruciform joint with continuous longitudinal member or full penetration welding.

Figure 16.4b represents the fillet welding under a bending moment. From the viewpoint of elastic design, the elastic section modulus Z_w of the deposited metal must equal the elastic section modulus Z_b of the plate. Each section modulus per unit width is expressed in Eqs. (16.2.3) and (16.2.4).

$$Z_b = \frac{t^2}{6} \tag{16.2.3}$$

$$Z_w = \left\{ \frac{(t+2w)^3}{12} - \frac{t^3}{12} \right\} \frac{2}{t+2w} \tag{16.2.4}$$

Then, equating these two section moduli,

$$\frac{t^2}{6} = \frac{(t+2w)^2}{6} - \frac{t^3}{6(t+2w)} \tag{16.2.5}$$

$$8\left(\frac{w}{t}\right)^3 + 12\left(\frac{w}{t}\right)^2 + 4\left(\frac{w}{t}\right) = 1 \tag{16.2.6}$$

$$w = 0.165t$$

$$l = 0.233t \tag{16.2.7}$$

On the basis of plastic design, the plastic section modulus M_w of a fillet weld should be equal to the plastic section modulus M_b of the plate. Each plastic section modulus is expressed as follows:

$$M_b = \frac{t^2}{4}\sigma_{yb} \tag{16.2.8}$$

$$M_w = w(w+t)\sigma_{yw} \tag{16.2.9}$$

where

σ_{yb}: yield stress of plate.
σ_{yw}: yield stress of welding metal.
σ_{yw} is usually greater than σ_{yb}, but considering the worst case where $\sigma_{yw} = \sigma_{yb}$ and equating Eqs. (16.2.8) and (16.2.9).

$$\frac{t^2}{4} = w(w+t)$$

$$\left(\frac{w}{t}\right)^2 + \left(\frac{w}{t}\right) = \frac{1}{4} \tag{16.2.10}$$

$$w = 0.21t$$

$$l = 0.296t \tag{16.2.11}$$

From both points of view, i.e. elastic design and plastic design, we can say that half of the leg length of a fillet weld under tension is much enough for the required leg length against the bending moment.

Typical examples of fillet welding against bending stress are observed in the bulkhead structure of tank boundaries. On the other hand, if fillet welding is used for connecting the web plate and face plate of a corrugated bulkhead structure of built-up type, the stress distribution of the fillet welding is different from the above one. Because the face plate of a corrugated bulkhead plays a role of extreme fiber based on beam theory, the fillet welding is subjected to tension or compression.

As discussed previously, different types of stress occur when the same fillet welding is applied to different kinds of structure. Therefore, the designer has to imagine the type and direction of the stress induced in the structural member and also think about possible failure modes of the structure due to the stress.

Figure 16.4c represents an example of a fillet welding under shearing force. In the case of a fillet weld against tension or in the case of a fillet weld against bending, the leg length can be rationally determined, since the plate stress directly flows through the fillet welding and is transmitted to the adjacent plate.

On the other hand, in the case of a fillet welding under shearing force, the discussion is slightly complicated. Cutting out a narrow slice, length Δx, of a beam under a shearing force including fillet welding in Fig. 16.4c, we obtain the equilibrium of forces in the following form:

$$2w\tau\Delta x = \Delta\sigma A_f$$

$$\therefore w = \frac{A_f\Delta\sigma}{2\tau\Delta x} \tag{16.2.12}$$

where

w: throat thickness of fillet welding
τ: shearing stress.
$\Delta\sigma$: change of nominal stress in face plate.
A_f: sectional area of face plate.
Considering the following relationship between bending moment and shearing force:

$$\frac{dM}{dx} = F \tag{16.2.13}$$

Then we can rewrite Eq. (16.2.12) into Eq. (16.2.14) as in the form:

$$w = \frac{A_f}{2\tau} \cdot \frac{F}{Z} \tag{16.2.14}$$

where

F: shearing force.
Z: section modulus of beam.

Introducing the sectional area of the web plate as A_w, the shearing force of the web is expressed as $F = A_w\tau$, then we find Eq. (16.2.15) under the condition that the average shear stress on the web is very large, reaching its allowable limit and taking the same value as the shear stress on the fillet weld.

$$w = \frac{A_f \cdot A_w}{2Z} \tag{16.2.15}$$

From the above Eq. (16.2.15), the throat thickness is a function not only of sectional web area but of sectional area of face plate and section modulus of beam.

In order to simplify Eq. (16.2.15), let us take advantage of the optimum design of beam as described in Sect. 8.4. In Table 8.1, if the theory of the optimum section of a beam is applied, the sectional area of a web plate A_w and sectional area of a face plate A_f can be determined by the section modulus Z. For instance, in case of a balanced girder, the following equations are derived:

$$\left.\begin{aligned} A_w &= 1.10\sqrt{Zt} \\ A_f &= 0.73\sqrt{Zt} \end{aligned}\right\} \tag{16.2.16}$$

Substituting Eq. (16.2.16) into Eq. (16.2.15), we obtain the following equation:

$$w = \frac{1.10 \times 0.73}{2} t = 0.4t \tag{16.2.17}$$

$$l = 0.565t \tag{16.2.18}$$

Consequently, the leg length of a fillet weld under shear can be simply decided by the plate thickness.

As mentioned earlier, the optimum design of a beam can uniquely provide the sectional shape, sectional moment of inertia and weight, once the required section modulus is given. In addition to that, the optimum design of a beam is helpful to treat several variables in terms of one parameter.

To summarize the above investigation, Fig. 16.5 illustrates the relation between the required leg length of a fillet weld and plate thickness in these three cases: fillet weld against tension Eq. (16.2.2), bending Eq. (16.2.7) or Eq. (16.2.11), and shearing Eq. (16.2.18). Requirement of NK's leg length, F1 and F2, for fillet welding are also plotted in the same figure. In order to compare directly the leg length requirement

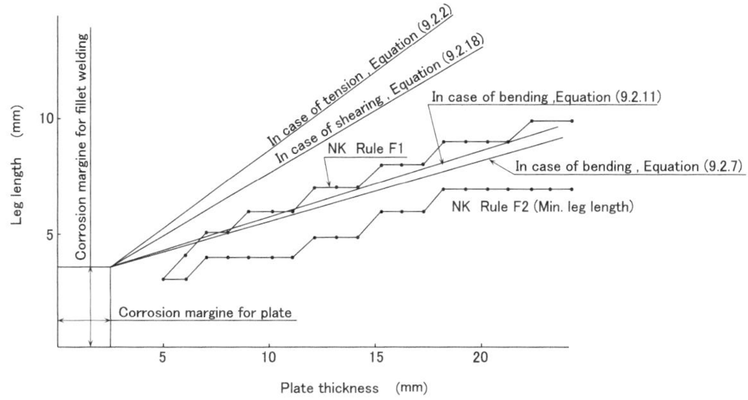

Fig. 16.5 Leg length of fillet weld

by the classification societies and the derived equations, the corrosion margin of a plate is assumed to be 2.5 mm and the corresponding leg length of a fillet weld as $2.5\text{mm} \times \sqrt{2} = 3.5\,\text{mm}$. For reference, the F1 requirement of leg length for a fillet weld by NK is normally applied to the following weld connections; rudder frames and rudder horn, bracket beneath main engine and inner plate of double bottom, tank boundary of water/oil-tight bulkhead.

From Fig. 16.5, it can be said that the leg length of the fillet weld under tension is the biggest. The leg length equations for bending (Eqs. (16.2.7) and (16.2.11)) agree quite well with the F1 requirement. The leg length equation for shearing (Eq. (16.2.18)) looks too demanding compared with the rule requirement. This is because it is assumed in the calculation that the maximum allowable shear stress is working on the web. Therefore, the leg length can be mitigated with respect to the fillet welding between the face plate and the web other than the span end locations where the shear stress on the web is critical.

The relation between the leg length of a fillet weld and deposit metal is indicated in Fig. 16.6, which shows that the quantity of deposit metal increases proportionally to the square of leg length. In order to decrease deposit metal and to save manpower for welding, the fillet weld with edge preparation is often used particularly for thicker plate more than 25 mm. In this case, 50° of groove is normally applied, as in the butt welding shown in Fig. 16.2.

Let us assume that the shape of deposit metal is an isosceles triangle having equal side length a as shown in Fig. 16.7, then the throat thickness must be half of the plate thickness t and it becomes

$$\left.\begin{aligned} x &= a\cos 50^\circ \\ w &= a\cos 25^\circ = \tfrac{t}{2} \end{aligned}\right\} \tag{16.2.19}$$

$$\therefore x = \frac{\cos 50^\circ}{2\cos 25^\circ} t = 0.35t$$

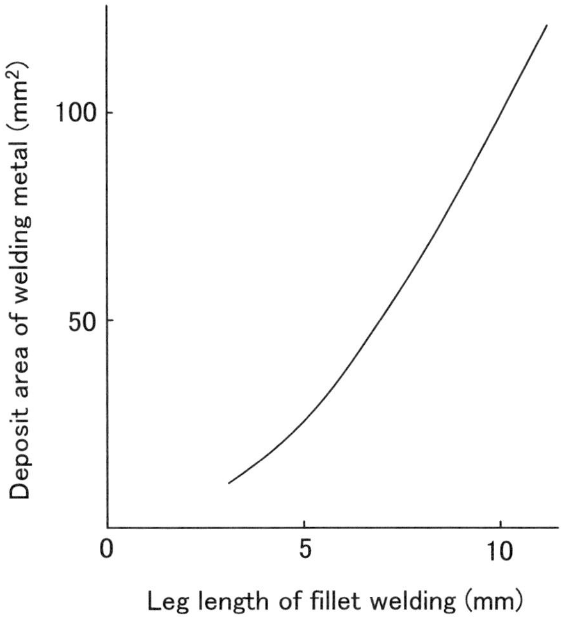

Fig. 16.6 Deposit of welding metal versus leg length of fillet weld

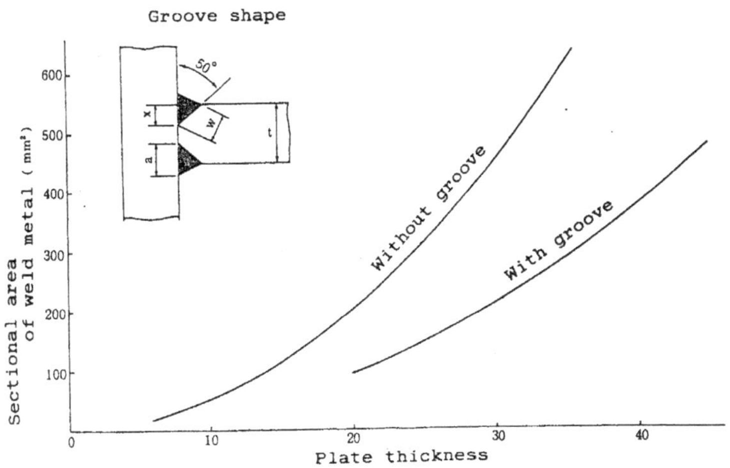

Fig. 16.7 Sectional area of fillet weld with and without groove

Namely, both edges of the plate should be cut by a 50° of groove with 0.35 t depth and the middle of the plate thickness must be left uncut for 0.3 t thickness.

Figure 16.7 represents the area of deposit metal of a fillet weld in both cases, with and without edge preparation. If the plate edge is prepared with a groove, then the area of weld deposit is less than half compared to that without a groove. In addition to that, the edge preparation is beneficial to reduce stress concentration due to smooth flow of stress. In critical areas with high tensile stress, it is recommended to apply full penetration welding ($x = t/2$). For example, IACS CSR requires full penetration welding to some specific critical joints such as the connection of vertical corrugated bulkhead to the lower hopper top plate.

16.3 Fillet Welding with Higher Strength Electrode

There are two effective ways to decrease the deposit metal of fillet welding; one is to provide edge preparation and the other is to carry out welding of mild steel plate by means of higher strength electrode. According to Fig. 16.6, we find that the weld deposit could be reduced to 77% if the leg length of fillet welding decreases from 8 to 7 mm. Furthermore, the leg length of 7 mm can be obtained by one pass welding, although the length of 8mm is achieved by two passes. For the sake of leg length reduction, the use of an electrode with higher strength than the base plate is effective. Taking σ_{BH} as the breaking stress of a higher strength electrode and σ_{BM} as that of a mild steel electrode, and l_H and l_M as the leg length of fillet welding, respectively, then the following equation can be obtained for the breaking condition:

$$\sigma_{BH} \cdot l_H = \sigma_{BM} \cdot l_M \tag{16.3.1}$$

Hence the leg length of fillet welding can be reduced to 80% using a higher strength electrode of 50 kgf/mm^2 class instead of a mild steel class of electrode.

Nevertheless, this equation gives an overestimated result, since the corrosion margin is neglected here. To take account of this margin, let us assume a is corrosion margin for both mild steel and higher strength steel of 50 kgf/mm^2 class, then Eq. (16.3.1) can be interpreted in the following form:

$$(l_H - 1.4a) = \frac{\sigma_{BM}}{\sigma_{BH}}(l_M - 1.4a) \tag{16.3.2}$$

Figure 16.8 shows a relation between l_H and l_M calculated by substituting $a = 2.5$mm into Eq. (16.3.2). It is found that 7 mm leg length with a higher strength electrode in 50 kgf/mm^2 class is enough to compensate 8 mm of length with mild steel electrode.

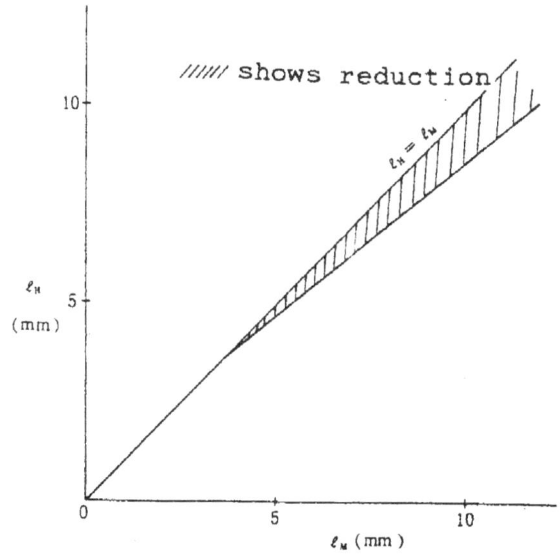

Fig. 16.8 Leg length reduction by high tensile electrode

16.4 Water Stopping Welding

When a riveted joint was applied to connect the plates, water flow through the gap between the lapped parts of plates was prevented by caulking the edge of the plate and the head of the rivets as illustrated in Fig. 16.9. In the same manner, the gap between the plate and stiffener still remains at the root face in the case of a fillet weld; therefore, it is necessary to provide water stopping welding to avoid the water flow in the gap.

Water stopping can be realized by filling the gap by welding at a location near the watertight boundary. Scallops with boxing fillet welding work for this purpose as shown in Fig. 16.9c. Another method is to perform a short full penetration welding in association with edge preparation of the stiffener as shown in Fig. 16.9d.

To confirm the water tightness of a fillet weld, air pressure is often applied to the compartment where the gap between plate and stiffener is interrupted by several water stopping welds. This way of confirmation can be performed when a block has been constructed and before the whole tank test, therefore, the final water tightness test can be done with full confidence.

16.5 Scallop and Serration

Scallops are to be provided to prevent the interference of welding lines, as shown in Fig. 16.10a. However, nowadays the designer is apt to eliminate scallops as much as possible for the following reasons:

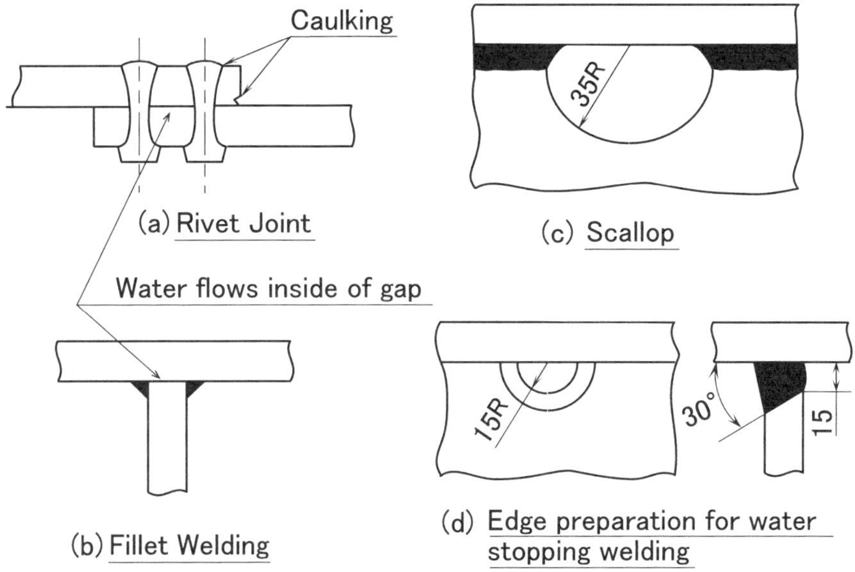

Fig. 16.9 Water stopping welding

(1) It was found that the interference of welding lines does not cause harmful damages neither to the strength or to the material deterioration.
(2) Scallops cause high stress concentration due to abrupt change of structure.
(3) Welding of scallops takes much time because of boxing fillet welding at both ends of a scallop.

When scallops are deleted, the designer must pay attention to the following two points:

(1) Reinforcement of a butt weld joining two plates should be well ground off in way of stiffener to make a smooth surface, which enables stiffeners to be welded on this plate.
(2) The corner of a web plate should be cut off in isosceles right-angled triangle where the corner directly touches a fillet weld of a stiffener as in Fig. 16.10b.

In the above two cases, the scallop works as an air hole or drain hole. If this scallop is eliminated, another air hole or drain hole must be provided to compensate for this scallop as illustrated in Fig. 16.10b.

Serration means an edge having a row of V-shapes like the teeth of a saw and it is applied to the edge of a stiffener with the intention of reducing weld length of the connection between stiffener and plate. In the case where two stiffeners are produced by cutting one channel bar or one I-section beam, the depth of the stiffener can be increased by providing serrations, i.e. the depth of a serrated stiffener becomes larger by a half depth of serration relative to that of non-serrated stiffener in Fig. 16.11. For this reason, serrated stiffeners were very popular until the 1955's. Nevertheless,

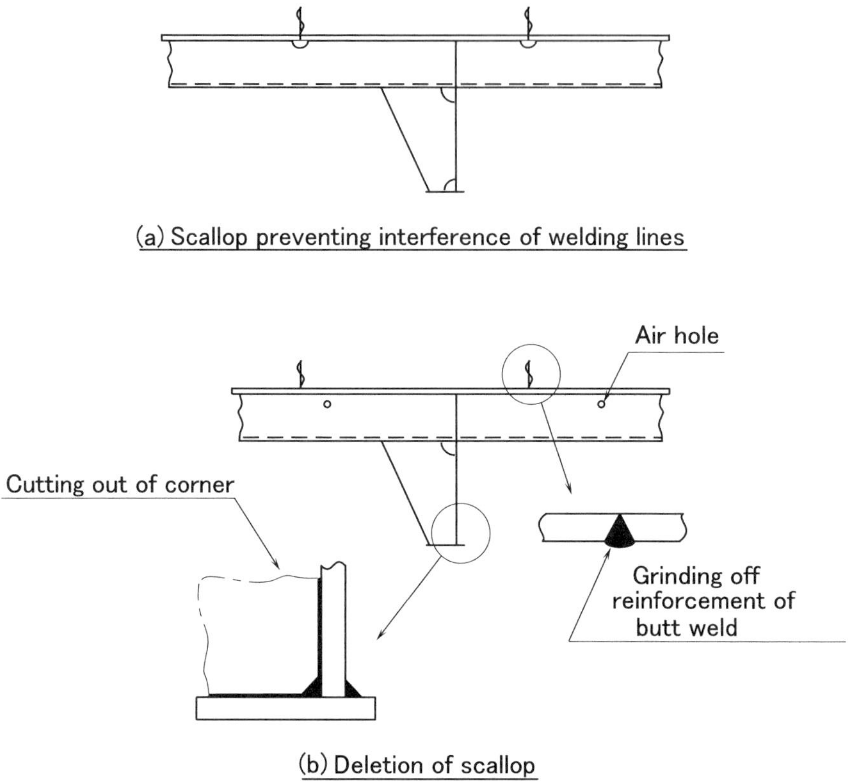

Fig. 16.10 Scallop preventing interference of welding line and its deletion

serrated stiffeners have disappeared together with scallops these days because of cracks due to high stress concentration and the difficulty for gravity welding and portable CO_2 semi-automatic welding due to boxing fillet welding.

16.6 Conversion of Butt Welding to Fillet Welding

Butt welding can be changed to fillet welding by inserting a plate in the athwart direction. Moreover, it is also converted to fillet welding by applying lap joint instead of butt joint.

The all welded ship *Shinwa-Maru*, which was constructed in Japan for the first time, was designed so as to be able to carry out all welding in the downward position or at least in the vertical position. The purpose of this design is to maintain good welding quality by changing difficult butt welding to easy fillet welding in the downward position, even when welding technique was not well established. This system is now called "plate insert method". In the design of *Shinwa-Maru*, the plate insert

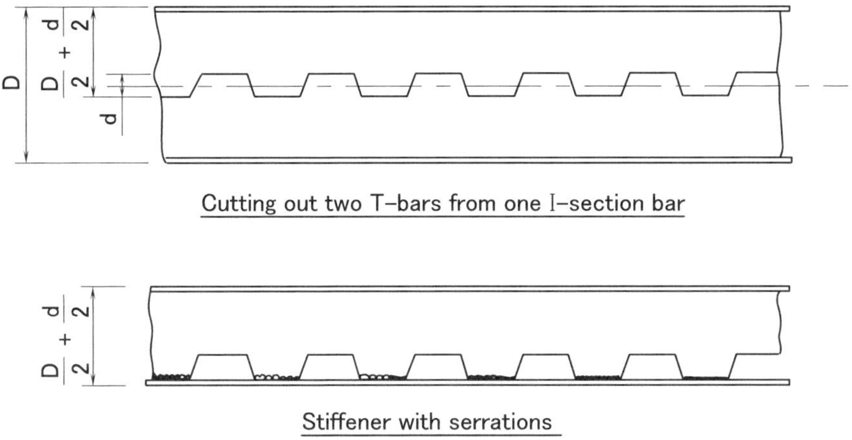

Fig. 16.11 Serration

arrangement is found in way of the connection between the bottom block and the side block, where the extension of the side longitudinal web is sandwiched by upper end of bilge shell plate and lower end of side shell plate as shown in Fig. 16.12. A similar arrangement is made for the joining between the bottom block and bulkhead block [97].

Nevertheless, nowadays horizontal butt welding can be performed more efficiently than fillet welding in a plate insert arrangement because butt welding can be done without any qualitative problem by the progress of welding technology, and in addition butt welding is carried out by automatic welding machines. We tend to overestimate the advantage of the plate insert arrangement, since that arrangement is able to replace difficult butt welding with easy fillet welding. However, we had better not

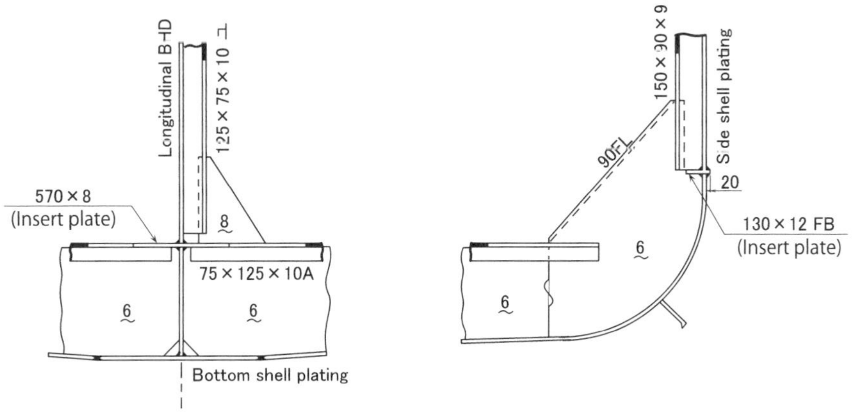

Fig. 16.12 Plate insert arrangement in *Shinwa-Maru*

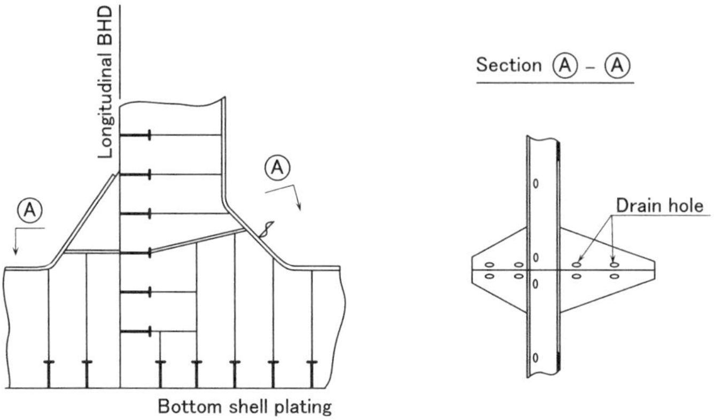

Fig. 16.13 Prevention of misalignment in plate insert arrangement

forget that the plate insert arrangement usually takes much more time for welding because we have to weld four lines of fillet weld instead of one butt weld.

Another important point in designing the plate insert arrangement is to prevent misalignment of abutting plates. It is easy to avoid misalignment, when the facing plates are welded near to the edge of sandwiched plate. But in cases where the two plates are joined in the middle of a wide sandwiched plate, it is not an easy task to arrange two plates to be in line with each other. To avoid misalignment in this case, it is recommended to provide drain holes to the sandwiched plate for the sake of easy adjustment as shown in Fig. 16.13.

Another significant point in the design of the plate insert arrangement is the need to be careful in determining the leg length of the fillet welding connecting web plates and insert plate. Although the leg length of fillet welding for shearing is applicable for the general case of a connection, a leg length $0.7t$ for tension should be given for the fillet welding near a face plate.

A corrugated bulkhead, in general, is constructed in such a way that the flat plate is bent in a Z shape and then the edges of the two bent plates are connected by butt welding as shown in Fig. 16.14a. This butt welding can be replaced with fillet welding by using two plates; i.e. flat plate and bent plate in a π shape. Let us compare the productivity of both types of corrugated bulkhead in Table 16.1, where the bulkhead consists of three corrugations. The corrugated bulkhead produced by fillet welding is more advantageous than the bulkhead by butt welding, from the view point of constructive efficiency. In addition to that, the corrugated bulkhead with fillet welding has such merit in designing that the designer can decide the spacing of corrugations freely because the width and thickness of the flat plates is independent of those of the bent plates in the π shape. In contrast, in the case of a butt welded bulkhead, the spacing of corrugation has restrictions, for the plate thickness must be the same for web plate and face plate.

If the corrugated bulkhead is divided into lower and upper parts, then they are fabricated by inserting a horizontal girder between them, which creates a plate insert

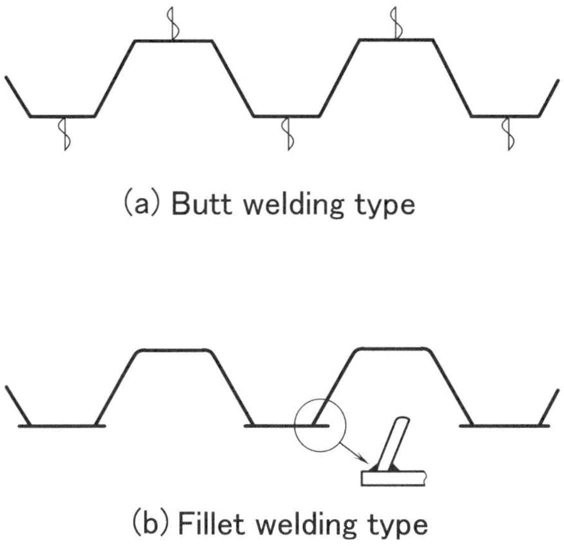

Fig. 16.14 Corrugated bulkhead

Table 16.1 Comparison of productivity of corrugated bulkhead

	Butt welding type	Fillet welding type
Number of plates	6	7
Number of welding lines	10	12
Number of back gougings	5	0
Number of press bendings	12	6*

* : 6 can be reduced to 3 by press bending with mould.

arrangement. It is difficult to avoid mis-alignment between lower and upper corrugated bulkhead, since the horizontal girder is wide. It is more difficult in the case of a butt welded bulkhead shown in Fig. 16.14a because one plate is bent twice and the accuracy of the corrugated shape cannot be expected. In contrast, in the case of a fillet welded bulkhead, the corrugated bending is carried out by one press using a press mould in the π shape, which helps to keep good accuracy of corrugation piece. Furthermore, as explained previously, the designer needs to provide drain holes on the web plate of the horizontal girder near corrugated structure.

An alternative idea to change butt welding to fillet welding is to make use of a lap joint. A lap joint used to be frequently adopted when the accuracy of the constructed block was not well controlled, since the lap joint allowed disagreement of connected edges. The welding time of a lap joint is half of that of a butt joint, therefore, lap welding is preferable from the view point of productivity. However, the lap joint gradually disappeared in way of structural members due to of the following reasons:

(1) Fatigue strength of lap joint is less than other joints.
(2) Lap joint applied to the internal members of a tanker causes corrosion quickly.
(3) Lapping of plates brings useless weight of steel.

This weight is in no way negligible for large tankers.

16.7 Long Intermittent Welding

There are two types of fillet welding; i.e. continuous welding and intermittent welding. Continuous fillet welding is applied to structural member in tanks or a member in exposed areas, since intermittent welding may causes rust in the gap between plate and stiffener. Hence intermittent welding is normally used for the welding of stiffeners in dry space such as accommodation spaces or the under deck space of cargo hold. Figure 16.15 represents an example of intermittent welding. Intermittent welding is carried out in the form of continuous welding of 27 length arranged in a parallel row or in a staggered row, and pitch p of the welding varies among 150 mm, 200 mm, 300 mm, etc. according to the required strength.

After developing the gravity process welding, long intermittent welding is taken into the welding practice to reduce the number of short continuous welding (75 mm) with the aid of long continuous welding (600 mm) of electrode used in gravity process welding, as shown in Fig. 16.15. Long intermittent welding seems to be curious, however, as fillet welding of continuous part transmits shearing force from web to plate and only one side of the welding is enough to support this shearing stress.

The intermitted welding is effective in reducing welding distortion in thin plates such as car decks of PCC. Some studies have been reported for several welding patterns theoretically and experimentally [98, 99].

16.8 Shrinkage of Deposit Metal

At the beginning of welding, the deposit metal is melting in more than 1,000°C, and then it starts solidification at the transformation temperature from liquid to solid, and finally it becomes solid steel at normal temperature. In this process, the deposit metal shrinks, which results in a tensile stress inside the deposit metal and compressive stress in the base metal around the deposit. These stresses are called "residual stress".

Let us roughly estimate this residual stress by assuming that the thermal expansion coefficient of steel is 10^{-5} and the change of temperature is 1,000°C. Then, the order of strain becomes 10^{-2} and it generates stress of 2,060 MPa by multiplying Young's modulus. This simple calculation overestimates the stress because the effect of yielding is neglected. However, we can easily imagine by this rough example that stresses around the yield stress are induced in way of the deposit metal. A designer

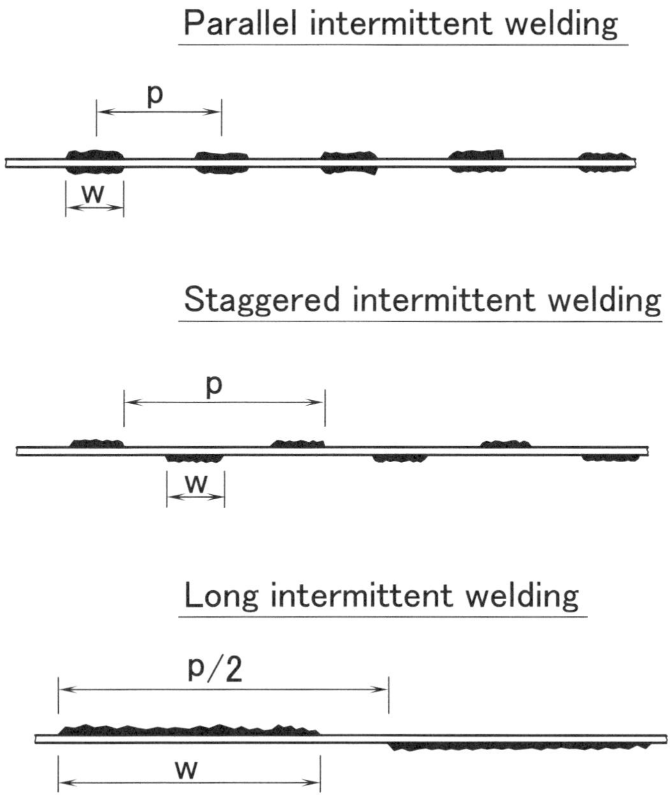

Fig. 16.15 Variable intermittent welding

should not forget the shrinking of this deposit metal, when he is in charge of detail welding design.

This shrinkage of deposit metal causes structural damages, as shown in Fig. 16.16. To prevent such welding damage, it is effective to extend the end of the sandwiched plate between the upper and lower plates as indicated by the chain line in Fig. 16.16.

16.9 One Side Welding

Both sides of a plate should be welded in butt welding as well as in fillet welding, in general. However, one side welding is inevitable where the space behind a plate is insufficient to permit welding. In such cases, it is necessary to execute complete fusion welding by using a backing metal strip.

One of the worst cases of one side welding is the case where a cover plate should be welded in the corner of a dead space. In this case, when the welding is carried out without backing, it is difficult to obtain good fusion. Figure 16.17 represents

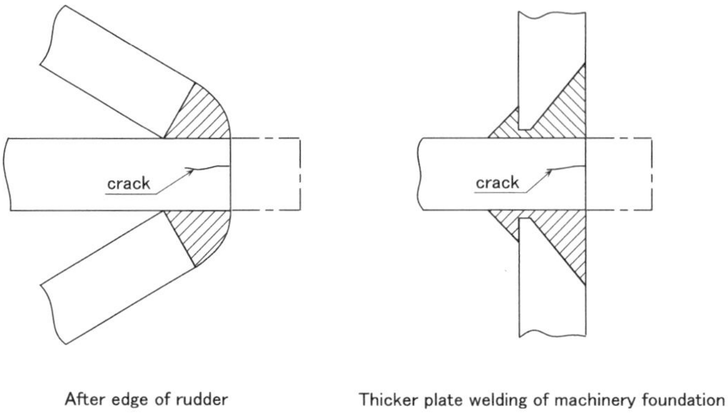

Fig. 16.16 Example of cracking due to welding metal shrinkage

an example of this case where the slant plates are provided at the lower ends of a corrugated bulkhead of a bulk carrier for the purpose of preventing the accumulation of grain cargo in the corners. While the welding of this slant plate is not so important in an ordinary cargo hold, it must be performed with the greatest care in a hold intended to carry water ballast or wet cargo such as oil in addition to dry cargo. If this slant plate is not perfectly tight, ballast water or cargo oil can seep into a space through imperfect welding, and then this liquid seeps out causing damage to any adjacent grain cargo.

These slant plates cause another structural damage; i.e. cracking, because adjacent slant plates cross at a point as shown in Fig. 16.17, which creates stress concentration at the cross point, requiring special attention to the welding quality and sufficient scantling when one side welding is employed.

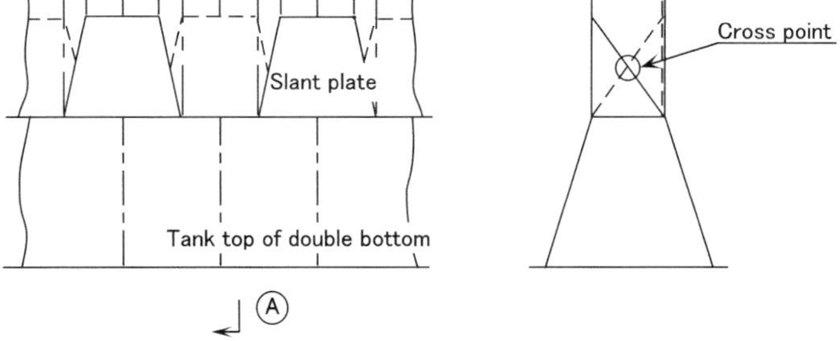

Fig. 16.17 Slant plate of lower corrugated bulkhead

Chapter 17
Fracture Control

17.1 Jack-Knifed Failure of the Liberty Ships

During World War II, USA was obliged to construct a large number of vessels urgently and the first all-welded ships, Liberty ships and T-2 Tankers, were produced according to an emergency ship building program. Some of these ships later broke completely in two, like a jack knife. Most of the failure occurred during the winter months. Failures occurred both when the ships were in rough sea conditions and when they were anchored at quay as shown in Fig. 17.1 [100]. The failure of these vessels gave a driving force to the study of brittle fracture and fracture mechanics. Brittle fracture is the phenomena where normally ductile mild steel becomes brittle in low temperature and the crack propagates very rapidly. The study of brittle fracture resulted in both improvement of assessing steel strength, and development of the design method. Hereafter, the failure of Liberty ships due to brittle fracture is explained and the fracture managing technology based on fracture mechanics is addressed.

At 11 pm on 16 January 1943, a few days after completing sea trials, the 152 m long T2 tanker the *Schenectady* broke in two amidships while lying at the outfitting pier in the construction yard in Portland, Oregon, USA. The temperature of the harbor water was about 4 °C and the conditions were still. The air temperature was approximately –3 °C and the winds were light.

The failure was sudden and accompanied by a report that it was heard a mile away. The fracture extended through the deck, the sides of the hull, the longitudinal bulkheads and the bottom girders. The vessel jack-knifed, hinging on the bottom plate which had remained intact. The central part of the ship rose clear of the water, so no flooding of the hull through the fracture occurred.

The *Schenectady* was built by the Kaiser Company as part of the huge World War II emergency shipbuilding program. This program produced 2,580 Liberty ships, 414 Victory ships and 530 T2 tankers over the years 1941–1946.

The failure of the *Schenectady* initiated on the deck between two bulkheads. A defective weld existed in a region of stress concentration arising at a design detail.

Y. Okumoto et al., *Design of Ship Hull Structures*,
https://doi.org/10.1007/978-3-032-12500-2_17

Fig. 17.1 Collapse of T2 tanker *Schenectady* [100]

The nominal tensile stress in the deck was calculated to be 68 N/mm^2. Poor welding procedures caused the catastrophe also.

A broad research program was undertaken to find the causes of these failures and to prescribe the remedies for their future prevention. In addition to the above research, other research was aimed at gaining a better understanding of the mechanism of brittle fracture and fracture in general.

Brittle fracture is quite dangerous, because normally ductile mild steel becomes brittle in low temperature and the crack propagates very rapidly at the rate of 1,500–2,000 m/s causing sudden fracture of the vessel as shown in Fig. 17.2 [101].

In order to avoid failure due to brittle fracture, the following measures are taken on the basis of much research;

(1) The absorbed energy in the Charpy V notch test correlates well with the observed crack initiation, propagation and arrest behavior of the ship' steel. Hence the Charpy V notch test is standardized in such a way that the absorbed energy of the ship's steel is required according to the specified temperature.
(2) High quality steels, more ductile steels, are developed by modifying alloy elements, grain size, deoxidation methods and normalizing heat treatments.
(3) More proper ship design and welding methods are implemented to avoid sharp defect in welded part, because these defects often initiate the crack.

Fig. 17.2 Photograph of typical brittle fracture surface [101]

17.2 Fracture Mechanics

17.2.1 Principles

Although the traditional design criteria of hull structures are generally based on the tensile strength, yield strength, and buckling strength, they are insufficient when there is a likelihood of cracks; especially in the case of the wide application of high strength steels, weight savings by detail stress analysis using Finite Element Method, and the development of refrigerated cargo ships such as LPG and LNG carriers which are exposed to low temperatures.

If preexisting flaws or high stress concentrations are present, especially in the vicinity of welding beads, the structures have a low crack resistance, that is, low fracture toughness, and the residual strength will be low. In such cases, the fracture may be brittle, and the application of fracture mechanics has been used for the evaluation of hull strength recently.

Figure 17.3 [102] shows the overall disciplines of fracture mechanics from the transgranular/intergranular fractures to engineering applications. Recently, "fail-safe design" has been widely applied. The criteria allow for the occurrence of small cracks and crack growth, if they are under the allowable values, provided that the residual strength is controlled to maintain the safety of the structure. Hence, the occurrence, propagation, and transition to the brittle fractures of fatigue cracks be well investigated, and the allowable stress, residual strength, allowable flaws, and material selection must be considered during the structural design.

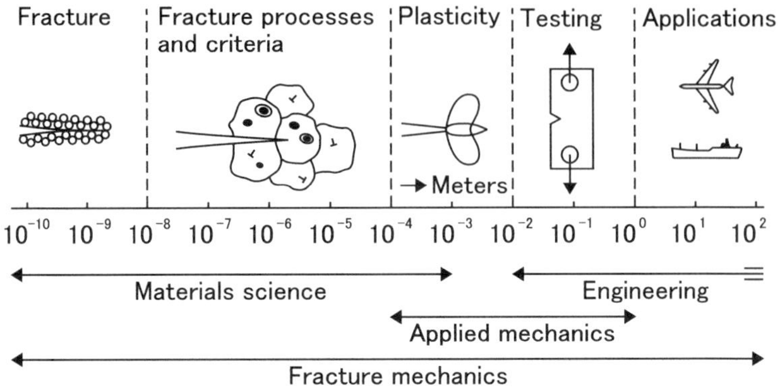

Fig. 17.3 Broad fields of fracture mechanic

17.2.2 Linear Fracture Mechanics

A crack in a solid is categorized in three modes as shown in Fig. 17.4. In these, mode I, opening mode, in which the displacement of the cracked surfaces is perpendicular to the plane of crack, is the most fundamental and important. In mode II, sliding mode, the displacement of the cracked surfaces is in the plane of the crack and perpendicular to the leading edge of the crack. Mode III, tearing mode, is caused by out-of-plane shear.

In mode I, in-plane crack tip stress is expressed as follows for an infinite plate subject to uniform tension (Fig. 17.5).

$$\sigma_{ij} = \frac{K_I}{\sqrt{2\pi r}} f_{ij}(\theta) \tag{17.2.1}$$

where

σ_{ij}: stress acting on a plate element $dxdy$ at a distance r from crack tip and at θ from the crack plane.

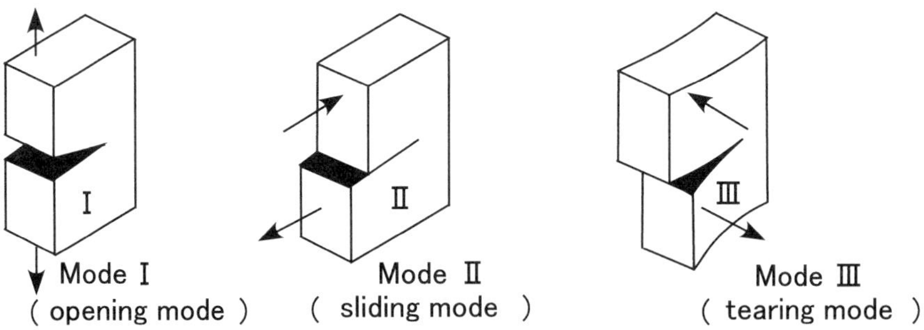

Fig. 17.4 Three modes of loading

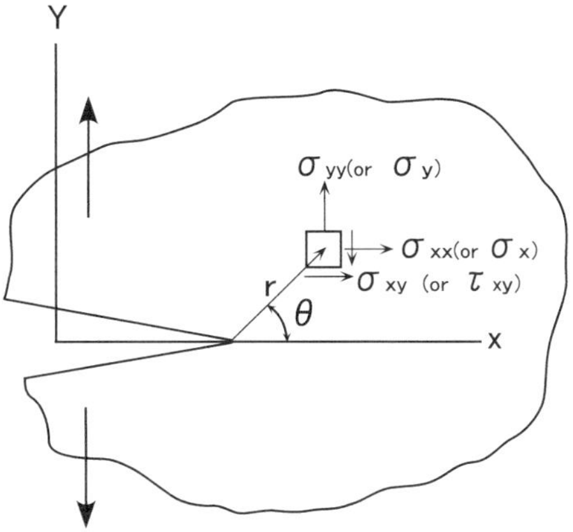

Fig. 17.5 Crack in arbitrary body

K_I: stress intensity factor for mode I.

The stresses for an area close to the crack tip as shown in Fig. 17.6 are shown in these equations:

$$\sigma_x = \sigma\sqrt{\frac{\pi a}{2\pi r}}\cos\frac{\theta}{2}\left[1 - \sin\frac{\theta}{2}\sin\frac{3\theta}{2}\right]$$

$$\sigma_y = \sigma\sqrt{\frac{\pi a}{2\pi r}}\cos\frac{\theta}{2}\left[1 + \sin\frac{\theta}{2}\sin\frac{3\theta}{2}\right] \tag{17.2.2}$$

$$\tau_x = \sigma\sqrt{\frac{\pi a}{2\pi r}}\sin\frac{\theta}{2}\cos\frac{\theta}{2}\cos\frac{3\theta}{2}$$

When $\theta = 0$, $\sigma_y = \frac{\sigma\sqrt{\pi a}}{\sqrt{2\pi r}} \equiv \frac{K_I}{\sqrt{2\pi r}}$

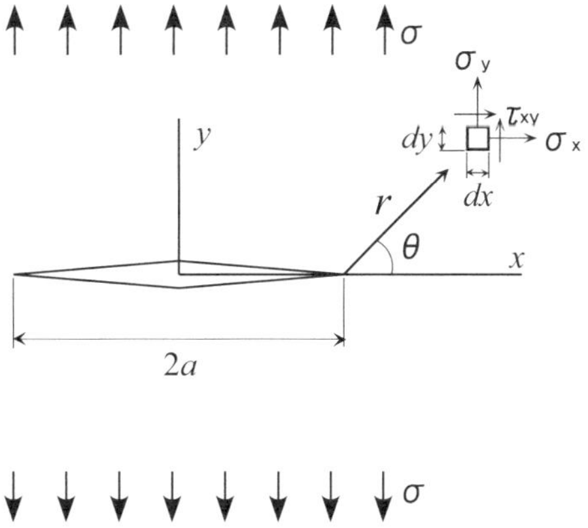

Fig. 17.6 Crack in infinite plate

Hence,

$$K_I = \sigma\sqrt{a\pi} \tag{17.2.3}$$

K_I is called the "stress intensity factor", which has the dimension of stress times the square root of the length and is defined only by the remote stress σ and the crack length.

For mode II and III, stress intensity factors are defined similarly. These values are applied for fractures under lower stress conditions such that the plastic zone is small compared to the size of the crack.

On the other hand, crack growth per unit thickness is written as:

$$G = \frac{\pi\sigma^2 a}{E'} = \frac{K^2}{E'} \tag{17.2.4}$$

where

$$E' = E \quad \text{for plane stress condition}$$

$$E' = \frac{E}{1-\nu^2} \quad \text{for plane strain condition}$$

E: Young's modulus.

ν: Poisson's ratio.

G is called "elastic energy release rate" per crack tip, or crack driving force. When G exceeds a certain critical value G_c, crack growth occurs.

$$G \geq G_c \tag{17.2.5}$$

G_c is called "critical energy released rate", and is generally determined by measurements.

Correspondingly, brittle fracture will occur when the K value exceeds a certain critical value K_c, which is called critical fracture toughness.

$$K \geq K_c \tag{17.2.6}$$

17.2.3 Non-linear Fracture Mechanics

When the plastic zone in the vicinity of a crack tip is not small, the crack opening displacement or crack-tip opening displacement (COD, CTOD) concept is applied; when the material at the crack tip reaches a maximum permissible plastic strain, crack extension takes place. The crack tip strain is related to COD, hence, crack

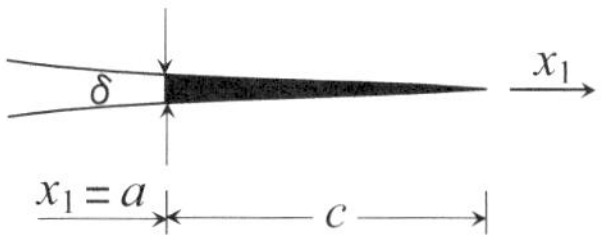

Fig. 17.7 COD δ in a thin plate

extension or fracture is assumed to occur when COD exceeds a critical value. COD is δ as shown in Fig. 17.7. This criterion is equivalent to K_c and G_c in the case of a small plastic arrangement.

In the case of Fig. 17.6, COD is written as

$$\delta = \frac{8\sigma_y a}{\pi E}\ln\left[\left(\cos\frac{\pi\sigma_\infty}{2\sigma_y}\right)^{-1}\right] \tag{17.2.7}$$

With decreasing σ_∞/σ_y (and hence c/a) this will asymptotically approach

$$\delta = \frac{K_I{}^2}{E\sigma_y} \tag{17.2.8}$$

17.2.4 Fracture Toughness

For fractures involving small-scale yielding, the Charpy test is usually applied to evaluate fracture toughness. The notched test piece is hit by a hammer at each temperature, and the energy loss is measured by the differences in hammer elevation before and after the hit. The test piece has usually V-notch in the center and is supported at both ends. Results of the V-notch Charpy impact tests are shown in Fig. 17.8. Fracture surface of V-notch Charpy impact tests is shown in Fig. 17.9.

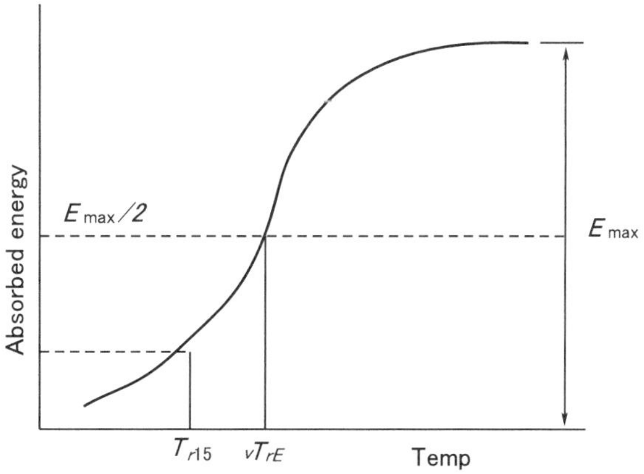

Fig. 17.8 Temperature curve of absorbed energy

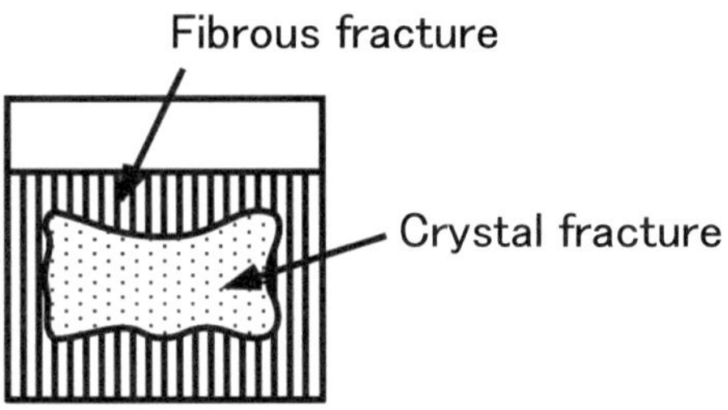

Fig. 17.9 Fracture of V notch Charpy specimen

The temperature at a half of maximum energy is called mean energy transition temperature ${}_vT_{rE}$, while such temperature that the areas of fibrous fracture and crystal fracture are same is called fracture appearance transition temperature ${}_vT_{rS}$. These two temperatures (${}_vT_{rE}$ and ${}_vT_{rS}$) are approximately the same. Furthermore, temperatures at which absorbed energy reaches 15 ft-lb (2.1 kgf-m) are said to be associated with the occurrence of brittle fracture, based on the experience of Liberty ships during World War II in USA.

17.2.5 Grade of Steel

Grades of steel materials are regulated by IACS unified rules. Here, the background of the application is introduced based on the NK rules [103, 104]. The principle is Eq. (17.2.6); the toughness K is taken to be greater than the critical fracture toughness. The model is shown in Fig. 17.10, in which a crack crosses at right angle to a weld bead. The length of the crack is estimated to be 240 mm at important places and 200 mm at others. The applied stresses are categorized as shown in Table 17.1. In addition, the residual stress is taken as being half the yield stress; then, the applied stress and K value are

$$\sigma = \sigma_d + 0.5\sigma_y \tag{17.2.9}$$

$$K = \left(\sigma_d + 0.5\sigma_y\right)\sqrt{\pi a} \tag{17.2.10}$$

where
σ_d: applied stress (Table 17.1).
σ_y: yield stress.
$2a$: crack length.

On the other hand, K_c is derived using the master curve of WES3003 using the V-notch Charpy impact test. The absorbed energy is expressed as follows as shown in Fig. 17.11.

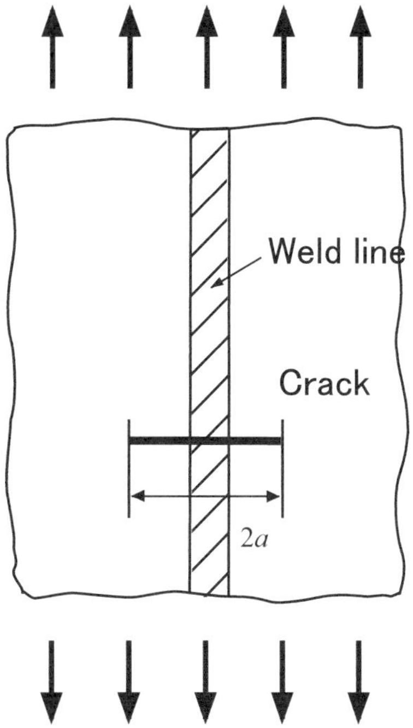

Fig. 17.10 Model with crack

Table 17.1 Applied stress, assumed crack length and K

Class	Mild steel (σ_y = 24 kgf/mm^2)				Higher tensile strength steel (σ_y = 32 kgf/mm^2)			
	σ_d (kgf/mm^2)	σ (kgf/mm^2)	a (mm)	K (kgf-mm$^{-3/2}$)	σ_d (kgf/mm^2)	σ (kgf/mm^2)	a (mm)	K (kgf-mm$^{-3/2}$)
I	4.8	16.8	100	298	6.2	22.2	100	393
II	12.0	24.0	100	425	15.4	31.4	100	557
III	14.4	26.4	100	468	18.5	34.5	100	611
IV	16.8	28.8	100	510	21.5	37.5	100	665
V	16.8	28.8	120	559	21.5	37.5	120	728

$$ {}_vE_T = \frac{1}{\sqrt{2\pi}} \int_{-\infty}^{T - {}_vT_{rE}} \exp\left\{ -\frac{1}{2} \left(\frac{T - {}_vT_{rE}}{20} \right)^2 \right\} dT \tag{17.2.11} $$

where

T: temperature of impact test (°C).

${}_vE_T$: absorbed energy (kgf · m).

${}_vT_{rE}$: fracture transition temperature (°C).

Using this master curve, the fracture transition temperature ${}_vT_{rE}$ can be obtained from the absorbed energy at an arbitrary temperature.

The brittle fracture temperature is determined from deep notch impact tests, for applied stresses of $\sigma_{y0}/2$ and crack lengths of 80 mm as follows

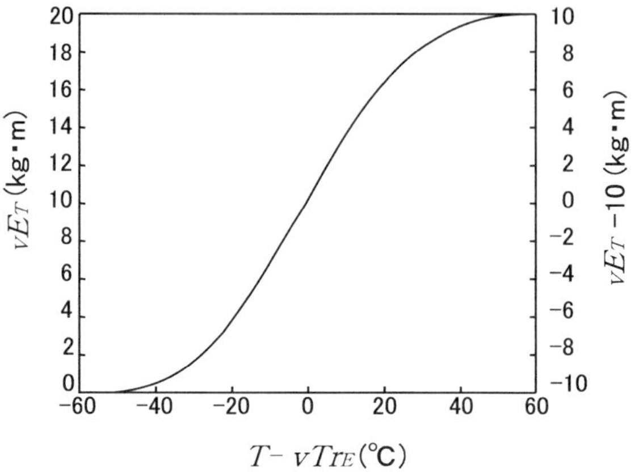

Fig. 17.11 Master curve of absorbed energy

$$ {}_{I}T_k = \left(0.00321\sigma_{y0} + 0.391\right)_{v}T_{rs} + 2.74\sqrt{t} - 5.44 \tag{17.2.12}$$

where

${}_{I}T_k$: Temperature of brittle fracture (°K).

${}_{v}T_{rS}$: transition temperature of 50% brittle fracture (°K).

σ_{y0}: nominal yield stress (kgf/mm^2).

t: plate thickness (mm).

Since ${}_{v}T_{rS}$ is almost same as ${}_{v}T_{rE}$, ${}_{I}T_k$ can be calculated from the above equation using ${}_{v}T_{rE}$ instead of ${}_{v}T_{rS}$. Finally, K_c is obtained from the following equation for a given temperature T_K(°K).

$$K_c = 3.81\sigma_{y0}\exp\left\{562\left(\frac{1}{{}_{I}T_K} - \frac{1}{T_K}\right)\right\} \tag{17.2.13}$$

Following this procedure, K_c values are obtained as shown in Table 17.2 using the specified absorbed energy of Charpy impact test at 0°C. Comparing the K value of Table 17.1 and K_c in Table 17.2, steel grades can be determined for each plate thickness and each stress class.

Table 17.2 K_c values at 0 °C

Grade	Plate thickness(mm)								
	10	15	20	25	30	35	40	45	50
KA	550	536	509	484	473	461	451	440	430
KB	630	597	580	550	536	522	509	484	473
KD	730	688	649	630	613	580	565	550	536
KE	1173	1089	1014	979	945	884	855	828	802
AH	715	679	646	630	601	587	573	560	548
DH	917	891	841	795	774	753	734	715	696
EH	1261	1178	1140	1059	1036	1004	974	945	917

Table 17.3 Material grade requirements of IACS unified rules for various material classes and plate thicknesses

Class	I		II		III	
Thickness (mm)	MS	HT	MS	HT	MS	HT
$t \leq 15$	A	AH	A	AH	A	AH
$15 < t \leq 20$	A	AH	A	AH	B	AH
$20 < t \leq 25$	A	AH	B	AH	D	DH
$25 < t \leq 30$	A	AH	D	DH	D	DH
$30 < t \leq 35$	B	AH	D	DH	E	EH
$35 < t \leq 40$	B	AH	D	DH	E	EH
$40 < t \leq 50$	D	DH	E	EH	E	EH

Table 17.3 shows the material grade requirements of IACS unified rules for various material classes and plate thicknesses [105]. The definition of the material classes (I ~ III) is given in detail in the rules for various structural members and their locations. Just for example, strength deck plating at corners of cargo hatch openings in bulk carriers is to be of Class III within 0.6 L amidships, and of Class II within rest of cargo region. In addition, for ships with length exceeding 250 m, shear strake at strength deck and stringer plate in strength deck are to be of grade E/EH within 0.4 L amidships, while bilge strake is to be of grade D/DH within 0.4 L amidships.

17.3 Fatigue Strength Design

17.3.1 Crack Propagation Calculation by Paris's Equation

In general, crack growth rate is considered to be governed by the K-value (stress intensity factor) range at the crack tip. The simplest application of fracture mechanics to fatigue strength design is the Paris's equation, expressed as follows:

$$\frac{da}{dN} = C(\Delta K)^m \tag{17.3.1}$$

where

da/dN: crack growth rate, or crack growth per cycle.

C, m: material constant.

ΔK: stress intensity factor range.

This equation is valid only within the region where crack growth is stable, and bounded at lower and upper extremes by ΔK_{th} and K_c. ΔK_{th} is the threshold stress intensity factor range, and the crack growth rate can be combined with Paris's equation as follows:

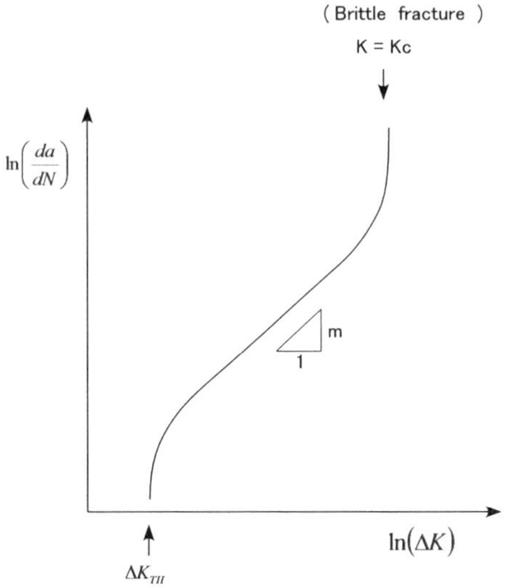

Fig. 17.12 Schematic crack growth rate curve

$$\begin{aligned}\frac{da}{dN} &= C(\Delta K)^m \qquad \Delta K > \Delta K_{th} \\ &= 0 \text{ (no crack)} \qquad \Delta K \le \Delta K_{th}\end{aligned} \tag{17.3.2}$$

Schematically, the crack growth rate curve is illustrated in Fig. 17.12.

Regarding the application of crack propagation analysis to a ship's structure, the IGC code for liquefied gas carriers requires this analysis for the independent Type B tanks, and there are many actual applications. However, recently such applications are spreading into conventional ships such as container ships, bulk carriers, and tankers. This makes it possible to evaluate the tolerance limit of undercut of the welding and the allowable values of initial crack defects or initial scratches. This improves the quality of hull structures rationally by integrating the design, workmanship, and inspection.

17.3.2 Fatigue Strength Design Considering Construction Tolerances [106]

Fatigue cracks are caused by variation in stresses at local points, and are influenced by structural stress concentration, construction tolerances, alignment, welding bead shape, as well as the exerted stress range and residual stress. In actual ship structure, some construction deviation is inevitable including angular distortion such as hungry horse mode and misalignment. Such construction deviations are controlled under construction standards such as JSQS, and it is considered that strength is warranted by the feedback from actual structural damage of ships in service.

Although a long history of shipbuilding proves that this system has functioned, simple standards such as JSQS do not accurately take into account the influence of design variations such as a wider application of higher tensile strength steel leading to increased nominal stress, or different structural configurations. By quantitatively evaluating the influence of construction tolerances, the quality in terms of fatigue strength can be enhanced.

(1) *Influence of construction tolerances to local stress:* Table 17.4 shows equations of local stress at the weld toe of several types of welding joints, and the stress concentration factor to be taken into account. In the case of boxing fillet welding and cruciform joints (non load transmitted), only the stress concentration factor at the weld toe (K_t) suffices, but in the case of butt joints and cruciform joints (load transmitted), the stress concentration factor due to misalignment (K_m) is also to be considered. In the case of fillet welding of skin plate or butt welding of skin plates, the stress range is to be divided into membrane stress and bending stress, and the stress concentration factor due to angular distortion such as hungry horse mode (K_d) and due to misalignment (K_m) is applied only to the membrane stress portion.

Table 17.5 shows the equations to derive each stress concentration factor [107, 108]. By applying such factors, the construction department can rationally control construction tolerances quantitatively depending on the location and exerted stresses.

(2) *Welding bead shape and K_t control:* As shown in the equation of K_t in Table 17.5, K_t is affected by the flank angle (θ) and the toe radius (ρ). It is readily understood that wide variation in the welding bead shape, or of such parameters as the flank angle and the toe radius, leads to a large probability of non-conformity. That is, if the variation in bead shape is large, the capacity distribution as shown in Fig. 1.6 in Part I becomes lower and more spread out, and the probability of fracture increases.

From this point of view, we adopt the idea of the K_t control method to control the stress concentration factor of weld toes. This means carrying out special control on bead shape for important places of high stress, and inspect and record them quantitatively. Figure 17.13 [109] shows the probability distribution of the K_t values of the beads of boxing fillet weld portions extracted at random. It is known that the welding quality is considerably enhanced through such special control.

One important aspect of K_t control is the rational agreement between design conditions and fabrication deviations. To warrant fatigue strength, actual bead shape (flank angle and toe radius) is to be measured and the resulting K_t is to be assessed such that it is within the allowable limits given by the design conditions.

Table 17.4 Local stress at weld toe

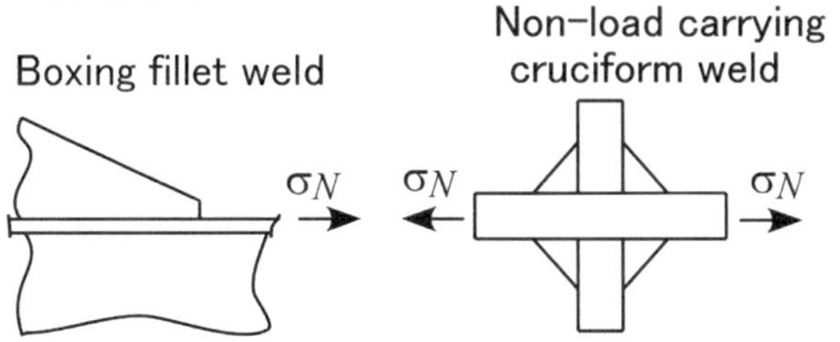

$$\sigma_L = K_t \sigma_N$$

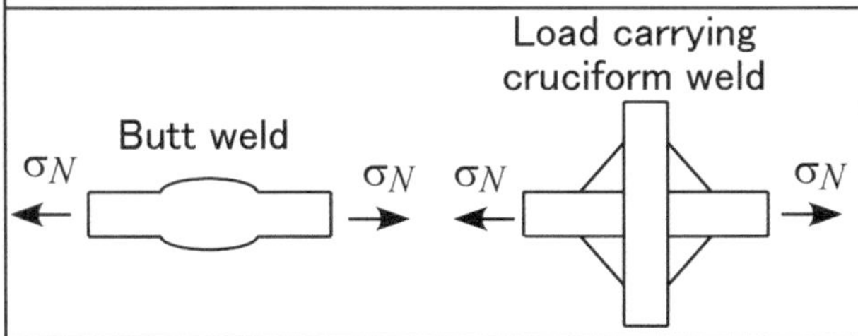

$$\sigma_L = K_t K_m \sigma_N$$

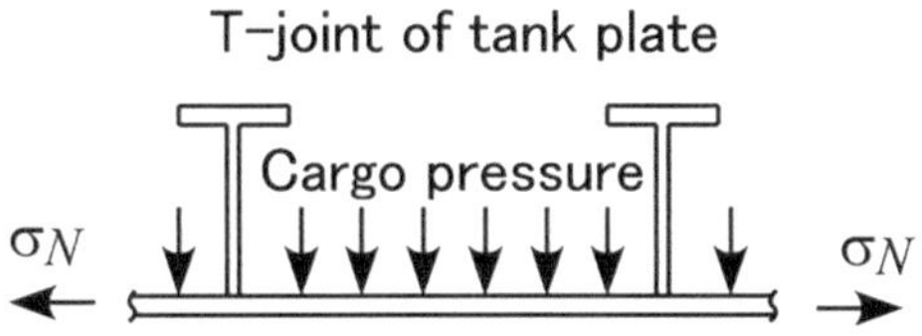

$$\sigma_L = K_t (\sigma_B + K_d \sigma_N)$$

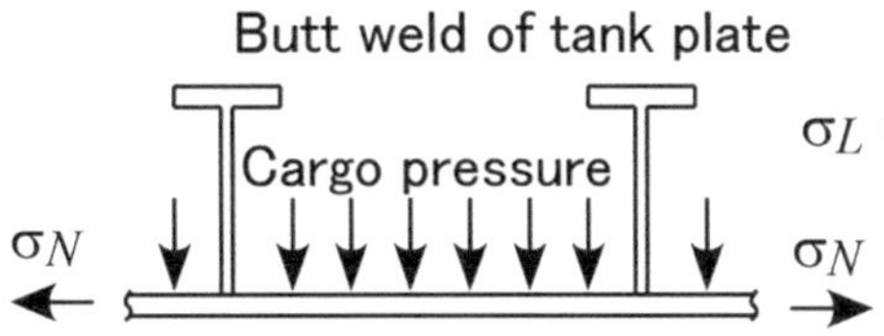

$$\sigma_L = K_t [\sigma_B + (K_d + K_m - 1) \sigma_N]$$

K_t : stress concentration factor at weld toe

K_m : stress concentration factor due to misalignment

K_d : stress concentration factor due to distortion

σ_L : local stress at weld toe

σ_N : nominal stress

σ_B : bending stress of tank plate

Table 17.5 Stress concentration factor and construction tolerances

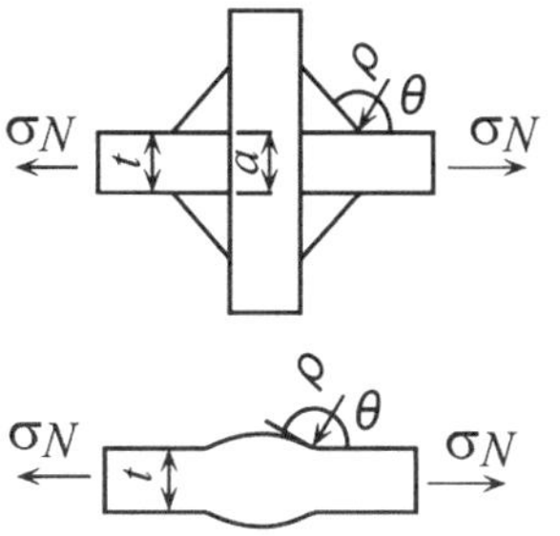

$$K_t = [1 + f(\theta)\{g(\rho) - 1\}]C\left(\frac{a}{t}\right)$$

θ : flank angle

ρ : toe radius

$f(\theta)$: term of the effect of flank angle [52]

$g(\rho)$: term of the effect of toe radius [52]

$C\left(\frac{a}{t}\right)$: term of the effect of root face [52]

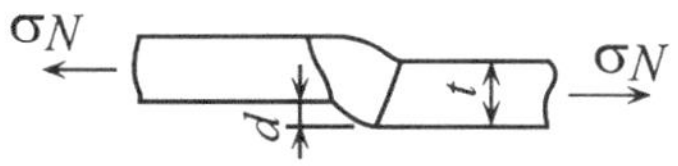

$$K_m = 1 + 3\frac{d}{t}$$

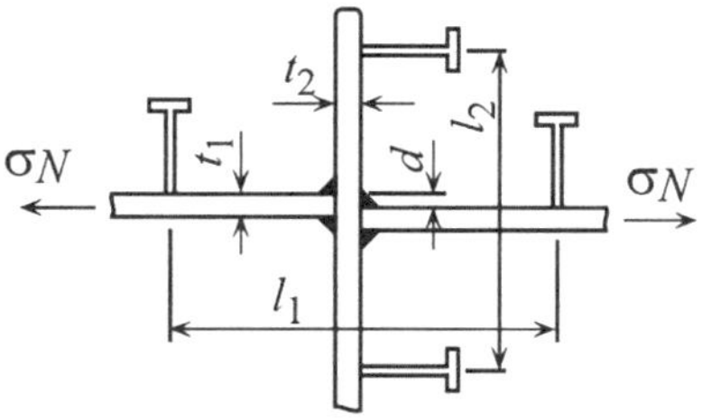

$$K_m = 1 + 3\frac{d}{t_1} \cdot \frac{1}{\frac{l_1}{l_2}\left(\frac{t_2}{t_1}\right)^2 + 1}$$

$$K_d = 1 + 6(1 - \nu^2)\frac{d}{t} \cdot \frac{\tanh\frac{m}{2}}{m}$$

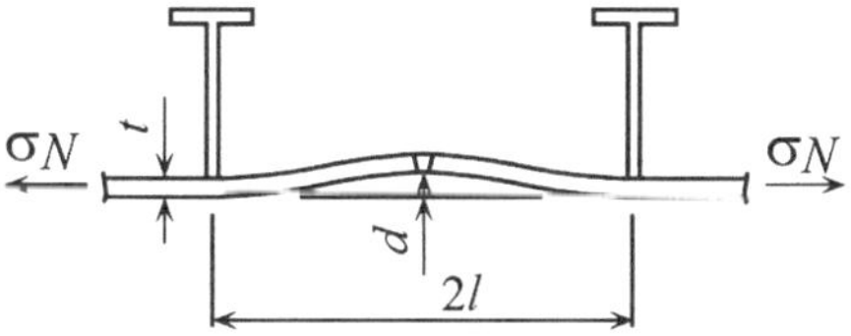

$$m = \sqrt{12(1 - \nu^2)\frac{\sigma_N}{E}\left(\frac{l}{t}\right)^2}$$

E : Young's modulus

ν : Poisson's ratio

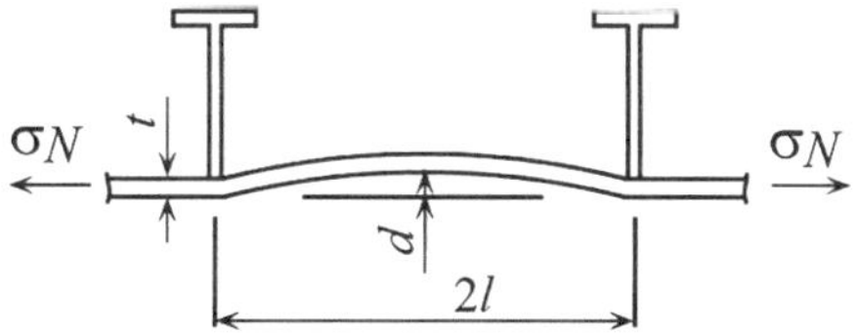

$$K_d = 1 + 6(1 - \nu^2)\frac{d}{t} \cdot 2\frac{m\sinh(m)\tanh\left(\frac{m}{2}\right) + m - \sinh(m)}{m^2\sinh m}$$

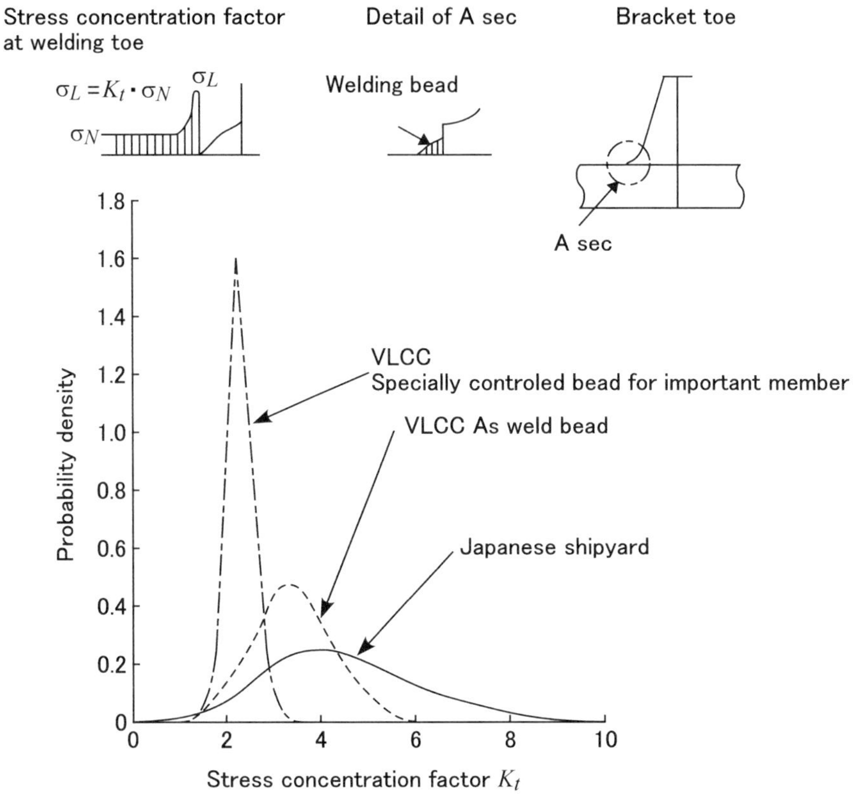

Fig. 17.13 Probability distribution of stress concentration factor K_t

Chapter 18
Hull Structural Vibration

18.1 Introduction

"Hull structural vibration is an old but a new problem" is often said by many managers. That is because the vibration of the hull structures caused serious problems in old times and still now brings new kinds of problems.

In addition, it is also frequently said "when some area vibrates heavily, vibration can be decreased by providing reinforcements to that part. Nevertheless, these reinforcements are likely to cause vibration in another area".

In this chapter, the change of boundary conditions for the vibration problem is mainly focused; both for social problems and ship structure problems. In the former case, for instance, the change of social and economic environments will bring different types of vibration problems. In the latter case, if reinforcements in the vibrating area are provided, they will change the boundary conditions of the surrounding areas. In other words, the reinforcements change the rigidity of the surrounding areas and their natural frequency as a result. This causes resonance between the exciting frequency and the natural frequency of the adjacent areas, which did not resonate with the exciting force before adding reinforcement.

In the case of forced vibration, there are three important relationships between the exciting force and the vibration response of a structure as follows:

(1) The structure always vibrates at the same frequency as the exciting frequency.
(2) The structure almost always vibrates in the same direction as in the exciting force.
(3) The vibration response of the structure is proportional to the magnitude of the exciting force.

In the following sections, we will explain two items: (a) basic features of hull structure vibration, (b) social, economic and physical changes of the environment affecting the boundary conditions of the hull structure vibration.

Y. Okumoto et al., *Design of Ship Hull Structures*,
https://doi.org/10.1007/978-3-032-12500-2_18

18.2 Basic Features of Hull Structure Vibration

Figure 18.1 illustrates the relationship between the basic vibration items. The figure shows that the basic vibration items are exciting forces, responses and criteria. Such basic items have three important parameters for each, i.e. direction, frequency, and amplitude. The vibration itself cannot be clearly defined, unless these three parameters "direction, frequency, and amplitude" are specified.

Let us explain the above matter by considering some examples.

Example 1 Let us suppose the following case, when we receive a report from a guarantee engineer on board a maiden voyage of a vessel saying that there is heavy vibration in the captain's room and that the captain complains seriously. Can we make any effective decisions or useful comments? We cannot because we are informed of nothing about the basic parameters of the vibration.

In contrast, when the report says that the deck plate of the captain's room vibrates vertically and that the frequency is 600 cpm (at the main engine shaft revolution 120 rpm) with single amplitude of 0.2 mm, what can we do? In such a case, we can easily investigate the reason for the vibration and can take effective countermeasures.

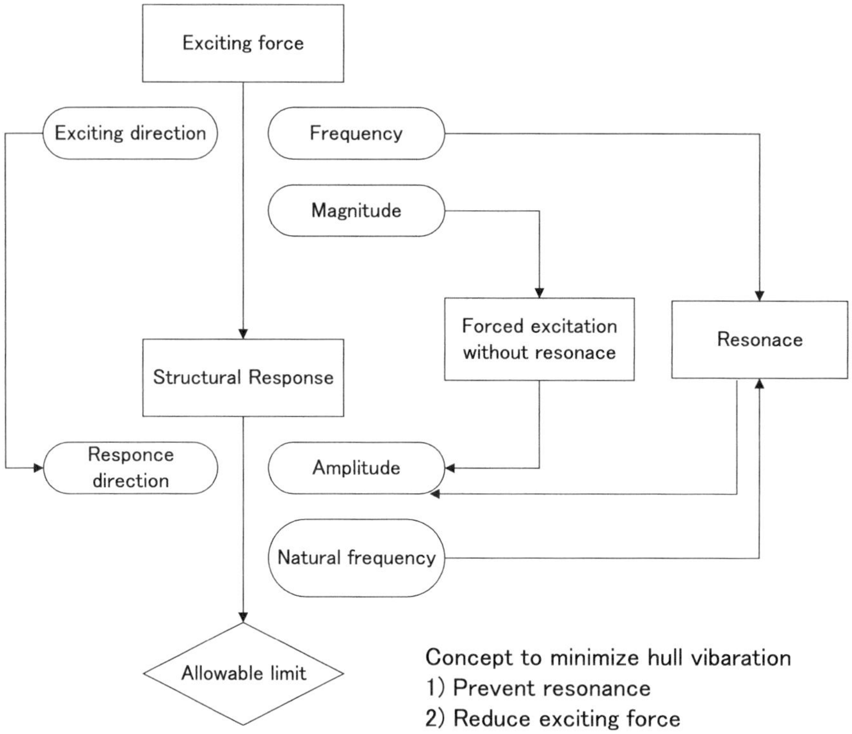

Fig. 18.1 Basic feature of hull structure vibration

- We can deduce from the report that the frequency of vibration is 5 times the shaft revolutions, hence, the vibration is probably caused by the exciting force of the propeller which has 5 blades.
- We can compare the vibration criteria, 9 mm/s for example, with the current vibration velocity of the deck 12.6 mm/s estimated from the frequency and the amplitude. Consequently, the result indicates some countermeasure is necessary.
- We can determine a suitable countermeasure, for the deck vibration occurs only in the captain's room and not over a wider area. This means we can solve the vibration trouble by providing reinforcements on the deck plate of the captain's room.
- We can ask the guarantee engineer on board to provide additional information; the distribution of vibration amplitude of deck plate, i.e. vibration level of deck girders and beams, the vibration level at the mid-point of panels surrounded by girders and beams. From these additional data, we can decide the most suitable countermeasure with confidence.

Example 2 Let us suppose we have a vibration problem, such that heavy vibration has appeared in a fresh water tank in the aft part of a vessel. In the first stage of investigation, judging from the vibration frequency which matched the exciting frequency due to the propeller, we have planned to take the following two countermeasures to reduce the vibration force of the propeller: (a) replacement of the ordinary propeller by a highly skewed propeller, (b) installation of the damping tank to decrease propeller surface forces. Since the vibration response decreases proportionally to the reduction of the exciting force, the countermeasures to reduce the propeller exciting force are expected to be effective.

However, the situation was found to be slightly different on this occasion, by careful observation of the measured data. From the data, we found heavy vibrations only in the panels of the tank, with the stiffeners of the tank scarcely vibrating. That is, in other words, the tank panels vibrate severely due to resonance with propeller excitation. Heavy vibration might be induced even by a small exciting force, if the tank panels resonate with the force. In such a case, it is best to avoid resonance from a practical point of view. To avoid resonance, there are two approaches: change the exciting frequency, or change the natural frequency of the structure. We thought the latter way is applicable for this case.

If the frequency of the exciting force agrees with the natural frequency of a structure, the structure is in resonance, and it will vibrate severely. That situation can be easily understood by considering that the structure absorbs the energy of the exciting force in resonance.

As discussed previously, the best way to prevent vibration is to avoid resonance. Thus, in planning a new vessel, it is necessary for designers to estimate natural frequencies of structural members and then to confirm them to avoid resonance with the frequencies of the exciting forces due to the propeller and main engine. It is, therefore, important in the design stage to estimate accurately the natural frequencies of structural members.

On the other hand, in the case that resonance is inevitable, an alternative way has recently been applied in designing. In this design method, firstly the response of the structure to resonance is estimated; secondly the response is compared with the allowable limit; then the design is accepted if the calculated result is within the limit. However, that judgment requires high accuracy of calculation, such as the estimation of the natural frequency and estimation of the exciting force. Therefore, we consider the response calculation method to be not as reliable as the resonance avoiding method, unless we accumulate much experience.

18.3 Overview of Ship Vibration

Ship structure vibration is mainly caused by the propeller and main engine. There are other vibration sources such as.

- impact force due to slamming which causes hull girder vibration, called "whipping"
- cyclic forces of machinery: pumps, generators, compressors, etc.

However, structural vibration due to the propeller or main engine is especially important, because the countermeasures for such vibration need a huge amount of expense and time compared to those of other vibration trouble. Therefore, we will focus on the vibration induced by the propeller and main engine.

Figure 18.2 illustrates the relationship between the exciting force and ship structural vibration, induced by the propeller and main engine. In the figure, an item boxed by a rectangle indicates an exciting force and an item circled by an ellipse shows a countermeasure suppressing the structural vibration which is induced by the exciting force.

Since the propeller is located at the aft part of a hull, the flow of the water is disturbed by the aft hull and the velocity of flow becomes non-uniform. As the propeller rotates in this non-uniform flow, the propeller creates fluctuating forces, which result in exciting forces acting in the ship structure. The exciting forces from propeller rotation in non-uniform flow can be called propeller shaft force. In addition, the propeller shaft force is usually considered as having two different components: one is bearing force which acts on the propeller shaft in an orthogonal direction, and the other is thrust force which works in the axial direction of the shaft.

On the other hand, the distance between the propeller blade and shell plating of a hull varies according to the propeller rotation. This leads to a change of pressure around the shell plating of the bottom structure. This kind of force is called a surface force, and it is independent of the shaft forces. A surface force is induced even in uniform flow because of the change of distance between the propeller blade and shell plate, while the shaft force happens only in non-uniform flow. The magnitude of the surface force is amplified by the existence of cavitation on a propeller blade; therefore, it is particularly important not to generate propeller cavitation to prevent vibration due to the surface force. The first step to decrease the surface force is

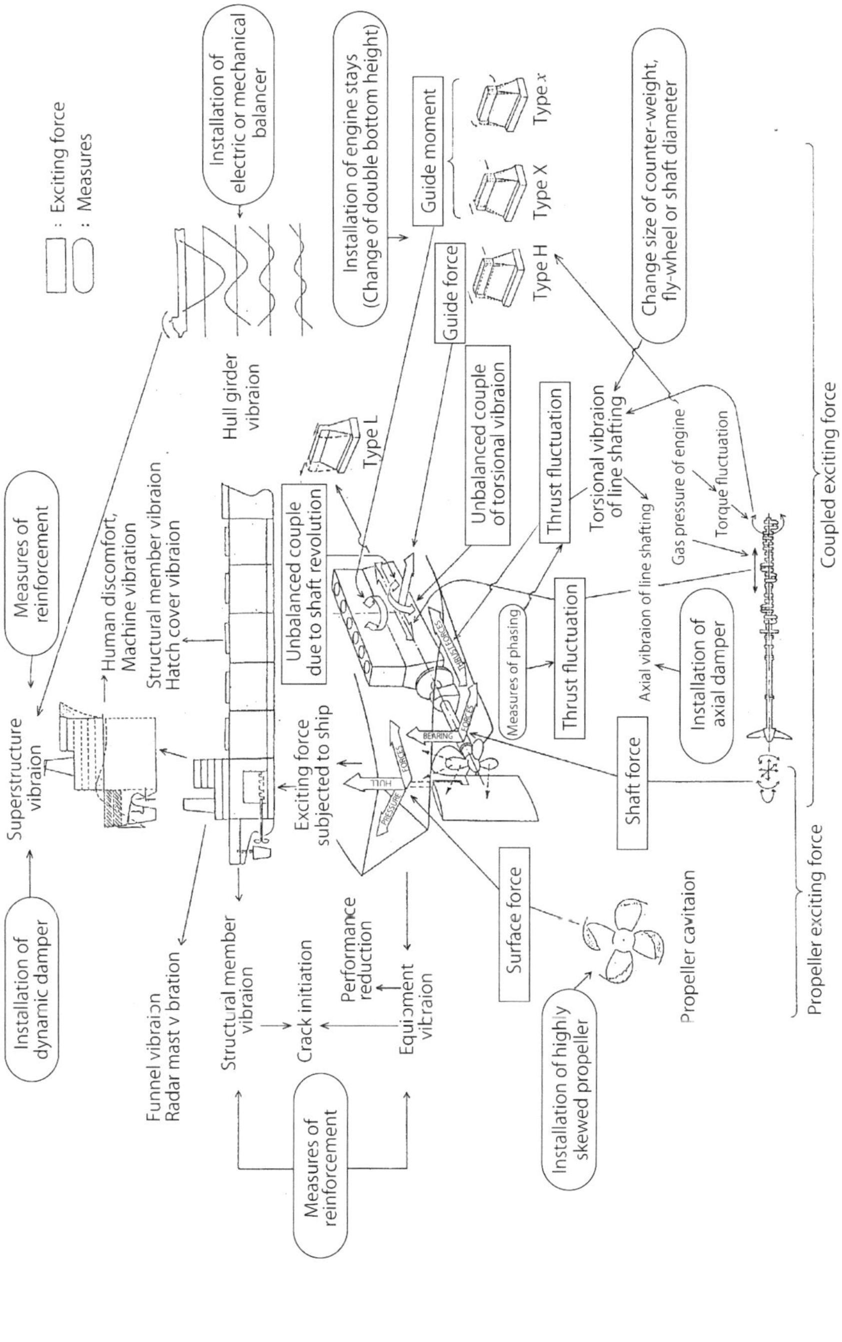

Fig. 18.2 OverView of ship vibration

to increase the distance between the propeller blade and hull surface as much as possible. The second step is to install a highly skewed propeller, which has a large skew-back angle of the blades.

Since a merchant vessel is normally driven by a diesel engine to minimize fuel consumption, the exciting forces of a diesel engine can be categorized by the following components according to the mechanism by which they are produced:

- Unbalanced forces or unbalanced moments induced by inertia forces due to the movement of pistons, etc.
- Guide forces or guide moments which are generated by the explosive pressure of gas. These are transmitted to the cylinder of main engine.
- Longitudinal exciting force which is induced by the inertia force of longitudinal deflection of the crankshaft due to gas pressure
- Fluctuation in thrust force which comes from torque variation in line shafting: Torque variation due to gas pressure causes torsional vibration of line shafting, including the propeller and propeller shaft, and torsional vibration of shafting causes cyclic fluctuation in flow velocity of propellers, which results in thrust fluctuation.

The magnitude of the unbalanced force and the unbalanced moment is so large, in general, that they can produce hull girder vibration if they resonate with the natural frequency of the hull girder vibration. Furthermore, hull girder vibration will bring longitudinal vibration to the superstructure. Therefore, it is normal in the design of line shafting to provide appropriate counterweights or a fly-wheel to the crankshaft which reduces the magnitude of the unbalanced forces or moments. Despite the above measures, in cases where considerable amount of unbalanced forces or moments still remain and it is predicted to cause hull girder vibration, an electric balancer is often installed at the aft end of the hull. The electric balancer generates an exciting force which can cancel hull girder vibration due to the unbalanced force or moment. An example of the effect of an electric balancer on the hull girder vibration is shown in Fig. 30.7 in Sect. 30.1.2, Part III.

Guide forces and moments are the exciting forces which induce lateral vibration in the main engine structure. Guide forces cause lateral vibration of the main engine structure which is called "H" type vibration. Guide moments can cause torsional vibration, that is "X" type, and it can also cause horizontal bending vibration, that is "x" type. When the natural frequency of the main engine structure resonates with the exciting frequency of the guide force or moment, the exciting force due to the guide force or moment is amplified extremely because the engine structure acts as a resonator. Such an increased exciting force is transmitted to the structural members in the engine room, and these may cause severe vibrations. To avoid the lateral vibration of main engine structure, it is a normal practice to provide engine stays at the top of the engine structure. An example of the effect of engine stays is shown in Fig. 30.16 in Sect. 30.1.3, Part III.

Once longitudinal vibration in a crankshaft happens, it causes a longitudinal exciting force on the crankshaft which is transmitted to the double bottom structure of the engine room via the thrust block of the main engine. This exciting force may cause a vibration problem in the superstructure or cause structural failures of panels, stiffeners, etc. in the area of the tanks. To prevent such vibration trouble, a longitudinal vibration damper is usually installed at the fore end of the main engine crankshaft. This damper, which is a kind of oil damper, can reduce the longitudinal vibration amplitude by absorbing the vibration energy. An example of the effect of a damper is shown in Fig. 30.12 in Sect. 30.1.3, Part III.

The fluctuating thrust force due to torsional vibration of the line shafting will cause longitudinal vibration of the main engine structure. If the natural frequency of the longitudinal vibration of the engine structure agrees with the frequency of the thrust fluctuation, the fluctuating thrust is magnified by this resonance, and this vibration is transmitted to structural members through the thrust block. In such cases, the installation of a torsional damper is a quite effective countermeasure to suppress torsional vibration of the line shafting.

18.4 Boundary Conditions of Hull Structure Vibration

Whenever a new type of vibration problem arises, it is caused by a change of the basic vibration items: (a) exciting force, (b) response of structure, (c) allowable criteria.

One of the above examples is seen in the vertical vibration problem of hull girder, which became troublesome in the latter half of the 1940's. This trouble is produced by the introduction of a diesel engine as the main engine, instead of a steam turbine driven engine, because the unbalanced moment of a diesel engine resulted in a large vibration response in the hull girder. The steam turbine engine had no such unbalanced moment.

The second example is a problem induced by the change of response. This trouble was due to the construction of the after-bridge ship, in accordance with the rationalization of design philosophy in the 1960s. The above after-bridge ship resulted in an increase in the height of the navigation bridge; consequently, the fore-aft vibration of bridge was highlighted.

Another example is recent severe requirements by the ship's crew for quiet habitability. In addition, the requirement of highly advanced electrical devices to keep reliable performance is also serious. These requirements generate severe criteria for vibration assessment. This matter can be considered as one of the social and technical changes in the environment which bring a change of boundary conditions for vibration.

Hull structural vibration is induced by the propeller and main engine, and these exciting sources acting on the hull may generate responses in the steel structure. Nowadays, we can estimate the exciting forces, the natural frequencies of the structure, and the structural responses, with the aid of the rapid progress of computers. In addition to this technical development, if we can predict what changes in the boundary conditions of hull vibration will be brought by changes in social and economic environments, we will be able to solve all vibration problems by providing suitable countermeasures in advance.

Some examples of the effects of physical boundary conditions are described below.

Example 1 Figure 18.3 illustrates a deck beam model of a pure car carrier (PCC). Figure 18.3a shows a deck beam fixed to a rigid wall at both ends and the natural frequency is calculated as 670 cpm; also, the natural frequency of a simply supported beam in Fig. 18.3b is estimated to be 300 cpm. Nevertheless, the measured natural frequency of the beam was 595 cpm as shown in Fig. 18.3d. Therefore, these estimated frequencies are very different from the measured one, and we must establish more reasonable boundary conditions for the beam; more flexible than those with fixed ends, but stiffer than those with simply supported ends. Hence, if we take into account the stiffness of the vertical frames supporting both ends of the beam, as shown in Fig. 18.3c, then the calculated natural frequency becomes 608 cpm. The calculated results show very good agreement with the measured frequency 595 cpm.

This indicates that even a simple beam model can be useful for estimating the natural frequency, if the boundary condition is considered in a reasonable way.

Example 2 The next example is a hatch cover vibration in a 60,000 DWT bulk carrier as shown in Fig. 18.4. The bulk carrier was fitted with a main engine of 5 cylinders and a propeller of 4 blades. The hatch covers of the vessel vibrated vertically at 110 rpm of main engine revolution, and the vibration amplitude was 0.4 mm at 550 cpm. We considered the vibration was caused by the resonance of the hatch cover with the exciting force of main engine, because the order of vibration was 5 times the main engine revolutions, which coincides with the number of 5 cylinders.

By modeling the hatch cover as a uniform beam simply supported at both ends, the natural frequency of the hatch cover can be obtained by the following equation:

$$f_{hc} = C\sqrt{\frac{I}{A}} \tag{18.4.1}$$

where

C: constant.

I: sectional moment of inertia of cross section.

A: sectional area.

As a countermeasure, the beam was reinforced by providing an additional web of 300 mm depth and a flange as indicated in Fig. 18.4a. From this increase in rigidity, we expected to increase I/A to twice the original value; consequently, the natural frequency of the hatch cover would become 1.41 $\left(=\sqrt{2}\right)$ times the original one. Despite our expectation, the natural frequency of about 550 cpm was found by measurements carried out after the reinforcement, which means the natural frequency was not increased by the reinforcement.

Fig. 18.3 Deck beam model of pure car carrier

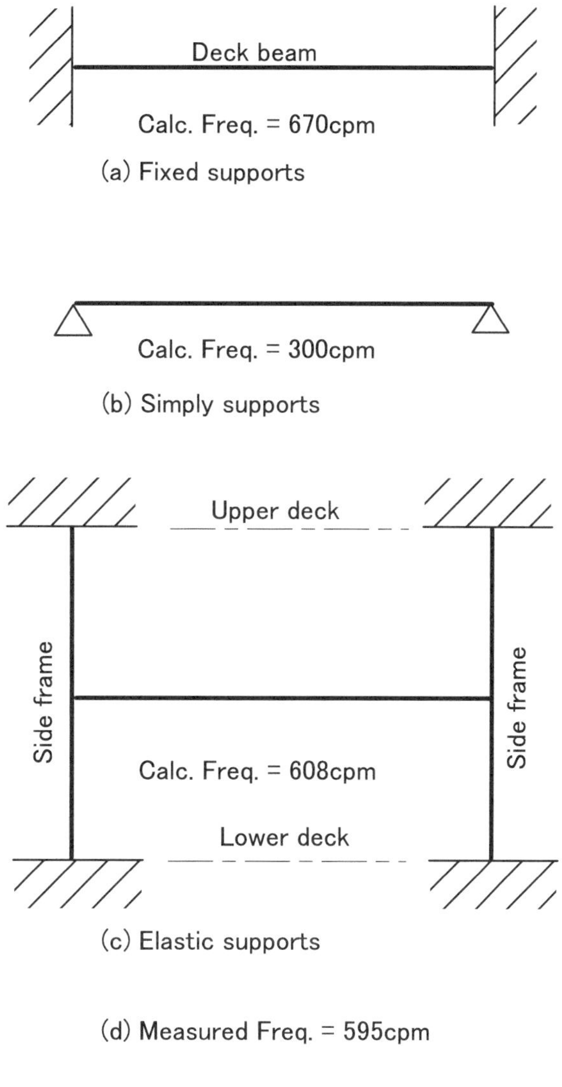

From this measured result, we found that the boundary condition of the hatch cover should be considered in a different way; the beam should be modeled not as a simply supported beam but as an elastically supported beam. Since the hatch cover is mounted on the hatch coaming on the top side tank, the outer edge of hatch cover may move vertically if the top side tank vibrates as shown in Fig. 18.4b. Hence, the natural frequency of an elastic supported beam is

$$\frac{1}{f_{coup}^2} = \frac{1}{f_{hc}^2} + \frac{1}{f_{tst}^2} \tag{18.4.2}$$

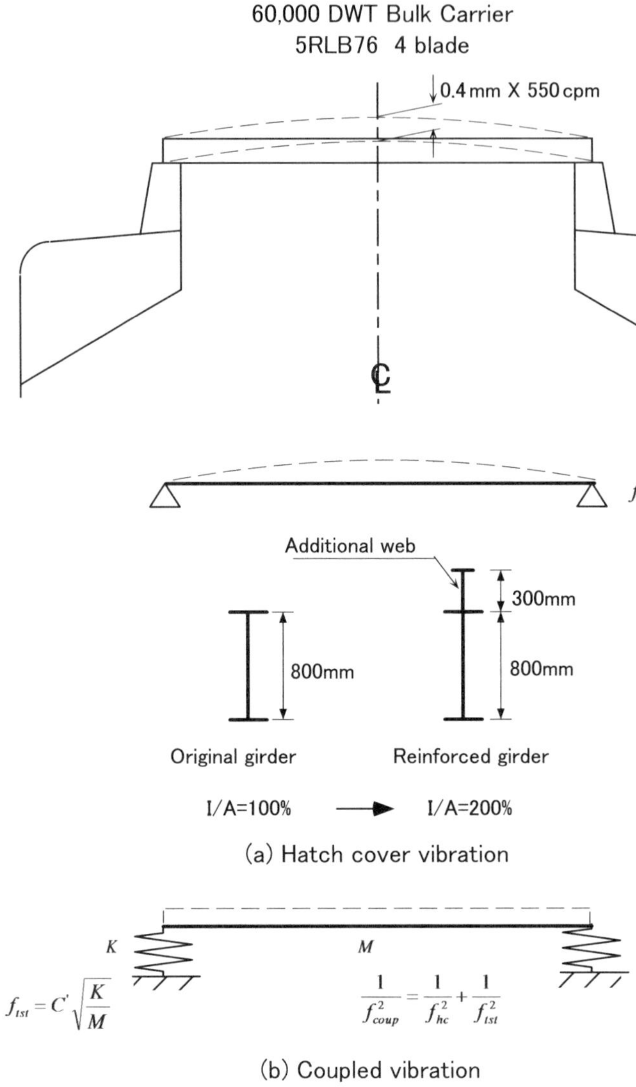

Fig. 18.4 Vertical vibration of hatch cover

where

f_{coup}: coupled natural frequency of hatch cover and top side tank.

f_{hc}: natural frequency of hatch cover itself.

f_{tst}: natural frequency of rigid hatch cover supported by equivalent spring of hatch coaming and top side tank.

From Eq. (18.4.2), in the case where f_{tst} is much smaller than f_{hc}, f_{coup} is governed by f_{tst} and f_{coup} cannot be increased by increasing f_{hc}. In this case, f_{coup} was 550 cpm. Therefore, if we assume the natural frequency of the hatch cover itself is $f_{hc} = 1,400$cpm, then f_{tst} becomes 600 cpm. By the reinforcement of the additional web, natural frequency of the hatch cover f_{hc} increases from 1,400 cpm to $1,400 \times 1.41$cpm, but conversely, natural frequency of the hatch cover foundation f_{tst} is reduced by the increase in the hatch cover weight due to the reinforcements. When the weight is assumed to increase by 13% by the reinforcement, f_{tst} reduces to 94% of the original and it becomes 564 cpm. Consequently, the coupled frequency becomes

$$\frac{1}{f_{coup}^2} = \frac{1}{(1,400 \times 1.41)^2} + \frac{1}{564^2}$$

$$\therefore f_{coup} = 542 \text{ cpm}$$

From the above calculation, we found that the coupled natural frequency decreases by 1.5% by providing the reinforcement to the hatch cover.

Example 2 shows that the boundary conditions of the hatch cover largely affect the natural frequency of the hatch cover.

Example 3 This example shows the vibration problem due to hydro-elastic interaction. The natural frequency of the upper deck structure of a tanker will change drastically, when the ballast water in the wing tank is touching the upper deck plate of the tank as shown in Fig. 18.5. In the case shown in the figure, the upper deck plate resonates with the exciting force of the propeller of 4 blades at 112 rpm of shaft revolution, because the natural frequency of the upper deck was 448 cpm, caused by the water contacting the deck plate. To avoid the resonance, the volume of ballast water was reduced to 98% of the tank volume, which resulted in a diminishing of the contact of ballast water with deck plate. By taking this countermeasure, the natural frequency of the deck plate becomes larger than the exciting frequency; consequently, the vibration due to resonance can be removed.

This example shows the ballast water changed the boundary conditions of the wing tank; water contact with the deck plate increases the weight of the deck plate.

18.5 Subsequent Boundary Condition Change of Hull Structure Vibration

As we have mentioned in Sect. 18.4, hull structure vibration takes place in a different form when the boundary conditions are changed. In this section, we will explain the subsequent situation of the boundary conditions together with the vibration problems induced by them.

The oil crisis in 1973 and 1977 led to a sudden rise in crude oil prices. Since that situation increased the proportion of fuel oil cost in the total running cost of

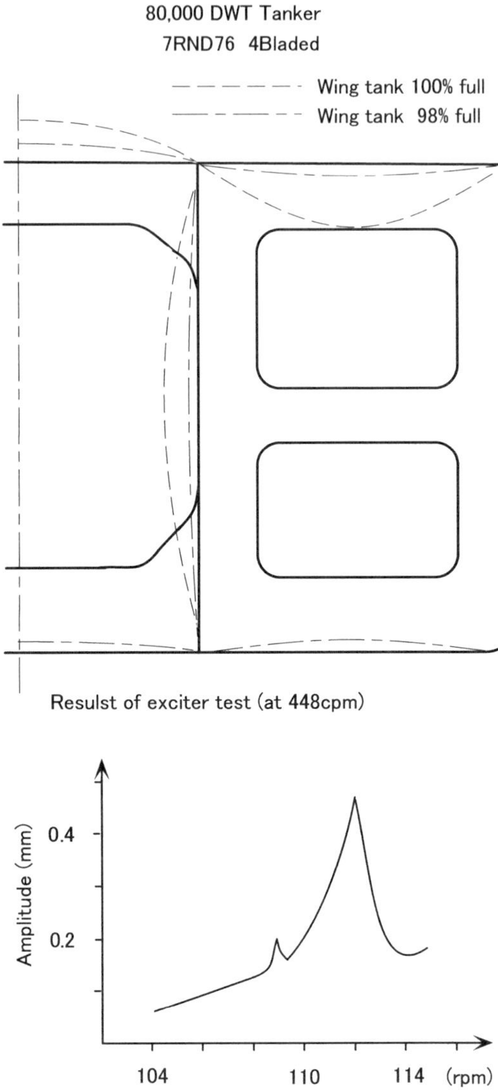

Fig. 18.5 Vertical vibration of upper deck of tanker

ships, necessary countermeasures were immediately taken in order to save energy. For the purpose of saving energy, ship owners, ship builders, and main engine makers have undertaken research and development. These efforts have reduced the fuel oil consumption, nowadays, to about a half of that prior to the oil crisis.

The energy saving efforts were achieved in the following three fields:

- Improvement of propulsive efficiency by the change of hull form
- Reduction of fuel oil consumption of main engine
- Improvement of heat efficiency by recovering the exhausted gas energy.

In addition to these, the application of higher strength steel to the hull structure contributed to energy savings because this brought about a decrease in the total hull steel weight. Furthermore, the adoption of self-polishing paint on the outer shell was successful.

Among these changes of the environment, the change of hull form, the improvement of propulsive efficiency, the reduction of fuel oil consumption, and the application of higher tensile strength steel have considerably affected hull vibration as explained below.

The change of hull form reduced the horsepower of the main engine required to produce the same deadweight and the same speed. In addition, the decrease in ship's service speed due to the rapid rise of fuel oil prices reduced the main engine power, which resulted in the decrease in engine exciting forces. For instance, although propeller revolution used to be 100 ~ 120 rpm in the case of a diesel driven vessel, and 80 ~ 90 rpm in the case of a turbine vessel, it became 70 ~ 80 rpm in diesel driven vessel, and in an extreme case even down to 60 rpm.

The reduction in propeller revolutions contributes to the reduction in flow velocity of propeller, and this prevents propeller cavitation, which means the reduction of propeller exciting forces.

To enhance propeller efficiency, the hull form design of the after body of the vessel was improved to produce a more uniform distribution of wake, leading to reduced propeller cavitation and propeller exciting forces.

Considering the above situation, both the improved hull form and improved propulsive efficiency have positive effects in preventing vibration problems. In contrast, efforts to reduce fuel oil consumption of a diesel engine have produced negative results from vibration viewpoint. The adoption of long stroke of main engines and the application of a small number of cylinders have caused a change in engine profile: larger height than length. Such a profile caused the longitudinal vibration of the main engine structure. The small number of cylinders resulted in increased unbalanced moment and bigger torque variation. The increased unbalanced moment caused problems of hull girder vibration and engine room double bottom vibration. The bigger torque variation induced variation in shaft revolution and thrust variation, causing local vibrations in the engine room. In the 1960's, 90% of the ship vibration problems was caused by propeller exciting forced, but due to the change in the boundary conditions as stated above, the dominant excitation source changed from a propeller to the main engine, and 70% of the ship vibration problem was caused by engine exciting forces in the 1980's.

Part III
Applications

Chapter 19
Hull Structure Arrangement

As explained in Chap. 1 of Part I the general arrangement of a ship is defined by the strength of the ship and many kinds of socio-economic requirements. The performance of the ship is closely related to the general arrangement. It used to be said that a ship designed by a designer with no experience in hull structure design may have deficiencies.

Hereunder we refer to the general arrangement from the viewpoint of hull structure design. In this design there are two main items, one of which is the tank arrangement not having excess longitudinal hull girder bending moment, and the other is the bulkhead arrangement, that is, the number of tanks or holds.

The former is well known and will be explained in Chap. 20 "Longitudinal Strength of Hull Girder". The latter is not so widely known and is explained here under as the hull structure arrangement.

19.1 Hold Arrangement

It has been a common view in the case of bulk carriers that a 5 holds arrangement can be applied up to 70,000 DWT, 7 holds up to 150,000 DWT and 9 or 11 holds for ships bigger than 150,000 DWT. Also there exists an opinion that a double hull side construction will be applied for ships bigger than 150,000 DWT. The biggest bulk carrier ever built is 250,000 DWT which has 9 holds and a single side hull construction.

How to decide the bulkhead arrangement, that is, the number of holds is a very important factor at the initial design stage but not yet clearly reported. Nowadays the hull structure is designed by being analyzed as a whole structure by computer. In this way the problem of how to decide the number of holds is submerged in the computer calculation process.

Y. Okumoto et al., *Design of Ship Hull Structures*,
https://doi.org/10.1007/978-3-032-12500-2_19

The main loads applied on the hull structure are buoyancy and weight which are forces in the vertical direction. Against these vertical loads, the vertical members such as side shell plates, transverse bulkheads and longitudinal bulkheads form the strength of hull structure. In other words, it can be said that the strength of the hull structure as a whole is maintained principally by the shearing strength of the side shell plates, transverse bulkheads and longitudinal bulkheads. Accordingly, for the larger ships, to increase the number of transverse bulkheads as well as increasing the thickness of side shell plates, transverse and longitudinal bulkheads plates are desirable. Also to construct a double hull structure is considered to be very effective.

In 1960's much damage was reported on the transverse rings in the wing tanks of large tankers. This was caused by the relative displacement between the longitudinal bulkhead and side shell plating due to the increased wing tank length. The swash bulkhead in long wing tanks was designed only for reducing the sloshing motion of free water surface in a tank and not for rigidity of the wing tank structure.

Figure 19.1 shows areas of buckling damage on a center line girder of one such 100,000 DWT tanker. In this case the center line girder was designed with the condition of being fixed at both ends by transverse bulkheads, one of which was a watertight bulkhead and the other was a swash bulkhead which had insufficient shear rigidity due to openings.

In the wing tank as well as in the center tank the strength members resisting the vertical loads should have a rigid frame work, otherwise the design condition (both ends fixed) of the primary supporting members will be faulty. By relaxing the both-ends-fixed condition there arises additional stresses in the primary supporting members which causes damage such as cracks and buckling.

19.2 Criteria of Design of Hull Structure Arrangement

19.2.1 Wing Tanks of Tankers

In the Rule Book of Nippon Kaiji Kyokai there was a regulation to limit the relative displacement between longitudinal bulkhead and side shell plate as follows:

Special consideration is to be paid to wing tank construction when the S-value as calculated by Eq. (19.2.1) exceeds 0.15.

$$S = \frac{h - 0.32d}{n_b k_b + n_s \eta_s k_s + n_t \eta_t k_t} \cdot \frac{a}{b} \cdot l \tag{19.2.1}$$

where

a: half breadth of center tank in m.

b: breadth of wing tank in m.

h: distance above top of keel to hatch opening top in center tank in m.

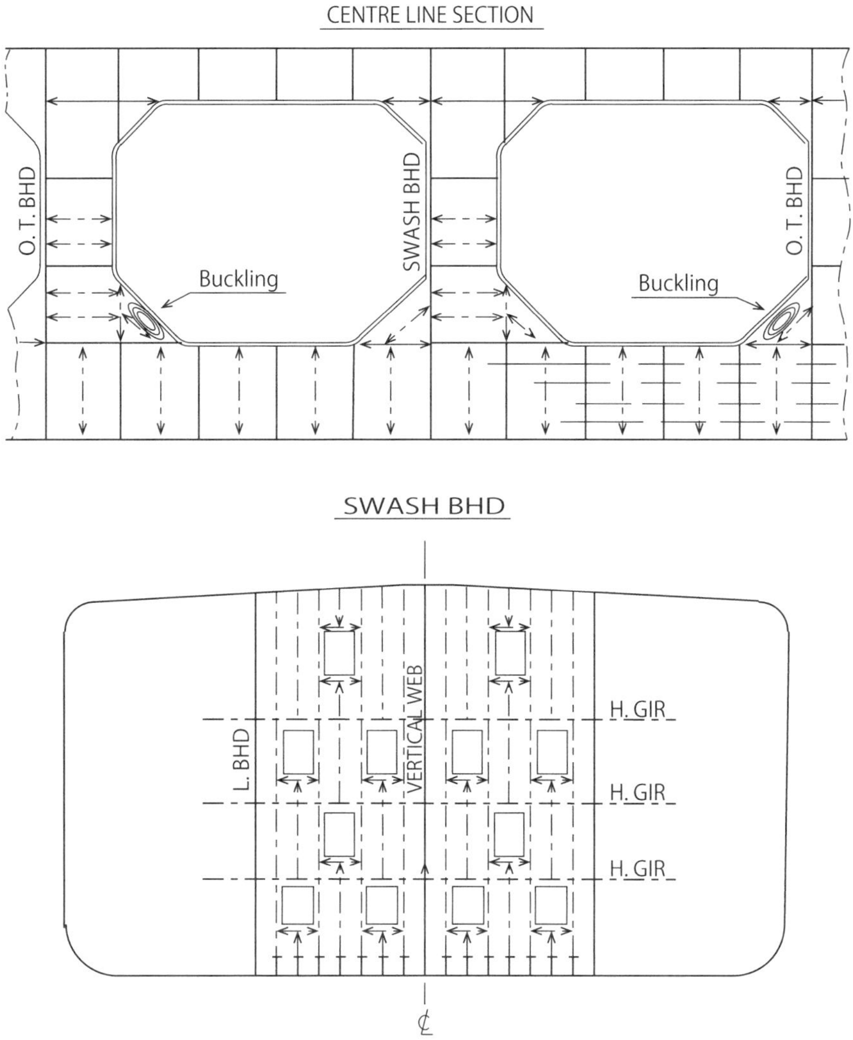

Fig. 19.1 Example of damage caused by lack of shear rigidity of swash bulkhead

d: full load draft in m.

l: center tank length between tight bulkheads in m.

n_b, n_s and n_t: numbers of transverse bulkheads, swash bulkheads and transverse rings in way of l respectively (in case of the ends of l, the number is 1/2).

η_s and η_t: coefficients depending on the opening ratio of swash bulkheads as shown in Table 19.1, for intermediate opening ratio coefficients delivered by interpolation.

k_b, k_s and k_t: values to be obtained by the following Eq. (19.2.2).

Table 19.1 Coefficient η_s, η_t due to opening ratio

Opening ratio (%)	0	5	10	20	30	40	50	60	70
η_s, η_t	1.00	0.95	0.80	0.55	0.35	0.23	0.15	0.10	0.06

$$81.0\frac{Dt}{\alpha b}\ (t\ \text{in mm}) \tag{19.2.2}$$

t: in calculation of k_b, mean thickness of the transverse bulkhead in the wing tank.

in calculation of k_s, mean thickness of the swash bulkhead in the wing tank.

in calculation of k_t, mean thickness of the transverse ring in the wing tank.

α: 1.0 for flat bulkheads, and for the corrugated bulkhead the values of following equations.

$$\text{in case of vertical corrugation}\ \alpha = \frac{\text{girthlength(breadthwise)}}{b}$$

$$\text{in case of horizontal corrugation}\ \alpha = \frac{\text{girthlength(depthwise)}}{D}$$

The numerator of Eq. (19.2.1) represents load W on the longitudinal bulkhead in the center-tank side in the floating condition with a shallow draft of $0.32d$ and with full ballast in the center tank. The denominator represents shear rigidity of wing tank construction which can be understood by the following deduction: 81.0 represents the shear modulus of steel in tonf/m · mm.

$$(n_b k_b + n_s \eta_s k_s + n_t \eta_t k_t)b = 81.0D\left(n_b\frac{t_b}{\alpha} + n_s \eta_s \frac{t_s}{\alpha} + n_t \eta_t k_t\right) \tag{19.2.3}$$

$$\text{Putting}\ n_b\frac{t_b}{\alpha} + n_s \eta_s \frac{t_s}{\alpha} + n_t \eta_t k_t = \sum t \tag{19.2.4}$$

$\sum t$ represents shear effective plate thickness.

Collecting the details above from Eqs. (19.2.1)–(19.2.4), the following Eq. (19.2.5) can be obtained.

$$S = \frac{W}{81.0 \times D\sum t} \leq 0.15 \tag{19.2.5}$$

where W in tonf, D in m, t in mm and 81.0 in tonf/m · mm. Accordingly S is non dimensional.

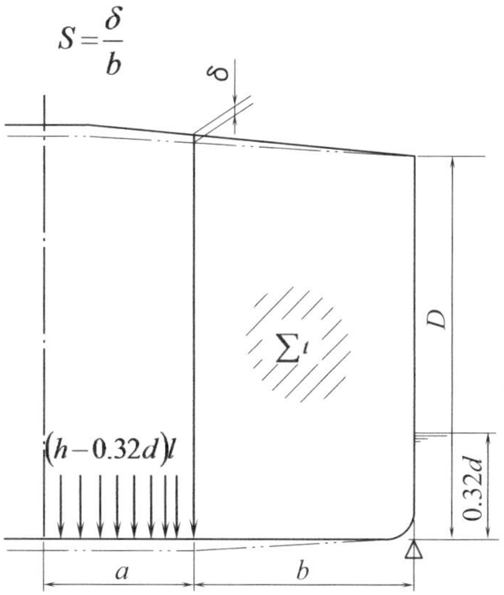

Fig. 19.2 Shearing deformation of VLCC wing tank

Because W is the load on the longitudinal bulkhead and $81.0D\sum t$ is the shear rigidity of the wing tank construction, S is the shearing strain of the wing tank as shown in Fig. 19.2.

Hereafter based on the above Nippon Kaiji Kyokai's regulation, a criterion for bulkhead arrangements of tankers is investigated.

Equation (19.2.5) can be modified to the following Eq. (19.2.6) assuming that the shear rigidity of a transverse ring is 10% of the shear rigidity of an oil tight bulkhead.

$$\frac{W}{D\left\{t_b + 0.1t_b\left(\frac{l}{s} - 1\right)\right\}} \leq 12.15 \tag{19.2.6}$$

where s: spacing of transverse ring

Furthermore, assuming $t_b \approx 0.7\left(3\sqrt{D} + 3.5\right)$ and $h \approx D$, Eq. (19.2.6) can be simplified into Eq. (19.2.7) as follows.

$$\frac{(D \quad 0.32d)al}{D \times 0.7\left(3\sqrt{D} + 3.5\right) \times \left(0.9 + 0.1\frac{l}{s}\right)} \leq 12.15 \tag{19.2.7}$$

By applying Eq. (19.2.7), the limits of a and l which are the main factors in determining the bulkhead arrangement can be obtained based on the principal dimensions of D and d.

For an example in Fig. 19.3 the limitation of length of the center tank for its half breadth is shown for a 200,000 DWT tanker which has $D = 26.3$ m, $d = 20.32$ m and $s = 4.8$ m. Actually this ship has a 48 m long center tank for a half breadth of 9.4 m, whereas Eq. (19.2.7) gives a 39 m center tank length for 9.4 m of center tank

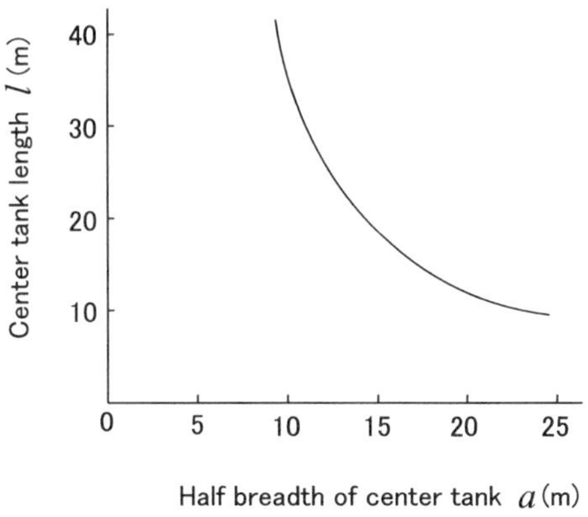

Fig. 19.3 Limitation for center tank size of VLCC

half breadth. From the viewpoint of shear deformation in the wing tank, the design of the hull structure arrangement of this ship will not be sufficiently strong.

19.2.2 Bulkhead Arrangement of Bulk Carriers

Let us apply the same principle of Sect. 19.2.1 to a bulk carrier. In the case of a bulk carrier the holds have the same breadth as the ship's breadth, hence the hold length provides the problem. Shear deformation on the transverse bulkhead is caused by the vertical load on the transverse bulkhead. For the loading condition in question, though the fully ballasted tank was considered in the ballast condition of tankers, there are two severe conditions in the case of the bulk carriers. One is an empty hold in the ballast condition and the other is a loaded hold in the fully loaded condition. Here the former case is investigated, that is, an empty hold in the ballast condition.

Assuming a transverse bulkhead supports $b/(b+l)$ of total load applied on the double bottom, the vertical load on a transverse bulkhead W is expressed by Eq. (19.2.8) as follows.

$$W = \frac{b}{b+l} \times 0.32d \times b \times l \tag{19.2.8}$$

where

b: breadth of double bottom in m.
l: length of hold in m.
d: full load draft in m.

Assuming the shear rigidity of a double bottom floor is 1% that of the transverse bulkhead, total shear rigidity G of a transverse bulkhead and floors is expressed by the following Eq. (19.2.9).

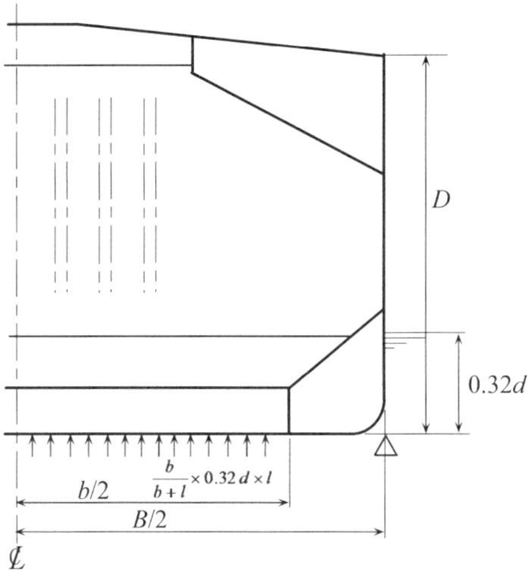

Fig. 19.4 Shearing deformation model of bulk carrier transverse bulkhead

$$G = 81.0D\left\{\frac{t_b}{\alpha} + 0.01 \times t_b\left(\frac{l}{s} - 1\right)\right\} \tag{19.2.9}$$

where

D: depth of ship in m.

t_b: mean thickness of bulkhead in mm.

α: correction factor for corrugated bulkhead, see Sect. 19.2.1

l: length of hold in m.

s: floor space in m.

Modeling these as shown in Fig. 19.4, the shearing strain S in the transverse bulkhead is expressed by Eq. (19.2.10).

$$S = \frac{W}{2G} \tag{19.2.10}$$

Furthermore assuming $b = 0.7B$, $\alpha = 1.8$ and $t_b = 0.7\left(3\sqrt{D} + 3.5\right)$, Eq. (19.2.11) can be derived from Eqs. (19.2.8)–(19.2.10) as follows.

$$S = \frac{B^2 dl}{722D(0.7B + l)\left(3\sqrt{D} + 3.5\right)\left(0.55 + 0.01\frac{l}{s}\right)} \tag{19.2.11}$$

In the case of a wing tank of a tanker, we imposed the limitation of $S \leq 0.15$ referring to Nippon Kaiji Kyokai's regulation, but in the case of bulk carriers we have no such reference. In such cases it is reasonable to refer to data of ships in service.

We calculated S values for two ships in service without any trouble. One is a 30,000 DWT ship with 5 holds and the other is a 130,000 DWT with 9 holds. The S values are 0.051 and 0.064 respectively, and we selected $S = 0.07$ as a criterion.

In the case of a 250,000 DWT bulk carrier, S values are 0.104 with 9 holds and 0.096 with 11 holds. It is recommended to have a stiffened flat plate or double hull transverse bulkhead instead of the current corrugated bulkhead.

19.3 Bulkhead Arrangement Beyond Cargo Hold

For large ships not only the structural arrangement in the vicinity of the cargo hold or tank but also the structure arrangement beyond cargo holds or tanks, such as bow and engine room construction, is important. In the vicinity of cargo tanks of a tanker, the side shell, longitudinal and transverse bulkheads provide the strength of the ship as a whole. However, at both ends of the tank part, to keep the strength continuity is difficult due to the absence of longitudinal bulkheads, hence a special attention should be paid.

19.3.1 Bow Construction Without Extended Longitudinal Bulkheads

There are usually two rows of longitudinal bulkheads in way of the cargo tanks of a tanker. Also forward of the collision bulkhead is usually the fore peak tank where no longitudinal bulkhead is necessary. In the case of a wide fore peak tank a center line swash bulkhead will be installed to avoid sloshing motion of the ballast water.

The volume of the fore peak tank will be quite considerable for large ships especially for a broad beam ship with a bulbous bow. The big volume of the fore peak tank causes great buoyancy in the fully loaded condition and a big load in the ballasted condition, while the fore peak tank is supported by the shear area of the side shell plate and the transverse bulkhead plate. In Fig. 19.5, volumes of fore peak tanks are plotted for the ship's depth. Assuming the volume V of the fore peak tank is proportional to the cube of the ship's depth D, Eq. (19.3.1) can be obtained as follows.

$$V = 0.4D^3 \tag{19.3.1}$$

The shearing force caused by the volume V of the fore peak tank is supported easily in the aft part of the collision bulkhead by side shell plates and longitudinal bulkhead plates. However, in the vicinity of the fore peak tank in some case it is hard for side shell plates to support the shearing force. In such cases a center line longitudinal bulkhead is to be fitted.

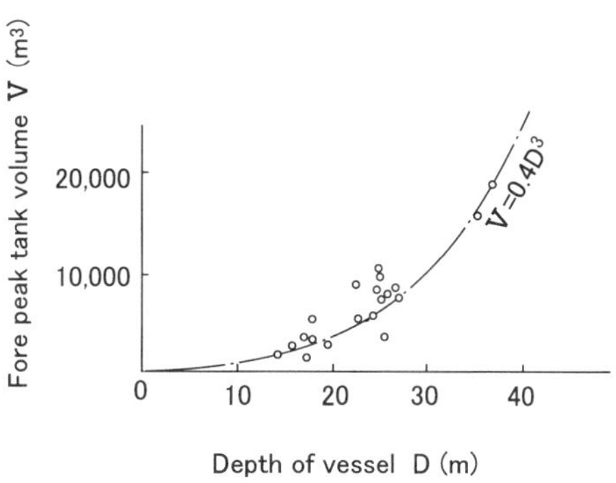

Fig. 19.5 Fore peak tank volume versus depth of vessel

19.3.2 Engine Room Construction Without Extended Longitudinal Bulkheads

Usually the two rows of longitudinal bulkheads in the cargo tank space are to be extended into the engine room as fuel oil tank bulkheads. But in the case of three rows of longitudinal bulkheads in the cargo tank, the center line longitudinal bulkhead is not able to be extended into the engine room. The shearing force supported by the center line longitudinal bulkhead is transmitted to side shell and side longitudinal bulkhead plates in the vicinity of engine room via the transverse bulkhead at the aft end of the cargo tank space.

To investigate the transmission of the shearing force an FEM calculation was carried out with a model as shown in Fig. 19.6. The results are shown in Fig. 19.7. The ship was a 600,000 DWT tanker of a trial design and has three rows of longitudinal bulkheads in the cargo tank space. The center line longitudinal bulkhead is extended aft to Fr.66 and two side longitudinal bulkheads are extended into the engine room. The model in Fig. 19.6 is fixed at Fr.71 and a uniform load of 10,000 tf is applied on half the breadth of the upper deck at Fr.28. The shearing stresses in the side shell, side and center line longitudinal bulkheads plates are calculated, then shearing forces are obtained by multiplying plate thickness by the calculated stresses. In Fig. 19.7, the supporting ratio of each member is shown.

From considerably forward of Fr.67, i.e. the aft end bulkhead of the aftermost tank, the shearing force supporting ratio of the longitudinal bulkheads decreases. In contrast, the shearing force supporting ratio of the side shell plates increases. The shearing force supporting ratio of the side longitudinal bulkheads also decreases from considerably forward of Fr.67, i.e. the aft end bulkhead of the tank part, which may be caused by reduced sectional area due to hull form.

Careful attention should be paid to avoid having troublesome deformation in the engine room structure caused by large buoyancy forces in the fully loaded condition, because there are many items such as main engine and main shafting for which alignment is very important. To avoid such troublesome deformation in the engine room structure, an effective bulkhead arrangement will be necessary.

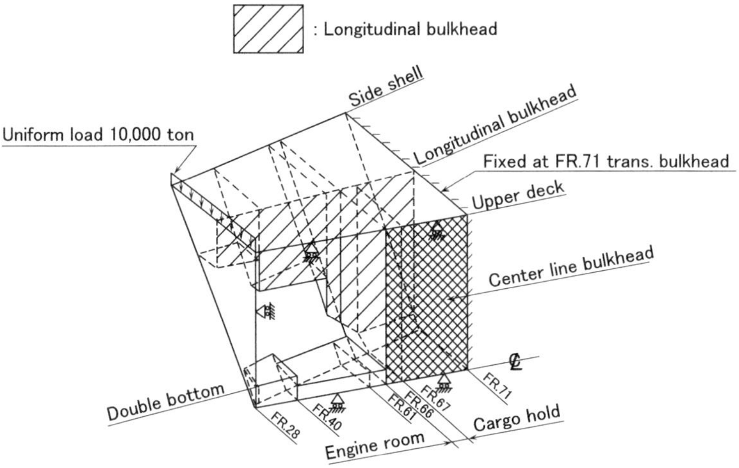

Fig. 19.6 Structural model of engine room fore part and cargo hold aft part

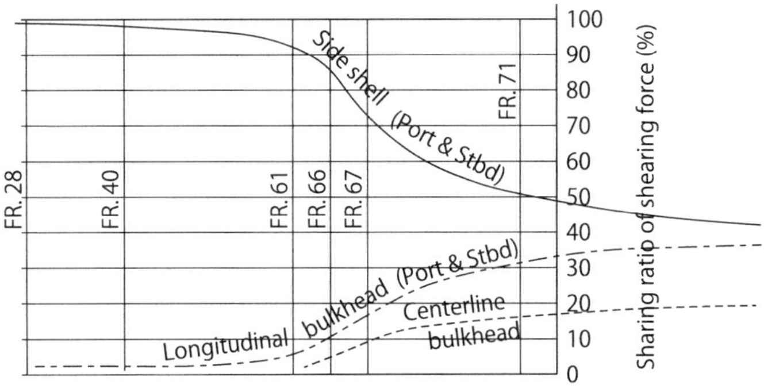

Fig. 19.7 Share of shearing force for each member

Chapter 20
Longitudinal Strength of Hull Girder

The longitudinal strength of the hull girder is one of the most important strength considerations in a ship, because lack of longitudinal strength may be a direct cause of the sinking of a ship. Damage such as jack knife deformation and breaking of the hull girder at midship are caused by the lack of longitudinal strength.

One example is the break-in-two accident of *Ryoyo Maru* as shown in Fig. 20.1 [110]. Generally speaking, damage due to the lack of longitudinal strength occurs mainly in older ships. Here the lack of longitudinal strength is caused by corroded deck plates and hence reduced thickness, and by reduced connections between deck plates and deck longitudinals because of corroded fillet weld beads.

Brittle fracture may happen in steels with low notch toughness because of defects caused by poor welding. Such damage to the longitudinal strength of hull girder can be avoided by anti-corrosion devices, good selection of materials, and good welding. It can be said that these are not structural design problems in the narrow sense but workmanship and maintenance problems.

In the design of hull structure arrangement, the vertical members such as side shells and bulkheads play a major role. On the other hand, in the design of the longitudinal strength of hull girders, the horizontal members at a greater distance from the neutral axis such as deck and bottom structures, will play the major role.

Basically, the classification societies' rules on longitudinal strength have been established based on extensive past experiences, and therefore, safety can be ensured as long as a ship is designed according to the rules. However, when the size or novelty of a ship transcends the scope of the past experiences, meticulous care should be taken of the integrity of the longitudinal strength. Anti-corrosion methods such as epoxy paint under the upper deck, notch tough materials, workmanship and maintenance should also be well controlled.

Hereunder some items are introduced on which the authors paid attention in the design of the longitudinal strength of hull girders.

Y. Okumoto et al., *Design of Ship Hull Structures*,
https://doi.org/10.1007/978-3-032-12500-2_20

Fig. 20.1 Hull girder collapse of *Ryoyo Maru* [110]

20.1 Review of Break-in-Two Accidents

In this section, history of break-in-two accidents of ships is reviewed referring to Sumi's review article [111] and others to understand learnt lessons in the past 100 years. Table 20.1 shows a chronology of break-in-two accidents, among which important accidents are explained hereafter.

In September 1901 during the maiden voyage from Vickers Armstrong Shipyard to Portsmouth, destroyer *HMS Cobra* broke in a sagging mode with the 3rd and 4th funnels colliding with each other and sank in a "middling rough" sea [111]. Just after the accident, an investigation committee was organized and the committee carried out static bending test on a similar destroyer *HMS Wolf* in a dry dock, and found that the measured deflection was much larger than the deflection calculated using beam theory. Root cause of this phenomenon was not made clear at that time, but almost a quarter century later, Hoffmann [112, 113] explained the reduction of the hull girder sectional moment of inertia due to the reduction of effective width of upper deck and shell plating after buckling of the transversely stiffened plate panels. Since then, the concept of effective width as explained in Sect. 8.1, Part II was established to account for the uneven distribution of stresses between rigid members.

Another early break-in-two accident such as that of *Schenectady* in 1943 was caused by brittle fracture. As explained in Sect. 17.1, Part II, thousands of cargo vessels were built over the period 1941–1946 in USA under the wartime emergency shipbuilding program as depicted in Fig. 1.2, Part I. To enable this unprecedented mass production, welding technique was boldly introduced all over the hull construction, but this caused many brittle facture accidents. US investigation committee and other worldwide researchers carried out wide-ranging and systematic research on

Table 20.1 Chronology of break-in-two accidents (revised based on [111])

Year	Ship name	Ship type	Root cause/failure mode	Hog/Sag
1901	Cobra	Destroyer	Buckling, ultimate strength	Sagging
1935	Hatsuyuki, Yugiri	Destroyer	Buckling, ultimate strength	Sagging
1943	Schnectady	Tanker	Brittle fracture	Hogging
1969	Bolivar Maru	Bulk	Failure of cargo hold	Unknown
1970	California Maru	Ore	Failure of side tank	Unknown
1976	Ryoyo Maru	Tanker	Corrosion, Wrong loading, ultimate strength	Sagging
1979	Kurdistan	Tanker	Brittle fracture	Sagging
1980	Energy concentration	Tanker	Wrong loading, ultimate strength	Hogging
1980	Onomichi Maru	Bulk	Fore part failure	Sagging
1997	Nakhodka	Tanker	Corrosion, ultimate strength	Sagging
1997	MSC Carla	Container	Deck fracture from doubling plates	Hogging
1999	Erika	Tanker	Corrosion, ultimate strength	Sagging
2002	Prestige	Tanker	Corrosion, ultimate strength	Sagging
2007	MSC Napoli	Container	Whipping, Bottom buckling, ultimate strength	Hogging
2013	MOL comfort	Container	Whipping, Bottom buckling, ultimate strength	Hogging

the root cause of this kind of accidents, and clarified the conditions causing brittle fracture, such as poor fracture toughness of materials, high stress intensity due to weld defect, low temperature, and tensile welding residual stresses. In addition, as a result of the application of welding instead of riveting all over the hull, there were no structural discontinuities where brittle cracks can be arrested. Thanks to all these research activities, rules were established to apply fracture mechanics to practical ship design and construction, and steel making and welding technologies progressed remarkably.

Major break-in-two accidents continued to occur especially for large bulk carriers. Accidents of *Bolivar Maru* (1969), *California Maru* (1970), and *Onomichi Maru* (1980) are considered to have been caused by abnormally high waves and inadequate ultimate strength of the hull girder in the forward part of the ship. Human errors also caused major accidents. In 1980, *Energy Concentration* broke her back during unloading operation of cargo oil in a harbor, caused by excessive hogging bending moment due to inappropriate unloading sequence. *Ryoyo Maru*'s accident shown in Fig. 20.1 is also considered to have been affected by improper weight distribution of water ballast. After these accidents, installation of a loading computer was mandated by the rules to ensure proper cargo loading to keep the still water bending moment and shearing force below the allowable limit.

In 1990's the focus shifted to the problem of substandard ships after many bulk carrier casualties and major oil spill accidents of tankers such as *Nakhodka* (1997), *Erika* (1999), and *Prestige* (2002). Excessive diminution of plate thickness was observed in many aged ships due to poor maintenance and caused serious structural failures. To cope with such situations, regulatory bodies and classification societies enacted stricter regulations step by step, such as enhanced survey program for bulk carriers and oil tankers (1993), Port State Control (PSC) (1996), Permanent Means of Access (PMA) for easy inspection and maintenance (2002), Performance Standard for the Protective Coatings (PSPC) (2006), Common Structural Rules for bulk carriers and oil tankers (CSR) (2006), and Goal-Based ship construction Standards (GBS) for bulk carriers and oil tankers (2010).

Up-sizing of container ship progressed remarkably as discussed in Sect. 5.3, Part I, especially after 1990's as depicted in Fig. 1.2, Part I. This was mainly caused by the expanding worldwide seaborne trade due to economic globalization and emergence of new industrial countries such as China. During the process of this up-sizing, we also experienced significant structural accidents.

In January 2007, *MSC Napoli* was sailing westward through the English Channel in a stormy weather of Beaufort scale of 10~11, when her hull girder collapsed suddenly at the forward end of the engine room as shown in Fig. 20.2 [114]. The transversely stiffened bottom plating buckled and collapsed, and the third assistant engineer witnessed water spraying from the long athwartship crack opening at the forward end of the engine room double bottom. The chief engineer stopped the main engine, and ordered all the personnel to leave the engine room. From the bridge wing, the master witnessed that the side plating just below the bridge wing was bulging and cracking, then decided to abandon the vessel. According to the investigation report [114], a combination of several factors has contributed to the hull girder collapse.

Engine room bottom plating is usually stiffened transversely due to the double bottom transverse floors arranged at every frame to support the heavy engine foundation (with an interval of 800 mm in case of MSC Napoli). Therefore, the plating is vulnerable to buckling against longitudinal bending. Meanwhile, because of the small block coefficient ($C_b = 0.609$ in case of *MSC Napoli*), the engine room forward bulkhead was located at a position closer to the midship (about $0.25L$ from the midship in case of *MSC Napoli*) than that of other ship types. In addition, because of the small C_b, the neutral axis at the engine room forward section was quite high, causing relatively larger bending stress at the bottom. As a result of the combination of all these factors, buckling strength of the bottom plating at the engine room forward was not sufficient, but the rules at that time did not require the buckling assessment outside the midship region ($-0.2L \sim 0.2L$ of midship). The engine room forward bulkhead was barely outside this region, and therefore, buckling assessment was not carried out in the design. After this accident, the classification societies' rules were amended to ensure buckling strength against longitudinal bending along the entire length of the hull. From this, we can learn that the classification societies' rules have a scope of application based on past experiences, and when designing ships that transcend the scope, it is not enough just to satisfy the rules, but meticulous consideration must be given based on the correct understanding of the structural behavior. Besides the

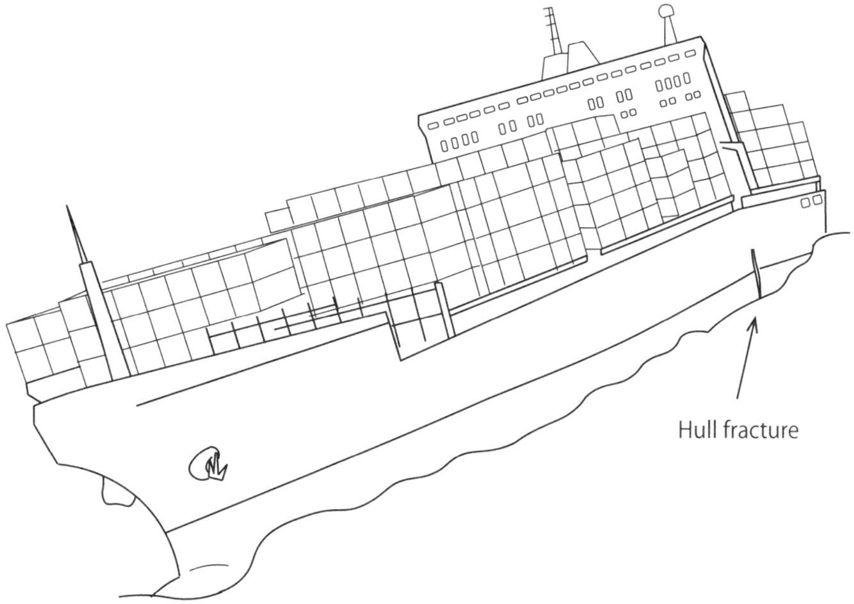

Fig. 20.2 Hull girder collapse of *MSC Napoli* [114]

buckling strength, it is reported that whipping load tends to become more significant in ships with finer hull form, and may have affected the collapse.

In June 2013, *MOL Comfort* fractured amidships as shown in Fig. 20.3 [115–117] while transiting the Indian Ocean from Singapore to Jeddah in a significant wave height of 5~6 m at a ship speed of 17 knots. Then, she split into 2 halves and sank. There was no loss of life thanks to the swift rescue efforts of the ships navigating the area and Indian disaster relief authorities.

The investigation report [117] pointed out the possibility of buckling collapse of bottom plating accelerated by the superposition of the double bottom local bending stress and the hull girder global bending stress. Actually, *MOL Comfort* had the

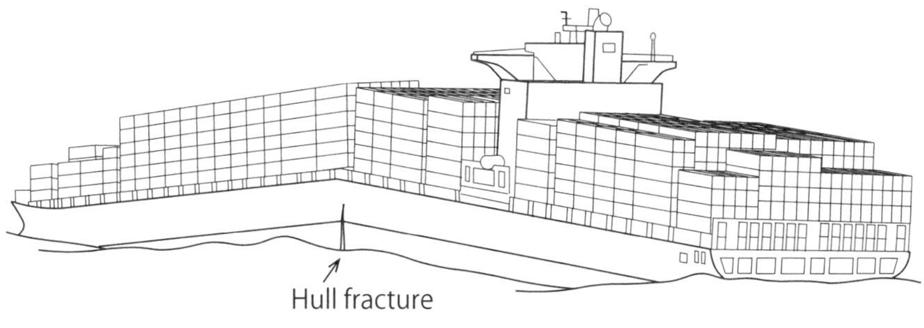

Fig. 20.3 Hull girder collapse of *MOL Comfort* [115]

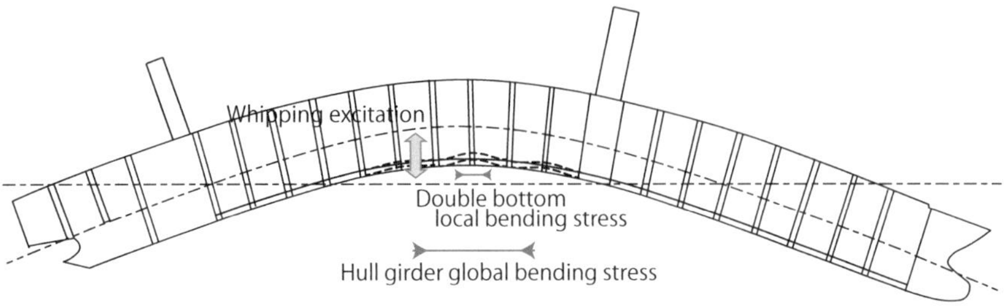

Fig. 20.4 Double bottom vibration excited by whipping

same level of the hull girder ultimate strength with other similar container ships when only the longitudinal members are taken into account using Smith's method, but a significant difference was observed when the combination of double bottom bending is considered using 3-D elasto-plastic finite element analysis. Due to this combination, hull girder ultimate strength was significantly decreased to about 70% of the value in case of pure bending. It was also pointed out that the whipping load has superimposed on the wave induced hull girder bending moment, and has affected the hull girder collapse.

The weak double bottom structure has further effects on the hull girder collapse. When whipping occurs, the double bottom is excited by the forced displacement by the whipping, and the double bottom bending stress at the bottom shell plating fluctuate with the same phase as the hull girder bending stress due to the whipping as shown in Fig. 20.4. According to numerical simulations and full-scale measurements of other container ships [118–120], the double bottom bending stress excited by the whipping reaches about one sixth of the hull girder bending stress induced by the whipping, which cannot be neglected when considering the effect of whipping on the hull girder ultimate strength. This effect is even greater when the double bottom is relatively weak. This effect was not taken into account in the investigation reports [115–117] but can explain the accidents better.

Another consideration is about the effect of the hull form. When C_b is large, consecutive three cargo holds may have a similar size of double bottom, and therefore, the model with the same section throughout the length can be employed without doubt (see Fig. 20.5a). However, when C_b is small as in the case of *MOL Comfort*, the double bottom in the fore and aft cargo holds becomes narrower in width and therefore, stronger. As a result, the boundary condition of the mid cargo hold changes as shown in Fig. 20.5b causing weaker ultimate strength. The investigation reports did not consider this effect, which further explains the accidents well. These increase in double bottom bending due to whipping or hull form reduces the hull girder ultimate strength by a few percent, and does not have a significant impact in isolation, but when they are combined with other deteriorating factors, they may cause serious consequences. Designers should always be aware of the difference from past established designs, and care for small differences in the design of novel ships.

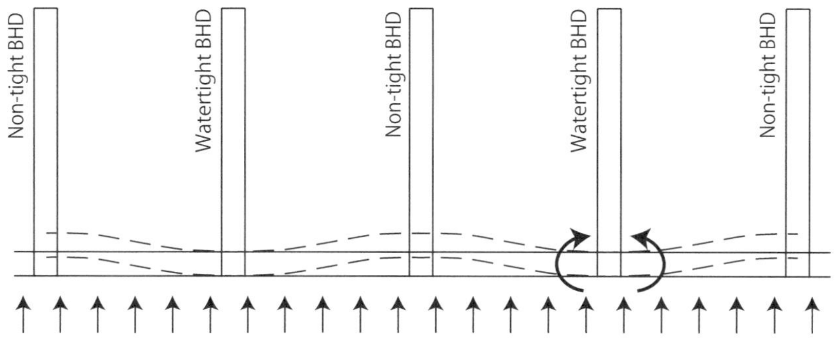

(a) Large C_b; same double bottom width throughout the length

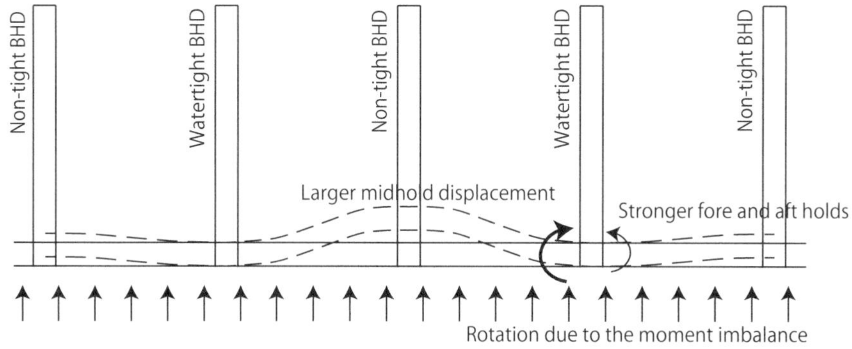

(b) Small C_b; narrow double bottom in fore and aft cargo holds

Fig. 20.5 Boundary condition of double bottom

After this accident, classification societies' rules were amended to explicitly consider the effect of the double bottom bending and whipping on the hull girder ultimate strength of large container ships. This is another example to show that the classification societies' rules are not necessarily perfect, and the designers should conduct independent considerations on the structural strength when designing novel ships transcending the past experiences. In association with the up-sizing of container ships, the breadth of the ship is becoming wider and wider and the double bottom is becoming weaker and weaker. One effective measure to strengthen the double bottom is to install a deep floor in way of the non-tight transverse bulkhead as shown in Fig. 20.6, which the authors used to apply to the container ships greater than about 8,000 TEU.

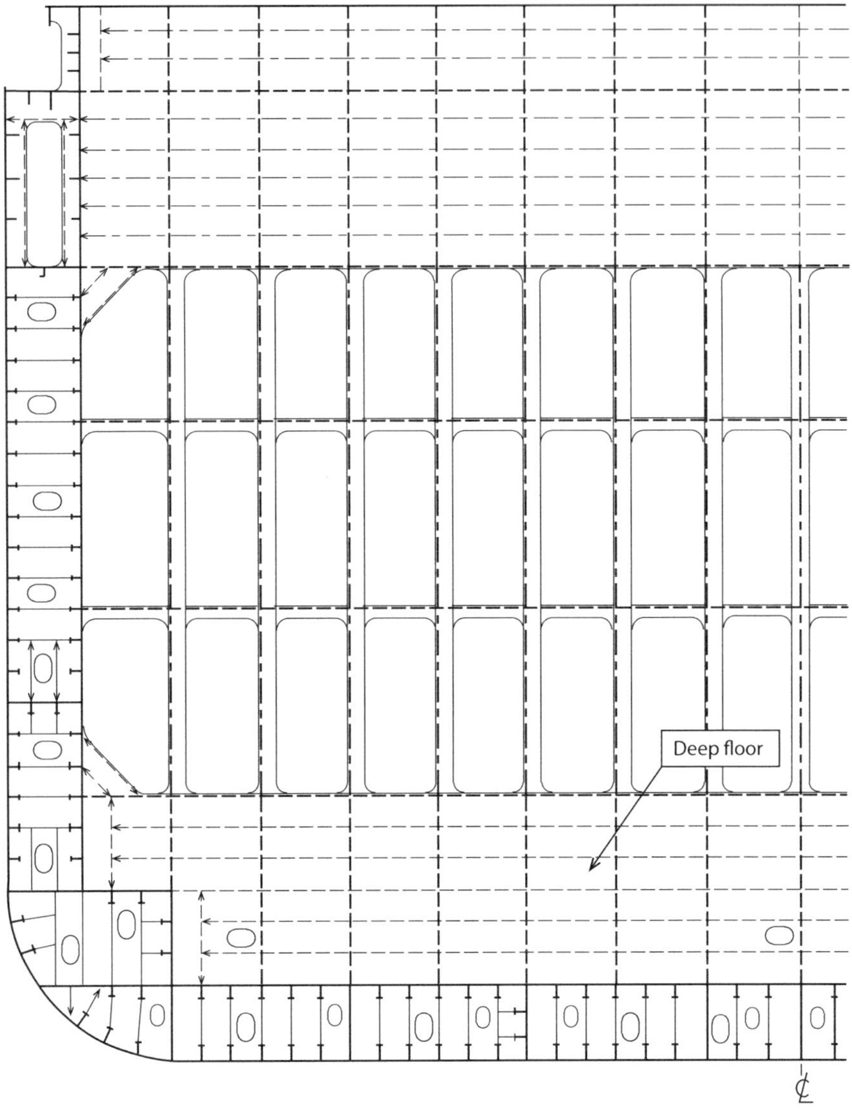

Fig. 20.6 Deep floor to reinforce double bottom

20.2 Allowable Stress for Longitudinal Strength

In the design of the longitudinal strength of hull girders, a total system for longitudinal strength design and direct calculation is becoming popular. These methods used to be applied in the designs of special ships such as super-size ships, high speed ships, unusual proportion ships, and ship with large deck opening. These methods were based on the response calculations in regular waves by strip theory, short term response predictions in irregular waves with sea spectrum, and long term response prediction with statistic sea conditions as shown in Sect. 2.4 of Part 1.

In the design of the longitudinal strength of hull girders, classification societies' rules have simple calculation methods, which give the required section modulus of the hull girder based on the bending moments in still water and in waves obtained from simple formula for the standard loading condition of the ship. There are two required section moduli, one is for wave bending moment M_w and the other is for sum of still water bending moment M_s and wave bending moment M_w. The former can be obtained from the ship length L, breadth B and block coefficient C_b irrespective of tank arrangement.

Here, an example of past classification rules is shown taking the rule of the Lloyd's Register of Shipping 1970 edition. The wave bending moment is given by the following Eq. (20.2.1).

$$M_w = \sigma_w C_1 L^2 B(C_b + 0.7) \times 10^{-3}\ \mathrm{tonf \cdot m} \tag{20.2.1}$$

where C_1 depends on ship length L.

The section modulus Z is taken as the bigger value obtained in the following Eqs. (20.2.2) and (20.2.3).

$$a)\, Z = C_1 L^2 B(C_b + 0.7)\ \mathrm{cm}^3 \tag{20.2.2}$$

$$b)\, Z = \frac{M_s + M_w}{\sigma_c} \times 10^3\ \mathrm{cm}^3 \tag{20.2.3}$$

σ_w in Eq. (20.2.1) means an allowable stress against the wave bending moment, and σ_c in Eq. (20.2.3) means an allowable stress against the total bending moment in still water M_s and in wave M_w which are shown in Table 20.2. σ_s in Table 20.2 means the allowable stress against the still water bending moment.

Table 20.2 Example of allowable stress for longitudinal strength (Unit in kgf/mm^2)

Ship	σ_c	Seagoing Service		Sheltered Water		Short Voyages	
		σ_s	σ_w	σ_s	σ_w	σ_s	σ_w
Tanker Ore oil carrier Ore carrier	16.40	6.4	10.0	11.4	5.0	8.4	8.0
Bulk carrier	18.15	8.15	10.0	13.15	5.0	10.15	8.0

Up-to-date rules such as IACS CSR have a similar system of equations, but different wave induced bending moment is given for hogging and sagging as shown in Eqs. (20.2.4) and (20.2.5) considering nonlinear effects. Equation (20.2.5) is basically identical to Eq. (20.2.1).

$$M_{w-hog} = 0.19 C_1 L^2 B C_b\ \mathrm{kN \cdot m} \tag{20.2.4}$$

$$M_{w-sag} = -0.11 C_1 L^2 B(C_b + 0.7) \text{ kN} \cdot \text{m} \quad (20.2.5)$$

Equation (20.2.3) is still valid in these rules, with $\sigma_c = 190$ MPa in case of CSR as explained in Sect. 3.3.2 of Part I. This allowable stress σ_c is larger than that in Table 20.1 because of the net scantling approach where the gross offered thickness reduced by half the corrosion addition must be used for the calculation of the section modulus. The formula in CSR corresponding to Eq. (20.2.2) is as shown in Eq. (20.2.6), from which σ_w is calculated to be 122 MPa, which is also larger than those in Table 20.1.

$$Z = 0.9 C_1 L^2 B(C_b + 0.7) \times 10^{-6} \text{ m}^3 \quad (20.2.6)$$

The required Z by Eq. (20.2.2) is for the wave bending moment and is the minimum requirement based on the principal dimensions, and the required Z by Eq. (20.2.3) is for the total bending moment in still water M_s and in waves M_w, which can be reduced by reducing the bending moment in still water M_s. In this case it is usual to decide the required Z from Eq. (20.2.3) for the total bending moment in still water M_s and in waves M_w so as not to exceed the required value given by Eq. (20.2.2) for the wave bending moment by adjusting the tank arrangement and the loading condition.

In Fig. 20.7 the maximum still water bending moments for each tank arrangement of a 600,000 DWT tanker (trial design) are shown. These are the values for the fully loaded condition. In the case of the ballast condition there is more flexibility than in the fully loaded condition in arranging the ballast water to produce a smaller bending moment.

In Fig. 20.7 three cases of the ballast tank arrangement are investigated:

case (1) No. 7 wing tanks.

case (2) No. 6 and 9 wing tanks.

case (3) No. 5 and 10 wing tanks.

For each arrangement the maximum still water bending moment is calculated in these three cases where the aft end bulkhead of the tank part is arranged at the A, B and C position as shown in Fig. 20.7. The maximum bending moments are all sagging moments. In case (3) where the ballast tanks are No. 5 and 10 wing tanks, bigger bending moment than in case (1) and case (2) is shown for all cases where the aft end bulkhead of the tank part is arranged at the A, B and C position. In case (1) and case (2) the maximum bending moments are equal for all cases where the aft end bulkhead of tank part is arranged at the A, B and the C position. The smallest bending moment is obtained in cases where the aft end bulkhead of tank part is arranged at the C position, while the biggest moment is obtained in case where the aft end bulkhead of tank part is arranged in the A position. This is a general rule and very important. To have a smaller still water bending moment, an arrangement with a shorter engine room and longer tank part is very effective.

Ballast tank arrangement	Bhd Arr. at aft.	Max. still water bending moment (kN-m)
1. No.7 Wing TK. Ballast 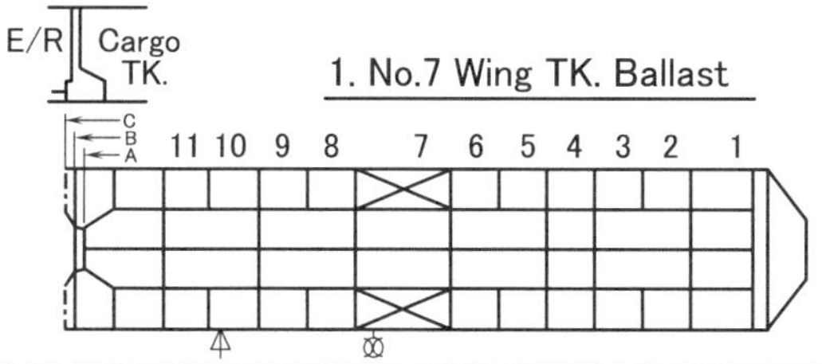	A	SAG. 1.44×10^7
	B	SAG. 1.27×10^7
	C	SAG. 1.20×10^7
2. No.6 & 9 Wing TK. Ballast 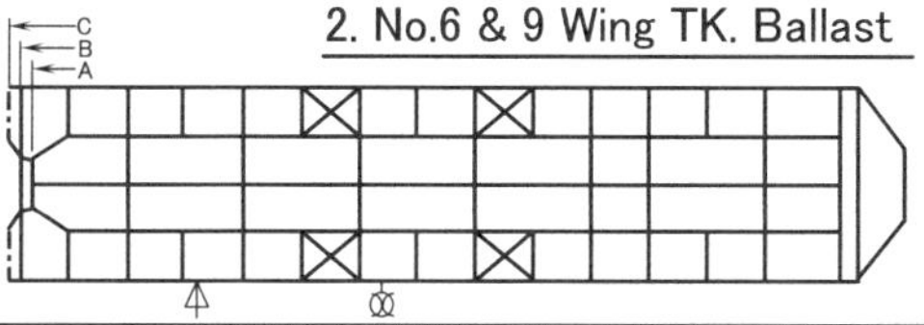	A	SAG. 1.44×10^7
	B	SAG. 1.27×10^7
	C	SAG. 1.20×10^7
3. No.5 & 10 Wing TK. Ballast 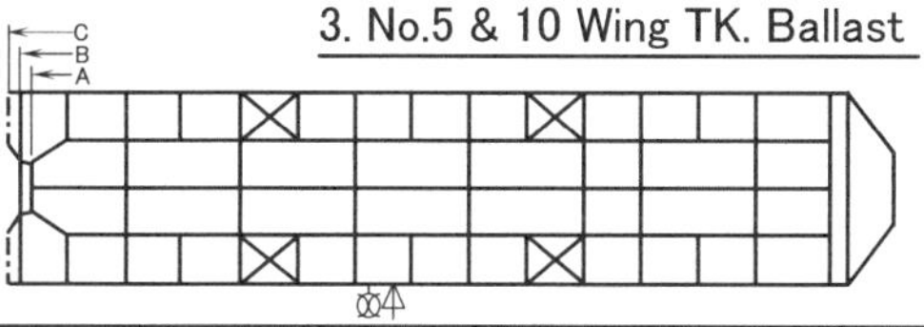	A	SAG. 1.85×10^7
	B	SAG. 1.61×10^7
	C	SAG. 1.56×10^7

denotes the position where max. bending moment occurs

Fig. 20.7 Still water bending moment in full load condition according to ballast tank arrangement

The authors suggest that in the design of a new ship the designer should have his/her own allowable stress values for the longitudinal bending moment based on the data of previous ships in service independent of the classification society rules. There is an owner who makes it a rule to have 10% bigger section modulus than the requirement of classification society rule for easy maintenance and longer life of the ship.

20.3 Position of Maximum Longitudinal Bending Moment

In Fig. 20.7 the positions of the maximum longitudinal bending moment are shown. It can be shown that the position of the maximum longitudinal bending moment in still water lies in some cases far from midship. Generally, the position of the maximum longitudinal bending moment in a wave lies at midship because the moment calculation is to be carried out in the condition where the crest or the hollow of the wave is at midship. However, in other conditions where the crest or hollow of the wave does not coincide with midship, the maximum longitudinal bending moment may not exceed that at midship.

The value and the position of the maximum longitudinal bending moment in a wave were investigated for the case where the crest or hollow of the wave does not coincide with midship. This investigation was carried out for a 600,000 DWT trial design tanker and the results are shown in Fig. 20.8. It can be shown from Fig. 20.8 that in this ship in the heavily ballasted condition the hogging moment is bigger than the sagging moment and the maximum longitudinal bending moment arises around S.S.6(1/2) and the value is about 85% of that in the fully loaded condition. On the other hand, in the fully loaded condition the sagging moment is bigger than the hogging moment and the maximum longitudinal bending moment rises to around S.S.4.

In this ship the above maximum bending moments arise within $0.4L$ of midship part, where the midship section modulus is kept constant and no problem is expected. However, it is recommended to check the longitudinal bending moment for cases where the crest or the hollow of the wave does not coincide with midship.

20.4 Rational Arrangement of Longitudinal Strength Members

20.4.1 Calculation of Section Modulus of Hull Girder

Hull girder section modulus can be calculated using the method explained in Sect. 8.3.1(4) of Part II. One example of hull girder section modulus calculation is shown in Fig. 20.9.

In this figure, the required section modulus for deck and bottom are specified at first. The position of an initial neutral axis is assumed tentatively by putting the distance from the neutral axis to deck and to bottom as Y'_{DK} and Y'_{BM}, respectively. Then, the sectional moment of inertia can be obtained by summing up the sectional moment of inertia for each member. In the summary table, B is the breadth of a member, t is the thickness, L is the vertical distance between the assumed neutral axis and the center of gravity of the member and H is the height of projecting length of the member. The distance from the true neutral axis to top or bottom can be derived by modifying the position of the neutral axis. Eventually, the obtained section modulus of the hull girder is compared with the required one and judgment is carried out as to whether the actual section modulus satisfies the requirement.

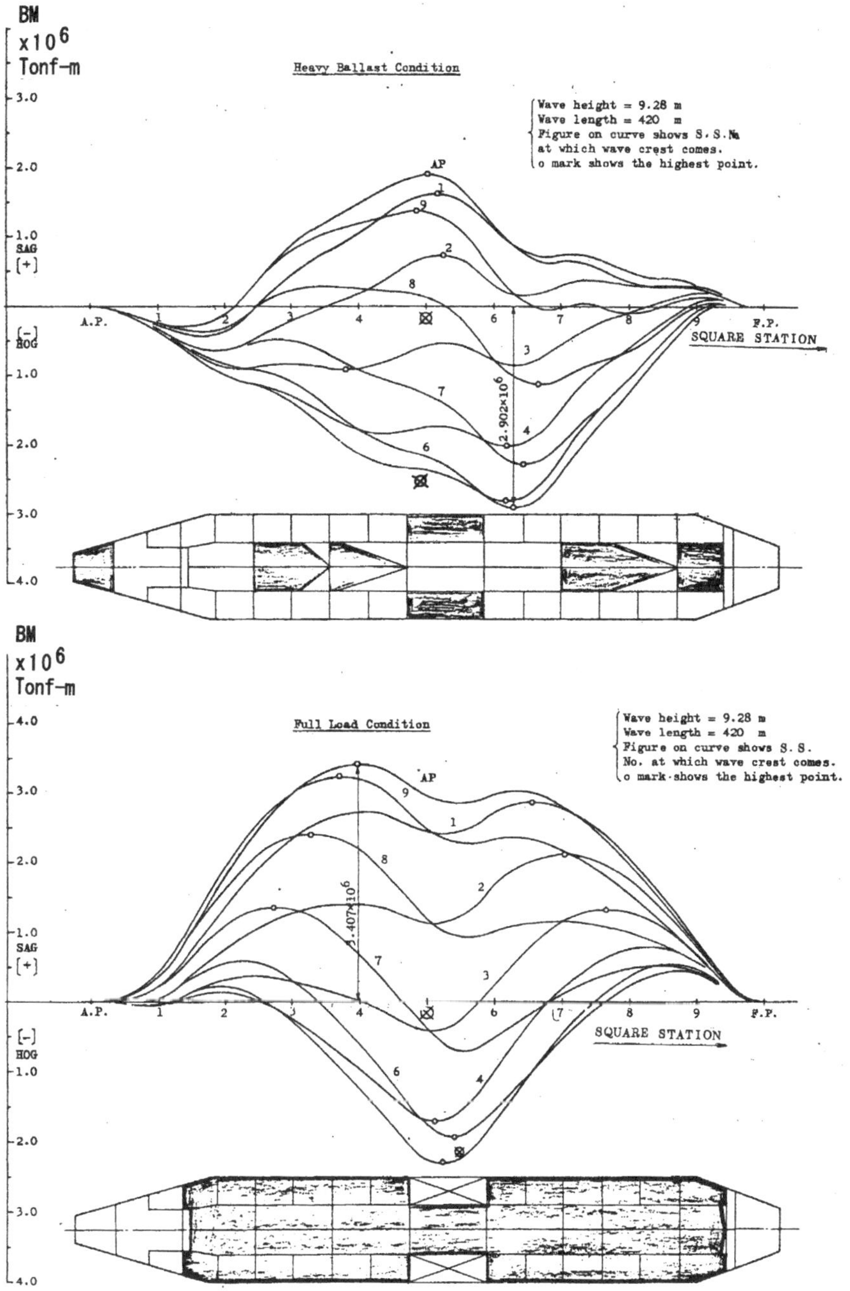

Fig. 20.8 Longitudinal bending moment in wave of which crest comes at each square station

1) Required section modulus

Z_{DK} = 6,749,200 cm^3

Z_{BM} = 6,749,200 cm^3

2) Assumed neutral axis

Y'_{DK} = 9.8 m

Y'_{BM} = 8.0 m

3) Summation of scantlings

Item	B (m)	T (mm)	A =BxT (mxmm)	L (m)	M =AxL (m^2xmm)	I =AxL^2 (m^3xmm)	H (m)	Self I (m^3xmm)
KLPLT1	1.095	13.0	14.235	-8.007	-113.97	912.5	0.00	0.00
BMPLT1	7.916	12.5	98.952	-7.796	-771.45	6014.4	0.42	1.45
BILGE1	4.712	12.5	58.905	-6.490	-382.28	2481.0	3.198	50.21
SSPLT1	5.38	12.5	67.25	-1.890	-127.10	240.2	5.380	162.21
:	:	:	:	:	:	:	:	:
DKLONG17F	0.141	9.0	1.267	9.730	12.33	120	0.141	0.00
DKLONG17W	0.09	9.19	0.827	9.655	7.990	77.1	0.00	0.00
Total			990.966		-492.4	41744.7		316.99

$E = \Sigma M / \Sigma A = -0.497$ (m)

Σ(I +self I)=42061.7

$\Sigma A \times E^2$= – 244.7

I* = 41817.0

4) Distance from true neutral axis to top and bottom

Y_{DK} = 9.8 + 0.497 = 10.297 m

Y_{BM} = 8.0 – 0.497 = 7.503 m

5) Comparison with required section modulus

For deck 2I*/Y_{DK} = 83634.0/10.297*1000 = 8122245.2 (cm^3) 120.344(%)

For bottom 2I*/Y_{BM} = 83634.0/ 7.503*1000 = 11146591.2 (cm^3) 165.154(%)

Fig. 20.9 An example of section modulus calculation

20.4.2 Rational Arrangement of Longitudinal Strength Members

At the beginning of this chapter, it is explained that most damages caused by the lack of longitudinal strength happen on older ships, and this can be avoided by anti-corrosion methods, selection of materials, and good welding. However, this damage can happen even on not so old ships.

The up-sizing of ships during the period 1960–1980 was achieved by means of three important technical innovations; rationalization of longitudinal strength, wider

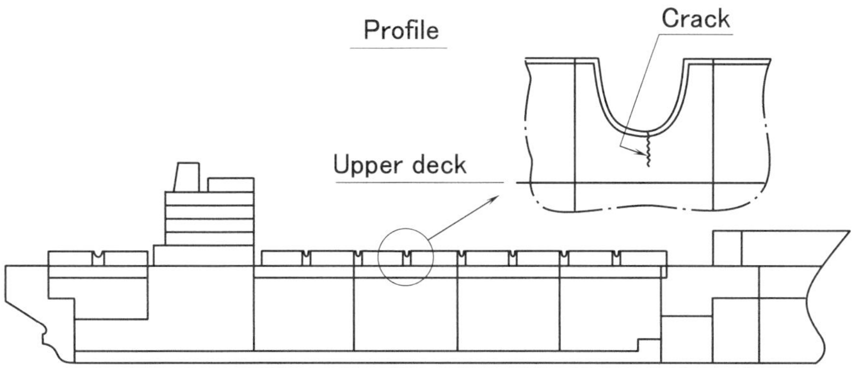

Fig. 20.10 Cracking of hatch side coaming of container vessel

spacing of frames and transverses, and increased tank length. The first of these was achieved by huge and energy requiring research on the generation of sea waves and ship motion in waves. By this research the hull girder bending moment could be estimated more accurately than the traditional calculation using $L/20$ wave height. As shown in Fig. 1.5 in Part I the required section modulus has decreased since 1950 down to 75% of the maximum [5].

The longitudinal strength design has become rational in this way, but this conversely means that the actual longitudinal bending stress has increased than before. In addition, wider application of higher strength steel has caused further increase in the stress level. Due to this increase in the stress level, it is becoming increasingly important to pay attention to the detail design with less stress concentration to prevent fatigue damage due to the longitudinal bending stress. An example of such damage is the crack on the continuous hatch side coaming of a container ship as shown in Fig. 20.10.

Before the rationalization of the longitudinal strength design, longitudinal strength, being the most important component of the ship strength, had the greatest margin of safety. On the other hand, small differences in longitudinal strength greatly affect the hull steel weight and hence ship's cost, which will be discussed in the next section. Therefore, every effort to reduce the scantling of the longitudinal members was done with reasonable excuse. Examples are shown in Fig. 20.11; one is a trunk section, and the other is a large camber section. In the calculation of the section modulus of the midship section, the sectional moment of inertia which includes all longitudinal members is divided by the distance from the neutral axis to the upper deck at the side. In this method it is beneficial to arrange longitudinal members as far from the neutral axis as possible.

Even in the case where the section modulus is calculated by dividing the sectional moment of inertia by the distance from the neutral axis to the farthest members, continuous members above upper deck, such as continuous hatch side coamings and bulwarks are effective in contributing to longitudinal strength as well as in saving

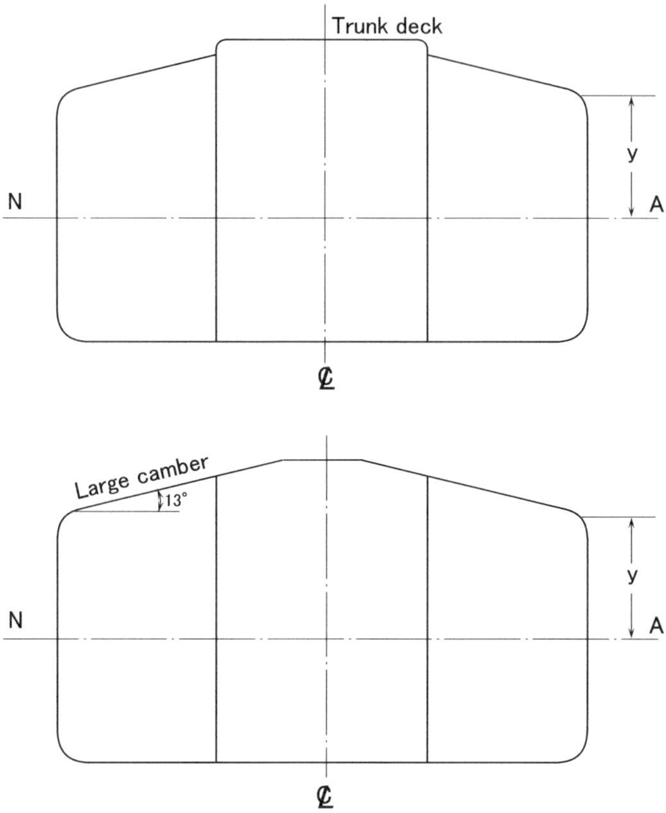

Fig. 20.11 Midship section of trunk deck type and large camber type

hull steel. An example is reported in which a continuous heavy bulwark reduced the longitudinal bending stress by 7–11% [121].

In the case of ordinary camber, the center line strake lies farther than other strakes from the neutral axis, and a thicker center line strake is preferable to keep the thickness of other strakes to a minimum.

It is current practice not to apply any reinforcement around the upper deck openings, provided the shape of the opening is ellipse with the longer axis in the longitudinal direction, more than twice the length of the shorter axis, and the length of shorter axis is not so big. However, if reinforcement is necessary around the upper deck opening, the reinforcement can be made minimum by keeping the strake which has the opening as thin as possible and increasing the thickness of the center line strake.

In general, by adding some sectional area to the midship section, the section modulus of the midship section can be increased, but when the addition is made at a small distance below the neutral axis, the section modulus at the deck decreases on the contrary. Conversely, by reducing material just below the neutral axis, an increase in the section modulus can be obtained. This is theoretically explained in Sect. 8.3.2 of Part II, and can be explained by the fact that putting material in that region brings bigger neutral axis movement than the increase in the sectional moment of inertia.

In an actual design the above phenomena can be put into effect by reducing the total sectional area of the plate and longitudinal stiffener below the neutral axis with narrower longitudinal stiffener spacing.

20.4.3 Longitudinal Member Arrangement and Hull Steel Weight

In Fig. 20.12 the weight ratio of the longitudinal strength members relative to the total hull steel weight is shown. The ratio is about 70–80% even though the value fluctuates depending on the ship's size and the type of ship. In Fig. 20.13 fluctuation in the weight of the longitudinal strength members for 1% change in the section modulus of the hull girder is shown. The absolute value of the increase or decrease of the weight of the longitudinal strength members for 1% change of the section modulus of the hull girder is considerable. Whereas the same 1% change of the section modulus of the hull girder will change the longitudinal bending stress by 1%, for example 200 MPa of the longitudinal bending stress comes to 202 or 198 MPa, which seems to be negligible.

To design a midship section which has exactly 100% of the required section modulus, adjusting the scantlings of many longitudinal members is a very important

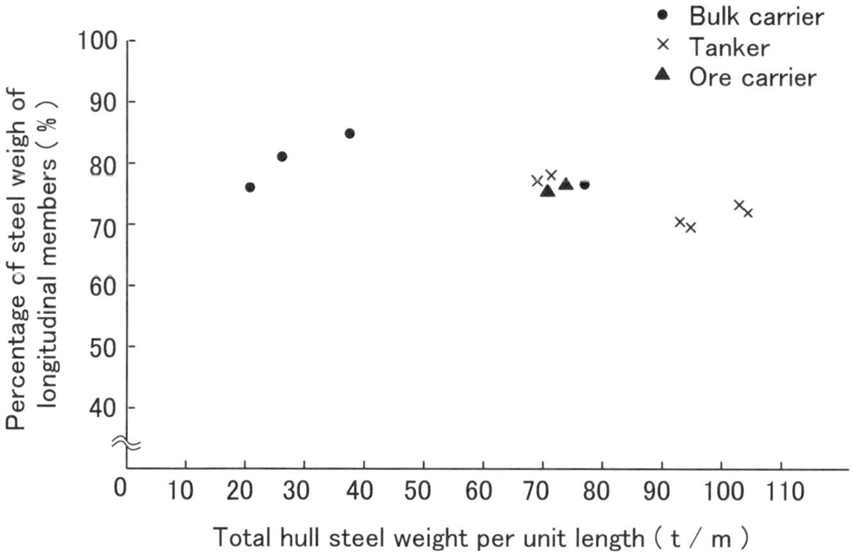

Fig. 20.12 Hull steel weight of longitudinal members

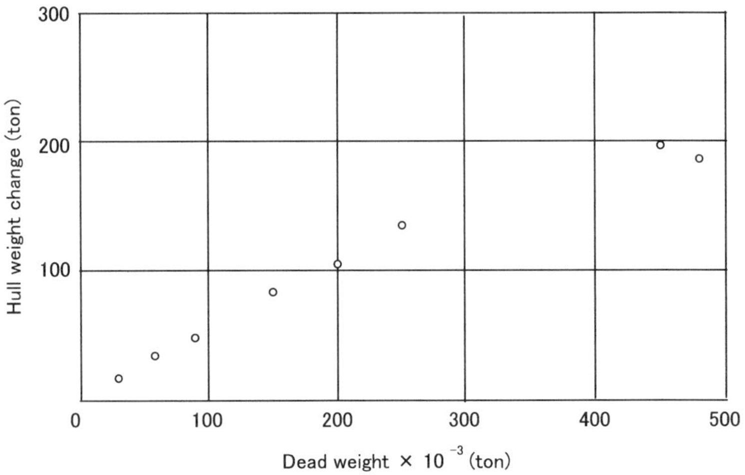

Fig. 20.13 Hul steel weight change due to 1% longitudinal I/y change

job at the initial design stage. The section modulus is obtained by dividing the moment of inertia by the distance from the neutral axis to the farthest member. It has been impossible to obtain exactly 100% of the required section modulus by adjusting the scantlings of a member in one action. Because the adjustment of the scantling of a member brings not only a change of moment of inertia but also a change of the position of the neutral axis. Thus it has to be done by a trial and error method. Practically it is done as follows:

The midship section modulus Z_0 is calculated from the longitudinal strength members using scantlings which are decided from a local strength viewpoint. Expressing the required section modulus by Z_r, in the case where $Z_0 \geq Z_r$ the scantling of these members can be applied as they are, but in the case where $Z_0 < Z_r$ the scantling of some member should be increased so as to have the condition of $Z_0 = Z_r$. In this case to increase the upper deck plate thickness which lies farthest from the neutral axis is simplest and effective.

The above story can be understood easily but actually some difficulties exist. For example, if the calculation result shows that the deck plate thickness should be 20.32 mm to have just the required section modulus, we cannot have the plate of 20.32 mm. We can have 20 mm or 20.5 mm thickness plate actually. The former gives insufficient section modulus and the latter gives excess section modulus. In such a case the 20.5 mm thickness plate will be used, giving 0.18 mm in excess thickness.

One solution to this problem is to apply a wider longitudinal spacing keeping the thickness of the deck plate as 20.5 mm. By applying wider longitudinal spacing, the number of longitudinal stiffeners can be reduced, and accordingly the sectional areas of the longitudinal stiffeners may be reduced. The thickness of the upper deck plate can be actually increased step by step by 0.5 mm, causing then the section modulus to also increases in a step function. By adjusting the number and the size of the deck longitudinal stiffeners, the effect of the step function may be relieved [122].

Another solution is to apply different thickness plates to the upper deck. In this case to apply a thicker plate to the center line strake keeping the same thickness for other plates, because the center line strake is further from the neutral axis than the other plates.

Comparing the above two solutions, the latter is more effective than the former from hull steel weight viewpoint and changing the spacing of the longitudinal stiffeners will cause many other modifications. However, in the case where different thickness plates for deck plating cannot be accepted from material handling, the former solution will be applied.

As shown in Fig. 20.13 a very small difference in the midship section modulus causes a big hull steel weight difference, and hence ship's cost difference. The section modulus should be just 100% as far as possible. A designer should take one of the above two ways with his own clear design philosophy.

20.5 Application of High Tensile Steel

After 1980's, thanks to the development of steel mill technology TMCP (Thermo-Mechanical Controlled Process), high tensile steel with good weldability has been widely applied in hull structure. The usage percent of high tensile steel is now 70–80% of total hull steel weight. High tensile steel has been applied in hull structures in the following two cases:

a. Cost reduction can be achieved.
b. To reduce plate thickness when the plate becomes too thick if the mild steel is employed. The thickness of 38 mm used to be a criterion, but recently 50 mm or more is acceptable thanks to the improved material and welding technology.

The cost reduction is evaluated in the following way. Hereafter HT means high tensile steel and MS means mild steel. And each item is described as follows.

Total weight of HT applied	W	ton
Reduction in weight due to HT application	ΔW	ton
Price of HT	P_{HT}	Yen/ton
Manufacturing cost of HT	L_{HT}	Yen/ton
Price of MS	P_{MS}	Yen/ton
Manufacturing cost of MS	L_{MS}	Yen/ton
Ships price per dead weight	P	Yen/ton

Using the above symbols, the following Eq. (20.5.1) means that the HT is more beneficial than MS.

$$W\{(P_{HT} + L_{HT}) - (P_{MS} + L_{MS})\} - \Delta W(P_{MS} + L_{MS}) - \Delta W \cdot P < 0 \quad (20.5.1)$$

In the case of TMCP steel L_{HT} is nearly equal to L_{MS}, then Eq. (20.5.1) can be simplified as follows.

$$W(P_{HT} - P_{MS}) - \Delta W(P_{MS} + L_{MS} + P) < 0 \tag{20.5.2}$$

$$\therefore \frac{\Delta W}{W} > \frac{P_{HT} - P_{MS}}{P_{MS} + L_{MS} + P} \tag{20.5.3}$$

$\Delta W/W$ depends on the yield strength of high tensile steel and specific design of each ship. In Fig. 20.14 the actual examples of $\Delta W/W$, where 315 MPahigh tensile steel is applied to high stress parts of longitudinal strength members, are shown. In the case where the value of the right side of Eq. (20.5.3), which depends on the steel price, manufacturing cost and ship's price, is smaller than $\Delta W/W$, high tensile steel can be applied economically.

For Ship A and Ship B in Fig. 20.14, $\Delta W/W$ is smaller than those of other ships, because these ships have greater depth than ordinary ships, and have excess section modulus for longitudinal strength with longitudinal members decided by local strength. In these ships the application of high tensile steel is not economical.

20.6 Stress Concentration on Longitudinal Strength Members

The crack shown in Fig. 20.10 is caused by cyclic high stress at a region of stress concentration. Although there are many discontinuous points in the hull structure and stress concentrations cannot be avoided, effort should be made to avoid high stress concentration.

The actual stress caused by the stress concentration is described by the following Eq. (20.6.1) [123].

$$S_a = f(S_n, K_d, K_w) \tag{20.6.1}$$

where

S_a: actual stress.

S_n: nominal stress due to scantling of member.

K_d: stress concentration factor for member's shape.

K_w: stress concentration factor for workmanship such as shape of weld bead, welding defects, misalignment of members, etc.

To avoid cracks it is important to have low S_a, which can be obtained by reducing S_n, K_d and K_w.

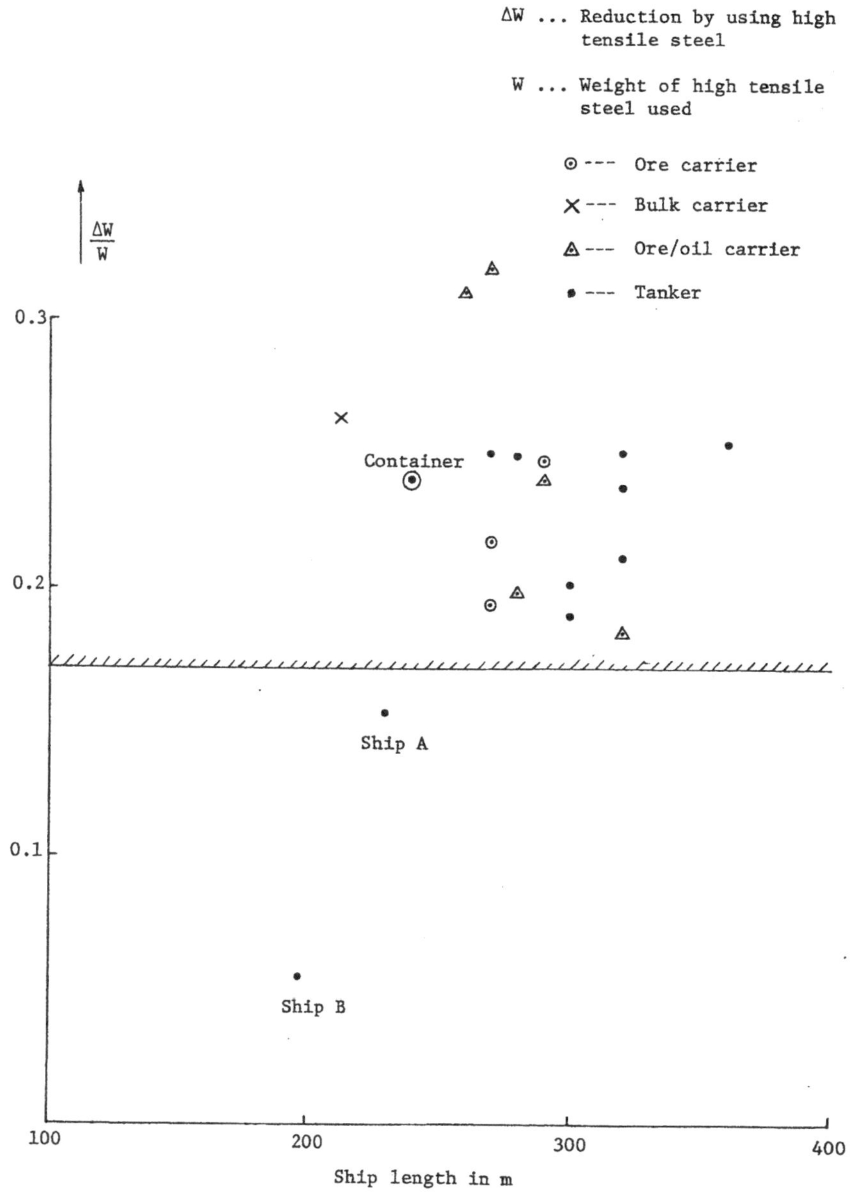

Fig. 20.14 Hull steel weight reduction due to use of high tensile steel

To reduce the nominal stress S_n, increasing the scantlings of the members is necessary. For a given tension or bending moment, a bigger sectional area or section modulus can result in a smaller nominal stress S_n. However, an increase in the scantlings means an increase in hull steel weight.

The stress concentration factor for workmanship K_w depends on the fabrication process and the welding techniques which are peculiar to the shipyard. The stress concentration factor for a member's shape K_d can be reduced by improving the member's shape without increasing the hull steel weight. For the improvement of the member's shape, the photo-elasticity technique had been applied in the past, and nowadays the FEM calculation is applied effectively.

Equation (20.6.1) indicates how important the design of the shapes of the structural members is in the detail design of the hull structure.

To avoid the crack shown in Fig. 20.10, an investigation of the shape of the hatch side coaming in the vicinity of the hatch end coamings was carried out [124]. In this coaming cut outs are necessary to pass pipes and to provide a passage. Accordingly, a continuous structure is not available. For several shapes of the cut-out, the stress concentration factor is calculated, then the best one was selected.

In Fig. 20.15 the FEM models and results are shown. The type "O" is the original shape which suffered from a crack. The types "A" and "D" have the same sill height as the type "O", and the types "B" and "E" have smaller sill heights than the type "O". The type "C" has highest sill height.

The calculations show the following facts.

1. Type "E" has a higher stress concentration because of smaller sill height.
2. Type "B" which has a smaller sill height than type "E" has a slight improvement because of smaller curvature.
3. Type "A" which has the original sill height and smaller curvature has stress concentration factor 7.5% less than the original type "O".
4. Type "C" shows a 15% improvement thanks to the higher sill height and smaller curvature.

The above investigation shows that a higher sill height and smaller curvature are effective in reducing the stress concentration in this case.

The crack shown in Fig. 20.10 happened only on the deck girder between three rows of hatches and not on the side coaming above the longitudinal bulkhead which has the same cut out as on the deck girder between the three rows of hatches. From this fact it can be said that the reduction of a girder's sectional area due to the cut-out can be a cause of cracking and the deck girders between the three rows of hatches effectively support the longitudinal bending moment.

Nowadays, due to the rapid up-sizing of container ships, longitudinal strength is becoming more and more critical, and higher strength steels up to the specified yield stress of 395 ~ 460 MPa are employed. Thus, it has become more important to ensure sufficient section modulus of the hull girder and to avoid stress concentration. As a result, it has become common practice to adopt longitudinally continuous hatch side coaming with extra thick plates for large container ships.

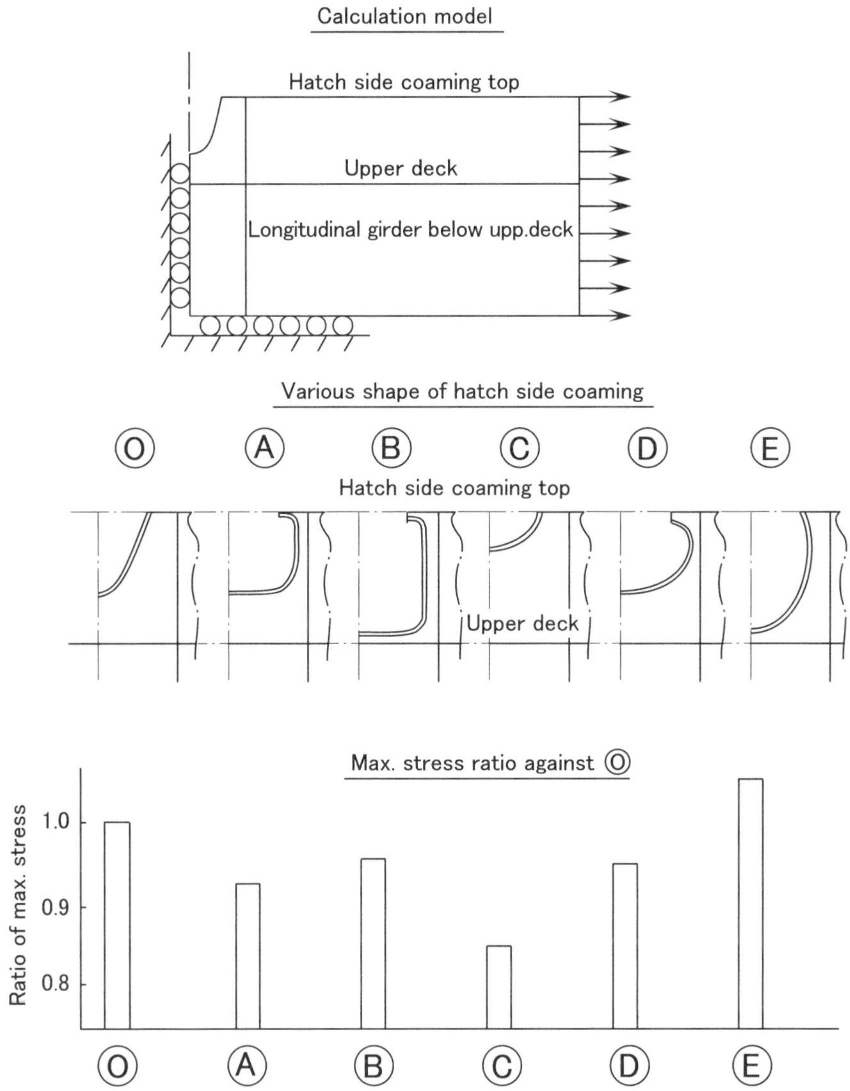

Fig. 20.15 Stress concentration due to hatch side coaming

20.7 Damage at the End of Long Attachment on Hull Girder

When a long longitudinally continuous member is attached on the hull girder, the hull girder bending stress flows into the attached member causing significant separating or pushing forces between the member and the hull girder at the ends of the attached member. This becomes even more hazardous with the stress concentration due to

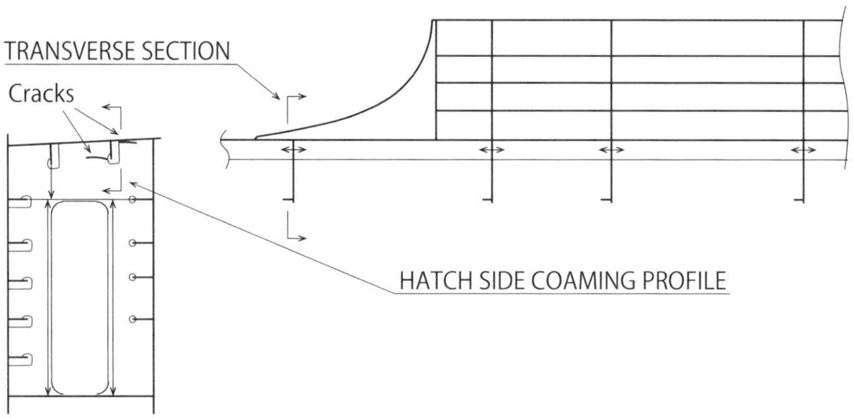

Fig. 20.16 Slot damage beneath the end bracket of continuous hatch side coaming

the abrupt change of the section at the ends of the attached member. Figure 20.16 shows an example of crack damage at a slot cutout for a deck longitudinal stiffener observed in the underdeck passage of a large container ship. There are hundreds of similar transverse webs and slot cutouts throughout the passage way, but the damage occurred only at this one frame. At first, it was difficult to identify the root cause of this damage, but finally it was found that the damaged transverse web was located just beneath the end bracket of a long continuous hatch side coaming. As a result, a large vertical force acted on the deck longitudinal stiffener from the end bracket causing large reaction force supported at the intersection between the deck longitudinal stiffener and the transverse web. The designer paid special attention to the sophisticated design of the end bracket with a long and soft shape and full penetration welding onto the deck plating. However, it seems that he did not care much for the structure beneath the end bracket. The separating force at the end of a long superstructure explained in Chap. 31 is caused by the same mechanism.

Similar phenomena occur when an additional longitudinal member is attached on the lower side of the upper deck plating. When deep girders and small deck longitudinals are longitudinally connected, an additional bending moment is exerted at the connection. Figure 20.17 shows an example of such an arrangement, where a crack was found at the toe of a bracket at the end of a partial deck girder under a crane post in a single hull tanker.

This structure consists of deck longitudinals, additional girders, and brackets. The additional girder is provided to support the load of the on-deck crane. In this case, it is usual to arrange scarfing brackets at both the ends of the additional girder to avoid an abrupt change of rigidity. However, the crack was found at the toe of these brackets.

The mechanism of this failure was studied and a new structural arrangement was proposed [125]. We assumed that this crack was caused by the longitudinal bending

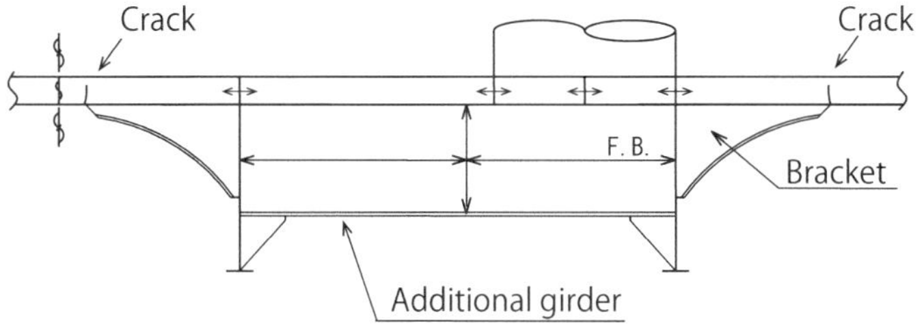

Fig. 20.17 Partial deck girder (with brackets)

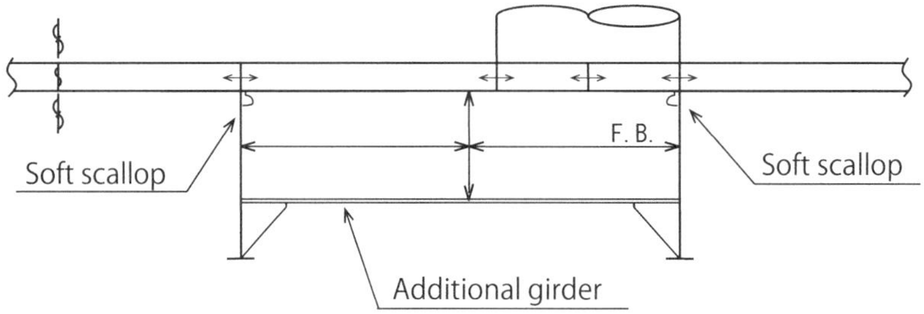

Fig. 20.18 Partial deck girder (without brackets)

moment of the hull girder, and we carried out a comparison of the original structure and the modified structure using finite element analysis. The original structure, which has brackets, is shown in Fig. 20.17, and the modified structure, which has no brackets, is shown in Fig. 20.18. The axial force to induce a nominal stress of 98 N/mm^2 is loaded. The results of this analysis are shown in Figs. 20.19a and b, respectively.

From the observation of Fig. 20.19a, it is found that:

1. The deck longitudinal is forced to deflect downward by 2.7 mm at the toe end of the bracket, due to rotation of the bracket caused by a shift in the position of the neutral axis between the deck longitudinal and the partial deck girder.
2. The maximum boundary stress on the face plate of the deck longitudinal is 190 N/mm^2, and the stress magnitude varies gradually along the length of the deck longitudinal. This fact indicates that the stress concentration at the toe of the bracket is not big thanks to the very soft shape of the bracket, but that the large stress is mainly caused by the additional bending moment exerted on the deck longitudinal due to the rotation of the bracket.

Based on this study, it was concluded that the cause of the damage was the additional bending moment exerted on the deck longitudinal due to the rotation of the bracket caused by the change in the position of the neutral axis between the deck longitudinal and the partial deck girder. Modification of the bracket to a softer shape cannot reduce the stress much, because the stress concentration at the toe end is already sufficiently small, as described above. To avoid the excessive additional

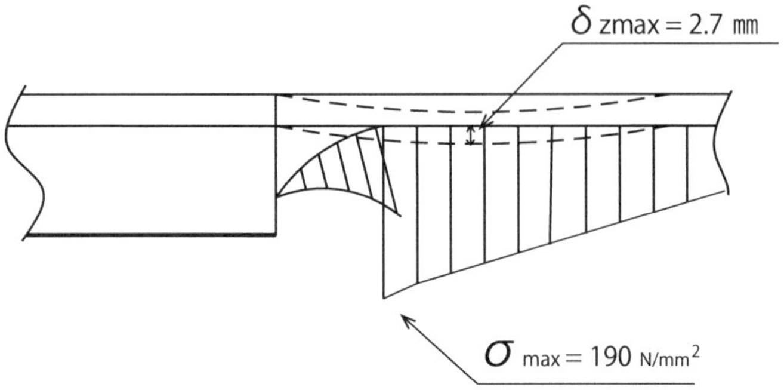

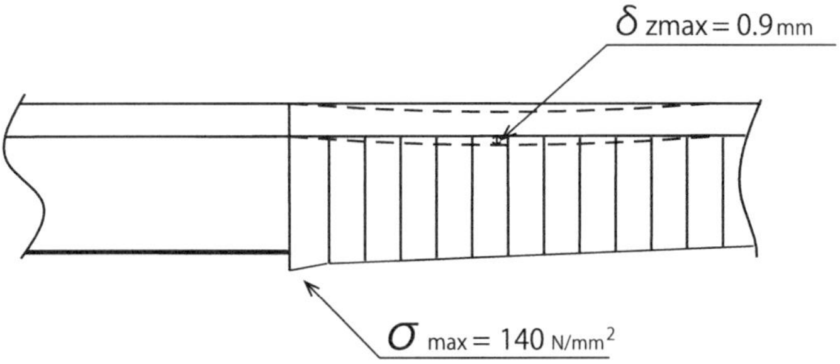

Fig. 20.19 Stress distribution along longitudinal stiffener

bending stress by the rotation of the bracket, the following two countermeasures are considered to be effective:

1. Decrease the additional bending moment acting on the deck longitudinal, by adding slightly less-deep girders at the fore and aft locations of the existing deep girder.
2. To create a structural arrangement to make the additional bending moment not to flow into the deck longitudinal.

The former countermeasure should actually work, but this solution requires much cost and weight. Then we studied the latter one and reached the bracket-less arrangement. From Fig. 20.19b, we can observe that:

1. By dispensing with the brackets, the downward deflection of the deck longitudinal was much improved (from 2.7 mm to 0.9 mm).
2. Maximum stress on the face plate of the deck longitudinal was reduced to 140 N/mm^2. A local stress concentration is also avoided by making the corner scallop of the partial deck girder softly rounded.

From this study, we can conclude that by dispensing with the brackets, the part of the downward force on the deck longitudinal due to the additional bending moment

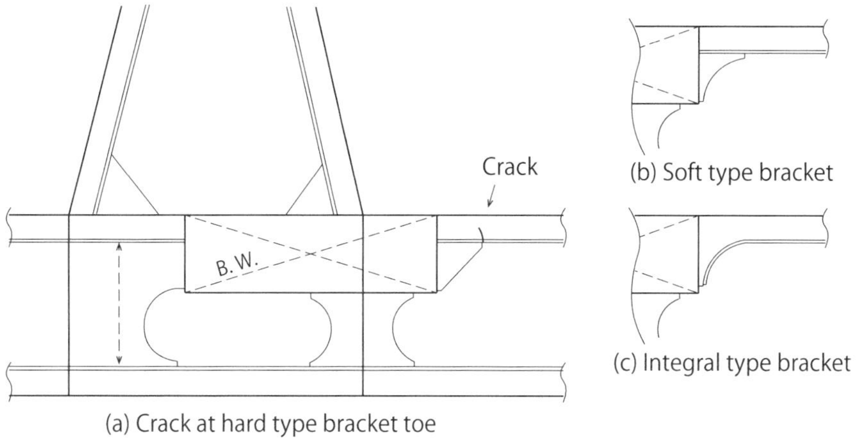

Fig. 20.20 Crack damage at the connection between inner bottom longitudinal and bilge well

was moved to the position of the transverse web, thus additional bending stress was considerably reduced because the acting point of the force is the same as the supported point.

This structural arrangement seems very strange at the beginning because common sense of structural designers is that an abrupt change of rigidity is to be avoided as far as practicable. Honestly speaking, some Classification Society's surveyors and ship owners commented about this bracket-less structure to add a bracket. However, by investigation about the stress flow, we can find some cases where adding reinforcing members sometimes deteriorates strength.

Similar damages caused by hull girder bending can occur in the bottom part of the ship as well. One typical example is the crack damage in way of the connection between inner bottom longitudinals and a bilge well as shown in Fig. 20.20. If this crack occurred in a relatively old ship, replacement of the bracket with a softer one as shown in Fig. 20.20b would suffice, but if it occurred much earlier after delivery, it would be necessary to change the stiffener end with an integral type bracket with a radiused face plate as shown in Fig. 20.20c.

20.8 Longitudinal Bending Stress in Fore & Aft Parts of Ship

In designing longitudinal strength members, firstly the scantlings of the members at midship, which are to be maintained in the vicinity of 0.4L midship, are decided so as to maintain the necessary section modulus and resist the midship bending moment, then the midship scantlings are tapered down to the scantlings at 0.1L from the ship's ends.

The above practice has been applied for a long time without any trouble. However the situation changed with the appearance of high speed ships and energy saving ships which have different hull forms from traditional ships. Damage was caused by the bending moment outside of the midship part. An example is that of a 10,000 DWT high speed cargo ship bent convex upwards at $(1/4)L$ from F.P. She suffered damage off South Africa in heavy weather. The damage looked like a folding jack knife and people called it "jack knife damage". This kind of damage has put into question the above traditional design practice, that is, the midship scantlings are to be tapered down to the scantlings at $0.1L$ ship's ends; this practice has been revised so as to keep the necessary section modulus at each section [126].

Actually a high speed container ship has a very fine bow form (refer to Fig. 5.3.1 in Part I), and it is necessary for the fore part of the $0.4L$ midship to have thicker plate than the midship part in order to have enough section modulus. In the case of a high speed ship with large bow flare, the slamming impact causes a big bending moment at the forward part in the sagging condition, which was proved by recent research using the strip method with consideration of elastic deformation of the hull structure [87–89].

The stern shape of the energy saving ships recently designed is quite different from the traditional ship. As shown in Fig. 5.6 in Part I, the so called "open bulbous stern" is so fine that the boss part has little connection to the main hull. In such a stern form, to keep the necessary section modulus requires heavy shell plates to be applied.

In a container ship, the semi-aft engine room arrangement used to be applied with many containers loaded on the upper deck aft of the engine room which caused a big hogging moment to occur in the vicinity of the open bulbous stern part. Attention should be paid to the buckling strength of the bottom shell plate in this area.

20.9 Hull Steel Weight Reduction Based on Ultimate Strength Criterion

In Fig. 20.12, it is indicated that the weight of the longitudinal strength members is about 70–80% of the total hull steel weight. Therefore, the required section modulus by the rules almost governs the hull steel weight. On the other hand, it is pointed out that the ultimate strength of existing hull girders is about 70% of the fully plastic moment of the section [127]. This means that in general hull girders collapse by buckling before a plastic hinge is formed at the section where the maximum bending moment acts upon.

From this, it can be said that the non-buckling design with existing scantling will bring greater ultimate strength by 43% (100/70 = 1.43). This means that if the ultimate strength can be employed as the hull girder strength criterion instead of the elastic design of the section modulus, 30% sectional area reduction of the longitudinal members can be expected keeping the same strength. In this case, the total hull steel weight can be reduced by 21–24% (70–80% $\times$ 0.3 = 21–24%).

Actual hull girder damage is governed by buckling in the compression side and cracks in the tension side. Then, improvement of buckling strength and stress concentration may directly contribute to the hull girder safety, which may ultimately lead to the optimum design with less hull steel weight.

Chapter 21
Transverse Strength of Ship

The longitudinal strength of a hull girder explained in Chap. 20 is the most important aspect of the ship strength, and its deficiency may cause the total loss of the ship. Accordingly, it has been designed so carefully that very little trouble has been reported. In the design of the longitudinal strength, the hull structure can be simply assumed to be a hull girder, and a sophisticated total system of the longitudinal strength calculation was developed. And now reliable design methods for the longitudinal strength are available.

On the other hand, the transverse strength of a ship is also important. Deficiency in the transverse strength will cause fatal damage as shown in Fig. 21.1. Both cases shown are damage to a tanker. One is an indent of a side shell caused by the buckling of a cross tie, and the other is bulging of the side shell construction caused by the loss of the connection between the shell plate and the transverse ring. Even in cases of big deformation, no crack in the shell plate was detected, except a small crack of a so called "flying bird" happened in the latter case.

The transverse strength members are statically indeterminate, being connected and affecting each other; therefore, it is not easy like in the case of the longitudinal strength members to find the response to the applied load. This is a reason why a total system for transverse strength is more difficult to apply.

The design of the hull structure arrangement explained in Chap. 19 very much affects the transverse strength. The shear deformation of a wing tank structure of a tanker causes secondary stress due to forced displacement on the structural members inside. The criteria for designing the hull structure arrangement in Chap. 19 are to avoid an excess secondary stress on the transverse strength members.

In this Chapter the allowable stress in the transverse members and some other considerations regarding their design will be explained.

Y. Okumoto et al., *Design of Ship Hull Structures*,
https://doi.org/10.1007/978-3-032-12500-2_21

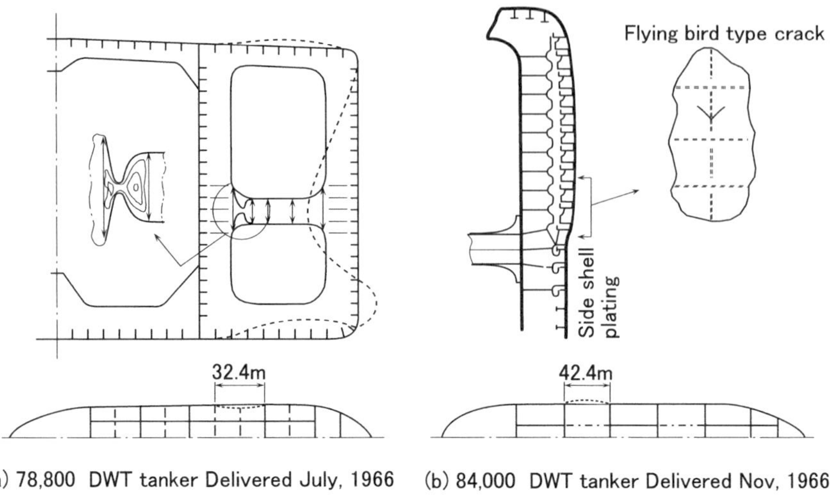

Fig. 21.1 Examples of failure due to lack of transverse strength

21.1 Allowable Stress for Transverse Strength

The allowable stress for transverse strength is given by each classification society together with a stress calculation method. As explained in Sect. 3.1.2, Part I, the load estimation, the response calculation, and the allowable limit are always to be connected together. For example, an allowable stress of a classification society can be applied only on the response calculated by the method specified by the classification society [128].

Hereunder the following example of load estimation, the response calculation and the allowable limit is introduced which was established in the 1970's by the authors when the increase in ship-size was so fast that the classification society's rules were not so effective.

The load is divided into two kinds, the long term and the short term. The former is for full loaded and ballast conditions in still water, and the latter is full loaded and ballast conditions in waves and the tank test condition.

The response is calculated by means of a three-dimensional frame model and a two-dimensional FEM model. The mesh division is shown in Fig. 21.2b which is better for the determination of the stress distribution along the face plate than that in Fig. 21.2a eliminating the irregularity at the mesh A. Firstly the cargo tank part of the tanker is modeled into a three-dimensional frame work and the shear deformation of the wing tanks is calculated with a draft corresponding to the trough of the wave. The model is supported at both ends of the side shell plates and longitudinal bulkheads. The wave height is that given by Nippon Kaiji Kyokai. Secondly a two-dimensional FEM calculation is carried out at the transverse ring where the shear deformation of the wing tank is the maximum. The draft at the subject transverse ring position is applied to the two-dimensional FEM.

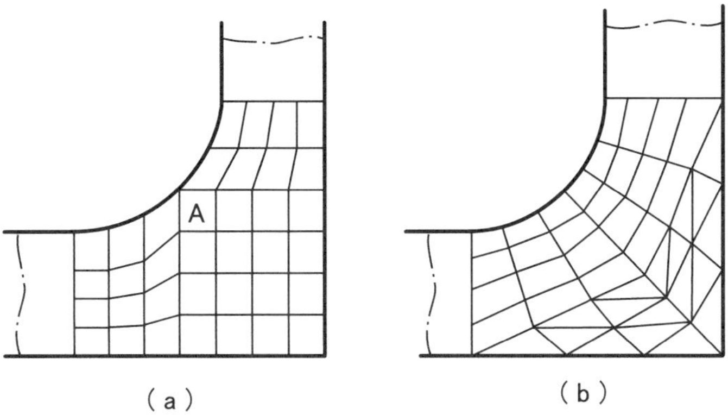

Fig. 21.2 Examples of FEM mesh subdivision

The six tankers shown in Table 21.1 are analyzed by the above method. These have a transverse main and no center girder construction system. Referring to the analyzed data, the allowable stress is decided as shown in Table 21.2 for each part shown in Fig. 21.3. The allowable stress is the maximum value analyzed on a ship in service without any structural trouble.

Table 21.1 Vessels in service analyzed for transverse strength

Vessel	Deadweight (t)	Principal dimensions (m)				Delivered year
		L	B	D	d	
A	80, 352	230	33.0	21.7	15.0	1965
B	150, 000	290	47.5	24.0	16.0	1966
C	372, 400	330	54.5	35.0	27.0	1971
D	210, 000	300	50.0	25.5	19.0	1971
E	251, 978	320	54.5	26.0	19.1	1971
F	477, 000	360	62.0	36.0	28.0	1973

Table 21.2 Allowable stress of transverse ring

Unit in MPa

		Long term	Short term	Location indicated by Fig.3.1.2
Web (von Mises stress)	High stress part	186	225	C,F,G,H,I,L,M,N,O
	Corner part	147	186	①～⑩,A,B,D,E,J,K,P
	Uniform part	118	147	1～12
Face (normal stress)	High stress part	186	225	A,B,C
	Corner part	127	157	D,E,F,G,H,I,J,K,L,M,N,O,P
	Uniform part	118	137	1～12

Fig. 21.3 Category of transverse strength member due to location

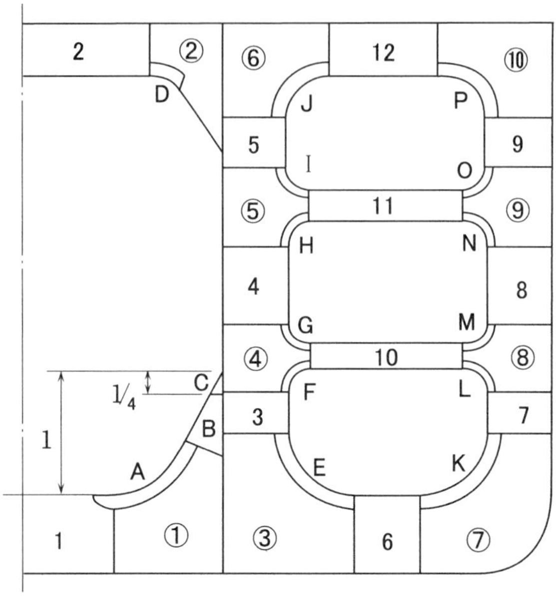

High stress zones as shown in Table 21.2 are to be designed carefully, because there are high or low stress concentrations in this zone. In Fig. 21.4 examples of cracks are also shown.

In Table 21.2 the allowable stress in the vicinity of the parallel part is lower than that of the corner part which seems unreasonable because the parallel part has no discontinuity, no stress concentration, and no reduction of effective breadth of the face plate. This is caused by the analytical method, namely the allowable stress was

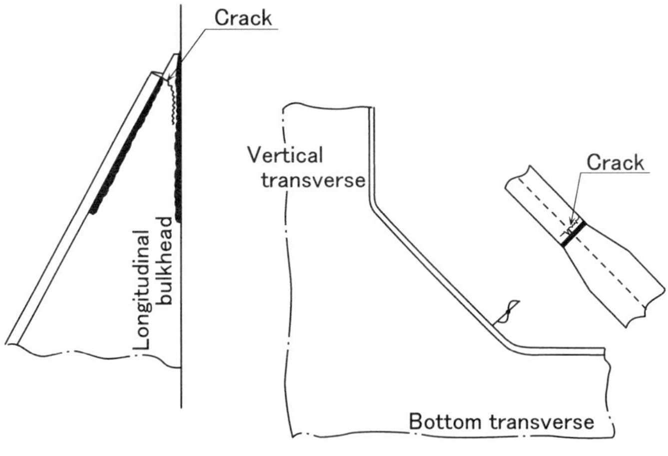

Fig. 21.4 Examples of crack in high stressed part

based on the data of previous ships in service and not theoretically. Theoretically the allowable stress in the vicinity of the parallel part can be higher than the corner part. In this matter, however, one important point should be kept in mind; the high stress in the vicinity of the parallel part of a transverse ring will cause a big deflection which causes additional stress in the secondary members supported by the transverse ring. The maximum deflection in this analysis occurred on the longitudinal bulkhead during tank testing with a fully loaded center tank and its value was about 1/1500 of the ship's depth. A deflection bigger than this value may cause problems for the secondary supporting members.

Nowadays, classification societies' rules specify allowable stress for transverse strength. For example, IACS CSR stipulates that the von Mises stress should not exceed the specified minimum yield stress (235 MPa for mild steel) in general for static plus dynamic load cases in combination with the net scantling approach as explained in Sect. 3.3.2(3), Part I. It should be emphasized again that this allowable stress is closely linked to the specified loading and analysis procedure including mesh size, model extent, boundary conditions, etc., and cannot be discussed in isolation.

21.2 Long Taper and Snake Head

In Sect. 21.1 it is mentioned that the allowable stress at the bracket toe of the bottom transverse is 225 MPa as a high stress zone, and a low stress concentration shape is necessary. For this purpose, a long taper and "snake head" are proposed [124].

In Fig. 21.5 a bottom transverse in the center tank of a tanker is shown. The bottom transverse has brackets at both ends where high shear stress and a large bending moment occur. In the face plate which lies farthest from the neutral axis high bending stress is generated. The face plate has a taper at its end at usually 15° on one side to allow for smooth stress flow at the end of the bracket. On this point an idea came to the mind of one of the authors, that is, a taper with a smaller angle than 15° will have a smoother stress flow, and it will require less material also.

In Fig. 21.5 a comparison is made of the stress distribution in the case of an ordinary taper and a long taper. The stress is calculated with the condition that both ends of the bottom transverse are fixed and have uniform bottom pressure. In the parallel part, the long tapered face plate shows slightly higher stress due to the slightly weaker fixation at the ends, but the difference is negligible. In contrast, at the corner of 1,500R, the ordinary tapered face plate shows a little higher stress, also due to the stronger fixation at the ends. Within the long tapered region where the sectional area of the face plate reduces, the long tapered face plate shows a higher stress as a matter of course, but finally at the extreme ends, the stress level of the long tapered face plate is about 1/2 of that of the ordinary tapered face plate, demonstrating a smooth stress flow in the long tapered face plate. The weight saving by the long taper is estimated to be about 4 tons in the case of a 480,000 DWT tanker which has 500 × 35mm face plates. The long tapered face plate is beneficial also with respect to the scrap ratio of the steel plate as explained in Sect. 9.2, Part II.

Fig. 21.5 Long tapered face plate of bottom transverse

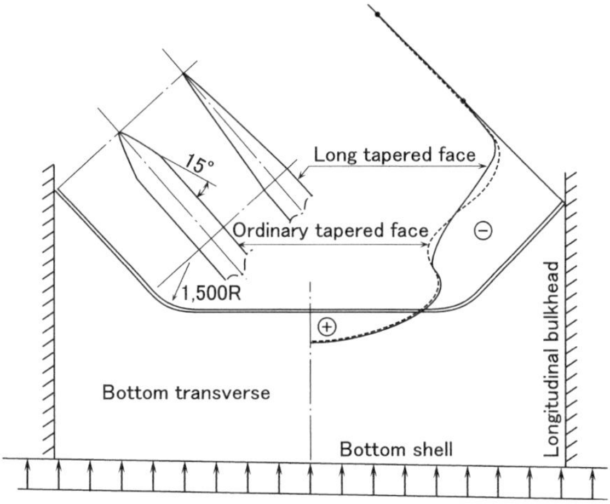

Even with the long tapered face plate, a high stress remains at the end of the bracket. To reduce this high stress a snake head bracket toe has been found effective; this was proposed and named by Dr. Ken-ichi Hattori. As shown in Fig. 21.6 the stress at the "snake head" is about 80% that of a circular bracket end, which used to be adopted.

Fig. 21.6 Maximum stress ratio against shape of bracket toe

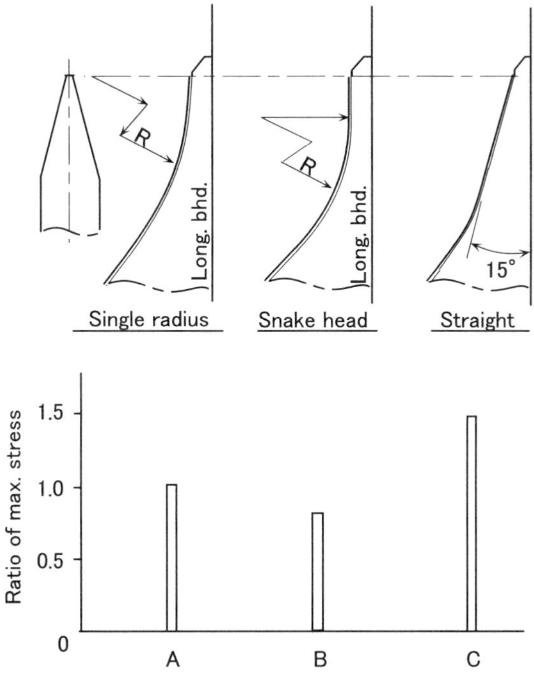

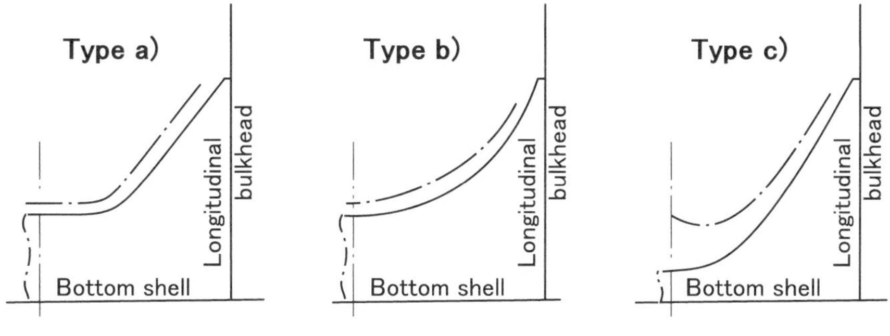

Fig. 21.7 Face plate stress due to shape of bottom transverse in center tank

By combining a long taper and a snake head, the crack as shown in Fig. 21.4 can be avoided with an allowable stress of 186 MPa for long term loads and 225 MPa for short term load.

21.3 Shape of Bottom Transverse in Center Tank

In Sect. 21.2 a long taper and snake head is discussed, keeping the section of the bottom transverse uniform except for the bracket part. In this section the stress distribution in the face plate of a bottom transverse with a non-uniform section is explained [129].

As shown in Fig. 21.7 three shapes were investigated (a) an ordinary shape (b) a large circle in way of bracket and (c) no uniform section with small depth at the center line. Type (c) is reasonable from a shearing force viewpoint which is zero at the center line and beneficial in the arrangement of low cargo pipes on the bottom transverse. The stress distribution in the face plate at a light draft condition with a fully loaded center tank is shown also in Fig. 21.7 from which it can be said that the stress distribution is quite similar for these three types except at the center line of Type (c).

21.4 Shape of Bottom Transverse in Wing Tank

At the connecting bracket of a bottom transverse in a wing tank and the vertical web on a longitudinal bulkhead, high bending stress in the face plate is often generated and cracks occur as shown in Fig. 21.4. The stress distribution was investigated for four types of bracket as shown in Fig. 21.8. Type (a) and (c) are circular types and Type (c) has a larger radius, and Type (b) and (d) are combination of circles and straight line, which are called "tray type", and Type (d) has a larger area [129].

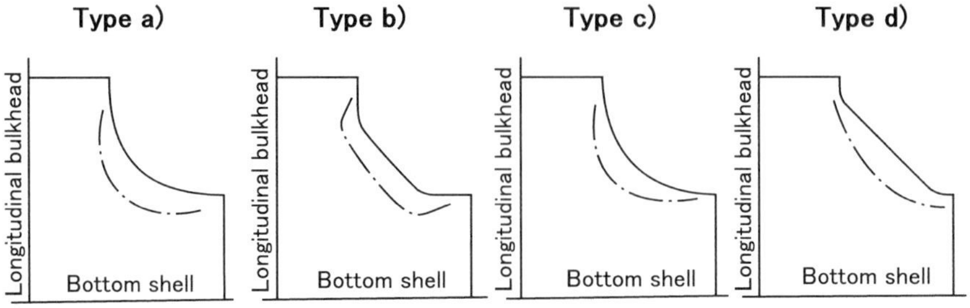

Fig. 21.8 Face plate stress due to shape of bottom transverse corner in wing tank

The calculated stress distribution in each face plate is shown in Fig. 21.8. As a matter of course, the stress is lower in the case of a larger radius of the circular face plate than that in the case of a smaller radius. Additionally, the tray type is better than the circular type.

The hull structure detail designer should keep in mind how important shape is. "Good appearance has good performance." and "good appearance covers seven flaws." are old Japanese proverbs.

21.5 Transverse Strength of Tanker

The transverse strength of a tanker is maintained by the transverse bulkheads and transverse rings. As explained in Chap. 19, the transverse bulkheads form a grillage structure with the longitudinal bulkheads and side shells, and form the basic strength of a ship. The transverse rings are supported by the grillage structure and prevent out-of-plane deflection of the grillage.

Hereunder the design of a transverse ring is explained.

21.5.1 Arrangement of Primary Supporting Members

Primary supporting members support stiffened panels composed of stiffeners and plates, and transmit the load into the bulkheads and shell plating which support the primary supporting members. When primary supporting members are arranged on a large rectangular stiffened panel, it is a common sense that it is most efficient to arrange the primary supporting members across the shorter span and transmit the load to the longer edges as shown in Fig. 21.9a.

As explained in Sect. 5.1, Part I, up-sizing of early oil tankers in 1950's and 1960's resulted in very deep tank with small width and length because the tank width and length were not increased in spite of the rapid increase in the ship size. An example can be seen in the rough general arrangement of *Universe Apollo* delivered in 1959,

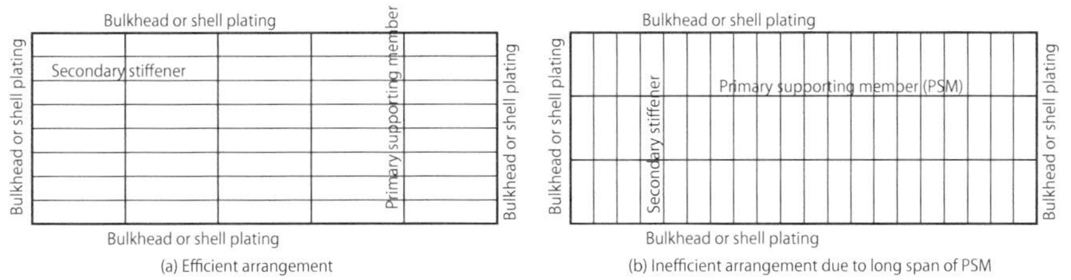

Fig. 21.9 Efficient arrangement of primary supporting members

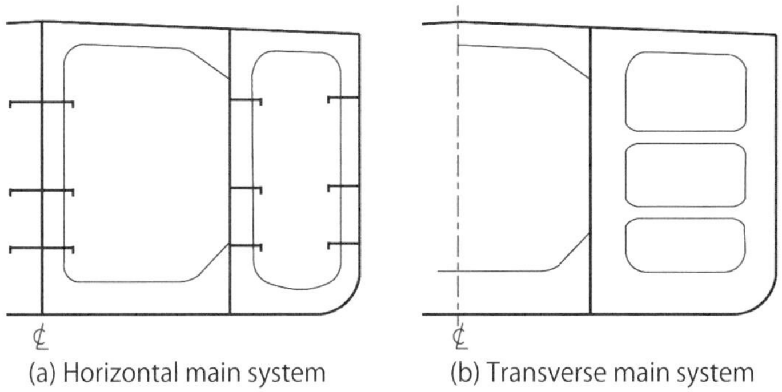

Fig. 21.10 Arrangement of primary supporting members

shown in Fig. 5.1, Part I. Therefore, $D > L, B$ where D, L, B are depth, length, and breadth of a tank, respectively. Because of this proportion, it was efficient to arrange primary supporting members across the shorter span, i.e., horizontally, and horizontal girders were adopted as shown in Fig. 21.10a. This structural arrangement is called the "horizontal main system".

Then, technological development enabled application of longer tanks up to length of 50 m or more, whose example can be seen in the rough general arrangement of *Terukuni Maru* delivered in 1971, shown in Fig. 5.1, Part I. This resulted in the tank proportion of $L > D > B$. In this case, the shorter spans are width and depth, and therefore, the transverse rings with struts as shown in Fig. 21.10b were more efficient and became the standard arrangement. This arrangement is called the "transverse main system".

As explained in Sect. 5.6, Part I, the grounding accident of *Exxon Valdez* triggered the mandatory application of double hull arrangement to oil tankers. When the single hull structure changed to the double hull structure, tank dimensions and proportion did not change as a whole. Therefore, the transverse main system was maintained, but new arrangement of cross ties appeared. In a single hull tanker structure, the rigidities of the vertical webs on the side and the longitudinal bulkhead were well balanced and they were effectively connected by cross ties. However, in the case of a double hull side structure, the rigidity of the vertical web on the side shell is

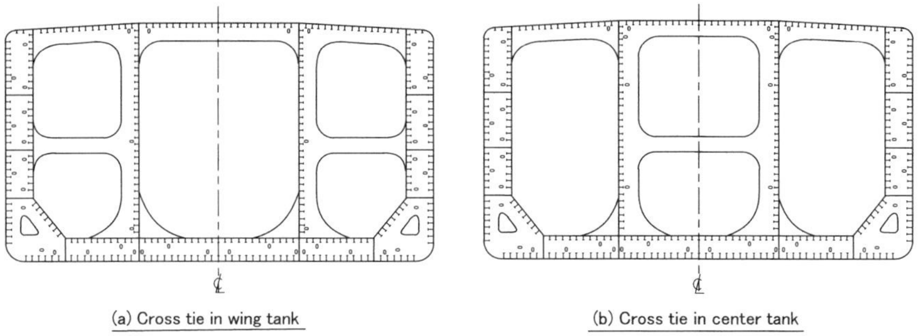

Fig. 21.11 Arrangement of cross tie of double hull tanker

far bigger than the rigidity of the vertical web on the longitudinal bulkhead. A trial calculation for a 300,000 DWT tanker indicates that the sectional moment of inertia of the vertical web on the longitudinal bulkhead is 28% of that of the vertical web on the side shell. The double side structure became so strong that it is no longer necessary to be supported by cross ties, and moreover, it is not so efficient to connect two members with different rigidities by a cross tie. Based on these backgrounds, a new arrangement with the cross tie in the center tank appeared as shown in Fig. 21.11b. In this arrangement a cross tie effectively connects the vertical webs of the same scantling on the longitudinal bulkheads. Similar discussions can be carried out based on the principle explained in Sect. 8.7 in Part II.

As explained in Sect. 20.1, amendment of the SOLAS convention was adopted to require permanent means of access (PMA) in 2002 to facilitate easy inspection and maintenance of hull structures. In this regulation, it is stated that the permanent means of access should as far as possible be integral to the structure of the ships, and therefore, deep horizontal girders were fitted inside the tanks as stages for easy access and inspection. This stage (horizontal girder) may work as a primary supporting member, and an idea was proposed to eliminate cross ties taking advantage of the horizontal girder supporting the longitudinal bulkhead structure as shown in Fig. 21.12 [130].

In this way, hull structural arrangement has changed to give the most efficient structure in response to the change of general arrangement and other social requirements. Hull structure designers should not be satisfied with existing designs, but should always pursue more efficient designs that suit the ever evolving socio-economic environment.

21.5.2 Cross Ties

At the beginning of the design of a transverse ring, the number of cross ties is to be decided, one cross tie, two cross ties, three cross ties or no cross ties. In the 1960's when a rapid increase in tanker size occurred, it was the common sense that the more cross ties the more safety, and the three cross ties arrangement was applied.

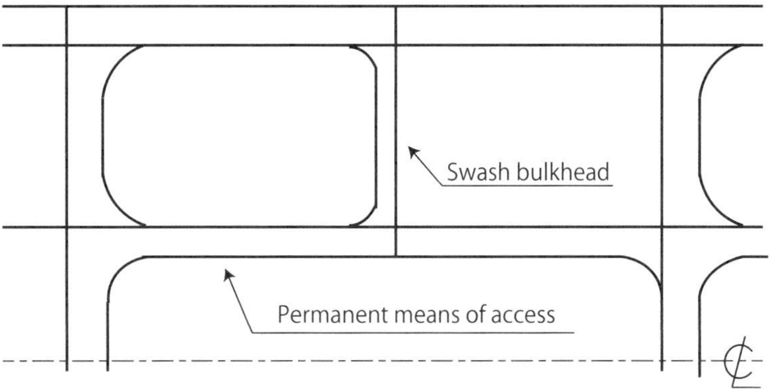

Fig. 21.12 Horizontal plan of cross-tieless tanker structure

The background of this common sense is that in the case of three cross ties, after one cross tie is broken, there remains still two cross ties which will maintain the transverse strength. However, this is not correct because the remaining ties cannot support the load which was taken by the three cross ties.

Then, it has become common practice to arrange two cross ties for big tankers and one cross tie for small ships. A 380,000 DWT tanker built in 1971 and a 480,000 DWT tanker in Fig. 21.13 built in 1973 had two cross ties, while a 78,800 DWT tanker built in 1966 had one cross tie as shown in Fig. 21.1.

The K type cross tie as shown in Fig. 21.14 was proposed and some European shipyards adopted it, but not the Japanese shipyards because of difficulty of erection work. From the strength and stiffness viewpoints, K type cross tie is effective in the

Fig. 21.13 Midship section of 480,000 DWT tanker

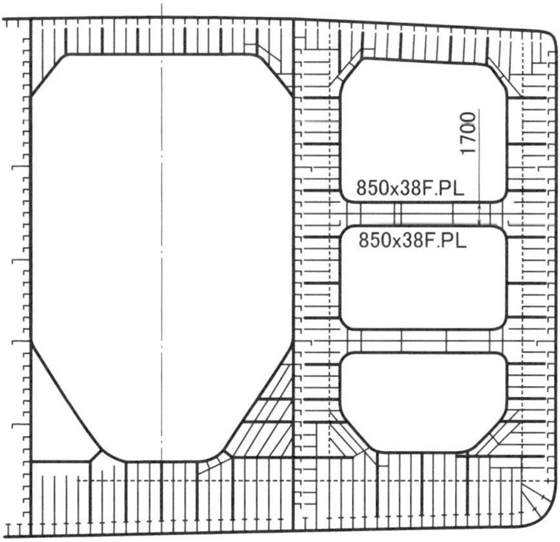

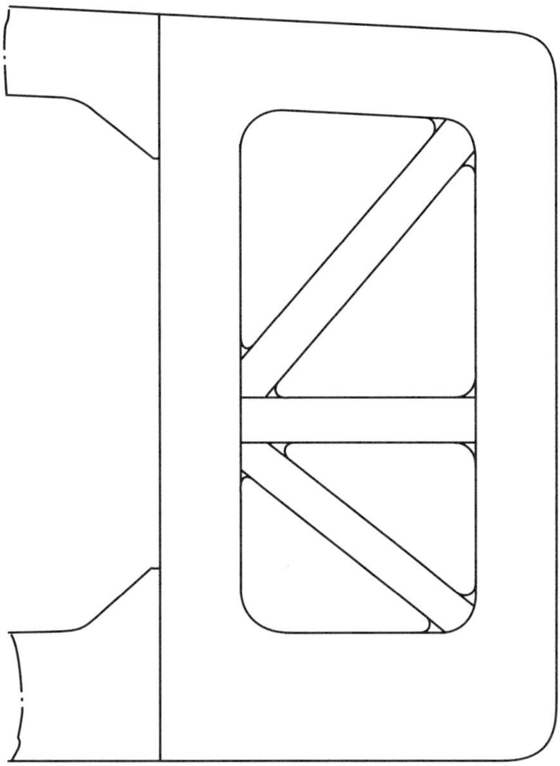

Fig. 21.14 Wing tank with K type cross tie

avoidance of shear deformation of the wing tank and also in increasing the strength of the vertical webs on the side shell and longitudinal bulkhead.

From the steel weight viewpoint there is little difference between one cross tie and two cross ties. One cross tie for a big ship means a longer unsupported span of vertical webs on the side shell and longitudinal bulkhead. This is a reason why the two cross tie arrangement is applied in big ships. No cross tie with deep bottom transverse is another arrangement, but it is inferior in terms of steel weight. The idea to connect the vertical webs on the side shell and longitudinal bulkhead is effective in increasing the strength and stiffness of these members.

The cross tie supports tension and compression as a pillar, and in addition to this it supports bending moments as a member of the transverse ring. The section of a cross tie has been "I" or "]" shape, but these shapes have unbalanced sectional moment of inertia around the x-axis and y-axis which is not suitable as pillars. Recently a well-balanced section shown in Fig. 21.15b has been applied. In the case of an "I" section, the sectional moment of inertia around the y-axis seems far less than that around the x-axis, accordingly the buckling strength around the y-axis is supposed to be far less than that around the x-axis. When the section as shown in Fig. 21.15b is applied, the sectional moment of inertia around the y-axis increases while that around the x-axis decreases, resulting in well-balanced proportion.

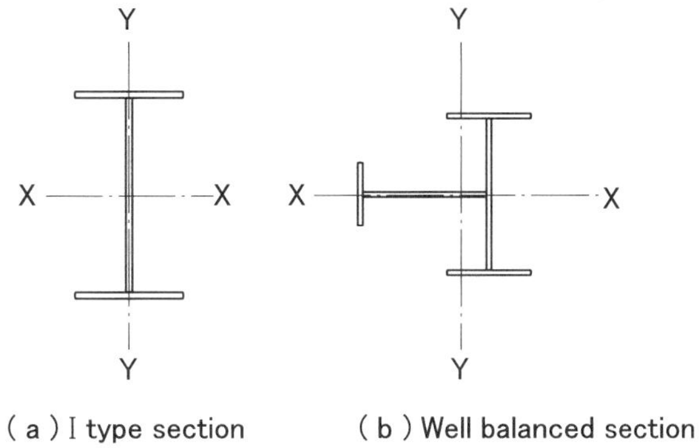

Fig. 21.15 Sectional shape of cross tie

21.5.3 Load Applied on Transverse Strength Members

The load applied on the transverse strength members is given by the rules of the classification societies such as 2.45 m head above the upper deck for the internal pressure of a fully loaded tank, external pressure corresponding to 1/3 of the fully loaded draft for the tank test condition, etc. The wave load is also given corresponding to the ship's size.

Hereunder the cross tie's strength is investigated based on a design standard established applying the above load and noting the buckling damage shown in Fig. 21.1.

Nippon Kaiji Kyokai [131] showed that the maximum variable pressure around the water line is 11.4 m at a probability level of $Q = 10^{-5}$ for ships over 200 m length. That around the bilge part is 5.1m, therefore the maximum pressure around the bilge part is (5.1 m + draft d). The above applied load is shown in Fig. 21.16, whose original formula is indicated in Fig. 2.3.1 of Sect. 2.3 "Transverse Strength Load" in Part I.

The axial force on a cross tie is given by Eq. (21.5.1).

$$P = h_c S l_i \tag{21.5.1}$$

where

h_c: load at the level of cross tie

S: spacing of side transverse.

l_i: distance between the two midpoints between the cross tie and the adjacent cross tie, deck transverse, or bottom transverse.

The design standard for a cross tie panel is given by Eqs. (21.5.2) and (21.5.3).

$$\frac{\sigma_{cr-p}}{\sigma_A} \geq 2.0 \tag{21.5.2}$$

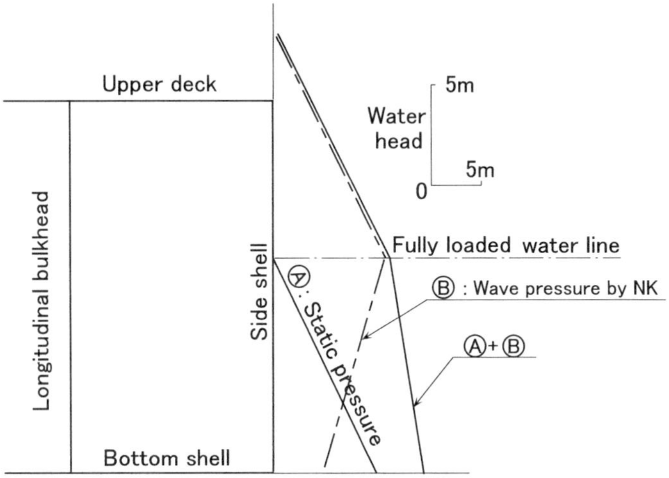

Fig. 21.16 Water pressure acting on side shell

$$\sigma_A = \frac{P}{A_C} \tag{21.5.3}$$

where

A_C: sectional area of cross tie

σ_{cr-p}: critical buckling stress by compression of panel surrounded by stiffeners in a boundary supported condition.

The design standard for a cross tie as a pillar is given by Eqs. (21.5.4) and (21.5.5). The critical buckling stress σ_{cr-c} of a cross tie as a pillar is calculated by means of Tetmajer's formula.

$$\sigma_{cr-c} = 304\left(1 - 0.00368\frac{l}{k}\right) \text{(MPa)} \tag{21.5.4}$$

$$\frac{\sigma_{cr-c}}{\sigma_A} \geq 2.0 \tag{21.5.5}$$

where

l: length of cross tie

k: minimum radius of gyration of cross tie

The analysis results are shown in Table 21.3. Ship B had buckling damage as shown in Fig. 21.1 and it can be seen that the safety factor for the panel of ship B is lower than 1.0. According to the above analysis a reinforcement was applied on the end panel of ship G which has a safety factor of 1.1.

Table 21.3 Safety factor of cross tie buckling

Vessel	Principal dimensions (m)					Safety factor		
	L	B	D	d	S	for pillar buckling	for panel buckling at middle of cross tie	for panel buckling at cross tie end
A	240	37.8	17.5	13.17	5.5	2.5	2.5	
B	232	38.0	17.8	13.01	5.4	2.1	0.8	0.8
C	290	47.5	24.0	16.00	4.5		3.1	2.0
D	230	33.3	19.5	13.29	4.6	2.5	2.2	
E	230	35.3	18.0	11.55	4.45	3.0	1.9	
F	245	39.0	23.0	16.33	4.05	3.2	1.8	
G	260	43.5	22.8	17.80	5.1	2.4	1.6	1.1
H	360	62.0	36.0	28.15	5.3		2.8	
I	360	62.0	36.0	28.15	5.3		2.1	
J	270	44.5	22.0	16.80	4.8	3.3	4.4	2.9
K	320	54.5	27.0	21.00	5.3	2.8	2.9	1.8

The above load is applied not only to the design of cross ties but also to the design of side shell vertical web frames. In Fig. 21.16 the variable pressure in waves as regulated by Nippon Kaiji Kyokai is also shown.

21.5.4 Inside Pressure in Wide Tanks

It is clear that the bulging damage in the side shell in the 84,000 DWT tanker shown in Fig. 21.1b was caused by inside pressure of the tank which was designed as a stabilizer tank where the longitudinal bulkheads had openings and the tank was considered to be a single tank spanning from the port side to the starboard side of the ship. The liquid in a tank works as a stabilizer and moves, for example to the starboard side when the ship is going to roll to the port side, and it hits the upper part of the starboard side shell. Moreover, in the fully loaded condition unexpected pressure will be caused on the upper part of the side shell by the ship's heel, as shown in Fig. 21.17.

From Fig. 21.17 the following relation can be derived:

$$BU = \frac{1}{2}a \cdot a\cot\theta \tag{21.5.6}$$

$$\therefore a = \sqrt{2BU\tan\theta} \tag{21.5.7}$$

$$c = B\tan\theta \tag{21.5.8}$$

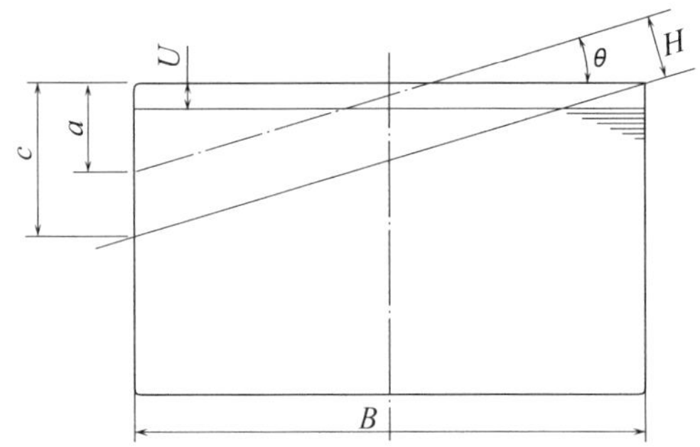

Fig. 21.17 Water head increase due to rolling

where

B: breadth of tank.

U: ullage.

θ: heel angle.

Accordingly, the increase of water head H above the upper deck at side is given by the following Eq. (21.5.9).

$$H = (c - a)\cos\theta = \left(B\tan\theta - \sqrt{2BU\tan\theta}\right)\cos\theta \tag{21.5.9}$$

In Fig. 21.18 the relation between B and H is shown in the case of $\theta = 30°$ and $U = 1$ m. In the case of an ordinary tank, H is more or less 2 or 3 m for about 15m of B; however, for a wider tank, H will be about 10 m.

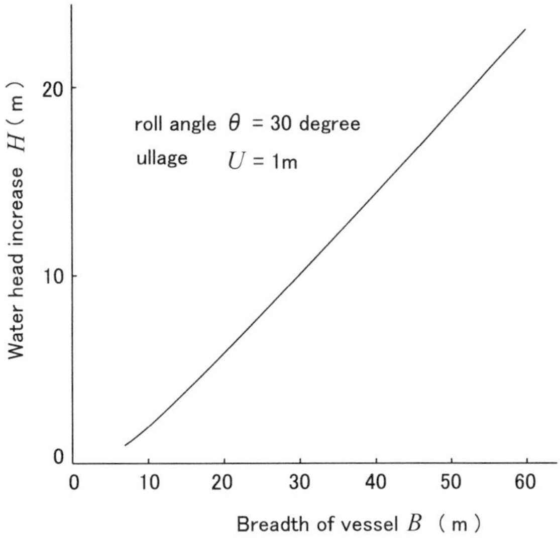

Fig. 21.18 Estimated water head increase according to ship's breadth

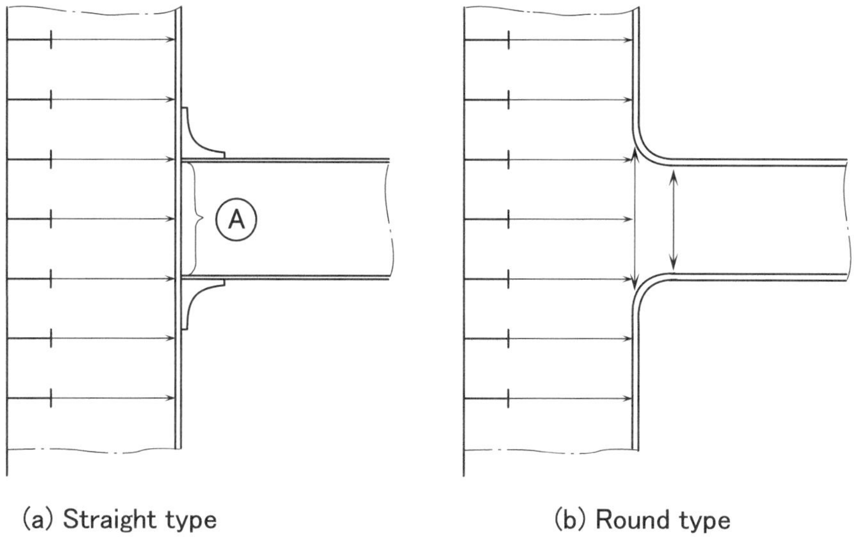

Fig. 21.19 Straight type and round type of cross tie

The ballast hold of a bulk carrier is in a similar condition and indent damage to a shoulder tank bottom will be caused by this behavior.

21.5.5 Straight Type and Circular Type Construction

From viewpoint of the fabrication of a transverse ring, much discussion has been held on which is better, straight type or circular type construction. Eventually in Japanese shipyards the circular type was adopted, on the other hand in European shipyards, the straight type is widely adopted. The straight type construction is better for labor saving and the circular type is better for weight saving and strength. New fabrication techniques such as NC cutting and automatic welding support the circular type construction.

A cross tie mainly plays the role of a pillar and the straight type cross tie seems to be reasonable in supporting the axial force. However, the circular type cross tie is lighter, because as shown in Fig. 21.19 the heavy face plate of the "A" part can be omitted.

21.5.6 Transverse Rings at Fore and Aft Parts of Tank

In the design of the transverse strength of the tank part of a tanker, the scantlings of each member are decided based on the midship shape and the same scantling is

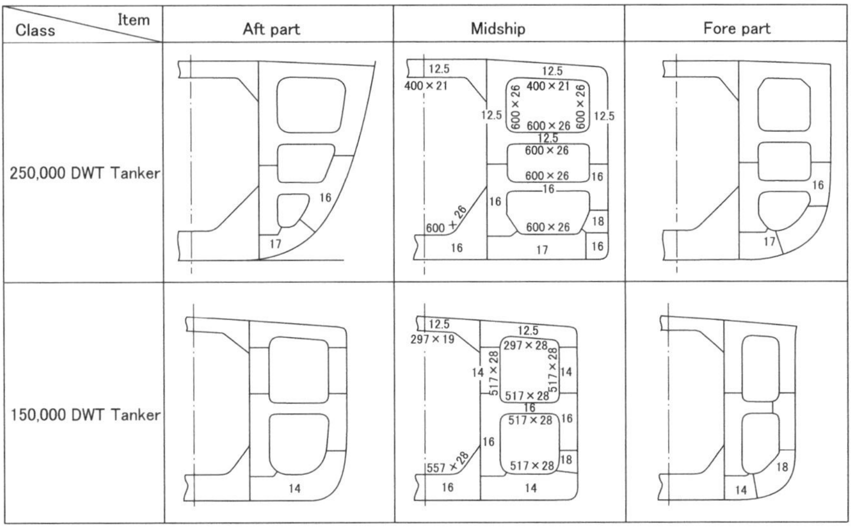

Fig. 21.20 Shape of transverse ring at fore, midship and aft

often applied to the fore and aft parts of a tank where the shape is different from the midship part. In Fig. 21.20 the shapes of transverse rings at the fore and aft parts of a 250,000 DWT and a 150,000 DWT single hull tanker are shown respectively. It seems that a remarkable material saving can be achieved if the proper design to meet the shapes is carried out considering the shorter span of primary supporting members.

In Sect. 21.5.1, it is explained that in a double hull tanker, the double side structure can stand by itself without cross ties thanks to its large rigidity. In the foremost cargo tank, the side shell plating shifts inward according to the hull shape and the inner hull plating is kept straight to ensure sufficient cargo tank volume as shown in Fig. 21.21a. Hence, the width of the double side structure may become extremely narrow as shown in Fig. 21.21b. Such a structure does not look feasible without the support by cross ties, but actually, the horizontal side stringers inside the double side structure are strong enough and effectively support the thin double side transverse webs thanks to the arch effect on the curved outer shell, converting the lateral pressure to the axial load (hoop stress) exerted on the side stringers as illustrated in the starboard side of Fig. 21.21a. That is, in the midship part, the side transverse web frames support the side stringers because the span of the side transverse web frames is much shorter than the side stringers, but conversely in the foremost cargo tank, the side stringers support the thin side transverse web frames. This kind of efficient structural arrangement can only be possible by carrying out full three-dimensional finite element analysis of the forward cargo tank structures and confirming the flow of forces in the structures.

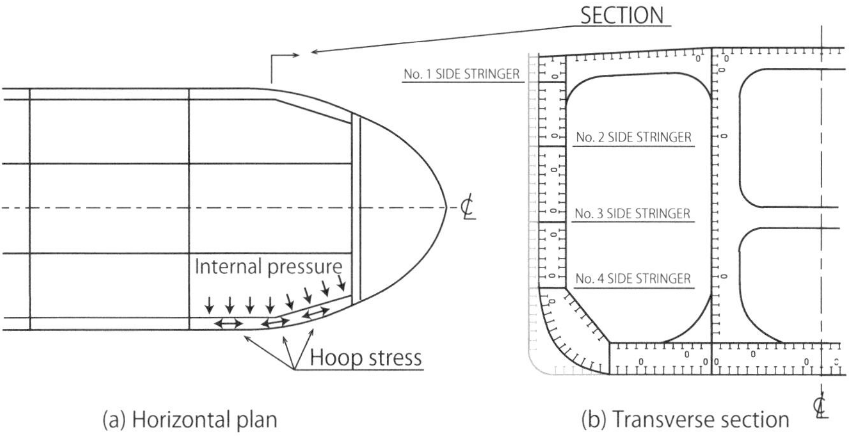

Fig. 21.21 Thin double side structure in the foremost tank

21.6 Transverse Strength of Ore Carrier

As stated in Sect. 5.2 of Part I, the ore carrier has a long and big ore hold in the middle of the ship which is sandwiched by ballast tanks. To achieve more efficient ore handling, less transverse bulkhead is fitted in the cargo hold and the hatch opening is bigger than that of an ordinary cargo ship. The transverse bulkhead is an important member in maintaining the transverse strength of a ship, and special care must be taken when designing a hull structure with a long hold. In such a case, the double bottom and decks between hatches play a major role in maintaining transverse strength. In this section the transverse strength members in the wing tank, knuckled part of longitudinal bulkhead, double bottom floor and deck between hatches are explained.

As shown in Fig. 21.22 the transverse bulkhead and the transverse ring in a wing tank suffer from compression in the fully loaded condition, and tension in the ballasted condition. This is also true for a transverse ring in a ballast wing tank of a tanker. However, for a transverse bulkhead the condition is a little different. As shown in Fig. 21.23 for the fully loaded condition the transverse bulkhead supports about twice the compression than in the case of a tanker. This is why a horizontal girder on a transverse bulkhead of an ore carrier is prone to buckling damage.

A knuckle is, in some cases, provided at the mid-level between the upper deck and double bottom on the longitudinal bulkheads of an ore carrier. In the design of the knuckle part and the radius part, attention should be paid to the out-of-plane force acting normal to the longitudinal bulkhead caused by in-plane force in the longitudinal bulkhead. Designers of land structures always provide reinforcement against the out-of-plane force on the plate caused by an in-plane force but the hull structure designers are so optimistic that they pay little attention to this point. Buckling damage happens at the radius part of a cross tie shown in Fig. 21.1a, and this is caused by the

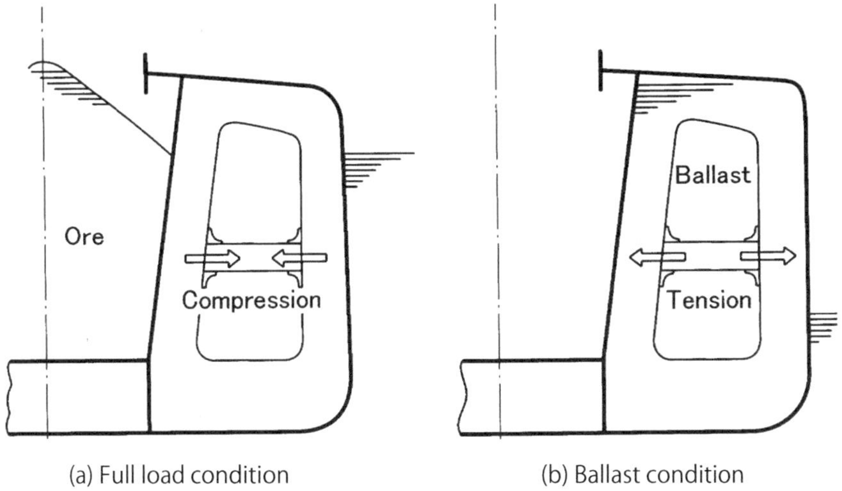

Fig. 21.22 Load on wing tank of ore carrier

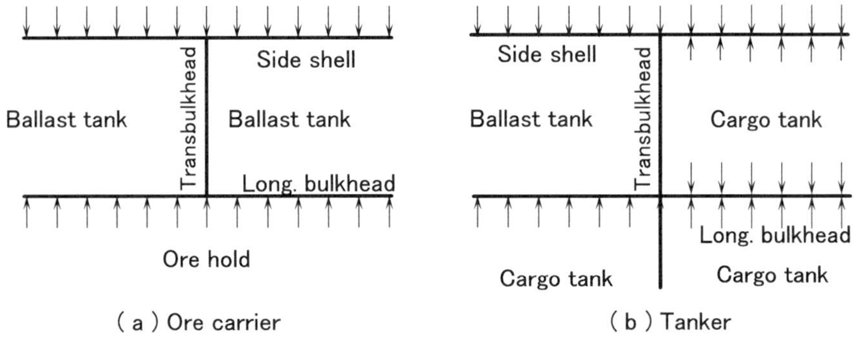

Fig. 21.23 Horizontal load acting on transverse bulkhead in full load condition

in-plane stress on the transverse web normal to the face plate, which is induced by the axial force on the radiused face plate. An example of a crack caused by this kind of stress working normal to the knuckled bulkhead induced by the in-plane force on the bulkhead is shown in Fig. 21.24. A countermeasure against this crack is to provide a longitudinal stiffener at the knuckled part to prevent the out-of-plane displacement of the plate.

In the design of an ore carrier, the double bottom can be designed with sufficient height because the required cargo hold volume is small due to the high specific gravity of iron ore. In this regard the design of transverse strength of an ore carrier is easier than that of a bulk carrier.

In the floor in a double bottom, slots are provided for the longitudinal stiffeners as well as in a transverse ring of a tanker or an ore carrier's wing tank. Around

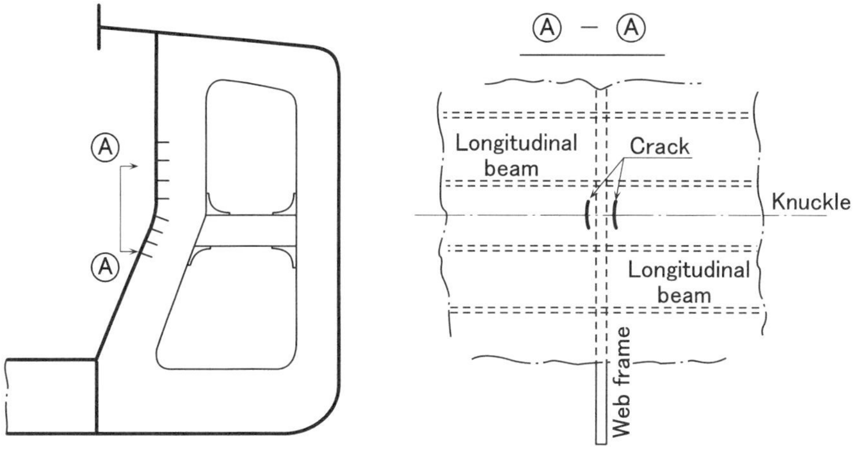

Fig. 21.24 Crack occurring at knuckled part in longitudinal bulkhead of ore carrier

the slots in a transverse ring of a tanker or an ore carrier's wing tank, many cracks occurred, which are explained in detail in Sect. 10.5, Part II. As explained in that section, cracks were seldom reported around the slots in the double bottom floors of ore carriers, bulk carriers, and cargo ships thanks to the small bending moment in the web stiffener, compared with the slots in a single hull structure as illustrated in Fig. 10.23, Part II.

In ore carriers a transverse force is generated in the deck between hatches. The force is compression in the case of a fully loaded condition, and tension in the case of a ballasted condition.

The forces can be estimated easily as follows. In a fully loaded condition a section of an ore carrier can be modeled as shown in Fig. 21.25. Applying bottom pressure, side pressure and weight of ore on the model the following equation can be obtained by the equilibrium of moments. Here the double bottom is more rigid than on upper deck between hatches.

$$l\left(\frac{dB}{2}\cdot\frac{B}{4}-W\cdot\frac{b}{2}\right)-F_mD=0 \qquad (21.6.1)$$

where

d: full load draft.

l: distance between centers of upper decks between hatches.

B: breadth of ship.

W: half of ore weight per unit length of ship.

b: half the breadth of the ore hold.

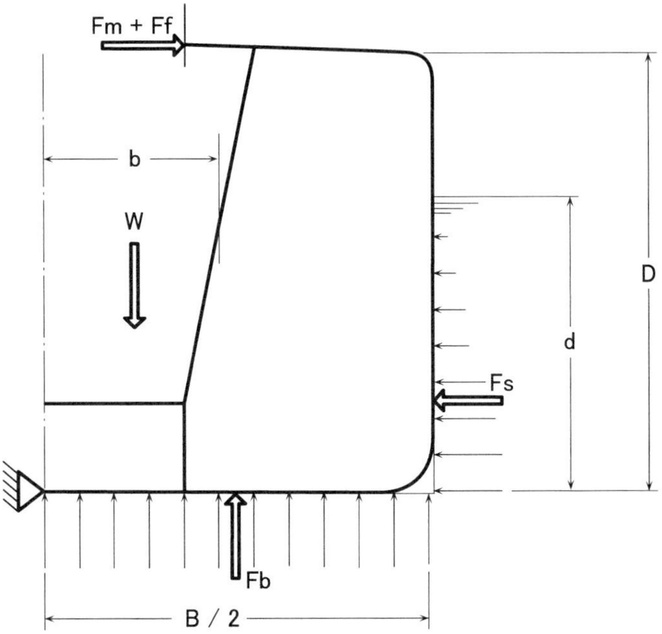

Fig. 21.25 Equilibrium of forces acting to transverse members of ore carrier

D: depth of ship.

F_m: force on upper deck between hatches due to bottom pressure and ore weight.

Expressing the force on the upper deck between hatches due to side pressure by F_f, Eq. (21.6.2) can be derived.

$$F_f D - l \cdot \frac{d^2}{2} \cdot \frac{d}{3} = 0 \tag{21.6.2}$$

From Eqs. (21.6.1) and (21.6.2) the total force in the upper deck between hatches $\left(F_m + F_f\right)$ can be obtained as shown by Eq. (21.6.3).

$$\left(F_m + F_f\right) = \frac{l}{2D}\left(\frac{dB^2}{4} - Wb + \frac{d^3}{3}\right) \tag{21.6.3}$$

The estimated compressive stresses in the upper decks between hatches, for large ore carriers, are shown in Table 21.4, and these were obtained assuming that the upper deck plate is 12 mm thick and the transverse beams of 300 × 90 × 11/16 inverted angle are fitted with 800 mm spaces. The critical buckling stress of the 800 × 4,000 × 12 rectangular plate with compressive force on the shorter edges is 152 MPa. Accordingly the safety factor for Ship A is 1.34, for Ship B is 1.17, and for Ship C is 2.54. In this case the design criteria of safety factor is taken to be the

Table 21.4 Compressive stress of upper deck between hatch openings of ore carrier

Vessel	Deadweight DW (ton)	Breadth B (m)	Depth D (m)	Draft d (m)	1/2 Breadth of cargo hold b (m)	Longitudinal distance between cargo hold centers l (m)	Ore weight (tonf / m)	Compressive stress (MPa)	Delivered year
A	226,300	52.0	23.45	18.1	13.6	41.6	907	113	1988
B	196,800	50.0	23.90	17.5	12.9	48.4	813	130	1985
C	197,100	50.0	23.30	17.8	14.3	39.6	830	60	1984

minimum value among the analyzed values for ships in service without any buckling, or the maximum value among the analyzed values for ships in service with buckling, if any.

21.7 Transverse Strength of Bulk Carrier

As explained in Sect. 5.2 in Part I, a bulk carrier has hopper tanks at both sides of the double bottom and shoulder tanks below the upper deck, whose torsional rigidity supports the double bottom and side frames. From this view point the distance between transverse bulkheads, that is the length of a hold, is important.

The shoulder tank and the hopper tank are connected by the side shell construction and have a strong resistance against the vertical force. Against the horizontal force the hopper tanks connected to the double bottom also have a strong resistance. The vertical force on the double bottom, such as water pressure on the bottom and the cargo weight, and the horizontal force on the hold frames on the side shell, causes torsional moments on the hopper and shoulder tanks, which are located at the supported ends of the double bottom and the hold frames on the side shell. The torsional rigidity of these parts is important in resisting the torsional moments. The torsional rigidity of a closed section is proportional to the square of the sectional area of the closed section, so the structural arrangement of small hopper and shoulder tanks, and of long cargo hold may have a transverse strength deficiency.

The torsional angle θ of a thin hollow tube is expressed by the following equation:

$$\theta = \frac{sTl}{4A^2Gt} \tag{21.7.1}$$

where

s: length of circumference passing through mid-points of plate thickness.

T: torsional moment

A: area surrounded by s

G: shear modulus of elasticity.

t: thickness of s

l: length of tube.

In the application of Eq. (21.7.1) to the hopper tank, l is to be half the length of a hold and T is to be a function of the ship breadth B, draft d and l. Here T is represented in Eq. (21.7.2) and θ is expressed by Eq. (21.7.3).

$$T \propto B^2 dl \tag{21.7.2}$$

$$\theta \propto \frac{sB^2 dl^2}{4A^2 Gt} \tag{21.7.3}$$

The structural arrangement of a bulk carrier is to be based on the standard θ which is derived by the analysis of θ in Eq. (21.7.3) for many ships in service.

The side shell construction connecting the hopper tank and the shoulder tank originally consisted of ordinary frames and web frames which were arranged in line with the transverse webs in the hopper tank and the shoulder tank. However, in the new design it is common to omit the web frames, and only the ordinary frames are retained. This is because as explained above the transverse strength of a bulk carrier is maintained by the torsional rigidity of the hopper and shoulder tanks. The upper and the lower ends of the side shell construction suffer from a forced rotational displacement from the hopper and shoulder tanks. Therefore, higher bending stress is generated in the deeper web frames, and the shallower ordinary frames are effective to mitigate the bending stress due to the forced displacement.

At the upper and the lower ends of the side frame, brackets should be fitted inside the shoulder tank and the hopper tank. There used to be no instruction to decide the scantling of these brackets in classification rules. Hereunder a suggestion to design these brackets is explained. As shown in Fig. 21.26a the bracket inside the hopper tank is supported by the side longitudinal A and the longitudinal stiffener B on the hopper plate; this resists the bending moment M at the lower end of the side frame. In this condition the equilibrium of the forces is expressed by Eq. (21.7.4).

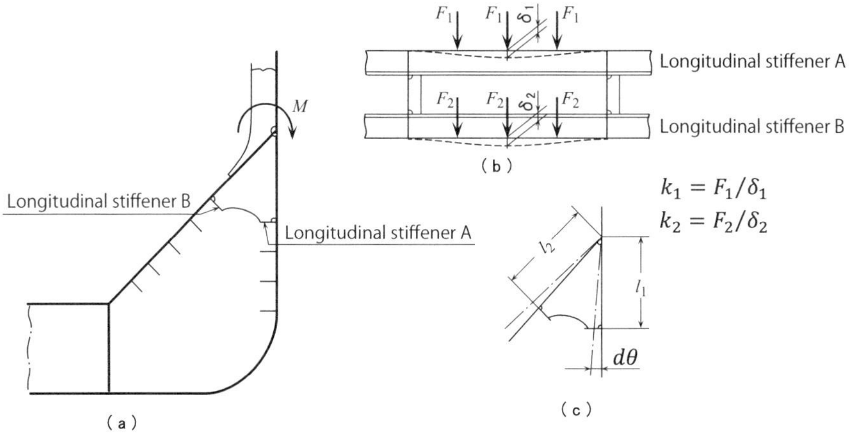

Fig. 21.26 Bracket supporting lower end of side frame

$$M = l_1F_1 + l_2F_2 \tag{21.7.4}$$

M can be expressed by $M = Z\sigma$ where Z is the section modulus of the side frame and σ is the design stress of the frame. F_1 and F_2 are the reactions from the longitudinal frame A and B, and can be obtained by limiting the stresses σ in the longitudinal frame A and B as shown in Fig. 21.26b. On the other hand, there exists a relation of Eq. (21.7.5) between F_1, F_2, l_1, and l_2 as shown in Fig. 21.26c, assuming the bracket is rigid.

$$d\theta = \frac{k_1}{l_1}F_1 = \frac{k_2}{l_2}F_2 \tag{21.7.5}$$

where k_1 and k_2 are spring constants of longitudinal frames.

According to the above consideration a reasonable design of the brackets at the lower and the upper ends of the side frames can be obtained and this depends on the size of the side frame. It should be noted that the scantling of the longitudinals A and B should be determined considering the forces F_1 and F_2 distributed along their length in addition to the lateral pressure acting on the longitudinals, and F_1 and F_2 are large when l_1 and l_2 are small (i.e. the bracket size is small).

An important point in designing the transverse strength of a bulk carrier is the connection of the double bottom to the hopper tank. Many cracks shown in Fig. 21.27 were reported during the era of the increasing size of bulk carriers. There are two ways of construction in this part, a straight welded type and a curved bending type, on which many discussions have been held from design and fabrication viewpoints.

It is easily understood that the crack shown in Fig. 21.27 is caused by the tensile stress due to the cargo weight on the double bottom. However in the smaller bulk carrier which has a narrow floor space in the double bottom no such crack has been reported even with a scallop in this part. From experience with such cracking, closing the scallop with a collar plate has been a common practice but for a narrow floor space up to about 2 m no collar plate arrangement is required.

At the connection of the double bottom to the hopper tank, either a straight welded type or a curved bending type has to be chosen from a fabrication view point as well as a strength view point. The latter requires a plate bending process but the fillet welding is half of the former. It is effective to fit a side girder around the midpart of the curve as shown in Fig. 21.28 to avoid movement perpendicular to the plate. Such movement caused cracks at the knuckled part of the longitudinal bulkhead of an ore carrier, as shown in Fig. 21.24. This alignment of the side girder is important to prevent structural damage from view point of strength, but it takes many labor hours to provide accurate alignment from the fabrication viewpoint.

In the case of the curved bending type, the effective breadth of the curved part relative to the bending moment at the end of the double bottom is reduced, for which some reinforcement to avoid out of plane displacement is necessary. Summing up the above discussion, the authors consider the curved type seems to have an advantage due to strength, in spite of the curved type not being so easily fabricated.

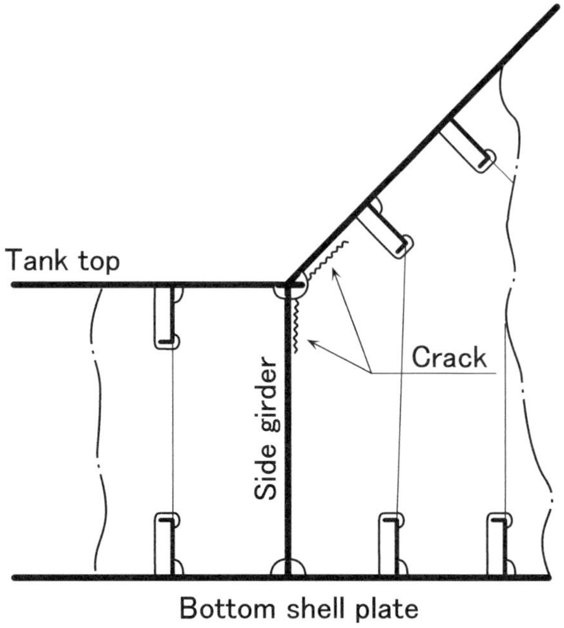

Fig. 21.27 Example of crack in double bottom end of bulk carrier

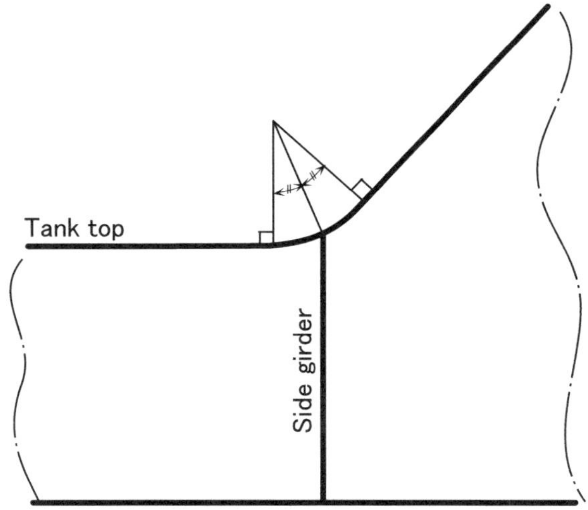

Fig. 21.28 Radiused bilge hopper knuckle construction

When designers discuss the strength of a structure, the load on the structure should be kept in the designer's mind. In the case of heavy cargo, alternate holds are loaded because homogeneous loading will bring the center of gravity lower causing a shorter rolling period. This is called alternate loading or jumping loading. Alternate loading is most important in bulk carriers.

Here as an example, homogeneous loading and alternate loading are compared for a 33,000 DWT bulk carrier. The loads on the double bottom are shown in Table 21.5. We can see that the total load for the alternate loading is 6–7 times of that for the homogeneous loading. The total load for the ballast condition is obtained assuming the draft is 32% of that of the fully loaded condition which is bigger than that for homogeneous loading. As explained above the load for the alternate loading is important for a bulk carrier double bottom which is supported by the torsional rigidity of the hopper tank.

Next, the in-plane load in the transverse bulkhead is to be considered. As shown in Fig. 21.29 the load on the double bottom is transmitted to the transverse bulkhead as an in-plane load, as well as to the side shell plates. The in-plane load in the transverse bulkhead W is expressed by the following Eq. (21.7.6):

$$W = \frac{a_1 w_1}{2}\alpha + \frac{a_2 w_2}{2}\alpha \tag{21.7.6}$$

where

α: double bottom load supporting ratio of transverse bulkhead.

Table 21.5 Applied load to double bottom of bulk carrier

Condition			Cargo load tonf / m²	Bottom pressure tonf / m²	Total load tonf / m²
Full load	Homogeneous loading		12.56	10.90	+ 1.66
	Alternate loading	Empty hold	0	10.90	- 10.90
		Loaded hold	23.36	10.90	+ 12.46
Ballast			0	3.49	- 3.49

+ : downward − : upward

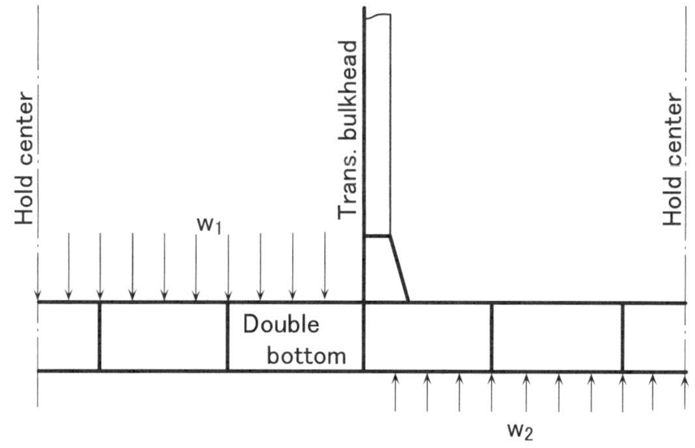

Fig. 21.29 Applied load to transverse bulkhead

Table 21.6 Load coefficient of transverse bulkhead of bulk carrier

		Load factor. tonf/m^2
Full load condition	Homogeneous load	+ 3.32
	Alternate load	+ 1.56
Ballast condition		− 6.98

a_i: area of double bottom No. i

w_i: load acting on double bottom No. i shown in Table 21.5.

Considering that the equation $a_1 = a_2$ holds in the midship part generally, Eq. (21.7.6) reduces to

$$W = \frac{a\alpha}{2}(w_1 + w_2) \tag{21.7.7}$$

where $a = a_1 = a_2$. In the homogeneous loading and the ballast condition $w_1 = w_2$. $(w_1 + w_2)$ is called the load factor for the transverse bulkhead and is shown in Table 21.6 which is obtained using the values in Table 21.5. It can be seen that the load factor for alternate loading is low because the load in the empty hold and the loaded hold cancel each other at the transverse bulkhead. According to the results shown in Table 21.6, the load factor is greatest in the ballast condition. This is the reason why we employed the ballast condition to assess the shearing strength of a transverse bulkhead in Sect. 19.2.2.

21.8 Transverse Strength of Container Ship

In the design and construction of the first generation of container ships, many broad and detailed assessments were made, especially on the torsional strength due to large hatch openings and on engine room construction due to high engine output. Regarding the transverse strength of a container ship, the hull structure designers seemed to be optimistic because containers are relatively light cargo and container ships have shallow draft even in the fully loaded condition, both resulting in smaller load on the double bottom structure than other ship types; container ships have many transverse bulkheads with intermediate partial bulkheads. In addition, the breadth of container ships was limited up to the Panamax size, i.e., 32.2 m breadth to be able to pass through the Panama Canal until around 1990. Thanks to this less severe conditions, very little damage on the transverse strength of container ships was reported.

However, prominent increase in the size of container ships is progressing nowadays as explained in Sect. 5.3, Part I. To increase the efficiency of transportation, many challenges are pursued, such as longer holds with less transverse bulkheads and optimized partial bulkhead construction. Furthermore, flexible double bottom structure adversely affects the hull girder ultimate strength as explained in Sect. 20.1. Along with these significant changes, more attention should be paid to the transverse strength of container ships.

Chapter 22
Torsional Strength

Following Chap. 20 "Longitudinal Strength of Hull Girder" and Chap. 21 "Transverse Strength of Ship", in this chapter torsional strength is explained referring to actual damages.

A ship, which has no large openings in its upper deck like a tanker, has enough torsional strength, and no damage caused by torsion has been reported. A bulk carrier and a container ship have large openings in the upper deck and a hull structure designer has to pay attention to the torsional strength.

As in the case of buckling, the torsional strength can be understood easily and clearly by means of a paper model. In Figs. 22.1 and 22.2, two paper models are shown. One is with a slit throughout the upper deck in a longitudinal direction, which has very low torsional rigidity, as shown in Fig. 22.1. The other is with large openings in the upper deck which shows tension and compression at the hatch corners as shown in Fig. 22.2. Actually damage such as buckling and cracking at the hatch corner are caused by torsion as explained in the following sections.

22.1 Structural Damage due to Torsion (Example No. 1)

A 750 TEU container vessel ($L \times B \times D \times d = 175.0 \times 25.2 \times 15.3 \times 9.724$ m, 28,000 HP) was sailing in the north pacific ocean from Oakland in North America to Kobe in Japan in the winter season, when she suffered heavy weather damage in the area of the bow structure. The damage was cracks in the starboard shell plate and buckling in the region of the upper deck, forecastle deck, and shell plating on the port and starboard side. After investigations, the reason for the damage was concluded to be that the bow structure was twisted counter-clockwise looking forward around the longitudinal axis of ship due to a large impact wave load on the large bow flare.

Y. Okumoto et al., *Design of Ship Hull Structures*,
https://doi.org/10.1007/978-3-032-12500-2_22

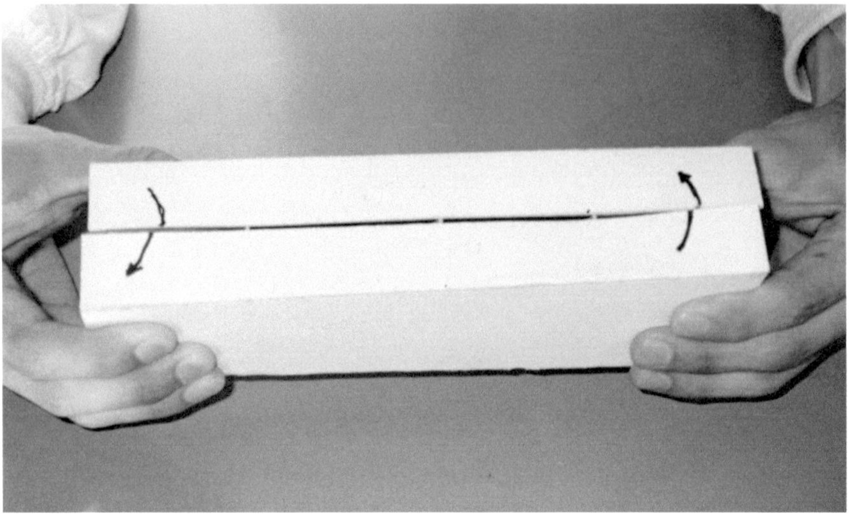

Fig. 22.1 Torsion of open rectangular section with a longitudinal slit

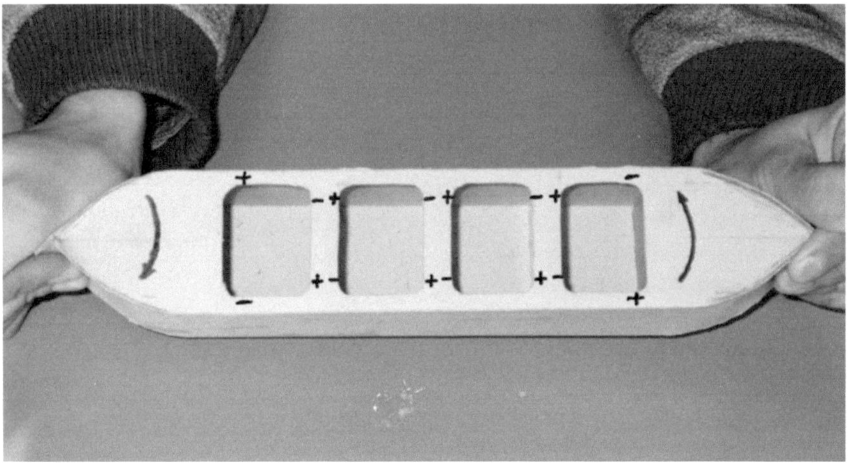

Fig. 22.2 Torsion of a paper model of a ship with a large opening

Figure 22.3 shows the port side shell plate where buckling occurred. The buckling was induced by a shearing force, judging from the fact that the direction of buckling is not straight upwards but inclined aft-wards. Accordingly, the damage was caused not by an upward force due to a wave impact load but by a counter-clockwise torsional moment around the longitudinal axis which was induced by a wave impact load acting on the starboard bow flare.

Figure 22.4 shows the deformed shape of the bow structure by FEM analysis at frame 154, in the forward-looking direction. The result shows that the cross section

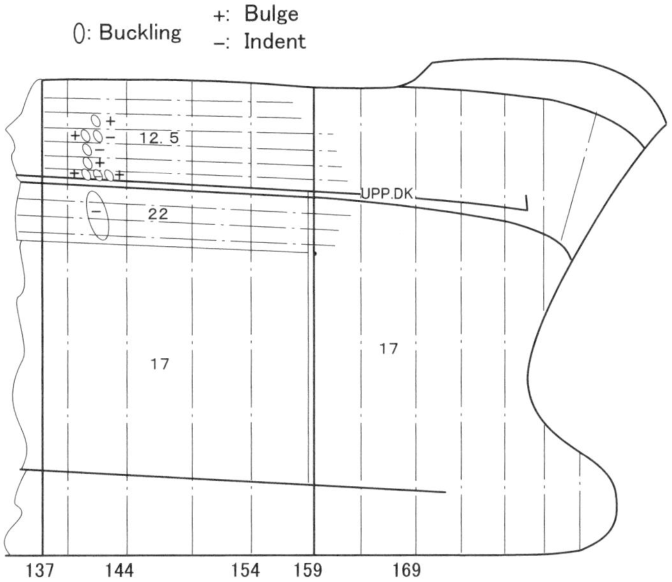

Fig. 22.3 Buckling of port side shell plating

was twisted counter-clockwise due to the wave impact load acting on the starboard bow flare.

Figure 22.5 illustrates the buckling of the forecastle deck plate of the subject vessel. A counter-clockwise torsional moment resulted in the buckling of the port side fore corner of the hatch opening and the starboard aft. We have to pay attention that the buckling of the forecastle deck might have been induced not only by torsional moments but also by horizontal bending moments which bend the bow structure in the port direction due to a starboard side wave impact load.

Fig. 22.4 Deformation of Fr. 154 cross section

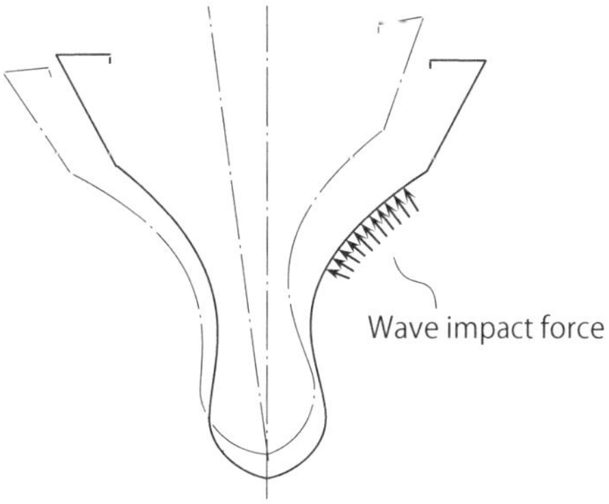

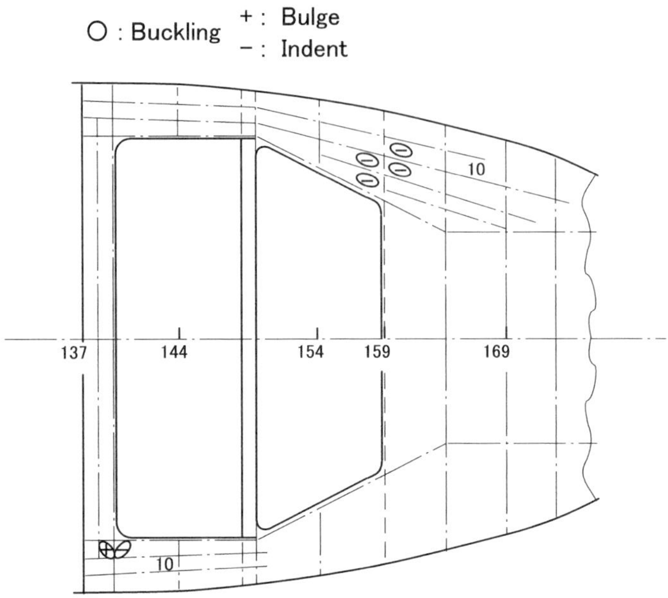

Fig. 22.5 Buckling of forecastle deck plating

22.2 Structural Damage due to Torsion (Example No. 2)

When a fully-loaded bulk carrier of 250,000 DWT with ore cargo was sailing from Brazil to Korea via the Cape of Good Hope, she encountered rough seas in the Indian Ocean and suffered structural damage to the deck area between hatch openings. According to the log data, the wind was force 8 grade on the Beaufort scale and the significant wave height was 7 m. The ship's speed was 7–8 knots and she proceeded with a rolling motion of $12-15^\circ$. The waves and winds were coming from ahead at 65° to starboard. Such heavy weather continued for 7 days and after the bad weather the above damage was found [132].

It happened in the return voyage of her maiden voyage and the wind class was around 3 Beaufort everyday during the maiden voyage except for the damage period. The distribution of the damage is shown in Fig. 22.6. There is no damage fore and aft of the No.6 hold, because No.6 is an empty hold and because the deck plate thickness fore and aft of the No.6 hold is thicker than that at the other holds. Figure 22.7 shows the scantlings of the deck plates where buckling occurred. Buckling took place in the transition area of 13mm to 25mm thick plate. From the distribution of buckling, it was deduced that the damage was caused by a torsional moment which created a counter-clockwise twisting at the bow and a clockwise twisting at stern. This deduction can be justified by the fact that the vessel encountered the winds and waves from 65° starboard ahead, and such winds and waves generate a torsional moment of a counter-clockwise twisting at the bow and a clockwise twisting at the stern.

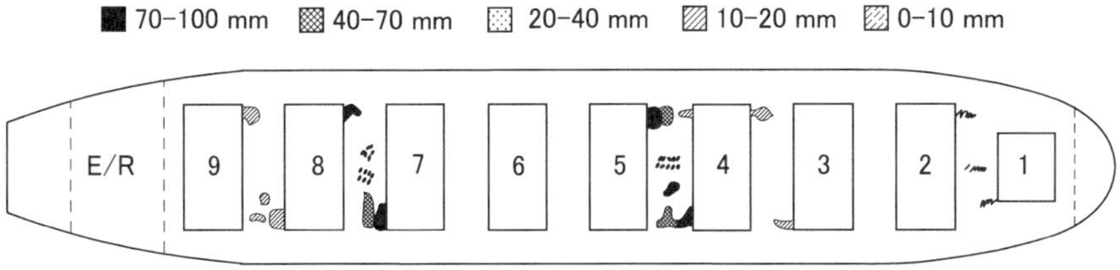

Fig. 22.6 Buckling of deck plating between hatches of 250,000 DWT bulk carrier

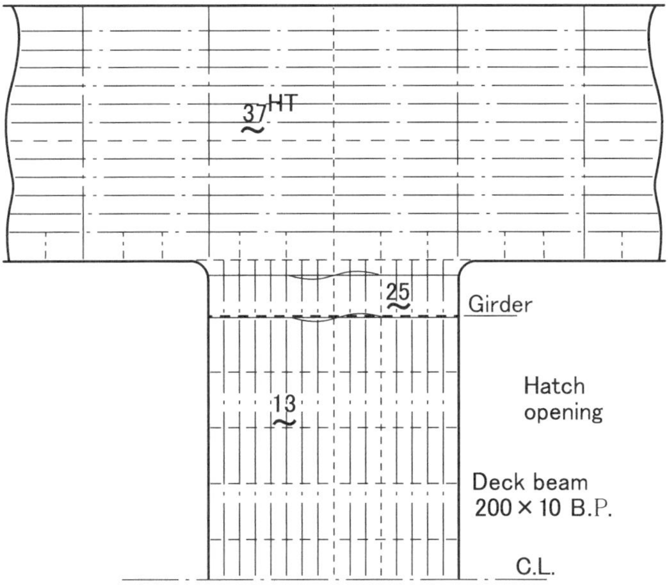

Fig. 22.7 Scantling of deck plating between hatches

This structural damage was investigated in detail and stress measurements on the vessel were carried out. Based on the analysis, the reinforcement of the deck structure was performed as shown in Fig. 22.8 and the following 4 recommendations were made:

(1) Since a bulk carrier is thought to have sufficient torsional strength, stress due to torsional deformation of the ship had been neglected. However, in cases where she carries ore in holds and in cases where large hatch openings whose breadth exceeds 50% of ship breadth are provided, special reinforcement is to be considered for deck structures against torsional deformation of the ship.

(2) In investigating the strength of the deck structure between the hatches, additional consideration is needed where the deck structure is subjected to bending due to a lateral load in the transverse bulkhead.

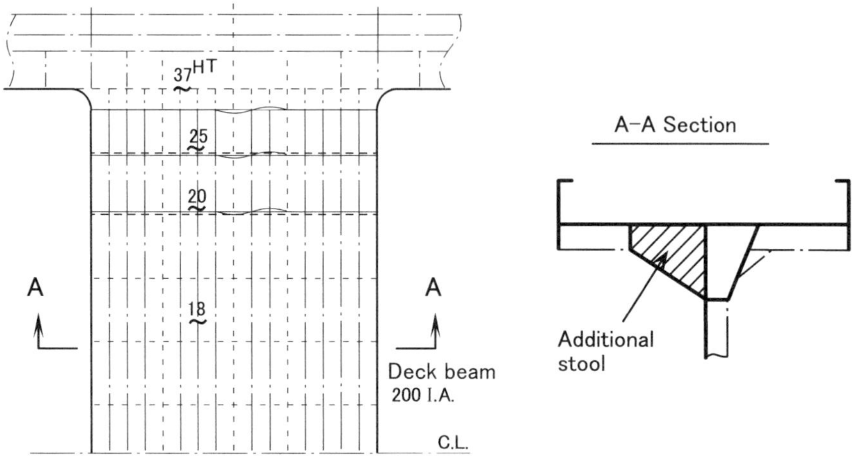

Fig. 22.8 Reinforcement for buckled deck plating between hatches

(3) The number of transverse bulkheads, that is the number of cargo holds, is to be carefully investigated, because this directly affects the compressive strength of the deck between hatches as well as the torsional strength of the ship.
(4) Homogeneous loading of ore cargo in holds is severe for the strength of the deck between hatches while alternate loading induces high stress in the double bottom. These deck and double bottom structures should be designed carefully taking account of the assumed worst loading conditions.

Chapter 23
Shell Structure

In Chaps. 19–22 of Part III, the structural design of a ship hull as a whole body is presented. Now let us explain the design for each part of the hull structure. The aim of the shell structure is to prevent water coming inside a hull, keeping the hull form. The shell structure supports generally a lateral pressure to the shell, and in some places, it supports in-plane forces such as tension and compression due to hull girder bending.

The thickness of the shell plate is very thin considering the huge structure of a ship. It can be said that the shell plate is not a plate but a membrane. The side shell plate of a VLCC of 300 m length is about 25 mm. This corresponds to 0.4 mm shell plate of a car of 5 m length, and 0.06 mm plate of a can of 10 cm diameter. Considering the scale effect, the ship's shell plate is very thin compared with other structures.

In the 1960's when ship size was increased, the shell plate thickness was not so increased. Hull structure designers were afraid of the shell plate thickness, and established a reasonable design method for the shell plate.

In this chapter a reasonable design method for the shell plate and examples of shell damage are explained.

23.1 Thickness of Shell Plates

The thickness of shell plates can be determined by the rules of classification societies. The shell plate endures not only lateral pressure caused by the water pressure but also in-plane loads caused by various deformation modes including hull girder bending and shearing, bending of primary supporting members, and bending of longitudinal stiffeners. However, classification society rules usually consider only lateral pressures and an uniaxial in-plane load due to the hull girder bending. This section

Y. Okumoto et al., *Design of Ship Hull Structures*,
https://doi.org/10.1007/978-3-032-12500-2_23

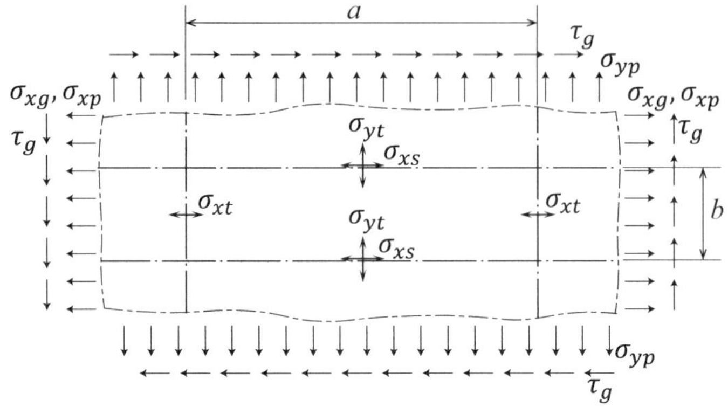

Fig. 23.1 Stress components acting on shell plates

describes a simplified strength assessment using von Mises stresses combining all the stress components acting on the shell plates.

The stresses acting in the shell plate are categorized as follows (see Fig. 23.1).

(1) longitudinal stress due to the hull girder bending: σ_{xg}
(2) shearing stress due to the hull girder bending: τ_g
(3) longitudinal and transverse stress due to the bending of primary supporting members and double bottom: σ_{xp}, σ_{yp}
(4) longitudinal stress due to the bending of longitudinal stiffeners: σ_{xs}
(5) plate bending stress due to the lateral water pressure: σ_{xt}, σ_{yt}

where the subscripts "*x*" and "*y*" designate the direction of the stress, and the subscripts "*g*", *p*", "*s*", and "*t*" denote the global, primary, secondary and tertiary stress, respectively.

Considering a rectangular plate supported by longitudinal stiffeners longitudinally and the transverse webs transversely, σ_{xg}, σ_{xp}, and σ_{xs} act longitudinally, and σ_{yp} and σ_{yt} act transversely in the mid-part along the longitudinal stiffeners, while σ_{xg}, σ_{xp} and σ_{xt} act longitudinally, and σ_{yp} acts transversely in the mid-part along the transverse webs. Of these stresses, σ_{yt} is generally the greatest because $a > b$. Hence, the stress in the mid-part along the longitudinal stiffeners should be considered.

The shearing stress τ_g mentioned above is negligible in the bottom shell plate, while τ_g in the side shell plate is calculated from the hull girder shearing force allowing for the share taken by the longitudinal bulkheads and the side shells. In this case, the von Mises equivalent stress σ_e at the mid-point along the longitudinal stiffeners is calculated for the plane stress condition as follows:

$$\sigma_e = \sqrt{\left(\sigma_{xg} + \sigma_{xp} + \sigma_{xs}\right)^2 - \left(\sigma_{xg} + \sigma_{xp} + \sigma_{xs}\right)\left(\sigma_{yp} + \sigma_{yt}\right) + \left(\sigma_{yp} + \sigma_{yt}\right)^2 + 3{\tau_g}^2} \tag{23.1.1}$$

The design criterion is that this equivalent stress is less than the yield stress of the material multiplied by the usage factor α.

$$\sigma_e \leq \alpha\sigma_Y \tag{23.1.2}$$

where σ_Y denotes the yield stress. As explained in Sect. 3.3.2(1) in Part I and Sect. 12.2 in Part II, α greater than 1.0 is usually allowed considering the plastic design.

Then, severe conditions should be selected for strength assessment as explained below taking a tanker hold part as an example.

(1) *Fully loaded condition*: In the fully loaded condition, the large water pressure in the empty tanks acts on the shell structure. Since the side longitudinals are bent inwards, σ_{xs} is a compressive stress. Hence, the hogging hull girder bending stress σ_{xg} should be considered to be combined with σ_{xs} in the bottom shell plate, and the sagging hull girder bending stress σ_{xg} should be considered to assess the plate strength in the upper side shell plate.

The strength assessment should be conducted at the sections where the hull girder bending moment or the shearing force becomes maximum. It is noted that the shearing stress becomes zero where the bending moment becomes maximum, while the bending moment is not zero where the shearing force becomes maximum.

(2) *Ballast condition*: In the ballast condition, a large water pressure in the ballast tanks acts on the shell structure from inside. Since the shell longitudinal is bent outwards, σ_{xs} is a tensile stress. Hence, the sagging hull girder bending stress σ_{xg} should be considered to be combined with σ_{xs} in the bottom shell plate, and the hogging hull girder bending stress σ_{xg} should be considered to assess the plate strength in the upper side shell plate.

The thickness of the shell plate in the above-mentioned calculations is applied usually in the 0.4 L (L: ship scantling length) midship region. However, it should not be overlooked that a large shearing stress due to hull girder bending may occur at the front part of the engine room and at the aft part of the fore peak tank. Examples of the shearing stresses calculated in the side shell of oil tankers at the aft part of the fore peak tank are shown in Fig. 23.2.

23.2 Shell at Bottom Forward

Due to the large increase in size of ships, dents of the shell plate in the bottom forward have arisen, because the distance from the pitching center to the forward shell is greater, and also the shape of the bottom forward region becomes flat. Figure 23.3 shows an example of the damage.

Where the draft is insufficient in rough seas, reinforcements are carried out in accordance with the classification rules. Since no reinforcement is usually carried

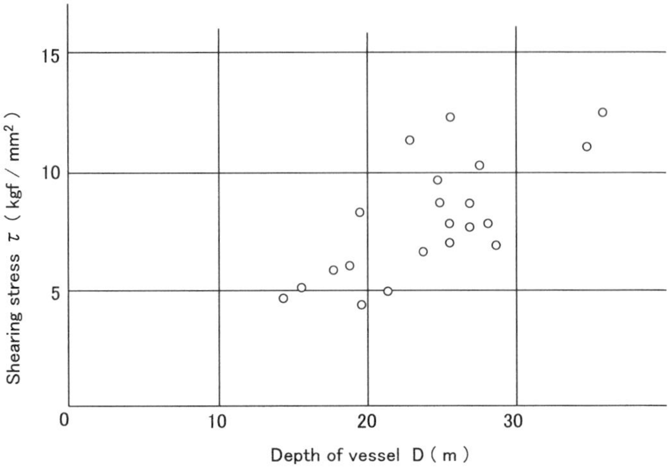

Fig. 23.2 Shearing stress of side shell around collision bulkhead

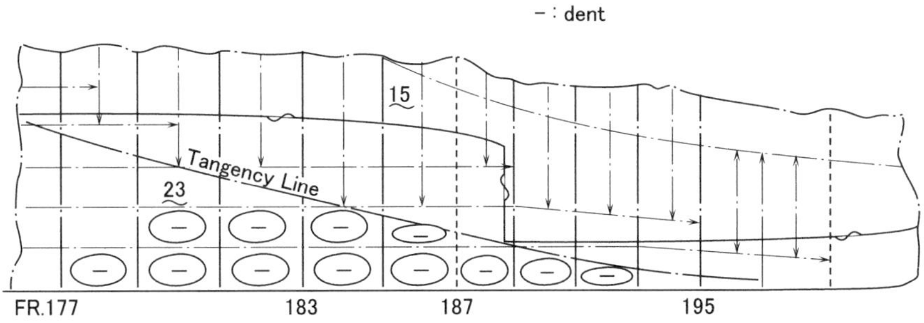

Fig. 23.3 Example of slamming damage

out where there is sufficient draft, the ship operator should know this condition and operate carefully in rough seas.

23.3 Shell at Bow Flare

Since most ships now have large bulbous bows, the bow flare in the vicinity of the bell mouth is made larger in order to avoid contact of the anchor with the shell. High speed ships also have a large bow flare for good seaworthiness. However when the bow flare is large, the pressure due to wave impact becomes large, and cracking and buckling will easily happen.

Figure 23.4 shows an example of a crack on the shell plate at the bow flare in a high-speed container ship. The crack started at Fr.151a and extended about 6.6

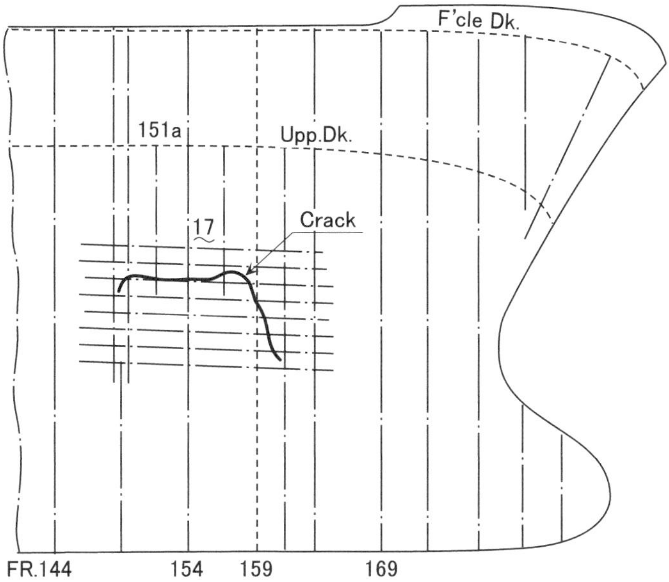

Fig. 23.4 Example of crack in bow flare

m horizontally and 3.5 m vertically. This damage occurred simultaneously with the structural damage due to torsion explained in Sect. 12.1. The ship had navigated in rough seas for 6 days before the accident; the scale of the wind force was over 10 for 11 h, and over 8 for 82 h; almost 60% of 6 days. It was reported that the ship had sailed at 9–12 knots in the center of a low atmospheric pressure region of 980 hPa with a 10–12 scale wind force for 7 h continuously.

The impact pressure due to the waves on the shell plate of the starboard side was calculated, based on ship motion in regular waves. The length of the waves was assumed to be 0.8× ship length ($\lambda/L = 0.8$) with an encounter angle of bow quartering seas ($\chi = 135°$). The maximum value is shown in Fig. 23.5, which includes two wave heights (H) of 12 m and 20 m. This shows that the pressure is proportional to the ship speed and that reducing speed is important to avoid damage due to wave impact.

Figure 23.6 shows also an example of bow damage. In this case, the water pressure to cause this damage was calculated to be about 100 m static head.

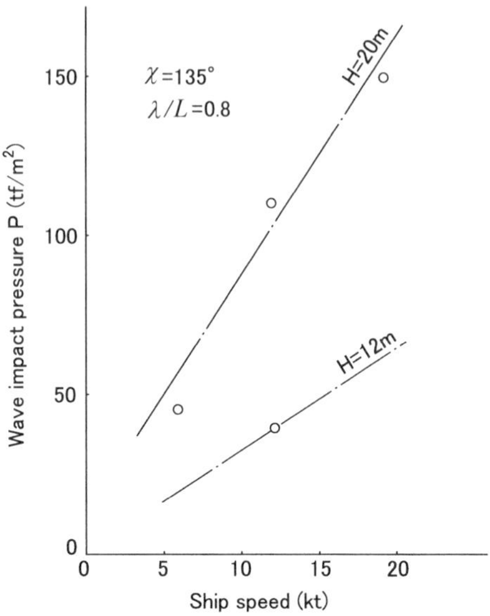

Fig. 23.5 Wave impact pressure of bow flare against ship speed

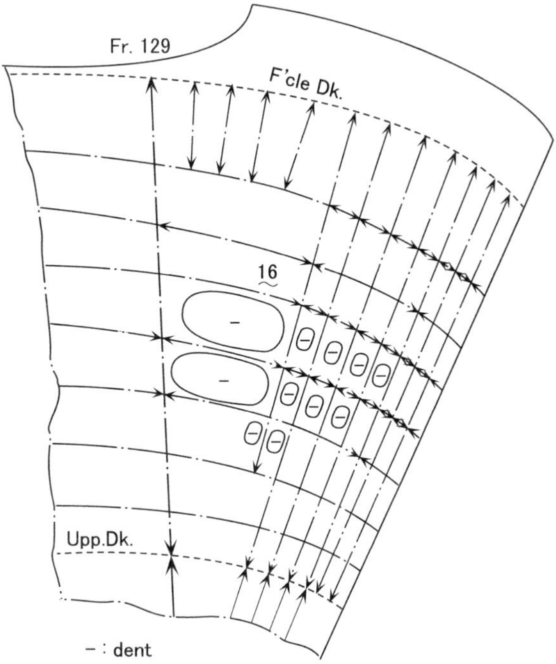

Fig. 23.6 Example of bow dent

23.4 Bilge Shell and Arch Effect

23.4.1 Unstiffened Bilge Shell Plating

It is preferable to have a large radius of bilge and a large rise-of-floor from the propulsion/resistance point of view, while smaller dimensions of both are desirable from the construction point of view. In the calculation of the ship's volume, Simpson's law used to be applied and smaller breadth of the baseline was beneficial to reduce the calculated ship's volume. In this regard, the rise-of-floor of 10 mm became a standard practice for large ships until around 1960's and 70's, allowing that the breadth of the keel plate was the breadth of the ship at the baseline. Bilge radius of 1,600 mm also became a standard because the radiused part could be made from only one strake with the width of 2,600 mm with an allowance of about 50 mm for fabrication.

It was thought that stiffening reinforcement seemed to be unnecessary for a part of a cylinder having a small radius like this thanks to the arch effect, and therefore, radiused bilge shell plate without longitudinal stiffeners appeared after structural analysis using FEM. At first it was of concern, but now this stiffener-less bilge shell plate has become a standard adopted in many ships, and no damage in this structure seems to have occurred.

When such an unstiffened bilge shell plate is employed, it is important to decide where to locate the first longitudinal stiffener adjacent to the bilge circle. In this regard, IACS CSR requires that a longitudinal is to be fitted at the bottom and at the side close to the position where the curvature of the bilge plate starts, with maximum distance of Δs_i generally not to be greater than one-third of the spacing s_i between the longitudinals, where Δs_i and s_i are as defined in Fig. 23.7. However, it is known that when a longitudinal stiffener is located very close to the position where the curvature of the bilge plate starts, the bending stress of the shell plating at the second longitudinal becomes greater than expected as a flat plate with a regular longitudinal spacing of s_i.

Now, let us explain the theoretical formulae to calculate the plate bending stress along an unstiffened bilge shell plate connected to infinitely continuous stiffened flat plate with regular spacings under uniform pressure p [133]. Figure 23.8 shows the simplified structural model employed to derive the formulae. For simplicity, the bottom and the side structures are assumed to be symmetrical about the axis

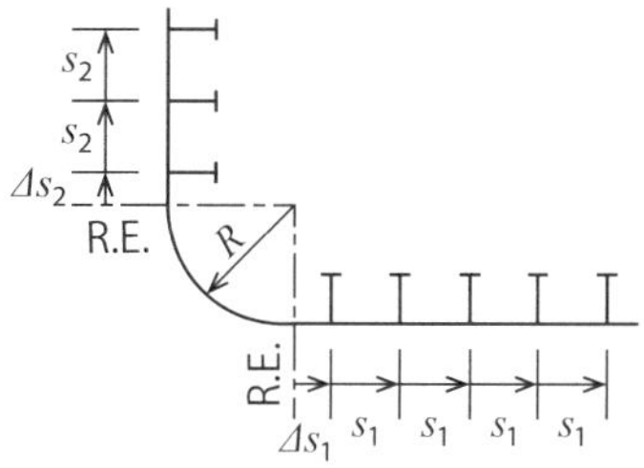

Fig. 23.7 Unstiffened bilge shell plate

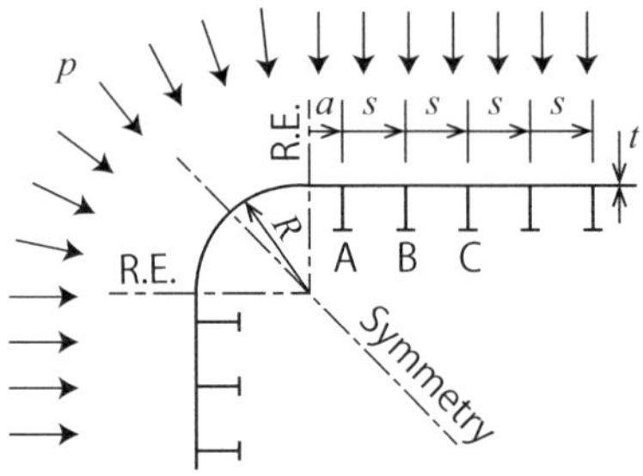

Fig. 23.8 Simplified structural model to derive plate bending stress

inclined by 45° passing through the center of the bilge circle. The shell plating of a uniform thickness of t is simply supported by longitudinal stiffeners. Then, the bending moment of the shell plating can be calculated using the following formulae, where M_A is the bending moment at the position A (first longitudinal); M_B at the position B (second longitudinal); and M_C at the position C (third longitudinal).

$$M_A = -M_{AL} - M_{\theta L} \tag{23.4.1}$$

$$M_B = -\frac{ps^2}{12} + \left(\sqrt{3} - 2\right) M_{\theta R} \tag{23.4.2}$$

$$M_C = -\frac{ps^2}{12} + \left(\sqrt{3} - 2\right)^2 M_{\theta R} \tag{23.4.3}$$

where

$$M_{AL} = \frac{EI_p}{R^2}\left(C_{1I} + \sqrt{2}C_{3I}\right) - \frac{\sqrt{2}EI_p}{R^3} C_{3I} a + \frac{pa^2}{2} \tag{23.4.4}$$

$$C_{1I} = -\frac{2R}{2\eta + \frac{\pi}{2}} \left\{ \frac{\sqrt{2}\Delta_2}{2\Delta_1}\left[2 + 2\eta - \eta^2 - 2\lambda_p(1+\eta)\right] - \frac{p}{E}\left[-\frac{a^3}{6I_p} + \frac{R}{t}\left(\frac{\pi}{4} + \eta\right)\right]\right\} \tag{23.4.5}$$

$$C_{3I} = \frac{\Delta_2}{\Delta_1} R \tag{23.4.6}$$

$$M_{\theta L} = -\frac{K_{AL}}{K_{AL} + K_{AR}}\left(M_{AL} - \frac{ps^2}{12}\right) \tag{23.4.7}$$

$$M_{\theta R} = -\frac{K_{AR}}{K_{AL} + K_{AR}}\left(M_{AL} - \frac{ps^2}{12}\right) \tag{23.4.8}$$

K_{AL} and K_{AR} are the rotational spring constants of the left-hand side (bilge shell plate side) and the right-hand side (flat stiffened plate side) of point A, respectively, and are expressed as follows:

$$K_{AL} = \frac{2EI_p}{\left(2\eta + \frac{\pi}{2}\right)R}\left\{1 + \frac{\sqrt{2}\Delta_3}{2\Delta_1}\left[\frac{\pi}{2} - 2 - \frac{\pi\eta}{2} - \eta^2 + 2\lambda_p(1+\eta)\right]\right\} \tag{23.4.9}$$

$$K_{AR} = 2\sqrt{3}\frac{EI_p}{s} \tag{23.4.10}$$

$$\begin{aligned}\Delta_1 = &\frac{1}{2\eta + \frac{\pi}{2}}\left\{4 + 8\eta - 4\eta^3 + \eta^4 - 2\lambda_p\left(2 + 4\eta + \eta^2 - \eta^3\right)\right\} \\ &- 1 - \frac{\pi}{2} - 2\eta + 2\eta^2 - \frac{2}{3}\eta^3\end{aligned} \tag{23.4.11}$$

$$\Delta_2 = \frac{\sqrt{2}p}{E}\left\{\frac{a^3(4 - 3\eta)}{24I_p} + \frac{2 + 2\eta - \eta^2}{2\eta + \frac{\pi}{2}}\left[-\frac{a^3}{6I_p} + \frac{R}{t}\left(\frac{\pi}{4} + \eta\right)\right]\right\} \tag{23.4.12}$$

$$\Delta_3 = \frac{\sqrt{2}\left(4 - \pi + \pi\eta + 2\eta^2\right)}{4\eta + \pi} \tag{23.4.13}$$

where

E: Young's modulus.

I_p: sectional moment of inertia of shell plating ($= t^3/12$)

$$\eta = a/R$$

$$\lambda_p = I_p/\left(tR^2\right)$$

According to Eqs. (23.4.1)–(23.4.3), bending stresses on the shell plating at longitudinal stiffener positions A, B and C can be calculated, and the results are shown in Fig. 23.9 taking a/s for the horizontal axis. Calculation conditions are given in Fig. 23.10. From Fig. 23.9, we can observe that when a/s is around 0.5, that is, when the first longitudinal is apart from the curvature end by about a half of the longitudinal spacing, the bending stresses at the three positions becomes equal with the best balance. Bending stress at point A becomes larger as a/s gets larger, while bending stress at point B becomes larger when a/s is smaller. When the first stiffener is located at the position where the curvature of the bilge plate starts ($a/s = 0$) as classification rules recommend, the bending stress at the longitudinal B becomes about 13% greater than expected as a flat plate with a regular longitudinal spacing. This effect is even greater when the bilge radius is increased.

In some cases, it is inevitable that the first longitudinal stiffener is located very close to the position where the curvature starts. Even in such cases, the bending stress at point B can be mitigated by arranging the second longitudinal B closer to the first longitudinal A. Even when $a/s = 0$, we can avoid the increase in the bending stress at point B by making the spacing between the first and the second longitudinals less than $0.85s$ [134].

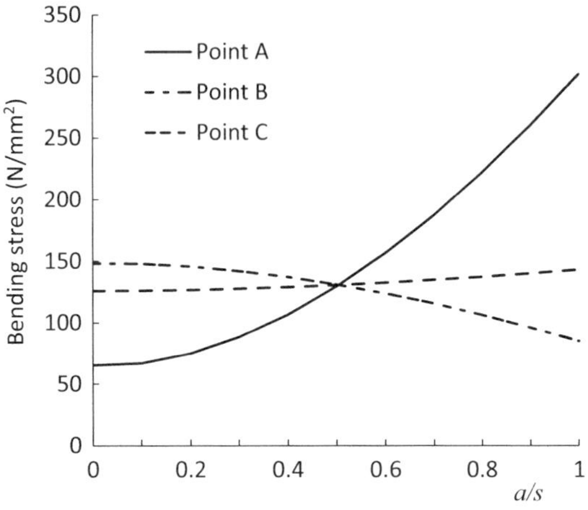

Fig. 23.9 Bending stress of shell plating at longitudinal stiffener positions

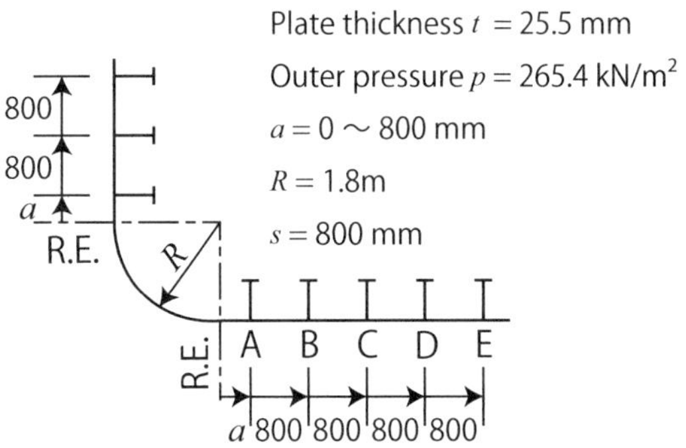

Fig. 23.10 Calculation conditions for unstiffened bilge shell plating

23.4.2 Stiffened Bilge Shell Plating

In general, bilge part of blunt ships, such as tankers and bulk carriers, is composed of unstiffened bilge plate. By contrast, fast ships, such as container ships and LNG carriers, have much larger bilge radii, and their bilge part is composed of stiffened bilge plate as shown in Fig. 23.11. In such a structural arrangement, longitudinal stiffeners are elastically supported effectively by the arch made of the curved shell plating. Thanks to this effect, the required section modulus of the longitudinal stiffeners can be reduced, although this effect is not considered in the major classification society rules.

The bending moment of the stiffener at its span end can be expressed as follows [135]:

$$M = -\frac{1}{12} psl^2 \kappa_a f_M \tag{23.4.14}$$

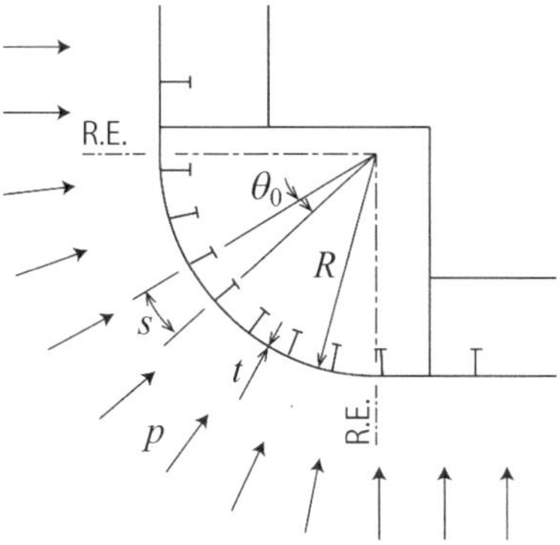

Fig. 23.11 Stiffened bilge shell plating

where

l: span length of the longitudinal stiffener

$$\kappa_a = \frac{1}{1 + \frac{1}{\lambda_p}\left\{\frac{1}{2}\left(1 + \frac{\theta_0}{2}\cot\frac{\theta_0}{2}\right)\frac{\theta_0}{2}\cot\frac{\theta_0}{2} + \frac{1}{2}\left(\frac{\theta_0}{2}\right)^2 - 1\right\}} \tag{23.4.15}$$

θ_0: included angle of two adjacent longitudinals as shown in Fig. 23.11

$$f_M = \frac{6(\sinh\lambda - \sin\lambda)}{\lambda^2(\sinh\lambda + \sin\lambda)} \tag{23.4.16}$$

$$\lambda = \sqrt[4]{\frac{\kappa_a}{4}\left(\frac{t}{R}\right)\left(\frac{s}{R}\right)\frac{l^4}{(1-\nu^2)I}} \tag{23.4.17}$$

ν: Poisson's ratio

I: sectional moment of inertia of the longitudinal stiffener with the attached plate

In Eq. (23.4.14), the part $psl^2/12$ is the bending moment at the span end for a longitudinal stiffener on a flat plate. Therefore, $\kappa_a f_M$ can be interpreted as the reduction ratio of the bending moment thanks to the elastic support of the curved shell plate. This reduction ratio $\kappa_a f_M$ is plotted in Fig. 23.12 for two cases: $\kappa_a = 0.35$ and $\kappa_a = 0.55$, taking λ in the horizontal axis. Regarding container ships with ordinary structural arrangement, κ_a ranges between 0.35 and 0.55, and λ ranges between 2.2 and 2.9. That is, the bending moment reduction falls into the hatched zone in Fig. 23.12, indicating that the required section modulus of the longitudinal stiffeners is usually less than a half of that required for the stiffeners on a flat plate. This effect may be beneficial in some cases in reducing the scantling of longitudinal stiffeners, but in reality, the parallel part of container ships is quite short, for example, only one block long, and it may not be so effective to reduce scantling within such a short length of a ship.

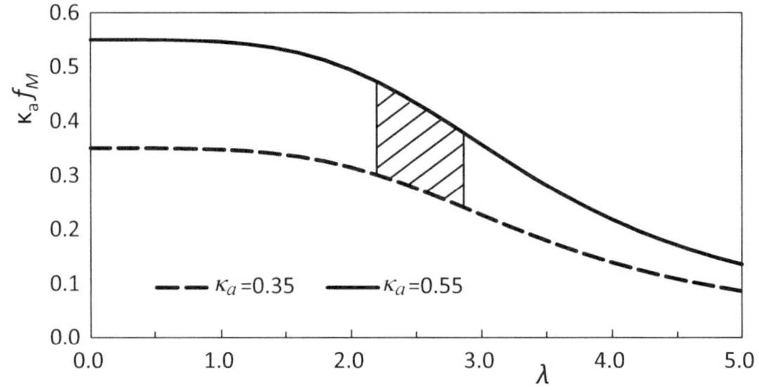

Fig. 23.12 Reduction ratio of stiffener bending moment on a curved shell

23.4.3 Benefits and Drawbacks of Arch Effect

As seen in the previous sections, cylindrical structures can support lateral pressure efficiently by balancing the lateral load by the membrane forces acting on the structure. Bilge shell plating is a typical example of such a structure. Cylindrical tanks and spherical tanks on liquefied gas carriers are another example. In addition, as explained in Fig. 21.21 in Sect. 21.5.6, large curved horizontal girders can support weak double hull structure in No. 1 C.O.T. of tanker structures thanks to the arch effect. There are many cylindrical shell areas and curved structures in the fore and aft structures. Since such a cylindrical shell is stronger than a flat plate under lateral force, in some areas the scantling can be reduced or the stiffener space can be wider, in accordance with each curvature. Further studies of the strength and vibration for such curved surfaces are now necessary for the hull designer in order to rationalize the hull structure design.

On the other hand, the arch effect is a double-edged sword and can cause damage in some cases. The bow flare failure of a high-speed vessel described in Sects. 23.3 and 27.2 was caused by a large hoop stress along the circumference of the bow flare due to wave impact loads acting on the shell plate of the bow flare, which is curved into a concave shape with a large radius. The rational arrangement of large curved horizontal girders in No. 1 C.O.T. of a tanker as mentioned above is possible because the double side structure is longitudinally framed, but the situation is different in the case of a bulk carrier. The single side structure of a bulk carrier is usually transversely framed as shown in Fig. 23.13, making the shell plate prone to buckling damages because of the large longitudinal hoop stresses due to the large radius of the shell plating. Figure 23.14 shows an example of such shell plate damages experienced on a cape-size bulk carrier, where shell plating between hold frames has collapsed due to the combination of external wave pressure and large compressive hoop stress. Two important factors are considered to have affected the damage as follows:

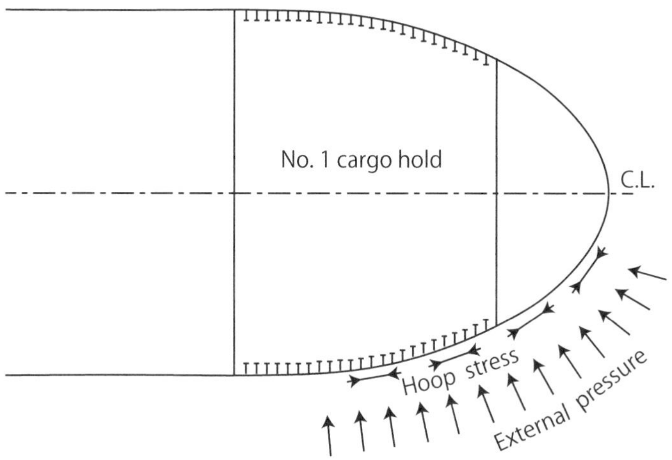

Fig. 23.13 Hold frames in No. 1 cargo hold of a bulk carrier

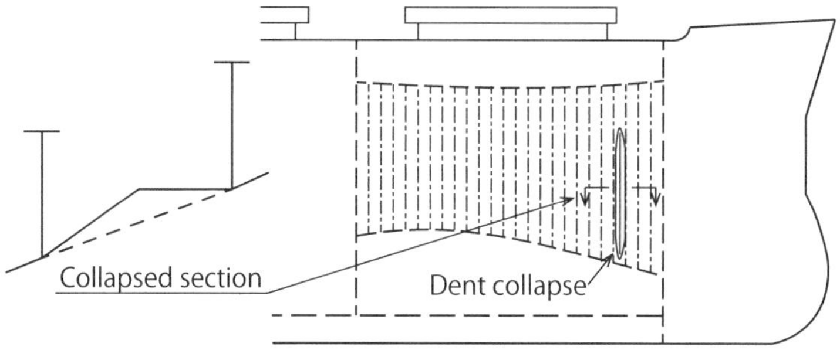

Fig. 23.14 Collapse of shell plate between hold frames in No. 1 cargo hold of a bulk carrier

(1) The shell plate thickness was thinner than that calculated using the actual stiffener spacing, which is larger than the frame space due to the horizontal inclination of the hull shape.
(2) Because of the long hold frames arranged in No. 1 cargo hold, the deflection of the hold frames was large causing large hoop stress.

To avoid this kind of shell damage, it is necessary to apply sufficient plate thickness and adequate hold frame rigidity corresponding to its span length.

Other examples of damages due to the arch effect are explained in various chapters of this book, such as:

(1) Crack damages along the welding between the curved faceplate and the web plate at the root of a strut of a tanker described in Sect. 11.6 of Part II.
(2) Crack damages at the knuckle part of a longitudinal bulkhead of an ore carrier described in Sect. 21.6.

(3) Crack damages at the intersection of the inner bottom plate and the bilge hopper slant plate of a bulk carrier in Sect. 21.7.

These damages were caused by the out-of-plane force of a curved or knuckled plate produced by the membrane forces acting on the plate, i.e., the same mechanism as the arch effect where the out-of-plane force is converted to and supported by membrane forces. Meticulous care must be taken when designing curved or knuckled structures subject to membrane forces.

23.5 Shell Near Stern Frame

For the shell in the vicinity of the stern frame, thicker plates are used in comparison with other shell plates. The stern frame with a G type rudder is an important structure in supporting the propeller, propulsive shaft, and rudder. The lower part contacts a keel block first during docking and transmits the force from the block to the hull.

Recent ship structures with a bulbous stern or open bulbous stern, associated with energy-saving, do not support the hull weight at docking, do not support the rudder as shown in Fig. 23.15, and only support the propeller shaft.

In the case of a structure with an open bulbous stern, increasing the thickness of the side shell in the vicinity of the casting is not desirable from the strength/vibration point of view, because it increases the weight at the end of a cantilever beam. If it is necessary to increase rigidity, increasing the plate thickness around part B not part A in Fig. 23.15 is effective. However, the ship breadth at B with a bulbous stern is wide already, hence the transverse rigidity seems to be strong enough.

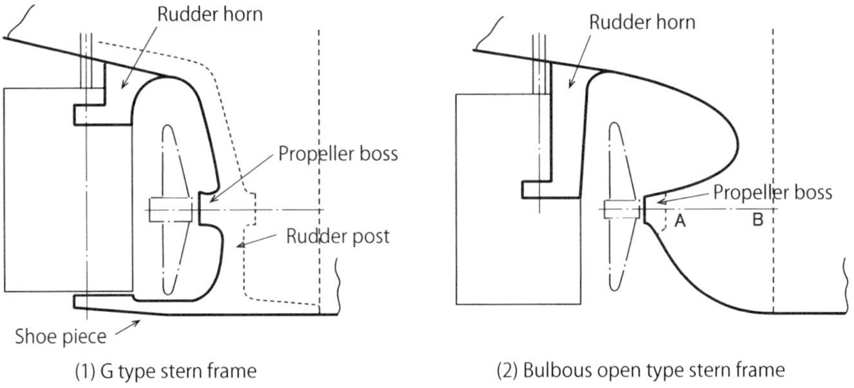

Fig. 23.15 Stern frame structure

23.6 Shell Damage

Since the shell structure protects a ship from exterior forces, it suffers easily from damage by waves and collision with floating objects. Considering that one of the important functions of shell plating is to prevent seawater flooding inside the hull, the damage is classified into 2 cases: with cracks and without cracks. When there are no cracks even if damage occurs on the shell, the shell structure keeps tightness and is fulfilling its function in this sense. The outflow of oil from a large tanker becomes a social problem from the environmental protection point of view, hence cracks in the shell cannot be allowed.

An example in Fig. 21.1a shows that the shell plate was dented in more than 3 m inside the hold by the buckling of a cross tie. But the cracking of the shell plate did not occur in this case fortunately. In the damage example shown in Fig. 21.1b, the shell plate was bent severely together with damage of the transverse webs. In this case the shell plate deformed 300mm outside convexly, but this deformation of shell plate did not cause cracks. A flying bird type crack occurred in the shell plate actually, but this was due to the propagation of the crack, which was generated in the web plate of the transverse web.

Figure 23.16 shows an example of damage to the bottom shell plate of a VLCC, which occurred by contact with the seabed when she navigated through the Malacca Strait in the fully loaded condition. Cracking did not occur. Such damages at grounding are shown in the reference [136].

Fig. 23.16 Dent of bottom shell by grounding

Fig. 23.17 Cracks of shell plate by floating object

Also in case of dent damage due to wave impact pressure, it is seldom associated with cracks as shown in Figs. 23.3 and 23.6. On the other hand, cracking of the shell will occur when large tensile hoop stress acts on the shell plate over a wide area caused by wave pressure, as shown in Fig. 23.4, and this cracking will start at a point having a welding defect.

Figure 23.17 shows damage with cracking which occurred due to lateral shearing forces by contact with a floating object. In addition, damage to the shell occurs often when the pushing force of a tugboat is excessive, or when the ship hits a pier at high speed.

Chapter 24
Bulkheads

Bulkheads which are installed to form the boundary of a compartment in which liquid cargo is to be loaded support a normal perpendicular force on the bulkhead plate. In addition to this, they support an in-plane force when they are installed as part of the frame work of the ship's whole structure, as explained in Chap. 19. Also, the longitudinal bulkheads of a tanker support the axial and shearing forces as well as the side shell plating, due to the longitudinal hull girder bending.

Especially in the case of a large ship, the in-plane force is important, and special attention is to be paid to the design of the swash bulkhead, in a long tank which is non-watertight and perforated.

A corrugated bulkhead is lighter than a stiffened plane bulkhead because it has no stiffeners but the plate itself plays the role of the stiffeners.

24.1 Horizontal Girders on Transverse Bulkheads (in Center Tank)

Here the measured behavior of a horizontal girder in a center tank of a tanker with a vertical stiffener arrangement is explained. Nowadays, thanks to the development of computer calculations such as matrix methods of structural analysis for frame and shell, on-board measurements have scarcely been carried out. However, such calculated results depend on the boundary conditions which can be understood only by on-board measurements. It is important for a designer to confirm the accuracy of a calculation by measurement.

At around the end of the 1960's, the horizontal girder on a transverse bulkhead was designed as a girder with both ends fixed, assuming the span points where the girder was fixed. With the increase in size of ships the span of the horizontal girder

Y. Okumoto et al., *Design of Ship Hull Structures*,
https://doi.org/10.1007/978-3-032-12500-2_24

was getting bigger and accordingly the question was raised whether span points for a smaller ship can be similarly applied to larger ships. Then measurements were carried out on board a 70,000 DWT tanker. The outline and the result of the measurements are shown in Fig. 24.1.

Assuming the horizontal girder has both ends elastically supported at the longitudinal bulkheads, the bending moment M at the point x (m) from the longitudinal bulkhead is presented by the following equation.

$$M = \mu \frac{wl^2}{12} - \frac{wl}{2}x + \frac{w}{2}x^2 \tag{24.1.1}$$

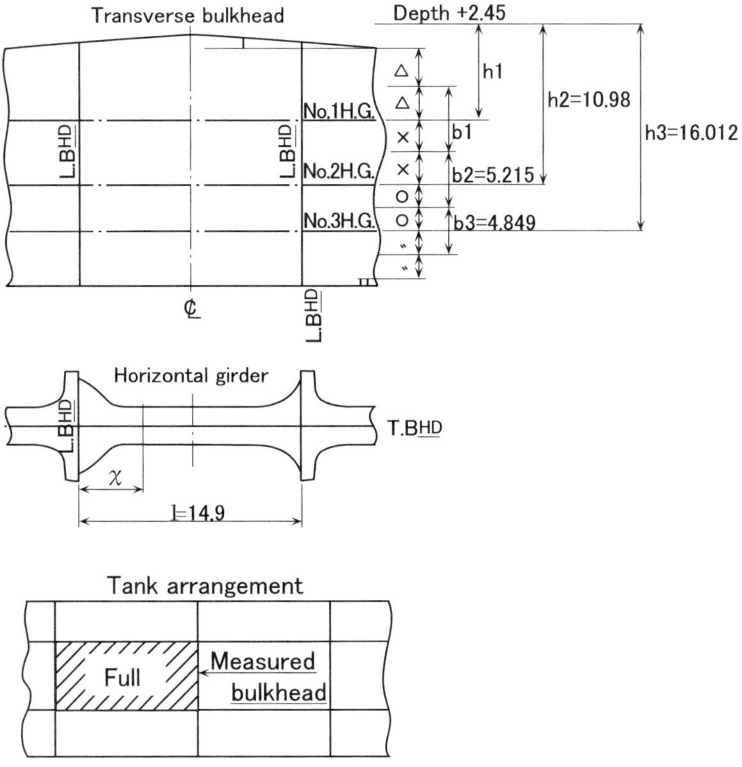

Measured results

	No.2H.G.	No.3H.G.
σ at ℄	15.5kgf/mm^2	11.0kgf/mm^2
x where σ = 0	3.45m	3.45m

Test head is depth+2.45m

Fig. 24.1 Example of stress measurement in horizontal girder of transverse bulkhead

where

μ: coefficient of degree of rigidity at ends.

l: distance between longitudinal bulkheads.

w: magnitude of applied uniform load.

From the measurements, the distance x where $M = 0$ and the moment at midspan $x = l/2$ can be obtained. Then, μ and w were derived using Eq. (24.1.1) as follows:

$\mu = 1.07$, $w = 1.54w_0$ for No. 2 horizontal girder.

$\mu = 1.07$, $w = 1.03w_0$ for No. 3 horizontal girder.

where w_0 is the nominal applied load, that is, the product hb of the water head h and the breadth of the applied load b on the horizontal girder, as shown in Fig. 24.1.

Theoretically, $\mu = 1.0$ in the case where a girder is fixed at the longitudinal bulkhead, and $\mu = 0$ in the case where a girder is simply supported at the longitudinal bulkhead. When the rotation of the span end is elastically supported, μ is expected to take a value varying between 0 and 1.0 corresponding to the rigidity of the end supports. However, the measured result shows that $\mu = 1.07$ which is out of the above range. This means that the span point (fixed point) lies a little inside of the center tank from the longitudinal bulkhead as shown in Fig. 24.2.

From Eq. (24.1.1), the bending moment at the midspan $M(l/2)$ is:

$$M\left(\frac{l}{2}\right) = \frac{wl^2}{24}(2\mu - 3) \tag{24.1.2}$$

At the span point, the magnitude of the bending moment must be twice as large as $-M(l/2)$. Therefore, equating Eqs. (24.1.1) and (24.1.2) multiplied by -2, the span point location can be found as follows:

$$\mu\frac{wl^2}{12} - \frac{wl}{2}x + \frac{w}{2}x^2 = -\frac{wl^2}{12}(2\mu - 3) \tag{24.1.3}$$

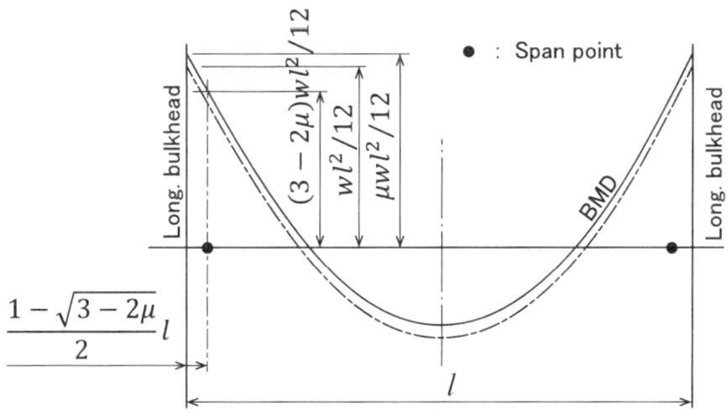

Fig. 24.2 Span point of horizontal girder of transverse bulkhead (measured result)

$$\therefore \frac{x}{l} = \frac{1 \pm \sqrt{3 - 2\mu}}{2} \tag{24.1.4}$$

When $\mu = 1.07$, $x/l = 0.0363$ is obtained.

From the above results it was found that the span point lies at a point 3.6% of span inside of the center tank from the longitudinal bulkhead. It was made clear from the measurements that despite the big brackets at both ends of the horizontal girder, the span points are quite close to the longitudinal bulkheads.

One more important item made clear by the measurement relates to the load supporting ratio of each horizontal girder. Putting $\mu = 1.0$, $\alpha = w/w_0$ is calculated for each horizontal girder. Then, $\alpha = 1.32$ for the No. 2 horizontal girder, and $\alpha = 0.89$ for the No. 3 horizontal girder are obtained. It can be understood that the load supporting ratio of the No. 2 horizontal girder (at the mid-depth of the tank) is bigger than that of the No. 1 and No. 3 horizontal girders, which are connected to the rigid deck or bottom structures by vertical bulkhead stiffeners. However, the value of $\alpha = 1.32$ for the No. 2 horizontal girder is more than common sense suggests.

Based on the above measured results, the following design standards for the horizontal girder in a center tank were established.

(1) The span is to be equal to the distance between the longitudinal bulkheads.
(2) The ratio α between the design load and the nominal load is to be:

1.35 for the middle girder in the case of a one or three horizontal girder arrangement.
1.15 for the uppermost horizontal girder in the case of a two or more horizontal girder arrangement.
1.0 for other horizontal girders.

Applying the above criteria on ships in service, the data shown in Table 24.1 were obtained. This kind of data is useful to establish allowable stresses for the design standards. The data show the differences between classification societies.

24.2 Horizontal Girder Arrangement on Bulkheads

The water pressure on a bulkhead of a tank increases with water depth. In the case of a vertical stiffener system, it is beneficial to arrange vertical stiffeners with an identical scantling throughout their length from viewpoint of procurement and fabrication. To achieve this condition, the horizontal girders which support the vertical stiffeners are best arranged so as to have an equal bending moment in each vertical stiffener span supported by the horizontal girders, deck and bottom. This leads to the spacing of the horizontal girders smaller near the bottom and wider near the deck as shown in Fig. 24.3.

Table 24.1 Calculated bending stress and shearing stress of horizontal girder of transverse bulkhead

SHIP	CLASS	σ (kgf / mm^2)			τ (kgf / mm^2)			τ_{cr} / τ			NOTE
		UPP.	MID.	LOW.	UPP.	MID.	LOW.	UPP.	MID.	LOW.	
A	NK	13.5	18.3	14.0							
B	NK	16.8	23.0	17.4							
C	AB	8.2	10.2	7.2							
D	NK	15.1	20.1	14.6							
E	NK	18.8	24.0	15.6							
F	NK	17.2	25.7	16.2							
G	AB NK	6.2	5.8	4.6	6.2	8.7	9.1	3.0	2.2	2.1	
H	NK	7.3	10.1	8.1	4.6	7.5	6.8	4.9	3.0	3.4	
I	NV				/	11.6	/		0.8		
J	AB				/	9.5	/		1.8		
K	NK				/	11.0	/		1.6		
L	NK				9.0	/	7.7	2.3		3.4	
M	AB				/	10.6	/		1.8		
N	AB				/	10.4	/		1.9		
O	AB				/	12.2	/		1.3		
P	LR				/	10.2	/		1.9		
Q	AB				/	12.6	/		1.8		
R	LR				7.1	11.0	11.7	2.4	1.6	1.5	
S	AB				6.1	8.1	6.3	3.9	2.9	3.8	

NOTE·····NK·····Nippon Kaiji Kyokai
AB·····American Bureau of Shipping
NV·····Det Norske Veritas
LR·····Lloyd´s Register of Shipping

First, let us find the rigorous optimum location of the horizontal girder in case only one horizontal girder is arranged, as shown in Fig. 24.4. λ denotes the ratio of the length of the vertical stiffener above the horizontal girder to the total length of the vertical stiffener. Linearly distributed load is applied on the bulkhead with w_U and w_L denoting the loading per unit length at the top and the bottom, respectively. The theorem of three moments is to be applied, assuming virtual spans outside the fixed boundary with infinite rigidity, and as a result, the bending moment at deck (M_U), at the horizontal girder (M_M), and at the bottom (M_L) are obtained as follows:

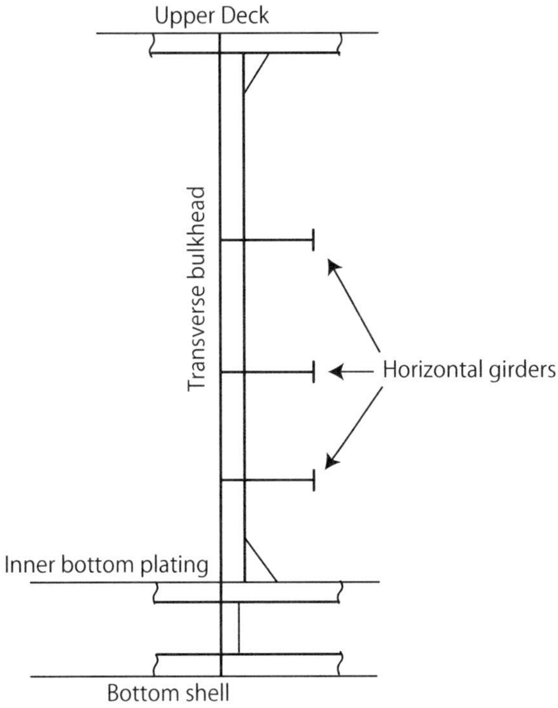

Fig. 24.3 Horizontal girder arrangement

$$\therefore \begin{pmatrix} M_U \\ M_M \\ M_L \end{pmatrix} = -\frac{l^2}{60} \begin{pmatrix} \frac{3+\lambda}{6} & -\frac{1}{3} & \frac{1-\lambda}{6} \\ -\frac{\lambda}{3} & \frac{2}{3} & -\frac{1-\lambda}{3} \\ \frac{\lambda}{6} & -\frac{1}{3} & \frac{4-\lambda}{6} \end{pmatrix} \begin{pmatrix} \lambda^2(15-7\lambda)w_U + 7\lambda^3 w_L \\ \left\{\lambda^3(15-8\lambda) + 8(1-\lambda)^4\right\}w_U + \left\{8\lambda^4 + (1-\lambda)^3(7+8\lambda)\right\}w_L \\ 7(1-\lambda)^3 w_U + (8+7\lambda)(1-\lambda)^2 w_L \end{pmatrix} \tag{24.2.1}$$

The bending moments at the three support points are plotted taking λ in the horizontal axis in Fig. 24.5. Figure 24.5(1) shows the case of uniformly distributed load ($w_U = w_L$), while Fig. 24.5(2) shows the case of triangular distribution ($w_U = 0$). In case of the uniformly distributed load, when the horizontal girder is located at the mid-depth ($\lambda = 0.5$), all the three bending moments becomes equal and minimum as a matter of course. Under the triangular pressure distribution, the curves of M_L and M_M cross at $\lambda = 0.605$, where the maximum of the absolute values of the three bending moments is minimized. When the horizontal girder is located higher than this ($\lambda < 0.605$), $|M_L|$ is maximum. When it is below this ($\lambda > 0.605$), $|M_M|$ is maximum.

This rigorous approach becomes quite cumbersome when the number of horizontal girders is more than one. Then, to simplify the problem, we now model the loading as uniformly distributed as shown in Fig. 24.6, where

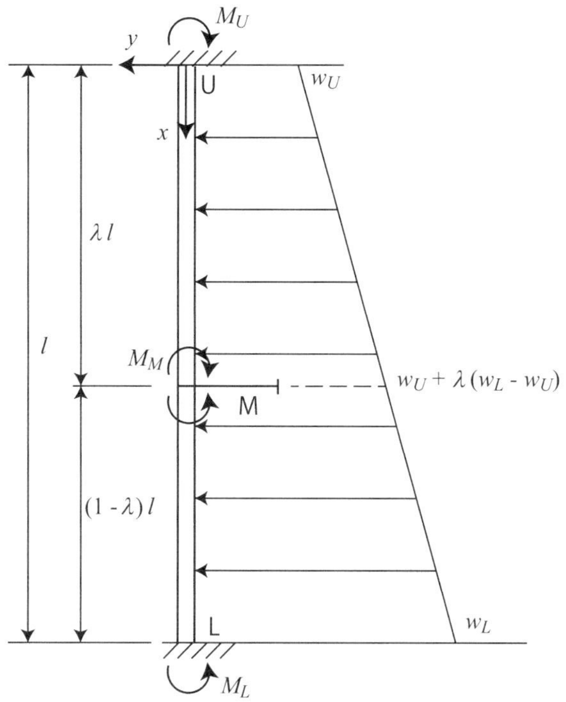

Fig. 24.4 Transverse bulkhead with one horizontal girder under linearly distributed load

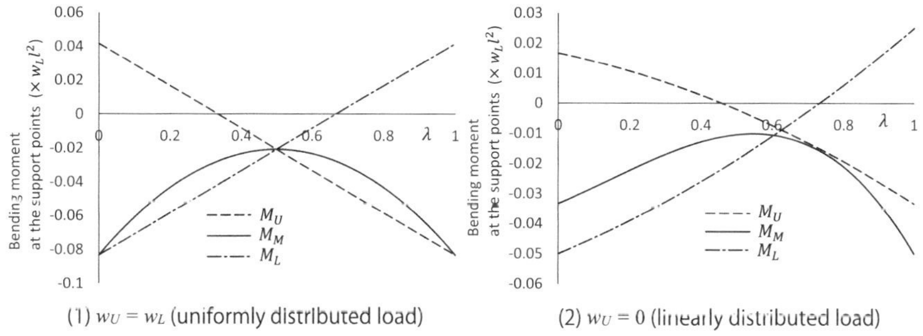

Fig. 24.5 Bending moment on vertical stiffener at support points (one horizontal girder)

$$w_{UM} = w_U + \frac{\lambda}{2}(w_L - w_U) \tag{24.2.2}$$

$$w_{LM} = w_U + \frac{1+\lambda}{2}(w_L - w_U) \tag{24.2.3}$$

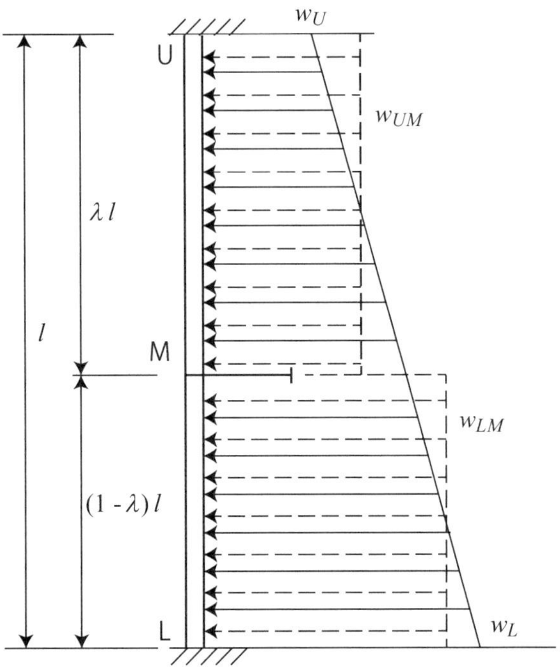

Fig. 24.6 Transverse bulkhead with one horizontal girder with uniform load approximation

This approximation is often used in the rules of classification societies. In this case, all the bending moments at the three supporting points U, M and L are identical when

$$w_{UM}(\lambda l)^2 = w_{LM}\{(1-\lambda)l\}^2 \tag{24.2.4}$$

Substituting Eqs. (24.2.2) and (24.2.3) into Eq. (24.2.4),

$$\left(-1+3\lambda-\lambda^2\right)w_U+\left(-1+\lambda+\lambda^2\right)w_L=0 \tag{24.2.5}$$

Now, we introduce a coefficient μ which represents the degree of uniformity of the distributed load ($w_U = \mu w_L$). $\mu = 1$ means completely uniform distribution while $\mu = 0$ means completely triangular distribution. Then, Eq. (24.2.5) reduces to:

$$\left(-1+3\lambda-\lambda^2\right)\mu-1+\lambda+\lambda^2=0$$

$$\therefore (1-\mu)\lambda^2+(1+3\mu)\lambda-1-\mu=0 \tag{24.2.6}$$

(1) *Uniform distribution* $\mu = 1$*:* Substituting $\mu = 1$ into Eq. (24.2.6), we can easily get $\lambda = 0.5$, which is same as the rigorous solution shown in Fig. 24.5(1).

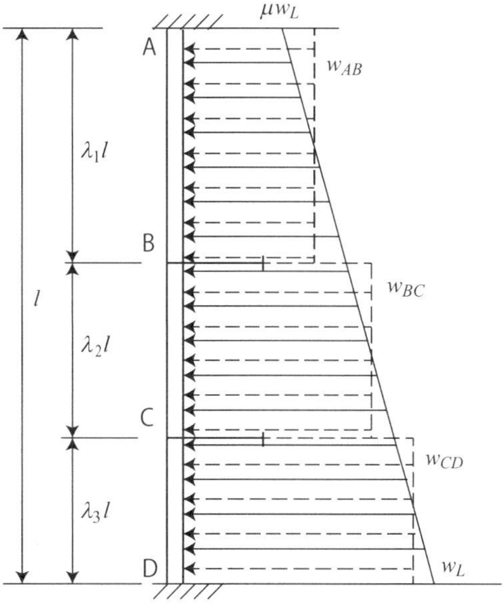

Fig. 24.7 Transverse bulkhead with two horizontal girders with uniform load approximation

(2) *Triangular distribution* $\mu = 0$*:* Substituting $\mu = 0$ into Eq. (24.2.6), the solution is:

$$\lambda = \frac{-1 + \sqrt{5}}{2} = 0.618 \tag{24.2.7}$$

which well approximates the rigorous solution of $\lambda = 0.605$ shown in Fig. 24.5(2).

When we have two horizontal girders as shown in Fig. 24.7, substituting.

$$w_{AB} = \left\{\mu + \frac{\lambda_1}{2}(1 - \mu)\right\} w_L \tag{24.2.8}$$

$$w_{BC} = \left\{\mu + \frac{2\lambda_1 + \lambda_2}{2}(1 - \mu)\right\} w_L \tag{24.2.9}$$

$$w_{CD} = \left\{\mu + \frac{1 + \lambda_1 + \lambda_2}{2}(1 - \mu)\right\} w_L \tag{24.2.10}$$

into

$$w_{AB}(\lambda_1 l)^2 = w_{BC}(\lambda_2 l)^2 = w_{CD}(\lambda_3 l)^2, \tag{24.2.11}$$

we can obtain the following governing equation.

$$\{2\mu + \lambda_1(1-\mu)\}\lambda_1^2 = \{2\mu + (2\lambda_1 + \lambda_2)(1-\mu)\}\lambda_2^2$$
$$= \{2\mu + (1 + \lambda_1 + \lambda_2)(1-\mu)\}\lambda_3^2 \quad (24.2.12)$$

From the first and the second terms of Eq. (24.2.12),

$$\left(\frac{\lambda_1}{\lambda_2}\right)^2 = \frac{\frac{\mu}{\lambda_1} + \left(1 + \frac{\lambda_2}{2\lambda_1}\right)(1-\mu)}{\frac{\mu}{\lambda_1} + \frac{1-\mu}{2}} \quad (24.2.13)$$

From the first and the third terms of Eq. (24.2.12),

$$\left(\frac{\lambda_1}{\lambda_3}\right)^2 = \frac{\frac{\mu}{\lambda_1} + \left(1 + \frac{\lambda_2}{\lambda_1} + \frac{\lambda_3}{2\lambda_1}\right)(1-\mu)}{\frac{\mu}{\lambda_1} + \frac{1-\mu}{2}} \quad (24.2.14)$$

Using these two equations, optimum arrangement of the horizontal girders can be found by identifying $\lambda_1, \lambda_2, \lambda_3$ which satisfy the condition $\lambda_1 + \lambda_2 + \lambda_3 = 1$ through an iterative method.

When the number of the horizontal girders is three introducing λ_4 in addition to $\lambda_1 \sim \lambda_3$, Eqs. (24.2.13) and (24.2.14) hold as they are, and in addition, the following equation holds with respect to the lowest span.

$$\left(\frac{\lambda_1}{\lambda_4}\right)^2 = \frac{\frac{\mu}{\lambda_1} + \left(1 + \frac{\lambda_2}{\lambda_1} + \frac{\lambda_3}{\lambda_1} + \frac{\lambda_4}{2\lambda_1}\right)(1-\mu)}{\frac{\mu}{\lambda_1} + \frac{1-\mu}{2}} \quad (24.2.15)$$

The optimum positions of horizontal girders depending on the pressure distribution pattern are plotted in Fig. 24.8. We can observe that the girders are equally spaced when the load is uniformly distributed ($\mu = 1$), while the girders are located lower as the top head decreases toward $\mu = 0$. The span of the uppermost stiffener is 1.5 or 2.0 times longer than the span of the lower stiffener when the load distribution is close to triangular. This kind of design chart is useful in initial design stage.

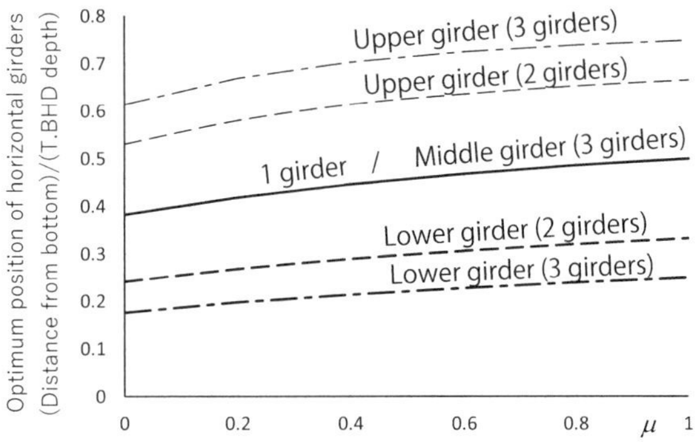

Fig. 24.8 Design chart for horizontal girder arrangement

24.3 Vertical Stiffeners on Transverse Bulkheads

In the design of a vertical stiffener on a transverse bulkhead which is supported at one side or both sides by horizontal girders, not only the water pressure but also the displacement of the horizontal girders needs to be considered. The stiffener at the lowest span is most adversely affected by this forced displacement because of the large relative displacement between the lowest horizontal girder and the rigid bottom, smaller span and larger depth of the stiffener. Figure 24.9a shows the additional stress in the vertical stiffener caused by the displacement of the horizontal girder, and Fig. 24.9b shows an example of the damage caused by the additional stress.

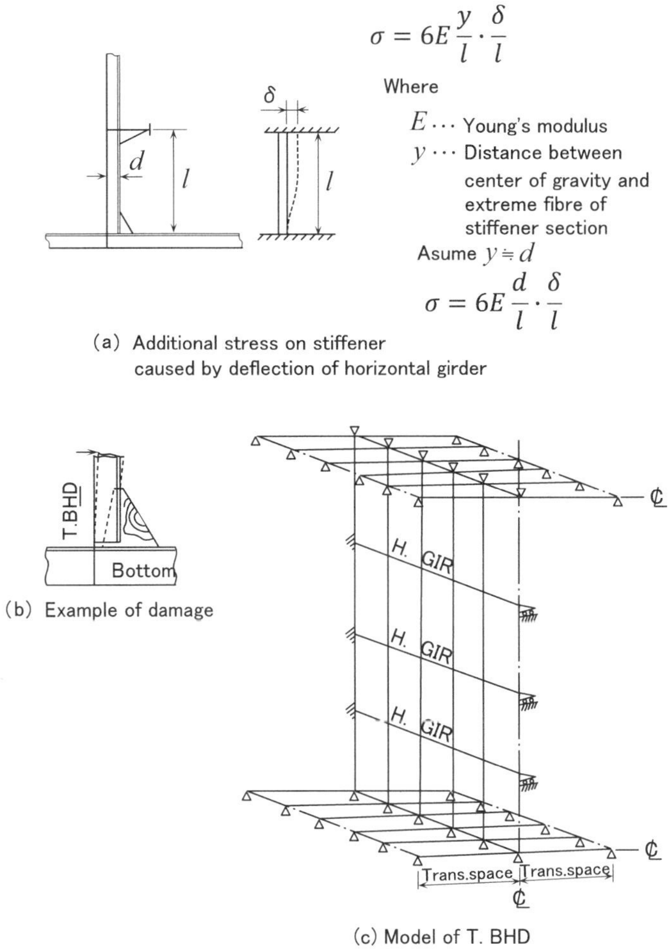

Fig. 24.9 Additional stress of stiffener due to deflection of horizontal girder and example of damage

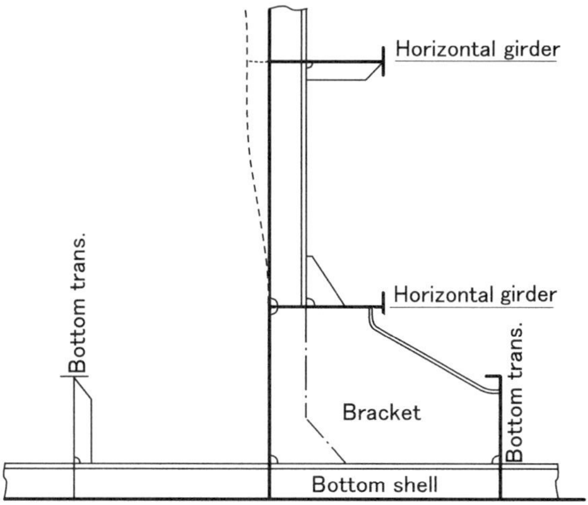

Fig. 24.10 Large bracket reducing deformation of horizontal girder

In the design of a vertical stiffener on a transverse bulkhead of a large ship, the displacement of the horizontal girders is first calculated, including the effect of the stiffness of the vertical stiffeners, and the scantling of the vertical stiffener is then decided considering the displacement of the horizontal girders. An example of the calculation model is shown in Fig. 24.9c.

In the case of a large displacement of the lowest horizontal girder, a large bracket connecting the horizontal girder to the adjacent bottom transverse is to be fitted to reduce the displacement, as shown in Fig. 24.10. In this case, attention needs to be given to the design of the vertical stiffener between the lowest horizontal girder and the second one from the bottom, because the relative displacement between them is considerable as shown by a dashed line in Fig. 24.10.

24.4 Swash Bulkheads

At the beginning of the era in large increase in ship size, much research was carried out and rationalization was achieved with the fruits of the research. One special fruit was the increase of the tank length. For a 30,000 DWT tanker the cargo oil tank length was 9 m or 12 m. With the increase in size of tankers a longer tank length was applied, for example from 15 m up to 45 m. In such long tanks one or two swash bulkheads were provided to avoid high dynamic liquid pressure on the transverse bulkhead due to the movement of cargo oil or water ballast. The swash bulkhead is non-tight and has uniform thickness of 10 mm or 12 mm with openings as shown in Table 24.2. Such an arrangement is good for avoiding liquid movement but has low shear rigidity to assist the frame work of the hull structure or to support the primary supporting members. Figure 19.1 shows damage caused by the lack of rigidity of a swash bulkhead.

Table 24.2 Opening shape of swash bulkhead, opening area ratio, and weight ratio

Ship	Principal dimensions (m)			Class	Type of swash bhd	Opening area ratio		Swash bulkhead weight ① (T/m²)		Tight bulkhead weight ② (T/m²)		Weight ratio ①/②	
	L	B	D			Center tank	Wing tank	Center tank	Wing tank	Center tank	Wing tank	Center tank	Wing tank
A	320.0	54.5	27.0	NK	F	0.52	0.35	0.10	0.13	0.37	0.33	0.27	0.39
B	190.0	28.2	17.0	NK	F	0.66	0.34	0.06	0.09	0.21	0.21	0.28	0.43
C	250.4	41.6	23.3	AB	E	0.17	0.38	0.17	0.13	0.34	0.28	0.50	0.46
D	320.0	54.5	27.0	AB	E	0.16	0.35	0.21	0.13	0.38	0.33	0.55	0.39
E	300.0	50.5	27.0	AB	E	0.14		0.21		0.35		0.60	
F	320.0	54.5	26.0	AB	E	0.17	0.30	0.21	0.16	0.36	0.33	0.58	0.48
G	360.0	62.0	36.0	AB	E	0.19		0.25		0.51		0.49	
H	260.0	43.5	22.8	AB	G	0.30	0.30	0.20	0.13	0.31	0.28	0.65	0.47
I	300.0	50.0	27.0	NK	C	0.14	0.28	0.22	0.19	0.36	0.35	0.61	0.54
J	300.0	50.0	27.0	NK	D	0.29	0.28	0.17	0.19	0.36	0.35	0.47	0.55
K	320.0	54.5	27.0	AB	B	0.13	0.27	0.22	0.16	0.36	0.33	0.61	0.49
L	307.0	48.2	25.5	AB	D	0.49	0.32	0.11	0.14	0.34	0.31	0.32	0.45
M	330.0	54.5	35.0	AB	D	0.27	0.28	0.19	0.16	0.40	0.35	0.48	0.46
N	330.0	54.5	35.0	AB	A	0.11	0.08	0.24	0.22	0.40	0.35	0.60	0.63
O	260.0	43.5	22.8	NK	A	0.14	0.12	0.22	0.18	0.30	0.25	0.73	0.72
P	230.0	32.2	10.7	AB	G		0.29		0.12		0.24		0.50

Type of swash bulkhead

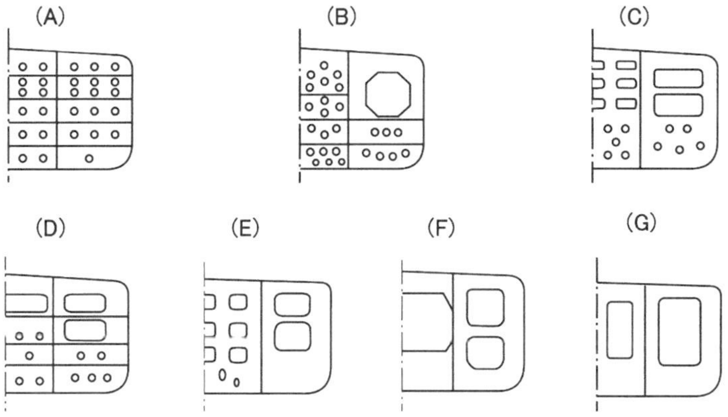

In a swash bulkhead, openings of many kinds of shape are cut out. However, the shape of the cut-out does not affect the shear rigidity too much. The shear rigidity depends mainly upon the area ratio of the opening. Figure 24.11 shows the relation between the shear rigidity and the area ratio of the opening [137], and Fig. 24.12 shows the relation between the opening area ratio and the weight ratio of a swash bulkhead relative to a tight bulkhead. From Fig. 24.12 it can be seen that the swash bulkhead without opening has 70% of the weight of the tight bulkhead which means the shear rigidity of the swash bulkhead will be about 70% of that of a tight bulkhead, even in the case of no opening. Sometimes an ore carrier or an ore/oil carrier is

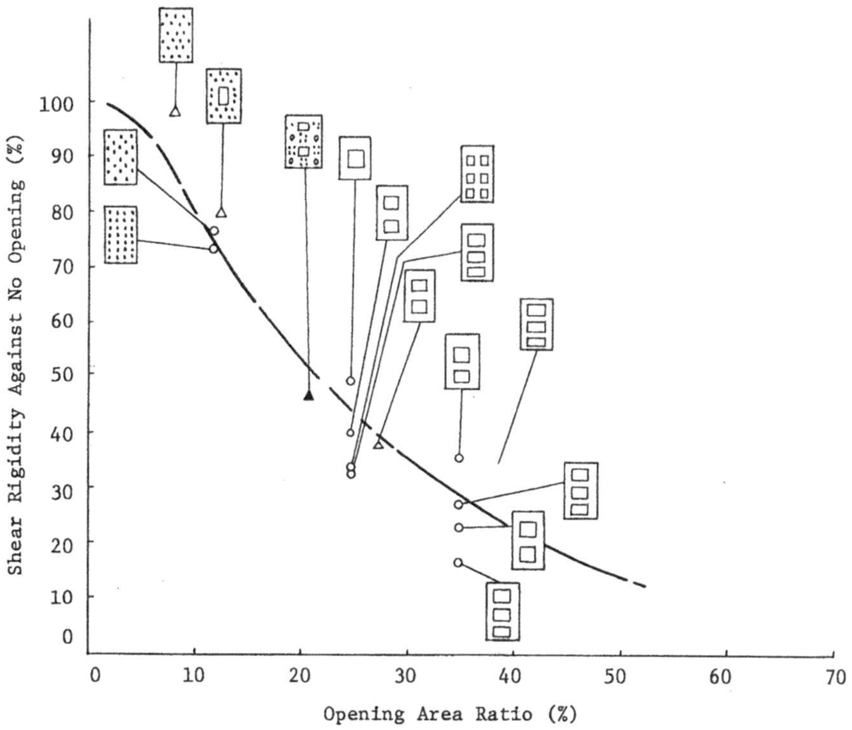

Fig. 24.11 Shear rigidity and opening area ratio of swash bulkhead

designed with an arrangement that has a swash bulkhead only in a wing tank and no bulkhead in the center tank at the same frame line. In this case the horizontal girders on the swash bulkhead support compressive force in the same way as the cross ties of an ordinary transverse ring, as explained in Sect. 21.5. Attention should be given not to cause any buckling.

24.5 Horizontal Stiffeners on Transverse Bulkheads

Nowadays a transverse bulkhead is constructed with vertical stiffeners, but previously horizontal stiffeners were used. In the case of a horizontal stiffener system, the horizontal stiffeners were not arranged at the same level as the longitudinal stiffeners on the longitudinal bulkhead, or on the side shell and no connection was provided between them. The ends of the horizontal stiffeners were snipped. The horizontal stiffeners were supported by vertical webs which were arranged so as to have an equal bending moment on the horizontal stiffener at the supporting points. To achieve this equal bending moment, the span length of the end span should be shorter than other

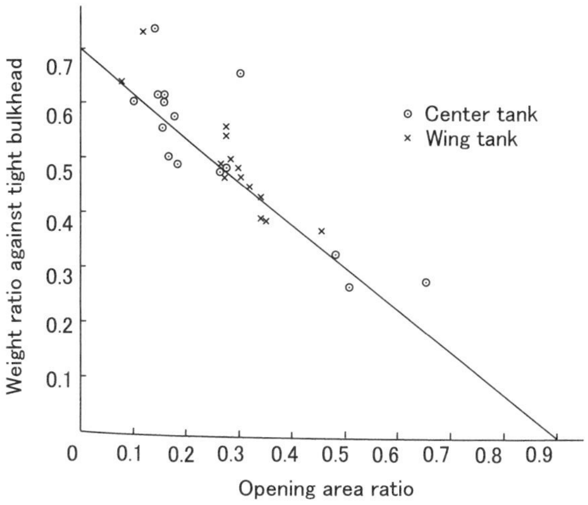

Fig. 24.12 Weight ratio of swash bulkhead versus opening area

regular span length by the ratio of $\alpha = \sqrt{2/3} = 0.816$ as shown in Fig. 24.13. This is theoretically explained in Sect. 8.6.3 in Part II.

The horizontal stiffener system was applied on 40,000 or 50,000 DWT tankers at the beginning of the era of the large increase in size of ships. The problem was corrosion at the snip end as shown in Fig. 24.13.

A transverse bulkhead supports the water pressure corresponding to the water head, accordingly the lower part is designed to be stronger than the upper part.

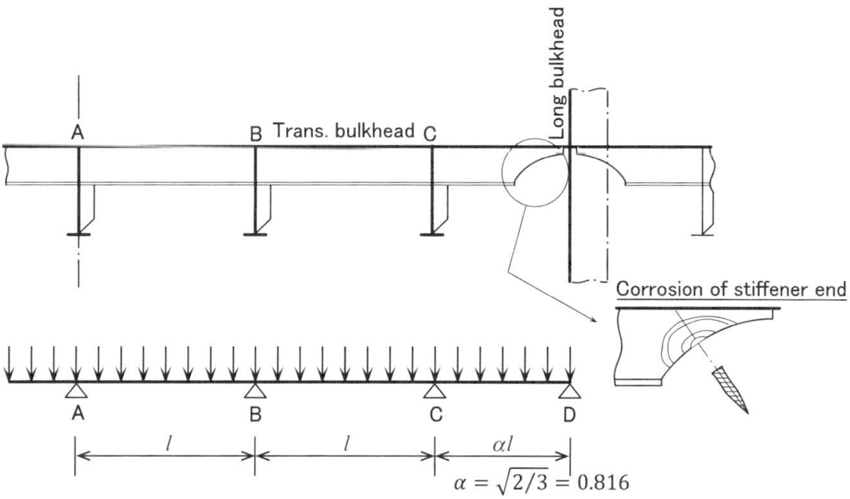

Fig. 24.13 Horizontal stiffener on transverse bulkhead

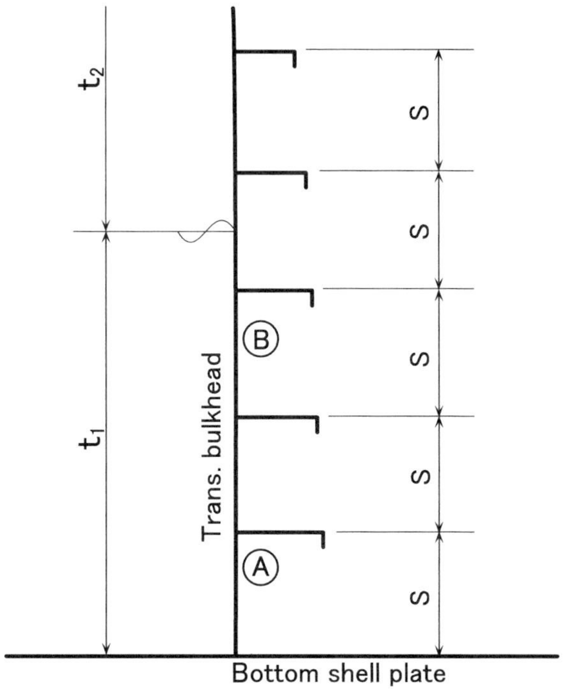

Fig. 24.14 Horizontal stiffener arrangement on transverse bulkhead

As shown in Fig. 24.14 where the horizontal stiffeners are arranged to have equal spacing, the lower stiffener has the bigger scantling corresponding to the water head. In addition to this, the lower bulkhead plate is required to be thicker than the upper one; however, in one strake of plating, the thickness is uniform. In Fig. 24.14, the A and B parts have the same thickness; therefore, the B part which supports less water pressure than the A part, has excess strength. The excess strength is caused by the equal thickness of the plate and the equal space arrangement of the horizontal stiffeners, and can be avoided by applying non-uniform thickness or unequal spacing.

The thickness t of the bulkhead plate is given by earlier Nippon Kaiji Kyokai's rule as follows:

$$t = 3.6s\sqrt{h} + 3.5 \tag{24.5.1}$$

And the section modulus Z of the horizontal stiffener is given as follows:

$$Z = 7cshl^2 \tag{24.5.2}$$

where

h: water head.

s: stiffener space.

l: span of stiffener.

c: coefficient (in case of bracket connection $c = 0.7$).

In the case of an equal space arrangement of the stiffeners, the required thickness of the bulkhead plate is proportional to $\sqrt{h}$ at each stiffener space. Therefore, it is beneficial if the thickness can be changed within one strake of the bulkhead plating. Existing steel mill technology enables such a non-uniform thickness plate such as the tapered thickness plate or longitudinally profiled steel plate explained in Sect. 5. 7 in Part I; however, in a shipyard some difficulty in the material control will appear. Another possibility is unequal spacing of stiffeners with uniform plate thickness, which may be more practical. Equation (24.5.1) can be modified as follows.

$$s = \frac{t - 3.5}{3.6\sqrt{h}} \tag{24.5.3}$$

Putting the lowest stiffener space as s_1; the next space as s_2 and so on, and putting the corresponding water head as h_1, h_2 and so on, the following equation can be obtained.

$$s_i = \frac{t - 3.5}{3.6\sqrt{h_{i-1} - s_{i-1}}} \tag{24.5.4}$$

where

$$s_1 = \frac{t - 3.5}{3.6\sqrt{h_1}}$$

$$i = 2, 3, 4, \cdots$$

As an example, s_i is calculated putting $t = 15\text{mm}$ and $h = 16\text{m}$, and the result is shown in Table 24.3. In this stiffener arrangement the required section modulus of the stiffener is given by the following equation.

$$Z_i = 7cl^2 \cdot \frac{1}{2}(s_i + s_{i+1})(h_{i-1} - s_i) \tag{24.5.5}$$

In the case of $c = 0.7$, $l = 4.5$ m and $h_0 = 16$ m, Z_i is calculated and shown in Table 24.4 in which the weight of the stiffener per unit length (kg/m) is also shown. The weight was calculated on the condition that the stiffener has an optimum section shape, according to Table 8.4.1 of Part II assuming an intermediate section with the web thickness of 12 mm.

The total length of s_1 to s_6 is 5,156 mm which corresponds to an equal spacing of 859.3 mm multiplied by 6. In this case the bulkhead plate thickness is 15.87 mm according to Eq. (24.5.1), and the required section modulus Z_i is calculated for each water head according to Eq. (24.5.2); this is shown in Table 24.5.

Table 24.3 Required stiffener space keeping constant panel thickness—Eq. (24.5.3)

i	S_i (mm)	S_i / S_1
1	799	1.000
2	819	1.025
3	842	1.054
4	868	1.086
5	897	1.123
6	931	1.165
	total 5,156	

Table 24.4 Required section modulus of stiffener keeping constant panel thickness—Eq. (24.5.5)

i	Z_i (cm^3)	W (kg/m)	Z_i / Z_1
1	1,220	52.0	1.000
2	1,186	51.3	0.972
3	1,149	50.5	0.942
4	1,110	49.6	0.910
5	1,068	48.7	0.875
6	1,023	47.7	0.839
		total 299.7	

Table 24.5 Required section modulus of stiffener keeping constant stiffener space

i	Z_i (cm^3)	W (kg/m)	Z_i / Z_1
1	1,291	53.5	1.000
2	1,218	52.0	0.943
3	1,144	50.4	0.886
4	1,071	48.8	0.830
5	998	47.1	0.773
6	925	45.3	0.716
		total 297.1	

Comparing Tables 24.4 with 24.5, it can be found that in the case of the unequal space arrangement the difference between the required section modulus of the adjacent stiffeners is smaller than that in the equal space situation, which is good for material management. In addition to the above, the total weight of plate and stiffeners of the 5,156 mm breadth and 1 m length panel is 906.8 kg/m for the unequal space arrangement, and 939.59 kg/m for the equal space arrangement.

The above suggests that a profit can be produced by applying a non-uniform instead of a uniform structure [44].

24.6 Minimum Thickness of Longitudinal Bulkhead Plates

The main role of a longitudinal bulkhead is to support the water pressure as well as a transverse bulkhead. However, in large ships some examples of buckling caused by an in-plane shearing force have occurred. The longitudinal bulkhead is one of the longitudinal strength members and supports the longitudinal bending moment which causes a tensile or compressive stress in the top and the bottom strakes of the longitudinal bulkhead. However, buckling damage caused by compressive stress has been scarcely reported because of the thicker top and bottom strakes where high compressive stress is generated. On the other hand, for plates around the neutral axis of the hull girder, where the shearing force is maximum, the minimum thickness for the water pressure is usually applied. This is a reason why the buckling is caused by the shearing force and not by the bending moment. An example of the buckling is shown in Fig. 24.15. In this case the thickness at the neutral axis is 14 mm and above the neutral axis it is 13 mm and 12 mm, below the neutral axis 15 mm and 16 mm. The buckling occurred on the 14 mm thick plate around the neutral axis and on the 13 mm and the 12 mm thick plates above the neutral axis. No buckling occurred on the 15 mm and the 16 mm thick plates below the neutral axis. The buckling pattern has an oblique axis which means the buckling is caused by the shearing force.

To avoid such buckling damage, the minimum thickness of a longitudinal bulkhead is to be decided based on the shearing force. The shearing force at a transverse section of a tanker is supported by the longitudinal bulkheads and the side shells. Disregarding the relative displacement between the longitudinal bulkhead and the side shell, the supporting ratios satisfy the following equation in case of the arrangement of two longitudinal bulkheads.

$$F = 2(\alpha_L F + \alpha_S F) \tag{24.6.1}$$

where F is shearing force; α_L and α_S are supporting ratio coefficients for the longitudinal bulkhead and the side shell, respectively. Derivation of α_L and α_S is explained in Sect. 24.7. According to the above expression, the shearing force F_L to be supported by the longitudinal bulkhead is given by the following equation.

$$F_L = \alpha_L F \tag{24.6.2}$$

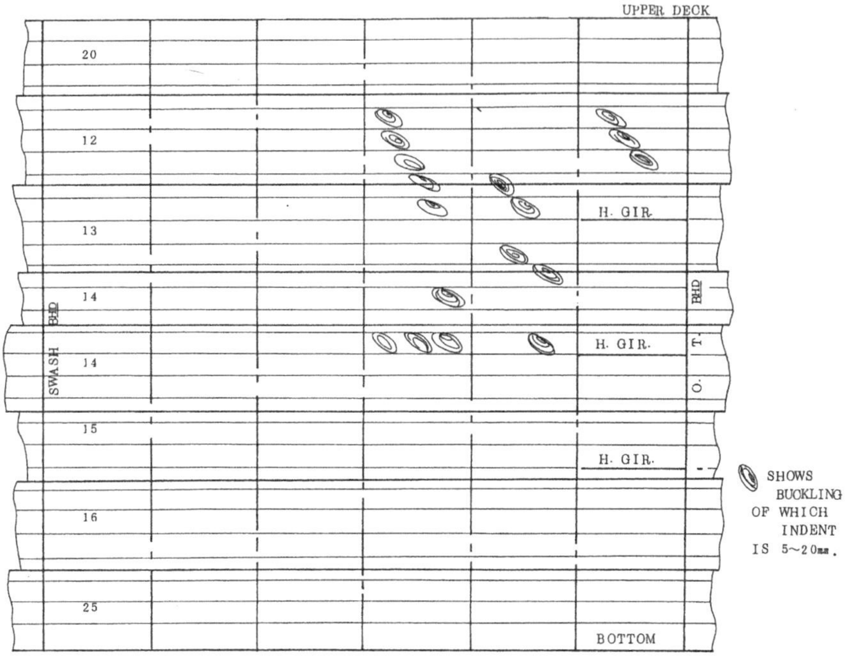

Fig. 24.15 Example of buckling on longitudinal bulkhead

The maximum shearing stress in the longitudinal bulkhead τ_{max} appears in the thinnest plate and is expressed by the following equation.

$$\tau_{max} = \frac{F_L}{D \cdot t_L} = \frac{\alpha_L F}{D \cdot t_L} \propto \frac{\alpha_L \Delta}{D \cdot t_L} \tag{24.6.3}$$

where

D: depth of ship.

t_L: thinnest plate thickness of longitudinal bulkhead.

Δ: full load displacement.

Table 24.6 and Fig. 24.16 show the relationship between τ_{max}, which is represented by $\alpha_L \Delta/(D \cdot t_L)$, and the ship length in service.

Figure 24.16 shows that a bigger τ_{max} exists for bigger ships. This comes from the assumption that the maximum shearing force is proportional to the full load displacement.

From Fig. 24.16, Ship J has larger maximum shearing stress than other ships and was considered insufficient. The ship was in the design stage, and the designed thickness was 15 mm for the longitudinal bulkhead and 22 mm for the side shell. To have a similar value of $\alpha_L \Delta/(D \cdot t_L)$ as that for ships in service without any damage,

Table 24.6 Maximum shearing stress of longitudinal bulkhead

Ship	L	B	D	d	Δ	a	b	$\frac{b}{a+b}$	t_L	t_s	α_{L0}	$\frac{\alpha_L}{\alpha_{L0}}$	α_L	$\frac{\Delta \cdot \alpha_L}{D \cdot t_L}$
A	360.0	62.0	36.0	28.00	547,300	10.34	20.66	0.666	14.0	26.0	0.175	1.213	0.212	230.5
B	330.0	54.5	35.0	27.00	424,890	9.40	17.85	0.655	14.0	26.0	0.175	1.215	0.213	184.4
C	320.0	54.5	27.0	21.00	312,435	9.40	17.85	0.655	14.0	23.0	0.189	1.215	0.230	190.0
D	300.0	50.0	27.0	20.70	266,469	8.46	16.54	0.662	14.0	24.0	0.184	1.210	0.223	157.1
E	279.6	44.5	24.7	18.88	201,063	8.10	14.15	0.636	14.0	23.5	0.187	1.220	0.228	132.4
F	230.0	40.0	18.0	13.50	102,154	7.20	12.80	0.640	12.5	22.0	0.181	1.218	0.221	100.2
G	230.0	33.0	21.7	15.15	94,560	7.45	9.05	0.548	9.5	20.0	0.161	1.220	0.196	90.1
H	230.0	35.5	18.0	12.46	84,551	7.65	10.00	0.567	10.0	20.0	0.167	1.222	0.204	95.7
I	236.0	38.0	17.2	12.05	90,356	8.50	10.50	0.553	10.5	21.0	0.167	1.220	0.203	101.7
J	340.0	68.0	29.0	22.60	426,234	11.28	22.72	0.668	15.0	22.0	0.203	1.210	0.245	240.3
K	320.0	54.5	28.0	21.00	312,360	11.75	15.50	0.569	14.0	24.5	0.182	1.220	0.222	176.8
L	278.8	44.5	24.5	17.90	191,412	9.00	13.25	0.596	14.0	23.0	0.189	1.220	0.231	128.8
M	190.0	28.2	17.0	12.60	55,183	4.80	9.30	0.660	11.5	18.5	0.192	1.220	0.234	66.0

Note : L : length of ship
B : breadth of ship
D : depth of ship
d : draft of ship
Δ : full load displacement
a, b : See Fig. 24.17
t_L : min. thickness of longitudinal bulkhead
t_s : thickness of side shell
α_{L0} : See Eq. (24.7.3)

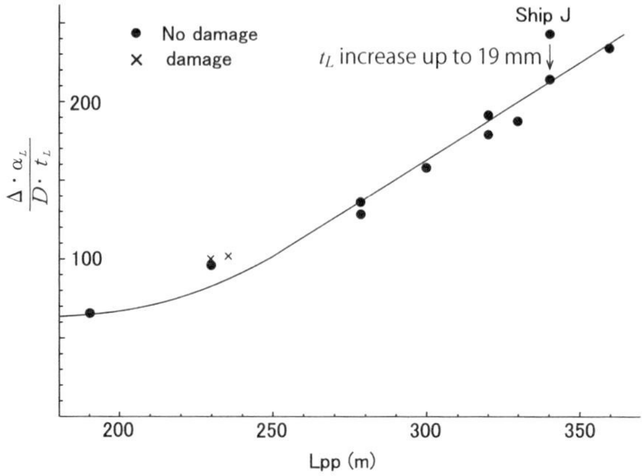

Fig. 24.16 Maximum shearing stress of longitudinal bulkhead versus ship length

the designed thickness of 15 mm for the longitudinal bulkhead is to be increased up to 19 mm keeping 22 mm for the side shell thickness. In another way the value of $\alpha_L \Delta/(D \cdot t_L)$ can be reduced by increasing the side shell plate. In this case the designed thickness of 22 mm for the side shell will be increased up to 30 mm, keeping 15 mm for the longitudinal bulkhead.

For reinforcement against the shearing force due to the longitudinal hull girder bending, it is usual to increase the side shell plate thickness. However as explained above the longitudinal bulkhead also supports the shearing force, so it is recommended to apply an efficient countermeasure, considering the supporting ratio coefficients for the longitudinal bulkhead and the side shell.

Actually the side shell plate is more important than the longitudinal bulkhead, and the ratio of the longitudinal bulkhead thickness t_L to the side shell thickness t_S will realistically fall between 0.55 and 0.80 ($0.55 \le t_L/t_S \le 0.80$).

24.7 Sharing Ratio of Shearing Force

The sharing ratio of the shearing force by the longitudinal bulkhead α_L and the side shell α_S can be obtained by the following equations [138].

$$\alpha_L + \alpha_S = \frac{1}{2} \tag{24.7.1}$$

$$\alpha_L = K\alpha_{L0} \tag{24.7.2}$$

$$\alpha_{L0} = \frac{t_L}{2(t_L + t_S)} \tag{24.7.3}$$

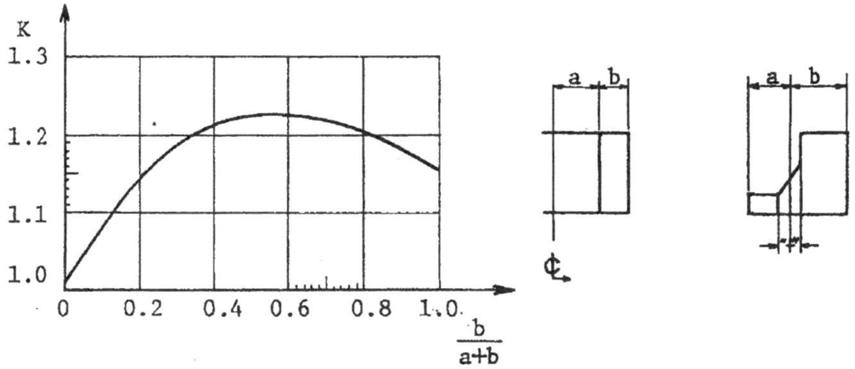

Fig. 24.17 Sharing coefficient of shear force on longitudinal bulkhead

K depends upon the distance between the longitudinal bulkhead and the center line and is given in Fig. 24.17. Figure 24.17 is effective in the case of a two-longitudinal bulkhead arrangement.

The sharing ratio of the shearing force by the longitudinal bulkhead is decided by the distance between the longitudinal bulkhead and the center line, and the minimum thickness of the longitudinal bulkhead. *K* has a maximum value of 1.23 where $b/(a + b)$ is around 0.5, where the longitudinal bulkhead lies at a quarter of the ship breadth from the side shell. *K* decreases corresponding to the distance between the longitudinal bulkhead and the center line down to 1.15 where the longitudinal bulkhead lies at the center line. On the other hand, *K* decreases also in cases of a closer position of the longitudinal bulkhead to the side shell, down to 1.0 where the longitudinal bulkhead comes to the side shell position. The value 1.0 means that the effect of the longitudinal bulkhead becomes the same as the side shell.

α_{L0} increases with the thickness of the longitudinal bulkhead. This will be the natural providence that the stronger one supports more load. This rate of increase gradually reduces when the thickness of the longitudinal bulkhead reaches that of the side shell plate. From this viewpoint, a criterion of $0.55 \le t_L/t_s \le 0.80$ stated in the previous section can be said to be practical.

In the case of a three longitudinal bulkheads arrangement, an example of the sharing ratio is shown in Figs. 24.18 and 19.7. This was calculated for a 600,000 DWT tanker which was not built actually. Figure 24.18 shows that the center line longitudinal bulkhead supports greater shearing force than each of the side longitudinal bulkheads. The longitudinal bulkheads of a two-longitudinal bulkhead arrangement, and even the side longitudinal bulkheads in the case of a three-longitudinal bulkhead arrangement can be extended into the engine room to maintain the structural continuity. However, the center line longitudinal bulkhead cannot be extended into the engine room. On the other hand, in the forward part the side longitudinal bulkheads are not extended into the fore peak tank but the centerline longitudinal bulkhead is. In such cases some countermeasures are necessary to maintain the structural continuity.

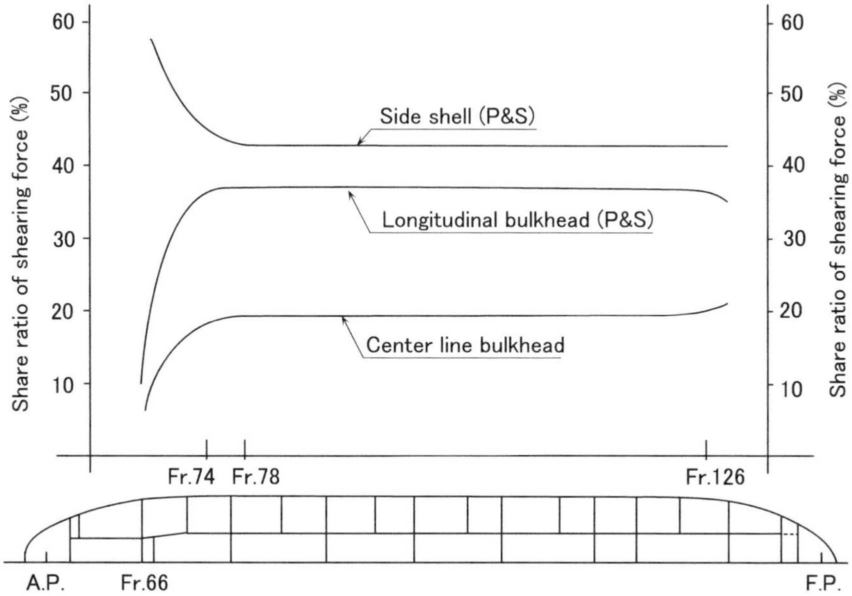

Fig. 24.18 Share ratio of shearing force of a tanker with three longitudinal bulkheads

In Sect. 19.3 "Bulkhead Arrangement Beyond Cargo Hold" some attention to maintain the structural continuity at the ends of the cargo hold were introduced. Special attention should be paid to the fact that the shearing force supporting ratio of the longitudinal bulkhead begins to decrease well away from the ends of the longitudinal bulkhead.

24.8 Corrugated Bulkheads

In a corrugated bulkhead the bulkhead plate itself resists the lateral load, such as water pressure, with a wave shaped section as shown in Fig. 24.19. This section describes how the shape and scantlings are determined.

Fig. 24.19 Plan of corrugated bulkhead

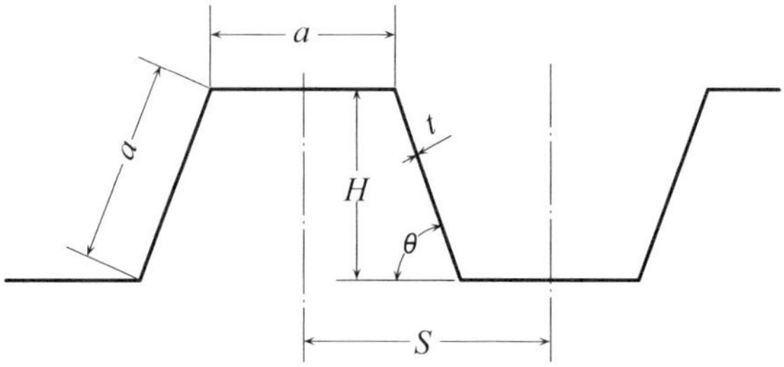

A corrugated bulkhead consists of parallel plates, and sloping plates which connect the parallel plates. Both support the water pressure equally and can have the same plate thickness provided the breadth is equal. In such a case the corrugated bulkhead can be made from one plate by bending. The strength calculation is to be carried out using the half breadth S of a wave as shown in Fig. 24.19. Assuming the breadth of the parallel plate and the sloping plate are both a and the distance between the parallel plates is H, the following equations are obtained:

$$a \sin\theta = H \tag{24.8.1}$$

$$a \cos\theta = S - a \tag{24.8.2}$$

$$Z = \frac{2atH}{3} \tag{24.8.3}$$

where

Z: section modulus of corrugated section.

t: thickness of corrugated plate.

t and Z can be calculated using Eqs. (24.5.1) and (24.5.2) from the earlier Nippon Kaiji Kyokai's rules. Now, let us express the equations introducing factors of proportionality k and m as shown in Eqs. (24.8.4) and (24.8.5), considering only the net thickness excluding corrosion margin for the plate thickness.

$$t = 3.6a\sqrt{h}(+3.5) \equiv ka \left(k = 3.6\sqrt{h}\right) \tag{24.8.4}$$

$$Z = 7cSh'l^2 \equiv mS \left(m = 7ch'l^2\right) \tag{24.8.5}$$

where

h'. water head at midspan of corrugated bulkhead

h: water head at bottom of corrugated bulkhead

Now five Eqs. (24.8.1), (24.8.2), (24.8.3), (24.8.4), and (24.8.5) have six unknowns of a, θ, H, S, Z and t, while t, h', h, c and l are given. So they can be solved by adding one more condition for which it is suitable to use the condition that the sectional area per unit breadth ($2at/S$) should be minimum.

From Eqs. (24.8.3), (24.8.4) and (24.8.5), the following equation is derived:

$$mS = \frac{2akaH}{3}$$

$$\therefore S = \frac{2ka^2}{3m}H \tag{24.8.6}$$

Furthermore, from Eqs. (24.8.6), (24.8.1) and (24.8.2), the following equation is derived.

$$S = \frac{2ka^2}{3m} a\sin\theta = a(1 + \cos\theta)$$

$$\therefore a^2 = \frac{3m(1 + \cos\theta)}{2k\sin\theta} \tag{24.8.7}$$

From Eq. (24.8.7), a value of a can be decided corresponding to θ, and H and S can be decided accordingly. Finally the optimum θ to have the minimum sectional area per unit breadth $(2at/S)_{min}$ can be obtained.

The sectional shapes of the corrugated bulkheads are shown in Fig. 24.20 for θ equals 30, 45, 60, 75 and 90°. In this calculation an assumption of $l = 4.5$m, $h' = 13.75$m and $h = 16$m is applied. In Fig. 24.20 the relation between θ and the sectional area per unit breadth $(2at/S)$ is also shown. In the case of $\theta = 65°$, the sectional area per unit breadth $(2at/S)$ becomes minimum. The sectional shape of the corrugated bulkhead for $\theta = 65°$ is also shown in Fig. 24.20. It seems well balanced. The calculated values for θ equals 30, 45, 60, 75 and 90° are shown in Table 24.7.

To compare the weight of a corrugated bulkhead with that of a flat plate bulkhead with vertical stiffeners, the scantlings are calculated. Assuming S, the stiffener spacing in the case of the flat plate bulkhead, is 0.67 m, the plate thickness excluding corrosion margin and the section modulus of the stiffener are obtained as follows.

$$t = 3.6S\sqrt{h} = 3.6 \times 0.67\sqrt{16} = 9.6 \text{ mm}$$

$$Z = 7cSh'l^2 = 7 \times 0.7 \times 0.67 \times 13.75 \times 4.5^2 = 914 \text{ cm}^3$$

The sectional area A of the stiffener is 54.78 cm^2 when the stiffener is designed with optimum sectional proportion for the section modulus of 914 cm^3. Therefore, the total sectional area of the plate and the stiffener is as follows.

$$\frac{St + A}{S} = \frac{67 \times 0.96 + 54.78}{0.67} = 177.8 \text{ cm}^2/\text{m}$$

In the case of a corrugated bulkhead of $\theta = 65°$, the weight per unit length is 95.4 cm^2/m which is 54% of the weight of the flat plate bulkhead. In this comparison no horizontal girder is included. The big weight difference is caused by the plate thickness which is decided by S, which is 9.6 mm in the case of the flat plate bulkhead. On the other hand, in the case of the corrugated bulkhead it is decided by a, giving a thickness of 6.8 mm.

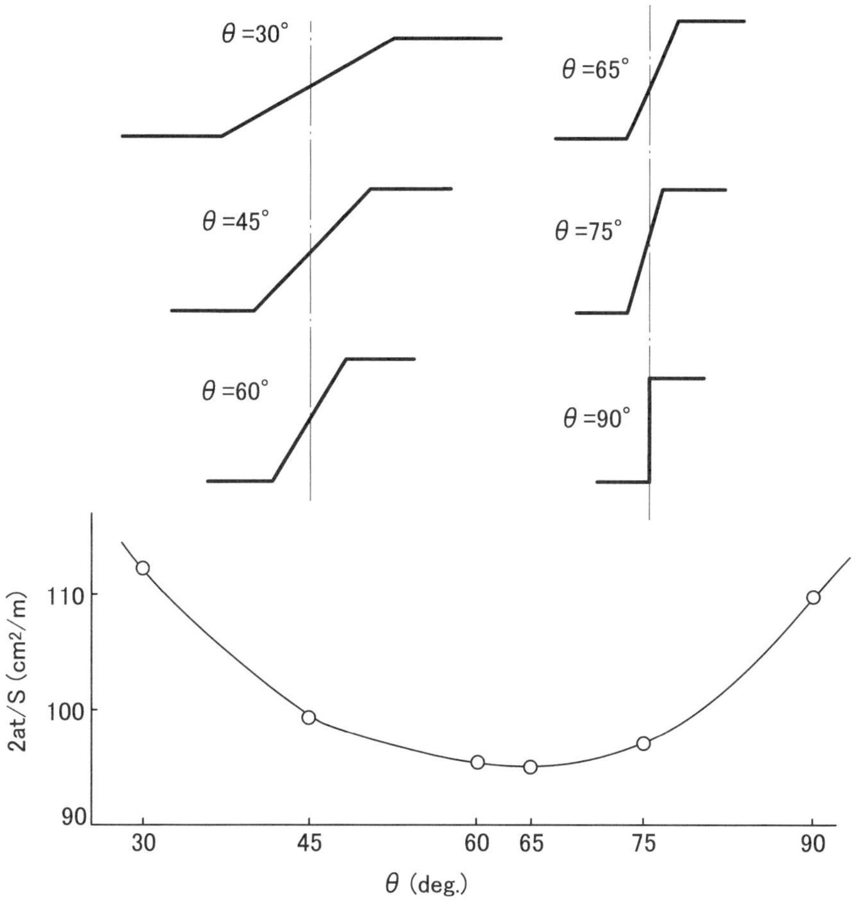

Fig. 24.20 Sectional area per unit length of corrugated bulkhead versus inclination of web plate

24.9 Horizontal Girders on Corrugated Bulkheads

One important item in the design of a horizontal girder on a corrugated bulkhead is the fact that the corrugated plate has little rigidity in the transverse direction and does not work as the extreme fiber of the girder. In the early days of the corrugated bulkhead, without paying attention to this fact, horizontal girders were provided on one side of the corrugated bulkhead as shown in Fig. 24.21. As a result, cracks occurred at the discontinuous points as shown in Fig. 24.21. In these cases, the scallops at the knuckled points made it more prone to the fatigue damage.

Table 24.7 Principal dimensions of corrugated bulkhead

θ (deg)	30	45	60	65	75	90
a (cm)	73	59	50	47	43	38
H (cm)	37	42	43	43	42	38
t (cm)	1.05	0.85	0.72	0.68	0.61	0.55
S (m)	1.36	1.01	0.75	0.67	0.54	0.38
$\frac{2at}{S}\left(\frac{\text{cm}^2}{\text{m}}\right)$	112.7	99.3	96.0	95.4	97.1	110.0

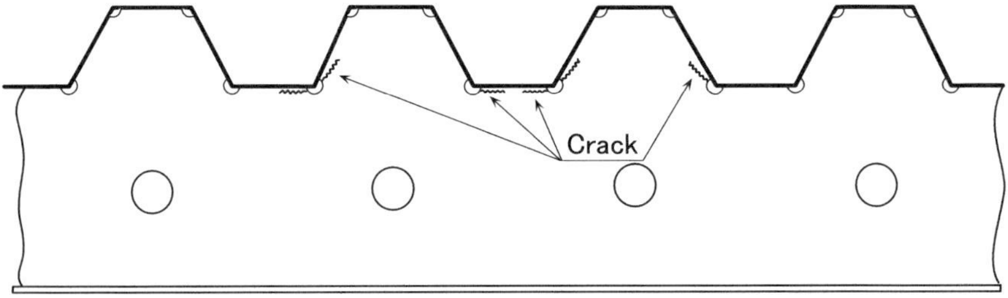

Fig. 24.21 Example of damage occurred in horizontal girder of corrugated bulkhead

To compensate for such a deficiency, a horizontal girder is provided on both sides of the corrugated bulkhead. In the era of the large increase in size of tankers, the horizontal girder was fitted to one side of the corrugated bulkhead for a 20,000 DWT or a 30,000 DWT tanker, on both sides for a 50,000 DWT or a 60,000 DWT tanker, and for bigger tankers than these, no corrugated bulkhead was fitted.

In the case where a horizontal girder is fitted on both sides of the corrugated bulkhead, the corrugated plate has to be divided into upper and lower parts, and welded to the girder, because it is difficult to cut the horizontal girder accurately into the wave shape. In such a case, attention should be paid to avoid misalignment between upper and lower parts of the corrugated plate. Especially for the parallel part of the corrugated bulkhead the alignment is more important than that for sloping plate, because the parallel part works in the same way as the extreme fiber of the bulkhead vertical stiffener and supports high tensile and compressive stress. To avoid misalignment of the corrugated plates, the drain hole arrangement shown in Fig. 24.22 is effective. In addition to the above, the horizontal girder on the corrugated bulkhead

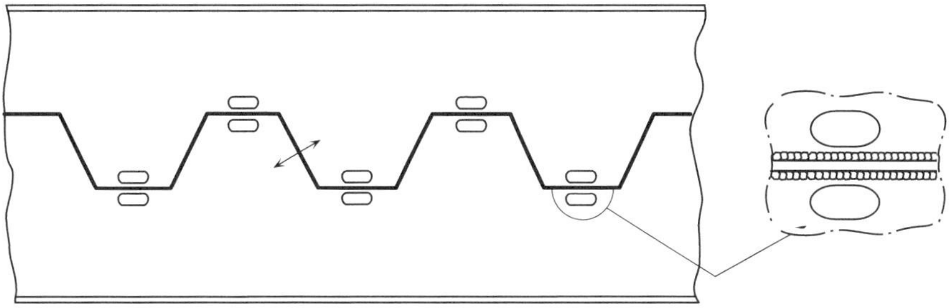

Fig. 24.22 Prevention of misalignment between upper and lower corrugated plates

is to be designed to support the axial force without the bulkhead plate because the corrugated plate has little rigidity in the transverse direction.

24.10 Stiffness of Corrugated Bulkheads Against In-Plane Loads

During the era of the large increase in size of tankers, corrugated bulkheads were not fitted but plane ones were used for tankers bigger than about 60,000 DWT. This might be due to the designer's feeling that it may be dangerous to apply it in a big ship where such corrugated bulkheads would have little in-plane rigidity.

However, the economic pressure to improve transportation efficiency forced hull structure designers to apply the corrugated bulkhead even for 200,000 DWT bulk carriers. For a bulk carrier or an ore carrier no horizontal girder is fitted on the transverse bulkhead to allow for good cargo handling. For big bulk carriers or big ore carriers, double plated transverse bulkheads had been fitted because the deep vertical stiffeners interfere with the cargo handling. A double-plated bulkhead has a disadvantage from the view point of not only hull steel weight but also hold capacity. From this view point the corrugated bulkhead has big advantages, and its application in big ships has been getting progressively popular.

Here, apart from a designer's feeling, some engineering analysis has been carried out on the in-plane rigidity of corrugated bulkheads. In-plane loads are divided into two types, the shearing load and the direct load. The shearing rigidity of the bulkhead is important in supporting the hull structure frame work as explained in Sect. 19.1. The shearing rigidity of a corrugated bulkhead is estimated roughly as $1/\alpha$ of a plane bulkhead, as explained in Sect. 19.2, where α is the ratio of the girth length of a corrugated bulkhead and the length of a plane bulkhead.

More reasonably the shearing rigidity of a corrugated bulkhead can be calculated by FEM [139]. This calculation was carried out on the corrugated bulkhead of a bulk carrier which had a depth of 10 m and a breadth of 31 m as shown in Fig. 24.23. Fixing the bulkhead at the side shell and putting a concentrated load vertically at each corrugated corner, the deflection at the center line of the bulkhead is calculated. The total load was 900 t and a symmetrical boundary condition was applied along the centerline. The calculated deflection at the center line was 17.1 mm.

Then, let us check this deflection through rough estimation. In case of the corrugation shape as shown in Fig. 24.23, $\alpha = 2$ which means that the shear rigidity of the corrugated bulkhead is 1/2 of that of the plane bulkhead. In the case of a plane plate of breadth 15.5 m, depth 10.0 m, thickness 12 mm with a total 900 t uniform load, the shearing force diagram is a triangle shape and the deflection is calculated as follows:

$$F = F_0 - \frac{F_0}{l}x = 900 - \frac{900}{15.5}x \tag{24.10.1}$$

$$\frac{dy}{dx} = \frac{F}{GA} = \frac{F}{810 \times 1.2 \times 1000} \tag{24.10.2}$$

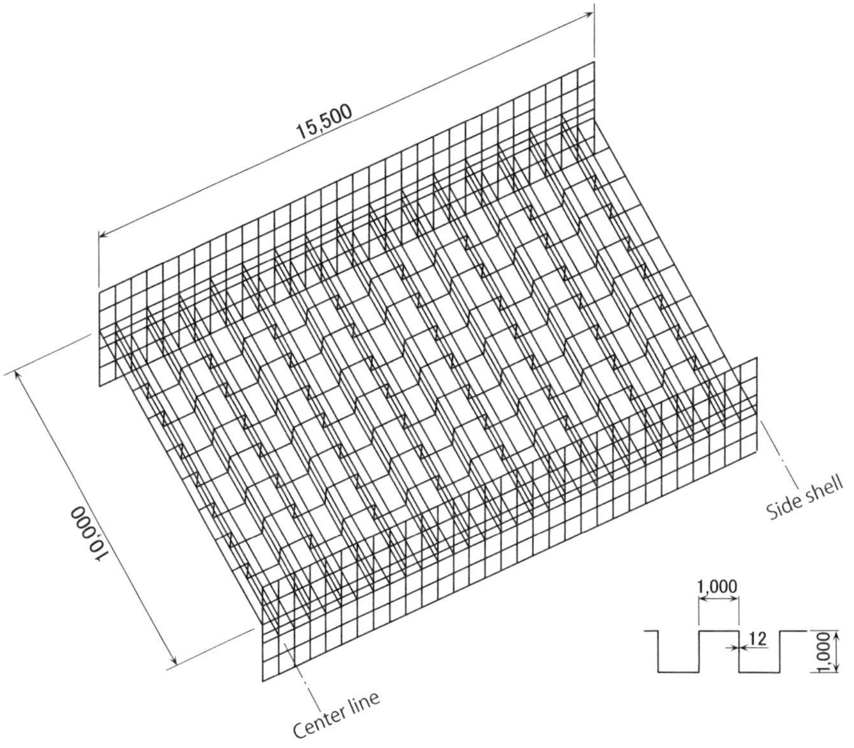

Fig. 24.23 FEM model of corrugated bulkhead

Putting Eq. (24.10.1) into Eq. (24.10.2) and integrating from the side shell to the centerline, that is, from $x = 0$ to $x = 15.5$, then the deflection y at the center line can be obtained.

$$y = \int_0^l \frac{F}{GA} dx = \frac{F_0 l}{2GA} = \frac{900 \times 15.5}{2 \times 810 \times 1.2 \times 1000} = 0.0072 \text{ m} \quad (24.10.3)$$

The deflection at the centerline for the plane bulkhead is 7.2 mm, and by applying the α correction, that for the corrugated bulkhead will be 14.4 mm. This is a little smaller than that by the FEM calculation but it can be accepted for practical use.

It seems that the rigidity of a corrugated bulkhead against the transverse direct load, tension, and compression is very small. This is calculated by taking out a unit breadth of the corrugated bulkhead and modeling it as shown in Fig. 24.24.

Assuming that the sloping plate is deformed by bending, the deformation δ_c is roughly calculated as follows.

$$P = \frac{12EI}{H^3}\delta_c = \frac{12Ebt_c{}^3}{12H^3}\delta_c \quad (24.10.4)$$

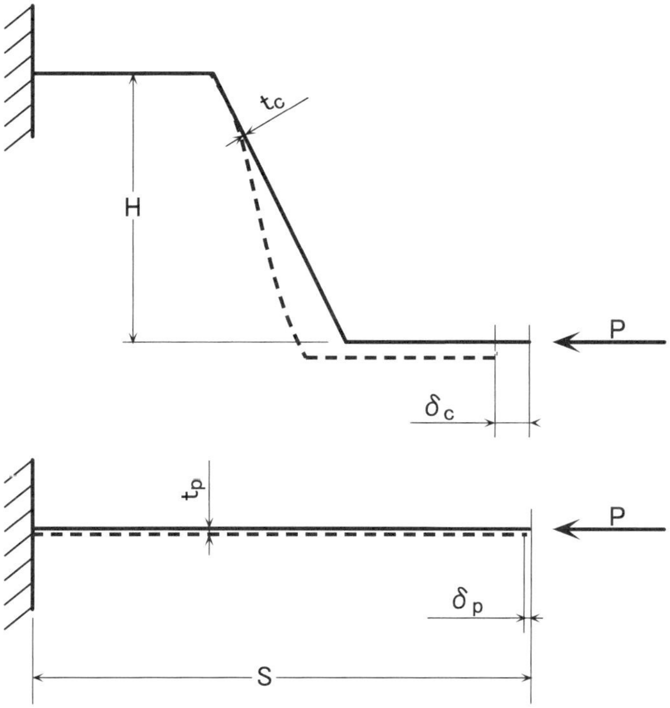

Fig. 24.24 Analysis model of corrugated and plane bulkheads subjected to horizontal force

where

P: load applied.

t_c: thickness of the corrugated bulkhead.

b: breadth of the corrugated bulkhead taken out for calculation.

E: Young's modulus.

On the other hand in the case of the plane bulkhead, as shown in Fig. 24.24, assuming that the plate of the S part is contracted by compression, the deformation δ_p is calculated as follows:

$$P = \frac{t_p b E}{S} \delta_p \tag{24.10.5}$$

Putting Eqs. (24.10.4) and (24.10.5) equal, the ratio of δ_c and δ_p for the equal load P can be obtained as follows:

$$\frac{\delta_c}{\delta_p} = \frac{H^3 t_p}{S t_c{}^3} \tag{24.10.6}$$

For a corrugated bulkhead with a slope angle of $\theta = 65°$ as explained in Sect. 24.8, the ratio δ_c/δ_p is calculated as:

$$\frac{\delta_c}{\delta_p} = \frac{H^3 t_p}{S t_c{}^3} = \frac{43^3 \times 0.96}{67 \times 0.68^3} = 3623 \tag{24.10.7}$$

Compared with a plane bulkhead, a corrugated bulkhead deforms more than 3,000 times under an in-plane load, that is, the rigidity of a corrugated bulkhead against an in-plane load is negligibly small. Therefore, the transverse corrugated bulkhead of a bulk carrier usually has a flat part adjacent to the side shell as shown in Fig. 24.25.

In the case of a chip carrier, longitudinal horizontal girders are often attached on the side shell to support vertical side frames. The horizontal girders are connected to and supported by the transverse corrugated bulkhead. In this construction, the end moment of the horizontal girder is an out-of-plane load which is supported by the vertical stiffeners fitted on the end flat part of the transverse corrugated bulkhead. On the other hand, the end reaction force of the horizontal girder is an in-plane load which is supported by the flat end-part of the transverse corrugated bulkhead.

Fig. 24.25 Side structure of corrugated bulkhead

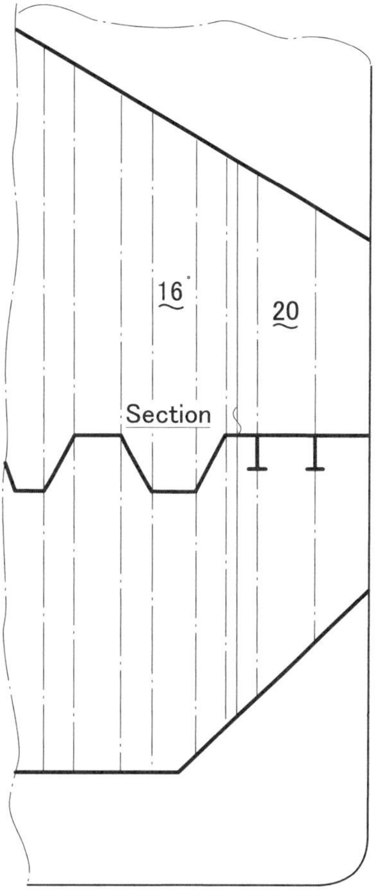

Chapter 25
Deck Structure

The deck receives a lateral external force as well as an in-plane force in relation to the longitudinal strength and the transverse strength, same as the shell and the bulkhead. Tensile and compressive stresses occur longitudinally on the upper deck due to the bending of the hull girder. For cargo ships, and container ships, tensile and compressive stresses also occur transversely on the cross deck between the hatch openings regarding transverse strength and torsional strength.

The strength characteristics are different between a full-width upper deck, such as that of a tanker, and a narrow upper side deck, such as that of a container ship. They are also different for the upper deck, the 2nd deck, and the 3rd deck.

Though the lateral loads on the shell and the bulkhead are assumed to be uniformly distributed, the deck loads are not always uniform depending on the cargo, and are often concentrated or locally distributed loads.

25.1 Stress Concentration at Hatch Corners

25.1.1 General

A stress concentration occurs near the hatch corners in the upper deck of a container ship, bulk carrier, and general cargo ship due to the hull girder bending and torsion. The maximum value of the stress is located at a point of 15–20° inside the circle from the end of the radius in the case of longitudinal bending, and the stress concentration factor depends on the ratio of (corner radius)/(hatch width).

Since the stress concentration factor is 3 for a circle and 2 for an ellipse (when the long axis is in the longitudinal direction) in the case of a hole in an infinitely wide plate, the latter is preferable for a hatch corner where longitudinal bending dominates. The stress distribution around the corner is obtained from the photo-elasticity tests,

Y. Okumoto et al., *Design of Ship Hull Structures*,
https://doi.org/10.1007/978-3-032-12500-2_25

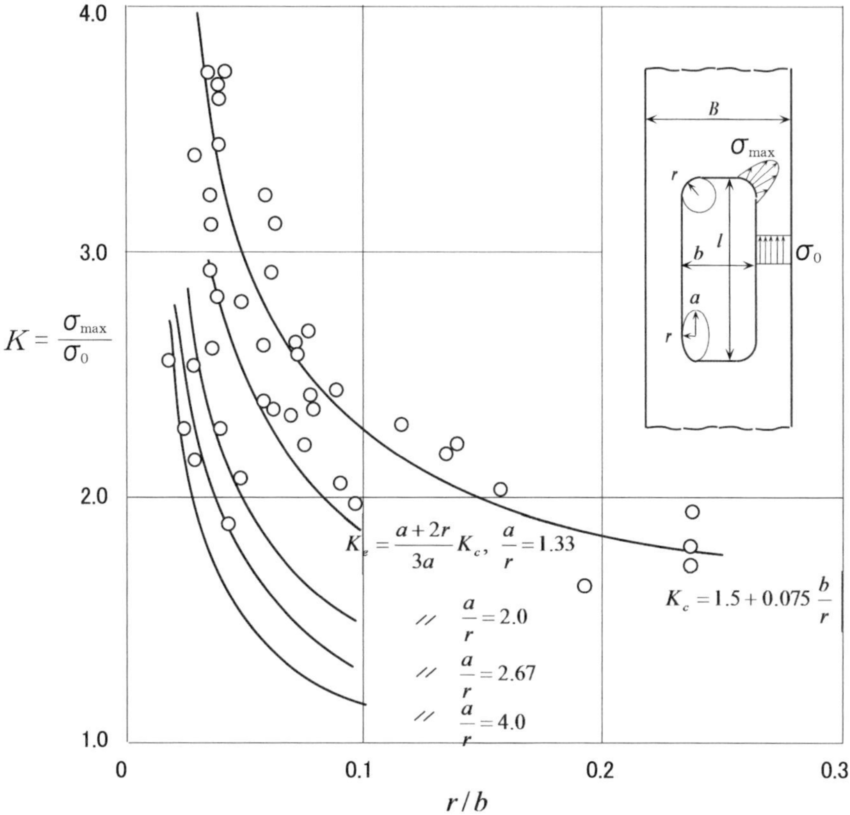

Fig. 25.1 Stress concentration factor around hatch corner

measurements using strain gauges, and also FEM analysis. Figure 25.1 shows these results [23]. The stress concentration factors are

$$K_c = 1.5 + 0.075\frac{b}{r} \quad \left(0.025 \le \frac{r}{b} \le 0.25\right)$$

$$K_e = \frac{a + 2r}{3a} K_c \quad \left(0.025 \le \frac{r}{b} \le 0.10, \quad \frac{a}{r} < 4\right) \qquad (25.1.1)$$

where

K_c: stress concentration factor of a circular hatch corner.

K_e: stress concentration factor of an elliptic hatch corner.

r: radius of circle or half length of shorter axis of ellipse.

NK	LRS
1. Circular corner with a radius of 600 mm or more 2. Double curvature $R_1 \geq 900$ mm $R_2 \geq 300$ mm $\theta \geq 45°$	1. Ellipse or parabola $a \geq \frac{b_L}{20}$ where 305 mm $\leq a \leq$ 610 mm b_L : width of hatch opening $b \geq 2a$

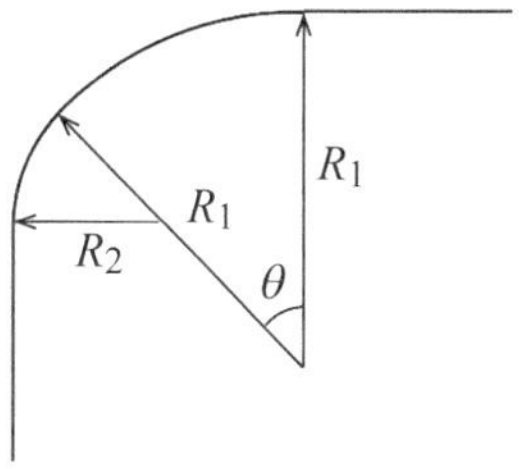

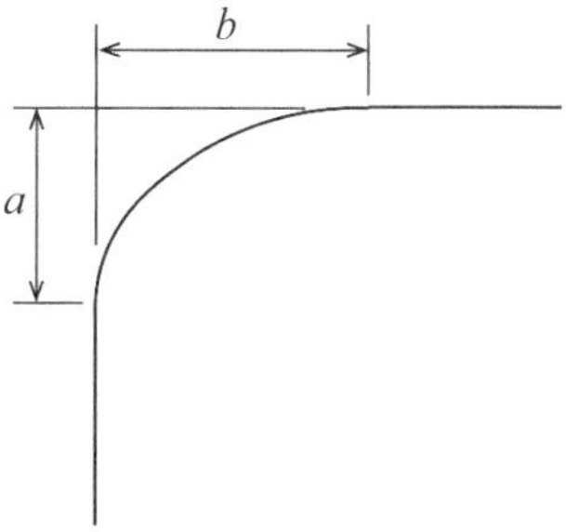

Fig. 25.2 Shape of hatch corner

b: hatch breadth.

a: half length of longer axis of ellipse.

Earlier classification society rules used to provide recommended hatch corner shapes as shown in Fig. 25.2 including large radius circle, double curvature, elliptic shape, or parabolic shape, where stress mitigation for hull girder vertical bending is mainly considered.

For the hatch corner of a container ship, the cell guide angle which holds containers is located near the corner, therefore a small radius is preferable to keep the hold space wider. On the other hand, the container ship has less torsional strength in the hull girder, because the ship has a smaller deck area, including both the side deck and the cross deck, due to the wide hatch. The stress concentration due to torsion is added.

The stress concentration at the hatch corner due to torsion is shown in Fig. 25.3. Since angular deformation between the side deck and the cross deck occurs, higher stress arises at the hatch corner. The position of maximum stress will be 45° to the side deck, while in the case of the longitudinal bending the position of maximum stress is 15 − 20° away from 45° inside the side deck as shown in Fig. 25.4.

The shape of the hatch corner should be carefully designed at the forward and aft hatches, because the stress levels are larger there considering the stress flow as shown in Fig. 25.5.

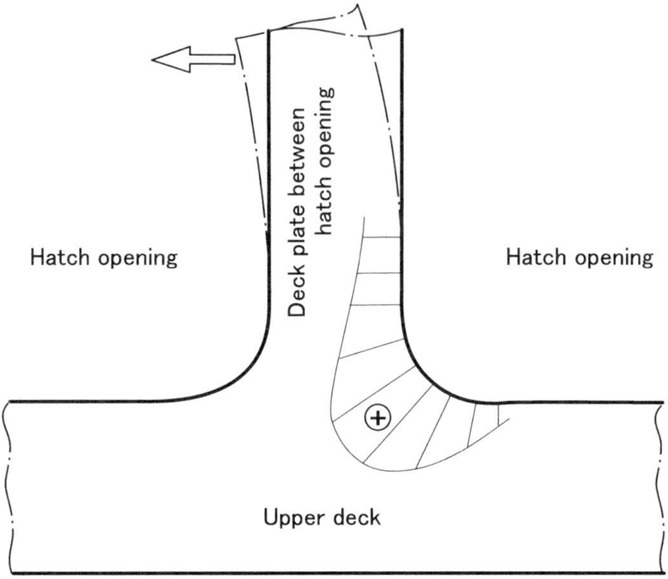

Fig. 25.3 Stress concentration around hatch opening corner due to torsion

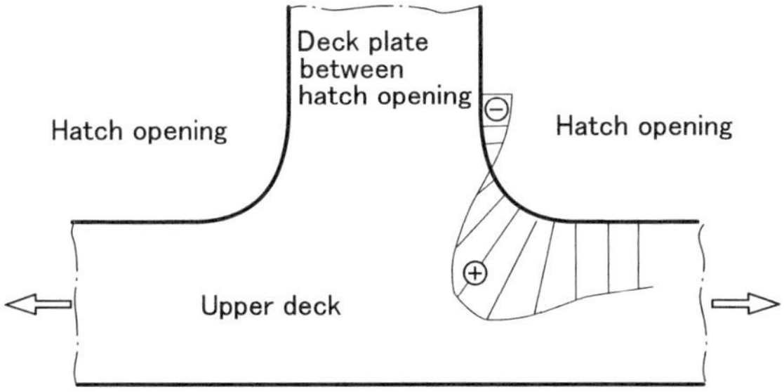

Fig. 25.4 Stress concentration around hatch opening corner due to tension

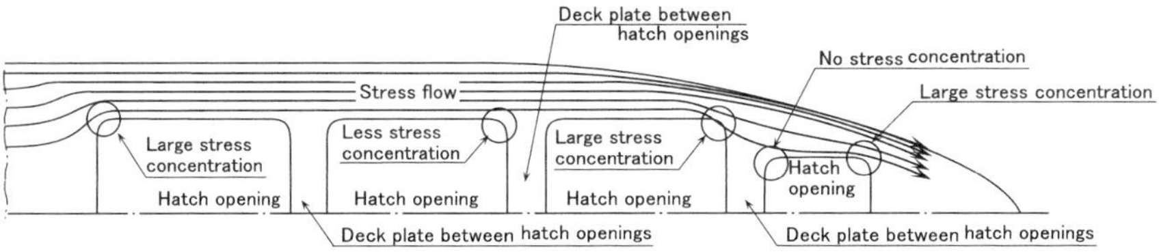

Fig. 25.5 Stress flow of upper deck due to longitudinal bending

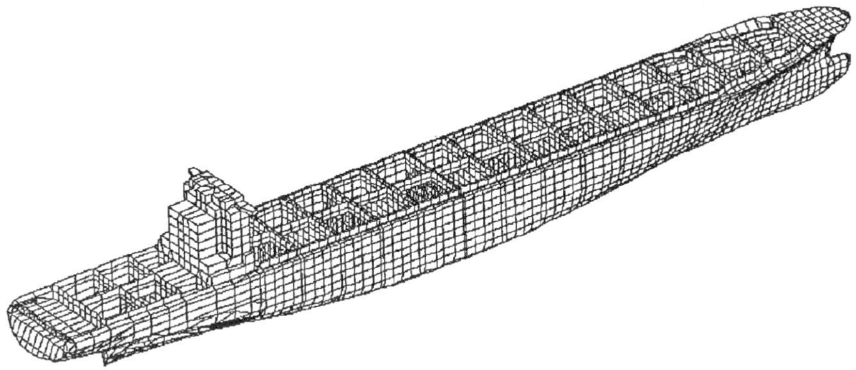

Fig. 25.6 Bird's-eye view of finite element model of container ship

25.1.2 Contour Shape Optimization of Container Ship Hatch Corners

Because the stress of a hatch corner of a container ship is influenced by several factors as explained in the previous section, the optimum shape that minimizes the stress level or stress variation is not necessarily similar to that shown in Fig. 25.2. In this section, we describe an example of contour shape optimization of a container ship hatch corner [140].

Figure 25.6 shows a bird's eye view of the finite element mesh configuration for the container ship concerned, which is one of the PANAMAX container ships of 4,000TEU class. Because this container ship has longitudinally continuous hatch side coaming, hatch corners on the coaming top level as well as on the upper deck level are studied.

(1) *Hatch corners on upper deck level:* Table 25.1 shows a combination of design parameters (independent variables) and constraints. Under the constraint that the clearance between the hatch corner contour and the cell guide is to be not less than 20 mm, *L*, *R*1 and *R*2 were optimized. The results are also shown in Table 25.1. Figure 25.7 shows the shape optimized for torsion and boundary stress distribution under torsional loads in oblique seas.

It is observed that we could obtain 15% less stress by optimization for vertical bending, and 20% less stress by optimization for torsion, compared with the double radius shape (900R + 300R). The optimized shape for vertical bending is quite good for vertical bending, but does not have good performance for torsion. On the other hand, the optimized shape for torsion has become similar to a single radius contour shape, and has quite good performance for both vertical bending and torsion. We can say that in this position, torsion is dominant and the optimized shape for torsion is to be selected.

Table 25.1 Results of hatch corner optimization at upper deck level

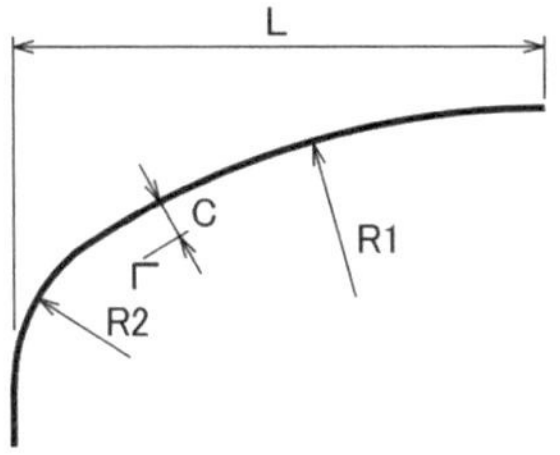

C : Clearance to cell guide (≧20)
(unit : mm)

	Design parameters			Maximum stress range (N/mm²)	
	L	R1	R2	Longitudinal bending	Torsion
900R + 300R	580	900	300	447	654
Optimum shape for longitudinal bending	660	1020	180	379	710
Optimum shape for torsion	510	1140	450	453	521

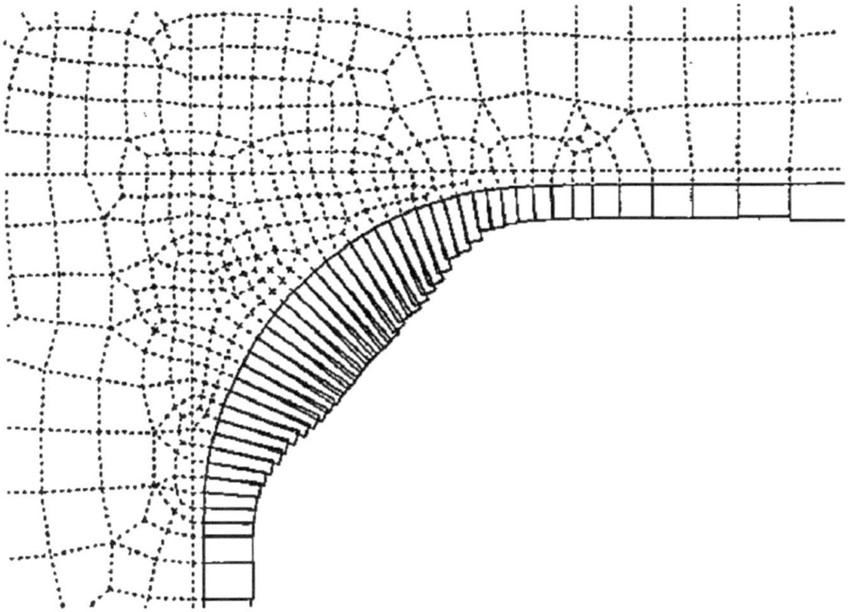

Fig. 25.7 Boundary stress distribution of hatch corner at upper deck level - shape optimized for torsion; torsional load applied

After the advent of post-Panamax container ships, to enhance container stowage efficiency, arrangement without between-hatch deck girders has become popular. In the case of such structural configuration, the inertia force of loaded containers in the longitudinal (fore-aft) direction acts directly on the cross deck structure. We can easily imagine that the maximum stress around the hatch corner due to this load would be rather closer to the centerline, than 45°. The designers should always be free from any fixed idea and should endeavor to seek optimum structure under changing circumstances.

(2) *Hatch corners on hatch coaming top:* Table 25.2 shows combination of design parameters (independent variables) and constraints on the coaming top level. Under the constraint that the athwartship distance between the inside end of radius *R*2 and the outside end of radius *R*1 is to be not greater than 100mm, *L*, *R*1 and *R*2 were optimized. The results are also shown in Table 25.2. Figure 25.8 shows the shape optimized for vertical bending and boundary stress distribution under vertical bending load in head seas.

It is observed that we could obtain 30% less stress by optimization for vertical bending, and 21% less stress by optimization for torsion, compared with the single radius of 100 mm. In this case, the optimized shape for vertical bending has quite good performance for both vertical bending and torsion, and the optimized shape is characterized by very large *R*1 and small *R*2. It is interesting that under the exactly same loading conditions, the optimum shape on the upper deck nearly becomes a single radius, and the optimum shape on the coaming top became very biased double radius.

This is considered to be due to the relation of rigidity between the cross deck strip and side structure. On the upper deck level, the side structure is wider than the cross

Table 25.2 Results of hatch corner optimization at coaming top level

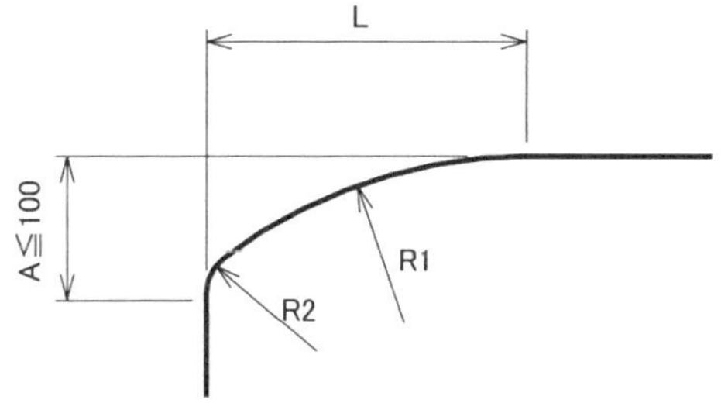

Unit : mm

	Design parameters			Maximum stress range (N/mm^2)	
	L	R1	R2	Longitudinal bending	Torsion
Single radius (100R)	100	100	100	754	474
Optimum shape for longitudinal bending	230	370	30	526	457
Optimum shape for torsion	190	250	30	580	373

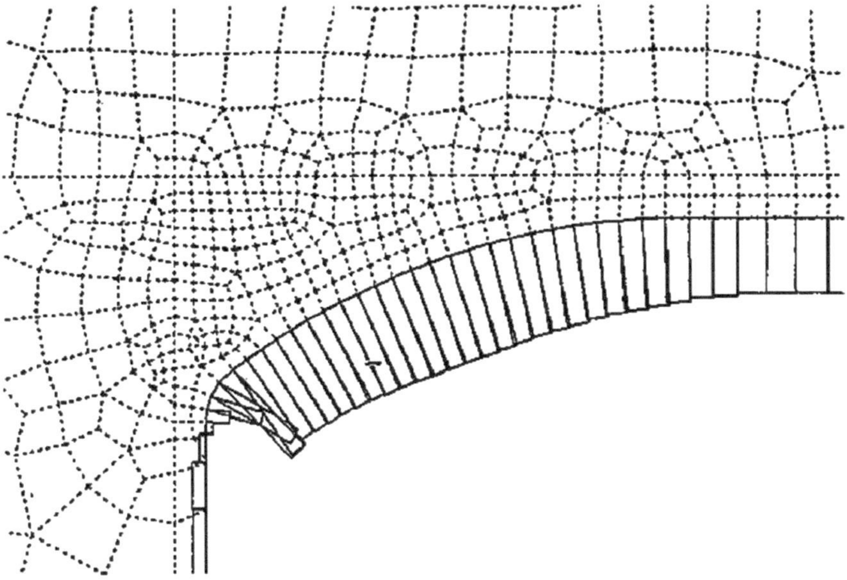

Fig. 25.8 Boundary stress distribution of hatch corner at coaming top level - shape optimized for vertical bending; vertical bending load applied

deck strip, and on the coaming top level, the side structure is narrower than the cross deck strip. Longitudinal bending stress is larger on the coaming top, whereas the bending moment of the cross deck strip due to torsion is supported mainly on upper deck level due to the strong rigidity of side structures by the upper deck level. These facts contributed to the optimized shape of each level.

25.2 Deck Strength for Locally Distributed Loads

Car decks and inner bottoms which support the wheel loads of vehicles or forklift trucks and steel coils are treated as a calculation of panels with locally distributed lateral load, not uniformly distributed loads such as water pressure. In this case, plastic theory is usually applied.

The steel coil is a product of thin steel plates, reeled as a cylindrical shape, which is produced by the roll process using a hot strip mill. The steel coils are loaded on the deck and the double bottom of a liner, a general cargo ship, or a bulk carrier, using the wooden dunnage as shown in Fig. 25.9. Though the arrangement of the wooden dunnage varies, the severest case for the plate strength seems to be the case where the load acts in the panel center as shown in Fig. 25.10. In order to simplify the calculation, intermittent loads on the center line of the panel are assumed to be distributed uniformly on the centerline as illustrated in Fig. 25.11, and roof shape collapse of a deck panel is considered on the basis of plastic theory. The work done by the external load and the work done by the plastic collapse are shown in Table 25.3 respectively [141]. As both values are the same,

$$P = \frac{8M_p\{a + b^2/(a-c)\}}{2b-d} \quad \text{for simply supported panel} \tag{25.2.1}$$

$$P = \frac{16M_p\{a + b^2/(a-c)\}}{2b-d} \quad \text{for fully clamped panel} \tag{25.2.2}$$

where

$$M_p = t^2\sigma_y/4$$

t: plate thickness

σ_y: yield stress.

Then, the collapse load is estimated for the panel according to the panel size and the print area of the local load.

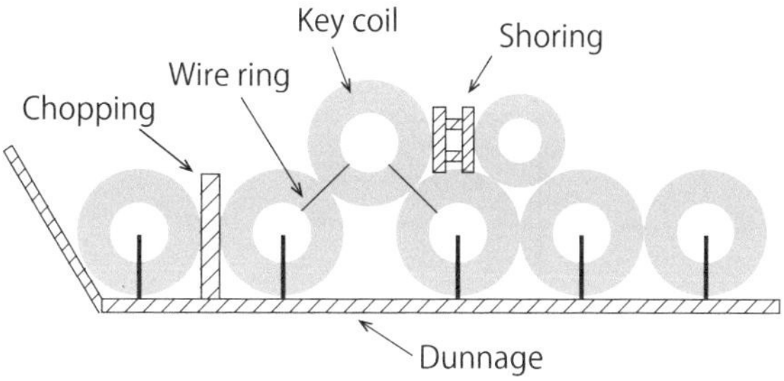

Fig. 25.9 Loading of hot coil

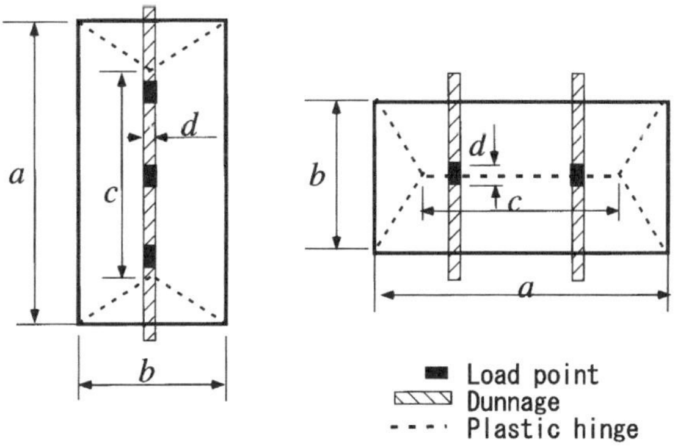

Fig. 25.10 Deck strength for hot coil loading

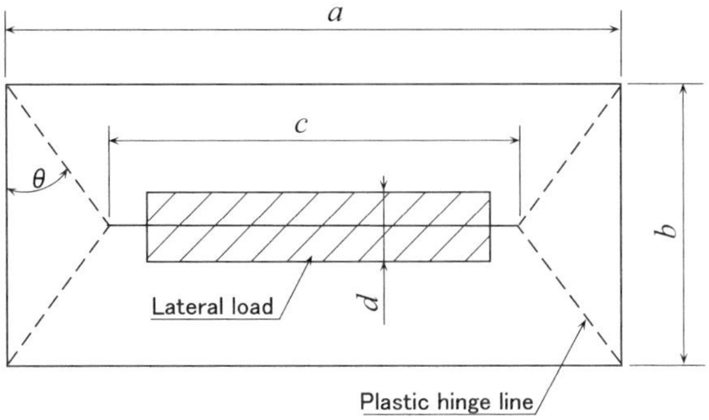

Fig. 25.11 Rectangular plate subjected to band width lateral load

Table 25.3 Internal work and external work due to applied load

U Inernal work	4 Edges simply supported	$4\left(\tan\theta+\frac{1}{\tan\theta}+\frac{c}{b}\right)M_p\delta$
	4 Edges clamped	$4\left(\tan\theta+\frac{2}{\tan\theta}+\frac{a+c}{b}\right)M_p\delta$
W External work		$\left(1-\frac{d}{2b}\right)P\delta$
Symbols :	$\tan\theta=\frac{a-c}{b}$	P : total load
	$M_p=\frac{4\sigma_y}{t^2}$	σ_y : yield stress
	δ : deflection of plastic hinge line c	

Next, let us consider a car deck sustaining a wheel load as shown in Fig. 25.12. A wheel load R_2 of 25 tons is acting in the center of the panel through double tires whose print areas are 350 × 420 mm each. The FEM calculation model is shown in Fig. 25.13, and the three types of load distribution are assumed as shown in Fig. 25.14. Figure 25.15 shows the results of the deflection, which are similar for the three types of loads. Case c, for the uniformly distributed load including the clearance between two print areas, is practical in evaluating the strength.

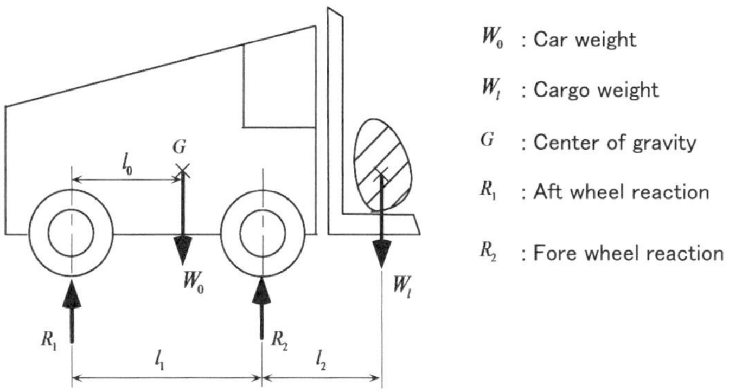

Fig. 25.12 Wheel load by forklift

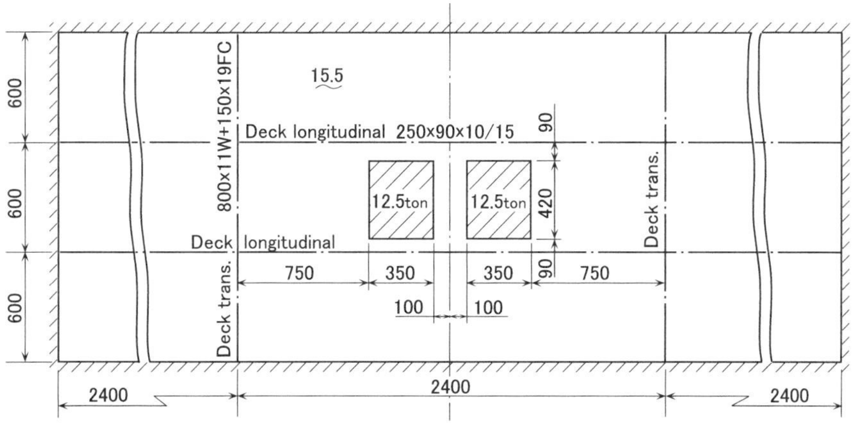

Fig. 25.13 Strength calculation model of deck plate subjected to wheel loads

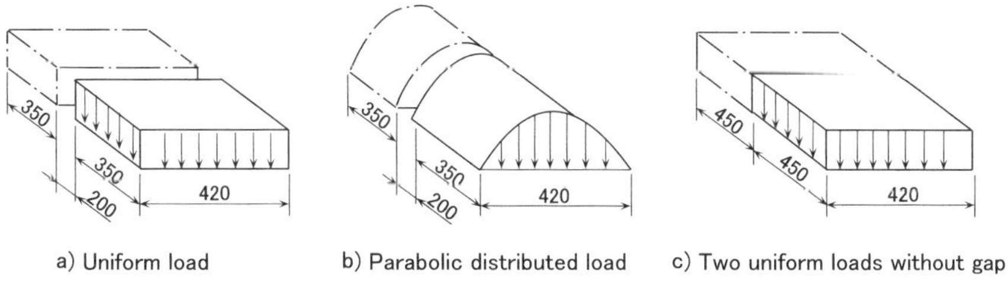

Fig. 25.14 Type of wheel load

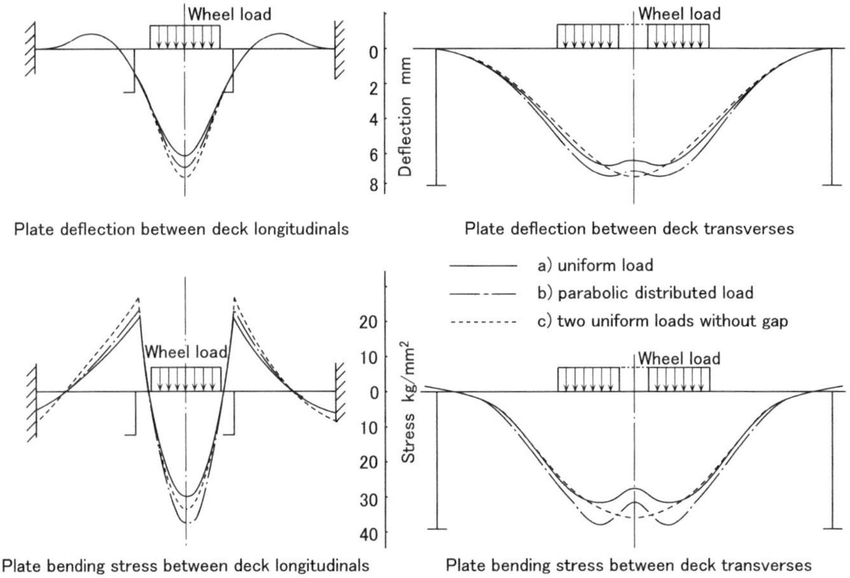

Fig. 25.15 Calculated results of deck plate deflection and bending moment under wheel loads

25.3 Damage to Deck Structure

A crack near a hatch corner will be seen often, especially at No.1 hold and the aftmost hold. Such cracks are caused by stress concentration by the hull girder bending and torsion. Figures 25.16 and 25.17 show examples of such a crack at the hatch corner of No.1 hold of a bulk carrier and the aftermost hatch before the engine room of a container ship respectively. Cracks also occur at the ends of hatch side coamings, because of the stress concentration due to the longitudinal force. Vertical tensile force separating the hatch side coaming and the upper deck acts at the ends of the hatch side coaming as explained in Sect. 20.7.

The collapse of the whole ship body often occurs in old ships, because of a lack of buckling strength due to the reduced thickness of plate suffering from corrosion. Painting maintenance is extremely important to maintain strength and for long life.

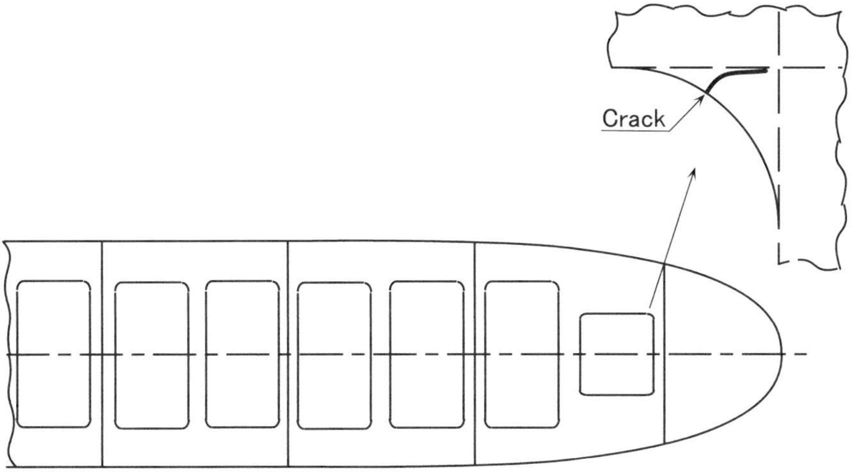

Fig. 25.16 Crack around foremost hatch corner

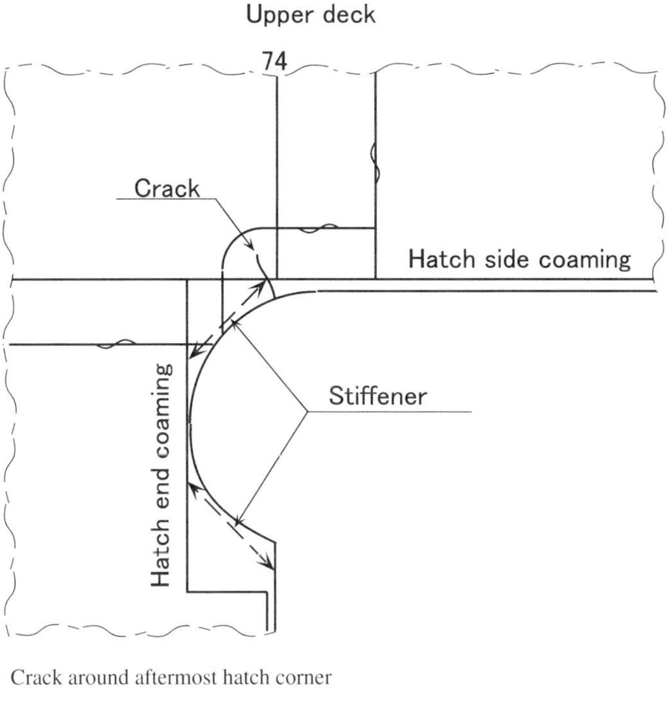

Fig. 25.17 Crack around aftermost hatch corner

Chapter 26
Double Hull Structure

The double hull structure has been applied to the bottom of cargo ships in general, and also to the newer types of ship such as container ships and LNG carriers. Furthermore, it was regulated that a large tanker should be provided with the double hull structure for the bottom and the side shell early in 1990's. The double hull structure of a cargo ship can prevent the water coming into the hold when the outer shell is wrecked by grounding and collision. For a container ship the double side hull structure is to be applied to maintain torsional strength which is reduced by the large hatch openings. A small container ship is designed with a single hull structure. These double hull compartments are used as ballast tanks.

The double side hull structure of an LNG carrier was originally for ballast tanks but works to some extent to protect the LNG tank from damage due to collision. The double side hull structure of a nuclear powered ship is designed to protect the nuclear reactor from damage due to a collision. The double side hull structure of a large bulk carrier is designed to support the shearing force. A famous 54,000 DWT bulk carrier, *Bolivar Maru*, had a double side hull structure with 5 long-holds arrangement, while the typical number of cargo holds for a ship of this size is seven.

An OBO (Ore/Bulk/Oil) ship also has a double side hull structure without stiffeners inside the hold for easy cleaning. Double side hull and bottom structures in large tankers are for the prevention of oil outflow during collision or grounding.

As explained above, there are many aims in applying the double hull structure, and the double hull structure has to be designed to meet these aims. In this chapter, the followings are explained: the structural system of a double hull structure, a comparison with the single hull structure, and how to design the double hull structure to meet the aims.

Y. Okumoto et al., *Design of Ship Hull Structures*,
https://doi.org/10.1007/978-3-032-12500-2_26

26.1 Structural System of Double Hull Structure

The ideas explained in Part II, Chap. 13 "Stiffened Panel" can be applied to the double hull structure. As shown in Fig. 26.1 in the case of the single hull structure, a plate and a stiffener always act together as one beam. On the other hand, in the case of the double hull structure, the top and the bottom parts act together, or separately depending on the boundary conditions. In the former case the merits of the double hull can be enjoyed, but in the latter case the structure is that of two beams and only has twice the strength of one side. Therefore, for the top and the bottom parts to act together, shearing members are to be provided between the top and the bottom plates instead of stiffeners. Hitachi Zosen Co. Ltd. in Japan applied such a structure system to a product carrier named Epoch Mark II as shown in Fig. 26.2, and Burmeister & Wain shipyard in Denmark also applied it to a 70,000 DWT OBO ship named BCT70 [142].

This web structure system has no structural member perpendicular to the web (usually girders are provided, but no floors) and therefore it has little rigidity in that direction. Accordingly special attention is necessary to avoid any deformation of

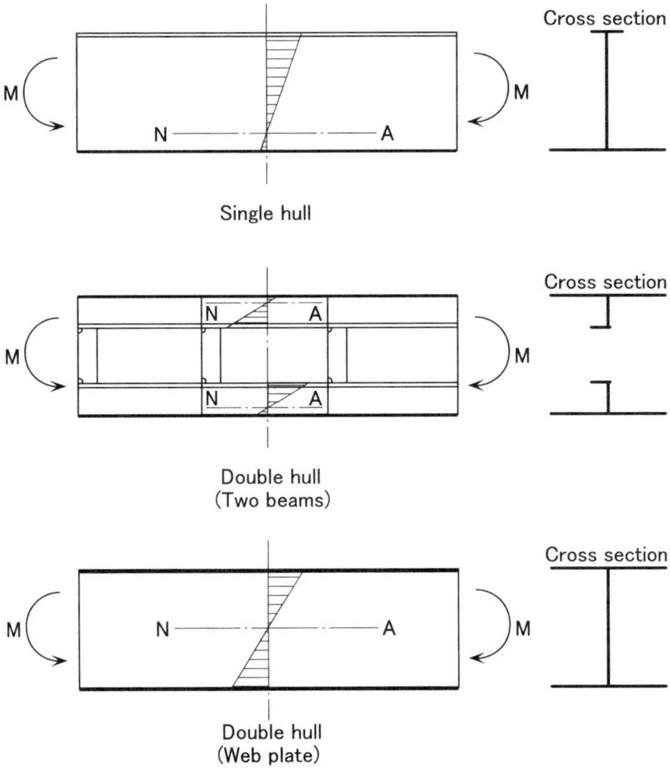

Fig. 26.1 Bending stress distribution of single hull and double hull structures

Fig. 26.2 Examples of web structure system in double hull

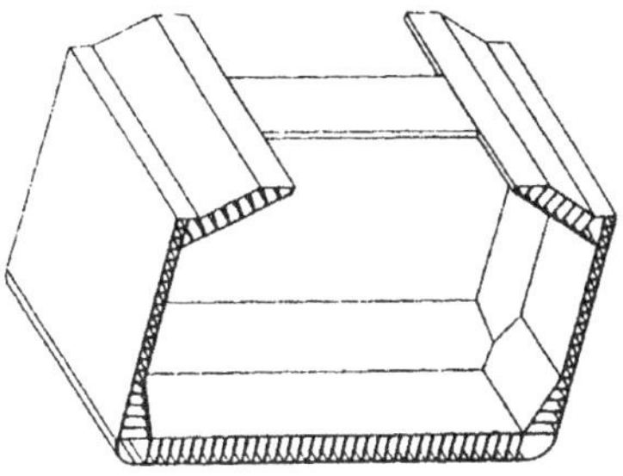

a) Double hull structure in BCT70

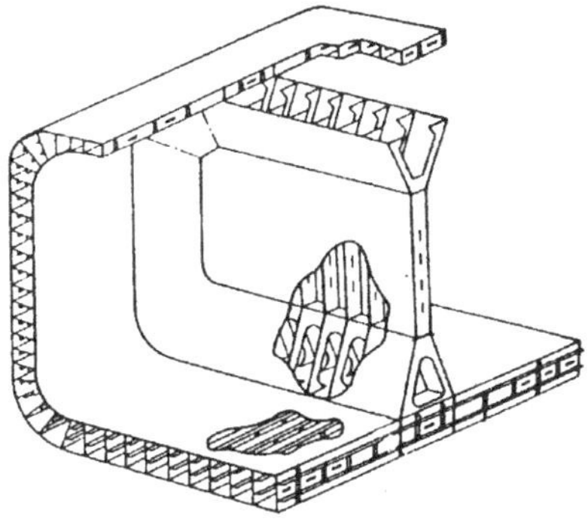

b) Double hull structure in EPOCH MARK II

the blocks during construction and transportation. Because of the longitudinal web arrangement, it is efficient to have shorter distances between the transverse bulkheads. A tank arrangement with wider width and shorter length is better for the application of the web structure system. In this system the structural members are arranged in one direction and there is no cross member, which brings a big advantage during production. Assembling can be done by running automatic welders in one direction only.

26.2 Double Hull Structure and Single Hull Structure

Many hull structural members support bending moment. Here the comparison of a double and single hull structure from the viewpoint of bending is explained. As shown in Fig. 26.1, the double hull structure is divided into two kinds: the double-beam system and the web structure system. In the case of the double-beam system, the plate has low bending stress similar to that of single hull structure. In the case of the web structure system the plate becomes a part of the extreme fiber of the structure and has high bending stress. In Sect. 23.1 the method of deciding the shell plate thickness is explained. Using the nomenclature in Sect. 23.1, only small σ_{xs} (the longitudinal stress due to the bending of longitudinal stiffeners) works in the single hull structure or double-beam system, while relatively larger σ_{xp} (the longitudinal stress due to the bending of the double hull) dominates in the web structure system.

In addition, when a fillet welding is employed to the double hull plating, the sum of the welding throat thickness is to be equal to the plate thickness as similar to that for the corrugated bulkhead.

The web structure system is more rigid than the double-beam system of equal depth. In other words, the web structure system is more effective than the double-beam system with higher bending stress in the top and bottom plates.

In the 1960's, with the increase in size of single hull tankers many cracks occurred at the intersection between the longitudinal stiffeners and the transverse webs as explained in Sect. 10.5.1, part II. They were called cracks around the slot. These were caused because the transverse web and the connecting flat bar between the longitudinal stiffener and the transverse web, did not increase as the original size, even though the load transmitted through this connection became greater because of the wider spacing and the greater draft of the larger ships. However, such cracking around slot has never been observed in the double bottom of a cargo ship even in the double bottom of a large bulk carrier.

The reason is, as explained in Fig. 10.23b in Part II, the effect of the lateral movement of the web stiffener. The lateral movement of the web stiffener causes a bending stress at the connection of the flat bar. In the case of the double hull structure, such large lateral movement does not appear and the countermeasures against a crack around the slot are less important than for a single hull structure.

26.3 Examples of Double Hull Structures

As already explained, there are many aims in applying the double hull structure, and the double hull structure is to be designed to meet these aims. Here the structural system of the double hull for each kind of ship is explained. In Fig. 5.3 in Part I examples of double hull structures are shown.

26.3.1 Cargo Ships

The double bottom of a cargo ship including a bulk carrier has to have water tightness of the inner bottom plate even in the case of damage to the outer shell plate caused by grounding or collision. To reduce the effect of a damaged outer shell plate to the inner bottom plate, it is more efficient to have a weak connection between them. From this view point it may be reasonable for the thickness of the double bottom floors to have reduced scantlings.

A structural system which has a strut connecting the inner bottom and bottom longitudinals at the midpoint between the floors has been applied popularly in the double bottom of cargo ships. This system is good from the structural viewpoint, because by connecting the inner bottom and bottom longitudinals both of them can carry load cooperatively. However, from the emergency viewpoint it is a little dangerous; the strut may pierce the inner bottom plate in the case of grounding.

A large bulk carrier once grounded causing severe damages at the starboard side bilge in the area of the hold. It was important to note that the web ring in the hopper tank had little damage because the thickness of the web was heavy and the big stiffeners were fitted nearly perpendicular to the shell and the sloped hopper plates. The applied force on the bilge shell plate by the grounding was transmitted directly to the sloped hopper plate which was badly broken. This is a bad example of the double bottom structure of a cargo ship.

26.3.2 Tankers

The double hull of a tanker, in a similar way to that of a cargo ship, has to be provided with oil tightness in the inner hull plate, particularly in the case of damage caused by grounding or collision to the outer shell plate. In the case of the double bottom of a cargo ship, a weak connection between the outer shell and the inner hull plate is preferable so that the inner hull plate is not directly affected by a load applied to the outer shell plate. This principle can be applied to the double bottom of a tanker.

Due to the application of the double hull side structure on a tanker, there are two problems which are different from a single hull side structure; one is the rigidity unbalance between the vertical webs on the side shell and the longitudinal bulkhead, and the other is the sharing ratio of the shearing force supported by the longitudinal bulkhead. The former, the rigidity unbalance, was explained in Sect. 21.5.1 in combination with the new strut arrangement in the center tank. Here, the shearing force sharing ratio is discussed.

The sharing ratio of the shearing force supported by the longitudinal bulkhead is estimated by the method explained in Sect. 24.7. A trial calculation for a 300,000 DWT tanker indicates that in the case of the double hull side structure the sharing ratio of the shearing force supported by the longitudinal bulkhead is 69% of that for the single hull structure. This reduction can be utilized in design improvement.

26.3.3 Container Ships

A container ship has large openings in the upper deck, and a high warping stress is expected at the corners of the openings due to the lack of torsional rigidity. To keep the necessary torsional rigidity, the double hull side structure is usually applied as a torsion box in the upper part of a large container ship.

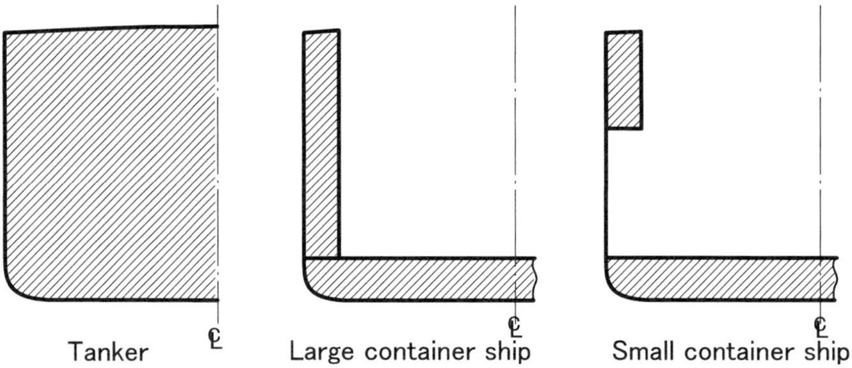

Fig. 26.3 Closed section of ships causing torsional rigidity

The torsional strength depends on the enclosed cross-sectional area of such plates, as shown in Fig. 26.3 by hatching. However, due to design limitations, a container ship cannot usually have enough cross-sectional area, and the designer has no other way but to increase the surrounding plate thickness.

26.3.4 Nuclear Ships

The aim of the double hull side structure of a nuclear ship is to protect the nuclear reactor and to prevent damage caused by collision with the other ships etc. Accordingly, enough strength against the collision is necessary. In this case, water tightness of the inner hull after the collision is not always required. Therefore, a strong connection between the outer and inner hull structures is preferable.

The collision damage is usually caused by the bow impact of a colliding ship. To prevent the bow from entering into the reactor room, heavy horizontal plates are fitted wherever possible. Also, an adequate distance between the outer and inner hull plates is necessary.

The nuclear ship "Mutsu" built in Japan ($L \times B \times D = 116 \times 19 \times 13.2\,\text{m}$) has a 4 m wide double side hull structure at both sides of the reactor room, and five horizontal heavy plates are fitted as shown in Fig. 26.4.

In the case of a tanker, preventing the oil outflow is of prime importance, and the oil tightness of the inner hull plate is most important. The design principles of double hull side structures of a tanker and a nuclear ship are similar but a little different.

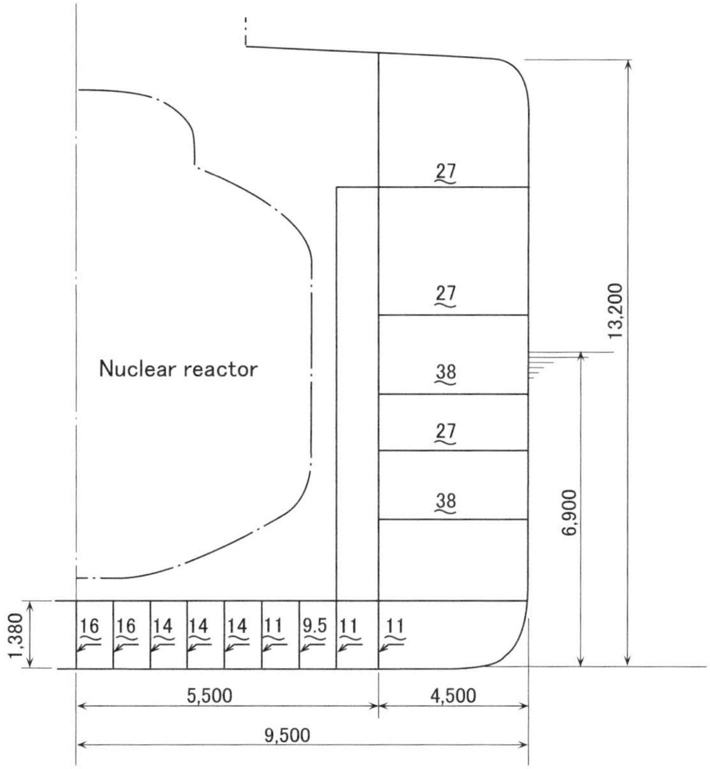

Fig. 26.4 Cross section of nuclear reactor room of nuclear vessel Mutsu

Chapter 27
Fore Construction

The fore construction is defined generally as the region from the collision bulkhead to the forefront of the ship. There are four important concepts in designing the fore construction. The first of them is that this portion collides with floating objects or other ships and also suffers from wave impact. The second is that the vertical acceleration depending on the pitching motion is large due to the long distance from the pitching center. The third is that a structural discontinuity area exists at the hold portion behind it. The last one is that the structural arrangement varies according to the ship body form.

27.1 Structural Arrangement

The transverse framing system has been generally adopted for the structure of a fore construction so far. The side stringer supports the frames at the ends. Since the ship side inclines toward the ship centerline, the transverse frames are fitted inclined to the shell plate, causing reduced bending rigidity of the transverse frames. If the frame is fitted perpendicularly to the shell, it is more effective from a strength point of view, but it is not preferable for assembly because the positioning of the frame is quite complex, especially in the area of a bulbous bow. As a result, a longitudinal framing system is now widely adopted, with the side longitudinals, the web frames, and the panting stringers as shown in Fig. 27.1.

The structural arrangement of the fore construction depends mainly on the bow shape. The main strength is maintained by the shell, the collision bulkhead, and the center line bulkhead.

Since the structure of the fore construction can be considered to be a cantilever beam fixed at the collision bulkhead to the main hull, the global strength such as bending, shearing, and torsional strength against the wave load should be considered as well as the local strength. In this point, attention should be paid to the fact that

Y. Okumoto et al., *Design of Ship Hull Structures*,
https://doi.org/10.1007/978-3-032-12500-2_27

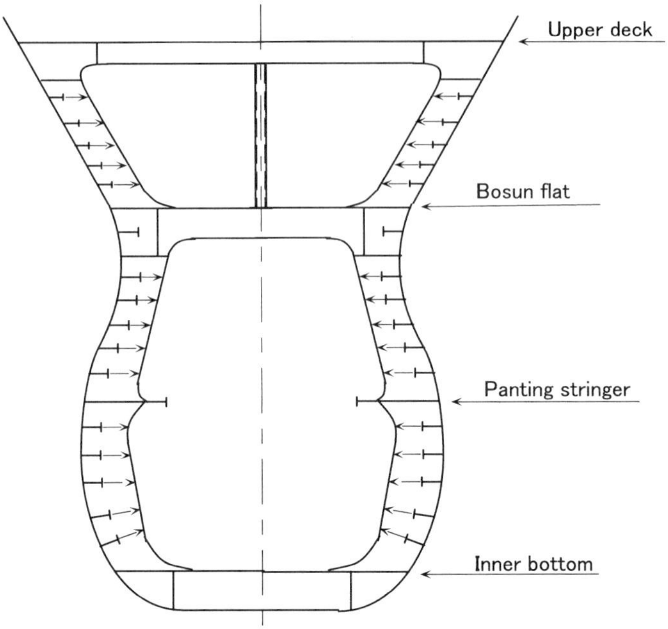

Fig. 27.1 Fore construction (Longitudinal framing system)

structural members inside the fore construction must deform according to the global deformation of the fore construction as a cantilever beam, in addition to the member's own deformation due to the local load acting on it. For example, secondary stress due to the forced displacement at the fore end should be studied for the centerline bottom girder as shown in Fig. 27.2.

Usually, a swash bulkhead is fitted at the center line for a wide forepeak tank. It is thought that the thickness of this plate is sufficient at 12 mm even for large ships, from the viewpoint of its purpose as a swash bulkhead. However, from the viewpoint of the whole structural design of the bow construction, the bulkhead is most important in maintaining the strength of the fore construction as a cantilever beam, especially for the shearing force, since a larger shearing stress acts on the bulkhead than on the side shell plating because of its shorter longitudinal girth length than the side shell plating, and therefore, buckling can easily occur.

27.2 Structure of Shell Construction

It is a common concept of current hull structure design that damage from a wave impact is not allowed though damage by collision against a floating object or another ship is inevitable in a sense and cannot be avoided solely by means of the structural

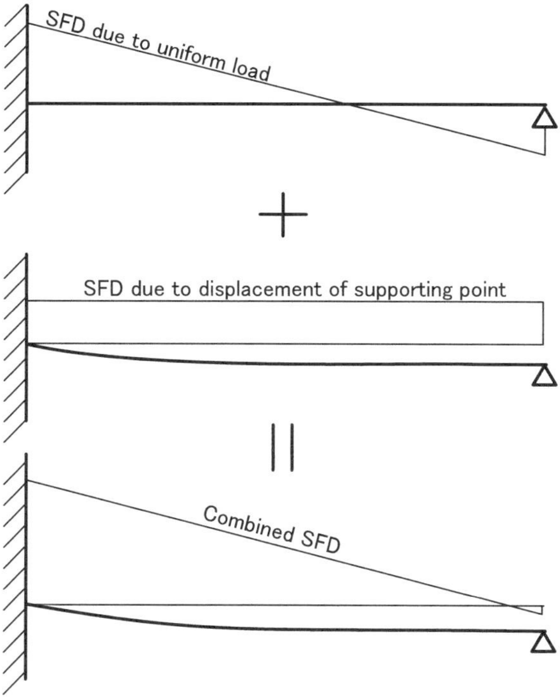

Fig. 27.2 Sharing forces on bottom centerline girder

design. Here, damage to the longitudinal frames and the web frames as a result of wave forces is introduced.

Figure 27.3 shows crack and buckling damages of the web frame on a container ship as shown in Fig. 23.4. Plate buckling occurred at the upper and lower parts of the web frame, and the web plate was cut in its mid-part, due to external water impact. The side shell plate was also cut at the upper part of L-23a. In this case, the shell plate seems to have a cylindrical form of approximately 6.5 m radius, which sustains the water pressure, and therefore a hoop stress of uniform distribution will be acting in the shell plate. The static water head, for which the hoop stress reaches to its rupture value, is estimated as 100 m in this case, as a result of detail calculation.

Figure 27.4 shows lateral buckling in the longitudinals at Fr.154-1/2. The flanges of the L-shaped angle have rotated backwards. The mechanism has already been stated in Sect. 8.8, Part II.

27.3 Vertical Acceleration Depending on Pitching

The 2nd characteristic of the fore construction is that it is far away from amidship. Therefore, the influence of the pitching is large. For a ship length of 300 m, a pitching angle of 1° corresponds to a vertical amplitude of 2.6 m at the fore end.

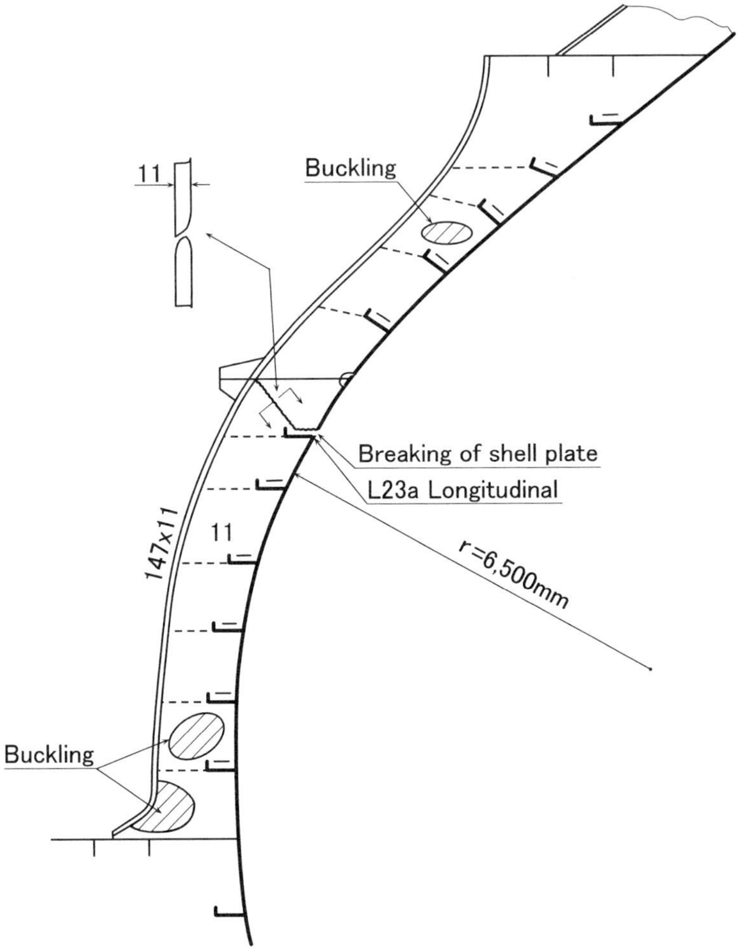

Fig. 27.3 Damages of web frame

The larger the pitching angle becomes, the bigger the vertical acceleration becomes. Hence, the water pressure on the bulkheads and the shell becomes larger when the fore peak tank is filled with ballast water. Such an additional water head due to the acceleration by pitching, as well as heaving, should be considered in the structural design of the fore construction.

Figure 27.5 shows the distribution of the vertical acceleration due to pitching and heaving along the ship length, from calculations using strip theory, for a container ship of 175 m length.

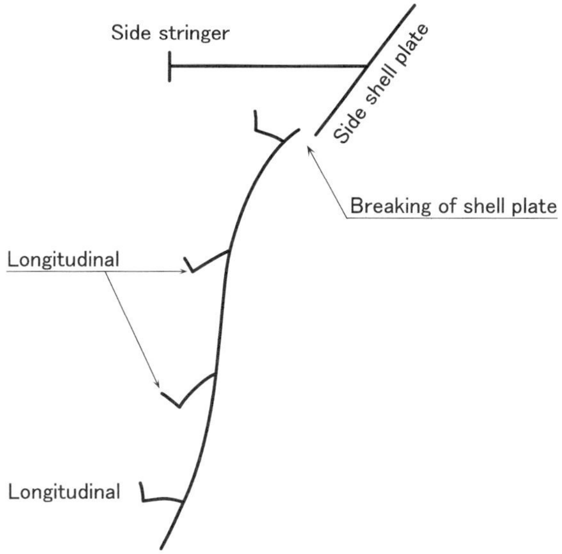

Fig. 27.4 Lateral bucking of longitudinals

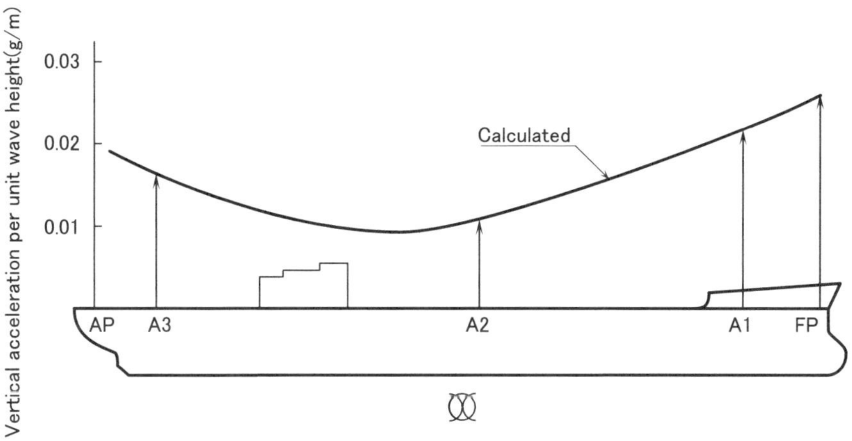

Fig. 27.5 Vertical acceleration due to ship motion

27.4 Deck Structure

The exposed deck located at the fore construction suffers from heavy wave loads, compared with other decks; it is said that the equivalent water head is 20 m.

There are three kinds of wave load: the wave impact on the ship bottom (bottom slamming), on the bow flare (bow flare slamming) and on the exposed deck (green water). They are related to each other, that is, a ship suffering from heavy bottom slamming and bow flare slamming has less green water, because the bottom slamming

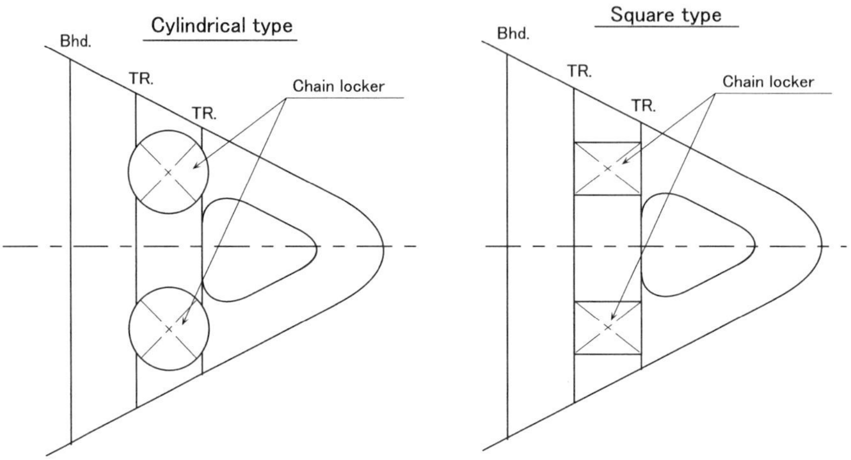

Fig. 27.6 Type of chain locker

and bow flare slamming resist the pitching motion, and also a large flare diverts waves away from the deck, as already seen in Fig. 2.7, Part I.

Recently, statistical analysis of ship motion and wave load can be done by using computer programs for each type of ship. Applying these may be an optimal design method for the fore construction.

The chain lockers are located in the fore construction. Although it is difficult to install them in high-speed ships because of their slender shape, a cylindrical type instead of a square type has been adopted for large ships because there is enough space, as illustrated in Fig. 27.6. In this type, the anchor chains are stored easily, and the stiffeners around the wall are not necessary. The standardization of the shape is also easy because it is an independent structural object. What is more, they act as a pillar between the upper deck and the lower deck.

27.5 Structural Continuity

The fore construction is a rather strong structure but the hold part following it is not so strong, hence structural continuity is very important.

For the transverse framing system in the hold of a cargo ship, the depth of the frames is ruled to be greater than or equal to 1/22 of the span in the region from the collision bulkhead to 0.15 L from FP, in order to prevent sudden structural change.

In the case where a horizontal girder is fitted inside the fore peak tank, a bracket is usually fitted in the hold part behind it. The shape should be rounded to prevent stress concentration and fatigue cracks, especially the aft end toe of the bracket.

Usually, the steel weight becomes less, and the production costs rise when the space between the longitudinal frames is narrow. The space is likely to be smaller in the fore peak tank from the strength viewpoint and to be larger in the hold from the production viewpoint. However, the continuity of the structure between the two regions needs to be considered.

27.6 Large Damage in Fore Construction

On September 26, 1935, a Japanese fleet encountered a violent typhoon during an exercise in the east of Aomori Prefecture Shiriyasaki. Two destroyers, the *Hatsuyuki* and the *Yugiri,* were cut in two in front of the bridge and then the fore part sunk. The Japanese navy set up a committee and investigated the accident as the 4th fleet incident in those days [143].

The loss of the *Bolivar Maru* on January 5, 1969, and the *California Maru* on February 9, 1970, were caused by the collapse of the fore part due to wave impact. Furthermore, the *Onomichi Maru* collapsed at the No.1 hold on December 30, 1980, and the remaining main hull sank after drifting. These three ships were large bulk carriers.

Such collapse of the fore part resulted from buckling of the upper deck after wave impact. Figure 27.7 shows the time history of the vertical wave bending moment obtained from a tank test of a container ship. It is clearly shown that additional vibration stress induced by wave shock and the bending stress in the sagging mode becomes large, when bow impact has occurred on the flare part.

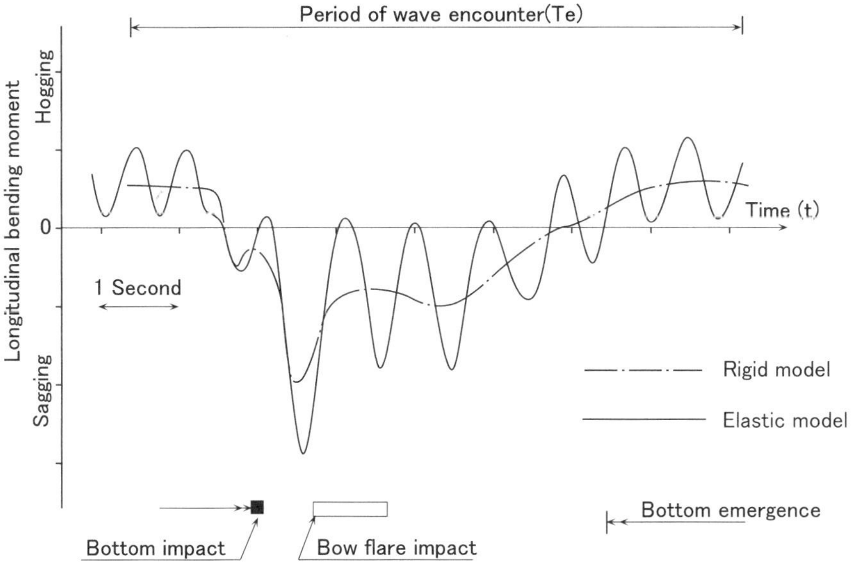

Fig. 27.7 Longitudinal bending moment due to wave impact

Fig. 27.8 Failure at fore part of Marcona Trader [144]

The *Marcona Trader* encountered such crucial accidents that the bottom and side shell of the fore region were torn off. Figure 27.8 shows the photograph at repair dock [144].

Chapter 28
Engine Room Construction

The most important item in the design of an engine room construction or a pump room construction of a tanker is to pay attention to the machinery installed there. Usually, the hull structure is to be designed based on stress not on deflection. In the case of a small ship the deflection is too small to cause any trouble to the machinery. However, in the case of a large ship the deflection of the hull structure causes some troubles.

Deformation in the turbine flat on which the turbines for the cargo oil pumps are installed causes a shaft misalignment between turbine and cargo oil pump. The deflection of a double bottom in the engine room badly affects the crank shaft of the main diesel engine and the shafting. It also badly affects, in a turbine ship, the alignment of the reduction gear and the shafting which causes non-uniform teeth contact between the bull gear and the pinion, as shown in Fig. 28.8. Excess bearing load on a shaft bearing sometimes causes damage to the bearing white metal.

Another important item concerning engine room construction is that the structure is also the foundation of the superstructure, especially in the case of an aft engine and an aft bridge arrangement. Inadequate foundation may cause some vibration problems. Arrangement of the engine room and the superstructure tends to be decided from the viewpoints of ship operation and habitability of the accommodation. With these considerations a poor foundation will result. A proverb says, "a ship designed by a designer who has no experience in the hull structure design will be malformed." This proverb will be true in the hull structure design of an engine room.

In the hull structure design of an engine room, the machinery arrangement is, of course, important, and besides this a hull structure designer should pay attention to the hull form in the area of the engine room. In the energy saving era after the oil crisis a hull form designers created very narrow forms in the area of the engine room to improve propulsion efficiency. A bow structure has been affected by the new hull form of a bulbous bow, and the engine room structure has been also affected by a bulbous stern. In these hull forms a big propeller is fitted at the end of a slender and weak cantilever.

Y. Okumoto et al., *Design of Ship Hull Structures*,
https://doi.org/10.1007/978-3-032-12500-2_28

In the hull structure design of an aft engine room which has a big main engine at the center, and which is situated near the propeller with an exciting force, more attention should be paid to anti-vibration design than in the case of a semi-aft or a midship engine room. A container ship built before the oil crisis had two diesel main engines of 34,800 BHP each and had a high speed of 25 knots. It is understood she had very little vibration during her sea trial, because of the semi-aft engine and the semi-aft bridge arrangement.

In this chapter the arrangement and the deformation of the engine and pump rooms are mainly explained.

28.1 Engine and Pump Rooms Arrangement

Here mainly the aft engine room arrangement is explained. The pump room arrangement is affected by the type of the pump, vertical or horizontal. As explained above in the case of a vertical shaft pump, the turbine flat is not to be connected to the pump room front bulkhead to avoid excessive deformation of the turbine flat.

Figure 28.1 shows the pump room arrangement of a 100,000 DWT tanker [145]. The turbine flat in the engine room is connected to the No. 3 horizontal girder on the pump room front bulkhead to reduce the scantling of the No. 3 horizontal girder. From the strength viewpoint this design will be OK. However, the water pressure in the aftermost center tank causes deflection in the turbine flat. Accordingly, the shaft connecting the pump and the turbine becomes misaligned.

Generally, a horizontal girder on a bulkhead is designed based on the stress and no consideration will be paid to the deflection. It is important for every naval architect to know the amount of deflection of a girder. The deflection of a girder is around 1/2,500 of the span. In this case the span of the No. 3 horizontal girder, that is the breadth of the center tank, is 15 m, accordingly the deflection will be about 6 mm which is big enough to cause the misalignment of the shaft.

In the machinery arrangement, the position of the boilers is important, i.e. whether to install them fore or aft. Originally, they were installed in the fore part of the engine room because the boiler tubes may crack due to vibration. The vibration level at the fore part of the engine room is less than that in the aft part. Now the improved wall-tube type boiler is tough enough to resist vibration, and they can be installed in any place in the engine room. It is common knowledge for a hull structure designer to install any machinery which may have trouble due to vibration at the fore part of the engine room.

Machinery which generates exciting force is best installed at the aft part of the engine room, because the aft part is narrow and the shorter span of the girder gives high local rigidity. In this view point a diesel generator is preferably installed at the aft part of the engine room.

The main engine is to be installed as far aft as possible to make the engine room length shorter. This is good for a rigid engine bed because the span of the double bottom floor is shorter at the aft part of the engine room. The container ship with

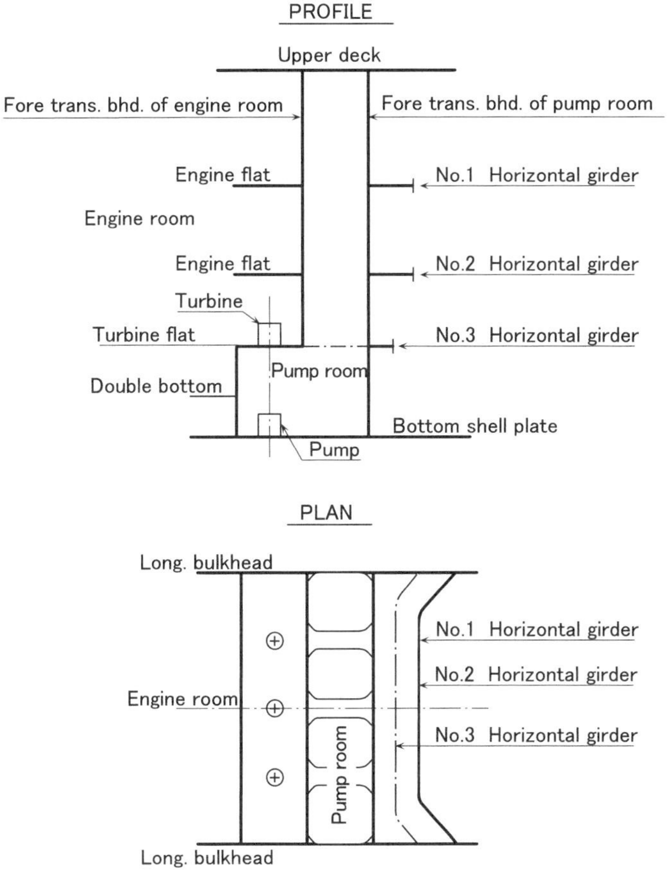

Fig. 28.1 Pump room arrangement of 100,000 DWT tanker

two 34,800 BHP engines, referred to above, has a semi-aft engine room arrangement which has a wider span of the double bottom floor in the engine room. In this case a center line longitudinal bulkhead is provided to support the wider span of the double bottom floor. In addition to the longitudinal center line bulkhead, the double bottom in the engine room is designed to be as high as possible to accommodate the raked propeller shaft. Then the rigidity of the double bottom, that is, the foundation of the main engines, seemed to be sufficient for the two Sulzer 12RND90 engines which were the most powerful in the world at that time.

In the design of the engine room, it will be the right road to first decide the height of the double bottom and the arrangement of the longitudinal and transverse bulkheads. Then the machinery, including the main engine, is to be arranged. At that stage the walls such as the front and the side walls of the superstructure are to be arranged over the bulkheads in the engine room. In this way vibration problems in the engine room, as well as in the superstructure, will be avoided.

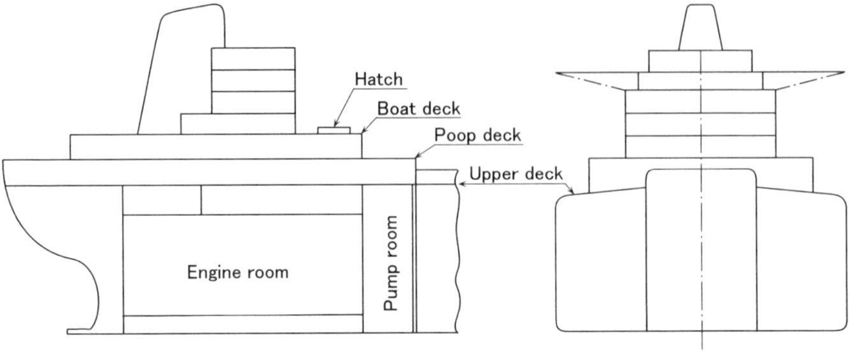

Fig. 28.2 Steel wall arrangement in engine room and deck house

The steel wall arrangement shown in Fig. 28.2 is for an 84,000 DWT tanker which suffered from a heavy vibration problem in the superstructure as well as in the engine room. The side walls of the superstructure are not aligned in a line and not on the longitudinal bulkhead in the engine room. The bridge front wall, the boat deck front wall, and the poop deck front wall are not in a plane. The fore and aft ends of the big superstructure, including the funnel, are not supported by the transverse bulkhead in the engine room. This arrangement was decided from the viewpoint of operation and habitability. For example, the replacement of a crankshaft can be performed through the hatch on the boat deck easily and not through the skylight as is usual, which is a little complicated. This is the reason why the front wall of the boat deck comes far forward from the bridge front wall. To separate the captain's and the owner's rooms from the crew space, a wider area under the boat deck is required which causes the side walls not to be in alignment.

As explained above, the steel wall arrangement in the engine room as well as in the superstructure should be decided not from the viewpoints of the operation and the habitability but from the viewpoint of the structure.

28.2 Rigidity Criteria in Engine Room Structure Design

Generally, the rigidity of a structure is expressed by the deflection of the structure. However, for a structure which may have a vibration problem such as an engine room, it is better to express its rigidity in terms of the natural frequency.

For a both-ends-fixed uniform beam, the deflection δ (cm) caused by its own weight is expressed by the following equation:

$$\delta = \frac{Wl^3}{384EI} \tag{28.2.1}$$

where

W: weight of beam (N).
l: span of beam (cm).
E: Young's modulus (N/cm^2).
I: sectional moment of inertia of beam (cm^4).

The natural frequency f (cpm) of the beam is expressed by the following equation:

$$f = \frac{60 \times 4.73^2}{2\pi} \sqrt{\frac{EIg}{l^3 W}} \tag{28.2.2}$$

where g is the acceleration of gravity (cm/s^2)

From Eqs. (28.2.1) and (28.2.2), the relation between δ (the deflection of beam due to its own weight) and f (lowest natural frequency of the beam) can be derived as follows:

$$f^2 = \frac{1.17 \times 10^5}{\delta} \tag{28.2.3}$$

The rigidity of a structure is usually shown in the form of the deflection or the natural frequency, and their relation can be expressed by Eq. (28.2.3)

The calculation method of the natural frequency of the double bottom, flat, etc. has been published in various articles [146], however the criteria for rigidity have been scarcely introduced. The authors consider it should be decided based on the designer's experience and his/her design philosophy.

For reference, here are some suggestions on how to decide the rigidity, that is, the natural frequency of the engine room structure.

28.2.1 Double Bottom in Engine Room

The vibration of the double bottom in an engine room is of fundamental one in the engine room. The main vibration modes in the double bottom in the engine room are shown in Fig. 28.3. The diesel main engine has several kinds of excitation which cause resonant vibration in the double bottom in the engine room when the vibration mode and natural frequency coincide with the exciting mode and frequency. For example, the exciting mode of the vertical unbalance force of the main engine corresponds to the fundamental vibration mode, and the exciting mode of the H-type lateral movement of the main engine corresponds to the transversely-unsymmetric vibration mode. These relations are also shown in Fig. 28.3.

As explained above each exciting force of the main engine corresponds to a vibration mode of the double bottom in the engine room. So the rigidity criterion should be that the natural frequency f_n of each mode of the double bottom vibration

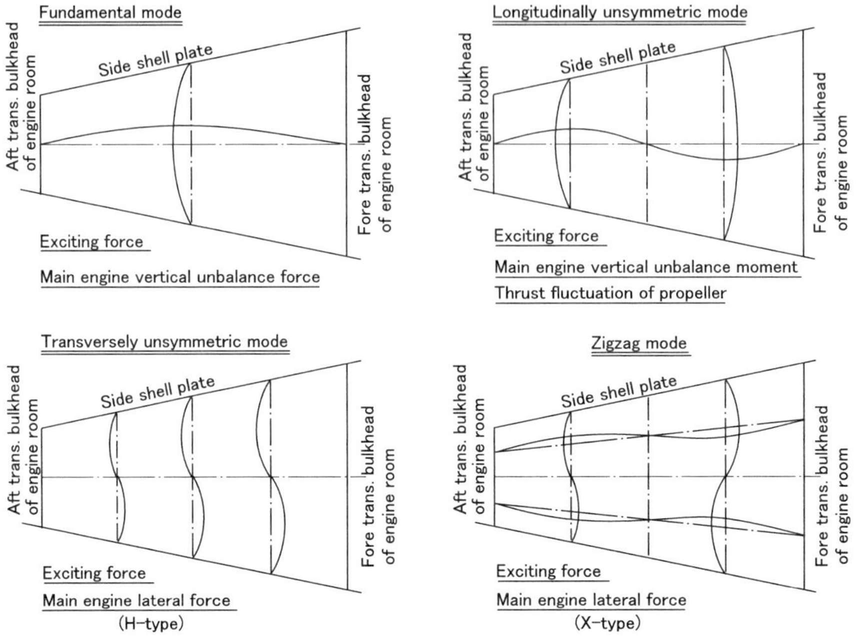

Fig. 28.3 Vibration mode of engine room double bottom (Aft engine room)

does not come close to the exciting frequency f_{ex} of the corresponding exciting mode. Assuming that the estimation allowance of the natural frequency of the structure is about $\pm 10\%$ and the resonance interval is also about $\pm 10\%$, the following relation can be obtained.

$$\frac{f_n}{f_{ex}} \leq 0.85, \quad \frac{f_n}{f_{ex}} \geq 1.15 \tag{28.2.4}$$

The above is for excitation of the main engine. The thrust variation among the propeller excitations causes a moment on the double bottom in the engine room through the thrust block, giving a similar effect to that of the main engine vertical unbalance moment. In this case, Eq. (28.2.4) is also effective by putting the propeller blade frequency as f_{ex} and the natural frequency of the double bottom longitudinal unsymmetrical mode vibration into f_n.

Equation (28.2.4) is derived based on the assumption that the estimation allowance of the natural frequency of the structure is about $\pm 10\%$. The figures of 0.85 and 1.15 depend on the estimation accuracy of the natural frequency of the structure. With improvement in the estimation accuracy, they can be modified.

28.2.2 Panel, Web, Stiffener etc.

For these local structural members, it is troublesome to apply the method explained in Sect. 28.2.1. To keep the natural frequency above some level is the practical way. These members are excited by the vibration displacement of the adjacent structures. As explained before, the vibration level in the aft part of the engine room is higher than that in the fore part. Therefore, it is reasonable to have severer criteria in the aft part than in the fore part.

To give an example, these members are to have natural frequencies of 1,500 cpm or higher in the aft part of the engine room, and 1,000 cpm or higher in the fore part of the engine room. In this case the natural frequency is calculated with the simply supported boundary condition with its own weight and added water if any.

28.3 Design of Structural Members in Engine Room

It is explained in Sect. 28.1 that in the design of the engine room, it is important to decide the height of the double bottom and the arrangement of the longitudinal and transverse bulkheads prior to the machinery arrangement. After fixing the bulkhead arrangement, i.e. the main frame work of the engine room, the structural members such as the engine flat and the web frame are designed.

The side shell in the engine room always supports external water pressure whose magnitude is large especially in the fully loaded condition. The water pressure is supported by the longitudinal frames and the load is transmitted to the web frames which are finally supported by the engine flats. From this viewpoint the engine flat is not only the foundation of the machinery but also an important strength member. In order to reduce the hull steel weight, the thickness of the engine flat reduced and finally the engine flat was sometimes constructed by a framework without a plate to support machinery foundations. Even in such a situation the hull structure designer should keep in mind that the engine flat is an important strength member to support the web frames in the engine room. Moreover, in the case of a narrow hull form at the aft part of the engine room, attention is to be paid to have enough breadth for the engine flat as shown in Fig. 28.4.

The engine flat is better arranged with an equal distance between the upper deck and the double bottom top so that the span of the web frame is equal, realizing a well-balanced structure.

The web frame is to support the water pressure from the side shell as a beam as well as supporting the water pressure from the bottom as a pillar. The web frame is compressed by water pressure from the bottom. Because of this contraction of the web frame in the engine room, the double bottom in the engine room which is supported by the web frame moves parallel upwards, and accordingly a deformation of the double bottom is caused. The deformation of the double bottom in the engine

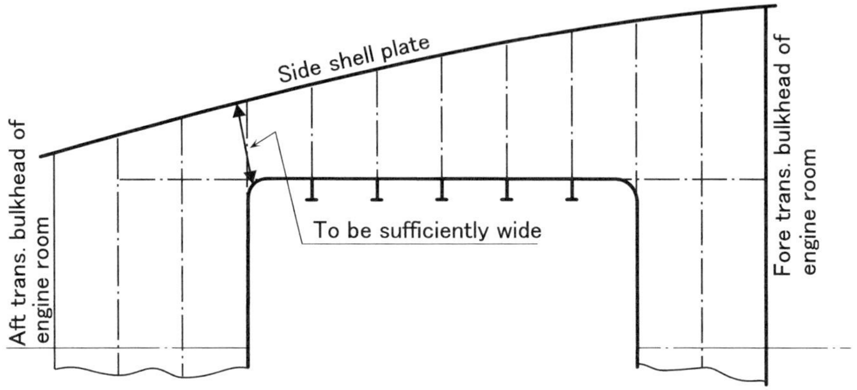

Fig. 28.4 Engine flat

room badly affects the main engine and the shaft alignment. However, the designer can control such deformation by adjusting the thickness of the web frame [147].

The pillar is also an important structural member in the engine room. It is popular to install H-section pillars below the engine casing at the web frame sections. These pillars which are installed apart from the center line, can be extended down to the double bottom in the engine room. However, in the aft part of the engine room the center line pillar cannot be extended down to the double bottom because of the propeller shaft. In this case the pillar is to be supported by a strong beam under the engine flat, as shown in Fig. 28.5. Such a structural arrangement is a kind of a misalignment and the pillar does not work well. One example is shown hereunder.

Assuming I is the sectional moment of inertia of the strong beam under the engine flat, l is the span of the strong beam and W is the concentrated load applied at the midspan of the strong beam, then the deflection δ of the strong beam is given by the following equation.

$$\delta = \frac{Wl^3}{192EI} + \frac{Wl}{4GA_w} \tag{28.3.1}$$

where

E: Young's modulus.
G: shear modulus of elasticity.
A_w: sectional area of web plate.

The first term of Eq. (28.3.1) shows the bending deflection and the second term shows the shear deflection.

The strong beam shown in Fig. 28.5 consists of a web of $1,000 \times 11$ mm and a face plate of 300×22 mm for both the upper and lower flanges, and the span l is 8 m. Then, the deflection δ can be calculated as follows.

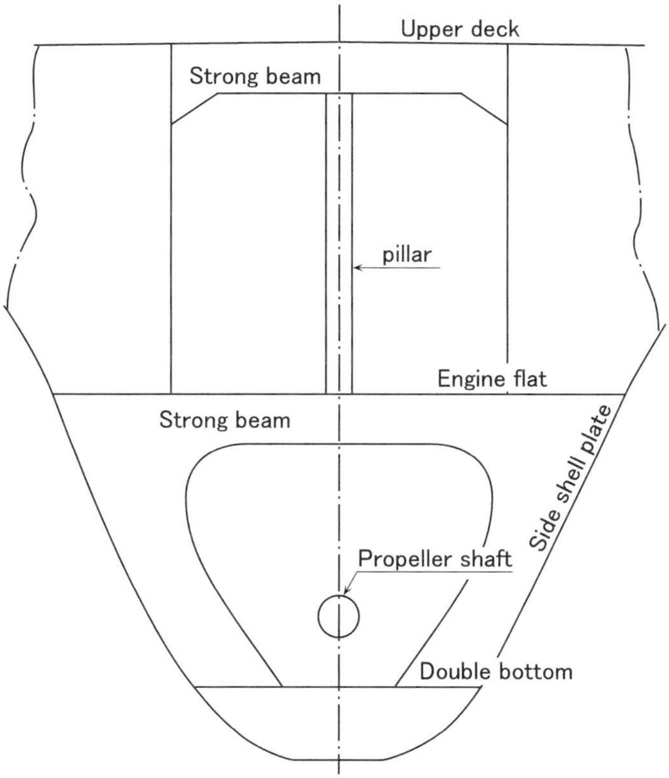

Fig. 28.5 Pillar installed on strong beam

$$
\begin{aligned}
\delta &= \frac{800^3\left(\text{cm}^3\right) \times W(\text{tonf})}{192 \times 2,100\left(\text{tonf/cm}^2\right) \times 421,667\left(\text{cm}^4\right)} \\
&\quad + \frac{800(\text{cm}) \times W(\text{tonf})}{4 \times 800\left(\text{tonf/cm}^2\right) \times 110\left(\text{cm}^2\right)} \\
&= (0.0030 + 0.0023)W(\text{cm}) \\
&= 0.0053W(\text{cm})
\end{aligned}
$$

The shear deflection is quite significant with the same order as the bending deflection because the strong beam is deep relative to the span and the web thickness is quite thin. Such a strong beam seems to have been designed paying more attention to the bending than to the shearing.

On the other hand, assuming the load W is supported by a pillar under the engine flat and on the double bottom, the shrinkage of the pillar Δh will be as follows.

$$\Delta h = \frac{hW}{EA} \tag{28.3.2}$$

where

A: sectional area of pillar.
h: length of pillar.

Instead of supporting the load W by a strong beam, if we assume an H-type pillar of $400 \times 400 \times 13/21$ between the engine flat and the double bottom shown in Fig. 28.5, the sectional area of the pillar is 215 cm^2, length h is 6 m, and the shrinkage Δh is as follows:

$$\Delta h = \frac{600(\text{cm}) \times W(\text{tonf})}{2,100(\text{tonf/cm}^2) \times 215(\text{cm}^2)} = 0.0013W(\text{cm})$$

Accordingly δ is about four times that of Δh which means that the strong beam cannot support the concentrated load from the pillar above as rigidly as the pillar on the double bottom.

28.4 Girders and Floors in Engine Room Double Bottom

It is a common practice to fit heavy engine girders just under the side plates of the main diesel engine structure. These heavy engine girders are often provided only in the region of the main diesel engine, but this practice must be examined carefully. In a similar way to a strong beam which supports a pillar at its midspan as explained in Sect. 28.3, the double bottom is a kind of a deep girder and its shear rigidity is important. So the plate thickness of the girder as well as the thickness of the floor is more important than the plate thickness of the tank top plate and the bottom shell plate.

In designing the girder and the floor arrangement in an engine room double bottom, the design philosophy of how to support the main engine is to be established, in addition to the engine maker's instructions and the classification society's rules.

Figure 28.6 shows a plan of an engine room double bottom of an aft engine ship. For the case where the distance between the fore end of the engine and the engine room front bulkhead is shorter than the distance between the fore end of the engine and the side shell, it is reasonable to extend the heavy engine girders to the engine room front bulkhead to support the main engine effectively. In this case, it would be regrettable to fit heavy engine girders only in the area of the main diesel engine. On the other hand, for the case where the distance between the fore end of the engine and the engine room front bulkhead is bigger, the floors supporting the fore part of the engine are to be thicker.

In both cases the shorter member is to be stiff which is understood in terms of the natural engineering philosophy.

There are many holes for access and easy fluid flow in the girders and floors, which impair the shear rigidity of the double bottom. The number of holes is to be limited and the holes should be opened in the areas where the shearing stress is low.

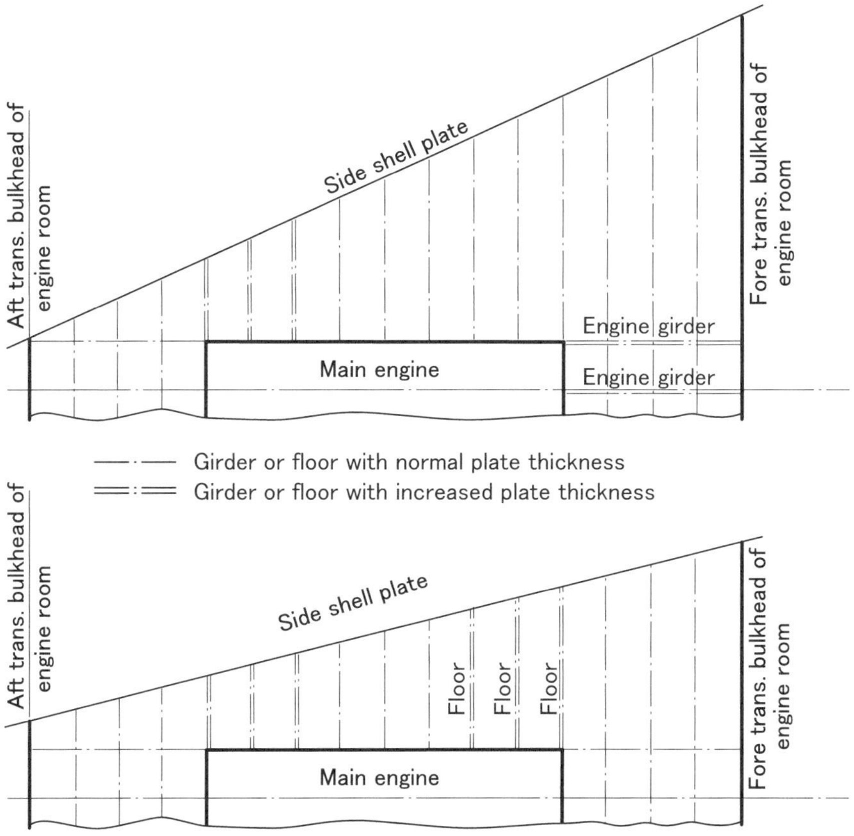

Fig. 28.6 Floor and girder arrangement of engine room double bottom

28.5 Problems Caused by Deflection of Engine Room Double Bottom

The large increase in size of a ship and their horse power in the era from 1950's to 1970's brought many important problems. One of them was the deflection of the engine room double bottom on which a high power main engine was installed. The profiles of such ships are shown in Fig. 28.7 [148].

According to the Ref. [148] the large increase in size is realized by increasing the length and the breadth of the ship. The depth of the ship was not increased so much. An increase in the breadth of the engine room double bottom followed an increase in ship breadth. The breadth of a 500,000 DWT tanker is about twice of that of a 36,000 DWT tanker. Supposing that the floor in the engine room double bottom is fixed at both ends by the side shell and the load does not change much, twice the breadth will bring 16 times the deflection of the double bottom.

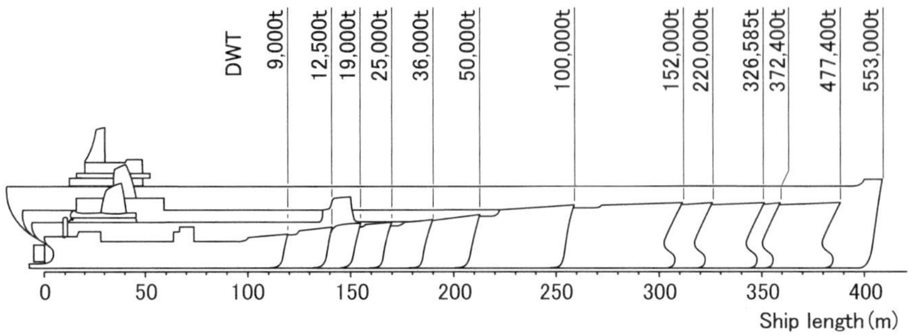

Fig. 28.7 History of increasing ship size

On the other hand, the diameter of the propeller shaft increased following the horse power increase. The shaft was designed for constant shearing stress, and the shaft revolutions was reduced to have higher propulsion efficiency. Accordingly, the torque became very big. The increase of horse power from 8,000 HP to 60,000 HP required a shaft with twice the diameter which had 16 times the bending rigidity.

The problems were caused by the deflection of the engine room double bottom which changed depending on the loading conditions, i.e. full load or ballast, the thrust variation at sailing, the wave load, and the thermal condition of the main engine. The rigid shaft is installed on a flexible hull structure, and this incompatibility has caused trouble at the connection parts between them, i.e. the shaft bearings.

The large increase in size of ships and their horse power caused much trouble at the shaft bearings and the reduction gear. This was because the propulsion machinery and the shafting were installed without paying any attention to the increased deflection of the engine room double bottom, which is the foundation for the propulsion machinery and the shafting. Recently high tensile steel has been applied widely in the hull structure which makes ships more flexible. The hull structure is mainly designed based on strength considerations. However, for some structural members, rigidity design or vibration design considerations are to be applied in addition to strength.

The trouble caused by the incompatibility between the rigidities of the hull and the machinery is divided into two: the damage to the shaft bearings and the damage to the reduction gear. The damage to the shaft bearings is caused by excess bearing load. Because of hull structure deformation an unbalance bearing load will appear, that is, some bearings support less load than the design load consequently the other bearings should support more load which destroys the bearings. In the case of a bearing which has no load, the unsupported span of the shaft will be nearly twice as long and the long span of the shaft may cause shaft lateral vibration which will destroy the bearing.

Damage to the reduction gear is caused by the non-uniform contact between the bull gear and the pinion which is caused by the deflection of the hull structure in the vicinity of the shafting, as shown in Fig. 28.8. The damage to the reduction gear shown is for a turbine ship, and in the case of a diesel ship, the bigger crank shaft

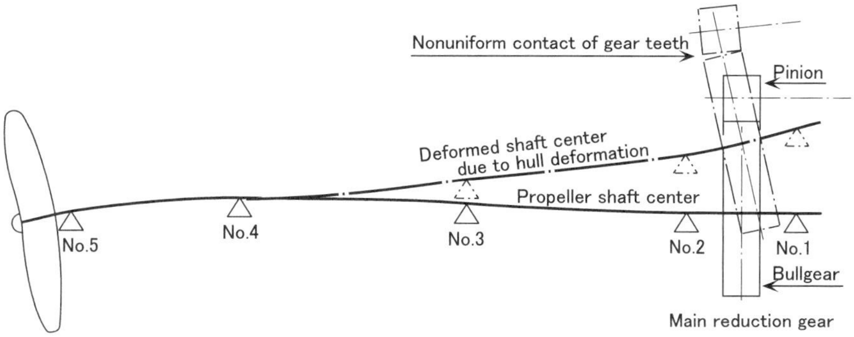

Fig. 28.8 Non-uniform contact of teeth of main reduction gear due to hull deformation

deflection, due to the engine room double bottom deflection, causes trouble such as an increase in the fillet stress in the crank shaft and damage to the bearing white metal.

One solution to this problem is to reduce the rigidity of the shafting by removing some shaft bearings, making the unsupported span longer.

Another solution is the rational alignment of the shafting. In this method shaft alignment is carried out so as not to have excess bearing load, and to settle the teeth contact in the reduction gears within the allowable limits, taking into account the hull structure deformation.

28.6 Deflection of Engine Room Double Bottom

For the rational alignment of the shafting considering the hull structure deformation, the hull structure deformation should be established by calculation. These days the final deflection mode can be accurately calculated by computer. However, to control the deflection it is better to estimate the deflection analytically.

The deflection of the engine room double bottom consists of the following three terms [147].

(1) Hull girder bending and shearing deflections in the vicinity of the engine room
(2) Deformation of the web frame which supports the engine room double bottom
(3) Bending and shearing deflection of the engine room double bottom itself

By obtaining the deflection of the engine room double bottom in this way, we can find which is the biggest component and how we can reduce the deflection effectively.

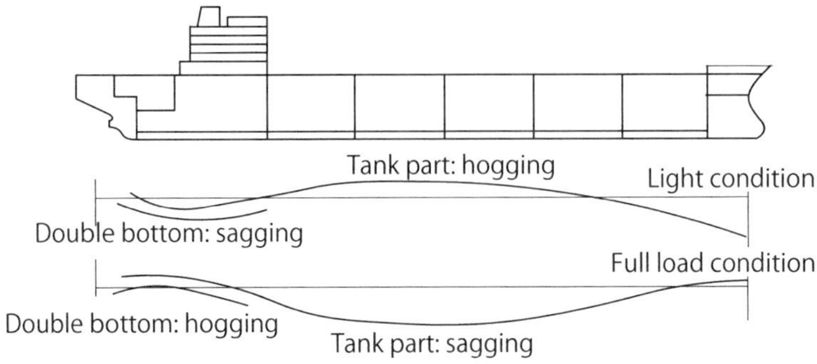

Fig. 28.9 Deflection of hull girder and double bottom in engine room

Moreover, by this analytical consideration the following question was clearly solved. It has been a well-known phenomenon that the deformation of the engine room double bottom of an aft engine tanker is convex upwards (hogging) in the fully loaded condition while the midship cargo tank part shows sagging, and in the ballast condition it is convex downwards (sagging) while the midship cargo tank part shows hogging as shown in Fig. 28.9. But nobody had explained this phenomenon clearly. In the above analytical consideration, it was found that shear deflection is dominant and its deflection is in the opposite direction to the bending deflection as explained in Sect. 15.5, Part II. Thus this long-continued question was clearly solved.

28.6.1 Bending and Shearing Deflection of Hull Girder in the Vicinity of Engine Room

Deflection in the engine room double bottom which affects the shaft alignment is the deviation from the straight line connecting the engine room fore and aft bulkheads. In the discussion of the deflection of the engine room double bottom, we can treat the hull girder in the vicinity of the engine room as a cantilever fixed at the forward section of the engine room front bulkhead, because the structure forward of the engine room front bulkhead does not affect the deflection of the engine room double bottom at all. Here the deflection is analyzed on the hull girder fixed at the midship section as shown in Fig. 28.10.

The deflection which causes trouble in the shafting, the reduction gear or the crank shaft is not that of the absolute value in the fully loaded or the ballast conditions, but rather the difference in the deflection caused by the load differences when in service, that is, the difference between the fully loaded and the ballast conditions. Based on this understanding, the load difference between the fully loaded and the ballast conditions is applied as the load on the cantilever model.

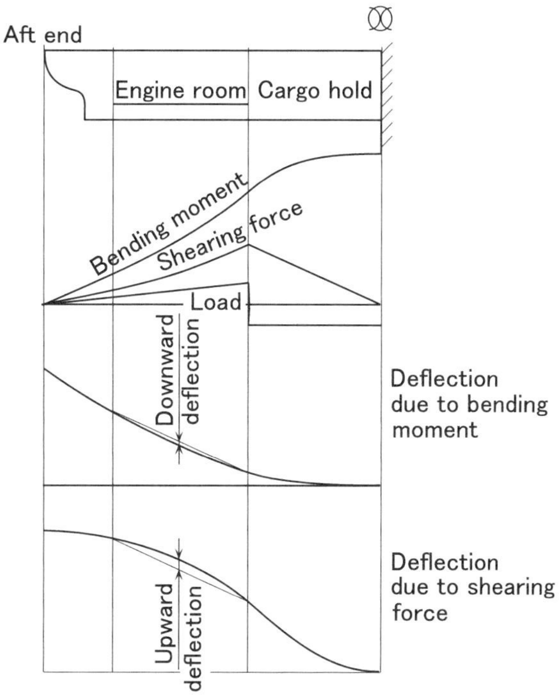

Fig. 28.10 Vertical deflection of hull girder

Assuming the ballast condition is the original condition, at the fully loaded condition upwards buoyancy caused by the draft increase is applied in the vicinity of the engine room; in contrast a downwards force caused by the cargo weight is applied in the vicinity of the cargo tank. Based on the above consideration, the load curve, the bending moment diagram, and the shearing force diagram were prepared as shown in Fig. 28.10. In Fig. 28.10 the deflections caused by the bending moment and the shearing force are also shown.

In Fig. 28.10, it is shown that the deviation from the straight line connecting the engine room fore and aft bulkheads is downwards for the bending deflection and it is upwards for the shearing deflection. The deviation caused by shearing is bigger than that by bending, because the hull girder cantilever is a deep girder in the engine room. Then, the total deviation comes upwards. This is the answer to the question raised in the previous discussion.

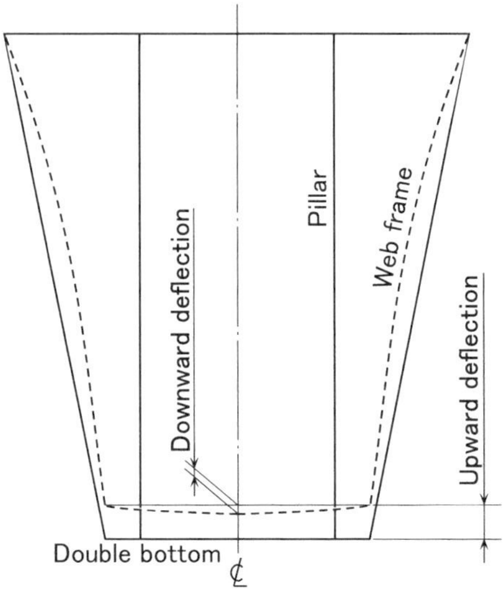

Fig. 28.11 Simplified deformation model of transverse ring of engine room

28.6.2 Deformation of Web Frame Which Supports Engine Room Double Bottom

The web Frame which supports the engine room double bottom is simplified and modeled as shown in Fig. 28.11. The deformed shape in the fully-loaded condition is shown in the broken line and the full line shows the original ballast condition.

As shown in Fig. 28.11 the web frames and the pillars are in compression because of the bottom water pressure and the double bottom moves upwards accordingly; also the center of the double bottom moves downwards as a result of side water pressure.

28.6.3 Bending and Shearing Deflections of Engine Room Double Bottom Itself

The engine room double bottom is a grillage structure consisting of floors and girders, and deforms upwards by the bottom pressure caused by the increased draft in the fully loaded condition.

The above deflection of the engine room double bottom is calculated for each component and the results are shown in Fig. 28.12. The ship is a 200,000 DWT turbine tanker of which the principal particulars are as follows:

$$L_{pp} = 300.0\,\text{m}$$

$$B_{mld} = 50.0\,\text{m}$$

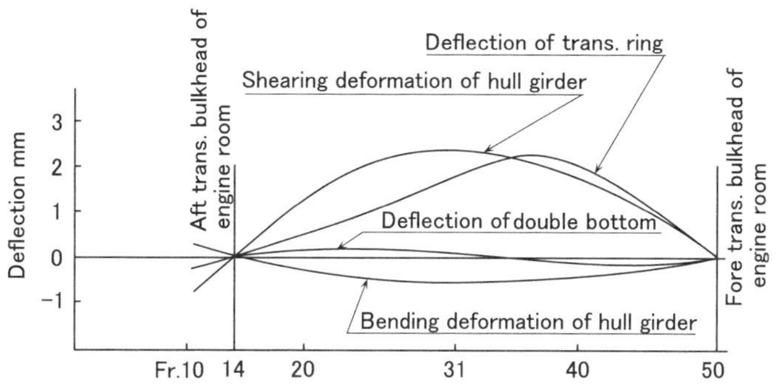

Fig. 28.12 Calculated example of deflection components of engine room double bottom

$$D_{mld} = 25.5\,\text{m}$$

$$d = 19.0\,\text{m}$$

Main engine: 33,000 SHP at 80 rpm

Speed: 16.4 knots

From Fig. 28.12 it can be found that the bending and the shearing deflections of the engine room double bottom itself are smaller than expected, and the main parts of the deflection come from deformation of the web frame, which supports the double bottom, and the hull girder bending and shearing deflections. Accordingly, increasing the double bottom height is not so effective in reducing the deflection of the engine room double bottom. On the other hand, as deflection by the hull girder shearing is dominant, special attention should be paid to the deflection of the engine room double bottom for a ship with small depth and with a considerably sloped side shell plating in way of the engine room.

Measured data for the engine room double bottom deflection at the reduction gear are shown in Fig. 28.13. It can be seen that the deflection is proportional to the draft increase.

28.7 Allowable Limit of Deflection of Engine Room Double Bottom

A 200,000 DWT turbine tanker is employed as an example in this section, and its shaft arrangement is shown in Fig. 28.14. In this shaft arrangement, the operating condition of each shaft bearing is as follows.

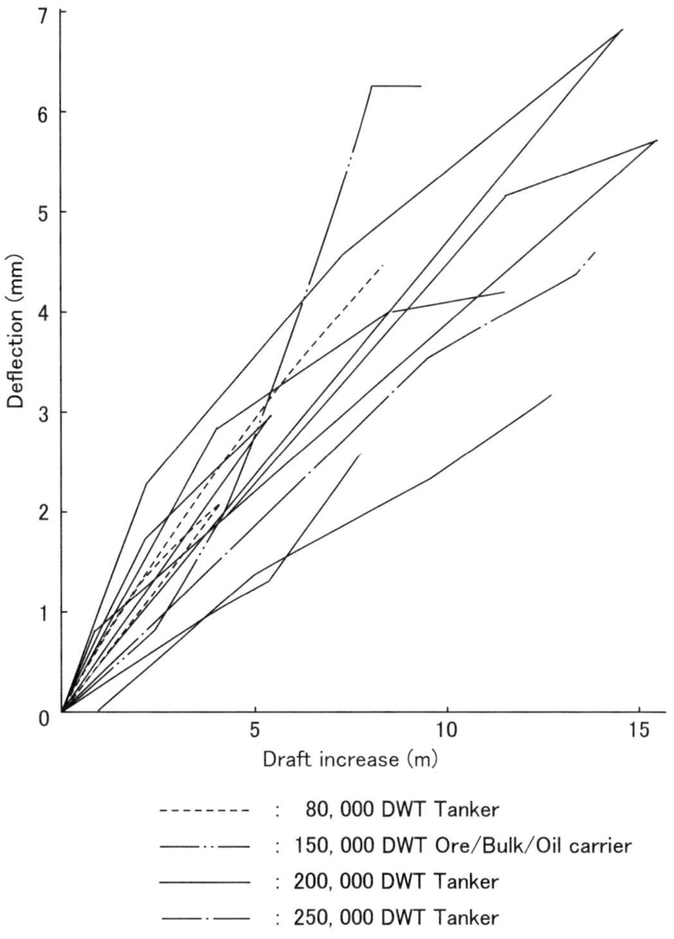

Fig. 28.13 Measured deflection of engine room double bottom at reduction gear

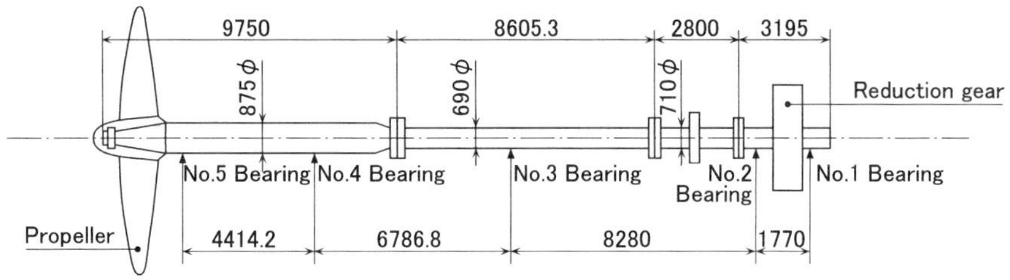

Fig. 28.14 Shaft arrangement of 200,000 DWT turbine tanker

(1) The bearing load on the No. 1 and 2 bearings is less than 46 tonf each and is positive (surface pressure is less than 10 kgf/cm^2).
(2) The load difference between the No. 1 and 2 bearings is less than 9.5 tonf (less than 20% of the total load).
(3) The bearing load of the No. 3, 4 and 5 bearings is less than 29, 27 and 89 tonf respectively (surface pressure is less than 5 kgf/cm^2) and is positive.

Assuming the bearing load on each bearing at the setting of the shaft with a light draft is R_1, R_2, R_3, R_4 and R_5 respectively and the bearing load on each bearing at the fully loaded condition is R'_1, R'_2, R'_3, R'_4 and R'_5 respectively, the above conditions are represented by the following equations.

$$\begin{aligned} 0 &< R'_1 = R_1 + I_{11}\Delta_1 + I_{12}\Delta_2 + I_{13}\Delta_3 \leq 46 \text{ tonf} \\ 0 &< R'_2 = R_2 + I_{21}\Delta_1 + I_{22}\Delta_2 + I_{23}\Delta_3 \leq 46 \text{ tonf} \\ 0 &< R'_3 = R_3 + I_{31}\Delta_1 + I_{32}\Delta_2 + I_{33}\Delta_3 \leq 29 \text{ tonf} \\ 0 &< R'_4 = R_4 + I_{41}\Delta_1 + I_{42}\Delta_2 + I_{43}\Delta_3 \leq 27 \text{ tonf} \\ 0 &< R'_5 = R_5 + I_{51}\Delta_1 + I_{52}\Delta_2 + I_{53}\Delta_3 \leq 89 \text{ tonf} \\ -9.5 \text{ tonf} &\leq R'_1 - R'_2 \leq 9.5 \text{ tonf} \end{aligned} \tag{28.7.1}$$

where

I_{ij}: influence coefficient.
Δ_i: deviation of each bearing from straight line connecting No. 4 and 5 bearing.

The influence coefficient I_{ij} is the variation in the bearing load (kgf) at the bearing i when the bearing j is moved in a vertical direction by the unit length (mm), and it depends on the rigidity of the shaft and the arrangement of the bearings. In the case of a 200,000 DWT turbine tanker, the influence coefficient is given in Table 28.1, and the bearing load on each bearing at the setting of the shaft, with a light draft, is given in Table 28.2.

Table 28.1 Coefficient I_{ij}

No. j	No. i				
	1	2	3	4	5
1	39,450	−50,130	14,700	−6,010	1,980
2	−50,130	64,480	−20,880	9,740	−3,210
3	14,700	−20,880	14,480	−15,390	7,080
4	−6,010	9,740	−15,390	25,000	−13,340
5	1,980	−3,210	7,080	−13,340	7,490

No. i: Supporting point where supporting force of bearing changes
No. j: Supporting point where 1 mm of vertical displacement is applied

Table 28.2 Supporting force for bearing and surface pressure

Supporting point No.	Item	
	Supporting force for bearing (kgf)	Surface pressure (kgf/cm^2)
1	23,520	5.1
2	23,880	5.1
3	23,580	4.0
4	16,680	3.0
5	68,020	3.8

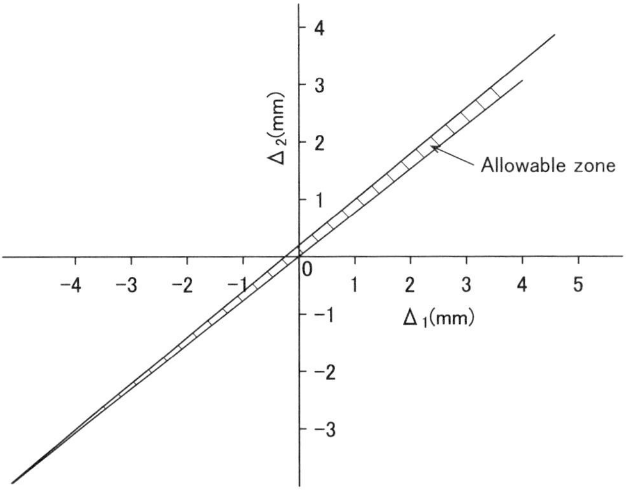

Fig. 28.15 Allowable deflection of engine room double bottom from shaft arrangement viewpoint

By putting these values in Tables 28.1 and 28.2 into Eq. (28.7.1), the allowable limits of Δ_1, Δ_2 and Δ_3 can be obtained. The allowable limits of Δ_1 and Δ_2 in the case of $\Delta_3 = -0.25$ mm is shown in Fig. 28.15. It can be said that the allowable limits of Δ_1 and Δ_2 in absolute values are moderately large, provided they have the relationship given in Fig. 28.15.

28.8 Control of Deflection of Engine Room Double Bottom

The phenomena of and the allowable limit of the deflection of the engine room double bottom have been made clear. So now how the deflection of the engine room double bottom is controlled, is explained. The 200,000 DWT turbine tanker previously introduced is taken as a model ship again.

In this case there are five bearings, so the deviations of the No. 1, 2 and 3 bearings from the straight line connecting the No. 4 and 5 bearings are important. As shown in Fig. 28.15 the allowable limits of the deviation of the No. 1 and 2 bearings are to be given for the given value of the deviation of the No. 3 bearing. For example in the case of Δ_3 is -0.25 mm, the combination of $\Delta_1 = -1.6$ mm and $\Delta_2 = -1.2$ mm is within the allowable limit, but the combination of $\Delta_1 = -1.6$ mm and $\Delta_2 = -0.8$ mm is outside of the allowable limit, even though the absolute value of the deviation is small. In this case having more deflection of the engine room double bottom at the No. 1 bearing and less deflection at the No. 2 bearing is satisfactory. For example $\Delta_1 = -1.4$ mm and $\Delta_2 = -1.0$ mm is satisfactory.

For the above control, adjustment of the thickness of the web plate which supports the engine room double bottom seems to be effective and practical.

Generally, in the case of trouble with the shafting and the reduction gear, it is recommended, and this has been applied, to increase the double bottom height in the vicinity of the shaft bearings and the reduction gear. However, as shown in Fig. 28.12 in an ordinary aft engine ship, the deflection of the engine room double bottom itself is very small and the increase of double bottom height will not solve the problem. A fine adjustment, giving a bigger deflection at the No. 1 bearing and smaller one at the No. 2 bearing as explained above, cannot be achieved by a uniform increase of the double bottom height.

Another way to control the deflection of the engine room double bottom is through adjustment of the hull girder deflection. As explained in Sect. 28.6.1 hull girder deflections by bending and by shearing have opposite direction. So there is a possibility to cancel them. Usually deflection by shearing is bigger than that by bending. Then, if we could increase the bending deflection of the hull girder in the engine room so that it cancels the shearing deflection, the effect of the hull girder deflection would disappear.

Natural phenomena are so well organized that a concentrated load at the end of a cantilever causes bending deflection only. The shearing force is constant along the length and no shearing deflection (deviation from a straight line) is caused. In an actual ship the ballast water in the aft peak tank can play the role of this concentrated load. The above case is outlined in Fig. 28.16.

28.9 Sea Chest in Engine Room Double Bottom

Cracks were found at the plate edges of a sea chest just after the sea trial of a newly built ship. The stiffeners around the sea chest box were not fitted in this case, although they were required by the design standards which were established by a senior designer of the company as shown in Fig. 28.17.

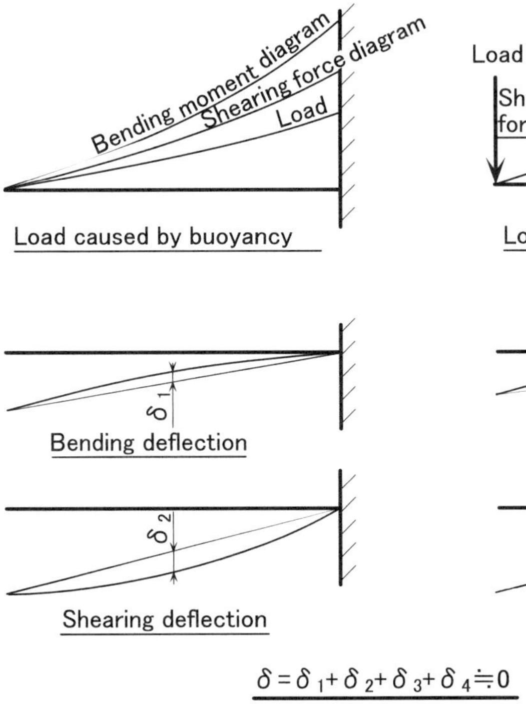

Fig. 28.16 Reduction of hull girder deflection

The trouble which was experienced by the company senior more than 30 years ago, appeared again. However, this time a young designer did not follow the instructions prepared by the senior. The senior, who instructed to fit stiffeners around the sea chest, by considering that the crack was caused by the vibration of the sea chest panel, was excellent. And the young designer was excellent too because he wanted to make clear the phenomena applying the latest technology. He solved the problem clearly by means of measurements on board the actual ship, the model test in a cavitation tunnel, and theoretical calculations.

The crack was caused by the vibration of the sea chest panel induced by the vortex of the incoming water flow. The resonance of the frequency of the induced vortex and the natural frequency of the sea chest panel, caused the heavy vibration. Accordingly, the problem was solved by the accurate estimation of the frequency of the induced vortex and the natural frequency of the sea chest panel.

The measurements of the frequency of the induced vortex for several shapes of sea chest opening, and of the natural frequency of a steel box filled with water using an exciter were carried out, and a design standard was established [149].

Engine room double bottom structure

6. Reinforcement of sea chest

Continuous horizontal tie band is to be fitted round the sea chest.

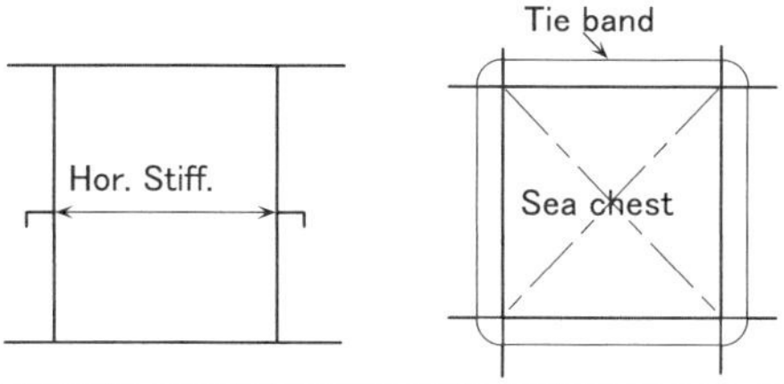

7. Bilge well of L.O.S.T. (bilge hat)

Attention is to be paid not to miss the well.

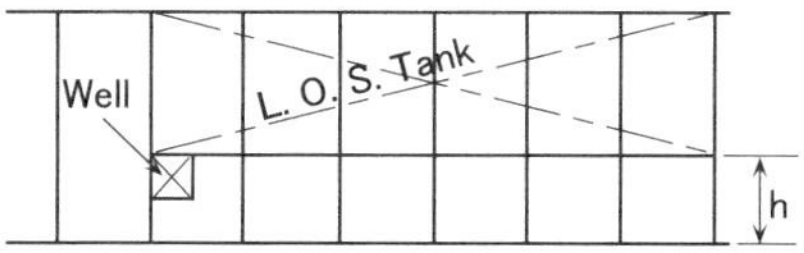

8. Cofferdam

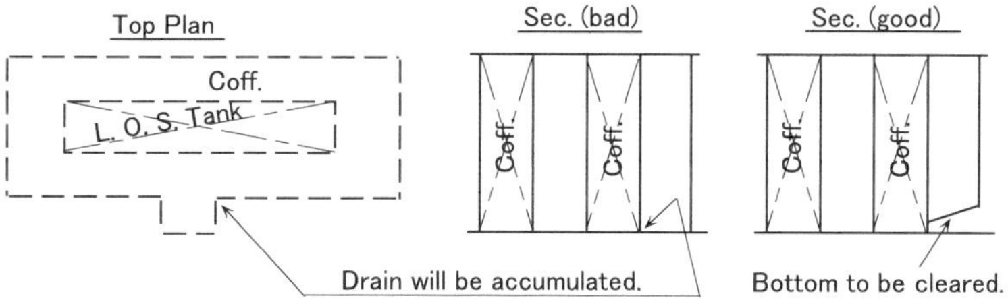

As shown above, the arrangement of extended cofferdam from top to bottom is not good, because drain will be accumulated at the front part of the extended part and a drain course is to be provided inside the cofferdam.

Fig. 28.17 An example of design standards of engine room double bottom structure

Chapter 29
Stern Construction and Stern Frame

The stern structure has the following special features not found in other structures:

- It is located near the propeller which is a vibration source.
- It supports the propeller and rudder.

That is, the stern must be designed so as not to allow excessive deflection which would disturb the smooth rotation of propeller and rudder, as well as keeping the stress in the stern within allowable limits. In general, design criteria based on allowable deflection requires much heavier scantlings for structural members, relative to that based on allowable stress.

Stern structure and engine room structure have several rotating machines inside them, therefore, they must be designed from the view point of allowable stress and also from allowable deflection. It was already mentioned that it is important in the design of an engine room that the structure is designed so as not to cause misalignment of machines or pumps, due to large deflection in the structure. In the same manner, much attention must be paid to restrict excessive deflection in the stern structure.

If we calculate the stress of the stern construction or stern frame by assuming proper external load, we find that the stress level of these members is almost one tenth those of the other structures. That is because the scantlings of these members are decided not by allowable stress criteria but by allowable deflection.

As explained in Sect. 28.7, in order to progress a design based on allowable deflection, allowable limits must be clarified first, and then the stiffness of structural members is determined so that the deflection of the member is lower than the critical limit. However, for the stern construction, stern frame and rudder, it is regretted that the allowable criteria of deflection has not been completely established at this stage.

Establishment of the following criteria is essential in designing more reasonable structures:

- allowable stress to prevent cracking or buckling
- allowable amplitude to prevent vibration
- allowable deflection to prevent machinery damage

Y. Okumoto et al., *Design of Ship Hull Structures*,
https://doi.org/10.1007/978-3-032-12500-2_29

29.1 Aft Peak Tank Construction

We normally call the after part of a vessel behind the aft peak bulkhead the stern structure, whose lower part is often utilized as an aft peak tank and upper part is for the steering gear room.

In the design of the stern structure, it is an important matter to choose whether a longitudinal framing system or a transverse framing system should be provided. Furthermore, when a transverse framing system has been selected, we should then decide which is best, a transverse system constituted by solid floors or transverse system by built-up frames.

Although it is a recent tendency to apply the longitudinal system in a stern structure, the transverse framing system was used formerly, in particular, the transverse system by built-up frames.

For example, soon after the time when one of the authors joined IHI, that was the middle of the 1950's, a 32,000 DWT tanker was planned to be built by IHI, which was a size larger than the 20,000 DWT tanker that we usually designed in those days. The owner gave us several structural drawings drawn by ISHERWOOD, a British company, which indicated that the aft peak tank was fabricated by a solid floor system. Considering that a fluctuating pressure induced by the propeller acts on the outer shell plating of the aft peak tank, it is reasonable to provide solid floors to the tank structure because they support the outer shell very rigidly. We were deeply impressed that such a rigid structure was used in the aft peak tank even for the larger size of vessel. Nevertheless, we had also some pity on the worker who had to perform tasks in the narrow space surrounded by solid floors; the floor space was 610 mm.

On the other hand, these solid floors are effective in preventing the vibration of the outer shell plating, but they may cause another vibration problem: vibration of the solid floor plate itself. Once the aft peak is filled with ballast water, the natural frequency of the floor plate decreases due to the increase of water weight, that is the effect of added mass of water. This makes the solid floor prone to vibration. Hence, we thought it would be helpful to arrange several lightning holes in the floor plating to reduce the weight of added mass.

In the case of the longitudinal framing system, longitudinal stiffeners are supported by transverse web frames, which means that slot openings are inevitable at the intersection between longitudinals and web frames. Such slots were recognized as starting points for cracks, because they create a stress concentration at the intersection. For this reason, we hesitated to apply a longitudinal framing system to the stern structure before the design of the 32,000 DWT tanker. Of course, regarding the application of a longitudinal system to the aft peak tank, we were well aware of the advantage of easy construction and of steel weight saving, while we knew the disadvantage of high stress concentration.

At the beginning of the stern structure design of the vessel, one young designer proposed to take the longitudinal framing system. "All slot openings had better be covered by collar plates, as long as there are problems because of high stress concentration caused by openings." It was a great idea to apply the longitudinal system to the stern structure, like the egg of Columbus.

29.2 Vibration of Stern Structure

The magnitude of the hull girder vibration is largest at the aft end of the stern part, in general. If we stand near the flag pole of the aft end during sea trials, we can experience the largest vibration amplitude. In other words, even when the vibration amplitude of other structures is greater than that of the stern, we will be able to reduce it to less than the stern by providing suitable reinforcement.

29.2.1 Vibration of Stern Overhang

Figure 29.1 shows a profile of the stern structure. As the length l of the overhang and its breadth b become larger, the natural frequency of the overhang decreases and it may cause a vibration problem.

The estimation method of the natural frequency of a stern overhang is explained hereafter. In this method, firstly the stern overhang is regarded as a uniform cantilever beam; secondly the natural frequency is obtained for the shearing deformation of the

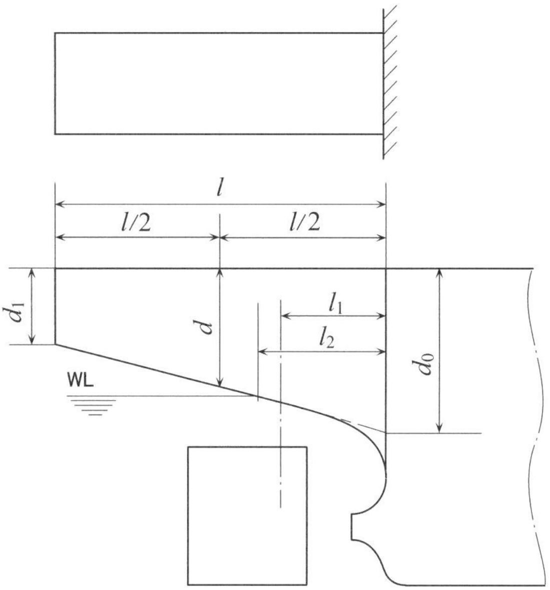

Fig. 29.1 Definition of stern overhang

beam, and finally it is modified by taking account the effect of bending deformation and of non-uniformity of the cross section of the beam. In addition to that, the weight of rudder and added mass of water is considered. The natural frequency is expressed by the following equations:

$$f = f_0 \times K_b \times K_v \tag{29.2.1}$$

where

$$f_0 = \frac{15}{l}\sqrt{\frac{gGA_s}{w}} \tag{29.2.2}$$

$$w = w_H + w_R + w_W$$

w_H: average hull steel weight per unit length.
w_R: equivalent weight of rudder per unit length.

$$w_R = \frac{W_R}{l} \times C_R$$

W_R: weight of rudder.
C_R: equivalent coefficient of rudder as indicated in Fig. 29.2.
w_W: equivalent weight of added mass of water per unit length.

$$w_W = \overline{w}_W \times C_W$$

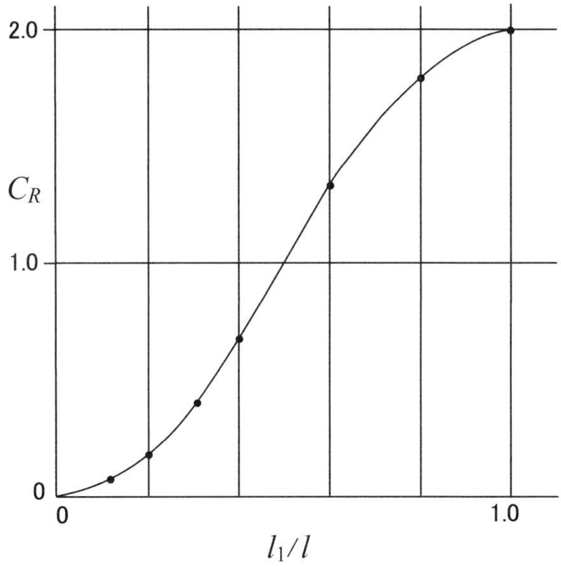

Fig. 29.2 Equivalent coefficient of rudder weight

$$\overline{w}_W = \frac{1}{2}\gamma_W \times \pi \times b^2$$

C_W: equivalent coefficient of added mass of water as indicated in Fig. 29.3.
γ_W: weight of sea water per unit volume.
l_2: length of overhang attaching to sea water.
b: breadth of water plane at the mid-point of l_2.
G: shear modulus of elasticity.
A_s: effective shear area at the middle point of l.
K_b: modification factor to include bending deformation as indicated in Fig. 29.4.
I: sectional moment of inertia at the middle point of l.

Fig. 29.3 Equivalent coefficient of added water mass

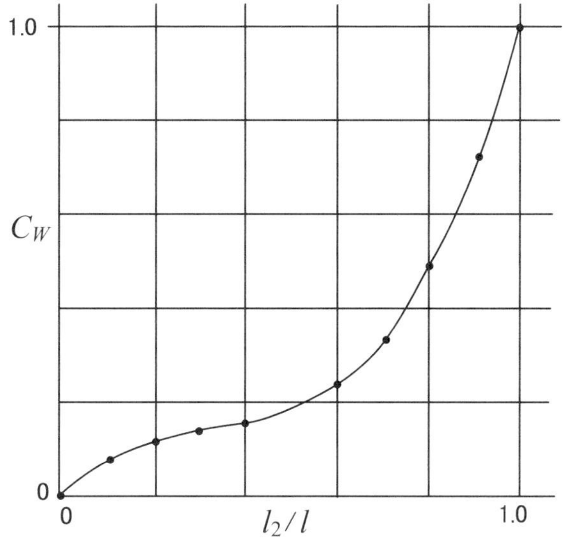

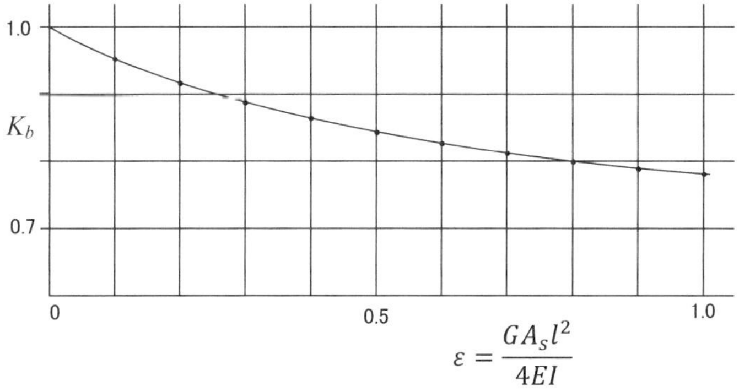

Fig. 29.4 Modification factor of bending

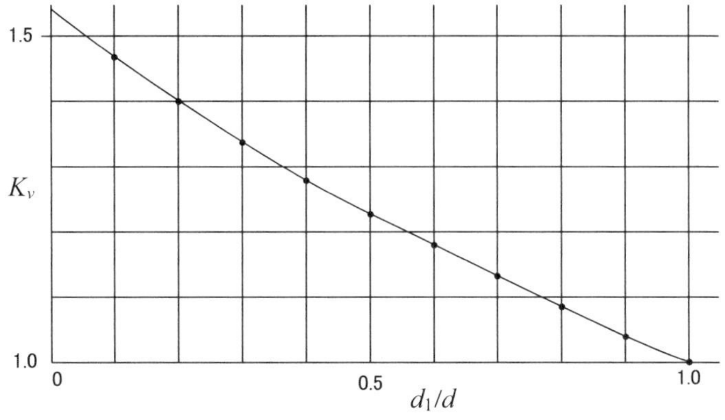

Fig. 29.5 Modification factor of cross section change

K_v: modification factor to include the change of cross section as indicated in Fig. 29.5.

In the case that the natural frequency obtained by Eq. (29.2.1) is close to the exciting frequency due to the propeller, resonance should be avoided by the following measures:

- Provide additional longitudinal bulkheads
- Increase plate thickness of the outer shell plating

29.2.2 Transverse Vibration of Stern Bossing of a Single Screw Vessel

Vibration of the bossing is induced by the vibration of the hull structure around bossing and of the propeller including the propulsive shaft. The vibration of the hull structure is affected by the movement of shaft and propeller through the spring effect of the lubricating oil film in the stern tube bearing and intermediate bearing. Hence the natural frequency of the bossing is usually calculated by means of FEM. We will explain the correlation between the exact result from the FEM calculations and the result obtained by simplified three-dimensional frame analysis. The model of the FEM calculations is illustrated in Fig. 29.6. An accurate natural frequency for the bossing can be obtained by the FEM model, however, making such FEM models takes much cost and time.

Therefore, a simplified calculation method is proposed with the aid of a three dimensional frame model as shown in Fig. 29.7. The stern bossing is modeled as co-supported beams connected to the propeller and propulsive shaft by an oil film spring bearing. The co-supported beams are composed of two groups of beams: the vertical cantilever beams, whose upper ends are fixed and which represent the transverse

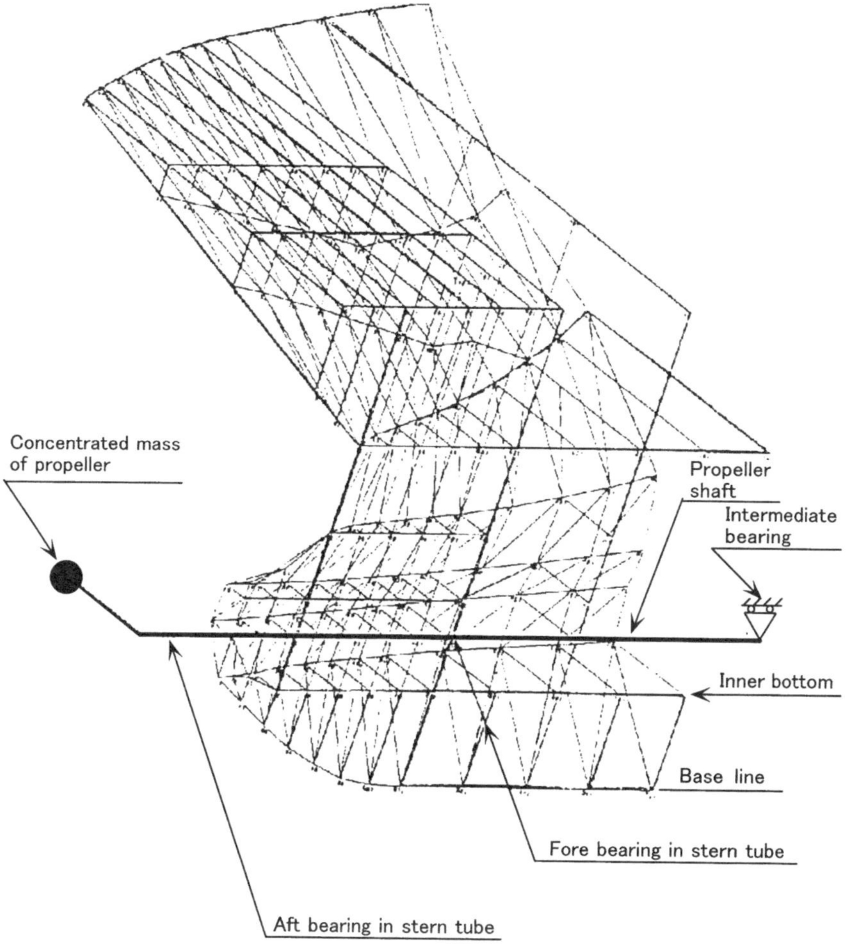

Fig. 29.6 3 Dimensional FEM model of bossing

members of the hull structure and stern frame, and the longitudinal cantilever beams with fixed fore ends represents the longitudinal members of the hull structure.

Table 29.1 demonstrates the calculated results obtained by each structural model. The simplified three-dimensional frame model can give only the natural frequencies in air, and the obtained natural frequency is a little higher than that of the FEM analysis. Therefore, it must be calibrated by correlation between the frequency in air obtained by the simplified frame analysis and that in water obtained by FEM analysis. The correction factor of the natural frequency due to water and analysis method becomes 309/363=0.85.

Consequently, when the natural frequency of the bossing in air is obtained by a simplified three dimensional frame model, multiplying it by 0.85 gives the natural frequency in water.

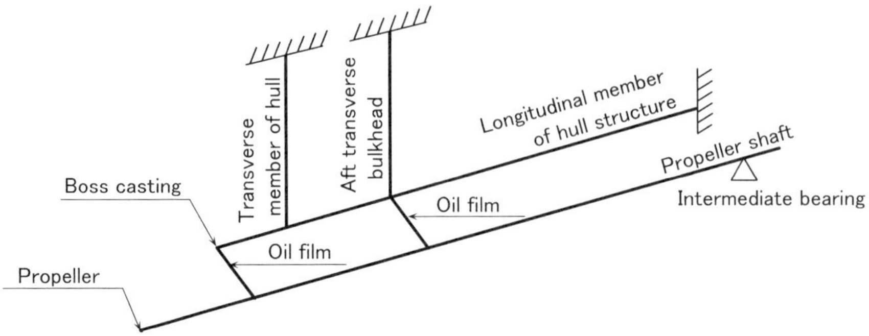

Fig. 29.7 Simplified frame analysis model of stern bossing

Table 29.1 Natural frequency of vertical vibration of bossing (single screw)

	Natural frequency		Note
	In air (cpm)	In water (cpm)	
3D FEM model with plate elements	330	309	Including propeller shaft
	530	487	Omitting propeller shaft
3D frame analysis model	363	309	Assuming natural freq. in water to be 0.85 times of natural freq. in air
Measurement	–	295	

29.2.3 *Vertical Vibration of Twin Bossing in Twin Screw Vessel*

In a vessel with twin screws, the vertical vibration of bossing may appear in two different vibration modes. One is a symmetrical mode, that is, port-side and starboard-side bossing vibrate simultaneously in a vertical direction, and the other is an unsymmetrical mode where the port-side bossing moves downward when the starboard-side moves upward, and vice versa. Strictly speaking, each bossing moves in a circular shape in the latter case. The symmetrical mode is caused by the vertical exciting force, while the unsymmetrical mode responds to transverse forces.

As for the calculation for vertical vibration of twin bossings, the method suggested by Ayling is well known and is still utilized for design purposes [150]. He assumed the bossings to be cantilever beams fixed to the main hull structure and derived the natural frequency of the symmetrical mode N as follows;

$$N = \phi_2 \cdot \frac{60}{2\pi}\sqrt{\frac{3E(I_1 + I_2)g}{(W_1 + W_2)l^3}} \text{ (cpm)} \tag{29.2.3}$$

where

ϕ_2: experimental modification factor of inclined angle of bossing as indicated in Fig. 29.8.

I_1, I_2: sectional moment of inertia of cast steel section and fabricated steel plate, respectively, at the base of the bossing (detail is explained in Fig. 29.9).

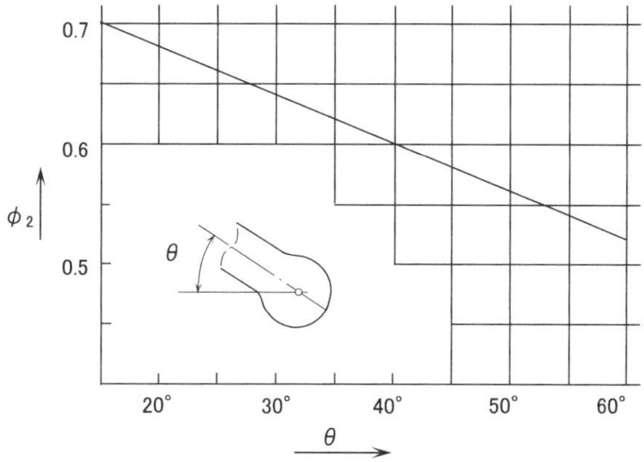

Fig. 29.8 Modification factor of raised angle of stern bossing

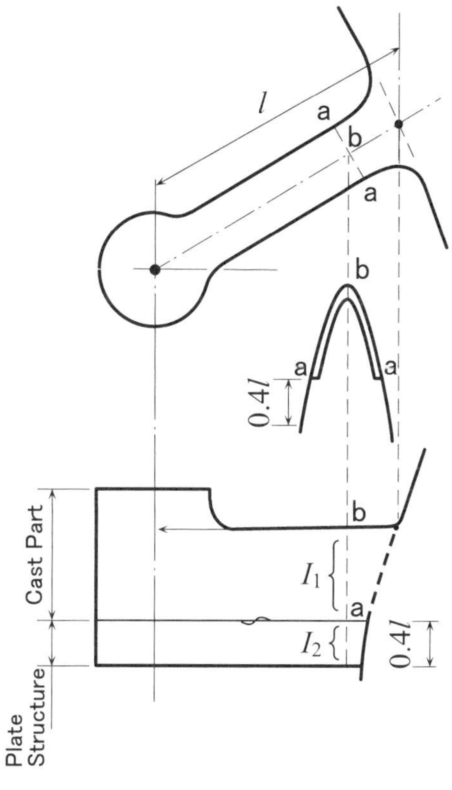

Fig. 29.9 Definition of stern bossing

W_1: propeller weight.
W_2: weight of bossing and propeller shaft.
l: distance between propeller shaft center and hull structure as shown in Fig. 29.9.
E: Young's modulus.
g: acceleration of gravity.

Moreover, the natural frequency of the unsymmetrical mode is reported to be approximately 85% of that of the symmetrical one. In order to confirm the above report, an excitation test was carried on a 1,800 TEU container carrier which was designed to have two diesel driven main engines and twin bossings to realize 34,800 HP. According to the measured results, the natural frequency of the symmetrical mode was 505 cpm and that for the unsymmetrical mode was 437 cpm, therefore, the ratio of the unsymmetrical frequency against the symmetrical one was proved to be 86.5%.

29.3 Stern Frame

The structure of the stern frame has changed drastically in response to the variation of rudder type and the change of stern shape.

In the case of a G type stern frame, not only the weight of the propeller and rudder but also the reaction force induced by keel blocks in dry dock must be supported, because the vessel touches the keel blocks firstly at the stern frame. In the design of a G type stern frame, the cast steel part around the propeller, called the propeller post, is easily designed for it acts as a pillar for the compressive load due to the reaction force from the keel blocks. In contrast, it is difficult to maintain sufficient strength in the shoe piece, because it works as a long cantilever beam if it is subjected to a reaction force at its aft end. To avoid this, the shoe piece must be designed so as to have a slight inclination which is greater than the maximum trim angle at dry docking as illustrated in Fig. 29.10.

As previously discussed in Sect. 23.5, in the case where the vessel has an open type of bulbous stern as indicated in Fig. 23.15, no reaction force is transmitted to the propeller post at dry docking. Consequently, the thickness of the lower part of the propeller post is considered to be sufficient as long as it has the same thickness

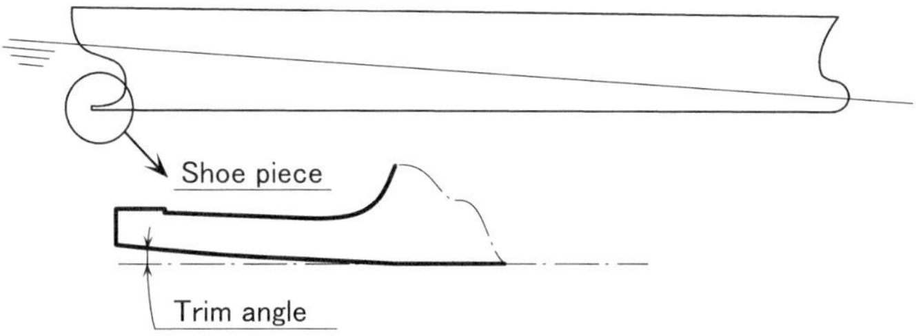

Fig. 29.10 Inclination of shoe piece bottom

as the side shell plate. If the plate thickness of this part is increased, it results in an increase of bossing weight, which acts as additional weight at the aft end of the cantilever beam. Therefore, such a design is not preferable in order to prevent any vibration problems.

By providing inclination to the shoe piece bottom, the reaction force can be avoided. Hence, the shoe piece is subjected to only a horizontal force due to the rudder. To investigate the strength of the shoe piece under such loading conditions, it is modeled as a cantilever beam subjected to a concentrated load at its end, and in this case the deflection at the end becomes of great concern.

Let us compare the deflection of a beam in the following two cases:

- Case A: shoe piece is modeled as a cantilever beam which has a triangular plan
- Case B: shoe piece is modeled as a uniform cantilever beam which has a straight form

Provided that the depth of the shoe piece b is the same in both cases, the total weight is also the same as shown in Fig. 29.11. In Case A, if the breadth at the fixed end is d_0, the breadth at the position whose distance is x from the fixed end is given by $(l-x)d_0/l$ and the bending moment at that section is $W(l-x)$. Thus, the deflection can be obtained by integrating the curvature according to the following process;

$$\frac{d^2y}{dx^2} = \frac{M}{EI} = \frac{W(l-x)12l^3}{Ebd_0^3(l-x)^3} = \frac{12Wl^3}{Ebd_0^3(l-x)^2} \tag{29.3.1}$$

Integrating the above curvature, we find the slope of the beam,

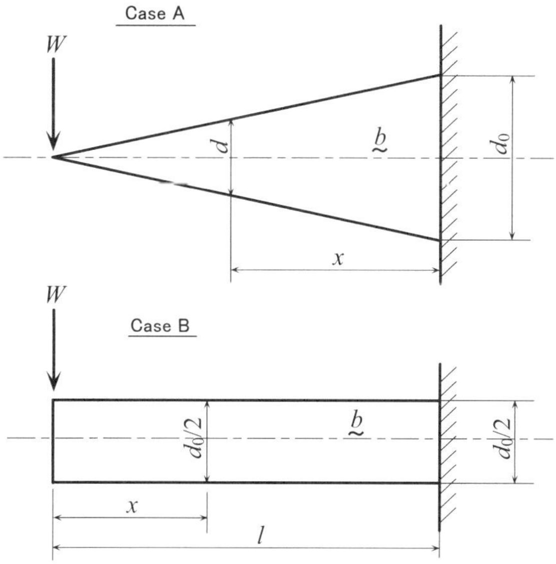

Fig. 29.11 Cantilever beams with triangular plane and with uniform plane

$$\frac{dy}{dx} = \frac{12Wl^3}{Ebd_0^3} \int (l-x)^{-2} dx = \frac{12Wl^3}{Ebd_0^3} (l-x)^{-1} + C \tag{29.3.2}$$

At the fixed end $x = 0$, the slope of deflection must become zero, that is $dy/dx = 0$; therefore, the arbitrary constant C must satisfy

$$C = -\frac{12Wl^3}{Ebd_0^3} \times \frac{1}{l}$$

Then the deflection can be expressed by introducing an arbitrary constant C_1,

$$y = \frac{12Wl^3}{Ebd_0^3} \left[\int (l-x)^{-1} dx - \int \frac{dx}{l} \right] = \frac{12Wl^3}{Ebd_0^3} \left[-\log(l-x) - \frac{x}{l} \right] + C_1$$

An arbitrary constant C_1 can be determined by the boundary condition that the deflection must be zero, i.e. $y = 0$, at the fixed end, i.e. $x = 0$.

$$C_1 = \frac{12Wl^3}{Ebd_0^3} \log l$$

Thus, we obtain the deformation form,

$$\therefore y = \frac{12Wl^3}{Ebd_0^3} \left[-\log(l-x) - \frac{x}{l} + \log l \right] = \frac{12Wl^3}{Ebd_0^3} \left[-\frac{x}{l} - \log\left(1 - \frac{x}{l}\right) \right] \tag{29.3.3}$$

In Case B, the deflection of the uniform beam is expressed by Eq. (29.3.4), taking the origin of the x-coordinate at the free edge.

$$y = \frac{12Wl^3}{Ebd_0^3} \times \frac{8}{3} \left(1 - \frac{3x}{2l} + \frac{x^3}{2l^3} \right) \tag{29.3.4}$$

The deformed shape of the beam is shown in Fig. 29.12 for both cases where the triangle beam and the uniform beam are loaded by the same magnitude of force. According to the figure, the triangular shape of the cantilever beam creates a large deflection around the tip of the beam. Considering the above, the rational design of the shoe piece is to have a triangular shape from the fixed end to $0.7l$ distance, and a uniform shape from $0.7l$ to the tip end.

Although the shape of the tip end of the shoe piece used to be designed in such a way that it had a teardrop section similar to the horizontal section of the rudder as shown in Fig. 29.13a, b, it has been changed to a blunt shape as shown in Fig. 29.13c. This is because it is unreasonable to provide such a sophisticated streamlined form to the tip of the shoe piece while it is in the turbulent flow induced by the propeller. In spite of the above discussion, the forward part of the shoe piece from the propeller position should be rounded at the corners as shown in the section

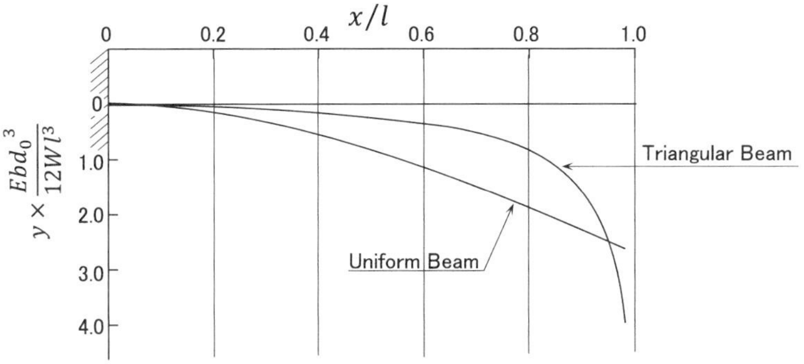

Fig. 29.12 Deformation of triangular beam and uniform beam

drawing in Fig. 29.13, because rectangular corners will generate vortices which may damage the propeller.

A fabricated type of shoe piece which consists of a few steel plates was widely used, since it is neither heavier nor as expensive as a steel casting. However, even in the case of applying a steel casting to the shoe piece, the steel weight can become approximately the same as a fabricated type by providing a ∩ shape in cross section in which the steel around the neutral axis of the horizontal bending is reduced. In addition to the above fact, a casting is more convenient than fabricated steel for the forming of a large corner radius to prevent the vortex problem.

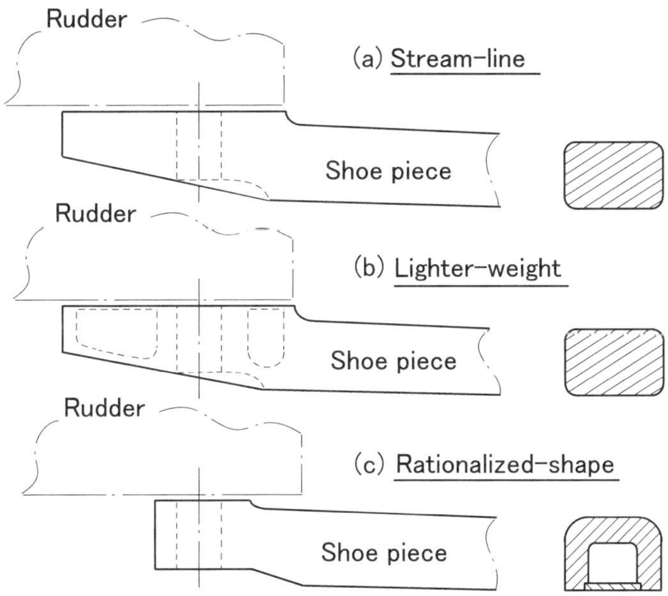

Fig. 29.13 Change of shoe piece tip

In the case of the stern frame without a shoe piece, the upper part of the stern frame, that is called the rudder horn, is important, because it must support every kind of the force induced by the rudder without the help of a shoe piece. As the rudder horn becomes larger, the bending moment created at the base of the rudder horn increases. In contrast, the force acting on the lower end of rudder horn decreases because the intermediate rudder bearing inside the hull can share the load acting on the upper part of the rudder.

The bending moment of the rudder horn is so big at the base of the horn that the rudder horn is usually designed to penetrate the outer shell plating and to extend to the upper flat or deck.

There are two types of rudder horn: welded type of rudder horn fabricated from steel plates, and cast-steel type. In case of the cast-steel type, the rudder horn is like a cylinder whose horizontal cross section has a D shape.

The stern frame is very important not only from the view point of strength but also from that of the construction schedule, for it takes a longer time to produce cast steel than ordinary steel parts. Therefore, the designer must pay careful attention to the production schedule of cast steel so as not to disturb the construction schedule.

In spite of these points, the authors suffered from serious trouble relating to the stern frame. Let us explain the valuable lessons learned from our experiences, which may benefit other structural designers:

The stern frame must be annealed by heating up to 650 ± 25 °C, for both fabricated type and casting type. The designer has to check if the size of the stern frame is within the capacity of the heating furnace. If the whole stern frame cannot be stored in the furnace, it must be divided into a few parts which are annealed individually and then connected to each other; being annealed by local heating. The designer has to take it for granted that he/she describes the above procedure of annealing very clearly in the drawing so as not to cause confusion in production.

Non-destructive tests of a stern frame are sometimes troublesome, in particular for radiographic tests utilizing X-rays. In some cases, it needs a special instrument for inspection due to the large thickness of cast steel. Therefore prior to the examination process, the designer must investigate the capability of the X-ray equipment.

The designer needs to prepare a repair plan for structural defects. For example, in almost all cases cast steel has internal defects. Cracks sometimes appear on the surface of the thicker steel plates during the bending process of the plate. This is often treated as a matter for the production engineer, but we think that the designer must go and see the defects himself to make a proper judgement to see if it is possible to repair the defects or not, and to determine a suitable repair method. The designer is apt to stay in his design office and to make decisions without going out to see the defects. However, we have to emphasize that the structural designer must go and see them himself in order to gather exact information to make correct decisions.

Chapter 30
Vibration Prevention

When we consider the structural strength, first we calculate the structural responses to the external forces and then we judge whether the result is within the allowable limits which are determined according to the data from existing vessels. In the same manner, when we design a structure from a vibration aspect, we first calculate the structural response to the exciting force and then we judge the result considering the allowable limits which are determined from human discomfort, structural damage, or the lifetime of the machinery.

Considering structural strength, an external force is predominantly induced by the sea conditions, which vary widely according to the change of sea state. Therefore, it is quite difficult to obtain the external forces correctly. For this reason, judgement must be carried out on the basis of calculations by assuming the external force appropriately.

In contrast with the above strength design, in the design of structural vibration, the exciting force can be estimated correctly. In spite of this advantage, the calculated structural response is not so accurate in many cases. This is because there are many unknown factors in the estimation of damping, added water mass, etc., and consequently they mislead the calculated response.

In the structural strength, it is impossible for the designer to decrease the external force. As a result, the purpose of the structural design is to reduce the structural response to the assumed external force.

In contrast, in vibration design there is no vibration of the structure as long as the exciting forces do not exist. Hence the designer must aim to reduce the exciting forces as low as possible in order to achieve a good vessel which has low vibration. Consequently, it is more important to pay attention to decreasing the exciting force than to calculating exactly the vibration response.

In this section, we are not willing to explain exact calculation methods for estimating damping etc. but we introduce experimental methods to allow the right decisions for vibration. The development of this method [151] was based on the vibration

Y. Okumoto et al., *Design of Ship Hull Structures*,
https://doi.org/10.1007/978-3-032-12500-2_30

measurements on many actual vessels together with experience of heavy vibrations of superstructures and vibrations of in-tank structures.

30.1 Exciting Forces

For the exciting forces, the frequency and magnitude are important. In a situation where there are two exciting forces of the same frequency, the phase difference between them is an important factor. The exciting forces on ship vibration come from the propeller and diesel engine. It is important to recognize that these exciting forces are magnified in the ship structure by resonator. For example, where there is resonance of axial vibration of the diesel engine crankshaft, the exciting force which comes from the diesel engine is transmitted to the hull and superstructure after being magnified by the resonator of the crankshaft, with the result that the superstructure then vibrates heavily [152, 153].

From the resonance viewpoint, the frequency of the exciting force is important, and in Fig. 30.1 the range of frequencies of exciting forces on a ship structure are shown. The frequency of propeller exciting force is mainly blade frequency (revolutions of propeller shaft multiplied by number of propeller blades) and sometimes twice that. The frequencies of diesel engine exciting forces are 1st, 2nd, 3rd orders, etc. (Nth order means N times propeller shaft revolutions). First and second order unbalanced forces or moments which sometimes cause hull girder flexural vibration are well known. Higher order exciting forces are transmitted to the hull structure after being magnified by resonators, such as crankshaft or engine structure. The magnitude of the higher order exciting force is usually too small to cause heavy vibration on a ship structure by itself. In spite of that, where the natural frequency of the crankshaft or engine structure vibration coincides with a certain order of the exciting force, it becomes a resonator and causes heavy vibration in the ship structure. In Fig. 30.1 the ranges of natural frequencies of ship structures are shown. Countermeasures are necessary where both the ranges of excitation and response overlap.

Usually, the propeller exciting force is treated as two forces: surface force and bearing force. The former depends very much on propeller aperture; recent research has discovered that a cavitating propeller generates far bigger surface force than a non cavitating propeller. The latter depends on the wake field, jN-th, $j(N - 1)$-th, and $j(N + 1)$-th order components of wake distribution (N is number of propeller blades). Recommended sizes of the propeller aperture have been proposed by many researchers, and these recommended values seem to be practical in avoiding excess surface force. From the authors' experience it can be said that in the ships with these recommended propeller aperture sizes the exciting force of the propeller is not a problem.

As already stated, where there are two exciting forces of the same frequency the phase difference between them is important. The cases are, for example.

(a) 6-bladed propeller and Sulzer 9RTA

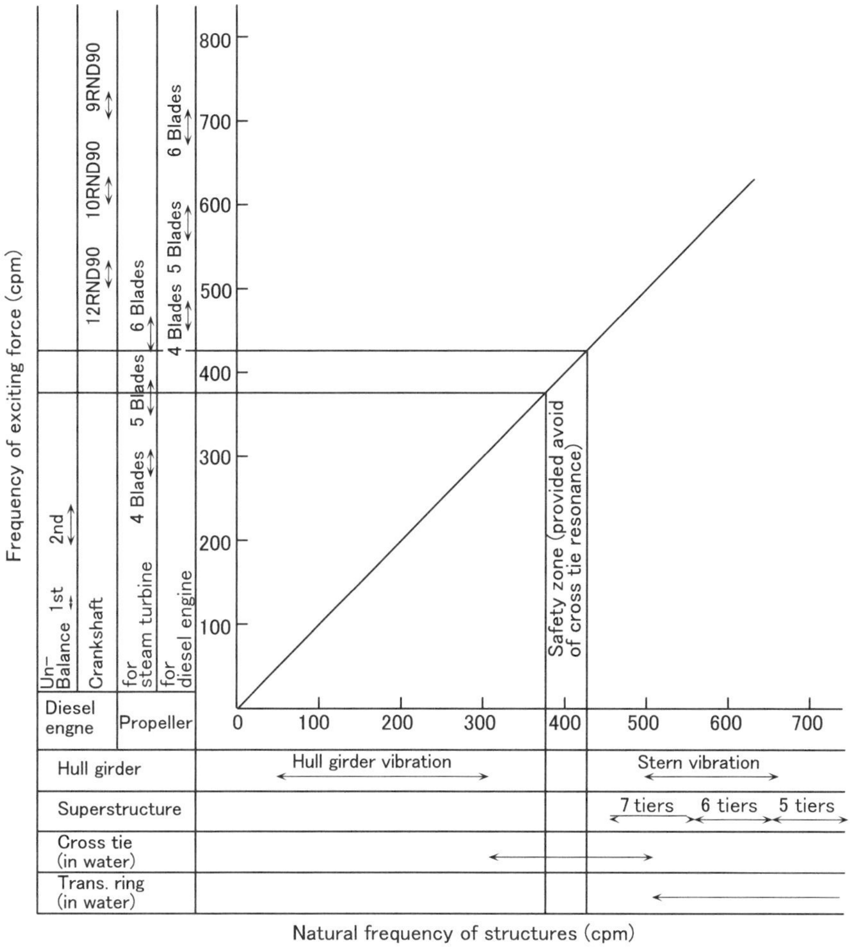

Fig. 30.1 Natural frequency of hull structure and frequency of exciting force

(b) 5-bladed propeller and Sulzer 10RTA
(c) 4-bladed propeller and Sulzer 12RTA

In these cases, the optimum propeller fitting angle must be investigated to cancel the exciting forces.

30.1.1 Magnitude of Propeller Excitation

The relation between the size of the propeller aperture and the vertical propeller exciting force has been studied by many researchers. A sample of their results is shown in Fig. 30.2. These results are obtained by comparing the ship's aft end vertical vibration amplitude when sailing and during exciter tests with the exciter installed just above the propeller. From Fig. 30.2 the magnitude of the exciting force is about 4–5% that of the total thrust in the case of a large aperture. The frequency of the propeller exciting force is equal to the blade frequency. The thrust of a 250,000 DWT tanker is 200 tonf, so the exciting force of the propeller will be about 10 tonf in this case [152].

Figure 30.3 shows the results of an exciter test on a 250,000 DWT tanker in which the exciter was fitted at the aft end of the ship. The lower figure represents the fore and aft vibration in the superstructure, and the upper is for the vertical vibration at the aft end of the ship. An exciting force of 10 tonf cannot cause significant vibration except in cases of resonance. It is important to avoid resonance in connection with the exciting force of the propeller to prevent ship vibration.

Many studies have been made to estimate the magnitude of the propeller exciting force [154]. Here the authors introduce Fujino's method developed for the initial design stage [155]. The important component of the exciting force is the vertical one, and Fujino's method uses this. Propeller excitation is divided into bearing force F_B and surface force F_S and expressed by the following.

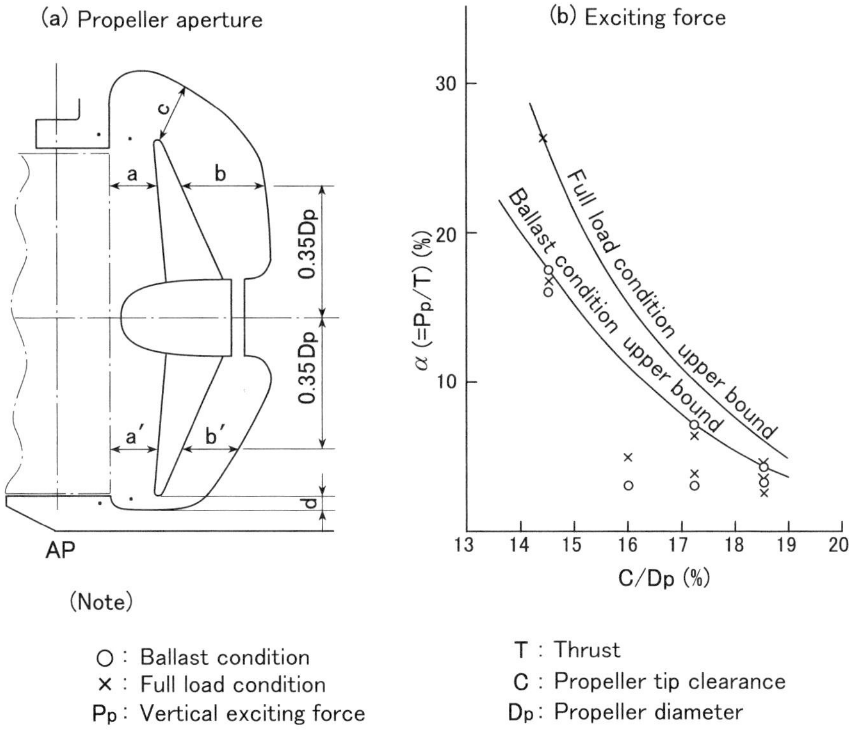

Fig. 30.2 Propeller aperture and exciting force

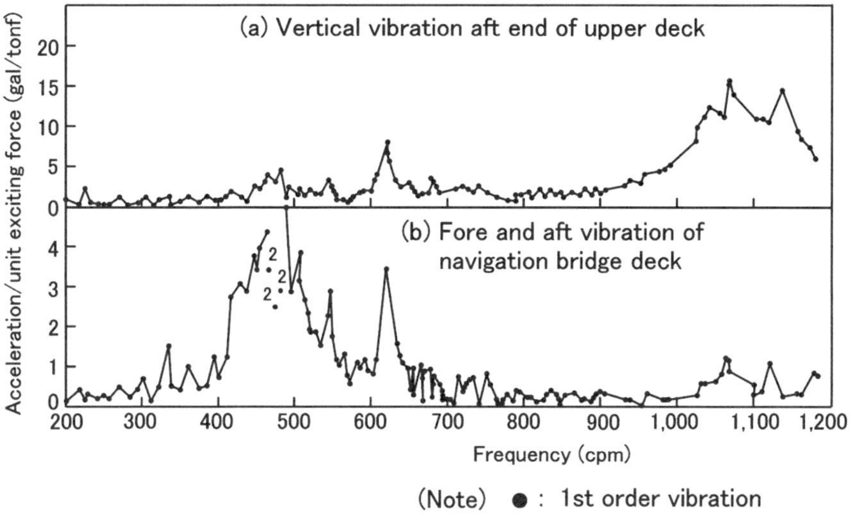

Fig. 30.3 Resonance curve of hull girder vertical vibration and bridge fore and aft vibration (250,000 DWT tanker)

$$F_B = K_B \times \frac{SHP}{10 \times N \times D_P} \tag{30.1.1}$$

$$F_S = 0.0212 \times \frac{K_{CAV}}{Z-1} \times \frac{SHP}{N \times D_P} \times \frac{0.1 + C/D_P}{(C/D_P)^{\frac{3}{2}}} \tag{30.1.2}$$

where

SHP: shaft horsepower (PS).

N: revolution of propeller shaft (rpm).

D_P: propeller diameter (m).

Z: number of propeller blades.

C: tip clearance (m).

K_B: 0.6 for 4-bladed propeller, 0.8 for 5-bladed propeller, 1.0 for 6-bladed propeller.

K_{CAV}: correction factor by cavitation (see Fig. 30.4).

In Fig. 30.4, parameters σ_0, k_T are as follows:

$$\sigma_0 = 19.6 \times \frac{d - h_0 + 9.707}{\left(\frac{N}{60} \times D_P \times J\right)} \tag{30.1.3}$$

where

k_T: thrust coefficient.

d: draft (m).

h_0: height of shaft center above base line (m).

J: advance ratio of the propeller.

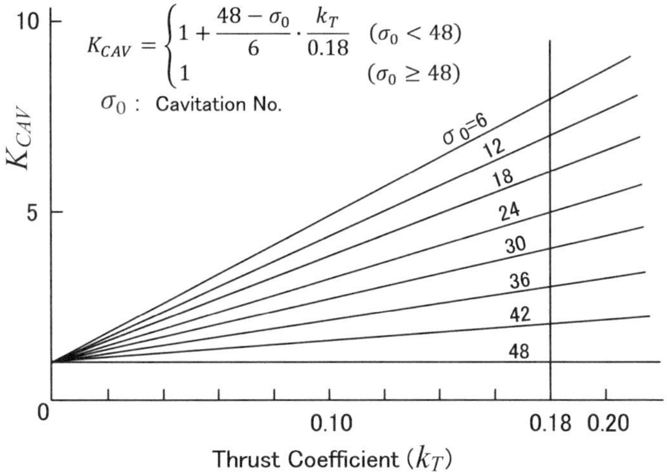

Fig. 30.4 Correction factor KCAV for cavitation

Finally, the vertical exciting force of the propeller is expressed as the sum of F_B and F_S. The response of the ship structure depends basically on the displacement of the ship. Figure 30.5 shows the relation between displacement of the ship and the sum of F_B and F_S for actual ships in service. In the case of a new design, the designer should check the propeller exciting force using Fujino's method. When the calculated propeller exciting force is less than the track record shown in Fig. 30.5, the propeller will not cause vibration problems as long as the resonance is avoided. Regarding the propeller exciting force, it is important for the designer to keep in mind that the effect of cavitation on the surface force will be far bigger than expected. By applying a skew propeller, the propeller exciting force can be reduced remarkably.

30.1.2 Magnitude of Diesel Engine Excitation

In Table 30.1 the magnitude of the first and second order exciting forces caused by the unbalanced inertia force of the moving parts of a Sulzer diesel engine are shown. In any case, the vertical and horizontal unbalanced forces are zero because of equal crank angle and equal moving mass of each cylinder, while the unbalanced moments remain. In Fig. 30.1 it can be seen that the first and second order unbalanced moments can provoke resonance with the hull girder flexural vibration. The natural frequency of hull girder flexural vibration can be estimated with reasonable accuracy and if resonance is anticipated, the following countermeasures are to be applied. The natural frequency of the hull girder flexural vibration depends on hull girder stiffness, displacement and weight distribution. To increase the hull girder stiffness much steel is required but displacement or weight distribution can be changed in

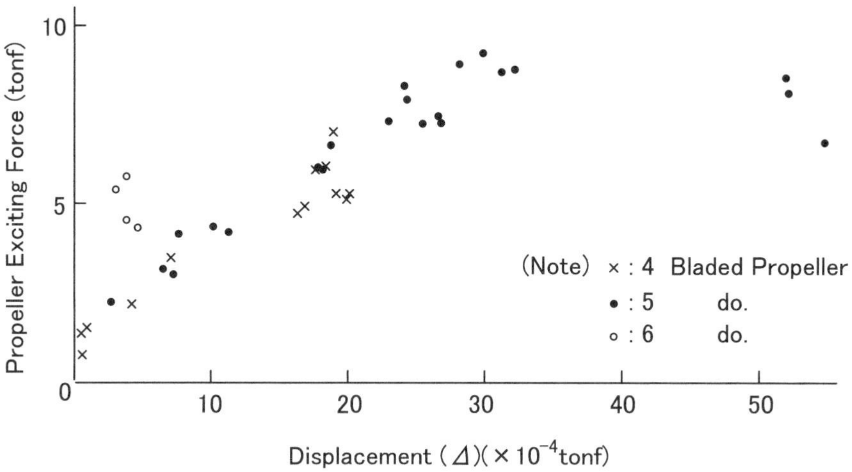

Fig. 30.5 Propeller exciting force and displacement

some cases without any expense. One way to prevent resonance is to settle the critical displacements or critical weight distributions to be avoided where hull girder flexural resonant vibration happens in a similar way to the barred range of propeller shaft revolution from the viewpoint of propeller shaft torsional resonant vibration.

However, the owners usually do not want to have such limitations. For unbalanced moments of the first order it is usual practice to put a counterweight on the flywheel to adjust any vertical or horizontal unbalanced moments keeping the sum of them almost unchanged. For example, where vertical resonant vibration is anticipated, the vertical unbalanced moment is to be decreased by the use of a counterweight, but this leads to the increase in the horizontal unbalanced moment. In this case it is important to confirm that no horizontal resonance exists. For a second order unbalance, it is usual practice to install a balancer to cancel the engine unbalance. The balancer consists of two unbalanced counterweights and a rotating mechanism which rotates twice as fast as the shaft. The rotation is regulated mechanically (chain or gear) or electrically (synchro-motor or pulse-motor). Usually a 6-cylinder diesel engine has a large unbalanced moment of the second order vertically, and often it causes resonance on Panamax type ships. In Fig. 30.6, one example of the effect of a balancer installed on a Panamax type ship is shown.

The balancer is a kind of vibration exciter so attention must be paid to prevent overbalance; an adjustable balancer is desirable.

30.1.3 Magnification of Exciting Force by Resonator

Where there is a resonator, exciting energy is accumulated and magnified on the way through the hull structure, largely affecting ship vibration. For example, resonances

Table 30.1 Unbalanced moment of SULZER engine

Engine type		Revolution at MCR (rpm)	Unbalanced moment (kN-m)		
Cylinder No.			Vertical		Horizontal
			1st order	2nd order	1st order
5	RTA48T	124	71	1,034	88
6			0	719	0
7			43	209	53
8			142	0	177
4	RTA52	130	330	860	330
5			110	1,070	110
6			0	740	0
7			60	220	60
8			210	0	210
5	RTA58T	103	141	1,878	149
6			0	1,306	0
7			84	379	88
8			282	0	297
4	RTA62	109	540	1,460	590
5			170	1,820	190
6			0	1,270	0
7			100	370	110
8			340	0	380
4	RTA72	94	910	2,330	910
5			290	2,900	290
6			0	2,020	0
7			170	590	170
8			100	0	100
5	RTA84T	74	328	4,186	482
6			0	2,981	0
7			195	865	287
8			128	0	188
9			350	1,498	514
4	RTA84C	102	1,405	3,719	2,374
5			453	4,613	747
6			0	3,221	0
7			276	935	439
8			230	0	358
9			516	1,581	799
10			242	1,112	370

(continued)

Table 30.1 (continued)

Engine type		Revolution at MCR (rpm)	Unbalanced moment (kN-m)		
Cylinder No.			Vertical		Horizontal
			1st order	2nd order	1st order
11			284	1,559	306
12			0	0	0
4	RTA84M	81	825	3,392	2,051
5			268	4,223	645
6			0	2,937	0
7			122	853	422
8			137	0	310
9			312	1,442	689
10			146	1,014	321
12			0	0	0
6	RTA96C	100	0	6,495	0
7			541	1,885	557
8			605	0	671
9			1,868	2,120	1,919
10			366	2,311	375
11			1,296	1,701	1,334
12			0	0	0

of the propeller shaft, crankshaft and diesel engine structure very often cause heavy vibration of the hull structure and superstructure; even resonances of the funnel or radar mast affect superstructure vibration.

(a) Lateral vibration of the propeller shaft: It is reported that the resonant lateral vibration of the propeller shaft with a propeller exciting force affects hull structure vibration. In the case of a big tanker, the authors seldom observed resonant vibration of the propeller shaft in the lateral direction, because the natural frequency of lateral vibration of the propeller shaft is far higher than the propeller blade frequency. One example is shown in Fig. 30.7. In the case of a container ship which has a narrow structure around the stern bossing, the natural frequency of the lateral vibration of the propeller shaft is low enough to cause resonance with the propeller exciting force.

In the case of nontraditional alignment of shafting, for example, on the occasion of straight alignment, shaft bearings do not work correctly in some cases because of the hull structure deformation caused by draft change. In such cases the span of the shaft becomes longer, and sometimes the natural frequency of the shafting becomes low enough to cause resonant vibration with the propeller blade frequency. To prevent resonant lateral vibration of the shafting, rational shaft alignment taking account of hull structure deformation caused by draft change is most important.

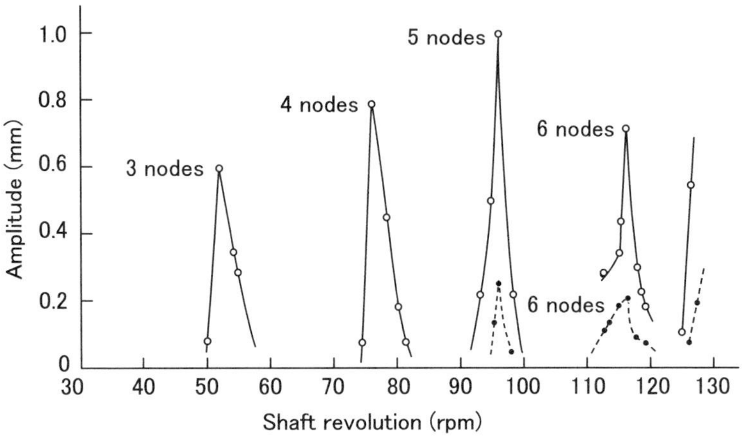

(Note) L × B × D × d (m) : 213 × 32.2 × 17.8 × 11.888

Main engine : SULZER 6RD90

Electric balancer : 5tonf (max.)

—○—: 2nd order (without balancer)

---•---: 2nd order (with balancer)

Fig. 30.6 Effect of balancer (vertical vibration at aft end in ballast condition)

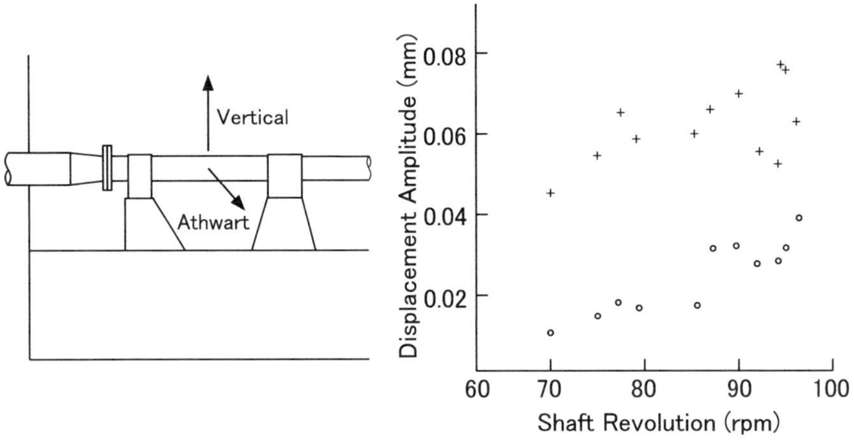

(Note) Ship: O/B/O

Dead Weight Capacity: 156,000 DWT

Main Engine: Steam Turbine (27,500 PS × 98 rpm)

Propeller: 5-Bladed Propeller

+: Athwart Vibration (5th Order)

○: Vertical Vibration (5th Order)

Fig. 30.7 Lateral vibration of propeller shaft

(b) Axial vibration of diesel engine crankshaft: The natural frequency of axial vibration of the shafting system including crankshaft, propeller shaft and propeller lies between 500–800 cpm. Therefore, there is a possibility of resonant vibration with a 5th–7th order component of cylinder pressure variation of the diesel engine in the normal revolution range of 110–120 rpm. In Fig. 30.8 natural frequencies of axial vibration of shafting systems of Sulzer RND engines are shown.

Where there is resonance of the axial vibration of the shafting system, fore and aft exciting force is transmitted to the hull structure and superstructure after being magnified by the resonator of the shafting system; this sometimes causes heavy vibration of the superstructure. In Fig. 30.9 fore and aft vibration of the superstructure is shown together with axial vibration of the crankshaft; the superstructure vibration increases as the shafting vibration increases. The authors estimated the magnified exciting force by resonance to be about several 10 tonf, which is considerably bigger than the propeller exciting force explained in Sect. 30.1.

To prevent resonant axial vibration of the shafting system, including diesel engine crankshaft, a detuner or damper is usually installed at the fore end of the crankshaft. In Fig. 30.10 the damper is illustrated. In Fig. 30.11 the effect of the damper on the

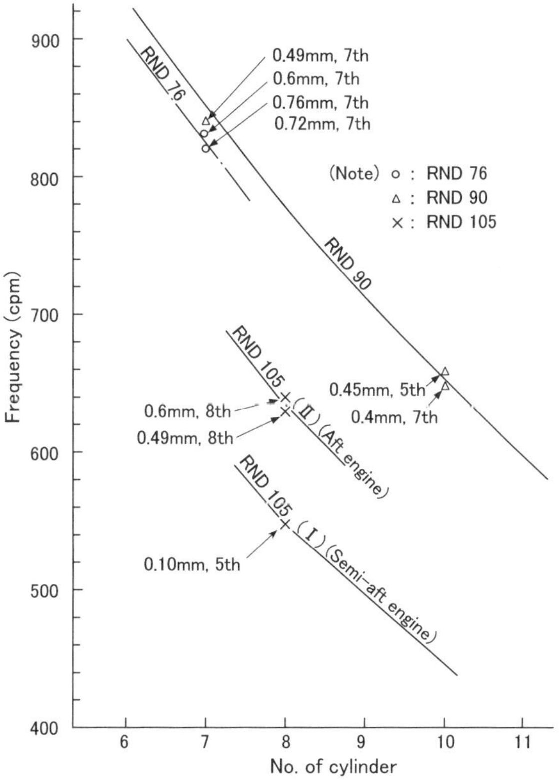

Fig. 30.8 Natural vibration of crankshaft axial vibration (at sea trial)

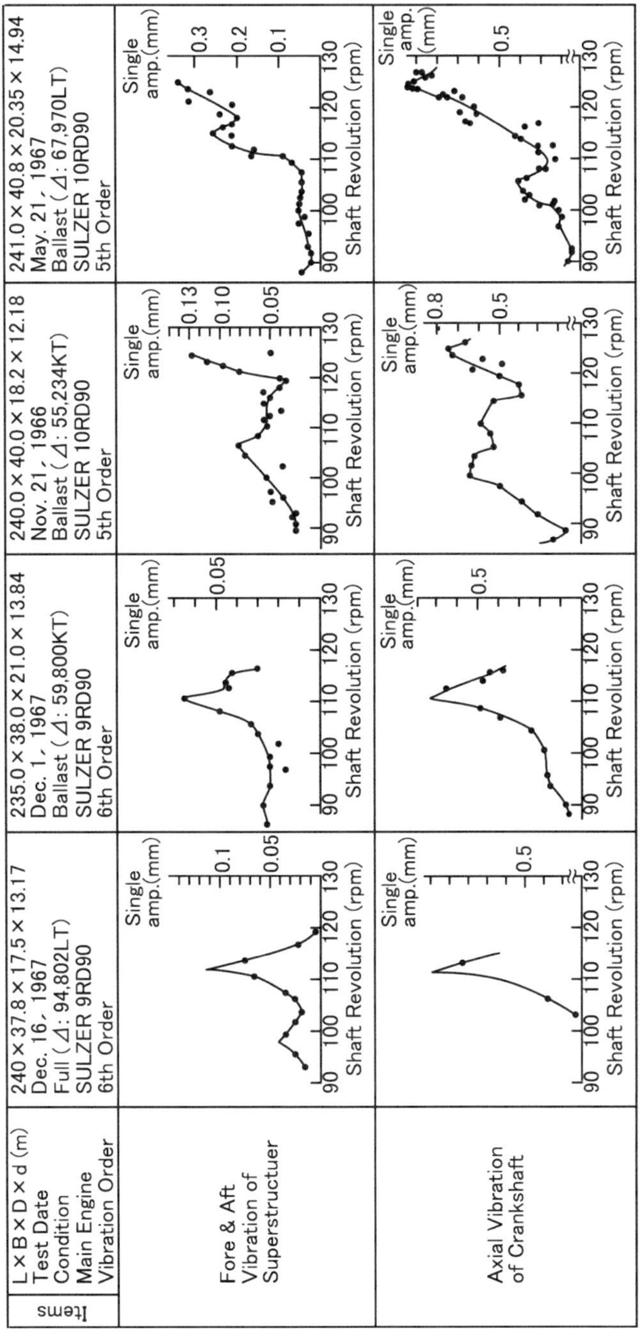

Fig. 30.9 Fore and aft vibration of superstructure and axial vibration of crankshaft

Fig. 30.10 Detail of damper

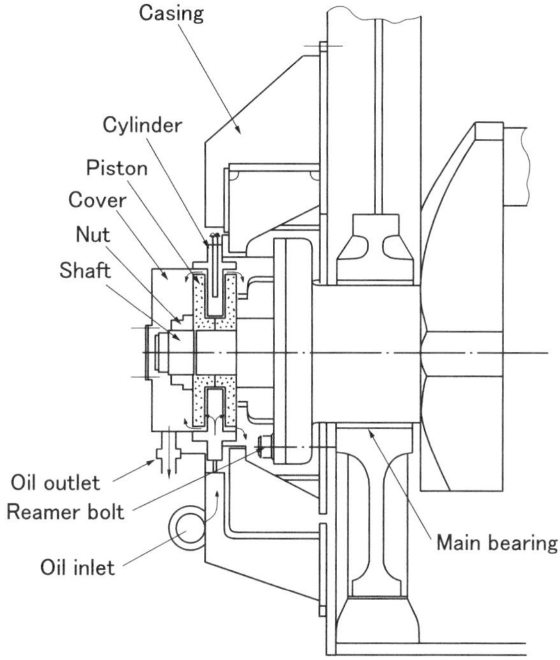

axial vibration of the crankshaft and fore and aft vibration of the superstructure are shown. It can be seen that the effect of the damper is remarkable.

To decrease the axial vibration of the crankshaft of a diesel engine, improvement of the firing order was studied. In case of the Sulzer 10RND90 engine, the 5th order fore and aft vibration of the crankshaft at normal revolutions can be eliminated by changing the firing order from P-7 (1, 9, 5, 6, 2, 10, 4, 3, 8, 7) to N-8 (1, 8, 7, 3, 5, 9, 4, 2, 10, 6).

In the case where resonance of the crankshaft axial vibration is foreseen, the best way to prevent crankshaft axial vibration is to install a damper at the fore end of the crankshaft.

(c) Torsional vibration of diesel engine crankshaft: Torsional vibration of the crankshaft causes both a fore and aft exciting force and a vertical and horizontal exciting force. The fore and aft exciting force is generated by contraction and elongation of the shaft due to torsion, and the effect is similar to the axial vibration of the crankshaft explained above in (b). In records of fore and aft vibration of the crankshaft measured at the fore end of the crankshaft, the resonant points of torsional vibration of the crankshaft appear clearly.

The mechanism by which a vertical exciting force is generated by torsional vibration is shown in Fig. 30.12. Because of torsional vibration, the moving parts vibrate; the vertical component of that vibration is the vertical exciting force. The order of vibration is ± 1 of the order of the torsional vibration, because the moving parts vibrate

Fig. 30.11 Effect of damper

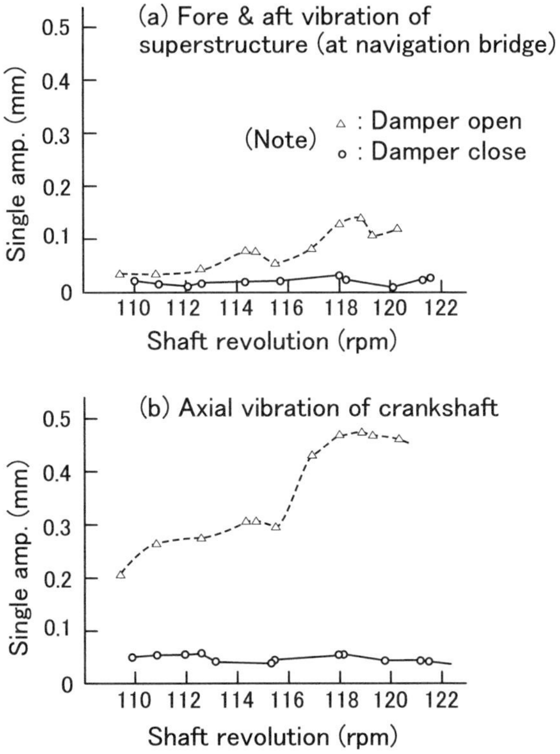

while rotating. This is an important phenomenon when considering the frequency of the exciting force.

In a similar way, a horizontal force is generated by torsional vibration. In Fig. 30.13 some examples of vibration caused by torsional vibration are shown. The records show that the double bottom is excited by 10 and 12th order frequency excitation generated by 11th order crankshaft torsional vibration. To prevent torsional resonant vibration, it is usual practice to install additional weight to shift the natural frequency away from the frequency of the exciting force.

(d) Lateral vibration of diesel engine structure: The exciting force, magnified by lateral vibration of the diesel engine structure, is transmitted to the ship structure as vertical vibration of the engine room double bottom, and also through rolling stays if fitted. The most well-known modes of lateral vibration of diesel engine structure are H, X, and *x* type vibrations, and each mode of vibration has a natural frequency corresponding to the particular engine. Some countermeasures are necessary when installing an engine which is expected to have resonance around the normal operating revolutions. In Fig. 30.14 the relationship between the number of cylinders and the natural frequency of lateral vibration of the engine structure is shown.

To prevent lateral vibration of the engine structure, it is usual practice to install stays. In Fig. 30.15 the effect of stays is shown. Oil cylinder dampers must be provided

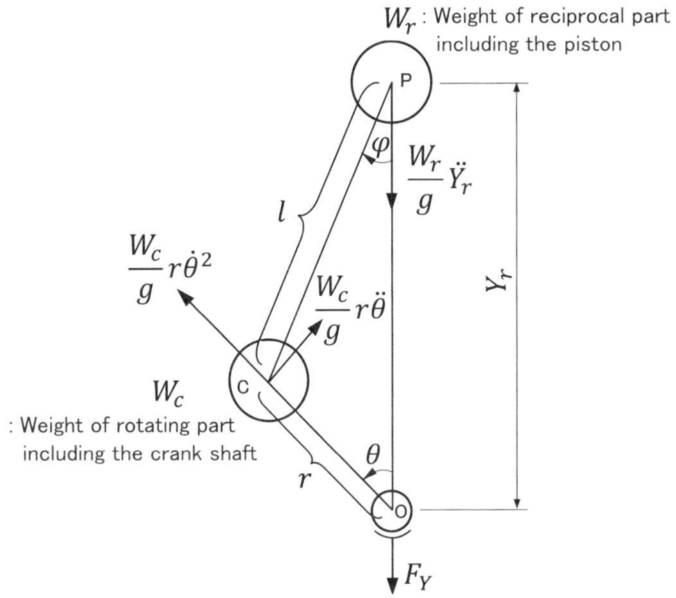

$$\begin{cases} F_Y = \dfrac{W_r}{g}\ddot{Y}_r - \dfrac{W_c}{g}r\dot{\theta}^2\cos\theta - \dfrac{W_c}{g}r\ddot{\theta}\sin\theta \\ Y_r = r\cos\theta + l\cos\varphi \\ \sin\varphi = \dfrac{r}{l}\sin\theta \end{cases}$$

By torsional vibration, θ oscillates with the amplitude of β

$$\theta = \omega_0 t + \beta\sin(n\omega_0 t + \alpha_n)$$

$$\therefore F_Y = -\frac{W_r + W_c}{g}r{\omega_0}^2\Big[\cos\omega_0 t + \beta_n\left(1 + \frac{n}{2}\right)\cos\{(n+1)\omega_0 t + \alpha_n\} + \beta_n\left(1 - \frac{n}{2}\right)\cos\{(n-1)\omega_0 t + \alpha_n\}\Big]$$

Fig. 30.12 Mechanism of vertical vibration force caused by torsional vibration of crankshaft

to absorb deflection in the engine room side shell structures where the amount of deflection is considerable. By changing the firing order, the exciting force of the lateral vibration of the engine structure can be reduced. On the Sulzer 10RND90 engine, changing the firing order from P-7 (1, 9, 5, 6, 2, 10, 4, 3, 8, 7) to N-8 (1, 8, 7, 3, 5, 9, 4, 2, 10, 6) caused the vanishing of the resonant point of the 5th order X mode vibration around normal operating revolutions.

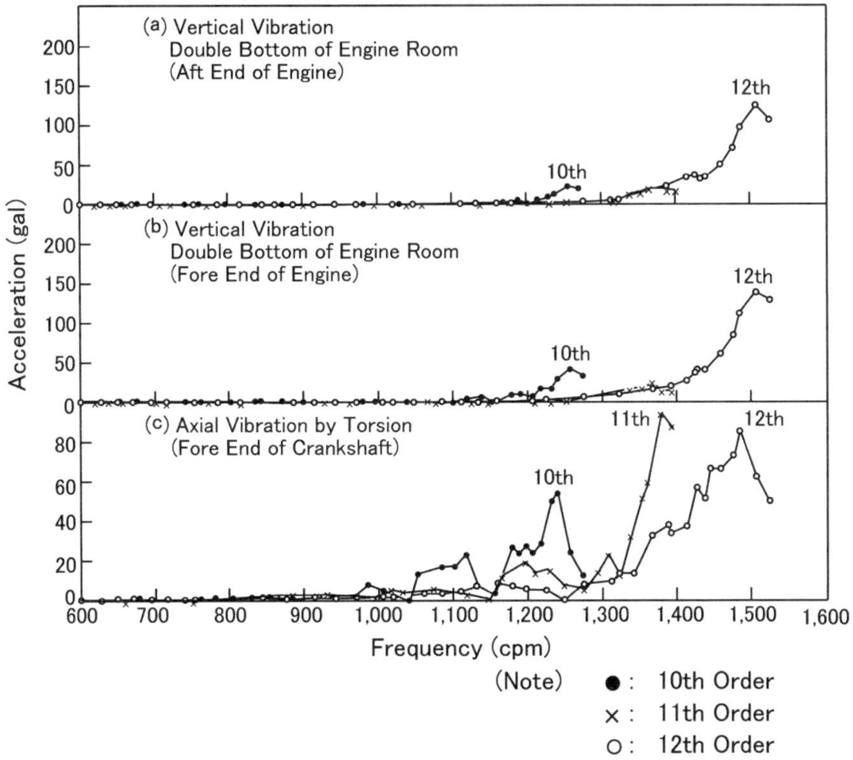

Fig. 30.13 Engine room double bottom vibration caused by torsional vibration of crankshaft (90,000 DWT tanker) (resonance curve)

(e) Fore and aft vibration of diesel engine structure: The exciting force magnified by the fore and aft vibration of the diesel engine structure is transmitted to the ship structure through the engine room double bottom. The natural frequency of the fore and aft vibration of the engine structure lies in the range of 400–600 cpm for the Sulzer RND type engine, and where it has resonance with propeller thrust variation, the vibration becomes substantial. In Fig. 30.16 the resonance curve of this vibration for the Sulzer 8RND105 type engine is shown.

To prevent fore and aft vibration of the engine structure, stays must be installed just as for lateral vibration. Furthermore, it is possible to increase the stiffness of the engine structure to change the natural frequency of the fore and aft vibration of the engine structure.

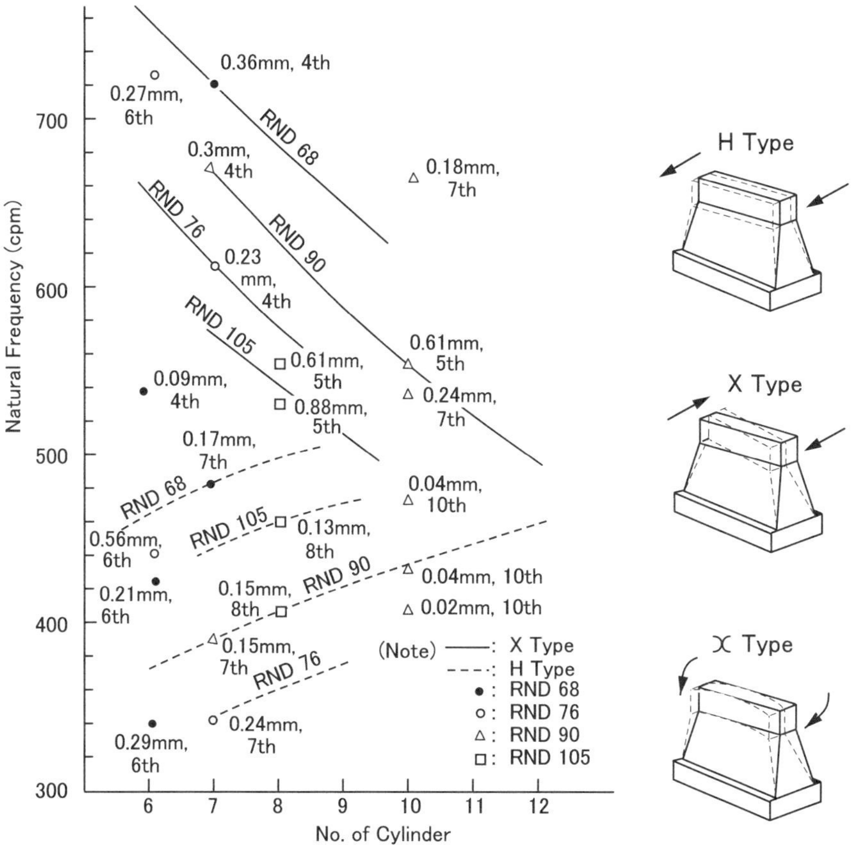

Fig. 30.14 Natural frequency of lateral vibration of SULZER engine

30.1.4 Cancellation of Exciting Force

Exciting forces can be cancelled by selecting a propeller fitting phase angle where two or more exciting forces of the same frequency exist. A typical example of cancellation is cancellation of the propeller thrust variation and axial vibration of the diesel engine crankshaft.

In the case of the Sulzer engine, axial vibrations of the crankshaft of the 6th order for the 9RND90 and 5th order for the 10RND90 have resonance around normal operating revolutions. By this rationale, the exciting forces magnified by these resonances can be cancelled by the 6th order thrust variation of a 6-bladed propeller, or 5th order thrust variation of a 5-bladed propeller, if the optimum fitting angle of the propeller is selected. This technique is called "phasing". In Fig. 30.17 an example of phasing for a Sulzer 9RND90 engine and a 6-bladed propeller is shown.

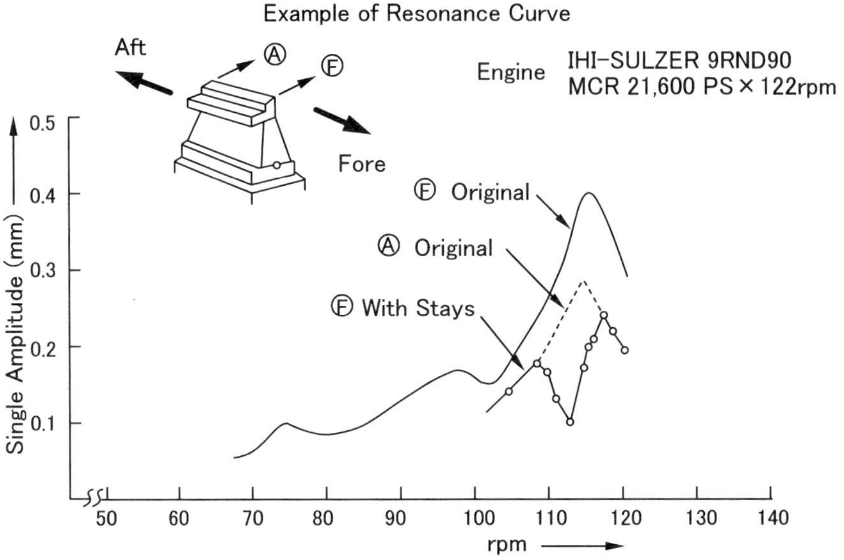

Ship	Main Engine	① Original Amplitude	② Amplitude with Stays	② / ①	Effects on Hull Vibration
105,000 LT Tanker	B & W 984-VT2BF-180	(Single) 0.18 mm	0.08	44 %	No
128,000 LT Tanker	B & W 1284-VT2BF-180	0.14	0.06	43	Engine room double bottom vertical vibration reduced
74,100 LT O/B/O	B & W 984-VT2BF-180	0.10	0.10	100	No
63,000 LT Tanker	SULZER 9RND90	0.22	0.09	40	No
750 TEU Container Ship	SULZER 8RND105	1.00	0.50	50	Trans. vibration of upper engine flat and navigation deck reduced

Fig. 30.15 Effect of stays for lateral vibration of main engine

30.1.5 Reduction of Main Engine Exciting Force by Elastic Mounting

It is important to reduce vibration and maintain comfortable habitability in vessels and this is particularly serious for passengers in luxury cruise ships. Although there are several forces inducing vibration problems in ships, one of the predominant

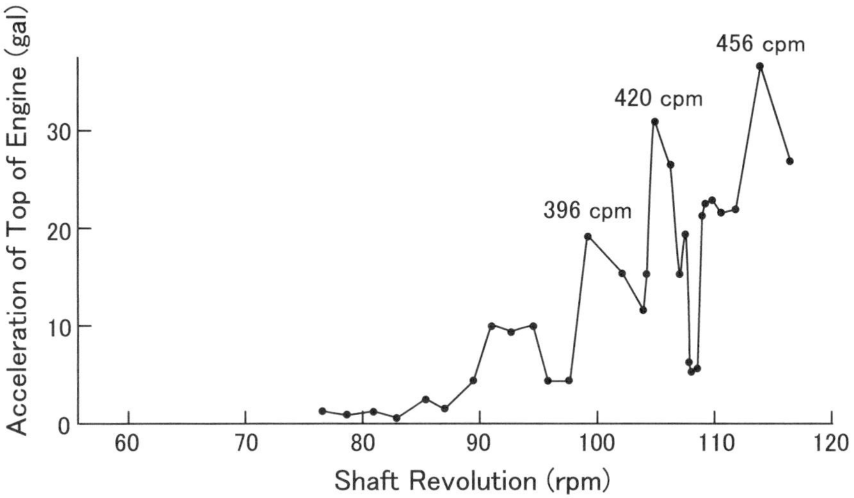

Fig. 30.16 Fore and aft vibration of diesel engine (SULZER 8RND105) (resonance curve)

vibration sources is its main engines. If the vibration forces due to the main engine are decreased considerably, this can contribute to a reduction in vibration trouble.

A luxury cruise ship of 660 passengers was built by IHI in 1990. The vessel was installed with two medium size engines, SEMT 12PC2-6V, which were elastically mounted [156] on engine beds by elastic components such as rubber pieces sandwiched by steel plates as illustrated in Fig. 30.18. Furthermore, a flexible coupling system was applied to the line shafting of the main engine to permit large movement of the engine. That elastic mounting system was designed to isolate the transmission of vibration forces from the main engine to the engine bed, and it was the first time for us to provide the system to a large cruise ship.

The system has proved to be effective according to the measurements carried out on sea trials. In the measurements, the elastic mounting system was effective in allowing the elastic components to function and, after that, it was turned off by using rigid components. The results showed that the system can suppress the vibration level in the hull structure to a satisfactory lower level, as shown in Fig. 30.19. The system was especially effective for the higher-order exciting forces. In the case of the 4.5th order of vibration, the structural response can be reduced to one fourth of the original level by installing the elastic mounting system.

30.2 Prevention of Ship Vibration

In Sect. 30.1, the authors explained the frequency of exciting force and how to reduce the magnitude of the exciting force. Reducing the magnitude of the exciting force is an important part of the prevention of ship vibration. Avoiding resonance is another

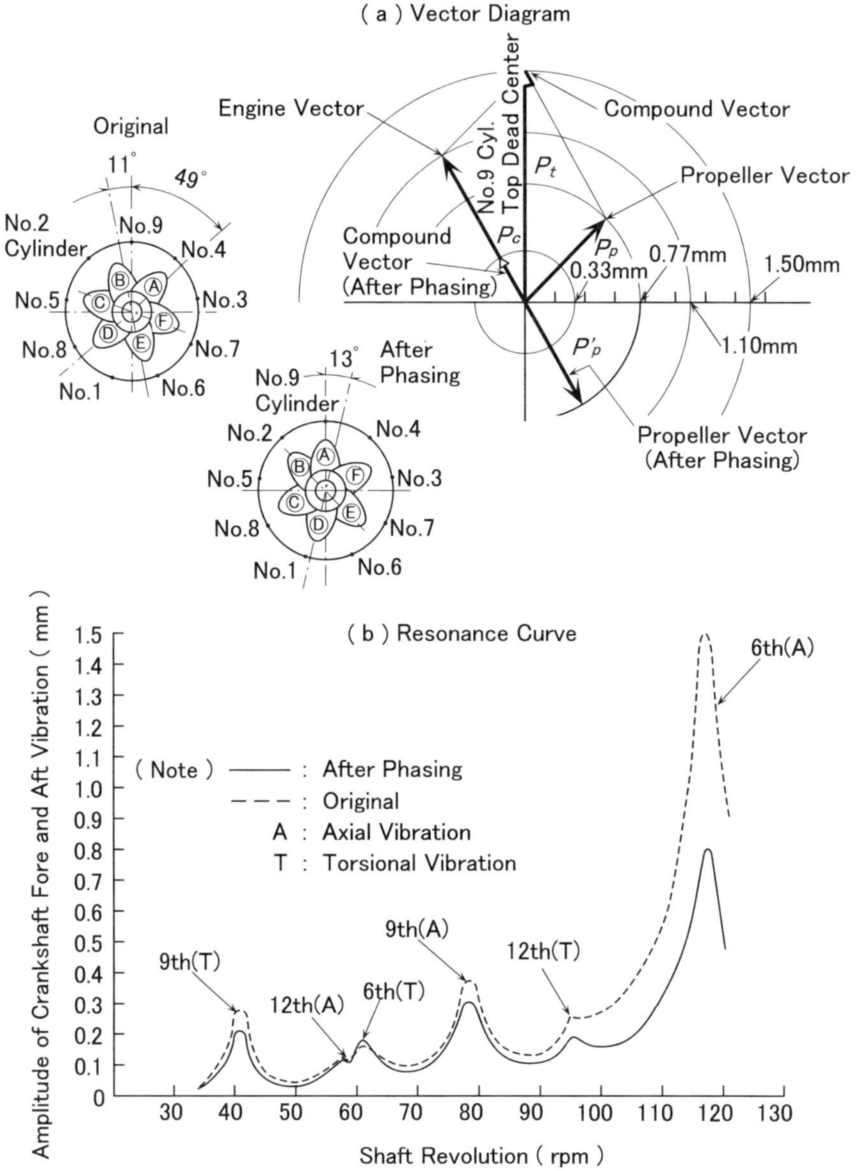

Fig. 30.17 Example of phasing

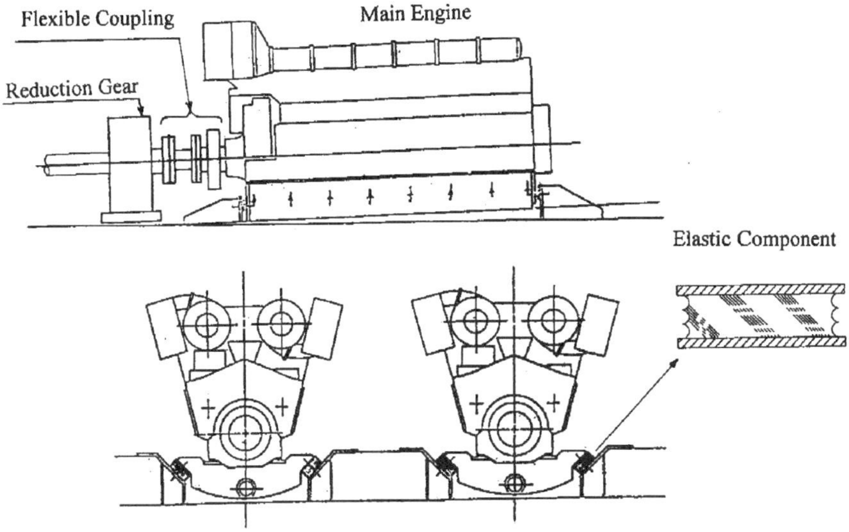

Fig. 30.18 Arrangement of elastically mounted main engine

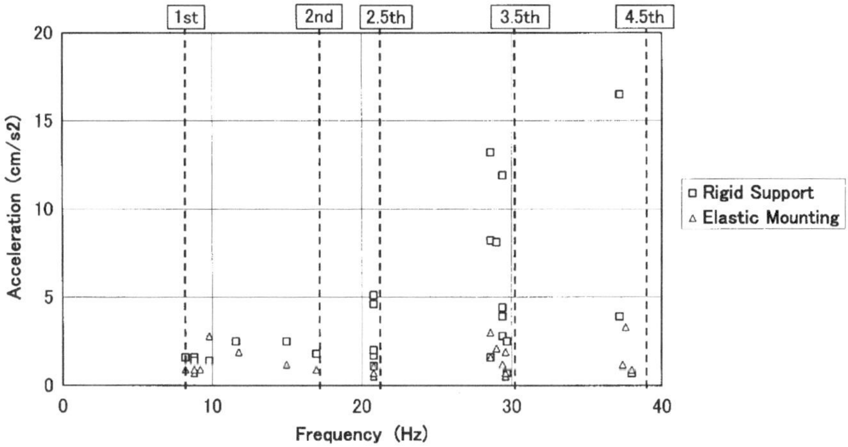

Fig. 30.19 Result of vibration measurement of No. 5 deck (Elastic component on/off test)

important element in the prevention of ship vibration. In this section the authors explain how to avoid resonance. The crucial procedure in avoiding resonance is to estimate the natural frequency of the ship structure correctly.

30.2.1 Flexural Vibration of Hull Girder

The calculation of the natural frequency of flexural vibration of the hull girder is done by using experimental formulae based on bending theory for 2–3 nodes of vibration, and on shear theory for 3 nodes and higher vibration. The experimental coefficients are determined from the analysis of measurements on actual ships. The results of calculations of the natural frequency of flexural vibration of the hull girder are accurate enough for practical design of up to 4–5 nodes of vibration vertically and horizontally. The exciting force, which can produce resonance with flexural vibration of the hull girder, is the diesel engine 1st order and 2nd order unbalanced moments.

Where it is foreseen that the flexural vibration of a hull girder at ballast condition may be resonant with the diesel engine 1st or 2nd order unbalanced moment, it is usual practice to avoid resonance by adjusting the volume of the ballast in the case of tankers, and adjusting the volume of the ballast in the aft peak tank in the case of cargo ships. It is reported that by filling the aft peak tank of a 40,000 DWT bulk carrier, the natural frequency of a 5-node vertical vibration comes down to 220 cpm from 230 cpm, and this 10 cpm of change is big enough to avoid resonance with a diesel engine unbalanced moment at normal operating revolutions. However, where the resonance is foreseen in the full load condition, and even at the ballast condition in cases where the owners do not want to have any operating limit, there is no way to avoid resonance. To increase the stiffness of the hull girders requires a large amount of steel, and to change the normal operating revolutions of a diesel engine results in a decrease of engine efficiency. In this case, a countermeasure to decrease the unbalanced moment of the diesel engine as stated in Sect. 30.1.2 should be applied instead of attempting to avoid resonance.

The above explanation is for lower frequency vibration which may cause resonance with diesel engine 1st and 2nd order unbalanced moments. Let us proceed to higher frequency vibration which may cause resonance with propeller excitation. In Fig. 30.20 an example of a resonance curve of flexural vibration of a hull girder is shown. In the frequency range below 190 cpm, we can observe apparent resonances of hull girder vibration with 2, 3, 4 or more nodes, and above 430 cpm the response is large even though no remarkable resonance exists. The authors call the former the frequency zone of hull girder vibration and the latter the frequency zone of stern vibration. Between the hull girder vibration zone and stern vibration zone there is an intermediate zone where the vibration is relatively small. This fact was confirmed on many ships by the exciter test and by numerical simulation of hull girder flexural vibration.

To prevent resonance of hull girder flexural vibration with propeller exciting force, it is important to decide the number of propeller blades so as to have the blade frequency come into this intermediate zone. In Fig. 30.21 the relationship between ship size and the intermediate zone is shown. For example, the intermediate zone is 300 ± 60 cpm for a ship $L_{pp} = 300$m, and 500 ± 60 cpm for $L_{pp} = 150$m. In the case of a VLCC which has 300 m length and 80 rpm propeller shaft revolution, a 4 bladed

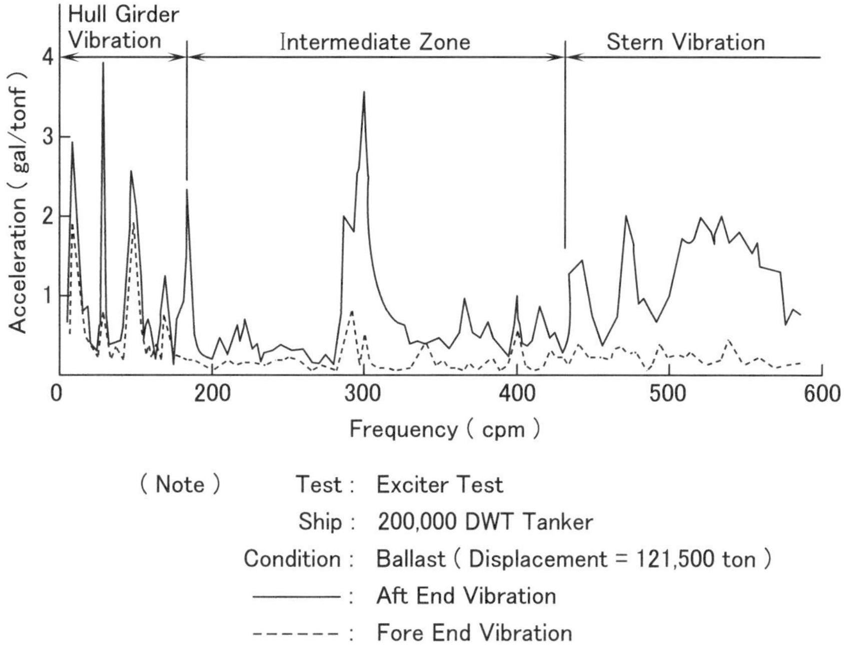

Fig. 30.20 Hull girder vibration

propeller is recommended to prevent resonance of hull girder flexural vibration with propeller exciting force.

30.2.2 Vibration of Superstructure

Since the aft bridge and aft engine arrangement was first applied in ship design, superstructure vibration has been one of the important problems in ship structure research. The aft part of the ship is close to the propeller which generates exciting forces and has less stiffness than the midship part. Also, diesel engines of large horse power are installed in the aft part of ships where the hull girder stiffness decreases and the aft bridge is higher to provide a better view [153].

None of these trends help to improve the superstructure vibration but only make matters worse. With a higher superstructure the natural frequency of the fore and aft vibration reduces so that there is the possibility of resonance from the exciting forces of the propeller or the diesel engine. Among the vibrations of a superstructure, the most important one is fore and aft vibration, and the authors explain below the calculation method to determine the natural frequency of this fore and aft vibration and countermeasures to prevent resonance.

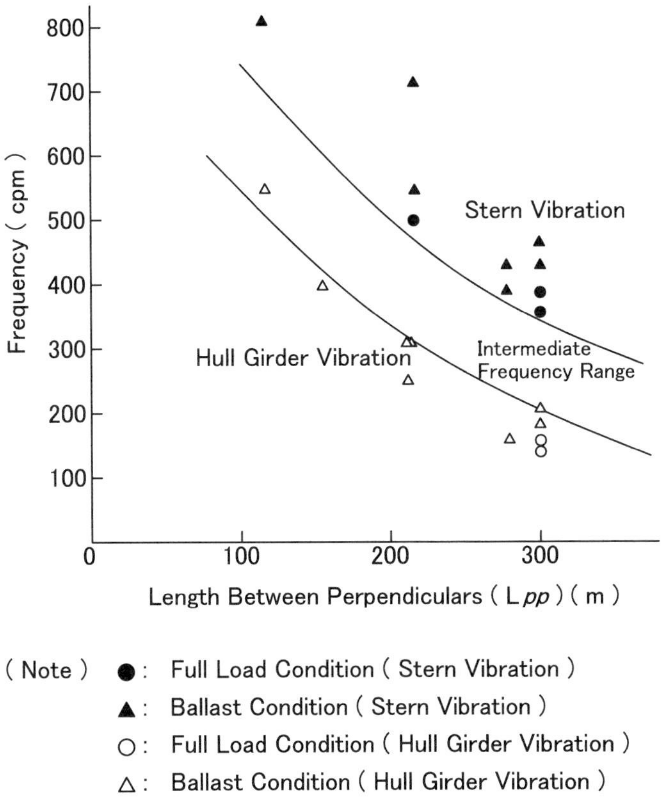

Fig. 30.21 Ship size and intermediate frequency range

The computational method to determine the natural frequency of the superstructure fore and aft vibration has been studied for many years by means of finite element analyses. The accuracy of the results is not yet adequate, and they require more labor and expense than simplified methods. Therefore, designers nowadays apply a simplified method based on measured data. In Fig. 30.22 it is explained that the fore and aft vibration of the superstructure consists of rotational vibration and shear vibration; the natural frequency of superstructure fore and aft vibration depends on the height of each type of superstructure. Based on this principle in Fig. 30.23, the relationship between the natural frequency and the height of the superstructure for A–C type arrangements is shown.

The exciting force, which can cause resonance with this fore and aft vibration, is propeller exciting force and diesel engine crankshaft axial vibration. Propeller excitation has a blade frequency, and the Sulzer RTA diesel engine has a 7th order frequency for a 7-cylinder, 6th order for a 12-cylinder, 5th order for a 11-cylinder and 4th order for a 12-cylinder engine. Where it is foreseen that resonance from a propeller exciting force may happen, the easiest and most effective method is to change the

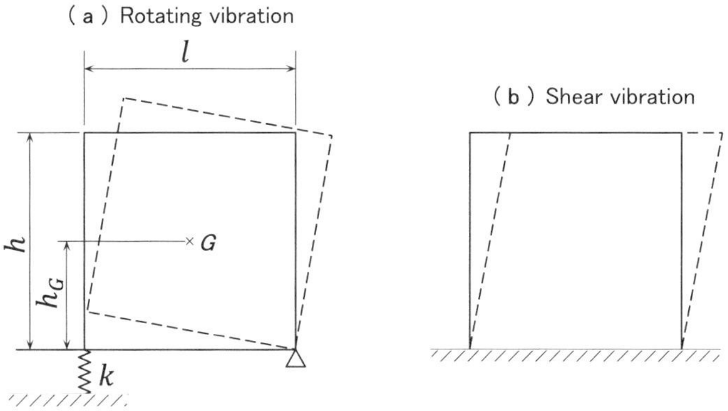

(Note) Natural frequency of rotating vibration: N_R

$$N_R = \frac{60}{2\pi}\sqrt{\frac{kl^2 g}{W{h_G}^2}} \quad \cdots\cdots (1)$$

Natural frequency of shear vibration:

$$N_S = \frac{\pi}{2}\cdot\frac{60}{2\pi}\sqrt{\frac{k'GAg}{hW}} \quad \cdots\cdots (2)$$

Weight of superstructure: $W \propto l \times h \times b$
Spring constant: $k \propto b$
Shear rigidity: $k'AG \propto tl \propto l$

b : Breadth of superstructure
t : Thickness of longitudinal wall (const.)

$$\therefore N_R \propto \frac{1}{h}\sqrt{\frac{l}{h}} \quad \cdots\cdots (3)$$

$$N_S \propto \frac{1}{h}\sqrt{\frac{1}{b}} \quad \cdots\cdots (4)$$

Natural frequency of superstructure: N

$$\frac{1}{N^2} = \frac{1}{{N_R}^2} + \frac{1}{{N_S}^2} \quad \cdots\cdots (5)$$

Fig. 30.22 Natural frequency of superstructure fore and aft vibration

number of propeller blades. But when it cannot be changed for some reason, the natural frequency of the superstructure fore and aft vibration must be changed to prevent resonance. Where resonance is foreseen with diesel engine crankshaft axial vibration, an axial damper must be installed at the fore end of the crankshaft, as stated in Sect. 30.1.3(b), or the natural frequency of the superstructure fore and aft vibration must be changed.

In this way, there are some cases where the natural frequency of the superstructure fore and aft vibration must be changed to prevent resonance. As shown in Fig. 30.24 the natural frequency N is affected by the natural frequency of rotational vibration

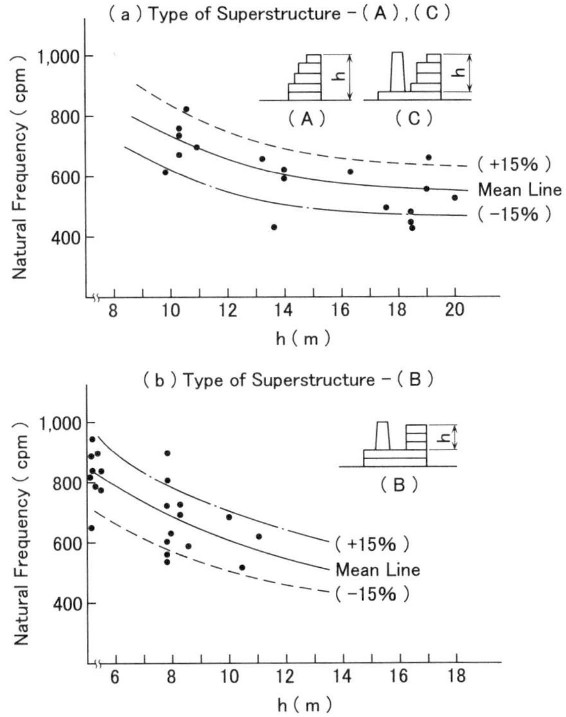

Fig. 30.23 Estimated natural frequency of superstructure fore and aft vibration

N_R and the natural frequency of shear vibration N_S. Equation (5) in Fig. 30.22 shows that where either N_R or N_S is far greater than the other, to increase the greater value is not effective for increasing N. In Fig. 30.24 the above condition is shown. Where N_R is low, that is, the spring constant k is small and the rigidity of the foundation is weak, to increase N_S, i.e., the shear rigidity of the superstructure, does not bring higher natural frequency N. With a reasonable value of N_R, an increase in shear rigidity becomes effective in increasing the natural frequency N.

Calculated values of N_R and N_S of typical ships are shown in Table 30.2. Table 30.2 shows that in general N_R is lower than N_S. Therefore, to increase the natural frequency of a superstructure fore and aft vibration, N_R must be increased, which means increasing the spring constant k in Eq. (1) of Fig. 30.22.

30.2.3 Active Mass Damper for Superstructure Vibration

The vibration amplitude of a superstructure has a large scatter, because much of the vibration is induced by the propeller and the propeller exciting forces fluctuate according to the change of flow velocity into the propeller. Moreover, the vibration frequency of a superstructure does not always have a single frequency but may have

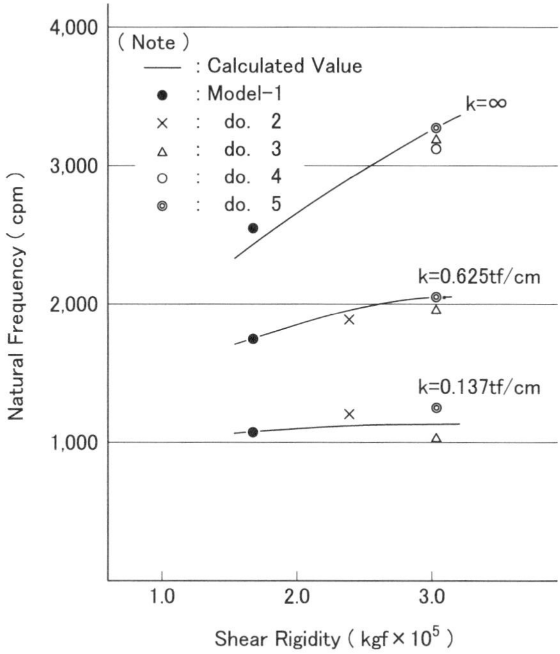

Fig. 30.24 Shear rigidity and natural frequency of superstructure

several frequencies because there are several exciting sources in a vessel. Therefore, the vibration of a superstructure cannot be cancelled by installing a balancer. A more sophisticated instrument is required.

In order to suppress such vibration, an active mass damper has been developed [157] as shown in Fig. 30.25. An active mass damper can effectively reduce the vibration of fluctuating amplitude and frequency induced by several vibration sources such as the main engine and the propeller. These vibration characteristics were confirmed by measurements on a 150,000 DWT tanker. Figure 30.26 shows the results of experiments on the active mass damper installed on the top of a superstructure. From the figure, the active mass damper reduced the vibration of the superstructure especially in the resonance peak area. The vibration due to an unbalanced moment (2nd order) decreased to one sixth of its original level, the vibration derived from the propeller (5th order) exciting force decreased to one half, and the vibration induced by the main engine exciting force (6th order) reduced to one fifth, by an active mass damper.

30.2.4 Vibration of In-Tank Structures

The natural frequency of the in-tank structure of a tanker lies in the range of the propeller blade frequency and the frequency of the diesel engine exciting force, and therefore there is a possibility of resonant vibration. Moreover, in case resonance

Table 30.2 Calculated value of N_R and N_S

Ship	Calculated Value (cpm)			Measured Value (cpm)	Type of Superstructure
	N_R	N_S	N		
A	466	881	441	570	A
B	452	1,115	448	610	A
C	337	791	331	535	A
D	570	1,340	561	560	D
E	815	1,129	707	658	D
F	853	1,142	731	740	D

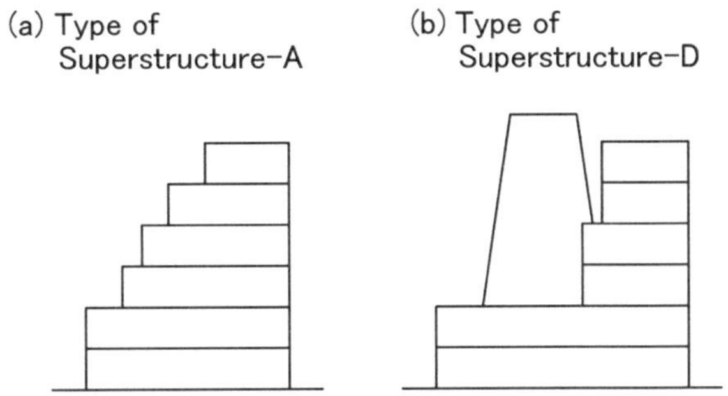

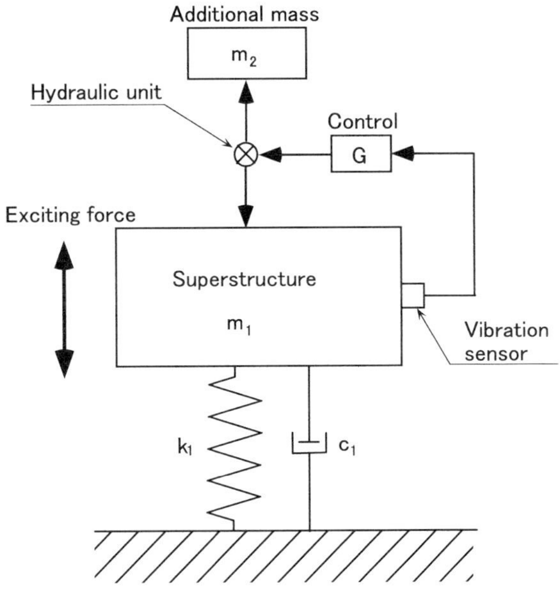

Fig. 30.25 Model of active mass damper on superstructure

Fig. 30.26 Effect of active mass damper for longitudinal vibration of superstructure

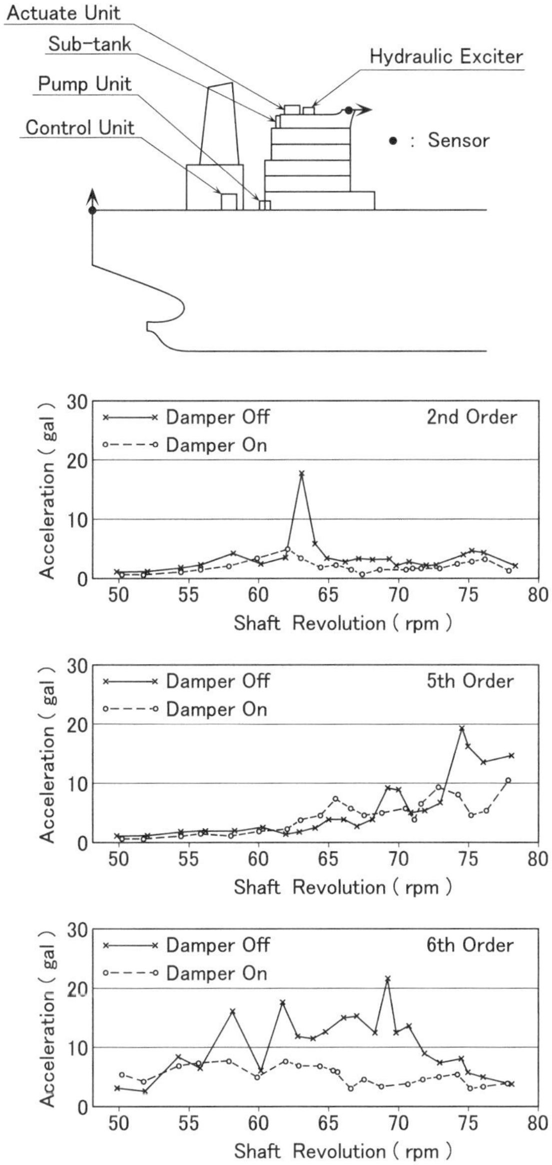

happens, cracks may be generated throughout the tank structure by heavy vibration because the structure is similar throughout the tank. Therefore, it is important to calculate the natural frequency of the in-tank structure to prevent resonance.

Recently the methods for calculating the natural frequency of in-tank structures have been improved, providing accurate values for practical design.

30.2.5 Calculation Methods of Natural Frequency of In-Tank Structures

(1) Simple formula: As shown in Fig. 30.27, the vibration of the web frame can be assumed to consist of panel vibration surrounded by stiffeners for which the natural frequency is f_P, the vibration of stiffeners for which the natural frequency is f_S, and the vibration of face plate for which natural frequency is f_{FB}. The calculation method for the natural frequency of a web frame is derived using f_P, f_S, and f_{FB} [158].

The natural frequency of face plate vibration f_{CFB} which is affected by the existence of panel vibration or stiffener vibration is described as follows.

$$f_{CFB} = C_{FB} C_L K_{LF} K_W \cdot \frac{428}{B^2} \cdot 10^5 \sqrt{\frac{1.2}{0.2 + \rho} \cdot \frac{I_B^*}{B l_B}} \tag{30.2.1}$$

where

C_{FB}: frequency reduction ratio affected by f_S or f_P.
C_L: frequency reduction ratio affected by longitudinal stiffener vibration.
K_{LF}: correction factor for stiffness of longitudinal stiffener and face plate.
K_W: correction factor for span (L) of web frame.
f_W: natural frequency of web frame (cpm).
f_B: natural frequency of bracket (cpm).
ρ: specific gravity of liquid (1.0 for seawater or oil, 0 for air).
l_B: space between brackets (cm).
I_B^*: equivalent sectional moment of inertia of bracket including stiffener (cm^4).
B: depth of web frame (cm).

The natural frequency of stiffener vibration f_{CS} which is affected by the existence of face plate vibration or panel vibration is described as follows:

$$f_{CS} = C_S C_{HS} \cdot \frac{483}{B^2} \cdot 10^5 \sqrt{\frac{K I_S}{A_S + \left(t_W + \rho C_S'' B\right) s}} \tag{30.2.2}$$

where

C_S: frequency reduction ratio affected by f_{FB} or f_P.
C_{HS}: frequency increase ratio affected by horizontal stiffener.
K: 1.0 for symmetrical stiffener, 0.85 for asymmetrical stiffener
I_S: sectional moment of inertia of stiffener (cm^4).
s: spacing of longitudinal stiffeners (cm).
t_W: plate thickness of web frame (cm).
A_S: sectional area of stiffener (cm^2).
C_S'': coefficient corresponding to the aspect ratio of the panel

$$\left(= \frac{0.08}{\sqrt{1 + (B/l_B)^2}} \right)$$

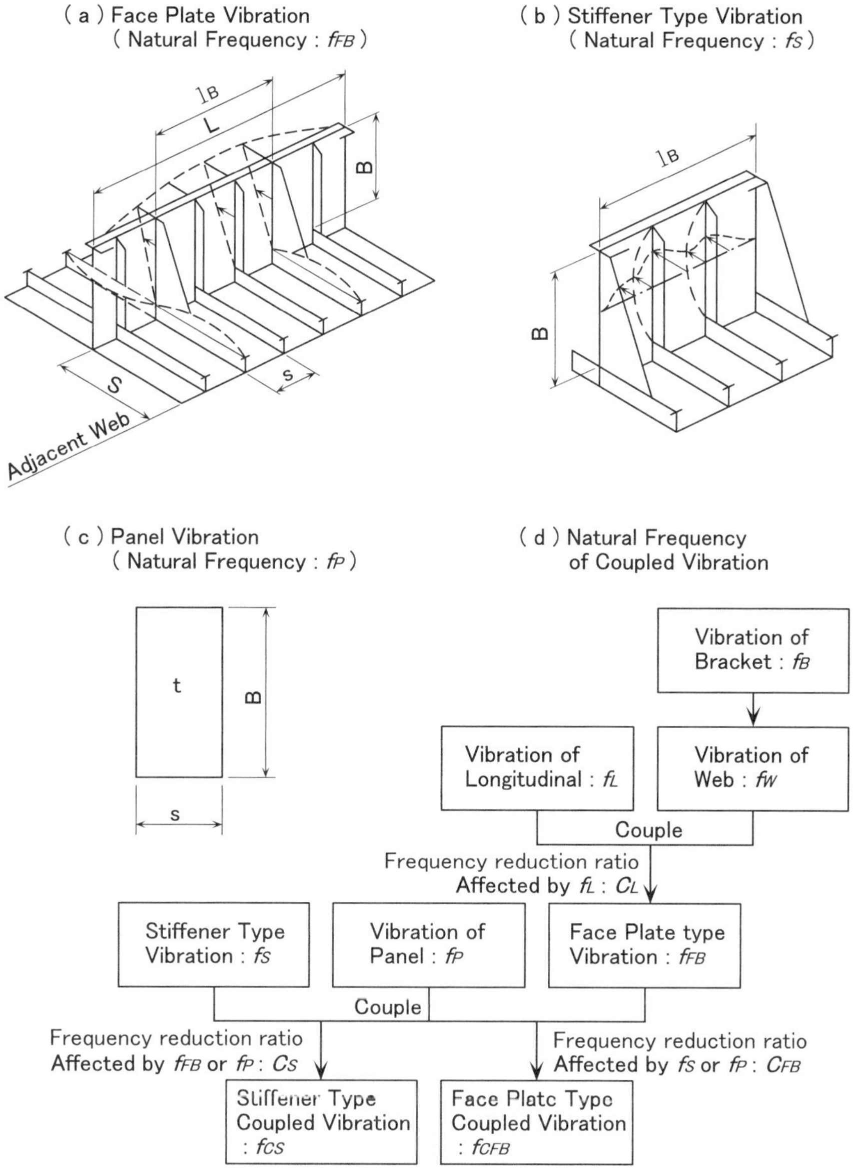

Fig. 30.27 Natural frequency of web frame

The natural frequency of a panel surrounded by stiffeners is described as follows:

$$f_P = \frac{934}{\left(C''_P s\right)^2} \cdot 10^2 \cdot \frac{t_W}{\sqrt{1 + \rho C''_P s / t_W}} \tag{30.2.3}$$

where

C''_P: coefficient corresponding to the aspect ratio of a panel

$$\left(= \frac{0.08}{\sqrt{1 + (s/B)^2}} \right)$$

The ratio between the natural frequencies calculated in the above formulae and frequencies of exciting forces coming from propeller or diesel engine N_{CAL}/N_{EX} is an important parameter in any attempt to avoid the resonant vibration of in-tank structure. In Fig. 30.28 the relationship between N_{CAL}/N_{EX} and the cracks caused by vibration are shown. No cracks occurred in the range where N_{CAL}/N_{EX} is over 1.1. According to this experience it can be stated that to prevent the cracks caused by vibration, the natural frequencies calculated by the above method must be higher than the frequencies of the exciting forces by 10% or more. In association with this vibration of web frames in a tank, cracks occur generally at the locations of stress concentration such as bracket toes attached to the longitudinal stiffeners. Therefore, in addition to the countermeasure stated above, application of soft-shaped bracket toe is effective to reduce fatigue damages due to vibration.

(2) Computer analysis: The calculated value of the natural frequency of a structure depends very much on the boundary conditions, which are difficult to assume correctly. To overcome this difficulty, the calculation for the whole structure has been

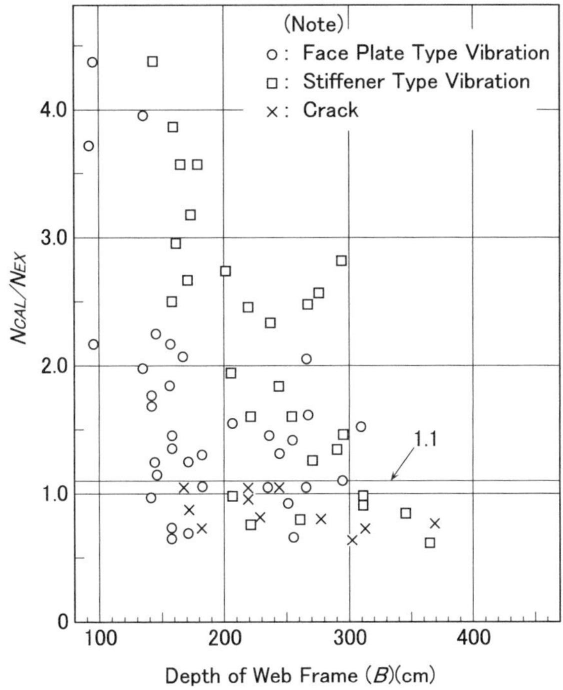

Fig. 30.28 Resonance of web and cracking

done by means of a computer. In the case of an in-tank structure, the whole transverse ring or half of it which takes symmetrical condition into consideration, is analyzed.

In Fig. 30.29 an example of this type of calculation is shown. In this case, the lowest natural frequency is for torsional vibration of the cross tie, the next is bending vibration of the tie, deck transverse vibration, vertical web vibration etc. Between the cross tie bending vibration, for which the natural frequency is 440 cpm, and the deck transverse vibration of 680 cpm, there is no resonant frequency, and above 680 cpm, many resonant frequencies exist densely. Accordingly, it can be stated that the range of 440–680 cpm is a safety zone and it is better to have a frequency of exciting force in this range.

Generally, it is difficult to keep the natural frequency of a cross tie higher than the frequency of the propeller and engine exciting forces. Therefore, designers used

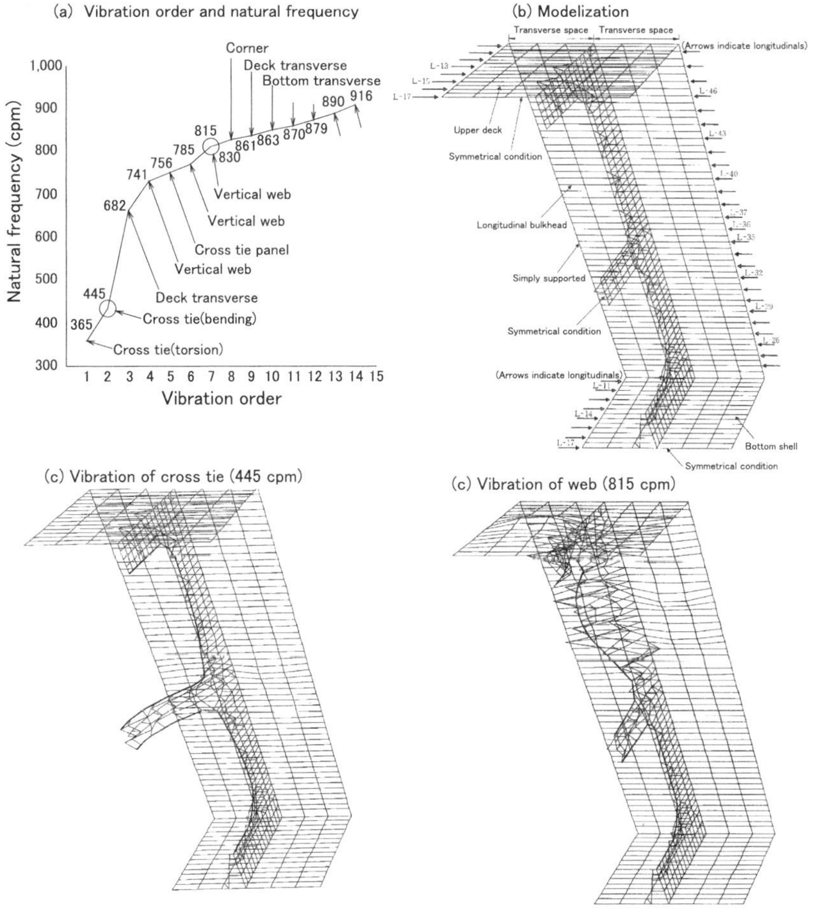

Fig. 30.29 Frequency and mode of transverse ring vibration

to design cross tie scantlings to have a natural frequency lower than, and the natural frequency of the web frame higher than the propeller blade frequency and frequency of diesel engine forces.

Chapter 31
Superstructure

The superstructure, which is not so long in comparison with the main hull structure, does not deflect in the same way as the main hull does, since the longitudinal stress of the main hull is not completely transmitted to the superstructure. In contrast, when the length of the superstructure is large enough, the longitudinal stress is fully transferred from the main hull to the superstructure, since the superstructure will act as part of the hull girder in absorbing hull girder bending.

The latter case is important for the strength of a superstructure because the longitudinal bending stress flowing into the superstructure may cause structural damage. Therefore, the strength of a vessel with a long superstructure is mainly discussed in this chapter.

31.1 Example of Damage to Long Superstructures

Figure 31.1 represents an example of structural damages in a car ferry with a long superstructure. Structural damage is found (a) in the corner of door opening, (b) in the connecting part between the slant end wall of the upper-most deck and the lower deck and (c) in the connection area of the pillar and deck. The damage (a) shows the cracks initiating from the free edge of a door opening corner in the diagonal direction, (b) the crack extending along the welding bead joining the lower deck plate and the slant plate of the upper deck, and (c) the crack in deck plate beneath the doubling plate of a pillar.

Y. Okumoto et al., *Design of Ship Hull Structures*,
https://doi.org/10.1007/978-3-032-12500-2_31

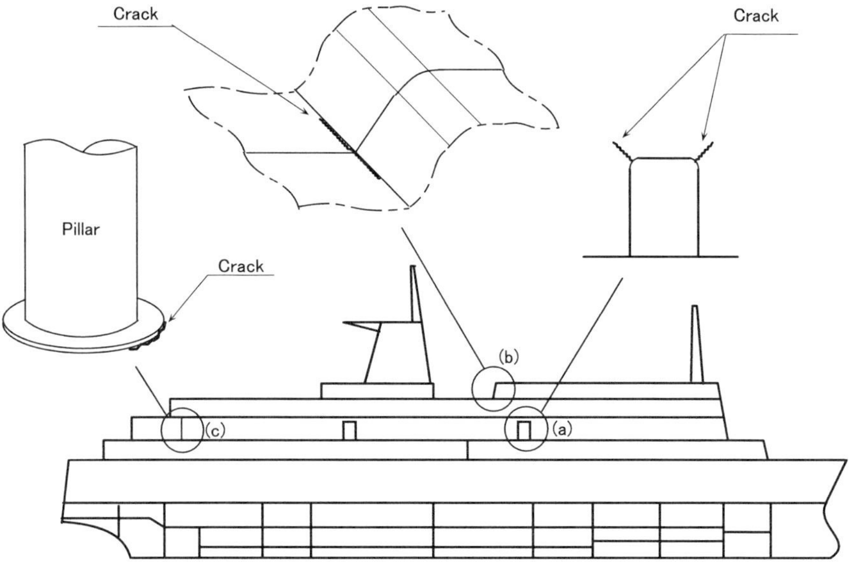

Fig. 31.1 Example of damage in car ferry with long superstructure

31.2 Interaction of Superstructures and Main Hull

Longitudinal bending stress is not completely transmitted from the main hull to the superstructure, in cases where the superstructure is comparatively short, such as a short deckhouse. In this case, the superstructure deforms in a different way from the main hull as shown in Fig. 31.2. From this figure, when the main hull is subjected to a longitudinal bending moment M, a shearing force T in the longitudinal direction is induced in the interface between the upper surface of the main hull and the lower surface of the superstructure. Reaction forces q also exist in the vertical direction. The shearing force T acting on a superstructure is likely to generate a bending deformation with a curvature in the opposite direction to the main hull bending. Therefore, the reaction of vertical force q results in a separating force at the end of the superstructure, if a hogging moment is applied to the main hull. For this reason, the end of a superstructure may suffer from cracks due to a separation force, when the length of the superstructure is short. In order to prevent such structural failures by the separation force, it is important to provide transverse bulkheads in the main hull in line with the end wall of the superstructure.

On the other hand, in the case of a vessel with a long superstructure, the superstructure and the main hull deflect in one united beam against a longitudinal bending moment. Hence the separation force does not appear, but longitudinal bending stress is induced in the superstructure.

The distribution of longitudinal bending stress in the superstructure and the main hull is described in Fig. 31.3, depending on the length of the superstructure versus that

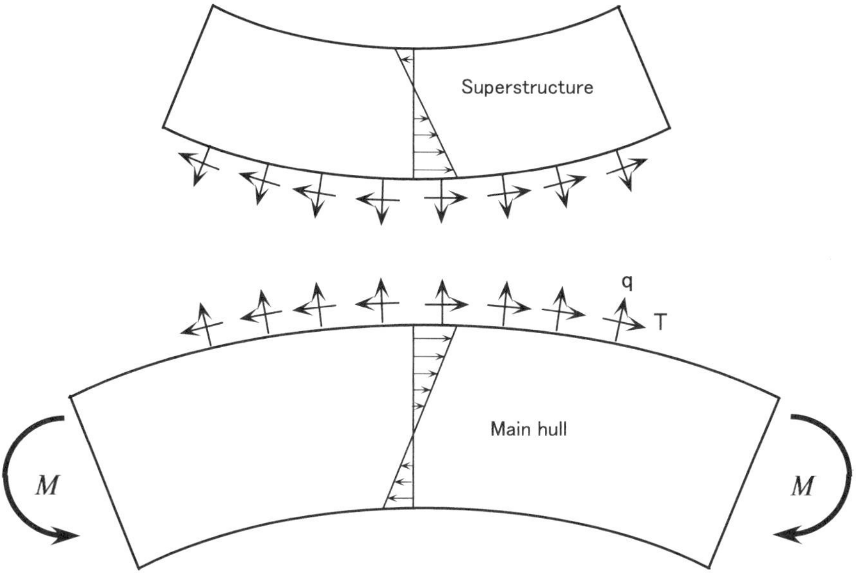

Fig. 31.2 Connecting force between main hull and superstructure due to longitudinal bending

of the main hull. In case A, where the superstructure is relatively long compared to the main hull, the longitudinal bending varies in a linear form, because both structures act as one beam and are bent simultaneously by the bending moment. Case B, where the superstructure is short, shows the bending stress distribution has a sharp change at the interface of the main hull and superstructure, because the sign of curvature in the superstructure is opposite to that in the main hull. Case C is an intermediate case of Case A and Case B. Based on several investigations, whether the superstructure is long or short can be defined by the length ratio of the superstructure against the main hull exceeding $15 \sim 25\%$ or not.

31.3 Magnitude of Longitudinal Bending Stress

To demonstrate what magnitude of longitudinal bending stress flows into a long superstructure, finite element analysis was carried out for a cruise ship of 22,600 GT, whose finite element model is shown in Fig. 31.4. Figure 31.4a represents the outside view of the portside of a whole FEM model, and Fig. 31.4b is the inside view from afterward. The length of this vessel is $L_{pp} = 155$m and the length of the superstructure is 140 m, hence, the length ratio of the superstructure is approximately 90%. This vessel has 7 decks as illustrated in Fig. 31.5, in which No. 4 deck is a strength deck running through from fore to aft end. The superstructure consists of No. 5 to No. 7 decks. Cross-hatching area in Fig. 31.4a means public space between No. 3 and

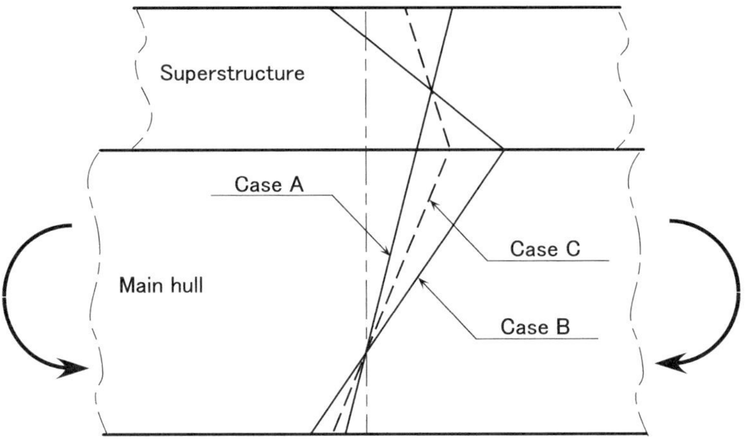

Fig. 31.3 Image of longitudinal bending stress distribution

No. 4 deck where a series of large windows are provided in the side shell to obtain a fine view for passengers. From the view point of strength, these large openings may affect the distribution of longitudinal bending stress, because shearing rigidity is reduced largely in this area. Therefore, to confirm this effect of large openings on the longitudinal bending stress distribution, two FEM models were prepared, i.e. a model with window openings and a model without windows. A loading case was selected to give maximum still water bending moment, and then concentrated forces were applied to the nodes on the side shell at the No. 2 deck level in such a way that the difference in the shearing forces corresponds to the concentrated force.

The calculated longitudinal stress distribution in each deck at the side was obtained as shown in Fig. 31.5. In Fig. 31.5, the broken line denotes the bending stress when the window openings are included in the model; the dotted line represents that when windows are ignored, and the straight line is obtained by beam theory by assuming the superstructure and the main hull work together as one beam against the bending moment. From this figure, we find that the bending stress distribution in the case of no windows is almost linear. Although the bending stress increases around No. 3 deck when the window openings are considered, the stress distribution can be regarded as straight in the superstructure. This stress increase at No. 3 deck is caused by the reduction of shear rigidity to 20% due to window openings. As explained previously, the critical length of the superstructure is 15 $\sim$ 25% of the main hull length. According to the above investigation, the superstructure of this vessel is long enough to transmit the longitudinal bending stress effectively, even though several large openings are provided in the side shell.

That brings us back to the question of what magnitude of longitudinal bending stress will flow into the superstructure. On the basis of the regulations specified by the Classification Society, only longitudinal strength members below the strength deck are taken into account in the section modulus of the hull girder; it is assumed other members in the superstructure, above the strength deck, are non-effective for

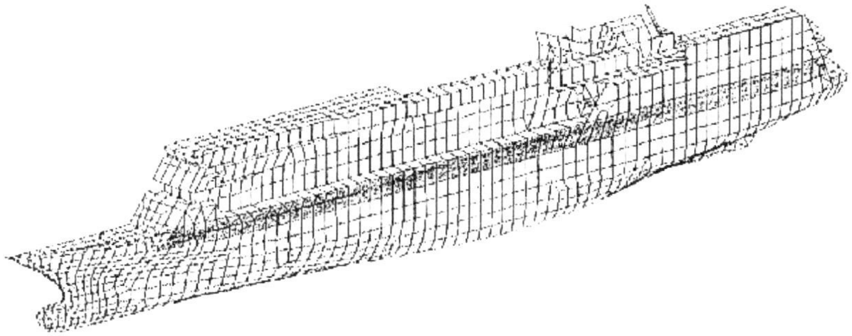

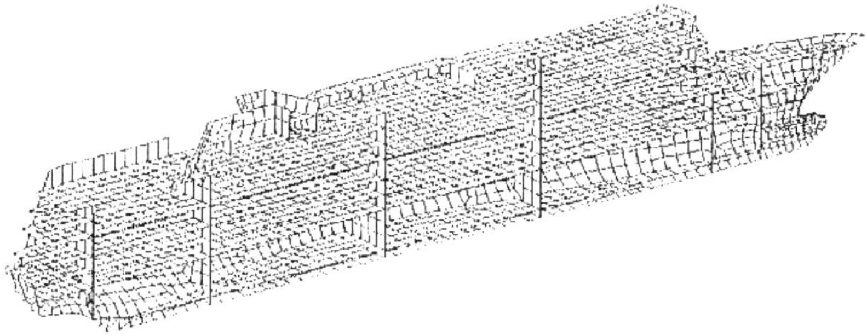

Fig. 31.4 Finite element model of cruise ship

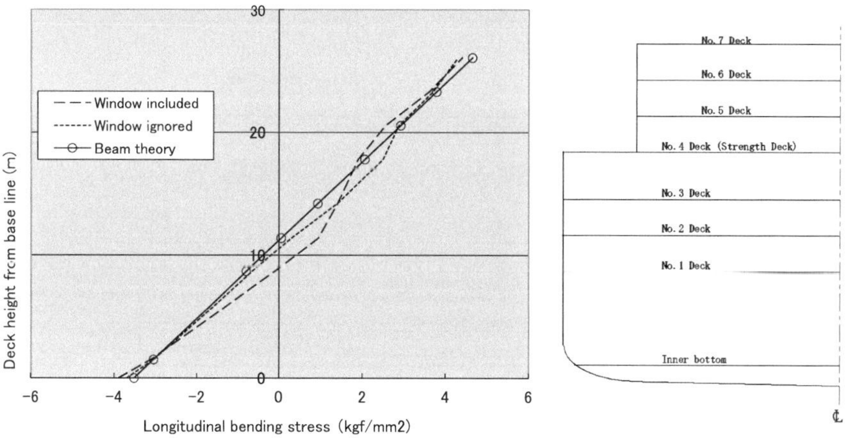

Fig. 31.5 Longitudinal bending stress distribution of cruise ship

longitudinal strength. However, as mentioned earlier, structural members in a superstructure contribute to the longitudinal bending strength together with the members in the main hull, where the superstructure is comparatively long. Therefore, the section modulus of the main hull is expected to increase by considering that members in the superstructure are also effective.

Let us take up this point by calculating the section modulus of the vessel in Fig. 31.4 in two cases: section modulus of main hull (a) with superstructure and (b) without superstructure. Table 31.1 indicates the comparison of section moduli: Case-A is the section modulus of the main hull only and Case-B is the section modulus of hull girder with superstructure. From this table, although the moment of inertia increases by 1.83 times by including the superstructure, the distance from the neutral axis to the uppermost deck increases by 1.62 times simultaneously; consequently the section modulus becomes only 1.13 times larger. Therefore, the designer should be careful about the strength of the superstructure, when it is a long superstructure, because the No. 7 deck stress reaches approximately the same level as that of No. 4 strength deck stress, required by the rules.

Table 31.1 Section modulus of hull girder with or without superstructure

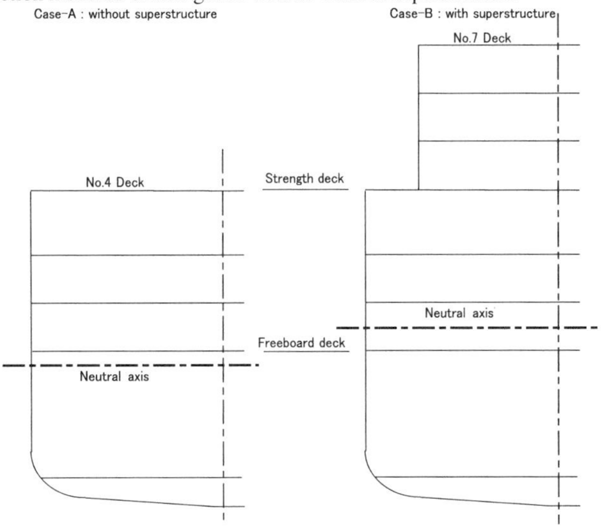

	Case-B / Case-A
Moment of inertia I	1.83
Distance from N.A. to upper most deck y_{DK}	1.62
Section modulus I/y_{DK}	1.13

31.4 Prevention of Structural Failures

In this section, several countermeasures are explained to prevent structural failures due to longitudinal stress in a superstructure.

31.4.1 Structural Discontinuity

As has been investigated above, longitudinal stress may be induced in the superstructure which is as large as the maximum stress in the main hull. For this reason, structural failures such as cracks may be easily caused by structural discontinuities due to stress concentrations as shown in Fig. 31.1.

Since a relative displacement of the upper and lower decks occurs in the vicinity of the connection between the lower deck plate and the end wall of the upper deck, as illustrated in Fig. 31.6, a flexible connection is sometimes applied to prevent cracks by absorbing this relative displacement. This figure shows a flexible type of connection and it is one of the available countermeasures to avoid structural failures at the discontinuous parts of a superstructure. Another solution is to apply a sufficiently strong structure to the discontinuous connection in order to reduce stress by reducing the relative displacement. Which type of solution should be taken depends on the design philosophy and the situation.

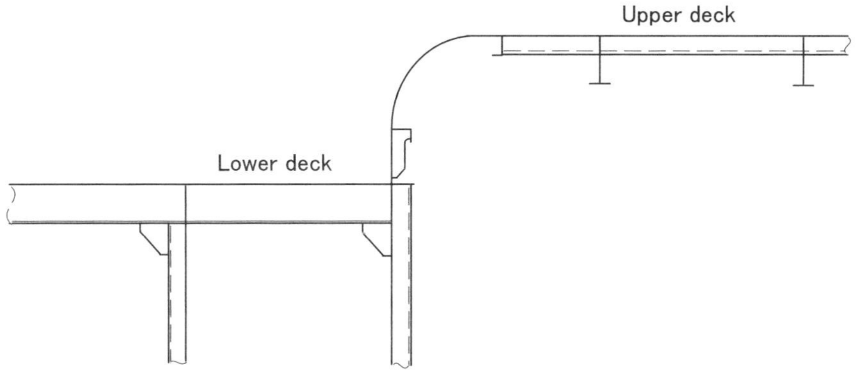

Fig. 31.6 Flexible connection at discontinuous part in long superstructure

31.4.2 Round Shape of Side Wall Opening Corner

Access openings are usually provided in steel walls in a superstructure and in some cases a negative rounded shape is applied to opening corners to overcome the interference between a corner edge and door frame, as indicated in Fig. 31.7. In these cases, cracks are apt to initiate from the free edge of the negative corner in the diagonal direction due to a combination of stress concentration at the round corner and the high longitudinal bending stress.

Cracks were found in a car ferry at the corners of these openings, and an FEM calculation was conducted to investigate the stress concentration around the corner. The structural model was a plane model as indicated by the cross-hatching area in Fig. 31.7 and was subjected to uniform tension, because longitudinal bending brings uniform tension in a longitudinal side wall. In order to confirm the stress concentration according to the corner shape, 4 types of door opening corner were modeled as illustrated in Fig. 31.8. Model CR1 is a damaged opening corner with a negative radius of 20R. Model CR2 indicates a normal corner with positive 200R and there is no damage in this type, however, the radius is so large that it is expensive to form such a door frame by bending work. Model CR3 is a negative radius corner with 60R and Model CR3 + S is an improved CR3 model by adding a horizontal stiffener.

The results are summarized in Table 31.2. From this table, the stress concentration factor of Model CR1, damaged, is 4.9, whereas the factor of Model CR2, with no

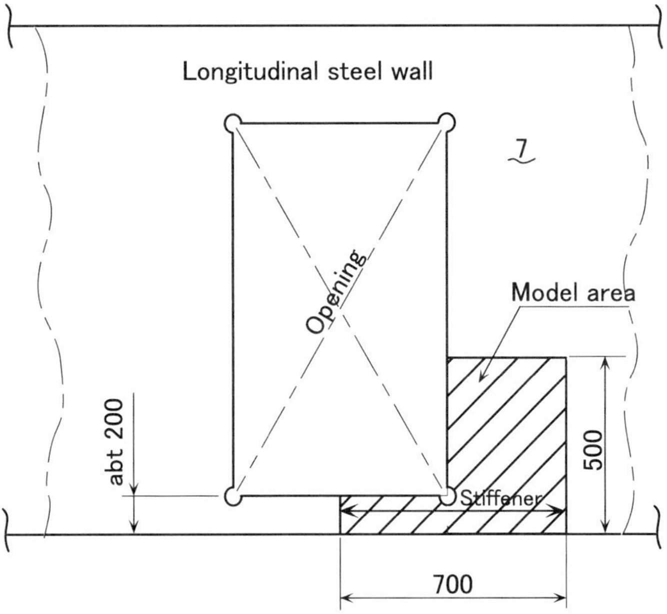

Fig. 31.7 Model area for stress concentration of door opening corner

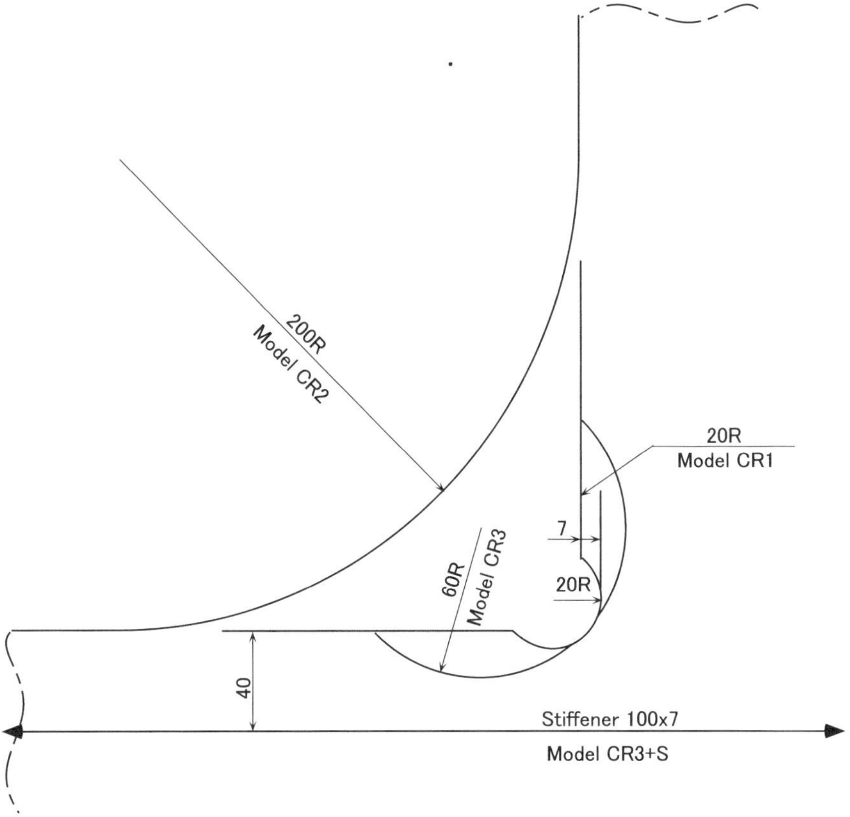

Fig. 31.8 Profiles of door opening corner

damage, is 2.4. In Model CR3 it is 3.9. On the other hand, in Model CR3 + S the stress concentration factor is 2.8, which is almost similar to that of Model CR2.

As a result of this investigation, adopting the shape of Model CR3 + S can prevent cracks.

31.4.3 Buckling

Regarding vessels with large superstructures such as a car ferry and a cruise ship, the superstructure is designed so as to minimize steel weight because of severe stability restrictions. To satisfy this requirement, the deck plate thickness is thin in comparison with other merchant vessels, which may result in buckling of deck plates under a sagging moment. Hence a longitudinal framing system is preferable to transverse framing in view of the buckling strength.

Table 31.2 Stress concentration around door opening corner

Model	Corner shape	Stress concentration factor	Note
CR1	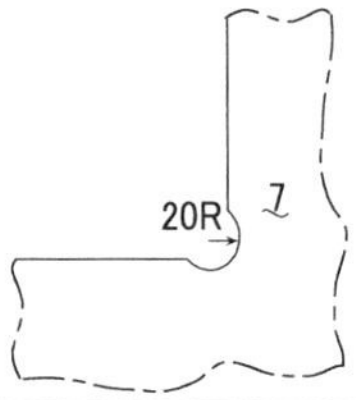	4.9	Damage
CR2	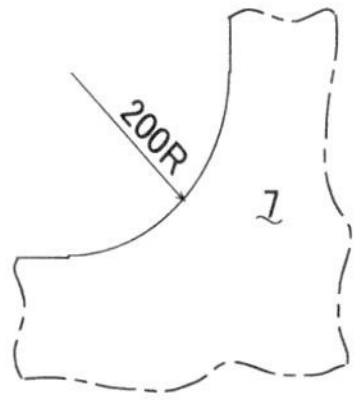	2.4	No damage
CR3	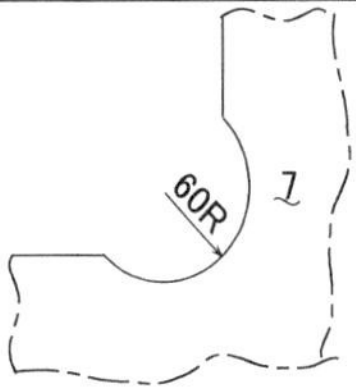	3.9	
CR3+S	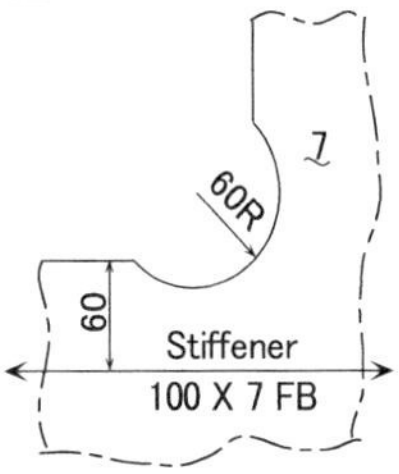	2.8	

Calculation model :

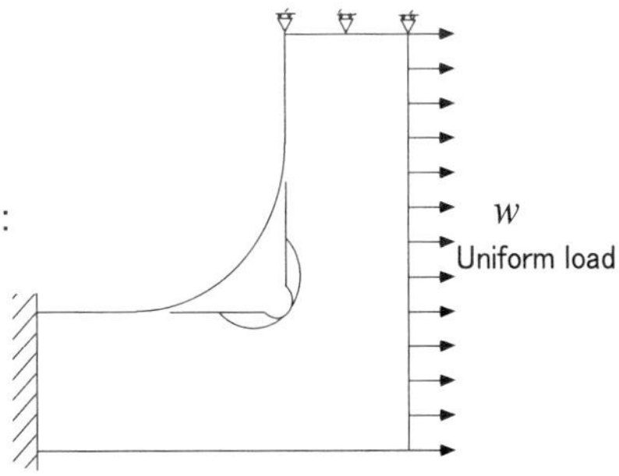

However, even when longitudinal stiffening is applied to the deck plating, the buckling strength of a deck plate should be investigated thoroughly, since the critical buckling strength differs depending on the calculation method for thin plates. For example, let us estimate the critical buckling stress of the deck plate in Fig. 31.9 by an analytical method first. Using Eq. (3.5.22) of Chap. 3 of Part I, we obtain the critical buckling stress as follows under the simply supported boundary condition:

$$\sigma_{cr} = \frac{\pi^2 E}{12(1-\nu^2)}\left(\frac{t}{b}\right)^2 K\eta = 4.85\ \text{kgf/mm}^2$$

where

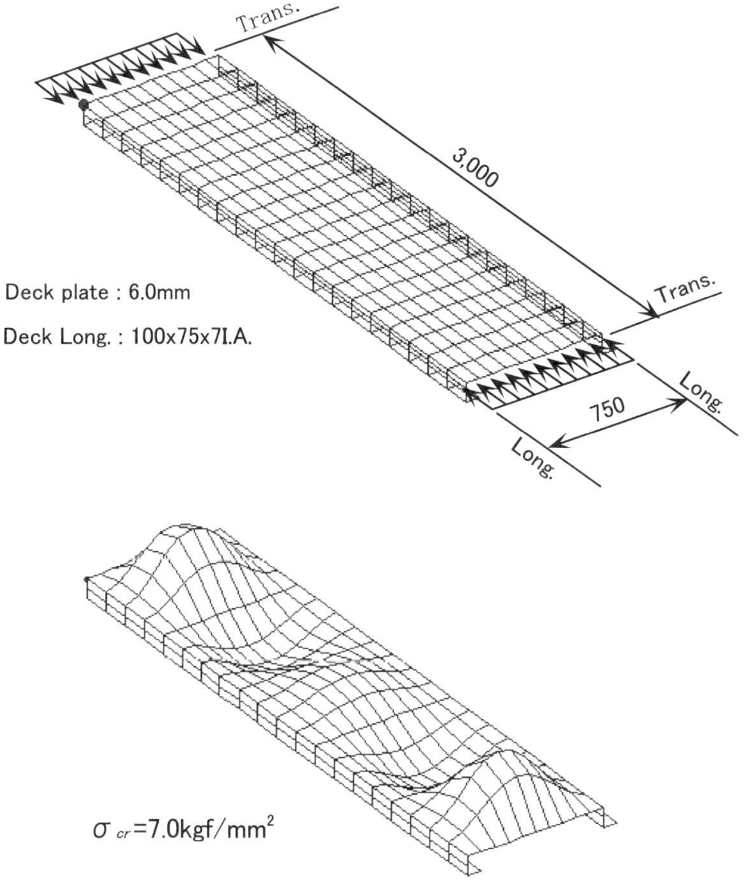

Fig. 31.9 Buckling strength of deck plate due to longitudinal bending stress

$$b = 750\ \text{mm}$$

$$K = 4$$

$$E = 21{,}000\ \text{kgf/mm}^2$$

$$t = 6\ \text{mm}$$

$$\nu = 0.3$$

$$\eta = 1 \text{ for } \sigma_e = 4.85 < \sigma_Y/2$$

On the other hand, the critical buckling stress calculated by FEM analysis is $\sigma_{cr} = 7.0\ \text{kgf/mm}^2$, which is 1.4 times greater than the analytical result. This difference is derived from the torsional rigidity of the deck longitudinals modeled in the FEM, affecting the increase of buckling stress. The difference becomes large in cases where the rigidity of plate bending is less than the torsional rigidity of the stiffener.

As discussed above, it is important to evaluate the buckling strength accurately in particular for thin deck plate structures.

31.4.4 Expansion Joints

For the purpose of reducing the longitudinal bending stress flowing into a superstructure, expansion joints are sometimes applied to the superstructure. An expansion joint is a slit provided in the side wall and decks of a superstructure to permit relative movement between adjacent superstructures as illustrated in Fig. 31.10 [159]. This slit is protected by a cover plate from the weather and has one edge welded to the side wall vertically with the other side free for sliding. Also the bottom of the slit is covered with a face plate to reduce the stress concentration around the free edge.

The expansion joint is effective in reducing longitudinal bending stress flowing into a superstructure. However, from the fatigue-strength view point, it behaves like a notch cut in a hull girder. Special attention should be paid regarding the shape of the bottom of the expansion joint [160].

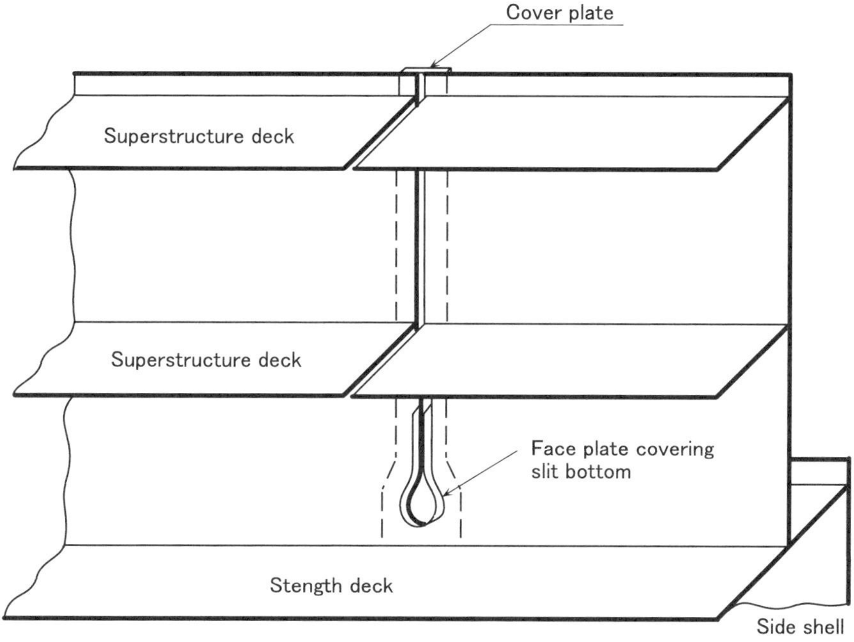

Fig. 31.10 Expansion joint applied to superstructure

References

1. M. Zako, Application of stochastic finite element method program to safety evaluation of storage tank. Ishikawajima-Harima Eng. Rev. No. 3 **22**, 143–147 (1982). (in Japanese)
2. X. Chen, Y. Kawamura, T. Okada, Stochastic finite element method based on response surface methodology considering uncertainty in shape of structures, *Proceedings of the 13th International Symposium on Practical Design of Ships and Other Floating Structures* (PRADS 2016) (2016)
3. X. Chen, Y. Kawamura, T. Okada, Solution of stochastic eigenvalue problem by improved stochastic inverse power method (I-SIPM). J. Mar. Sci. Technol. **23**, 814–834 (2018)
4. N.K. Kyokai, *Explanation of Revision of Rules and Regulations* (1964). (in Japanese)
5. N.K. Kyokai, *Explanation of Revision of Rules and Regulations and Other Technical Rules* (1972). (in Japanese)
6. M. Mano, Safety and reliability of welded structures—examples of marine structure design. Mach. Design, No. 9 **23**, 50–55 (1979). (in Japanese)
7. No. 134 Research Group: Research report on allowable stress of hull structure members, Japan Ship Research Association (1974). (in Japanese)
8. J.B. Caldwell, "Structural reliability", Colloquium on Reliability as a Design Parameter in Marine Reactor System
9. H. Okamura, H. Itagaki, *Probabilistic Treatment of Strength*. Baifukan Co., Ltd. (1979). (in Japanese)
10. Y. Kumakura, T. Nagano, Y. Okumoto, K. Tanida, Stress and motion measurements on the container ship, "Japan Ace" in service. IHI Eng. Rev. **8**(2), 47–67 (1975)
11. N. Salvesen, E.O. Tuck, O. Faltinsen, Ship motions and sea loads. Trans. SNAME **78**, 250–287 (1970)
12. J. Fukuda, "Statistic Prediction of Ship Response", Society of Naval Architects of Japan, Symposium on Ship Manoeuvrability, pp. 99–120 (1969). (in Japanese)
13. Y. Tomita, S. Tozawa: "Wave Load, High Tensile Steel and Fatigue Strength", Society of Naval Architects of Japan, Symposium on Ship's Strength and Failure Prevention, pp. 81–128 (1993). (in Japanese)
14. H. Kusumoto, H. Yu Takeda, K. Yashima, Y.U. Yamazaki, K. Iida, Ship hull structure design by DLA (Dynamic Loading Approach). IHI Eng. Rev. **27**(2), 61–65 (1994)
15. H. Mitsuyasu, F. Tasai, T. Suhara, S. Mizuno, M. Ohkusu, T. Honda, K. Rikiishi, Observations of the directional spectrum of ocean waves using a cloverleaf buoy. J. Phys. Oceanogr. **5**, 750–760 (1975)
16. S. Taira: *Strength of Materials*. Ohmsha, Ltd. (1990). (in Japanese)
17. Y. Okumoto, A. Ando, Y. Niiro, K. Hiyoku: Computer-aided Engineering System for Hull Structure—"FRESCO-S", ICCAS'88 China, pp. 79–90 (1988)

Y. Okumoto et al., *Design of Ship Hull Structures*,
https://doi.org/10.1007/978-3-032-12500-2

18. M. Yoshiki, T. Kanazawa, *Comprehensive Material Strength.* p. 96, Yokendo (in Japanese) (1959)
19. International Association of Classification Societies: "Common Structural Rules for Bulk Carriers and Oil Tankers" (2024)
20. The Japan Welding Society: "Welding Handbook, 4th edition", p. 46. (in Japanese)
21. W. Muckle, *Strength of Ships' Structures. Edward Arnold* (Publishers) Ltd. London (1967)
22. K. Terasawa (ed.), *Strength of Ship Hull Structure.* Kaibundo, (in Japanese) (1974)
23. The Kansai Society of Naval Architects: "Handbook of Ship Design, 4th Edition", Kaibundo Shuppan (1983). (in Japanese)
24. N. Jones, Plastic behavior of ship structure. Transactions of SNAME **84**, 115–139 (1976)
25. S. Timoshenko, *Vibration Problems in Engineering.* Van Nostrand (1950)
26. T. Matsudaira, *Basic Theory of Vibration.* Gendaikogakusha (1973). (in Japanese)
27. Y. Takeuchi (ed.), Pandect of design engineering. Fuji Thecno Syst., 842 (1996). (in Japanese)
28. Y. Okumoto, Y. Takeda, K. Hiyoku, Modern Hull Structure Design System "COSMOS", ICCAS'85 Trieste, Italy, pp. 275–284 (1985)
29. K. Kitamura, Studies on optimization of ship structure (3rd report)—optimum design of longitudinal members of tanker. J. Soc. Naval Architects Japan, No. **132**, 335–348 (1972). (in Japanese)
30. T. Okada, I. Neki, Utilization of genetic algorithm for optimizing the design of ship hull structures. Naval Archit. Ocean. Eng. **31**, 71–83 (1994)
31. A. Abe, T. Okada, I. Neki, Development of optimum design system for container ship. IHI Eng. Rev. **32**(4), 260–266 (1992). (in Japanese)
32. T. Okada, M. Toyoda, E. Kobayashi, Practical initial design optimization of large container ships. *Proceedings of PRADS2007 Houston*, USA, 574–579 (2007)
33. M. Toyoda, *A Study on Structural Design of Large Container Ship, Doctoral Dissertation*, The University of Tokyo (2009). (in Japanese)
34. M. Toyoda, H. Kusumoto, T. Okada, E. Kobayashi: Advanced structural design of containerships to achieve overall safety and economy. Design & Operation of Container Ship, The Royal Institution of Naval Architects, pp. 31–41 (2006)
35. T. Nagashima, F. Niwa, H. Hirano, T. Okada, Application of automatic nesting to ship hulls using the FINEST algorithm. IHI Eng. Rev. **38**(1), 38–44 (2005)
36. T. Nagashima, S. Matsuoka, H. Tawara, F. Niwa, T. Okada, Automatic nesting of ship hulls using the FINEST algorithm (2nd Report)—application to plate size determination. IHI Eng. Rev. **39**(1), 12–17 (2006)
37. International Maritime Organization: "International Code for the Construction and Equipment of Ships Carrying Liquefied Gases in Bulk" (2016)
38. O. Baba, Y. Okumoto, A. Abe, Welding of thicker steel joints of container ships. J. Ship Prod. **23**(1), 46–52 (2007)
39. M. Toyoda, N. Kiji, Y. Nakajima, T. Okada, Y. Nakanishi, K. Inose, Safety of mega container ship focusing on brittle crack initiation and arrest behavior of heavy thickness plate. *The ASME 27th International Conference on Offshore Mechanics and Arctic Engineering*, OMAE2008, OMAE2008–57375 (2008)
40. T. Okada, Y. Takeda, T. Maeda, On board measurement of stresses and deflections of a Post-Panamax containership and its feedback to rational design. Mar. Struct. **19**, 141–172 (2006)
41. M. Toyoda, H. Kusumoto, T. Okada, E. Kobayashi, Advanced structural design of containerships to achieve overall safety and economy. Design & Operation of Container Ship, The Royal Institution of Naval Architects, pp. 31–41 (2006)
42. M. Mano, On recent ship vibration. Bulletin Soc. Naval Archit. Japan, No. 692, pp. 75–83 (1987). (in Japanese)
43. Japan Ship Research Association: "40 Years History of Japan Ship Research Association" (1993). (in Japanese)
44. T. Nagano, O. Ushirokawa, M. Mano, Y. Yamada, Introduction of non-uniformity into ship hull structure. *Proceedings of PRADS'87 Trondheim*, 1098–1105 (1987)

45. H. Yajima, T. Yao, M. Yamamoto, A study on buckling / plastic collapse behavior of steel plate having surface layers with ultra fine grain microstructure. *Proceedings of TEAM'97* Singapore (1997)
46. Nippon Steel Corporation: "NSC's Steel Plates for Marine Structures", Cat. No. EXE350, p. 104 (1974)
47. Study Group of Shipbuilding Textbook: "Fundamentals of Merchant Ship Design", Seizandou, p. 168 (2001). (in Japanese)
48. International Association of Classification Societies: UR W-14 Rev. 3 (2021)
49. Y. Ishii, H Tamura, *General of Welding Engineering*, Kyouritsu-shuppan, p. 106 (1976). (in Japanese)
50. H. Kihara, *Modern Welding Engineering*. Ohmsya, p. 147 (1974). (in Japanese)
51. International Association of Classification Societies: UR W-13 Rev. 7 (2021)
52. Y. Okumoto, Dimensional error of assembled blocks. J. Ship Prod. **17**(1), 8–15 (2001)
53. N. Yamamoto, Studies on the generation and progression of corrosion in structural hull members. ClassNK Techn. Bull. **15**, 43–49 (1997)
54. Y. Okumoto, Residual stress on ship hull structure. *Proceedings of TEAM'97* Singapore, pp. 290–303 (1997)
55. H. Togawa, *Guide for Finite Element Method,* Saiensu-sha Co., Ltd. (1979). (in Japanese)
56. R. Kamichika, H. Ohya, Stress analysis (2)—application of FEM stress analysis to actual structures. J. Japan Welding Soc. **66**(3), 147–150 (1997). (in Japanese)
57. D. Faulkner, A review of effective plating for use in the analysis of stiffened plating in bending and compression. J. Ship Res. **19**(1), 1–17 (1975)
58. Lloyd's Register of Shipping: "Rules and Regulations for the Classification of Ships Part 3" (1996)
59. Pandect of Hull Structure Strength, Commemoration of Retirement of Prof. Yoshiki (1970), pp. 21–24 (in Japanese).
60. M. Mano, Calculation of girder's face plates. Harima Zosen Techn. Rev. **7**, 35–40 (1958). (in Japanese)
61. T. Okada, K. Watanabe, Y. Kawamura, Y. Tanaka, N. Osawa, H. Murayama, *Strength of Materials—Fundamentals of Structural Strength Design,* Asakura Publishing Co., Ltd. (2021). (in Japanese)
62. Ships Design Department, Aioi, Some developments in design details of hull structure. Ishikawajima-Harima Eng. Rev. **8**(39), 33-41 (1968). (in Japanese)
63. A. Nitta, Plastic strength of hybrid steel beams subjected to bending and axial thrust. J. Zosen Kiokai **116**, 69–77 (1964). (in Japanese)
64. D. Zhang, Z. Xu, T. Okada, Y. Kawamura, G. Hayakawa, K. Ishibashi, H. Koyama, A study on the scantling formulae of a stiffener against lateral pressure under the simultaneous action of axial load. Mar. Struct. **95**, 103595 (2024)
65. T. Okada, Z. Xu, Y. Kawamura, G. Hayakawa, Simple formulation of shear stress effect on the bending strength of stiffeners of ship hull structures. *Proceedings of the 37th Asian-Pacific Technical Exchange and Advisory Meeting on Marine Structures* (TEAM 2024), Kure, Japan, pp. 540–549 (2024)
66. Lloyd's Register of Shipping: "Klasserfaren heter Betraffande Skrovskador panyare Och Storre Typer av Fartyg" (1971)
67. Y. Ochi, T. Nagano, K. Yamaguchi, H. Kimura, H. Kashima, Strength around slot for penetration of longitudinals. Ishikawajima-Harima Eng. Rev. **11**(6), 463–474 (1971). (in Japanese)
68. T. Okada, Y. Kawamura, Strength evaluation of intersection between stiffeners and primary supporting members considering the effect of shear force on the primary member web. Mar. Struct. **59**, 25–46 (2018)
69. L. Gu, T. Okada, L. Xia, Y. Kawamura, Strength evaluation of intersection between stiffeners and primary supporting members in double hull structure. Mar. Struct. **74**, 102789 (2020)
70. N.K. Kyokai, On damage of side longitudinals of VLCC's. Report presented to Study Group on High Tensile Steel Hull Structure (1991). (in Japanese)

71. Y. Nakajima, N. Iino, I. Neki, O. Ushirokawa, M. Maeda, H. Sasajima, Fatigue strength design of ship structure. IHI Engineering Review **26**(2), 54–61 (1993)
72. K. Yuasa, *Strength of Materials*. Sankaidou Book Store (1942). (in Japanese)
73. Technical Research Committee, The West-Japan Society of Naval Architects: "Research and Investigation on Hull Structure Damage" (1970). (in Japanese)
74. S.P. Timoshenko, S. Woinowsky-Krieger, *Theory of Plates and Shells* (Second Edition). McGRAW-HILL (1959)
75. T. Fujii, K. Uchino, Strength of panels of plating loaded beyond the elastic limits. Ishikawajima-Harima Eng. Rev. **2**(4), 84–93 (1962). (in Japanese)
76. R. Umezawa, Y. Naruse, T. Okada, Y. Kawamura, K. Ishibashi, H. Koyama, A Study on scantling formulae of plate members due to lateral pressure under the effect of axial loads. Mar. Struct. **80**, 103099 (2021)
77. N.K. Kyokai, Rules for the survey and construction of steel ships. Part C (2023)
78. M. Yamada, T. Okada, Y. Naruse, Y. Kawamura, G. Hayakawa, K. Ishibashi, H. Koyama, Influence of plate aspect ratio on the axial load effect on the plate strength against lateral pressure. *Practical Design of Ships and Other Floating Structures: Proceedings of the 15th International Symposium*, PRADS 2022, October 9–13, Dubrovnik, Croatia, 145–158 (2022)
79. T. Nitta, T. Wada, M. Mano, Some development in design details of hull structure—report no. 2. Ishikawajima-Harima Eng. Rev. **9**(4), 395–402 (1969). (in Japanese)
80. M. Mano, M. Takata, I. Ogura, H. Okamoto, H. Matsumoto, Some applications of the finite element method to hull structure—2nd Report. Ishikawajima-Harima Eng. Rev. **12**(3), 250–258 (1972). (in Japanese)
81. NKK, Low Carbon Equivalent Steel of 36 kgf/mm^2 Class of Yield Stress Produced by NCT Process. NKK Technical Report (in Japanese)
82. J. Yagi, W. Yasukawa, On the minimum weight construction of single plated flat grillage under lateral loads. J. Kansai Soc. Naval Archit. Japan, No. 104, 31–37 (1961). (in Japanese)
83. K. Kitamura, Studies on optimization of ship structures (the 2nd Report)—optimum design of grillages. J. Soc. Naval Archit. Japan, No. 130, 259–268 (1971). (in Japanese)
84. J. Nagasawa, Y. Honma, On the minimum weight of orthogonally stiffened plates under lateral pressure. J. Soc. Naval Archit. Japan, No. 127, 129–141 (1970). (in Japanese)
85. S.P. Timoshenko, J.N. Goodier, *Theory of Elasticity (Third Edition)*. McGRAW-HILL (1982)
86. A. Nanba, K. Tomiyoshi, M. Nakayama, S. Miyazaki, Y. Kitamura, Initial deflection of hull. J. Kansai Soc. Naval Archit. Japan, No. 137, 6–11 (1970). (in Japanese)
87. Y. Yamamoto, M. Fujino, T. Fukasawa, Motion and longitudinal strength of a ship in head sea and the effects of non-linearities. J. Soc. Naval Archit. Japan, No. 143, 179–187 (1978). (in Japanese)
88. Y. Yamamoto, M. Fujino, T. Fukasawa, Motion and longitudinal strength of a ship in head sea and the effects of non-linearities (2nd Report). J. Soc. Naval Archit. Japan, No. 144, 179–187 (1978). (in Japanese)
89. Y. Yamamoto, M. Fujino, T. Fukasawa, Motion and longitudinal strength of a ship in head sea and the effects of non-linearities (3rd Report). J. Soc. Naval Archit. Japan, No. 145, 63–70 (1979). (in Japanese)
90. M. Mano, Convenient method for calculation of ship's deflection. Harima Zosen Techn. Rev. No. 10, 58–66 (1959). (in Japanese)
91. M. Mano, Happiness of designer. Bull. Soc. Naval Archit. Japan, No. 613, 395–400 (1980). (in Japanese)
92. Y. Yamaguchi, Study on fatigue strength of hull structural members (No. 2). Report of NK's Res. Inst. **2**, 157–179 (1970). (in Japanese)
93. The Shipbuilding Research Association of Japan: "Research on Fatigue Strength of Ship Structure", The 200th Research Committee 2nd Sub Committee, Report No. 86 (1977). (in Japanese)
94. Transverse Strength Sub-WG, SNAJ, Determination of leg length of fillet welding (1). Bull. Soc. Naval Archit. Japan, No. 595, 29–34 (1979). (in Japanese)

95. Transverse Strength Sub-WG, SNAJ, Determination of leg length of fillet welding (2). Bull. Soc. Naval Archit. Japan, No. 597, 118–123 (1979). (in Japanese)
96. Transverse Strength Sub-WG, SNAJ, Determination of leg length of fillet welding (3). Bull. Soc. Naval Archit. Japan, No. 598, 170–174 (1979). (in Japanese)
97. Y. Matsuyama, Y. Shibata, On all-welded vessel. Zosen-kyokai zassan, No. 267, 16–26 continued to p. 14 (1946). (in Japanese)
98. Y. Okumoto, M. Sano, H. Murakawa, S. Rashed, Reduction of welding distortion based on theoretical prediction using inherent deformation method. ISOPE 2011, Maui, pp. 138–144 (2011)
99. Y. Okumoto, H. Murakawa, S. Rashed, Y. Katayama, M. Sano, M. Yoshimura, Study to reduce straightening work on PCC car deck. Report of Kindai University Fundamental Technol. Next Gen. Res. Inst. **8**, 7–13 (2008). (in Japanese)
100. Wikimedia Commons, File:TankerSchenectady.jpg. Available from: https://commons.wikimedia.org/wiki/File:TankerSchenectady.jpg. Accessed on 14th April 2025
101. Khalifeh, H.D. Manesh, A. Dehghan, Fracture behavior of structural steels under earthquake dynamic loadings. IntechOpen, 2024, Available from: https://doi.org/10.5772/intechopen.1004017. Accessed on 14th April 2025
102. David Broek, *Elementary Engineering Fracture Mechanics (fourth revised edition).* Martinus Nijhoff Publishers, p. 8 (1986)
103. N.K. Kyokai, Commentary on the revisions of the rules on the grade of steel in the rules for the survey and construction of steel ships Part C in 1981. Trans. Nippon Kaiji Kyokai **178**, 2–8 (1982). (in Japanese)
104. H. Yajima, M. Tada, S. Noda, H. Kitada, K. Ikegami, N. Takahashi, A suggested method of ship hull plating fracture toughness assessment. Trans. The West-Japan Soc. Naval Archit. No. 75, pp. 200–211 (1988). (in Japanese)
105. International Association of Classification Societies: UR S-6 Rev. 9 (2018)
106. A. Abe, Study of Structural Design Method of SPB Tank. Doctoral dissertation of Tokyo Univ., pp. 103–136 (2000). (in Japanese)
107. O. Ushirokawa, E. Nakayama, Stress concentration factor of welded joints. Ishikawajima-Harima Eng. Rev. **23**(4), 351–355 (1983). (in Japanese)
108. Y. Akita, T. Maeda, T. Yada, K. Sakai, Effect of angular distortion in welded joints on brittle fracture initiation. IIW Doc. X-569–70 (1970)
109. T. Okada, A. Abe, O. Ushirokawa, A. Itoh, M. Itoh, Y. Nakajima, Ship structure design based on direct calculation. IHI Eng. Rev. **29**(3), 101–108 (1996)
110. Japan Archives, https://jaa2100.org/. Accessed July, 2025
111. Y. Sumi, Structural safety of ships developed by lessons learned from the 100-year history of break-in-two accidents. Mar. Struct. **64**, 481–491 (2019)
112. G.H. Hoffmann, Analysis of Sir John Biles's experiments on H.M.S. Wolf in the light of Pietzker's theory. Trans. INA, 67, pp. 41–65 (1925)
113. G.H. Hoffmann, The effective I of H.M.S. Wolf. Trans. INA, 70, pp. 152–172 (1928)
114. Marine Accident Investigation Branch: "Report on the Investigation of the Structural Failure of MSC Napoli English Channel on 18 January 2007", Report No 9/2008 (2008)
115. Committee on Large Container Ship Safety: "Interim Report of Committee on Large Container Ship Safety", Maritime Bureau of Japan's Ministry of Land, Infrastructure, Transport and Tourism (2013)
116. Committee on Large Container Ship Safety: "Final Report of Committee on Large Container Ship Safety", Maritime Bureau of Japan's Ministry of Land, Infrastructure, Transport and Tourism (2015)
117. The Investigative Panel on Large Container Ship Safety: "Investigation Report on Structural Safety of Large Container Ships", ClassNK (2014)
118. Y. Kawasaki, T. Okada, H. Kobayakawa, I. Amaya, T. Miyashita, T. Nagashima, I. Neki, Strength evaluation of containerships based on dynamic elastic response calculation of hull girder, 2nd report—influence of hull girder rigidity and correlation between double bottom bending and hull girder bending. J. Japan Soc. Naval Archit. Ocean Eng. **25**, 191–203 (2017). (in Japanese)

119. Y. Kawasaki, T. Okada, H. Kobayakawa, I. Amaya, T. Miyashita, T. Nagashima, I. Neki, A study on forced vibration of double bottom structure due to whipping on an ultra large container ship. *Proceedings of the ASME 2017 36th International Conference on Ocean, Offshore and Arctic Engineering (OMAE2017)*, Paper No. OMAE2017–61149 (2017)
120. J. Chen, T. Okada, Y. Kawamura, X. Chen, T. Mitsuyuki, A study on whipping related double bottom response as well as its statistical characteristics based on full-scale measurements. J. Mar. Sci. Technol. **26**(3), 846–860 (2021)
121. N.V. Barabanov, On replacement of floating bulwarks for ones attached to sheer strake. OMAE, Houston, Texas (1988)
122. H. Ohtsubo, M. Kitamura, Structural design of mid-ship section by expert system. J. Soc. Naval Archit. Japan, No. 164, 267–272 (1988). (in Japanese)
123. J. Yagi, Y. Tomita, Y. Fujimoto, Fatigue strength criteria for ship structures—fatigue strength criterion for longitudinal and transverse strength members. J. Soc. Naval Archit. Japan, No. 147, 211–218 (1980). (in Japanese)
124. M. Mano, Detail design of hull structure. PRADS ’83 Tokyo & Seoul, pp. 525–530 (1983)
125. S. Tamai, F. Niwa, M. Toyoda, T. Okada, Practical consideration on detailed design of ship structure. *Proceedings of TEAM2001 Jochiwon* (2001)
126. J. McCallum, The strength of fast cargo ships. Trans. Royal Inst. Naval Archit. **117**, 1–18 (1975)
127. A. Thayamballi, L. Kutt, Y.N. Chen, Advanced strength and structural reliability assessment of the ship’s hull girder. Adv. Marine Struct., Elsevier Applied Science Publishers (1986)
128. M. Mano, A reliable design of hull structure. *First International Symposium on Ship Reliability*, Varna (1985)
129. Y. Ogawa, Structural system in IHI (Ishikawajima-Harima Heavy Industries Co., Ltd.) of 480,000 DWT Tanker. *Symposium on Structure and Strength of Large Ship* (Sponsored by the Society of Naval Architects of Japan, pp. 119–136 (1974). (in Japanese)
130. C.I. Ha, N. Mun, M.S. Kim, M.S. Kim, Development of VLCC without the Cross-tie between Longitudinal Bulkheads (“Cross-tieless Cargo Tank of VLCC”), TSCF2013, Shanghai (2013)
131. N.K. Kyokai, Commentary on the revisions of the rules for the survey and construction of steel ships and inspection guidelines in 1974, Trans. Nippon Kaiji Kyokai, No. 150, pp. 3–41 (1975). (in Japanese)
132. D.P. Hong, O.H. Kim, Y.S. Lee, *Analysis of Structural Damage of a Large Ore/Coal Carrier* (SNAME STAR Symposium, Philadelphia, 1987)
133. T. Okada, T. Toyama, Y. Kawamura, Theoretical strength assessment of unstiffened bilge shell plating and some considerations on rule prescriptions. J. Marine Sci.Technol. **22**(1), 85–100 (2017)
134. T. Okada, T. Toyama, Y. Kawamura, Theoretical study on structural arrangement to control strength of unstiffened bilge shell plating. *Proceedings of the 13th International Symposium on Practical Design of Ships and Other Floating Structures (PRADS2016)*, 4–8 September, 2015, Copenhagen, Denmark (2016)
135. T. Okada, J. Lin, Y. Kawamura, Strength of curved shell plating of ships under hoop stress—2nd report verification and considerations through finite element analysis. Conf. Proc. Japan Soc. Naval Archit. Ocean Eng. **21**, 423–428 (2015). (in Japanese)
136. Y. Okumoto, Study on Ship Hull Structure at Grounding Condition. Doctoral dissertation of Hiroshima University, pp. 18–46 (1986). (in Japanese)
137. Y. Akita, S. Shimizu, K. Kitamura, On the shear rigidity of swash bulkhead in wing tank for large tankers. J. Soc. Naval Archit. Japan, No. 123, 144–152 (1968). (in Japanese)
138. M. Yamakoshi, K. Kagawa, Shear deformation of wing tanks. Trans. of the West-Japan Society of Naval Architects, No.29, pp. 83–95 (1965). (in Japanese)
139. S. Kurose, Shear Rigidity of Corrugated Bulkhead. Internal Report of Tsuneishi Shipbuilding Co. Ltd. (1988). (in Japanese)
140. I. Neki, T. Okada, Optimum structural design using genetic algorithm and finite element method. J. Soc. Naval Archit. Japan **177**, 327–338 (1995). (in Japanese)

141. No. 166 Committee: The Report of the Structural Strength of Ships Loading Steel Coil, Shipbuilding Research Association of Japan (1977). (in Japanese)
142. The Naval Architect, "BCT70: heralding a new standard family of Burmeister & Wain bulk carriers", The Naval Architect, pp. E175–E180 (1991)
143. Society of the Naval Architects of Japan: "Shipbuilding History in Showa Era", Vol. 1, Hara Book Store.
144. H. Yajima, Lessons to learn from examples of damages to large steel structures. Trans. West-Japan Soc. Naval Archit. **90**, 245–251 (1995). (in Japanese)
145. M. Mano, "Strength and Vibration Problems in the Area of Hull Fittings" No.12 Summer Seminar New Shipbuilding Technology, Society of the Naval Architects of Japan (1986). (in Japanese)
146. Y. Ochi, K. Tanida, R. Fujino, Prevention of engine room vibration—1st report an engine room double bottom—. Ishikawajima-Harima Eng. Rev. **16**(6), 655–666 (1976). (in Japanese)
147. S. Nakashima, M. Mano, Study on deflection of engine room double bottom. Ishikawajima-Harima Eng. Rev. **11**(5), 394–400 (1971). (in Japanese)
148. G.C. Volcy, "Experience with Marine Engineering System over the Last Thirty Years" Andrew Laing Lecture 1983-NECIES Newcastle upon tyne
149. N. Okamura, H. Sasajima, K. Tanida, Vortex induced heavy vibration in enclosed hull cavities. Trans. Royal Inst. Naval Archit. **130** (1988)
150. P.W. Ayling, P.W. Knaggs, Propeller bossing vibration. Trans. Royal Inst. Naval Archit. **107**, 57–81 (1965)
151. M. Mano, Y. Ochi, K. Fujii, Prevention and remedy of ship vibration. IHI Eng. Rev. **10**(3), 25–41 (1977)
152. M. Mano, F. Takano, K. Furuya, Investigation of ship vibration and its application. Ishikawajima-Harima Eng. Rev. **1**(1), 39–48 (1961). (in Japanese)
153. M. Mano, Vibrations des superstructures des Navires. Bureau Veritas Bull. Techn. **53**(3) (1971)
154. M. Yamaguchi, Experimental investigation on propeller exciting forces. Ishikawajima-Harima Eng. Rev. **16**(6), 648–654 (1976). (in Japanese)
155. R. Fujino, Propeller exciting force. *Proceedings of 5th TEAM, Jangseungpo*, Korea (1991)
156. Y. Takeda, H. Kusumoto, Elastic mounting system of main engine of a cruise ship. *Proceedings of TEAM'98* Kanazawa (1998)
157. Y. Takeda, H. Kusumoto, H. Sasajima, K. Tanida, Y. Koike, N. Takahashi, T. Iwanami, Development of active mass damper for ships by hydraulic control. J. Soc. Naval Archit. Japan **171**, 289–298 (1992). (in Japanese)
158. Y. Ochi, T. Nagano, Y. Yoshida, Measurements and design standard of vibration of web frame in a tank—2nd report design standard. Ishikawajima-Harima Eng. Rev. **15**(3), 390–407 (1975). (in Japanese)
159. Robert Taggart (Editor): "Ship Design and Construction", SNAME (1980)
160. H.W. Stapel, A.W. Vredeveldt, J.M.J. Journee, W. de Koning, Fatigue damage in the expansion joints of SS Rotterdam. *Proceedings of the 7th International Symposium PRADS'98*, Hague, the Netherlands, pp. 905–912 (1998)

Index

Y. Okumoto et al., *Design of Ship Hull Structures*,
https://doi.org/10.1007/978-3-032-12500-2

MIX
Papier aus verantwortungsvollen Quellen
Paper from responsible sources
FSC® C105338

If you have any concerns about our products,
you can contact us on
ProductSafety@springernature.com

In case Publisher is established outside the EU,
the EU authorized representative is:
Springer Nature Customer Service Center GmbH
Europaplatz 3, 69115 Heidelberg, Germany

Printed by Libri Plureos GmbH
in Hamburg, Germany